Springer Series in Operations Research and Financial Engineering

Editors:
Thomas V. Mikosch   Sidney I. Resnick   Stephen M. Robinson

Springer Series in Operation Research and Financial Engineering

Jorge Nocedal   Stephen J. Wright

Numerical Optimization

Second Edition

 Springer

Jorge Nocedal
EECS Department
Northwestern University
Evanston, IL 60208-3118
USA
nocedal@eecs.northwestern.edu

Stephen J. Wright
Computer Sciences Department
University of Wisconsin
1210 West Dayton Street
Madison, WI 53706–1613
USA
swright@cs.wisc.edu

Series Editors:
Thomas V. Mikosch
University of Copenhagen
Laboratory of Actuarial Mathematics
DK-1017 Copenhagen
Denmark
mikosch@act.ku.dk

Sidney I. Resnick
Cornell University
School of Operations Research and
    Industrial Engineering
Ithaca, NY 14853
USA
sirl@cornell.edu

Stephen M. Robinson
Department of Industrial and Systems
    Engineering
University of Wisconsin
1513 University Avenue
Madison, WI 53706–1539
USA
smrobins@facstaff.wise.edu

Mathematics Subject Classification (2000): 90B30, 90C11, 90-01, 90-02

ISBN-10: 1-4939-3711-1   ISBN-13: 978-1-4939-3711-0
DOI 10.1007/978-0-387-40065-5

Printed on acid-free paper.

springer.com

To Sue, Isabel and Martin
and
To Mum and Dad

Contents

Preface

This is a book for people interested in solving optimization problems. Because of the wide (and growing) use of optimization in science, engineering, economics, and industry, it is essential for students and practitioners alike to develop an understanding of optimization algorithms. Knowledge of the capabilities and limitations of these algorithms leads to a better understanding of their impact on various applications, and points the way to future research on improving and extending optimization algorithms and software. Our goal in this book is to give a comprehensive description of the most powerful, state-of-the-art, techniques for solving continuous optimization problems. By presenting the motivating ideas for each algorithm, we try to stimulate the reader's intuition and make the technical details easier to follow. Formal mathematical requirements are kept to a minimum.

Because of our focus on continuous problems, we have omitted discussion of important optimization topics such as discrete and stochastic optimization. However, there are a great many applications that can be formulated as continuous optimization problems; for instance,

finding the optimal trajectory for an aircraft or a robot arm;

identifying the seismic properties of a piece of the earth's crust by fitting a model of the region under study to a set of readings from a network of recording stations;

designing a portfolio of investments to maximize expected return while maintaining an acceptable level of risk;

controlling a chemical process or a mechanical device to optimize performance or meet standards of robustness;

computing the optimal shape of an automobile or aircraft component.

Every year optimization algorithms are being called on to handle problems that are much larger and complex than in the past. Accordingly, the book emphasizes large-scale optimization techniques, such as interior-point methods, inexact Newton methods, limited-memory methods, and the role of partially separable functions and automatic differentiation. It treats important topics such as trust-region methods and sequential quadratic programming more thoroughly than existing texts, and includes comprehensive discussion of such "core curriculum" topics as constrained optimization theory, Newton and quasi-Newton methods, nonlinear least squares and nonlinear equations, the simplex method, and penalty and barrier methods for nonlinear programming.

The Audience

We intend that this book will be used in graduate-level courses in optimization, as offered in engineering, operations research, computer science, and mathematics departments. There is enough material here for a two-semester (or three-quarter) sequence of courses. We hope, too, that this book will be used by practitioners in engineering, basic science, and industry, and our presentation style is intended to facilitate self-study. Since the book treats a number of new algorithms and ideas that have not been described in earlier textbooks, we hope that this book will also be a useful reference for optimization researchers.

Prerequisites for this book include some knowledge of linear algebra (including numerical linear algebra) and the standard sequence of calculus courses. To make the book as self-contained as possible, we have summarized much of the relevant material from these areas in the Appendix. Our experience in teaching engineering students has shown us that the material is best assimilated when combined with computer programming projects in which the student gains a good feeling for the algorithms—their complexity, memory demands, and elegance—and for the applications. In most chapters we provide simple computer exercises that require only minimal programming proficiency.

Emphasis and Writing Style

We have used a conversational style to motivate the ideas and present the numerical algorithms. Rather than being as concise as possible, our aim is to make the discussion flow in a natural way. As a result, the book is comparatively long, but we believe that it can be read relatively rapidly. The instructor can assign substantial reading assignments from the text and focus in class only on the main ideas.

A typical chapter begins with a nonrigorous discussion of the topic at hand, including figures and diagrams and excluding technical details as far as possible. In subsequent sections, the algorithms are motivated and discussed, and then stated explicitly. The major theoretical results are stated, and in many cases proved, in a rigorous fashion. These proofs can be skipped by readers who wish to avoid technical details.

The practice of optimization depends not only on efficient and robust algorithms, but also on good modeling techniques, careful interpretation of results, and user-friendly software. In this book we discuss the various aspects of the optimization process—modeling, optimality conditions, algorithms, implementation, and interpretation of results—but not with equal weight. Examples throughout the book show how practical problems are formulated as optimization problems, but our treatment of modeling is light and serves mainly to set the stage for algorithmic developments. We refer the reader to Dantzig [86] and Fourer, Gay, and Kernighan [112] for more comprehensive discussion of this issue. Our treatment of optimality conditions is thorough but not exhaustive; some concepts are discussed more extensively in Mangasarian [198] and Clarke [62]. As mentioned above, we are quite comprehensive in discussing optimization algorithms.

Topics Not Covered

We omit some important topics, such as network optimization, integer programming, stochastic programming, nonsmooth optimization, and global optimization. Network and integer optimization are described in some excellent texts: for instance, Ahuja, Magnanti, and Orlin [1] in the case of network optimization and Nemhauser and Wolsey [224], Papadimitriou and Steiglitz [235], and Wolsey [312] in the case of integer programming. Books on stochastic optimization are only now appearing; we mention those of Kall and Wallace [174], Birge and Louveaux [22]. Nonsmooth optimization comes in many flavors. The relatively simple structures that arise in robust data fitting (which is sometimes based on the ℓ_1 norm) are treated by Osborne [232] and Fletcher [101]. The latter book also discusses algorithms for nonsmooth penalty functions that arise in constrained optimization; we discuss these briefly, too, in Chapter 18. A more analytical treatment of nonsmooth optimization is given by Hiriart-Urruty and Lemaréchal [170]. We omit detailed treatment of some important topics that are the focus of intense current research, including interior-point methods for nonlinear programming and algorithms for complementarity problems.

Additional Resource

The material in the book is complemented by an online resource called the NEOS Guide, which can be found on the World-Wide Web at

```
http://www.mcs.anl.gov/otc/Guide/
```

The Guide contains information about most areas of optimization, and presents a number of case studies that describe applications of various optimization algorithms to real-world problems such as portfolio optimization and optimal dieting. Some of this material is interactive in nature and has been used extensively for class exercises.

For the most part, we have omitted detailed discussions of specific software packages, and refer the reader to Moré and Wright [217] or to the Software Guide section of the NEOS Guide, which can be found at

```
http://www.mcs.anl.gov/otc/Guide/SoftwareGuide/
```

Users of optimization software refer in great numbers to this web site, which is being constantly updated to reflect new packages and changes to existing software.

Acknowledgments

We are most grateful to the following colleagues for their input and feedback on various sections of this work: Chris Bischof, Richard Byrd, George Corliss, Bob Fourer, David Gay, Jean-Charles Gilbert, Phillip Gill, Jean-Pierre Goux, Don Goldfarb, Nick Gould, Andreas Griewank, Matthias Heinkenschloss, Marcelo Marazzi, Hans Mittelmann, Jorge Moré, Will Naylor, Michael Overton, Bob Plemmons, Hugo Scolnik, David Stewart, Philippe Toint, Luis Vicente, Andreas Wächter, and Ya-xiang Yuan. We thank Guanghui Liu, who provided help with many of the exercises, and Jill Lavelle who assisted us in preparing the figures. We also express our gratitude to our sponsors at the Department of Energy and the National Science Foundation, who have strongly supported our research efforts in optimization over the years.

One of us (JN) would like to express his deep gratitude to Richard Byrd, who has taught him so much about optimization and who has helped him in very many ways throughout the course of his career.

Final Remark

In the preface to his 1987 book [101], Roger Fletcher described the field of optimization as a "fascinating blend of theory and computation, heuristics and rigor." The ever-growing realm of applications and the explosion in computing power is driving optimization research in new and exciting directions, and the ingredients identified by Fletcher will continue to play important roles for many years to come.

Jorge Nocedal   Stephen J. Wright
Evanston, IL   *Argonne, IL*

Preface to the Second Edition

During the six years since the first edition of this book appeared, the field of continuous optimization has continued to grow and evolve. This new edition reflects a better understanding of constrained optimization at both the algorithmic and theoretical levels, and of the demands imposed by practical applications. Perhaps most notably, new chapters have been added on two important topics: derivative-free optimization (Chapter 9) and interior-point methods for nonlinear programming (Chapter 19). The former topic has proved to be of great interest in applications, while the latter topic has come into its own in recent years and now forms the basis of successful codes for nonlinear programming.

Apart from the new chapters, we have revised and updated throughout the book, de-emphasizing or omitting less important topics, enhancing the treatment of subjects of evident interest, and adding new material in many places. The first part (unconstrained optimization) has been comprehensively reorganized to improve clarity. Discussion of Newton's method—the touchstone method for unconstrained problems—is distributed more naturally throughout this part rather than being isolated in a single chapter. An expanded discussion of large-scale problems appears in Chapter 7.

Some reorganization has taken place also in the second part (constrained optimization), with material common to sequential quadratic programming and interior-point methods now appearing in the chapter on fundamentals of nonlinear programming algorithms (Chapter 15) and the discussion of primal barrier methods moved to the new interior-point chapter. There is much new material in this part, including a treatment of nonlinear programming duality, an expanded discussion of algorithms for inequality constrained quadratic programming, a discussion of dual simplex and presolving in linear programming, a summary of practical issues in the implementation of interior-point linear programming algorithms, a description of conjugate-gradient methods for quadratic programming, and a discussion of filter methods and nonsmooth penalty methods in nonlinear programming algorithms.

In many chapters we have added a Perspectives and Software section near the end, to place the preceding discussion in context and discuss the state of the art in software. The appendix has been rearranged with some additional topics added, so that it can be used in a more stand-alone fashion to cover some of the mathematical background required for the rest of the book. The exercises have been revised in most chapters. After these many additions, deletions, and changes, the second edition is only slightly longer than the first, reflecting our belief that careful selection of the material to include and exclude is an important responsibility for authors of books of this type.

A manual containing solutions for selected problems will be available to bona fide instructors through the publisher. A list of typos will be maintained on the book's web site, which is accessible from the web pages of both authors.

We acknowledge with gratitude the comments and suggestions of many readers of the first edition, who sent corrections to many errors and provided valuable perspectives on the material, which led often to substantial changes. We mention in particular Frank Curtis, Michael Ferris, Andreas Griewank, Jacek Gondzio, Sven Leyffer, Philip Loewen, Rembert Reemtsen, and David Stewart.

Our special thanks goes to Michael Overton, who taught from a draft of the second edition and sent many detailed and excellent suggestions. We also thank colleagues who read various chapters of the new edition carefully during development, including Richard Byrd, Nick Gould, Paul Hovland, Gabo Lopéz-Calva, Long Hei, Katya Scheinberg, Andreas Wächter, and Richard Waltz. We thank Jill Wright for improving some of the figures and for the new cover graphic.

We mentioned in the original preface several areas of optimization that are not covered in this book. During the past six years, this list has only grown longer, as the field has continued to expand in new directions. In this regard, the following areas are particularly noteworthy: optimization problems with complementarity constraints, second-order cone and semidefinite programming, simulation-based optimization, robust optimization, and mixed-integer nonlinear programming. All these areas have seen theoretical and algorithmic advances in recent years, and in many cases developments are being driven by new classes of applications. Although this book does not cover any of these areas directly, it provides a foundation from which they can be studied.

Jorge Nocedal   Stephen J. Wright
Evanston, IL   *Madison, WI*

CHAPTER *1*

Introduction

People optimize. Investors seek to create portfolios that avoid excessive risk while achieving a high rate of return. Manufacturers aim for maximum efficiency in the design and operation of their production processes. Engineers adjust parameters to optimize the performance of their designs.

Nature optimizes. Physical systems tend to a state of minimum energy. The molecules in an isolated chemical system react with each other until the total potential energy of their electrons is minimized. Rays of light follow paths that minimize their travel time.

Optimization is an important tool in decision science and in the analysis of physical systems. To make use of this tool, we must first identify some *objective*, a quantitative measure of the performance of the system under study. This objective could be profit, time, potential energy, or any quantity or combination of quantities that can be represented by a single number. The objective depends on certain characteristics of the system, called *variables* or *unknowns*. Our goal is to find values of the variables that optimize the objective. Often the variables are restricted, or *constrained*, in some way. For instance, quantities such as electron density in a molecule and the interest rate on a loan cannot be negative.

The process of identifying objective, variables, and constraints for a given problem is known as *modeling*. Construction of an appropriate model is the first step—sometimes the most important step—in the optimization process. If the model is too simplistic, it will not give useful insights into the practical problem. If it is too complex, it may be too difficult to solve.

Once the model has been formulated, an optimization algorithm can be used to find its solution, usually with the help of a computer. There is no universal optimization algorithm but rather a collection of algorithms, each of which is tailored to a particular type of optimization problem. The responsibility of choosing the algorithm that is appropriate for a specific application often falls on the user. This choice is an important one, as it may determine whether the problem is solved rapidly or slowly and, indeed, whether the solution is found at all.

After an optimization algorithm has been applied to the model, we must be able to recognize whether it has succeeded in its task of finding a solution. In many cases, there are elegant mathematical expressions known as *optimality conditions* for checking that the current set of variables is indeed the solution of the problem. If the optimality conditions are not satisfied, they may give useful information on how the current estimate of the solution can be improved. The model may be improved by applying techniques such as *sensitivity analysis*, which reveals the sensitivity of the solution to changes in the model and data. Interpretation of the solution in terms of the application may also suggest ways in which the model can be refined or improved (or corrected). If any changes are made to the model, the optimization problem is solved anew, and the process repeats.

MATHEMATICAL FORMULATION

Mathematically speaking, optimization is the minimization or maximization of a function subject to constraints on its variables. We use the following notation:

- x is the vector of *variables*, also called *unknowns* or *parameters*;

- f is the *objective function*, a (scalar) function of x that we want to maximize or minimize;

- c_i are *constraint* functions, which are scalar functions of x that define certain equations and inequalities that the unknown vector x must satisfy.

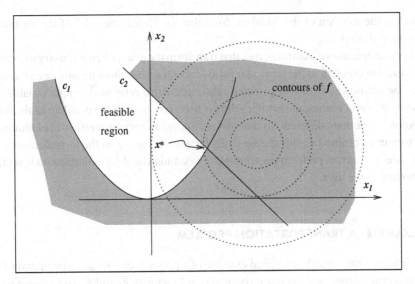

Figure 1.1 Geometrical representation of the problem (1.2).

Using this notation, the optimization problem can be written as follows:

$$\min_{x \in \mathbb{R}^n} f(x) \quad \text{subject to} \quad \begin{aligned} c_i(x) &= 0, \quad i \in \mathcal{E}, \\ c_i(x) &\geq 0, \quad i \in \mathcal{I}. \end{aligned} \tag{1.1}$$

Here $\mathcal{I}$ and $\mathcal{E}$ are sets of indices for equality and inequality constraints, respectively.

As a simple example, consider the problem

$$\min (x_1 - 2)^2 + (x_2 - 1)^2 \quad \text{subject to} \quad \begin{aligned} x_1^2 - x_2 &\leq 0, \\ x_1 + x_2 &\leq 2. \end{aligned} \tag{1.2}$$

We can write this problem in the form (1.1) by defining

$$f(x) = (x_1 - 2)^2 + (x_2 - 1)^2, \qquad x = \begin{bmatrix} x_1 \\ x_2 \end{bmatrix},$$

$$c(x) = \begin{bmatrix} c_1(x) \\ c_2(x) \end{bmatrix} = \begin{bmatrix} -x_1^2 + x_2 \\ -x_1 - x_2 + 2 \end{bmatrix}, \qquad \mathcal{I} = \{1, 2\}, \quad \mathcal{E} = \emptyset.$$

Figure 1.1 shows the contours of the objective function, that is, the set of points for which $f(x)$ has a constant value. It also illustrates the *feasible region*, which is the set of points satisfying all the constraints (the area between the two constraint boundaries), and the point

x^*, which is the solution of the problem. Note that the "infeasible side" of the inequality constraints is shaded.

The example above illustrates, too, that transformations are often necessary to express an optimization problem in the particular form (1.1). Often it is more natural or convenient to label the unknowns with two or three subscripts, or to refer to different variables by completely different names, so that relabeling is necessary to pose the problem in the form (1.1). Another common difference is that we are required to *maximize* rather than minimize f, but we can accommodate this change easily by *minimizing* $-f$ in the formulation (1.1). Good modeling systems perform the conversion to standardized formulations such as (1.1) transparently to the user.

EXAMPLE: A TRANSPORTATION PROBLEM

We begin with a much simplified example of a problem that might arise in manufacturing and transportation. A chemical company has 2 factories F_1 and F_2 and a dozen retail outlets $R_1, R_2, \ldots, R_{12}$. Each factory F_i can produce a_i tons of a certain chemical product each week; a_i is called the *capacity* of the plant. Each retail outlet R_j has a known weekly *demand* of b_j tons of the product. The cost of shipping one ton of the product from factory F_i to retail outlet R_j is c_{ij}.

The problem is to determine how much of the product to ship from each factory to each outlet so as to satisfy all the requirements and minimize cost. The variables of the problem are x_{ij}, $i = 1, 2$, $j = 1, \ldots, 12$, where x_{ij} is the number of tons of the product shipped from factory F_i to retail outlet R_j; see Figure 1.2. We can write the problem as

$$\min \sum_{ij} c_{ij} x_{ij} \tag{1.3a}$$

$$\text{subject to} \sum_{j=1}^{12} x_{ij} \le a_i, \quad i = 1, 2, \tag{1.3b}$$

$$\sum_{i=1}^{2} x_{ij} \ge b_j, \quad j = 1, \ldots, 12, \tag{1.3c}$$

$$x_{ij} \ge 0, \quad i = 1, 2, \quad j = 1, \ldots, 12. \tag{1.3d}$$

This type of problem is known as a *linear programming* problem, since the objective function and the constraints are all linear functions. In a more practical model, we would also include costs associated with manufacturing and storing the product. There may be volume discounts in practice for shipping the product; for example the cost (1.3a) could be represented by $\sum_{ij} c_{ij} \sqrt{\delta + x_{ij}}$, where $\delta > 0$ is a small subscription fee. In this case, the problem is a *nonlinear program* because the objective function is nonlinear.

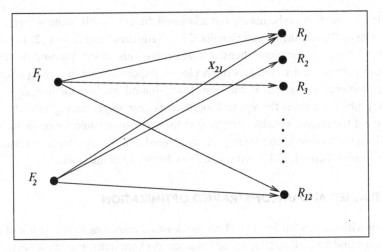

Figure 1.2 A transportation problem.

CONTINUOUS VERSUS DISCRETE OPTIMIZATION

In some optimization problems the variables make sense only if they take on integer values. For example, a variable x_i could represent the number of power plants of type i that should be constructed by an electicity provider during the next 5 years, or it could indicate whether or not a particular factory should be located in a particular city. The mathematical formulation of such problems includes integrality constraints, which have the form $x_i \in \mathbf{Z}$, where $\mathbf{Z}$ is the set of integers, or binary constraints, which have the form $x_i \in \{0, 1\}$, in addition to algebraic constraints like those appearing in (1.1). Problems of this type are called *integer programming* problems. If some of the variables in the problem are *not* restricted to be integer or binary variables, they are sometimes called *mixed integer programming* problems, or MIPs for short.

Integer programming problems are a type of *discrete optimization* problem. Generally, discrete optimization problems may contain not only integers and binary variables, but also more abstract variable objects such as permutations of an ordered set. The defining feature of a discrete optimization problem is that the unknown x is drawn from a a finite (but often very large) set. By contrast, the feasible set for *continuous optimization* problems—the class of problems studied in this book—is usually uncountably infinite, as when the components of x are allowed to be real numbers. Continuous optimization problems are normally easier to solve because the smoothness of the functions makes it possible to use objective and constraint information at a particular point x to deduce information about the function's behavior at all points close to x. In discrete problems, by constrast, the behavior of the objective and constraints may change significantly as we move from one feasible point to another, even if the two points are "close" by some measure. The feasible sets for discrete optimization problems can be thought of as exhibiting an extreme form of nonconvexity, as a convex combination of two feasible points is in general not feasible.

Discrete optimization problems are not addressed directly in this book; we refer the reader to the texts by Papadimitriou and Steiglitz [235], Nemhauser and Wolsey [224], Cook et al. [77], and Wolsey [312] for comprehensive treatments of this subject. We note, however, that continuous optimization techniques often play an important role in solving discrete optimization problems. For instance, the branch-and-bound method for integer linear programming problems requires the repeated solution of linear programming "relaxations," in which some of the integer variables are fixed at integer values, while for other integer variables the integrality constraints are temporarily ignored. These subproblems are usually solved by the simplex method, which is discussed in Chapter 13 of this book.

CONSTRAINED AND UNCONSTRAINED OPTIMIZATION

Problems with the general form (1.1) can be classified according to the nature of the objective function and constraints (linear, nonlinear, convex), the number of variables (large or small), the smoothness of the functions (differentiable or nondifferentiable), and so on. An important distinction is between problems that have constraints on the variables and those that do not. This book is divided into two parts according to this classification.

Unconstrained optimization problems, for which we have $\mathcal{E} = \mathcal{I} = \emptyset$ in (1.1), arise directly in many practical applications. Even for some problems with natural constraints on the variables, it may be safe to disregard them as they do not affect on the solution and do not interfere with algorithms. Unconstrained problems arise also as reformulations of constrained optimization problems, in which the constraints are replaced by penalization terms added to objective function that have the effect of discouraging constraint violations.

Constrained optimization problems arise from models in which constraints play an essential role, for example in imposing budgetary constraints in an economic problem or shape constraints in a design problem. These constraints may be simple bounds such as $0 \leq x_1 \leq 100$, more general linear constraints such as $\sum_i x_i \leq 1$, or nonlinear inequalities that represent complex relationships among the variables.

When the objective function and all the constraints are linear functions of x, the problem is a *linear programming* problem. Problems of this type are probably the most widely formulated and solved of all optimization problems, particularly in management, financial, and economic applications. *Nonlinear programming* problems, in which at least some of the constraints or the objective are nonlinear functions, tend to arise naturally in the physical sciences and engineering, and are becoming more widely used in management and economic sciences as well.

GLOBAL AND LOCAL OPTIMIZATION

Many algorithms for nonlinear optimization problems seek only a local solution, a point at which the objective function is smaller than at all other feasible nearby points. They do not always find the *global solution*, which is the point with lowest function value among *all* feasible points. Global solutions are needed in some applications, but for many problems they are difficult to recognize and even more difficult to locate. For *convex programming* problems, and more particularly for linear programs, local solutions are also global solutions. General nonlinear problems, both constrained and unconstrained, may possess local solutions that are not global solutions.

In this book we treat global optimization only in passing and focus instead on the computation and characterization of local solutions. We note, however, that many successful global optimization algorithms require the solution of many local optimization problems, to which the algorithms described in this book can be applied.

Research papers on global optimization can be found in Floudas and Pardalos [109] and in the *Journal of Global Optimization*.

STOCHASTIC AND DETERMINISTIC OPTIMIZATION

In some optimization problems, the model cannot be fully specified because it depends on quantities that are unknown at the time of formulation. This characteristic is shared by many economic and financial planning models, which may depend for example on future interest rates, future demands for a product, or future commodity prices, but uncertainty can arise naturally in almost any type of application.

Rather than just use a "best guess" for the uncertain quantities, modelers may obtain more useful solutions by incorporating additional knowledge about these quantities into the model. For example, they may know a number of possible scenarios for the uncertain demand, along with estimates of the probabilities of each scenario. *Stochastic optimization* algorithms use these quantifications of the uncertainty to produce solutions that optimize the *expected* performance of the model.

Related paradigms for dealing with uncertain data in the model include *chance-constrained optimization*, in which we ensure that the variables x satisfy the given constraints to some specified probability, and *robust optimization*, in which certain constraints are required to hold for all possible values of the uncertain data.

We do not consider stochastic optimization problems further in this book, focusing instead on *deterministic optimization* problems, in which the model is completely known. Many algorithms for stochastic optimization do, however, proceed by formulating one or more deterministic subproblems, each of which can be solved by the techniques outlined here.

Stochastic and robust optimization have seen a great deal of recent research activity. For further information on stochastic optimization, consult the books of Birge and Louveaux [22] and Kall and Wallace [174]. Robust optimization is discussed in Ben-Tal and Nemirovski [15].

CONVEXITY

The concept of convexity is fundamental in optimization. Many practical problems possess this property, which generally makes them easier to solve both in theory and practice.

The term "convex" can be applied both to sets and to functions. A set $S \in \mathbb{R}^n$ is a *convex set* if the straight line segment connecting any two points in S lies entirely inside S. Formally, for any two points $x \in S$ and $y \in S$, we have $\alpha x + (1 - \alpha)y \in S$ for all $\alpha \in [0, 1]$. The function f is a *convex function* if its domain S is a convex set and if for any two points x and y in S, the following property is satisfied:

$$f(\alpha x + (1 - \alpha)y) \leq \alpha f(x) + (1 - \alpha)f(y), \quad \text{for all } \alpha \in [0, 1]. \tag{1.4}$$

Simple instances of convex sets include the unit ball $\{y \in \mathbb{R}^n \mid \|y\|_2 \leq 1\}$; and any polyhedron, which is a set defined by linear equalities and inequalities, that is,

$$\{x \in \mathbb{R}^n \mid Ax = b, \ Cx \leq d\},$$

where A and C are matrices of appropriate dimension, and b and d are vectors. Simple instances of convex functions include the linear function $f(x) = c^T x + \alpha$, for any constant vector $c \in \mathbb{R}^n$ and scalar α; and the convex quadratic function $f(x) = x^T H x$, where H is a symmetric positive semidefinite matrix.

We say that f is *strictly convex* if the inequality in (1.4) is strict whenever $x \neq y$ and α is in the open interval $(0, 1)$. A function f is said to be *concave* if $-f$ is convex.

If the objective function in the optimization problem (1.1) and the feasible region are both convex, then any local solution of the problem is in fact a global solution.

The term *convex programming* is used to describe a special case of the general constrained optimization problem (1.1) in which

- the objective function is convex,

- the equality constraint functions $c_i(\cdot)$, $i \in \mathcal{E}$, are linear, and

- the inequality constraint functions $c_i(\cdot)$, $i \in \mathcal{I}$, are concave.

OPTIMIZATION ALGORITHMS

Optimization algorithms are iterative. They begin with an initial guess of the variable x and generate a sequence of improved estimates (called "iterates") until they terminate, hopefully at a solution. The strategy used to move from one iterate to the next distinguishes one algorithm from another. Most strategies make use of the values of the objective function f, the constraint functions c_i, and possibly the first and second derivatives of these functions. Some algorithms accumulate information gathered at previous iterations, while others use only local information obtained at the current point. Regardless of these specifics (which will receive plenty of attention in the rest of the book), good algorithms should possess the following properties:

- Robustness. They should perform well on a wide variety of problems in their class, for all reasonable values of the starting point.

- Efficiency. They should not require excessive computer time or storage.

- Accuracy. They should be able to identify a solution with precision, without being overly sensitive to errors in the data or to the arithmetic rounding errors that occur when the algorithm is implemented on a computer.

These goals may conflict. For example, a rapidly convergent method for a large unconstrained nonlinear problem may require too much computer storage. On the other hand, a robust method may also be the slowest. Tradeoffs between convergence rate and storage requirements, and between robustness and speed, and so on, are central issues in numerical optimization. They receive careful consideration in this book.

The mathematical theory of optimization is used both to characterize optimal points and to provide the basis for most algorithms. It is not possible to have a good understanding of numerical optimization without a firm grasp of the supporting theory. Accordingly, this book gives a solid (though not comprehensive) treatment of optimality conditions, as well as convergence analysis that reveals the strengths and weaknesses of some of the most important algorithms.

NOTES AND REFERENCES

Optimization traces its roots to the calculus of variations and the work of Euler and Lagrange. The development of linear programming n the 1940s broadened the field and stimulated much of the progress in modern optimization theory and practice during the past 60 years.

Optimization is often called *mathematical programming*, a somewhat confusing term coined in the 1940s, before the word "programming" became inextricably linked with computer software. The original meaning of this word (and the intended one in this context) was more inclusive, with connotations of algorithm design and analysis.

Modeling will not be treated extensively in the book. It is an essential subject in its own right, as it makes the connection between optimization algorithms and software on the one hand, and applications on the other hand. Information about modeling techniques for various application areas can be found in Dantzig [86], Ahuja, Magnanti, and Orlin [1], Fourer, Gay, and Kernighan [112], Winston [308], and Rardin [262].

CHAPTER 2

Fundamentals of Unconstrained Optimization

In unconstrained optimization, we minimize an objective function that depends on real variables, with no restrictions at all on the values of these variables. The mathematical formulation is

$$\min_{x} \; f(x), \tag{2.1}$$

where $x \in \mathbb{R}^n$ is a real vector with $n \geq 1$ components and $f : \mathbb{R}^n \to \mathbb{R}$ is a smooth function.

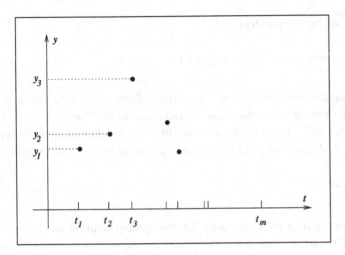

Figure 2.1 Least squares data fitting problem.

Usually, we lack a global perspective on the function f. All we know are the values of f and maybe some of its derivatives at a set of points $x_0, x_1, x_2, \ldots$. Fortunately, our algorithms get to choose these points, and they try to do so in a way that identifies a solution reliably and without using too much computer time or storage. Often, the information about f does not come cheaply, so we usually prefer algorithms that do not call for this information unnecessarily.

❑ EXAMPLE 2.1

Suppose that we are trying to find a curve that fits some experimental data. Figure 2.1 plots measurements $y_1, y_2, \ldots, y_m$ of a signal taken at times $t_1, t_2, \ldots, t_m$. From the data and our knowledge of the application, we deduce that the signal has exponential and oscillatory behavior of certain types, and we choose to model it by the function

$$\phi(t; x) = x_1 + x_2 e^{-(x_3-t)^2/x_4} + x_5 \cos(x_6 t).$$

The real numbers x_i, $i = 1, 2, \ldots, 6$, are the parameters of the model; we would like to choose them to make the model values $\phi(t_j; x)$ fit the observed data y_j as closely as possible. To state our objective as an optimization problem, we group the parameters x_i into a vector of unknowns $x = (x_1, x_2, \ldots, x_6)^T$, and define the residuals

$$r_j(x) = y_j - \phi(t_j; x), \qquad j = 1, 2, \ldots, m, \tag{2.2}$$

which measure the discrepancy between the model and the observed data. Our estimate of

x will be obtained by solving the problem

$$\min_{x \in \mathbb{R}^6} f(x) = r_1^2(x) + r_2^2(x) + \cdots + r_m^2(x). \tag{2.3}$$

This is a *nonlinear least-squares problem*, a special case of unconstrained optimization. It illustrates that some objective functions can be expensive to evaluate even when the number of variables is small. Here we have $n = 6$, but if the number of measurements m is large (10^5, say), evaluation of $f(x)$ for a given parameter vector x is a significant computation.

∎

Suppose that for the data given in Figure 2.1 the optimal solution of (2.3) is approximately $x^* = (1.1, 0.01, 1.2, 1.5, 2.0, 1.5)$ and the corresponding function value is $f(x^*) = 0.34$. Because the optimal objective is nonzero, there must be discrepancies between the observed measurements y_j and the model predictions $\phi(t_j, x^*)$ for some (usually most) values of j—the model has not reproduced all the data points exactly. How, then, can we verify that x^* is indeed a minimizer of f? To answer this question, we need to define the term "solution" and explain how to recognize solutions. Only then can we discuss algorithms for unconstrained optimization problems.

2.1   WHAT IS A SOLUTION?

Generally, we would be happiest if we found a *global minimizer* of f, a point where the function attains its least value. A formal definition is

A point x^* is a *global minimizer* if $f(x^*) \leq f(x)$ for all x,

where x ranges over all of $\mathbb{R}^n$ (or at least over the domain of interest to the modeler). The global minimizer can be difficult to find, because our knowledge of f is usually only local. Since our algorithm does not visit many points (we hope!), we usually do not have a good picture of the overall shape of f, and we can never be sure that the function does not take a sharp dip in some region that has not been sampled by the algorithm. Most algorithms are able to find only a *local* minimizer, which is a point that achieves the smallest value of f in its neighborhood. Formally, we say:

A point x^* is a *local minimizer* if there is a neighborhood $\mathcal{N}$ of x^* such that $f(x^*) \leq f(x)$ for all $x \in \mathcal{N}$.

(Recall that a neighborhood of x^* is simply an open set that contains x^*.) A point that satisfies this definition is sometimes called a *weak local minimizer*. This terminology distinguishes it from a strict local minimizer, which is the outright winner in its neighborhood. Formally,

> A point x^* is a *strict local minimizer* (also called a *strong local minimizer*) if there is a neighborhood $\mathcal{N}$ of x^* such that $f(x^*) < f(x)$ for all $x \in \mathcal{N}$ with $x \neq x^*$.

For the constant function $f(x) = 2$, every point x is a weak local minimizer, while the function $f(x) = (x - 2)^4$ has a strict local minimizer at $x = 2$.

A slightly more exotic type of local minimizer is defined as follows.

> A point x^* is an *isolated local minimizer* if there is a neighborhood $\mathcal{N}$ of x^* such that x^* is the only local minimizer in $\mathcal{N}$.

Some strict local minimizers are not isolated, as illustrated by the function

$$f(x) = x^4 \cos(1/x) + 2x^4, \qquad f(0) = 0,$$

which is twice continuously differentiable and has a strict local minimizer at $x^* = 0$. However, there are strict local minimizers at many nearby points x_j, and we can label these points so that $x_j \to 0$ as $j \to \infty$.

While strict local minimizers are not always isolated, it is true that all isolated local minimizers are strict.

Figure 2.2 illustrates a function with many local minimizers. It is usually difficult to find the global minimizer for such functions, because algorithms tend to be "trapped" at local minimizers. This example is by no means pathological. In optimization problems associated with the determination of molecular conformation, the potential function to be minimized may have millions of local minima.

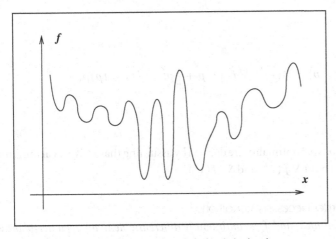

Figure 2.2 A difficult case for global minimization.

Sometimes we have additional "global" knowledge about f that may help in identifying global minima. An important special case is that of convex functions, for which every local minimizer is also a global minimizer.

RECOGNIZING A LOCAL MINIMUM

From the definitions given above, it might seem that the only way to find out whether a point x^* is a local minimum is to examine all the points in its immediate vicinity, to make sure that none of them has a smaller function value. When the function f is *smooth*, however, there are more efficient and practical ways to identify local minima. In particular, if f is twice continuously differentiable, we may be able to tell that x^* is a local minimizer (and possibly a strict local minimizer) by examining just the gradient $\nabla f(x^*)$ and the Hessian $\nabla^2 f(x^*)$.

The mathematical tool used to study minimizers of smooth functions is Taylor's theorem. Because this theorem is central to our analysis throughout the book, we state it now. Its proof can be found in any calculus textbook.

Theorem 2.1 (Taylor's Theorem).

Suppose that $f : \mathbb{R}^n \rightarrow \mathbb{R}$ is continuously differentiable and that $p \in \mathbb{R}^n$. Then we have that

$$f(x + p) = f(x) + \nabla f(x + tp)^T p, \tag{2.4}$$

for some $t \in (0, 1)$. Moreover, if f is twice continuously differentiable, we have that

$$\nabla f(x + p) = \nabla f(x) + \int_0^1 \nabla^2 f(x + tp)p \, dt, \tag{2.5}$$

and that

$$f(x + p) = f(x) + \nabla f(x)^T p + \tfrac{1}{2} p^T \nabla^2 f(x + tp)p, \tag{2.6}$$

for some $t \in (0, 1)$.

Necessary conditions for optimality are derived by assuming that x^* is a local minimizer and then proving facts about $\nabla f(x^*)$ and $\nabla^2 f(x^*)$.

Theorem 2.2 (First-Order Necessary Conditions).

If x^* is a local minimizer and f is continuously differentiable in an open neighborhood of x^*, then $\nabla f(x^*) = 0$.

PROOF. Suppose for contradiction that $\nabla f(x^*) \neq 0$. Define the vector $p = -\nabla f(x^*)$ and note that $p^T \nabla f(x^*) = -\|\nabla f(x^*)\|^2 < 0$. Because ∇f is continuous near x^*, there is a scalar $T > 0$ such that

$$p^T \nabla f(x^* + tp) < 0, \qquad \text{for all } t \in [0, T].$$

For any $\bar{t} \in (0, T]$, we have by Taylor's theorem that

$$f(x^* + \bar{t}p) = f(x^*) + \bar{t}p^T \nabla f(x^* + tp), \qquad \text{for some } t \in (0, \bar{t}).$$

Therefore, $f(x^* + \bar{t}p) < f(x^*)$ for all $\bar{t} \in (0, T]$. We have found a direction leading away from x^* along which f decreases, so x^* is not a local minimizer, and we have a contradiction. □

We call x^* a *stationary point* if $\nabla f(x^*) = 0$. According to Theorem 2.2, any local minimizer must be a stationary point.

For the next result we recall that a matrix B is positive definite if $p^T B p > 0$ for all $p \neq 0$, and positive semidefinite if $p^T B p \geq 0$ for all p (see the Appendix).

Theorem 2.3 (Second-Order Necessary Conditions).

If x^* is a local minimizer of f and $\nabla^2 f$ exists and is continuous in an open neighborhood of x^*, then $\nabla f(x^*) = 0$ and $\nabla^2 f(x^*)$ is positive semidefinite.

PROOF. We know from Theorem 2.2 that $\nabla f(x^*) = 0$. For contradiction, assume that $\nabla^2 f(x^*)$ is not positive semidefinite. Then we can choose a vector p such that $p^T \nabla^2 f(x^*) p < 0$, and because $\nabla^2 f$ is continuous near x^*, there is a scalar $T > 0$ such that $p^T \nabla^2 f(x^* + tp) p < 0$ for all $t \in [0, T]$.

By doing a Taylor series expansion around x^*, we have for all $\bar{t} \in (0, T]$ and some $t \in (0, \bar{t})$ that

$$f(x^* + \bar{t}p) = f(x^*) + \bar{t}p^T \nabla f(x^*) + \tfrac{1}{2}\bar{t}^2 p^T \nabla^2 f(x^* + tp) p < f(x^*).$$

As in Theorem 2.2, we have found a direction from x^* along which f is decreasing, and so again, x^* is not a local minimizer. □

We now describe *sufficient conditions*, which are conditions on the derivatives of f at the point z^* that guarantee that x^* is a local minimizer.

Theorem 2.4 (Second-Order Sufficient Conditions).

Suppose that $\nabla^2 f$ is continuous in an open neighborhood of x^* and that $\nabla f(x^*) = 0$ and $\nabla^2 f(x^*)$ is positive definite. Then x^* is a strict local minimizer of f.

PROOF.   Because the Hessian is continuous and positive definite at x^*, we can choose a radius $r > 0$ so that $\nabla^2 f(x)$ remains positive definite for all x in the open ball $\mathcal{D} = \{z \mid \|z - x^*\| < r\}$. Taking any nonzero vector p with $\|p\| < r$, we have $x^* + p \in \mathcal{D}$ and so

$$f(x^* + p) = f(x^*) + p^T \nabla f(x^*) + \tfrac{1}{2} p^T \nabla^2 f(z) p$$
$$= f(x^*) + \tfrac{1}{2} p^T \nabla^2 f(z) p,$$

where $z = x^* + tp$ for some $t \in (0, 1)$. Since $z \in \mathcal{D}$, we have $p^T \nabla^2 f(z) p > 0$, and therefore $f(x^* + p) > f(x^*)$, giving the result.   □

Note that the second-order sufficient conditions of Theorem 2.4 guarantee something stronger than the necessary conditions discussed earlier; namely, that the minimizer is a *strict* local minimizer. Note too that the second-order sufficient conditions are not necessary: A point x^* may be a strict local minimizer, and yet may fail to satisfy the sufficient conditions. A simple example is given by the function $f(x) = x^4$, for which the point $x^* = 0$ is a strict local minimizer at which the Hessian matrix vanishes (and is therefore not positive definite).

When the objective function is convex, local and global minimizers are simple to characterize.

Theorem 2.5.

When f is convex, any local minimizer x^* is a global minimizer of f. If in addition f is differentiable, then any stationary point x^* is a global minimizer of f.

PROOF.   Suppose that x^* is a local but not a global minimizer. Then we can find a point $z \in \mathbb{R}^n$ with $f(z) < f(x^*)$. Consider the line segment that joins x^* to z, that is,

$$x = \lambda z + (1 - \lambda) x^*, \qquad \text{for some } \lambda \in (0, 1]. \tag{2.7}$$

By the convexity property for f, we have

$$f(x) \le \lambda f(z) + (1 - \lambda) f(x^*) < f(x^*). \tag{2.8}$$

Any neighborhood $\mathcal{N}$ of x^* contains a piece of the line segment (2.7), so there will always be points $x \in \mathcal{N}$ at which (2.8) is satisfied. Hence, x^* is not a local minimizer.

For the second part of the theorem, suppose that x^* is not a global minimizer and choose z as above. Then, from convexity, we have

$$
\nabla f(x^*)^T (z - x^*) = \frac{d}{d\lambda} f(x^* + \lambda(z - x^*)) \,|_{\lambda=0} \quad \text{(see the Appendix)}
$$
$$
= \lim_{\lambda \downarrow 0} \frac{f(x^* + \lambda(z - x^*)) - f(x^*)}{\lambda}
$$
$$
\leq \lim_{\lambda \downarrow 0} \frac{\lambda f(z) + (1 - \lambda) f(x^*) - f(x^*)}{\lambda}
$$
$$
= f(z) - f(x^*) < 0.
$$

Therefore, $\nabla f(x^*) \neq 0$, and so x^* is not a stationary point. $\qquad\square$

These results, which are based on elementary calculus, provide the foundations for unconstrained optimization algorithms. In one way or another, all algorithms seek a point where $\nabla f(\cdot)$ vanishes.

NONSMOOTH PROBLEMS

This book focuses on smooth functions, by which we generally mean functions whose second derivatives exist and are continuous. We note, however, that there are interesting problems in which the functions involved may be nonsmooth and even discontinuous. It is not possible in general to identify a minimizer of a general discontinuous function. If, however, the function consists of a few smooth pieces, with discontinuities between the pieces, it may be possible to find the minimizer by minimizing each smooth piece individually.

If the function is continuous everywhere but nondifferentiable at certain points, as in Figure 2.3, we can identify a solution by examing the *subgradient* or *generalized*

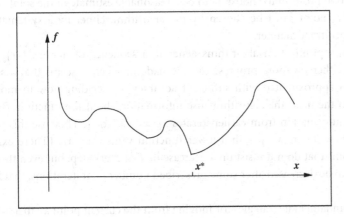

Figure 2.3 Nonsmooth function with minimum at a kink.

gradient, which are generalizations of the concept of gradient to the nonsmooth case. Nonsmooth optimization is beyond the scope of this book; we refer instead to Hiriart-Urruty and Lemaréchal [170] for an extensive discussion of theory. Here, we mention only that the minimization of a function such as the one illustrated in Figure 2.3 (which contains a jump discontinuity in the first derivative $f'(x)$ at the minimum) is difficult because the behavior of f is not predictable near the point of nonsmoothness. That is, we cannot be sure that information about f obtained at one point can be used to infer anything about f at neighboring points, because points of nondifferentiability may intervene. However, minimization of certain special nondifferentiable functions, such as

$$f(x) = \|r(x)\|_1, \qquad f(x) = \|r(x)\|_\infty \qquad (2.9)$$

(where $r(x)$ is a vector function), can be reformulated as smooth constrained optimization problems; see Exercise 12.5 in Chapter 12 and (17.31). The functions (2.9) are useful in data fitting, where $r(x)$ is the residual vector whose components are defined in (2.2).

2.2 OVERVIEW OF ALGORITHMS

The last forty years have seen the development of a powerful collection of algorithms for unconstrained optimization of smooth functions. We now give a broad description of their main properties, and we describe them in more detail in Chapters 3, 4, 5, 6, and 7. All algorithms for unconstrained minimization require the user to supply a starting point, which we usually denote by x_0. The user with knowledge about the application and the data set may be in a good position to choose x_0 to be a reasonable estimate of the solution. Otherwise, the starting point must be chosen by the algorithm, either by a systematic approach or in some arbitrary manner.

Beginning at x_0, optimization algorithms generate a sequence of iterates $\{x_k\}_{k=0}^\infty$ that terminate when either no more progress can be made or when it seems that a solution point has been approximated with sufficient accuracy. In deciding how to move from one iterate x_k to the next, the algorithms use information about the function f at x_k, and possibly also information from earlier iterates $x_0, x_1, \ldots, x_{k-1}$. They use this information to find a new iterate x_{k+1} with a lower function value than x_k. (There exist *nonmonotone* algorithms that do not insist on a decrease in f at every step, but even these algorithms require f to be decreased after some prescribed number m of iterations, that is, $f(x_k) < f(x_{k-m})$.)

There are two fundamental strategies for moving from the current point x_k to a new iterate x_{k+1}. Most of the algorithms described in this book follow one of these approaches.

TWO STRATEGIES: LINE SEARCH AND TRUST REGION

In the *line search* strategy, the algorithm chooses a direction p_k and searches along this direction from the current iterate x_k for a new iterate with a lower function value. The distance to move along p_k can be found by approximately solving the following one-dimensional minimization problem to find a step length α:

$$\min_{\alpha > 0} \ f(x_k + \alpha p_k). \tag{2.10}$$

By solving (2.10) exactly, we would derive the maximum benefit from the direction p_k, but an exact minimization may be expensive and is usually unnecessary. Instead, the line search algorithm generates a limited number of trial step lengths until it finds one that loosely approximates the minimum of (2.10). At the new point, a new search direction and step length are computed, and the process is repeated.

In the second algorithmic strategy, known as *trust region*, the information gathered about f is used to construct a *model function* m_k whose behavior near the current point x_k is similar to that of the actual objective function f. Because the model m_k may not be a good approximation of f when x is far from x_k, we restrict the search for a minimizer of m_k to some region around x_k. In other words, we find the candidate step p by approximately solving the following subproblem:

$$\min_{p} m_k(x_k + p), \qquad \text{where } x_k + p \text{ lies inside the trust region.} \tag{2.11}$$

If the candidate solution does not produce a sufficient decrease in f, we conclude that the trust region is too large, and we shrink it and re-solve (2.11). Usually, the trust region is a ball defined by $\|p\|_2 \leq \Delta$, where the scalar $\Delta > 0$ is called the trust-region radius. Elliptical and box-shaped trust regions may also be used.

The model m_k in (2.11) is usually defined to be a quadratic function of the form

$$m_k(x_k + p) = f_k + p^T \nabla f_k + \tfrac{1}{2} p^T B_k p, \tag{2.12}$$

where f_k, ∇f_k, and B_k are a scalar, vector, and matrix, respectively. As the notation indicates, f_k and ∇f_k are chosen to be the function and gradient values at the point x_k, so that m_k and f are in agreement to first order at the current iterate x_k. The matrix B_k is either the Hessian $\nabla^2 f_k$ or some approximation to it.

Suppose that the objective function is given by $f(x) = 10(x_2 - x_1^2)^2 + (1 - x_1)^2$. At the point $x_k = (0, 1)$ its gradient and Hessian are

$$\nabla f_k = \begin{bmatrix} -2 \\ 20 \end{bmatrix}, \qquad \nabla^2 f_k = \begin{bmatrix} -38 & 0 \\ 0 & 20 \end{bmatrix}.$$

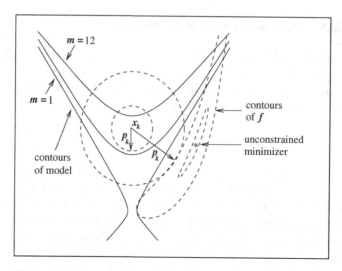

Figure 2.4 Two possible trust regions (circles) and their corresponding steps p_k. The solid lines are contours of the model function m_k.

The contour lines of the quadratic model (2.12) with $B_k = \nabla^2 f_k$ are depicted in Figure 2.4, which also illustrates the contours of the objective function f and the trust region. We have indicated contour lines where the model m_k has values 1 and 12. Note from Figure 2.4 that each time we decrease the size of the trust region after failure of a candidate iterate, the step from x_k to the new candidate will be shorter, and it usually points in a different direction from the previous candidate. The trust-region strategy differs in this respect from line search, which stays with a single search direction.

In a sense, the line search and trust-region approaches differ in the order in which they choose the *direction* and *distance* of the move to the next iterate. Line search starts by fixing the direction p_k and then identifying an appropriate distance, namely the step length α_k. In trust region, we first choose a maximum distance—the trust-region radius Δ_k—and then seek a direction and step that attain the best improvement possible subject to this distance constraint. If this step proves to be unsatisfactory, we reduce the distance measure Δ_k and try again.

The line search approach is discussed in more detail in Chapter 3. Chapter 4 discusses the trust-region strategy, including techniques for choosing and adjusting the size of the region and for computing approximate solutions to the trust-region problems (2.11). We now preview two major issues: choice of the search direction p_k in line search methods, and choice of the Hessian B_k in trust-region methods. These issues are closely related, as we now observe.

SEARCH DIRECTIONS FOR LINE SEARCH METHODS

The steepest descent direction $-\nabla f_k$ is the most obvious choice for search direction for a line search method. It is intuitive; among all the directions we could move from x_k, it is the one along which f decreases most rapidly. To verify this claim, we appeal again to Taylor's theorem (Theorem 2.1), which tells us that for any search direction p and step-length parameter α, we have

$$f(x_k + \alpha p) = f(x_k) + \alpha p^T \nabla f_k + \tfrac{1}{2}\alpha^2 p^T \nabla^2 f(x_k + tp)p, \quad \text{for some } t \in (0, \alpha)$$

(see (2.6)). The rate of change in f along the direction p at x_k is simply the coefficient of α, namely, $p^T \nabla f_k$. Hence, the unit direction p of most rapid decrease is the solution to the problem

$$\min_p \; p^T \nabla f_k, \qquad \text{subject to } \|p\| = 1. \tag{2.13}$$

Since $p^T \nabla f_k = \|p\| \, \|\nabla f_k\| \cos\theta = \|\nabla f_k\| \cos\theta$, where θ is the angle between p and ∇f_k, it is easy to see that the minimizer is attained when $\cos\theta = -1$ and

$$p = -\nabla f_k / \|\nabla f_k\|,$$

as claimed. As we illustrate in Figure 2.5, this direction is orthogonal to the contours of the function.

The *steepest descent method* is a line search method that moves along $p_k = -\nabla f_k$ at every step. It can choose the step length α_k in a variety of ways, as we discuss in Chapter 3. One advantage of the steepest descent direction is that it requires calculation of the gradient ∇f_k but not of second derivatives. However, it can be excruciatingly slow on difficult problems.

Line search methods may use search directions other than the steepest descent direction. In general, any *descent* direction—one that makes an angle of strictly less than $\pi/2$ radians with $-\nabla f_k$—is guaranteed to produce a decrease in f, provided that the step length

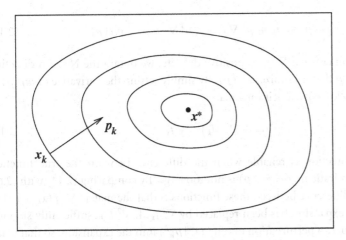

Figure 2.5 Steepest descent direction for a function of two variables.

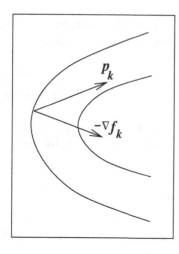

Figure 2.6
A downhill direction p_k.

is sufficiently small (see Figure 2.6). We can verify this claim by using Taylor's theorem. From (2.6), we have that

$$f(x_k + \epsilon p_k) = f(x_k) + \epsilon p_k^T \nabla f_k + O(\epsilon^2).$$

When p_k is a downhill direction, the angle θ_k between p_k and ∇f_k has $\cos \theta_k < 0$, so that

$$p_k^T \nabla f_k = \|p_k\| \, \|\nabla f_k\| \cos \theta_k < 0.$$

It follows that $f(x_k + \epsilon p_k) < f(x_k)$ for all positive but sufficiently small values of ϵ.

Another important search direction—perhaps the most important one of all—is the *Newton direction*. This direction is derived from the second-order Taylor series approximation to $f(x_k + p)$, which is

$$f(x_k + p) \approx f_k + p^T \nabla f_k + \tfrac{1}{2} p^T \nabla^2 f_k p \stackrel{\text{def}}{=} m_k(p). \tag{2.14}$$

Assuming for the moment that $\nabla^2 f_k$ is positive definite, we obtain the Newton direction by finding the vector p that minimizes $m_k(p)$. By simply setting the derivative of $m_k(p)$ to zero, we obtain the following explicit formula:

$$p_k^N = -\left(\nabla^2 f_k\right)^{-1} \nabla f_k. \tag{2.15}$$

The Newton direction is reliable when the difference between the true function $f(x_k + p)$ and its quadratic model $m_k(p)$ is not too large. By comparing (2.14) with (2.6), we see that the only difference between these functions is that the matrix $\nabla^2 f(x_k + tp)$ in the third term of the expansion has been replaced by $\nabla^2 f_k$. If $\nabla^2 f$ is sufficiently smooth, this difference introduces a perturbation of only $O(\|p\|^3)$ into the expansion, so that when $\|p\|$ is small, the approximation $f(x_k + p) \approx m_k(p)$ is quite accurate.

The Newton direction can be used in a line search method when $\nabla^2 f_k$ is positive definite, for in this case we have

$$\nabla f_k^T p_k^N = -p_k^{N^T} \nabla^2 f_k p_k^N \leq -\sigma_k \| p_k^N \|^2$$

for some $\sigma_k > 0$. Unless the gradient ∇f_k (and therefore the step p_k^N) is zero, we have that $\nabla f_k^T p_k^N < 0$, so the Newton direction is a descent direction.

Unlike the steepest descent direction, there is a "natural" step length of 1 associated with the Newton direction. Most line search implementations of Newton's method use the unit step $\alpha = 1$ where possible and adjust α only when it does not produce a satisfactory reduction in the value of f.

When $\nabla^2 f_k$ is not positive definite, the Newton direction may not even be defined, since $(\nabla^2 f_k)^{-1}$ may not exist. Even when it *is* defined, it may not satisfy the descent property $\nabla f_k^T p_k^N < 0$, in which case it is unsuitable as a search direction. In these situations, line search methods modify the definition of p_k to make it satisfy the descent condition while retaining the benefit of the second-order information contained in $\nabla^2 f_k$. We describe these modifications in Chapter 3.

Methods that use the Newton direction have a fast rate of local convergence, typically quadratic. After a neighborhood of the solution is reached, convergence to high accuracy often occurs in just a few iterations. The main drawback of the Newton direction is the need for the Hessian $\nabla^2 f(x)$. Explicit computation of this matrix of second derivatives can sometimes be a cumbersome, error-prone, and expensive process. Finite-difference and automatic differentiation techniques described in Chapter 8 may be useful in avoiding the need to calculate second derivatives by hand.

Quasi-Newton search directions provide an attractive alternative to Newton's method in that they do not require computation of the Hessian and yet still attain a superlinear rate of convergence. In place of the true Hessian $\nabla^2 f_k$, they use an approximation B_k, which is updated after each step to take account of the additional knowledge gained during the step. The updates make use of the fact that changes in the gradient g provide information about the second derivative of f along the search direction. By using the expression (2.5) from our statement of Taylor's theorem, we have by adding and subtracting the term $\nabla^2 f(x)p$ that

$$\nabla f(x + p) = \nabla f(x) + \nabla^2 f(x)p + \int_0^1 \left[\nabla^2 f(x + tp) - \nabla^2 f(x) \right] p \, dt.$$

Because $\nabla f(\cdot)$ is continuous, the size of the final integral term is $o(\|p\|)$. By setting $x = x_k$ and $p = x_{k+1} - x_k$, we obtain

$$\nabla f_{k+1} = \nabla f_k + \nabla^2 f_k (x_{k+1} - x_k) + o(\|x_{k+1} - x_k\|).$$

When x_k and x_{k+1} lie in a region near the solution x^*, within which $\nabla^2 f$ is positive definite, the final term in this expansion is eventually dominated by the $\nabla^2 f_k (x_{k+1} - x_k)$ term, and we can write

$$\nabla^2 f_k(x_{k+1} - x_k) \approx \nabla f_{k+1} - \nabla f_k. \tag{2.16}$$

We choose the new Hessian approximation B_{k+1} so that it mimics the property (2.16) of the true Hessian, that is, we require it to satisfy the following condition, known as the *secant equation*:

$$B_{k+1} s_k = y_k, \tag{2.17}$$

where

$$s_k = x_{k+1} - x_k, \qquad y_k = \nabla f_{k+1} - \nabla f_k.$$

Typically, we impose additional conditions on B_{k+1}, such as symmetry (motivated by symmetry of the exact Hessian), and a requirement that the difference between successive approximations B_k and B_{k+1} have low rank.

Two of the most popular formulae for updating the Hessian approximation B_k are the *symmetric-rank-one* (SR1) formula, defined by

$$B_{k+1} = B_k + \frac{(y_k - B_k s_k)(y_k - B_k s_k)^T}{(y_k - B_k s_k)^T s_k}, \tag{2.18}$$

and the *BFGS formula*, named after its inventors, Broyden, Fletcher, Goldfarb, and Shanno, which is defined by

$$B_{k+1} = B_k - \frac{B_k s_k s_k^T B_k}{s_k^T B_k s_k} + \frac{y_k y_k^T}{y_k^T s_k}. \tag{2.19}$$

Note that the difference between the matrices B_k and B_{k+1} is a rank-one matrix in the case of (2.18) and a rank-two matrix in the case of (2.19). Both updates satisfy the secant equation and both maintain symmetry. One can show that BFGS update (2.19) generates positive definite approximations whenever the initial approximation B_0 is positive definite and $s_k^T y_k > 0$. We discuss these issues further in Chapter 6.

The quasi-Newton search direction is obtained by using B_k in place of the exact Hessian in the formula (2.15), that is,

$$p_k = -B_k^{-1} \nabla f_k. \tag{2.20}$$

Some practical implementations of quasi-Newton methods avoid the need to factorize B_k at each iteration by updating the *inverse* of B_k, instead of B_k itself. In fact, the equivalent formula for (2.18) and (2.19), applied to the inverse approximation $H_k \overset{\text{def}}{=} B_k^{-1}$, is

$$H_{k+1} = \left(I - \rho_k s_k y_k^T\right) H_k \left(I - \rho_k y_k s_k^T\right) + \rho_k s_k s_k^T, \qquad \rho_k = \frac{1}{y_k^T s_k}. \qquad (2.21)$$

Calculation of p_k can then be performed by using the formula $p_k = -H_k \nabla f_k$. This matrix–vector multiplication is simpler than the factorization/back-substitution procedure that is needed to implement the formula (2.20).

Two variants of quasi-Newton methods designed to solve large problems—partially separable and limited-memory updating—are described in Chapter 7.

The last class of search directions we preview here is that generated by *nonlinear conjugate gradient methods*. They have the form

$$p_k = -\nabla f(x_k) + \beta_k p_{k-1},$$

where β_k is a scalar that ensures that p_k and p_{k-1} are *conjugate*—an important concept in the minimization of quadratic functions that will be defined in Chapter 5. Conjugate gradient methods were originally designed to solve systems of linear equations $Ax = b$, where the coefficient matrix A is symmetric and positive definite. The problem of solving this linear system is equivalent to the problem of minimizing the convex quadratic function defined by

$$\phi(x) = \tfrac{1}{2} x^T A x - b^T x,$$

so it was natural to investigate extensions of these algorithms to more general types of unconstrained minimization problems. In general, nonlinear conjugate gradient directions are much more effective than the steepest descent direction and are almost as simple to compute. These methods do not attain the fast convergence rates of Newton or quasi-Newton methods, but they have the advantage of not requiring storage of matrices. An extensive discussion of nonlinear conjugate gradient methods is given in Chapter 5.

All of the search directions discussed so far can be used directly in a line search framework. They give rise to the steepest descent, Newton, quasi-Newton, and conjugate gradient line search methods. All except conjugate gradients have an analogue in the trust-region framework, as we now discuss.

MODELS FOR TRUST-REGION METHODS

If we set $B_k = 0$ in (2.12) and define the trust region using the Euclidean norm, the trust-region subproblem (2.11) becomes

$$\min_{p} \ f_k + p^T \nabla f_k \qquad \text{subject to } \|p\|_2 \leq \Delta_k.$$

We can write the solution to this problem in closed form as

$$p_k = -\frac{\Delta_k \nabla f_k}{\|\nabla f_k\|}.$$

This is simply a steepest descent step in which the step length is determined by the trust-region radius; the trust-region and line search approaches are essentially the same in this case.

A more interesting trust-region algorithm is obtained by choosing B_k to be the exact Hessian $\nabla^2 f_k$ in the quadratic model (2.12). Because of the trust-region restriction $\|p\|_2 \leq \Delta_k$, the subproblem (2.11) is guaranteed to have a solution even when $\nabla^2 f_k$ is not positive definite p_k, as we see in Figure 2.4. The trust-region Newton method has proved to be highly effective in practice, as we discuss in Chapter 7.

If the matrix B_k in the quadratic model function m_k of (2.12) is defined by means of a quasi-Newton approximation, we obtain a trust-region quasi-Newton method.

SCALING

The performance of an algorithm may depend crucially on how the problem is formulated. One important issue in problem formulation is *scaling*. In unconstrained optimization, a problem is said to be *poorly scaled* if changes to x in a certain direction produce much larger variations in the value of f than do changes to x in another direction. A simple example is provided by the function $f(x) = 10^9 x_1^2 + x_2^2$, which is very sensitive to small changes in x_1 but not so sensitive to perturbations in x_2.

Poorly scaled functions arise, for example, in simulations of physical and chemical systems where different processes are taking place at very different rates. To be more specific, consider a chemical system in which four reactions occur. Associated with each reaction is a *rate constant* that describes the speed at which the reaction takes place. The optimization problem is to find values for these rate constants by observing the concentrations of each chemical in the system at different times. The four constants differ greatly in magnitude, since the reactions take place at vastly different speeds. Suppose we have the following rough estimates for the final values of the constants, each correct to within, say, an order of magnitude:

$$x_1 \approx 10^{-10}, \quad x_2 \approx x_3 \approx 1, \quad x_4 \approx 10^5.$$

Before solving this problem we could introduce a new variable z defined by

$$\begin{bmatrix} x_1 \\ x_2 \\ x_3 \\ x_4 \end{bmatrix} = \begin{bmatrix} 10^{-10} & 0 & 0 & 0 \\ 0 & 1 & 0 & 0 \\ 0 & 0 & 1 & 0 \\ 0 & 0 & 0 & 10^5 \end{bmatrix} \begin{bmatrix} z_1 \\ z_2 \\ z_3 \\ z_4 \end{bmatrix},$$

and then define and solve the optimization problem in terms of the new variable z. The

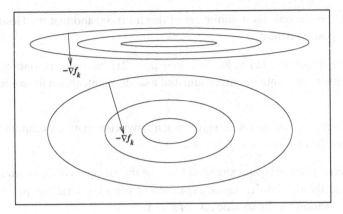

Figure 2.7 Poorly scaled and well scaled problems, and performance of the steepest descent direction.

optimal values of z will be within about an order of magnitude of 1, making the solution more balanced. This kind of scaling of the variables is known as *diagonal scaling*.

Scaling is performed (sometimes unintentionally) when the units used to represent variables are changed. During the modeling process, we may decide to change the units of some variables, say from meters to millimeters. If we do, the range of those variables and their size relative to the other variables will both change.

Some optimization algorithms, such as steepest descent, are sensitive to poor scaling, while others, such as Newton's method, are unaffected by it. Figure 2.7 shows the contours of two convex nearly quadratic functions, the first of which is poorly scaled, while the second is well scaled. For the poorly scaled problem, the one with highly elongated contours, the steepest descent direction does not yield much reduction in the function, while for the well-scaled problem it performs much better. In both cases, Newton's method will produce a much better step, since the second-order quadratic model (m_k in (2.14)) happens to be a good approximation of f.

Algorithms that are not sensitive to scaling are preferable, because they can handle poor problem formulations in a more robust fashion. In designing complete algorithms, we try to incorporate *scale invariance* into all aspects of the algorithm, including the line search or trust-region strategies and convergence tests. Generally speaking, it is easier to preserve scale invariance for line search algorithms than for trust-region algorithms.

✐ EXERCISES

✐ **2.1** Compute the gradient $\nabla f(x)$ and Hessian $\nabla^2 f(x)$ of the Rosenbrock function

$$f(x) = 100(x_2 - x_1^2)^2 + (1 - x_1)^2. \tag{2.22}$$

Show that $x^* = (1, 1)^T$ is the only local minimizer of this function, and that the Hessian matrix at that point is positive definite.

2.2 Show that the function $f(x) = 8x_1 + 12x_2 + x_1^2 - 2x_2^2$ has only one stationary point, and that it is neither a maximum or minimum, but a saddle point. Sketch the contour lines of f.

2.3 Let a be a given n-vector, and A be a given $n \times n$ symmetric matrix. Compute the gradient and Hessian of $f_1(x) = a^T x$ and $f_2(x) = x^T A x$.

2.4 Write the second-order Taylor expansion (2.6) for the function $\cos(1/x)$ around a nonzero point x, and the third-order Taylor expansion of $\cos(x)$ around any point x. Evaluate the second expansion for the specific case of $x = 1$.

2.5 Consider the function $f : \mathbb{R}^2 \to \mathbb{R}$ defined by $f(x) = \|x\|^2$. Show that the sequence of iterates $\{x_k\}$ defined by

$$x_k = \left(1 + \frac{1}{2^k}\right) \begin{bmatrix} \cos k \\ \sin k \end{bmatrix}$$

satisfies $f(x_{k+1}) < f(x_k)$ for $k = 0, 1, 2, \dots$. Show that every point on the unit circle $\{x \mid \|x\|^2 = 1\}$ is a limit point for $\{x_k\}$. Hint: Every value $\theta \in [0, 2\pi]$ is a limit point of the subsequence $\{\xi_k\}$ defined by

$$\xi_k = k \pmod{2\pi} = k - 2\pi \left\lfloor \frac{k}{2\pi} \right\rfloor,$$

where the operator $\lfloor \cdot \rfloor$ denotes rounding down to the next integer.

2.6 Prove that all isolated local minimizers are strict. (Hint: Take an isolated local minimizer x^* and a neighborhood $\mathcal{N}$. Show that for any $x \in \mathcal{N}$, $x \neq x^*$ we must have $f(x) > f(x^*)$.)

2.7 Suppose that $f(x) = x^T Q x$, where Q is an $n \times n$ symmetric positive semidefinite matrix. Show using the definition (1.4) that $f(x)$ is convex on the domain $\mathbb{R}^n$. Hint: It may be convenient to prove the following equivalent inequality:

$$f(y + \alpha(x - y)) - \alpha f(x) - (1 - \alpha) f(y) \leq 0,$$

for all $\alpha \in [0, 1]$ and all $x, y \in \mathbb{R}^n$.

2.8 Suppose that f is a convex function. Show that the set of global minimizers of f is a convex set.

✐ **2.9** Consider the function $f(x_1, x_2) = (x_1 + x_2^2)^2$. At the point $x^T = (1, 0)$ we consider the search direction $p^T = (-1, 1)$. Show that p is a descent direction and find all minimizers of the problem (2.10).

✐ **2.10** Suppose that $\tilde{f}(z) = f(x)$, where $x = Sz + s$ for some $S \in \mathbb{R}^{n \times n}$ and $s \in \mathbb{R}^n$. Show that

$$\nabla \tilde{f}(z) = S^T \nabla f(x), \qquad \nabla^2 \tilde{f}(z) = S^T \nabla^2 f(x) S.$$

(Hint: Use the chain rule to express $d\tilde{f}/dz_j$ in terms of df/dx_i and dx_i/dz_j for all $i, j = 1, 2, \ldots, n$.)

✐ **2.11** Show that the symmetric rank-one update (2.18) and the BFGS update (2.19) are scale-invariant if the initial Hessian approximations B_0 are chosen appropriately. That is, using the notation of the previous exercise, show that if these methods are applied to $f(x)$ starting from $x_0 = Sz_0 + s$ with initial Hessian B_0, and to $\tilde{f}(z)$ starting from z_0 with initial Hessian $S^T B_0 S$, then all iterates are related by $x_k = Sz_k + s$. (Assume for simplicity that the methods take unit step lengths.)

✐ **2.12** Suppose that a function f of two variables is poorly scaled at the solution x^*. Write two Taylor expansions of f around x^*—one along each coordinate direction—and use them to show that the Hessian $\nabla^2 f(x^*)$ is ill-conditioned.

✐ **2.13** (For this and the following three questions, refer to the material on "Rates of Convergence" in Section A.2 of the Appendix.) Show that the sequence $x_k = 1/k$ is not Q-linearly convergent, though it does converge to zero. (This is called *sublinear convergence*.)

✐ **2.14** Show that the sequence $x_k = 1 + (0.5)^{2^k}$ is Q-quadratically convergent to 1.

✐ **2.15** Does the sequence $x_k = 1/k!$ converge Q-superlinearly? Q-quadratically?

✐ **2.16** Consider the sequence $\{x_k\}$ defined by

$$x_k = \begin{cases} \left(\frac{1}{4}\right)^{2^k}, & k \text{ even,} \\ (x_{k-1})/k, & k \text{ odd.} \end{cases}$$

Is this sequence Q-superlinearly convergent? Q-quadratically convergent? R-quadratically convergent?

CHAPTER 3

Line Search Methods

Each iteration of a line search method computes a search direction p_k and then decides how far to move along that direction. The iteration is given by

$$x_{k+1} = x_k + \alpha_k p_k, \tag{3.1}$$

where the positive scalar α_k is called the *step length*. The success of a line search method depends on effective choices of both the direction p_k and the step length α_k.

Most line search algorithms require p_k to be a *descent direction*—one for which $p_k^T \nabla f_k < 0$—because this property guarantees that the function f can be reduced along this direction, as discussed in the previous chapter. Moreover, the search direction often has the form

$$p_k = -B_k^{-1} \nabla f_k, \qquad (3.2)$$

where B_k is a symmetric and nonsingular matrix. In the steepest descent method, B_k is simply the identity matrix I, while in Newton's method, B_k is the exact Hessian $\nabla^2 f(x_k)$. In quasi-Newton methods, B_k is an approximation to the Hessian that is updated at every iteration by means of a low-rank formula. When p_k is defined by (3.2) and B_k is positive definite, we have

$$p_k^T \nabla f_k = -\nabla f_k^T B_k^{-1} \nabla f_k < 0,$$

and therefore p_k is a descent direction.

In this chapter, we discuss how to choose α_k and p_k to promote convergence from remote starting points. We also study the rate of convergence of steepest descent, quasi-Newton, and Newton methods. Since the pure Newton iteration is not guaranteed to produce descent directions when the current iterate is not close to a solution, we discuss modifications in Section 3.4 that allow it to start from any initial point.

We now give careful consideration to the choice of the step-length parameter α_k.

3.1 STEP LENGTH

In computing the step length α_k, we face a tradeoff. We would like to choose α_k to give a substantial reduction of f, but at the same time we do not want to spend too much time making the choice. The ideal choice would be the global minimizer of the univariate function $\phi(\cdot)$ defined by

$$\phi(\alpha) = f(x_k + \alpha p_k), \quad \alpha > 0, \qquad (3.3)$$

but in general, it is too expensive to identify this value (see Figure 3.1). To find even a local minimizer of ϕ to moderate precision generally requires too many evaluations of the objective function f and possibly the gradient ∇f. More practical strategies perform an *inexact* line search to identify a step length that achieves adequate reductions in f at minimal cost.

Typical line search algorithms try out a sequence of candidate values for α, stopping to accept one of these values when certain conditions are satisfied. The line search is done in two stages: A bracketing phase finds an interval containing desirable step lengths, and a bisection or interpolation phase computes a good step length within this interval. Sophisticated line search algorithms can be quite complicated, so we defer a full description until Section 3.5.

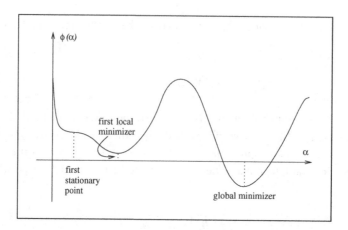

Figure 3.1 The ideal step length is the global minimizer.

We now discuss various termination conditions for line search algorithms and show that effective step lengths need not lie near minimizers of the univariate function $\phi(\alpha)$ defined in (3.3).

A simple condition we could impose on α_k is to require a reduction in f, that is, $f(x_k + \alpha_k p_k) < f(x_k)$. That this requirement is not enough to produce convergence to x^* is illustrated in Figure 3.2, for which the minimum function value is $f^* = -1$, but a sequence of iterates $\{x_k\}$ for which $f(x_k) = 5/k$, $k = 0, 1, \ldots$ yields a decrease at each iteration but has a limiting function value of zero. The insufficient reduction in f at each step causes it to fail to converge to the minimizer of this convex function. To avoid this behavior we need to enforce a *sufficient decrease* condition, a concept we discuss next.

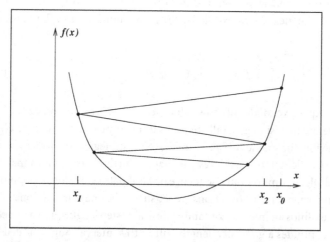

Figure 3.2 Insufficient reduction in f.

THE WOLFE CONDITIONS

A popular inexact line search condition stipulates that α_k should first of all give *sufficient decrease* in the objective function f, as measured by the following inequality:

$$f(x_k + \alpha p_k) \leq f(x_k) + c_1 \alpha \nabla f_k^T p_k, \tag{3.4}$$

for some constant $c_1 \in (0, 1)$. In other words, the reduction in f should be proportional to both the step length α_k and the directional derivative $\nabla f_k^T p_k$. Inequality (3.4) is sometimes called the *Armijo condition*.

The sufficient decrease condition is illustrated in Figure 3.3. The right-hand-side of (3.4), which is a linear function, can be denoted by $l(\alpha)$. The function $l(\cdot)$ has negative slope $c_1 \nabla f_k^T p_k$, but because $c_1 \in (0, 1)$, it lies above the graph of ϕ for small positive values of α. The sufficient decrease condition states that α is acceptable only if $\phi(\alpha) \leq l(\alpha)$. The intervals on which this condition is satisfied are shown in Figure 3.3. In practice, c_1 is chosen to be quite small, say $c_1 = 10^{-4}$.

The sufficient decrease condition is not enough by itself to ensure that the algorithm makes reasonable progress because, as we see from Figure 3.3, it is satisfied for all sufficiently small values of α. To rule out unacceptably short steps we introduce a second requirement, called the *curvature condition*, which requires α_k to satisfy

$$\nabla f(x_k + \alpha_k p_k)^T p_k \geq c_2 \nabla f_k^T p_k, \tag{3.5}$$

for some constant $c_2 \in (c_1, 1)$, where c_1 is the constant from (3.4). Note that the left-hand-side is simply the derivative $\phi'(\alpha_k)$, so the curvature condition ensures that the slope of ϕ at α_k is greater than c_2 times the initial slope $\phi'(0)$. This makes sense because if the slope $\phi'(\alpha)$

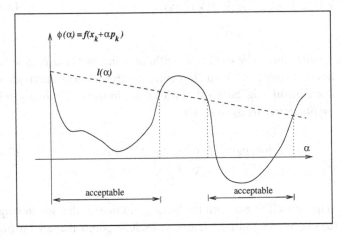

Figure 3.3 Sufficient decrease condition.

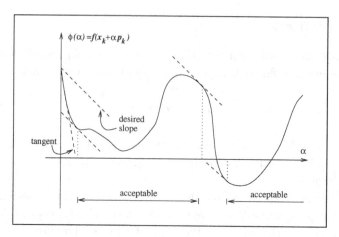

Figure 3.4 The curvature condition.

is strongly negative, we have an indication that we can reduce f significantly by moving further along the chosen direction.

On the other hand, if $\phi'(\alpha_k)$ is only slightly negative or even positive, it is a sign that we cannot expect much more decrease in f in this direction, so it makes sense to terminate the line search. The curvature condition is illustrated in Figure 3.4. Typical values of c_2 are 0.9 when the search direction p_k is chosen by a Newton or quasi-Newton method, and 0.1 when p_k is obtained from a nonlinear conjugate gradient method.

The sufficient decrease and curvature conditions are known collectively as the *Wolfe conditions*. We illustrate them in Figure 3.5 and restate them here for future reference:

$$f(x_k + \alpha_k p_k) \leq f(x_k) + c_1 \alpha_k \nabla f_k^T p_k, \tag{3.6a}$$

$$\nabla f(x_k + \alpha_k p_k)^T p_k \geq c_2 \nabla f_k^T p_k, \tag{3.6b}$$

with $0 < c_1 < c_2 < 1$.

A step length may satisfy the Wolfe conditions without being particularly close to a minimizer of ϕ, as we show in Figure 3.5. We can, however, modify the curvature condition to force α_k to lie in at least a broad neighborhood of a local minimizer or stationary point of ϕ. The *strong Wolfe conditions* require α_k to satisfy

$$f(x_k + \alpha_k p_k) \leq f(x_k) + c_1 \alpha_k \nabla f_k^T p_k, \tag{3.7a}$$

$$|\nabla f(x_k + \alpha_k p_k)^T p_k| \leq c_2 |\nabla f_k^T p_k|, \tag{3.7b}$$

with $0 < c_1 < c_2 < 1$. The only difference with the Wolfe conditions is that we no longer allow the derivative $\phi'(\alpha_k)$ to be too positive. Hence, we exclude points that are far from stationary points of ϕ.

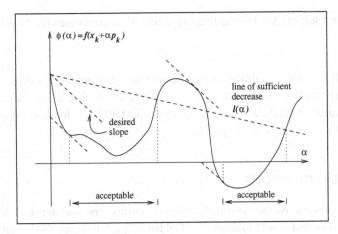

Figure 3.5 Step lengths satisfying the Wolfe conditions.

It is not difficult to prove that there exist step lengths that satisfy the Wolfe conditions for every function f that is smooth and bounded below.

Lemma 3.1.

Suppose that $f : \mathbb{R}^n \to \mathbb{R}$ is continuously differentiable. Let p_k be a descent direction at x_k, and assume that f is bounded below along the ray $\{x_k + \alpha p_k | \alpha > 0\}$. Then if $0 < c_1 < c_2 < 1$, there exist intervals of step lengths satisfying the Wolfe conditions (3.6) and the strong Wolfe conditions (3.7).

PROOF. Note that $\phi(\alpha) = f(x_k + \alpha p_k)$ is bounded below for all $\alpha > 0$. Since $0 < c_1 < 1$, the line $l(\alpha) = f(x_k) + \alpha c_1 \nabla f_k^T p_k$ is unbounded below and must therefore intersect the graph of ϕ at least once. Let $\alpha' > 0$ be the smallest intersecting value of α, that is,

$$f(x_k + \alpha' p_k) = f(x_k) + \alpha' c_1 \nabla f_k^T p_k. \tag{3.8}$$

The sufficient decrease condition (3.6a) clearly holds for all step lengths less than α'.

By the mean value theorem (see (A.55)), there exists $\alpha'' \in (0, \alpha')$ such that

$$f(x_k + \alpha' p_k) - f(x_k) = \alpha' \nabla f(x_k + \alpha'' p_k)^T p_k. \tag{3.9}$$

By combining (3.8) and (3.9), we obtain

$$\nabla f(x_k + \alpha'' p_k)^T p_k = c_1 \nabla f_k^T p_k > c_2 \nabla f_k^T p_k, \tag{3.10}$$

since $c_1 < c_2$ and $\nabla f_k^T p_k < 0$. Therefore, α'' satisfies the Wolfe conditions (3.6), and the inequalities hold strictly in both (3.6a) and (3.6b). Hence, by our smoothness assumption on f, there is an interval around α'' for which the Wolfe conditions hold. Moreover, since the term in the left-hand side of (3.10) is negative, the strong Wolfe conditions (3.7) hold in the same interval. □

The Wolfe conditions are scale-invariant in a broad sense: Multiplying the objective function by a constant or making an affine change of variables does not alter them. They can be used in most line search methods, and are particularly important in the implementation of quasi-Newton methods, as we see in Chapter 6.

THE GOLDSTEIN CONDITIONS

Like the Wolfe conditions, the *Goldstein conditions* ensure that the step length α achieves sufficient decrease but is not too short. The Goldstein conditions can also be stated as a pair of inequalities, in the following way:

$$f(x_k) + (1 - c)\alpha_k \nabla f_k^T p_k \leq f(x_k + \alpha_k p_k) \leq f(x_k) + c\alpha_k \nabla f_k^T p_k, \tag{3.11}$$

with $0 < c < 1/2$. The second inequality is the sufficient decrease condition (3.4), whereas the first inequality is introduced to control the step length from below; see Figure 3.6

A disadvantage of the Goldstein conditions vis-à-vis the Wolfe conditions is that the first inequality in (3.11) may exclude all minimizers of ϕ. However, the Goldstein and Wolfe conditions have much in common, and their convergence theories are quite similar. The Goldstein conditions are often used in Newton-type methods but are not well suited for quasi-Newton methods that maintain a positive definite Hessian approximation.

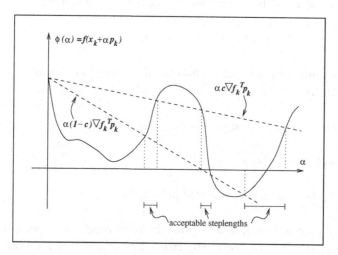

Figure 3.6   The Goldstein conditions.

SUFFICIENT DECREASE AND BACKTRACKING

We have mentioned that the sufficient decrease condition (3.6a) alone is not sufficient to ensure that the algorithm makes reasonable progress along the given search direction. However, if the line search algorithm chooses its candidate step lengths appropriately, by using a so-called *backtracking* approach, we can dispense with the extra condition (3.6b) and use just the sufficient decrease condition to terminate the line search procedure. In its most basic form, backtracking proceeds as follows.

Algorithm 3.1 (Backtracking Line Search).
 Choose $\bar{\alpha} > 0$, $\rho \in (0, 1)$, $c \in (0, 1)$; Set $\alpha \leftarrow \bar{\alpha}$;
 repeat until $f(x_k + \alpha p_k) \leq f(x_k) + c\alpha \nabla f_k^T p_k$
  $\alpha \leftarrow \rho\alpha$;
 end (repeat)
 Terminate with $\alpha_k = \alpha$.

In this procedure, the initial step length $\bar{\alpha}$ is chosen to be 1 in Newton and quasi-Newton methods, but can have different values in other algorithms such as steepest descent or conjugate gradient. An acceptable step length will be found after a finite number of trials, because α_k will eventually become small enough that the sufficient decrease condition holds (see Figure 3.3). In practice, the contraction factor ρ is often allowed to vary at each iteration of the line search. For example, it can be chosen by safeguarded interpolation, as we describe later. We need ensure only that at each iteration we have $\rho \in [\rho_{lo}, \rho_{hi}]$, for some fixed constants $0 < \rho_{lo} < \rho_{hi} < 1$.

The backtracking approach ensures either that the selected step length α_k is some fixed value (the initial choice $\bar{\alpha}$), or else that it is short enough to satisfy the sufficient decrease condition but not *too* short. The latter claim holds because the accepted value α_k is within a factor ρ of the previous trial value, α_k/ρ, which was rejected for violating the sufficient decrease condition, that is, for being too long.

This simple and popular strategy for terminating a line search is well suited for Newton methods but is less appropriate for quasi-Newton and conjugate gradient methods.

3.2 CONVERGENCE OF LINE SEARCH METHODS

To obtain global convergence, we must not only have well chosen step lengths but also well chosen search directions p_k. We discuss requirements on the search direction in this section, focusing on one key property: the angle θ_k between p_k and the steepest descent direction $-\nabla f_k$, defined by

$$\cos \theta_k = \frac{-\nabla f_k^T p_k}{\|\nabla f_k\| \|p_k\|}. \tag{3.12}$$

The following theorem, due to Zoutendijk, has far-reaching consequences. It quantifies the effect of properly chosen step lengths α_k, and shows, for example, that the steepest descent method is globally convergent. For other algorithms, it describes how far p_k can deviate from the steepest descent direction and still produce a globally convergent iteration. Various line search termination conditions can be used to establish this result, but for concreteness we will consider only the Wolfe conditions (3.6). Though Zoutendijk's result appears at first to be technical and obscure, its power will soon become evident.

Theorem 3.2.

Consider any iteration of the form (3.1), where p_k is a descent direction and α_k satisfies the Wolfe conditions (3.6). Suppose that f is bounded below in $\mathbb{R}^n$ and that f is continuously differentiable in an open set $\mathcal{N}$ containing the level set $\mathcal{L} \stackrel{\text{def}}{=} \{x : f(x) \leq f(x_0)\}$, where x_0 is the starting point of the iteration. Assume also that the gradient ∇f is Lipschitz continuous on $\mathcal{N}$, that is, there exists a constant $L > 0$ such that

$$\|\nabla f(x) - \nabla f(\tilde{x})\| \leq L\|x - \tilde{x}\|, \quad \text{for all } x, \tilde{x} \in \mathcal{N}. \tag{3.13}$$

Then

$$\sum_{k \geq 0} \cos^2 \theta_k \, \|\nabla f_k\|^2 < \infty. \tag{3.14}$$

PROOF.   From (3.6b) and (3.1) we have that

$$(\nabla f_{k+1} - \nabla f_k)^T p_k \geq (c_2 - 1)\nabla f_k^T p_k,$$

while the Lipschitz condition (3.13) implies that

$$(\nabla f_{k+1} - \nabla f_k)^T p_k \leq \alpha_k L \|p_k\|^2.$$

By combining these two relations, we obtain

$$\alpha_k \geq \frac{c_2 - 1}{L} \frac{\nabla f_k^T p_k}{\|p_k\|^2}.$$

By substituting this inequality into the first Wolfe condition (3.6a), we obtain

$$f_{k+1} \leq f_k - c_1 \frac{1 - c_2}{L} \frac{(\nabla f_k^T p_k)^2}{\|p_k\|^2}.$$

From the definition (3.12), we can write this relation as

$$f_{k+1} \leq f_k - c \cos^2 \theta_k \|\nabla f_k\|^2,$$

where $c = c_1(1 - c_2)/L$. By summing this expression over all indices less than or equal to k, we obtain

$$f_{k+1} \leq f_0 - c \sum_{j=0}^{k} \cos^2 \theta_j \| \nabla f_j \|^2. \tag{3.15}$$

Since f is bounded below, we have that $f_0 - f_{k+1}$ is less than some positive constant, for all k. Hence, by taking limits in (3.15), we obtain

$$\sum_{k=0}^{\infty} \cos^2 \theta_k \| \nabla f_k \|^2 < \infty,$$

which concludes the proof. □

Similar results to this theorem hold when the Goldstein conditions (3.11) or strong Wolfe conditions (3.7) are used in place of the Wolfe conditions. For all these strategies, the step length selection implies inequality (3.14), which we call the *Zoutendijk condition*.

Note that the assumptions of Theorem 3.2 are not too restrictive. If the function f were not bounded below, the optimization problem would not be well defined. The smoothness assumption—Lipschitz continuity of the gradient—is implied by many of the smoothness conditions that are used in local convergence theorems (see Chapters 6 and 7) and are often satisfied in practice.

The Zoutendijk condition (3.14) implies that

$$\cos^2 \theta_k \| \nabla f_k \|^2 \to 0. \tag{3.16}$$

This limit can be used in turn to derive global convergence results for line search algorithms.

If our method for choosing the search direction p_k in the iteration (3.1) ensures that the angle θ_k defined by (3.12) is bounded away from 90°, there is a positive constant δ such that

$$\cos \theta_k \geq \delta > 0, \quad \text{for all } k. \tag{3.17}$$

It follows immediately from (3.16) that

$$\lim_{k \to \infty} \| \nabla f_k \| = 0. \tag{3.18}$$

In other words, we can be sure that the gradient norms $\| \nabla f_k \|$ converge to zero, provided that the search directions are never too close to orthogonality with the gradient. In particular, the method of steepest descent (for which the search direction p_k is parallel to the negative gradient) produces a gradient sequence that converges to zero, provided that it uses a line search satisfying the Wolfe or Goldstein conditions.

We use the term *globally convergent* to refer to algorithms for which the property (3.18) is satisfied, but note that this term is sometimes used in other contexts to mean different things. For line search methods of the general form (3.1), the limit (3.18) is the strongest global convergence result that can be obtained: We cannot guarantee that the method converges to a minimizer, but only that it is attracted by stationary points. Only by making additional requirements on the search direction p_k—by introducing negative curvature information from the Hessian $\nabla^2 f(x_k)$, for example—can we strengthen these results to include convergence to a local minimum. See the Notes and References at the end of this chapter for further discussion of this point.

Consider now the Newton-like method (3.1), (3.2) and assume that the matrices B_k are positive definite with a uniformly bounded condition number. That is, there is a constant M such that

$$\|B_k\| \, \|B_k^{-1}\| \le M, \quad \text{for all } k.$$

It is easy to show from the definition (3.12) that

$$\cos \theta_k \ge 1/M \tag{3.19}$$

(see Exercise 3.5). By combining this bound with (3.16) we find that

$$\lim_{k \to \infty} \|\nabla f_k\| = 0. \tag{3.20}$$

Therefore, we have shown that Newton and quasi-Newton methods are globally convergent if the matrices B_k have a bounded condition number and are positive definite (which is needed to ensure that p_k is a descent direction), and if the step lengths satisfy the Wolfe conditions.

For some algorithms, such as conjugate gradient methods, we will be able to prove the limit (3.18), but only the weaker result

$$\liminf_{k \to \infty} \|\nabla f_k\| = 0. \tag{3.21}$$

In other words, just a subsequence of the gradient norms $\|\nabla f_{k_j}\|$ converges to zero, rather than the whole sequence (see Appendix A). This result, too, can be proved by using Zoutendijk's condition (3.14), but instead of a constructive proof, we outline a proof by contradiction. Suppose that (3.21) does not hold, so that the gradients remain bounded away from zero, that is, there exists $\gamma > 0$ such that

$$\|\nabla f_k\| \ge \gamma, \quad \text{for all } k \text{ sufficiently large.} \tag{3.22}$$

Then from (3.16) we conclude that

$$\cos \theta_k \rightarrow 0, \qquad\qquad (3.23)$$

that is, the entire sequence $\{\cos \theta_k\}$ converges to 0. To establish (3.21), therefore, it is enough to show that a subsequence $\{\cos \theta_{k_j}\}$ is bounded away from zero. We will use this strategy in Chapter 5 to study the convergence of nonlinear conjugate gradient methods.

By applying this proof technique, we can prove global convergence in the sense of (3.20) or (3.21) for a general class of algorithms. Consider *any* algorithm for which (i) every iteration produces a decrease in the objective function, and (ii) every mth iteration is a steepest descent step, with step length chosen to satisfy the Wolfe or Goldstein conditions. Then, since $\cos \theta_k = 1$ for the steepest descent steps, the result (3.21) holds. Of course, we would design the algorithm so that it does something "better" than steepest descent at the other $m - 1$ iterates. The occasional steepest descent steps may not make much progress, but they at least guarantee overall global convergence.

Note that throughout this section we have used only the fact that Zoutendijk's condition implies the limit (3.16). In later chapters we will make use of the bounded sum condition (3.14), which forces the sequence $\{\cos^2 \theta_k \|\nabla f_k\|^2\}$ to converge to zero at a sufficiently rapid rate.

3.3 RATE OF CONVERGENCE

It would seem that designing optimization algorithms with good convergence properties is easy, since all we need to ensure is that the search direction p_k does not tend to become orthogonal to the gradient ∇f_k, or that steepest descent steps are taken regularly. We could simply compute $\cos \theta_k$ at every iteration and turn p_k toward the steepest descent direction if $\cos \theta_k$ is smaller than some preselected constant $\delta > 0$. Angle tests of this type ensure global convergence, but they are undesirable for two reasons. First, they may impede a fast rate of convergence, because for problems with an ill-conditioned Hessian, it may be necessary to produce search directions that are almost orthogonal to the gradient, and an inappropriate choice of the parameter δ may cause such steps to be rejected. Second, angle tests destroy the invariance properties of quasi-Newton methods.

Algorithmic strategies that achieve rapid convergence can sometimes conflict with the requirements of global convergence, and vice versa. For example, the steepest descent method is the quintessential globally convergent algorithm, but it is quite slow in practice, as we shall see below. On the other hand, the pure Newton iteration converges rapidly when started close enough to a solution, but its steps may not even be descent directions away from the solution. The challenge is to design algorithms that incorporate both properties: good global convergence guarantees and a rapid rate of convergence.

We begin our study of convergence rates of line search methods by considering the most basic approach of all: the steepest descent method.

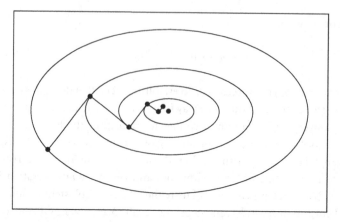

Figure 3.7   Steepest descent steps.

CONVERGENCE RATE OF STEEPEST DESCENT

We can learn much about the steepest descent method by considering the ideal case, in which the objective function is quadratic and the line searches are exact. Let us suppose that

$$f(x) = \tfrac{1}{2} x^T Q x - b^T x, \tag{3.24}$$

where Q is symmetric and positive definite. The gradient is given by $\nabla f(x) = Qx - b$ and the minimizer x^* is the unique solution of the linear system $Qx = b$.

It is easy to compute the step length α_k that minimizes $f(x_k - \alpha \nabla f_k)$. By differentiating the function

$$f(x_k - \alpha \nabla f_k) = \frac{1}{2}(x_k - \alpha \nabla f_k)^T Q (x_k - \alpha \nabla f_k) - b^T (x_k - \alpha \nabla f_k)$$

with respect to α, and setting the derivative to zero, we obtain

$$\alpha_k = \frac{\nabla f_k^T \nabla f_k}{\nabla f_k^T Q \nabla f_k}. \tag{3.25}$$

If we use this exact minimizer α_k, the steepest descent iteration for (3.24) is given by

$$x_{k+1} = x_k - \left(\frac{\nabla f_k^T \nabla f_k}{\nabla f_k^T Q \nabla f_k} \right) \nabla f_k. \tag{3.26}$$

Since $\nabla f_k = Qx_k - b$, this equation yields a closed-form expression for x_{k+1} in terms of x_k. In Figure 3.7 we plot a typical sequence of iterates generated by the steepest descent method on a two-dimensional quadratic objective function. The contours of f are ellipsoids whose axes lie along the orthogonal eigenvectors of Q. Note that the iterates zigzag toward the solution.

To quantify the rate of convergence we introduce the weighted norm $\|x\|_Q^2 = x^T Q x$. By using the relation $Q x^* = b$, we can show that

$$\tfrac{1}{2}\|x - x^*\|_Q^2 = f(x) - f(x^*), \tag{3.27}$$

so this norm measures the difference between the current objective value and the optimal value. By using the equality (3.26) and noting that $\nabla f_k = Q(x_k - x^*)$, we can derive the equality

$$\|x_{k+1} - x^*\|_Q^2 = \left\{ 1 - \frac{\left(\nabla f_k^T \nabla f_k\right)^2}{\left(\nabla f_k^T Q \nabla f_k\right)\left(\nabla f_k^T Q^{-1} \nabla f_k\right)} \right\} \|x_k - x^*\|_Q^2 \tag{3.28}$$

(see Exercise 3.7). This expression describes the exact decrease in f at each iteration, but since the term inside the brackets is difficult to interpret, it is more useful to bound it in terms of the condition number of the problem.

Theorem 3.3.

When the steepest descent method with exact line searches (3.26) is applied to the strongly convex quadratic function (3.24), the error norm (3.27) satisfies

$$\|x_{k+1} - x^*\|_Q^2 \leq \left(\frac{\lambda_n - \lambda_1}{\lambda_n + \lambda_1} \right)^2 \|x_k - x^*\|_Q^2, \tag{3.29}$$

where $0 < \lambda_1 \leq \lambda_2 \leq \cdots \leq \lambda_n$ are the eigenvalues of Q.

The proof of this result is given by Luenberger [195]. The inequalities (3.29) and (3.27) show that the function values f_k converge to the minimum f_* at a linear rate. As a special case of this result, we see that convergence is achieved in one iteration if all the eigenvalues are equal. In this case, Q is a multiple of the identity matrix, so the contours in Figure 3.7 are circles and the steepest descent direction always points at the solution. In general, as the condition number $\kappa(Q) = \lambda_n/\lambda_1$ increases, the contours of the quadratic become more elongated, the zigzagging in Figure 3.7 becomes more pronounced, and (3.29) implies that the convergence degrades. Even though (3.29) is a worst-case bound, it gives an accurate indication of the behavior of the algorithm when $n > 2$.

The rate-of-convergence behavior of the steepest descent method is essentially the same on general nonlinear objective functions. In the following result we assume that the step length is the global minimizer along the search direction.

Theorem 3.4.

Suppose that $f : \mathbb{R}^n \to \mathbb{R}$ is twice continuously differentiable, and that the iterates generated by the steepest-descent method with exact line searches converge to a point x^* at *which the Hessian matrix $\nabla^2 f(x^*)$ is positive definite. Let r be any scalar satisfying*

$$r \in \left(\frac{\lambda_n - \lambda_1}{\lambda_n + \lambda_1}, 1 \right),$$

where $\lambda_1 \le \lambda_2 \le \cdots \le \lambda_n$ are the eigenvalues of $\nabla^2 f(x^*)$. Then for all k sufficiently large, we have

$$f(x_{k+1}) - f(x^*) \le r^2 [f(x_k) - f(x^*)].$$

In general, we cannot expect the rate of convergence to improve if an inexact line search is used. Therefore, Theorem 3.4 shows that the steepest descent method can have an unacceptably slow rate of convergence, even when the Hessian is reasonably well conditioned. For example, if $\kappa(Q) = 800$, $f(x_1) = 1$, and $f(x^*) = 0$, Theorem 3.4 suggests that the function value will still be about 0.08 after one thousand iterations of the steepest descent method with exact line search.

NEWTON'S METHOD

We now consider the Newton iteration, for which the search is given by

$$p_k^N = -\nabla^2 f_k^{-1} \nabla f_k. \tag{3.30}$$

Since the Hessian matrix $\nabla^2 f_k$ may not always be positive definite, p_k^N may not always be a descent direction, and many of the ideas discussed so far in this chapter no longer apply. In Section 3.4 and Chapter 4 we will describe two approaches for obtaining a globally convergent iteration based on the Newton step: a line search approach, in which the Hessian $\nabla^2 f_k$ is modified, if necessary, to make it positive definite and thereby yield descent, and a trust region approach, in which $\nabla^2 f_k$ is used to form a quadratic model that is minimized in a ball around the current iterate x_k.

Here we discuss just the local rate-of-convergence properties of Newton's method. We know that for all x in the vicinity of a solution point x^* such that $\nabla^2 f(x^*)$ is positive definite, the Hessian $\nabla^2 f(x)$ will also be positive definite. Newton's method will be well defined in this region and will converge quadratically, provided that the step lengths α_k are eventually always 1.

Theorem 3.5.

Suppose that f is twice differentiable and that the Hessian $\nabla^2 f(x)$ is Lipschitz continuous (see (A.42)) in a neighborhood of a solution x^* at which the sufficient conditions (Theorem 2.4) are satisfied. Consider the iteration $x_{k+1} = x_k + p_k$, where p_k is given by (3.30). Then

(i) *if the starting point x_0 is sufficiently close to x^*, the sequence of iterates converges to x^*;*

(ii) *the rate of convergence of $\{x_k\}$ is quadratic; and*

(iii) *the sequence of gradient norms $\{\|\nabla f_k\|\}$ converges quadratically to zero.*

PROOF. From the definition of the Newton step and the optimality condition $\nabla f_* = 0$ we have that

$$
\begin{aligned}
x_k + p_k^{\text{N}} - x^* &= x_k - x^* - \nabla^2 f_k^{-1} \nabla f_k \\
&= \nabla^2 f_k^{-1} \left[\nabla^2 f_k(x_k - x^*) - (\nabla f_k - \nabla f_*) \right].
\end{aligned}
\tag{3.31}
$$

Since Taylor's theorem (Theorem 2.1) tells us that

$$
\nabla f_k - \nabla f_* = \int_0^1 \nabla^2 f(x_k + t(x^* - x_k))(x_k - x^*) \, dt,
$$

we have

$$
\begin{aligned}
&\left\| \nabla^2 f(x_k)(x_k - x^*) - (\nabla f_k - \nabla f(x^*)) \right\| \\
&= \left\| \int_0^1 \left[\nabla^2 f(x_k) - \nabla^2 f(x_k + t(x^* - x_k)) \right] (x_k - x^*) \, dt \right\| \\
&\leq \int_0^1 \left\| \nabla^2 f(x_k) - \nabla^2 f(x_k + t(x^* - x_k)) \right\| \, \|x_k - x^*\| \, dt \\
&\leq \|x_k - x^*\|^2 \int_0^1 Lt \, dt = \tfrac{1}{2} L \|x_k - x^*\|^2,
\end{aligned}
\tag{3.32}
$$

where L is the Lipschitz constant for $\nabla^2 f(x)$ for x near x^*. Since $\nabla^2 f(x^*)$ is nonsingular, there is a radius $r > 0$ such that $\|\nabla^2 f_k^{-1}\| \leq 2\|\nabla^2 f(x^*)^{-1}\|$ for all x_k with $\|x_k - x^*\| \leq r$. By substituting in (3.31) and (3.32), we obtain

$$
\|x_k + p_k^{\text{N}} - x^*\| \leq L \|\nabla^2 f(x^*)^{-1}\| \|x_k - x^*\|^2 = \tilde{L} \|x_k - x^*\|^2,
\tag{3.33}
$$

where $\tilde{L} = L \|\nabla^2 f(x^*)^{-1}\|$. Choosing x_0 so that $\|x_0 - x^*\| \leq \min(r, 1/(2\tilde{L}))$, we can use this inequality inductively to deduce that the sequence converges to x^*, and the rate of convergence is quadratic.

By using the relations $x_{k+1} - x_k = p_k^{\text{N}}$ and $\nabla f_k + \nabla^2 f_k p_k^{\text{N}} = 0$, we obtain that

$$
\begin{aligned}
\|\nabla f(x_{k+1})\| &= \|\nabla f(x_{k+1}) - \nabla f_k - \nabla^2 f(x_k) p_k^{\text{N}}\| \\
&= \left\| \int_0^1 \nabla^2 f(x_k + t p_k^{\text{N}})(x_{k+1} - x_k) \, dt - \nabla^2 f(x_k) p_k^{\text{N}} \right\| \\
&\leq \int_0^1 \left\| \nabla^2 f(x_k + t p_k^{\text{N}}) - \nabla^2 f(x_k) \right\| \, \|p_k^{\text{N}}\| \, dt \\
&\leq \tfrac{1}{2} L \|p_k^{\text{N}}\|^2 \\
&\leq \tfrac{1}{2} L \|\nabla^2 f(x_k)^{-1}\|^2 \|\nabla f_k\|^2 \\
&\leq 2L \|\nabla^2 f(x^*)^{-1}\|^2 \|\nabla f_k\|^2,
\end{aligned}
$$

proving that the gradient norms converge to zero quadratically. $\qquad\square$

As the iterates generated by Newton's method approach the solution, the Wolfe (or Goldstein) conditions will accept the step length $\alpha_k = 1$ for all large k. This observation follows from Theorem 3.6 below. Indeed, when the search direction is given by Newton's method, the limit (3.35) is satisfied—the ratio is zero for all k! Implementations of Newton's method using these line search conditions, and in which the line search always tries the unit step length first, will set $\alpha_k = 1$ for all large k and attain a local quadratic rate of convergence.

QUASI-NEWTON METHODS

Suppose now that the search direction has the form

$$p_k = -B_k^{-1} \nabla f_k, \tag{3.34}$$

where the symmetric and positive definite matrix B_k is updated at every iteration by a quasi-Newton updating formula. We already encountered one quasi-Newton formula, the BFGS formula, in Chapter 2; others will be discussed in Chapter 6. We assume here that the step length α_k is computed by an inexact line search that satisfies the Wolfe or strong Wolfe conditions, with the same proviso mentioned above for Newton's method: The line search algorithm will always try the step length $\alpha = 1$ first, and will accept this value if it satisfies the Wolfe conditions. (We could enforce this condition by setting $\bar{\alpha} = 1$ in Algorithm 3.1, for example.) This implementation detail turns out to be crucial in obtaining a fast rate of convergence.

The following result shows that if the search direction of a quasi-Newton method approximates the Newton direction well enough, then the unit step length will satisfy the Wolfe conditions as the iterates converge to the solution. It also specifies a condition that the search direction must satisfy in order to give rise to a superlinearly convergent iteration. To bring out the full generality of this result, we state it first in terms of a general descent iteration, and then examine its consequences for quasi-Newton and Newton methods.

Theorem 3.6.

Suppose that $f : \mathbb{R}^n \to \mathbb{R}$ is twice continuously differentiable. Consider the iteration $x_{k+1} = x_k + \alpha_k p_k$, where p_k is a descent direction and α_k satisfies the Wolfe conditions (3.6) with $c_1 \leq 1/2$. If the sequence $\{x_k\}$ converges to a point x^* such that $\nabla f(x^*) = 0$ and $\nabla^2 f(x^*)$ is positive definite, and if the search direction satisfies

$$\lim_{k \to \infty} \frac{\|\nabla f_k + \nabla^2 f_k p_k\|}{\|p_k\|} = 0, \tag{3.35}$$

then

(i) *the step length $\alpha_k = 1$ is admissible for all k greater than a certain index k_0; and*

(ii) *if $\alpha_k = 1$ for all $k > k_0$, $\{x_k\}$ converges to x^* superlinearly.*

It is easy to see that if $c_1 > 1/2$, then the line search would exclude the minimizer of a quadratic, and unit step lengths may not be admissible.

If p_k is a quasi-Newton search direction of the form (3.34), then (3.35) is equivalent to

$$\lim_{k \to \infty} \frac{\|(B_k - \nabla^2 f(x^*))p_k\|}{\|p_k\|} = 0. \tag{3.36}$$

Hence, we have the surprising (and delightful) result that a superlinear convergence rate can be attained even if the sequence of quasi-Newton matrices B_k does not converge to $\nabla^2 f(x^*)$; it suffices that the B_k become increasingly accurate approximations to $\nabla^2 f(x^*)$ *along the search directions* p_k. Importantly, condition (3.36) is *both necessary and sufficient* for the superlinear convergence of quasi-Newton methods.

Theorem 3.7.

Suppose that $f : \mathbb{R}^n \to \mathbb{R}$ is twice continuously differentiable. Consider the iteration $x_{k+1} = x_k + p_k$ (that is, the step length α_k is uniformly 1) and that p_k is given by (3.34). Let us assume also that $\{x_k\}$ converges to a point x^* such that $\nabla f(x^*) = 0$ and $\nabla^2 f(x^*)$ is positive definite. Then $\{x_k\}$ converges superlinearly if and only if (3.36) holds.

PROOF.   We first show that (3.36) is equivalent to

$$p_k - p_k^{\text{N}} = o(\|p_k\|), \tag{3.37}$$

where $p_k^{\text{N}} = -\nabla^2 f_k^{-1} \nabla f_k$ is the Newton step. Assuming that (3.36) holds, we have that

$$
\begin{aligned}
p_k - p_k^{\text{N}} &= \nabla^2 f_k^{-1}(\nabla^2 f_k p_k + \nabla f_k) \\
&= \nabla^2 f_k^{-1}(\nabla^2 f_k - B_k)p_k \\
&= O(\|(\nabla^2 f_k - B_k)p_k\|) \\
&= o(\|p_k\|),
\end{aligned}
$$

where we have used the fact that $\|\nabla^2 f_k^{-1}\|$ is bounded above for x_k sufficiently close to x^*, since the limiting Hessian $\nabla^2 f(x^*)$ is positive definite. The converse follows readily if we multiply both sides of (3.37) by $\nabla^2 f_k$ and recall (3.34).

By combining (3.33) and (3.37), we obtain that

$$\|x_k + p_k - x^*\| \le \|x_k + p_k^{\text{N}} - x^*\| + \|p_k - p_k^{\text{N}}\| = O(\|x_k - x^*\|^2) + o(\|p_k\|).$$

A simple manipulation of this inequality reveals that $\|p_k\| = O(\|x_k - x^*\|)$, so we obtain

$$\|x_k + p_k - x^*\| \le o(\|x_k - x^*\|),$$

giving the superlinear convergence result.   □

We will see in Chapter 6 that quasi-Newton methods normally satisfy condition (3.36) and are therefore superlinearly convergent.

3.4 NEWTON'S METHOD WITH HESSIAN MODIFICATION

Away from the solution, the Hessian matrix $\nabla^2 f(x)$ may not be positive definite, so the Newton direction p_k^N defined by

$$\nabla^2 f(x_k) p_k^N = -\nabla f(x_k) \tag{3.38}$$

(see (3.30)) may not be a descent direction. We now describe an approach to overcome this difficulty when a direct linear algebra technique, such as Gaussian elimination, is used to solve the Newton equations (3.38). This approach obtains the step p_k from a linear system identical to (3.38), except that the coefficient matrix is replaced with a positive definite approximation, formed before or during the solution process. The modified Hessian is obtained by adding either a positive diagonal matrix or a full matrix to the true Hessian $\nabla^2 f(x_k)$. A general description of this method follows.

Algorithm 3.2 (Line Search Newton with Modification).
　Given initial point x_0;
　for　$k = 0, 1, 2, \ldots$
　　　Factorize the matrix $B_k = \nabla^2 f(x_k) + E_k$, where $E_k = 0$ if $\nabla^2 f(x_k)$
　　　　　is sufficiently positive definite; otherwise, E_k is chosen to
　　　　　ensure that B_k is sufficiently positive definite;
　　　Solve $B_k p_k = -\nabla f(x_k)$;
　　　Set $x_{k+1} \leftarrow x_k + \alpha_k p_k$, where α_k satisfies the Wolfe, Goldstein, or
　　　　　Armijo backtracking conditions;
　end

Some approaches do not compute E_k explicitly, but rather introduce extra steps and tests into standard factorization procedures, modifying these procedures "on the fly" so that the computed factors are the factors of a positive definite matrix. Strategies based on modifying a Cholesky factorization and on modifying a symmetric indefinite factorization of the Hessian are described in this section.

Algorithm 3.2 is a practical Newton method that can be applied from any starting point. We can establish fairly satisfactory global convergence results for it, provided that the strategy for choosing E_k (and hence B_k) satisfies the *bounded modified factorization* property. This property is that the matrices in the sequence $\{B_k\}$ have bounded condition number whenever the sequence of Hessians $\{\nabla^2 f(x_k)\}$ is bounded; that is,

$$\kappa(B_k) = \|B_k\| \, \|B_k^{-1}\| \leq C, \quad \text{some } C > 0 \text{ and all } k = 0, 1, 2, \ldots. \tag{3.39}$$

If this property holds, global convergence of the modified line search Newton method follows from the results of Section 3.2.

Theorem 3.8.

Let f be twice continuously differentiable on an open set $\mathcal{D}$, and assume that the starting point x_0 of Algorithm 3.2 is such that the level set $\mathcal{L} = \{x \in \mathcal{D} : f(x) \leq f(x_0)\}$ is compact. Then if the bounded modified factorization property holds, we have that

$$\lim_{k \to \infty} \nabla f(x_k) = 0.$$

For a proof this result see [215].

We now consider the convergence rate of Algorithm 3.2. Suppose that the sequence of iterates x_k converges to a point x^* where $\nabla^2 f(x^*)$ is sufficiently positive definite in the sense that the modification strategies described in the next section return the modification $E_k = 0$ for all sufficiently large k. By Theorem 3.6, we have that $\alpha_k = 1$ for all sufficiently large k, so that Algorithm 3.2 reduces to a pure Newton method, and the rate of convergence is quadratic.

For problems in which ∇f^* is close to singular, there is no guarantee that the modification E_k will eventually vanish, and the convergence rate may be only linear. Besides requiring the modified matrix B_k to be well conditioned (so that Theorem 3.8 holds), we would like the modification to be as small as possible, so that the second-order information in the Hessian is preserved as far as possible. Naturally, we would also like the modified factorization to be computable at moderate cost.

To set the stage for the matrix factorization techniques that will be used in Algorithm 3.2, we will begin by assuming that the eigenvalue decomposition of $\nabla^2 f(x_k)$ is available. This is not realistic for large-scale problems because this decomposition is generally too expensive to compute, but it will motivate several practical modification strategies.

EIGENVALUE MODIFICATION

Consider a problem in which, at the current iterate x_k, $\nabla f(x_k) = (1, -3, 2)^T$ and $\nabla^2 f(x_k) = \text{diag}(10, 3, -1)$, which is clearly indefinite. By the spectral decomposition theorem (see Appendix A) we can define $Q = I$ and $\Lambda = \text{diag}(\lambda_1, \lambda_2, \lambda_3)$, and write

$$\nabla^2 f(x_k) = Q \Lambda Q^T = \sum_{i=1}^{n} \lambda_i q_i q_i^T. \tag{3.40}$$

The pure Newton step—the solution of (3.38)—is $p_k^N = (-0.1, 1, 2)^T$, which is not a descent direction, since $\nabla f(x_k)^T p_k^N > 0$. One might suggest a modified strategy in which we replace $\nabla^2 f(x_k)$ by a positive definite approximation B_k, in which all negative eigenvalues in $\nabla^2 f(x_k)$ are replaced by a small positive number δ that is somewhat larger than machine precision $\mathbf{u}$; say $\delta = \sqrt{\mathbf{u}}$. For a machine precision of 10^{-16}, the resulting matrix in our example is

$$B_k = \sum_{i=1}^{2} \lambda_i q_i q_i^T + \delta q_3 q_3^T = \text{diag}\left(10, 3, 10^{-8}\right), \tag{3.41}$$

which is numerically positive definite and whose curvature along the eigenvectors q_1 and q_2 has been preserved. Note, however, that the search direction based on this modified Hessian is

$$\begin{aligned} p_k = -B_k^{-1}\nabla f_k &= -\sum_{i=1}^{2} \frac{1}{\lambda_i} q_i \left(q_i^T \nabla f_k\right) - \frac{1}{\delta} q_3 \left(q_3^T \nabla f(x_k)\right) \\ &\approx -\left(2 \times 10^8\right) q_3. \end{aligned} \tag{3.42}$$

For small δ, this step is nearly parallel to q_3 (with relatively small contributions from q_1 and q_2) and quite long. Although f decreases along the direction p_k, its extreme length violates the spirit of Newton's method, which relies on a quadratic approximation of the objective function that is valid in a neighborhood of the current iterate x_k. It is therefore not clear that this search direction is effective.

Various other modification strategies are possible. We could flip the signs of the negative eigenvalues in (3.40), which amounts to setting $\delta = 1$ in our example. We could set the last term in (3.42) to zero, so that the search direction has no components along the negative curvature directions. We could adapt the choice of δ to ensure that the length of the step is not excessive, a strategy that has the flavor of trust-region methods. As this discussion shows, there is a great deal of freedom in devising modification strategies, and there is currently no agreement on which strategy is best.

Setting the issue of the choice of δ aside for the moment, let us look more closely at the process of modifying a matrix so that it becomes positive definite. The modification (3.41) to the example matrix (3.40) can be shown to be optimal in the following sense. If A is a symmetric matrix with spectral decomposition $A = Q\Lambda Q^T$, then the correction matrix ΔA of *minimum Frobenius norm* that ensures that $\lambda_{\min}(A + \Delta A) \geq \delta$ is given by

$$\Delta A = Q \, \text{diag}\,(\tau_i) Q^T, \quad \text{with} \quad \tau_i = \begin{cases} 0, & \lambda_i \geq \delta, \\ \delta - \lambda_i, & \lambda_i < \delta. \end{cases} \tag{3.43}$$

Here, $\lambda_{\min}(A)$ denotes the smallest eigenvalue of A, and the Frobenius norm of a matrix is defined as $\|A\|_F^2 = \sum_{i,j=1}^{n} a_{ij}^2$ (see (A.9)). Note that ΔA is not diagonal in general, and that the modified matrix is given by

$$A + \Delta A = Q(\Lambda + \text{diag}(\tau_i))Q^T.$$

By using a different norm we can obtain a *diagonal* modification. Suppose again that A is a symmetric matrix with spectral decomposition $A = Q\Lambda Q^T$. A correction matrix

ΔA with minimum Euclidean norm that satisfies $\lambda_{\min}(A + \Delta A) \geq \delta$ is given by

$$\Delta A = \tau I, \qquad \text{with} \qquad \tau = \max(0, \delta - \lambda_{\min}(A)). \tag{3.44}$$

The modified matrix now has the form

$$A + \tau I, \tag{3.45}$$

which happens to have the same form as the matrix occurring in (unscaled) trust–region methods (see Chapter 4). All the eigenvalues of (3.45) have thus been shifted, and all are greater than δ.

These results suggest that both diagonal and nondiagonal modifications can be considered. Even though we have not answered the question of what constitutes a good modification, various practical diagonal and nondiagonal modifications have been proposed and implemented in software. They do not make use of the spectral decomposition of the Hessian, since it is generally too expensive to compute. Instead, they use Gaussian elimination, choosing the modifications indirectly and hoping that somehow they will produce good steps. Numerical experience indicates that the strategies described next often (but not always) produce good search directions.

ADDING A MULTIPLE OF THE IDENTITY

Perhaps the simplest idea is to find a scalar $\tau > 0$ such that $\nabla^2 f(x_k) + \tau I$ is sufficiently positive definite. From the previous discussion we know that τ must satisfy (3.44), but a good estimate of the smallest eigenvalue of the Hessian is normally not available. The following algorithm describes a method that tries successively larger values of τ. (Here, a_{ii} denotes a diagonal element of A.)

Algorithm 3.3 (Cholesky with Added Multiple of the Identity).
  Choose $\beta > 0$;
  if $\min_i a_{ii} > 0$
      set $\tau_0 \leftarrow 0$;
  else
      $\tau_0 = -\min(a_{ii}) + \beta$;
  end (if)
  for $k = 0, 1, 2, \ldots$
      Attempt to apply the Cholesky algorithm to obtain $LL^T = A + \tau_k I$;
      if the factorization is completed successfully
          stop and return L;
      else
          $\tau_{k+1} \leftarrow \max(2\tau_k, \beta)$;
      end (if)
  end (for)

The choice of β is heuristic; a typical value is $\beta = 10^{-3}$. We could choose the first nonzero shift τ_0 to be proportional to be the final value of τ used in the latest Hessian modification; see also Algorithm B.1. The strategy implemented in Algorithm 3.3 is quite simple and may be preferable to the modified factorization techniques described next, but it suffers from one drawback. Every value of τ_k requires a new factorization of $A + \tau_k I$, and the algorithm can be quite expensive if several trial values are generated. Therefore it may be advantageous to increase τ more rapidly, say by a factor of 10 instead of 2 in the last **else** clause.

MODIFIED CHOLESKY FACTORIZATION

Another approach for modifying a Hessian matrix that is not positive definite is to perform a Cholesky factorization of $\nabla^2 f(x_k)$, but to increase the diagonal elements encountered during the factorization (where necessary) to ensure that they are sufficiently positive. This *modified Cholesky* approach is designed to accomplish two goals: It guarantees that the modified Cholesky factors exist and are bounded relative to the norm of the actual Hessian, and it does not modify the Hessian if it is sufficiently positive definite.

We begin our description of this approach by briefly reviewing the Cholesky factorization. Every symmetric positive definite matrix A can be written as

$$A = LDL^T, \tag{3.46}$$

where L is a lower triangular matrix with unit diagonal elements and D is a diagonal matrix with positive elements on the diagonal. By equating the elements in (3.46), column by column, it is easy to derive formulas for computing L and D.

❏ EXAMPLE 3.1

Consider the case $n = 3$. The equation $A = LDL^T$ is given by

$$
\begin{bmatrix} a_{11} & a_{21} & a_{31} \\ a_{21} & a_{22} & a_{32} \\ a_{31} & a_{32} & a_{33} \end{bmatrix}
=
\begin{bmatrix} 1 & 0 & 0 \\ l_{21} & 1 & 0 \\ l_{31} & l_{32} & 1 \end{bmatrix}
\begin{bmatrix} d_1 & 0 & 0 \\ 0 & d_2 & 0 \\ 0 & 0 & d_3 \end{bmatrix}
\begin{bmatrix} 1 & l_{21} & l_{31} \\ 0 & 1 & l_{32} \\ 0 & 0 & 1 \end{bmatrix}.
$$

(The notation indicates that A is symmetric.) By equating the elements of the first column, we have

$$
\begin{aligned}
a_{11} &= d_1, \\
a_{21} &= d_1 l_{21} \quad \Rightarrow \quad l_{21} = a_{21}/d_1, \\
a_{31} &= d_1 l_{31} \quad \Rightarrow \quad l_{31} = a_{31}/d_1.
\end{aligned}
$$

Proceeding with the next two columns, we obtain

$$
\begin{aligned}
a_{22} &= d_1 l_{21}^2 + d_2 & \Rightarrow \quad d_2 &= a_{22} - d_1 l_{21}^2, \\
a_{32} &= d_1 l_{31} l_{21} + d_2 l_{32} & \Rightarrow \quad l_{32} &= \left(a_{32} - d_1 l_{31} l_{21} \right) / d_2, \\
a_{33} &= d_1 l_{31}^2 + d_2 l_{32}^2 + d_3 & \Rightarrow \quad d_3 &= a_{33} - d_1 l_{31}^2 - d_2 l_{32}^2.
\end{aligned}
$$

$\square$

This procedure is generalized in the following algorithm.

Algorithm 3.4 (Cholesky Factorization, LDL^T Form).

$\quad$ **for** $\quad j = 1, 2, \ldots, n$

$\qquad c_{jj} \leftarrow a_{jj} - \sum_{s=1}^{j-1} d_s l_{js}^2;$

$\qquad d_j \leftarrow c_{jj};$

$\qquad$ **for** $\quad i = j + 1, \ldots, n$

$\qquad\qquad c_{ij} \leftarrow a_{ij} - \sum_{s=1}^{j-1} d_s l_{is} l_{js};$

$\qquad\qquad l_{ij} \leftarrow c_{ij} / d_j;$

$\qquad$ **end**

$\quad$ **end**

One can show (see, for example, Golub and Van Loan [136, Section 4.2.3]) that the diagonal elements d_{jj} are all positive whenever A is positive definite. The scalars c_{ij} have been introduced only to facilitate the description of the modified factorization discussed below. We should note that Algorithm 3.4 differs a little from the standard form of the Cholesky factorization, which produces a lower triangular matrix M such that

$$
A = MM^T. \tag{3.47}
$$

In fact, we can make the identification $M = LD^{1/2}$ to relate M to the factors L and D computed in Algorithm 3.4. The technique for computing M appears as Algorithm A.2 in Appendix A.

If A is indefinite, the factorization $A = LDL^T$ may not exist. Even if it does exist, Algorithm 3.4 is numerically unstable when applied to such matrices, in the sense that the elements of L and D can become arbitrarily large. It follows that a strategy of computing the LDL^T factorization and then modifying the diagonal after the fact to force its elements to be positive may break down, or may result in a matrix that is drastically different from A.

Instead, we can modify the matrix A during the course of the factorization in such a way that all elements in D are sufficiently positive, and so that the elements of D and L are not too large. To control the quality of the modification, we choose two positive parameters δ and β, and require that during the computation of the jth columns of L and D in Algorithm 3.4 (that is, for each j in the outer loop of the algorithm) the following bounds be satisfied:

$$d_j \geq \delta, \qquad |m_{ij}| \leq \beta, \quad i = j+1, j+2, \ldots, n, \qquad (3.48)$$

where $m_{ij} = l_{ij}\sqrt{d_j}$. To satisfy these bounds we only need to change one step in Algorithm 3.4: The formula for computing the diagonal element d_j in Algorithm 3.4 is replaced by

$$d_j = \max\left(|c_{jj}|, \left(\frac{\theta_j}{\beta}\right)^2, \delta \right), \quad \text{with} \ \ \theta_j = \max_{j<i\leq n} |c_{ij}|. \qquad (3.49)$$

To verify that (3.48) holds, we note from Algorithm 3.4 that $c_{ij} = l_{ij}d_j$, and therefore

$$|m_{ij}| = |l_{ij}\sqrt{d_j}| = \frac{|c_{ij}|}{\sqrt{d_j}} \leq \frac{|c_{ij}|\beta}{\theta_j} \leq \beta, \quad \text{for all } i > j.$$

We note that θ_j can be computed prior to d_j because the elements c_{ij} in the second **for** loop of Algorithm 3.4 do not involve d_j. In fact, this is the reason for introducing the quantities c_{ij} into the algorithm.

These observations are the basis of the modified Cholesky algorithm described in detail in Gill, Murray, and Wright [130], which introduces symmetric interchanges of rows and columns to try to reduce the size of the modification. If P denotes the permutation matrix associated with the row and column interchanges, the algorithm produces the Cholesky factorization of the permuted, modified matrix $PAP^T + E$, that is,

$$PAP^T + E = LDL^T = MM^T, \qquad (3.50)$$

where E is a nonnegative diagonal matrix that is zero if A is sufficiently positive definite. One can show (Moré and Sorensen [215]) that the matrices B_k obtained by this modified Cholesky algorithm to the exact Hessians $\nabla^2 f(x_k)$ have bounded condition numbers, that is, the bound (3.39) holds for some value of C.

MODIFIED SYMMETRIC INDEFINITE FACTORIZATION

Another strategy for modifying an indefinite Hessian is to use a procedure based on a symmetric indefinite factorization. Any symmetric matrix A, whether positive definite or not, can be written as

$$PAP^T = LBL^T, \qquad (3.51)$$

where L is unit lower triangular, B is a block diagonal matrix with blocks of dimension 1 or 2, and P is a permutation matrix (see our discussion in Appendix A and also Golub and

Van Loan [136, Section 4.4]). We mentioned earlier that attempting to compute the LDL^T factorization of an indefinite matrix (where D is a *diagonal* matrix) is inadvisable because even if the factors L and D are well defined, they may contain entries that are larger than the original elements of A, thus amplifying rounding errors that arise during the computation. However, by using the block diagonal matrix B, which allows 2×2 blocks as well as 1×1 blocks on the diagonal, we can guarantee that the factorization (3.51) always exists and can be computed by a numerically stable process.

❏ **Example 3.2**

The matrix

$$A = \begin{bmatrix} 0 & 1 & 2 & 3 \\ 1 & 2 & 2 & 2 \\ 2 & 2 & 3 & 3 \\ 3 & 2 & 3 & 4 \end{bmatrix}$$

can be written in the form (3.51) with $P = [e_1, e_4, e_3, e_2]$,

$$L = \begin{bmatrix} 1 & 0 & 0 & 0 \\ 0 & 1 & 0 & 0 \\ \frac{1}{9} & \frac{2}{3} & 1 & 0 \\ \frac{2}{9} & \frac{1}{3} & 0 & 1 \end{bmatrix}, \quad B = \begin{bmatrix} 0 & 3 & 0 & 0 \\ 3 & 4 & 0 & 0 \\ 0 & 0 & \frac{7}{9} & \frac{5}{9} \\ 0 & 0 & \frac{5}{9} & \frac{10}{9} \end{bmatrix}. \tag{3.52}$$

Note that both diagonal blocks in B are 2×2. Several algorithms for computing symmetric indefinite factorizations are discussed in Section A.1 of Appendix A.   ❐

The symmetric indefinite factorization allows us to determine the *inertia* of a matrix, that is, the number of positive, zero, and negative eigenvalues. One can show that the inertia of B equals the inertia of A. Moreover, the 2×2 blocks in B are always constructed to have one positive and one negative eigenvalue. Thus the number of positive eigenvalues in A equals the number of positive 1×1 blocks plus the number of 2×2 blocks.

As for the Cholesky factorization, an indefinite symmetric factorization algorithm can be modified to ensure that the modified factors are the factors of a positive definite matrix. The strategy is first to compute the factorization (3.51), as well as the spectral decomposition $B = Q \Lambda Q^T$, which is inexpensive to compute because B is block diagonal (see Exercise 3.12). We then construct a modification matrix F such that

$$L(B + F)L^T$$

is sufficiently positive definite. Motivated by the modified spectral decomposition (3.43), we choose a parameter $\delta > 0$ and define F to be

$$F = Q \operatorname{diag}(\tau_i) Q^T, \quad \tau_i = \begin{cases} 0, & \lambda_i \geq \delta, \\ \delta - \lambda_i, & \lambda_i < \delta, \end{cases} \quad i = 1, 2, \ldots, n, \quad (3.53)$$

where λ_i are the eigenvalues of B. The matrix F is thus the modification of minimum Frobenius norm that ensures that all eigenvalues of the modified matrix $B + F$ are no less than δ. This strategy therefore modifies the factorization (3.51) as follows:

$$P(A + E)P^T = L(B + F)L^T, \quad \text{where } E = P^T LFL^T P.$$

(Note that E will not be diagonal, in general.) Hence, in contrast to the modified Cholesky approach, this modification strategy changes the entire matrix A, not just its diagonal. The aim of strategy (3.53) is that the modified matrix satisfies $\lambda_{\min}(A + E) \approx \delta$ whenever the original matrix A has $\lambda_{\min}(A) < \delta$. It is not clear, however, whether it always comes close to attaining this goal.

3.5   STEP-LENGTH SELECTION ALGORITHMS

We now consider techniques for finding a minimum of the one-dimensional function

$$\phi(\alpha) = f(x_k + \alpha p_k), \quad (3.54)$$

or for simply finding a step length α_k satisfying one of the termination conditions described in Section 3.1. We assume that p_k is a descent direction—that is, $\phi'(0) < 0$—so that our search can be confined to positive values of α.

If f is a convex quadratic, $f(x) = \frac{1}{2}x^T Q x - b^T x$, its one-dimensional minimizer along the ray $x_k + \alpha p_k$ can be computed analytically and is given by

$$\alpha_k = -\frac{\nabla f_k^T p_k}{p_k^T Q p_k}. \quad (3.55)$$

For general nonlinear functions, it is necessary to use an iterative procedure. The line search procedure deserves particular attention because it has a major impact on the robustness and efficiency of all nonlinear optimization methods.

Line search procedures can be classified according to the type of derivative information they use. Algorithms that use only function values can be inefficient since, to be theoretically sound, they need to continue iterating until the search for the minimizer is narrowed down to a small interval. In contrast, knowledge of gradient information allows us to determine whether a suitable step length has been located, as stipulated, for example, by the Wolfe conditions (3.6) or Goldstein conditions (3.11). Often, particularly when x_k is close to the solution, the very first choice of α satisfies these conditions, so the line search need not be invoked at all. In the rest of this section, we discuss only algorithms that make use of derivative information. More information on derivative-free procedures is given in the notes at the end of this chapter.

All line search procedures require an initial estimate α_0 and generate a sequence $\{\alpha_i\}$ that either terminates with a step length satisfying the conditions specified by the user (for example, the Wolfe conditions) or determines that such a step length does not exist. Typical procedures consist of two phases: a *bracketing phase* that finds an interval $[\bar{a}, \bar{b}]$ containing acceptable step lengths, and a *selection phase* that zooms in to locate the final step length. The selection phase usually reduces the bracketing interval during its search for the desired step length and interpolates some of the function and derivative information gathered on earlier steps to guess the location of the minimizer. We first discuss how to perform this interpolation.

In the following discussion we let α_k and α_{k-1} denote the step lengths used at iterations k and $k-1$ of the optimization algorithm, respectively. On the other hand, we denote the trial step lengths generated during the line search by α_i and α_{i-1} and also α_j. We use α_0 to denote the initial guess.

INTERPOLATION

We begin by describing a line search procedure based on interpolation of known function and derivative values of the function ϕ. This procedure can be viewed as an enhancement of Algorithm 3.1. The aim is to find a value of α that satisfies the sufficient decrease condition (3.6a), without being "too small." Accordingly, the procedures here generate a decreasing sequence of values α_i such that each value α_i is not too much smaller than its predecessor α_{i-1}.

Note that we can write the sufficient decrease condition in the notation of (3.54) as

$$\phi(\alpha_k) \leq \phi(0) + c_1 \alpha_k \phi'(0), \tag{3.56}$$

and that since the constant c_1 is usually chosen to be small in practice ($c_1 = 10^{-4}$, say), this condition asks for little more than descent in f. We design the procedure to be "efficient" in the sense that it computes the derivative $\nabla f(x)$ as few times as possible.

Suppose that the initial guess α_0 is given. If we have

$$\phi(\alpha_0) \leq \phi(0) + c_1 \alpha_0 \phi'(0),$$

this step length satisfies the condition, and we terminate the search. Otherwise, we know that the interval $[0, \alpha_0]$ contains acceptable step lengths (see Figure 3.3). We form a quadratic approximation $\phi_q(\alpha)$ to ϕ by interpolating the three pieces of information available—$\phi(0)$, $\phi'(0)$, and $\phi(\alpha_0)$—to obtain

$$\phi_q(\alpha) = \left(\frac{\phi(\alpha_0) - \phi(0) - \alpha_0 \phi'(0)}{\alpha_0^2} \right) \alpha^2 + \phi'(0)\alpha + \phi(0). \tag{3.57}$$

(Note that this function is constructed so that it satisfies the interpolation conditions $\phi_q(0) = \phi(0)$, $\phi_q'(0) = \phi'(0)$, and $\phi_q(\alpha_0) = \phi(\alpha_0)$.) The new trial value α_1 is defined as the minimizer of this quadratic, that is, we obtain

$$\alpha_1 = -\frac{\phi'(0)\alpha_0^2}{2\left[\phi(\alpha_0) - \phi(0) - \phi'(0)\alpha_0\right]}. \tag{3.58}$$

If the sufficient decrease condition (3.56) is satisfied at α_1, we terminate the search. Otherwise, we construct a *cubic* function that interpolates the four pieces of information $\phi(0)$, $\phi'(0)$, $\phi(\alpha_0)$, and $\phi(\alpha_1)$, obtaining

$$\phi_c(\alpha) = a\alpha^3 + b\alpha^2 + \alpha\phi'(0) + \phi(0),$$

where

$$\begin{bmatrix} a \\ b \end{bmatrix} = \frac{1}{\alpha_0^2 \alpha_1^2 (\alpha_1 - \alpha_0)} \begin{bmatrix} \alpha_0^2 & -\alpha_1^2 \\ -\alpha_0^3 & \alpha_1^3 \end{bmatrix} \begin{bmatrix} \phi(\alpha_1) - \phi(0) - \phi'(0)\alpha_1 \\ \phi(\alpha_0) - \phi(0) - \phi'(0)\alpha_0 \end{bmatrix}.$$

By differentiating $\phi_c(x)$, we see that the minimizer α_2 of ϕ_c lies in the interval $[0, \alpha_1]$ and is given by

$$\alpha_2 = \frac{-b + \sqrt{b^2 - 3a\phi'(0)}}{3a}.$$

If necessary, this process is repeated, using a cubic interpolant of $\phi(0)$, $\phi'(0)$ and the two most recent values of ϕ, until an α that satisfies (3.56) is located. If any α_i is either too close to its predecessor α_{i-1} or else too much smaller than α_{i-1}, we reset $\alpha_i = \alpha_{i-1}/2$. This safeguard procedure ensures that we make reasonable progress on each iteration and that the final α is not too small.

The strategy just described assumes that derivative values are significantly more expensive to compute than function values. It is often possible, however, to compute the directional derivative simultaneously with the function, at little additional cost; see Chapter 8. Accordingly, we can design an alternative strategy based on cubic interpolation of the values of ϕ and ϕ' at the two most recent values of α.

Cubic interpolation provides a good model for functions with significant changes of curvature. Suppose we have an interval $[\bar{a}, \bar{b}]$ known to contain desirable step lengths, and two previous step length estimates α_{i-1} and α_i in this interval. We use a cubic function to interpolate $\phi(\alpha_{i-1})$, $\phi'(\alpha_{i-1})$, $\phi(\alpha_i)$, and $\phi'(\alpha_i)$. (This cubic function always exists and is unique; see, for example, Bulirsch and Stoer [41, p. 52].) The minimizer of this cubic in $[\bar{a}, \bar{b}]$ is either at one of the endpoints or else in the interior, in which case it is given by

$$\alpha_{i+1} = \alpha_i - (\alpha_i - \alpha_{i-1}) \left[\frac{\phi'(\alpha_i) + d_2 - d_1}{\phi'(\alpha_i) - \phi'(\alpha_{i-1}) + 2d_2} \right], \tag{3.59}$$

with

$$d_1 = \phi'(\alpha_{i-1}) + \phi'(\alpha_i) - 3\frac{\phi(\alpha_{i-1}) - \phi(\alpha_i)}{\alpha_{i-1} - \alpha_i},$$
$$d_2 = \text{sign}(\alpha_i - \alpha_{i-1}) \left[d_1^2 - \phi'(\alpha_{i-1})\phi'(\alpha_i) \right]^{1/2}.$$

The interpolation process can be repeated by discarding the data at one of the step lengths α_{i-1} or α_i and replacing it by $\phi(\alpha_{i+1})$ and $\phi'(\alpha_{i+1})$. The decision on which of α_{i-1} and α_i should be kept and which discarded depends on the specific conditions used to terminate the line search; we discuss this issue further below in the context of the Wolfe conditions. Cubic interpolation is a powerful strategy, since it usually produces a quadratic rate of convergence of the iteration (3.59) to the minimizing value of α.

INITIAL STEP LENGTH

For Newton and quasi-Newton methods, the step $\alpha_0 = 1$ should always be used as the initial trial step length. This choice ensures that unit step lengths are taken whenever they satisfy the termination conditions and allows the rapid rate-of-convergence properties of these methods to take effect.

For methods that do not produce well scaled search directions, such as the steepest descent and conjugate gradient methods, it is important to use current information about the problem and the algorithm to make the initial guess. A popular strategy is to assume that the first-order change in the function at iterate x_k will be the same as that obtained at the previous step. In other words, we choose the initial guess α_0 so that $\alpha_0 \nabla f_k^T p_k = \alpha_{k-1} \nabla f_{k-1}^T p_{k-1}$, that is,

$$\alpha_0 = \alpha_{k-1} \frac{\nabla f_{k-1}^T p_{k-1}}{\nabla f_k^T p_k}.$$

Another useful strategy is to interpolate a quadratic to the data $f(x_{k-1})$, $f(x_k)$, and $\nabla f_{k-1}^T p_{k-1}$ and to define α_0 to be its minimizer. This strategy yields

$$\alpha_0 = \frac{2(f_k - f_{k-1})}{\phi'(0)}. \tag{3.60}$$

It can be shown that if $x_k \to x^*$ superlinearly, then the ratio in this expression converges to 1. If we adjust the choice (3.60) by setting

$$\alpha_0 \leftarrow \min(1, 1.01\alpha_0),$$

we find that the unit step length $\alpha_0 = 1$ will eventually always be tried and accepted, and the superlinear convergence properties of Newton and quasi-Newton methods will be observed.

A LINE SEARCH ALGORITHM FOR THE WOLFE CONDITIONS

The Wolfe (or strong Wolfe) conditions are among the most widely applicable and useful termination conditions. We now describe in some detail a one-dimensional search procedure that is guaranteed to find a step length satisfying the *strong* Wolfe conditions (3.7) for any parameters c_1 and c_2 satisfying $0 < c_1 < c_2 < 1$. As before, we assume that p is a descent direction and that f is bounded below along the direction p.

The algorithm has two stages. This first stage begins with a trial estimate α_1, and keeps increasing it until it finds either an acceptable step length or an interval that brackets the desired step lengths. In the latter case, the second stage is invoked by calling a function called **zoom** (Algorithm 3.6, below), which successively decreases the size of the interval until an acceptable step length is identified.

A formal specification of the line search algorithm follows. We refer to (3.7a) as the *sufficient decrease condition* and to (3.7b) as the *curvature condition*. The parameter $\alpha_{\max}$ is a user-supplied bound on the maximum step length allowed. The line search algorithm terminates with α_* set to a step length that satisfies the strong Wolfe conditions.

Algorithm 3.5 (Line Search Algorithm).
> Set $\alpha_0 \leftarrow 0$, choose $\alpha_{\max} > 0$ and $\alpha_1 \in (0, \alpha_{\max})$;
> $i \leftarrow 1$;
> **repeat**
> > Evaluate $\phi(\alpha_i)$;
> > **if** $\phi(\alpha_i) > \phi(0) + c_1\alpha_i\phi'(0)$ or $[\phi(\alpha_i) \geq \phi(\alpha_{i-1})$ and $i > 1]$
> > > $\alpha_* \leftarrow$ **zoom**(α_{i-1}, α_i) and **stop**;
> > Evaluate $\phi'(\alpha_i)$;
> > **if** $|\phi'(\alpha_i)| \leq -c_2\phi'(0)$
> > > set $\alpha_* \leftarrow \alpha_i$ and **stop**;
> > **if** $\phi'(\alpha_i) \geq 0$
> > > set $\alpha_* \leftarrow$ **zoom**(α_i, α_{i-1}) and **stop**;
> > Choose $\alpha_{i+1} \in (\alpha_i, \alpha_{\max})$;
> > $i \leftarrow i + 1$;
> **end (repeat)**

Note that the sequence of trial step lengths $\{\alpha_i\}$ is monotonically increasing, but that the order of the arguments supplied to the **zoom** function may vary. The procedure uses the knowledge that the interval (α_{i-1}, α_i) contains step lengths satisfying the strong Wolfe conditions if one of the following three conditions is satisfied:

(i) α_i violates the sufficient decrease condition;

(ii) $\phi(\alpha_i) \geq \phi(\alpha_{i-1})$;

(iii) $\phi'(\alpha_i) \geq 0$.

The last step of the algorithm performs extrapolation to find the next trial value α_{i+1}. To implement this step we can use approaches like the interpolation procedures above, or we can simply set α_{i+1} to some constant multiple of α_i. Whichever strategy we use, it is important that the successive steps increase quickly enough to reach the upper limit α_{max} in a finite number of iterations.

We now specify the function **zoom**, which requires a little explanation. The order of its input arguments is such that each call has the form $\textbf{zoom}(\alpha_{lo}, \alpha_{hi})$, where

(a) the interval bounded by α_{lo} and α_{hi} contains step lengths that satisfy the strong Wolfe conditions;

(b) α_{lo} is, among all step lengths generated so far and satisfying the sufficient decrease condition, the one giving the smallest function value; and

(c) α_{hi} is chosen so that $\phi'(\alpha_{lo})(\alpha_{hi} - \alpha_{lo}) < 0$.

Each iteration of **zoom** generates an iterate α_j between α_{lo} and α_{hi}, and then replaces one of these endpoints by α_j in such a way that the properties (a), (b), and (c) continue to hold.

Algorithm 3.6 (zoom).

  repeat

        Interpolate (using quadratic, cubic, or bisection) to find

           a trial step length α_j between α_{lo} and α_{hi};

        Evaluate $\phi(\alpha_j)$;

        if $\phi(\alpha_j) > \phi(0) + c_1\alpha_j\phi'(0)$ or $\phi(\alpha_j) \geq \phi(\alpha_{lo})$

           $\alpha_{hi} \leftarrow \alpha_j$;

        else

           Evaluate $\phi'(\alpha_j)$;

           if $|\phi'(\alpha_j)| \leq -c_2\phi'(0)$

               Set $\alpha_* \leftarrow \alpha_j$ and **stop**;

           if $\phi'(\alpha_j)(\alpha_{hi} - \alpha_{lo}) \geq 0$

               $\alpha_{hi} \leftarrow \alpha_{lo}$;

           $\alpha_{lo} \leftarrow \alpha_j$;

  end (repeat)

If the new estimate α_j happens to satisfy the strong Wolfe conditions, then **zoom** has served its purpose of identifying such a point, so it terminates with $\alpha_* = \alpha_j$. Otherwise, if α_j satisfies the sufficient decrease condition and has a lower function value than x_{lo}, then we set $\alpha_{lo} \leftarrow \alpha_j$ to maintain condition (b). If this setting results in a violation of condition (c), we remedy the situation by setting α_{hi} to the old value of α_{lo}. Readers should sketch some graphs to see for themselves how **zoom** works!

As mentioned earlier, the interpolation step that determines α_j should be safeguarded to ensure that the new step length is not too close to the endpoints of the interval. Practical line search algorithms also make use of the properties of the interpolating polynomials to make educated guesses of where the next step length should lie; see [39, 216]. A problem that can arise is that as the optimization algorithm approaches the solution, two consecutive function values $f(x_k)$ and $f(x_{k-1})$ may be indistinguishable in finite-precision arithmetic. Therefore, the line search must include a stopping test if it cannot attain a lower function value after a certain number (typically, ten) of trial step lengths. Some procedures also stop if the relative change in x is close to machine precision, or to some user-specified threshold.

A line search algorithm that incorporates all these features is difficult to code. We advocate the use of one of the several good software implementations available in the public domain. See Dennis and Schnabel [92], Lemaréchal [189], Fletcher [101], Moré and Thuente [216] (in particular), and Hager and Zhang [161].

One may ask how much more expensive it is to require the strong Wolfe conditions instead of the regular Wolfe conditions. Our experience suggests that for a "loose" line search (with parameters such as $c_1 = 10^{-4}$ and $c_2 = 0.9$), both strategies require a similar amount of work. The strong Wolfe conditions have the advantage that by decreasing c_2 we can directly control the quality of the search, by forcing the accepted value of α to lie closer to a local minimum. This feature is important in steepest descent or nonlinear conjugate gradient methods, and therefore a step selection routine that enforces the strong Wolfe conditions has wide applicability.

NOTES AND REFERENCES

For an extensive discussion of line search termination conditions see Ortega and Rheinboldt [230]. Akaike [2] presents a probabilistic analysis of the steepest descent method with exact line searches on quadratic functions. He shows that when $n > 2$, the worst-case bound (3.29) can be expected to hold for most starting points. The case $n = 2$ can be studied in closed form; see Bazaraa, Sherali, and Shetty [14]. Theorem 3.6 is due to Dennis and Moré.

Some line search methods (see Goldfarb [132] and Moré and Sorensen [213]) compute a direction of negative curvature, whenever it exists, to prevent the iteration from converging to nonminimizing stationary points. A direction of negative curvature p_- is one that satisfies $p_-^T \nabla^2 f(x_k) p_- < 0$. These algorithms generate a search direction by combining p_- with the steepest descent direction $-\nabla f_k$, often performing a curvilinear backtracking line search.

It is difficult to determine the relative contributions of the steepest descent and negative curvature directions. Because of this fact, the approach fell out of favor after the introduction of trust-region methods.

For a more thorough treatment of the modified Cholesky factorization see Gill, Murray, and Wright [130] or Dennis and Schnabel [92]. A modified Cholesky factorization based on Gershgorin disk estimates is described in Schnabel and Eskow [276]. The modified indefinite factorization is from Cheng and Higham [58].

Another strategy for implementing a line search Newton method when the Hessian contains negative eigenvalues is to compute a direction of negative curvature and use it to define the search direction (see Moré and Sorensen [213] and Goldfarb [132]).

Derivative-free line search algorithms include golden section and Fibonacci search. They share some of the features with the line search method given in this chapter. They typically store three trial points that determine an interval containing a one-dimensional minimizer. Golden section and Fibonacci differ in the way in which the trial step lengths are generated; see, for example, [79, 39].

Our discussion of interpolation follows Dennis and Schnabel [92], and the algorithm for finding a step length satisfying the strong Wolfe conditions can be found in Fletcher [101].

✎ EXERCISES

✎ **3.1** Program the steepest descent and Newton algorithms using the backtracking line search, Algorithm 3.1. Use them to minimize the Rosenbrock function (2.22). Set the initial step length $\alpha_0 = 1$ and print the step length used by each method at each iteration. First try the initial point $x_0 = (1.2, 1.2)^T$ and then the more difficult starting point $x_0 = (-1.2, 1)^T$.

✎ **3.2** Show that if $0 < c_2 < c_1 < 1$, there may be no step lengths that satisfy the Wolfe conditions.

✎ **3.3** Show that the one-dimensional minimizer of a strongly convex quadratic function is given by (3.55).

✎ **3.4** Show that the one-dimensional minimizer of a strongly convex quadratic function always satisfies the Goldstein conditions (3.11).

✎ **3.5** Prove that $\|Bx\| \geq \|x\|/\|B^{-1}\|$ for any nonsingular matrix B. Use this fact to establish (3.19).

✎ **3.6** Consider the steepest descent method with exact line searches applied to the convex quadratic function (3.24). Using the properties given in this chapter, show that if the initial point is such that $x_0 - x^*$ is parallel to an eigenvector of Q, then the steepest descent method will find the solution in one step.

✎   **3.7** Prove the result (3.28) by working through the following steps. First, use (3.26) to show that

$$\|x_k - x^*\|_Q^2 - \|x_{k+1} - x^*\|_Q^2 = 2\alpha_k \nabla f_k^T Q(x_k - x^*) - \alpha_k^2 \nabla f_k^T Q \nabla f_k,$$

where $\| \cdot \|_Q$ is defined by (3.27). Second, use the fact that $\nabla f_k = Q(x_k - x^*)$ to obtain

$$\|x_k - x^*\|_Q^2 - \|x_{k+1} - x^*\|_Q^2 = \frac{2(\nabla f_k^T \nabla f_k)^2}{(\nabla f_k^T Q \nabla f_k)} - \frac{(\nabla f_k^T \nabla f_k)^2}{(\nabla f_k^T Q \nabla f_k)}$$

and

$$\|x_k - x^*\|_Q^2 = \nabla f_k^T Q^{-1} \nabla f_k.$$

✎   **3.8** Let Q be a positive definite symmetric matrix. Prove that for any vector x, we have

$$\frac{(x^T x)^2}{(x^T Q x)(x^T Q^{-1} x)} \geq \frac{4\lambda_n \lambda_1}{(\lambda_n + \lambda_1)^2},$$

where λ_n and λ_1 are, respectively, the largest and smallest eigenvalues of Q. (This relation, which is known as the Kantorovich inequality, can be used to deduce (3.29) from (3.28).)

✎   **3.9** Program the BFGS algorithm using the line search algorithm described in this chapter that implements the strong Wolfe conditions. Have the code verify that $y_k^T s_k$ is always positive. Use it to minimize the Rosenbrock function using the starting points given in Exercise 3.1.

✎   **3.10** Compute the eigenvalues of the 2 diagonal blocks of (3.52) and verify that each block has a positive and a negative eigenvalue. Then compute the eigenvalues of A and verify that its inertia is the same as that of B.

✎   **3.11** Describe the effect that the modified Cholesky factorization (3.50) would have on the Hessian $\nabla^2 f(x_k) = \text{diag}(-2, 12, 4)$.

✎   **3.12** Consider a block diagonal matrix B with 1×1 and 2×2 blocks. Show that the eigenvalues and eigenvectors of B can be obtained by computing the spectral decomposition of each diagonal block separately.

✎   **3.13** Show that the quadratic function that interpolates $\phi(0)$, $\phi'(0)$, and $\phi(\alpha_0)$ is given by (3.57). Then, make use of the fact that the sufficient decrease condition (3.6a) is not satisfied at α_0 to show that this quadratic has positive curvature and that the minimizer satisfies

$$\alpha_1 < \frac{\alpha_0}{2(1 - c_1)}.$$

Since c_1 is chosen to be quite small in practice, this inequality indicates that α_1 cannot be much greater than $\frac{1}{2}$ (and may be smaller), which gives us an idea of the new step length.

✐ **3.14** If $\phi(\alpha_0)$ is large, (3.58) shows that α_1 can be quite small. Give an example of a function and a step length α_0 for which this situation arises. (Drastic changes to the estimate of the step length are not desirable, since they indicate that the current interpolant does not provide a good approximation to the function and that it should be modified before being trusted to produce a good step length estimate. In practice, one imposes a lower bound—typically, $\rho = 0.1$—and defines the new step length as $\alpha_i = \max(\rho\alpha_{i-1}, \hat{\alpha}_i)$, where $\hat{\alpha}_i$ is the minimizer of the interpolant.)

✐ **3.15** Suppose that the sufficient decrease condition (3.6a) is not satisfied at the step lengths α_0, and α_1, and consider the cubic interpolating $\phi(0)$, $\phi'(0)$, $\phi(\alpha_0)$ and $\phi(\alpha_1)$. By drawing graphs illustrating the two situations that can arise, show that the minimizer of the cubic lies in $[0, \alpha_1]$. Then show that if $\phi(0) < \phi(\alpha_1)$, the minimizer is less than $\frac{2}{3}\alpha_1$.

CHAPTER 4

Trust-Region Methods

Line search methods and trust-region methods both generate steps with the help of a quadratic model of the objective function, but they use this model in different ways. Line search methods use it to generate a search direction, and then focus their efforts on finding a suitable step length α along this direction. Trust-region methods define a region around the current iterate within which they *trust* the model to be an adequate representation of the objective function, and then choose the step to be the approximate minimizer of the model in this region. In effect, they choose the direction and length of the step simultaneously. If a step is not acceptable, they reduce the size of the region and find a new minimizer. In general, the direction of the step changes whenever the size of the trust region is altered.

The size of the trust region is critical to the effectiveness of each step. If the region is too small, the algorithm misses an opportunity to take a substantial step that will move it much closer to the minimizer of the objective function. If too large, the minimizer of the model may be far from the minimizer of the objective function in the region, so we may have to reduce the size of the region and try again. In practical algorithms, we choose the size of the region according to the performance of the algorithm during previous iterations. If the model is consistently reliable, producing good steps and accurately predicting the behavior of the objective function along these steps, the size of the trust region may be increased to allow longer, more ambitious, steps to be taken. A failed step is an indication that our model is an inadequate representation of the objective function over the current trust region. After such a step, we reduce the size of the region and try again.

Figure 4.1 illustrates the trust-region approach on a function f of two variables in which the current point x_k and the minimizer x^* lie at opposite ends of a curved valley. The quadratic model function m_k, whose elliptical contours are shown as dashed lines, is constructed from function and derivative information at x_k and possibly also on information accumulated from previous iterations and steps. A line search method based on this model searches along the step to the minimizer of m_k (shown), but this direction will yield at most a small reduction in f, even if the optimal steplength is used. The trust-region method steps to the minimizer of m_k within the dotted circle (shown), yielding a more significant reduction in f and better progress toward the solution.

In this chapter, we will assume that the model function m_k that is used at each iterate x_k is quadratic. Moreover, m_k is based on the Taylor-series expansion of f around

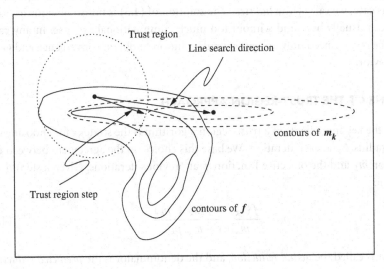

Figure 4.1   Trust-region and line search steps.

x_k, which is

$$f(x_k + p) = f_k + g_k^T p + \tfrac{1}{2} p^T \nabla^2 f(x_k + tp)p, \tag{4.1}$$

where $f_k = f(x_k)$ and $g_k = \nabla f(x_k)$, and t is some scalar in the interval $(0, 1)$. By using an approximation B_k to the Hessian in the second-order term, m_k is defined as follows:

$$m_k(p) = f_k + g_k^T p + \tfrac{1}{2} p^T B_k p, \tag{4.2}$$

where B_k is some symmetric matrix. The difference between $m_k(p)$ and $f(x_k + p)$ is $O\left(\|p\|^2\right)$, which is small when p is small.

When B_k is equal to the true Hessian $\nabla^2 f(x_k)$, the approximation error in the model function m_k is $O\left(\|p\|^3\right)$, so this model is especially accurate when $\|p\|$ is small. This choice $B_k = \nabla^2 f(x_k)$ leads to the trust-region Newton method, and will be discussed further in Section 4.4. In other sections of this chapter, we emphasize the generality of the trust-region approach by assuming little about B_k except symmetry and uniform boundedness.

To obtain each step, we seek a solution of the subproblem

$$\min_{p \in \mathbb{R}^n} m_k(p) = f_k + g_k^T p + \tfrac{1}{2} p^T B_k p \qquad \text{s.t. } \|p\| \le \Delta_k, \tag{4.3}$$

where $\Delta_k > 0$ is the trust-region radius. In most of our discussions, we define $\|\cdot\|$ to be the Euclidean norm, so that the solution p_k^* of (4.3) is the minimizer of m_k in the ball of radius Δ_k. Thus, the trust-region approach requires us to solve a sequence of subproblems (4.3) in which the objective function and constraint (which can be written as $p^T p \le \Delta_k^2$) are both quadratic. When B_k is positive definite and $\|B_k^{-1} g_k\| \le \Delta_k$, the solution of (4.3) is easy to identify—it is simply the unconstrained minimum $p_k^{\mathrm{B}} = -B_k^{-1} g_k$ of the quadratic $m_k(p)$. In this case, we call p_k^{B} the *full step*. The solution of (4.3) is not so obvious in other cases, but it can usually be found without too much computational expense. In any case, as described below, we need only an *approximate* solution to obtain convergence and good practical behavior.

OUTLINE OF THE TRUST-REGION APPROACH

One of the key ingredients in a trust-region algorithm is the strategy for choosing the trust-region radius Δ_k at each iteration. We base this choice on the agreement between the model function m_k and the objective function f at previous iterations. Given a step p_k we define the ratio

$$\rho_k = \frac{f(x_k) - f(x_k + p_k)}{m_k(0) - m_k(p_k)}; \tag{4.4}$$

the numerator is called the *actual reduction*, and the denominator is the *predicted reduction* (that is, the reduction in f predicted by the model function). Note that since the step p_k

is obtained by minimizing the model m_k over a region that includes $p = 0$, the predicted reduction will always be nonnegative. Hence, if ρ_k is negative, the new objective value $f(x_k + p_k)$ is greater than the current value $f(x_k)$, so the step must be rejected. On the other hand, if ρ_k is close to 1, there is good agreement between the model m_k and the function f over this step, so it is safe to expand the trust region for the next iteration. If ρ_k is positive but significantly smaller than 1, we do not alter the trust region, but if it is close to zero or negative, we shrink the trust region by reducing Δ_k at the next iteration.

The following algorithm describes the process.

Algorithm 4.1 (Trust Region).
Given $\hat{\Delta} > 0$, $\Delta_0 \in (0, \hat{\Delta})$, and $\eta \in \left[0, \frac{1}{4}\right)$:
for $k = 0, 1, 2, \ldots$
      Obtain p_k by (approximately) solving (4.3);
      Evaluate ρ_k from (4.4);
      if $\rho_k < \frac{1}{4}$
          $\Delta_{k+1} = \frac{1}{4}\Delta_k$
      else
          if $\rho_k > \frac{3}{4}$ and $\|p_k\| = \Delta_k$
              $\Delta_{k+1} = \min(2\Delta_k, \hat{\Delta})$
          else
              $\Delta_{k+1} = \Delta_k$;
      if $\rho_k > \eta$
          $x_{k+1} = x_k + p_k$
      else
          $x_{k+1} = x_k$;
end (for).

Here $\hat{\Delta}$ is an overall bound on the step lengths. Note that the radius is increased only if $\|p_k\|$ actually reaches the boundary of the trust region. If the step stays strictly inside the region, we infer that the current value of Δ_k is not interfering with the progress of the algorithm, so we leave its value unchanged for the next iteration.

To turn Algorithm 4.1 into a practical algorithm, we need to focus on solving the trust-region subproblem (4.3). In discussing this matter, we sometimes drop the iteration subscript k and restate the problem (4.3) as follows:

$$\min_{p \in \mathbb{R}^n} m(p) \overset{\text{def}}{=} f + g^T p + \tfrac{1}{2} p^T B p \qquad \text{s.t.} \quad \|p\| \leq \Delta. \qquad (4.5)$$

A first step to characterizing exact solutions of (4.5) is given by the following theorem (due to Moré and Sorensen [214]), which shows that the solution p^* of (4.5) satisfies

$$(B + \lambda I)p^* = -g \qquad (4.6)$$

for some $\lambda \geq 0$.

Theorem 4.1.

The vector p^* is a global solution of the trust-region problem

$$\min_{p \in \mathbb{R}^n} m(p) = f + g^T p + \tfrac{1}{2} p^T B p, \quad \text{s.t. } \|p\| \le \Delta, \tag{4.7}$$

if and only if p^* is feasible and there is a scalar $\lambda \ge 0$ such that the following conditions are satisfied:

$$(B + \lambda I) p^* = -g, \tag{4.8a}$$

$$\lambda(\Delta - \|p^*\|) = 0, \tag{4.8b}$$

$$(B + \lambda I) \quad \text{is positive semidefinite.} \tag{4.8c}$$

We delay the proof of this result until Section 4.3, and instead discuss just its key features here with the help of Figure 4.2. The condition (4.8b) is a complementarity condition that states that at least one of the nonnegative quantities λ and $(\Delta - \|p^*\|)$ must be zero. Hence, when the solution lies strictly inside the trust region (as it does when $\Delta = \Delta_1$ in Figure 4.2), we must have $\lambda = 0$ and so $Bp^* = -g$ with B positive semidefinite, from (4.8a) and (4.8c), respectively. In the other cases $\Delta = \Delta_2$ and $\Delta = \Delta_3$, we have $\|p^*\| = \Delta$, and so λ is allowed to take a positive value. Note from (4.8a) that

$$\lambda p^* = -B p^* - g = -\nabla m(p^*).$$

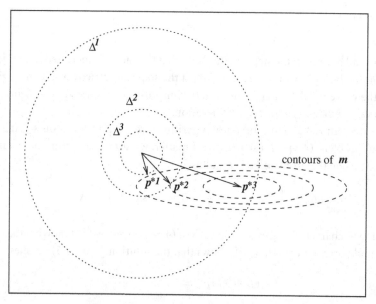

Figure 4.2   Solution of trust-region subproblem for different radii $\Delta^1, \Delta^2, \Delta^3$.

Thus, when $\lambda > 0$, the solution p^* is collinear with the negative gradient of m and normal to its contours. These properties can be seen in Figure 4.2.

In Section 4.1, we describe two strategies for finding *approximate* solutions of the subproblem (4.3), which achieve at least as much reduction in m_k as the reduction achieved by the so-called *Cauchy point*. This point is simply the minimizer of m_k along the steepest descent direction $-g_k$. subject to the trust-region bound. The first approximate strategy is the *dogleg method*, which is appropriate when the model Hessian B_k is positive definite. The second strategy, known as *two-dimensional subspace minimization*, can be applied when B_k is indefinite, though it requires an estimate of the most negative eigenvalue of this matrix. A third strategy, described in Section 7.1, uses an approach based on the conjugate gradient method to minimize m_k, and can therefore be applied when B is large and sparse.

Section 4.3 is devoted to a strategy in which an iterative method is used to identify the value of λ for which (4.6) is satisfied by the solution of the subproblem. We prove global convergence results in Section 4.2. Section 4.4 discusses the trust-region Newton method, in which the Hessian B_k of the model function is equal to the Hessian $\nabla^2 f(x_k)$ of the objective function. The key result of this section is that, when the trust-region Newton algorithm converges to a point x^* satisfying second-order sufficient conditions, it converges superlinearly.

4.1 ALGORITHMS BASED ON THE CAUCHY POINT

THE CAUCHY POINT

As we saw in Chapter 3, line search methods can be globally convergent even when the optimal step length is not used at each iteration. In fact, the step length α_k need only satisfy fairly loose criteria. A similar situation applies in trust-region methods. Although in principle we seek the optimal solution of the subproblem (4.3), it is enough for purposes of global convergence to find an approximate solution p_k that lies within the trust region and gives a *sufficient reduction* in the model. The sufficient reduction can be quantified in terms of the Cauchy point, which we denote by p_k^c and define in terms of the following simple procedure.

Algorithm 4.2 (Cauchy Point Calculation).
Find the vector p_k^s that solves a linear version of (4.3), that is,

$$p_k^s = \arg\min_{p \in \mathbb{R}^n} f_k + g_k^T p \qquad \text{s.t. } \|p\| \le \Delta_k; \tag{4.9}$$

Calculate the scalar $\tau_k > 0$ that minimizes $m_k(\tau p_k^s)$ subject to
    satisfying the trust-region bound, that is,

$$\tau_k = \arg\min_{\tau \ge 0} m_k(\tau p_k^s) \qquad \text{s.t. } \|\tau p_k^s\| \le \Delta_k; \tag{4.10}$$

Set $p_k^c = \tau_k p_k^s$.

It is easy to write down a closed-form definition of the Cauchy point. For a start, the solution of (4.9) is simply

$$p_k^s = -\frac{\Delta_k}{\|g_k\|}g_k.$$

To obtain τ_k explicitly, we consider the cases of $g_k^T B_k g_k \leq 0$ and $g_k^T B_k g_k > 0$ separately. For the former case, the function $m_k(\tau p_k^s)$ decreases monotonically with τ whenever $g_k \neq 0$, so τ_k is simply the largest value that satisfies the trust-region bound, namely, $\tau_k = 1$. For the case $g_k^T B_k g_k > 0$, $m_k(\tau p_k^s)$ is a convex quadratic in τ, so τ_k is either the unconstrained minimizer of this quadratic, $\|g_k\|^3/(\Delta_k g_k^T B_k g_k)$, or the boundary value 1, whichever comes first. In summary, we have

$$p_k^c = -\tau_k \frac{\Delta_k}{\|g_k\|}g_k, \tag{4.11}$$

where

$$\tau_k = \begin{cases} 1 & \text{if } g_k^T B_k g_k \leq 0; \\ \min\left(\|g_k\|^3/(\Delta_k g_k^T B_k g_k), 1\right) & \text{otherwise.} \end{cases} \tag{4.12}$$

Figure 4.3 illustrates the Cauchy point for a subproblem in which B_k is positive definite. In this example, p_k^c lies strictly inside the trust region.

The Cauchy step p_k^c is inexpensive to calculate—no matrix factorizations are required—and is of crucial importance in deciding if an approximate solution of the trust-region subproblem is acceptable. Specifically, a trust-region method will be globally

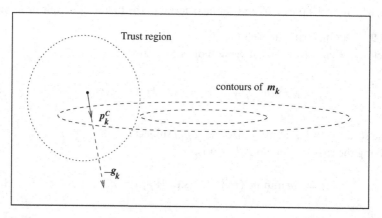

Figure 4.3   The Cauchy point.

convergent if its steps p_k give a reduction in the model m_k that is at least some fixed positive multiple of the decrease attained by the Cauchy step.

IMPROVING ON THE CAUCHY POINT

Since the Cauchy point p_k^C provides sufficient reduction in the model function m_k to yield global convergence, and since the cost of calculating it is so small, why should we look any further for a better approximate solution of (4.3)? The reason is that by always taking the Cauchy point as our step, we are simply implementing the steepest descent method with a particular choice of step length. As we have seen in Chapter 3, steepest descent performs poorly even if an *optimal* step length is used at each iteration.

The Cauchy point does not depend very strongly on the matrix B_k, which is used only in the calculation of the step length. Rapid convergence can be expected only if B_k plays a role in determining the *direction* of the step as well as its length, and if B_k contains valid curvature information about the function.

A number of trust-region algorithms compute the Cauchy point and then try to improve on it. The improvement strategy is often designed so that the full step $p_k^B = -B_k^{-1} g_k$ is chosen whenever B_k is positive definite and $\|p_k^B\| \leq \Delta_k$. When B_k is the exact Hessian $\nabla^2 f(x_k)$ or a quasi-Newton approximation, this strategy can be expected to yield superlinear convergence.

We now consider three methods for finding approximate solutions to (4.3) that have the features just described. Throughout this section we will be focusing on the internal workings of a single iteration, so we simplify the notation by dropping the subscript "k" from the quantities Δ_k, p_k, m_k, and g_k and refer to the formulation (4.5) of the subproblem. In this section, we denote the solution of (4.5) by $p^*(\Delta)$, to emphasize the dependence on Δ.

THE DOGLEG METHOD

The first approach we discuss goes by the descriptive title of the *dogleg method*. It can be used when B is positive definite.

To motivate this method, we start by examining the effect of the trust-region radius Δ on the solution $p^*(\Delta)$ of the subproblem (4.5). When B is positive definite, we have already noted that the unconstrained minimizer of m is $p^B = -B^{-1}g$. When this point is feasible for (4.5), it is obviously a solution, so we have

$$p^*(\Delta) = p^B, \qquad \text{when } \Delta \geq \|p^B\|. \tag{4.13}$$

When Δ is small relative to p^B, the restriction $\|p\| \leq \Delta$ ensures that the quadratic term in m has little effect on the solution of (4.5). For such Δ, we can get an approximation to $p(\Delta)$

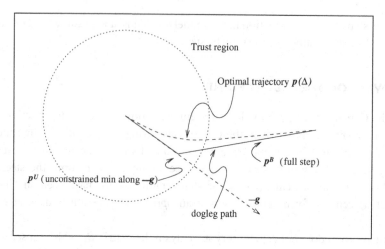

Figure 4.4 Exact trajectory and dogleg approximation.

by simply omitting the quadratic term from (4.5) and writing

$$p^*(\Delta) \approx -\Delta \frac{g}{\|g\|}, \qquad \text{when } \Delta \text{ is small.} \qquad (4.14)$$

For intermediate values of Δ, the solution $p^*(\Delta)$ typically follows a curved trajectory like the one in Figure 4.4.

The dogleg method finds an approximate solution by replacing the curved trajectory for $p^*(\Delta)$ with a path consisting of two line segments. The first line segment runs from the origin to the minimizer of m along the steepest descent direction, which is

$$p^{\mathrm{U}} = -\frac{g^T g}{g^T B g} g, \qquad (4.15)$$

while the second line segment runs from p^{U} to p^{B} (see Figure 4.4). Formally, we denote this trajectory by $\tilde{p}(\tau)$ for $\tau \in [0, 2]$, where

$$\tilde{p}(\tau) = \begin{cases} \tau p^{\mathrm{U}}, & 0 \le \tau \le 1, \\ p^{\mathrm{U}} + (\tau - 1)(p^{\mathrm{B}} - p^{\mathrm{U}}), & 1 \le \tau \le 2. \end{cases} \qquad (4.16)$$

The dogleg method chooses p to minimize the model m along this path, subject to the trust-region bound. The following lemma shows that the minimum along the dogleg path can be found easily.

Lemma 4.2.

Let B be positive definite. Then

(i) $\|\tilde{p}(\tau)\|$ *is an increasing function of* τ, *and*

(ii) $m(\tilde{p}(\tau))$ *is a decreasing function of* τ.

PROOF. It is easy to show that (i) and (ii) both hold for $\tau \in [0, 1]$, so we restrict our attention to the case of $\tau \in [1, 2]$. For (i), define $h(\alpha)$ by

$$
\begin{aligned}
h(\alpha) &= \tfrac{1}{2}\|\tilde{p}(1+\alpha)\|^2 \\
&= \tfrac{1}{2}\|p^{\mathrm{U}} + \alpha(p^{\mathrm{B}} - p^{\mathrm{U}})\|^2 \\
&= \tfrac{1}{2}\|p^{\mathrm{U}}\|^2 + \alpha(p^{\mathrm{U}})^T(p^{\mathrm{B}} - p^{\mathrm{U}}) + \tfrac{1}{2}\alpha^2\|p^{\mathrm{B}} - p^{\mathrm{U}}\|^2.
\end{aligned}
$$

Our result is proved if we can show that $h'(\alpha) \geq 0$ for $\alpha \in (0, 1)$. Now,

$$
\begin{aligned}
h'(\alpha) &= -(p^{\mathrm{U}})^T(p^{\mathrm{U}} - p^{\mathrm{B}}) + \alpha\|p^{\mathrm{U}} - p^{\mathrm{B}}\|^2 \\
&\geq -(p^{\mathrm{U}})^T(p^{\mathrm{U}} - p^{\mathrm{B}}) \\
&= \frac{g^T g}{g^T B g} g^T \left(-\frac{g^T g}{g^T B g} g + B^{-1} g \right) \\
&= g^T g \frac{g B^{-1} g}{g^T B g} \left[1 - \frac{(g^T g)^2}{(g^T B g)(g^T B^{-1} g)} \right] \\
&\geq 0,
\end{aligned}
$$

where the final inequality is a consequence of the Cauchy-Schwarz inequality. (We leave the details as an exercise.)

For (ii), we define $\hat{h}(\alpha) = m(\tilde{p}(1+\alpha))$ and show that $\hat{h}'(\alpha) \leq 0$ for $\alpha \in (0, 1)$. Substitution of (4.16) into (4.5) and differentiation with respect to the argument leads to

$$
\begin{aligned}
\hat{h}'(\alpha) &= (p^{\mathrm{B}} - p^{\mathrm{U}})^T(g + B p^{\mathrm{U}}) + \alpha(p^{\mathrm{B}} - p^{\mathrm{U}})^T B(p^{\mathrm{B}} - p^{\mathrm{U}}) \\
&\leq (p^{\mathrm{B}} - p^{\mathrm{U}})^T(g + B p^{\mathrm{U}} + B(p^{\mathrm{B}} - p^{\mathrm{U}})) \\
&= (p^{\mathrm{B}} - p^{\mathrm{U}})^T(g + B p^{\mathrm{B}}) = 0,
\end{aligned}
$$

giving the result. $\qquad\square$

It follows from this lemma that the path $\tilde{p}(\tau)$ intersects the trust-region boundary $\|p\| = \Delta$ at exactly one point if $\|p^{\mathrm{B}}\| \geq \Delta$, and nowhere otherwise. Since m is decreasing along the path, the chosen value of p will be at p^{B} if $\|p^{\mathrm{B}}\| \leq \Delta$, otherwise at the point of intersection of the dogleg and the trust-region boundary. In the latter case, we compute the appropriate value of τ by solving the following scalar quadratic equation:

$$
\|p^{\mathrm{U}} + (\tau - 1)(p^{\mathrm{B}} - p^{\mathrm{U}})\|^2 = \Delta^2.
$$

Consider now the case in which the exact Hessian $\nabla^2 f(x_k)$ is available for use in the model problem (4.5). When $\nabla^2 f(x_k)$ is positive definite, we can simply set $B = \nabla^2 f(x_k)$ (that is, $p^B = (\nabla^2 f(x_k))^{-1} g_k$) and apply the procedure above to find the Newton-dogleg step. Otherwise, we can define p^B by choosing B to be one of the positive definite modified Hessians described in Section 3.4, then proceed as above to find the dogleg step. Near a solution satisfying second-order sufficient conditions (see Theorem 2.4), p^B will be set to the usual Newton step, allowing the possibility of rapid local convergence of Newton's method (see Section 4.4).

The use of a modified Hessian in the Newton-dogleg method is not completely satisfying from an intuitive viewpoint, however. A modified factorization perturbs the diagonals of $\nabla^2 f(x_k)$ in a somewhat arbitrary manner, and the benefits of the trust-region approach may not be realized. In fact, the modification introduced during the factorization of the Hessian is redundant in some sense because the trust-region strategy introduces its own modification. As we show in Section 4.3, the exact solution of the trust-region problem (4.3) with $B_k = \nabla^2 f(x_k)$ is $(\nabla^2 f(x_k) + \lambda I)^{-1} g_k$, where λ is chosen large enough to make $(\nabla^2 f(x_k) + \lambda I)$ positive definite, and its value depends on the trust-region radius Δ_k. We conclude that the Newton-dogleg method is most appropriate when the objective function is convex (that is, $\nabla^2 f(x_k)$ is always positive semidefinite). The techniques described below may be more suitable for the general case.

The dogleg strategy can be adapted to handle indefinite matrices B, but there is not much point in doing so because the full step p^B is not the unconstrained minimizer of m in this case. Instead, we now describe another strategy, which aims to include directions of negative curvature (that is, directions d for which $d^T B d < 0$) in the space of candidate trust-region steps.

TWO-DIMENSIONAL SUBSPACE MINIMIZATION

When B is positive definite, the dogleg method strategy can be made slightly more sophisticated by widening the search for p to the entire two-dimensional subspace spanned by p^U and p^B (equivalently, g and $-B^{-1}g$). The subproblem (4.5) is replaced by

$$\min_p m(p) = f + g^T p + \tfrac{1}{2} p^T B p \quad \text{s.t. } \|p\| \le \Delta, \; p \in \text{span}[g, B^{-1}g]. \tag{4.17}$$

This is a problem in two variables that is computationally inexpensive to solve. (After some algebraic manipulation it can be reduced to finding the roots of a fourth degree polynomial.) Clearly, the Cauchy point p^C is feasible for (4.17), so the optimal solution of this subproblem yields at least as much reduction in m as the Cauchy point, resulting in global convergence of the algorithm. The two-dimensional subspace minimization strategy is obviously an extension of the dogleg method as well, since the entire dogleg path lies in $\text{span}[g, B^{-1}g]$.

This strategy can be modified to handle the case of indefinite B in a way that is intuitive, practical, and theoretically sound. We mention just the salient points of the handling of the indefiniteness here, and refer the reader to papers by Byrd, Schnabel, and Schultz (see [54] and [279]) for details. When B has negative eigenvalues, the two-dimensional subspace in (4.17) is changed to

$$\text{span}[g, (B + \alpha I)^{-1}g], \qquad \text{for some } \alpha \in (-\lambda_1, -2\lambda_1), \qquad (4.18)$$

where λ_1 denotes the most negative eigenvalue of B. (This choice of α ensures that $B + \alpha I$ is positive definite, and the flexibility in the choice of α allows us to use a numerical procedure such as the Lanczos method to compute it.) When $\|(B + \alpha I)^{-1}g\| \leq \Delta$, we discard the subspace search of (4.17), (4.18) and instead define the step to be

$$p = -(B + \alpha I)^{-1}g + v, \qquad (4.19)$$

where v is a vector that satisfies $v^T (B + \alpha I)^{-1}g \leq 0$. (This condition ensures that $\|p\| \geq \|(B + \alpha I)^{-1}g\|$.) When B has zero eigenvalues but no negative eigenvalues, we define the step to be the Cauchy point $p = p^c$.

When the exact Hessian is available, we can set $B = \nabla^2 f(x_k)$, and note that $B^{-1}g$ is the Newton step. Hence, when the Hessian is positive definite at the solution x^* and when x_k is close to x^* and Δ is sufficiently large, the subspace minimization problem (4.17) will be solved by the Newton step.

The reduction in model function m achieved by the two-dimensional subspace minimization strategy often is close to the reduction achieved by the exact solution of (4.5). Most of the computational effort lies in a single factorization of B or $B + \alpha I$ (estimation of α and solution of (4.17) are less significant), while strategies that find nearly exact solutions of (4.5) typically require two or three such factorizations (see Section 4.3).

4.2   GLOBAL CONVERGENCE

REDUCTION OBTAINED BY THE CAUCHY POINT

In the preceding discussion of algorithms for approximately solving the trust-region subproblem, we have repeatedly emphasized that global convergence depends on the approximate solution obtaining at least as much decrease in the model function m as the Cauchy point. (In fact, a fixed positive fraction of the Cauchy decrease suffices.) We start the global convergence analysis by obtaining an estimate of the decrease in m achieved by the Cauchy point. We then use this estimate to prove that the sequence of gradients $\{g_k\}$ generated by Algorithm 4.1 has an accumulation point at zero, and in fact converges to zero when η is strictly positive.

Our first main result is that the dogleg and two-dimensional subspace minimization algorithms and Steihaug's algorithm (Algorithm 7.2) produce approximate solutions p_k of the subproblem (4.3) that satisfy the following estimate of decrease in the model function:

$$m_k(0) - m_k(p_k) \geq c_1 \|g_k\| \min \left(\Delta_k, \frac{\|g_k\|}{\|B_k\|} \right), \qquad (4.20)$$

for some constant $c_1 \in (0, 1]$. The usefulness of this estimate will become clear in the following two sections. For now, we note that when Δ_k is the minimum value in (4.20), the condition is slightly reminiscent of the first Wolfe condition: The desired reduction in the model is proportional to the gradient and the size of the step.

We show now that the Cauchy point p_k^c satisfies (4.20), with $c_1 = \frac{1}{2}$.

Lemma 4.3.

The Cauchy point p_k^c satisfies (4.20) with $c_1 = \frac{1}{2}$, that is,

$$m_k(0) - m_k(p_k^c) \geq \frac{1}{2} \|g_k\| \min\left(\Delta_k, \frac{\|g_k\|}{\|B_k\|} \right). \tag{4.21}$$

Proof.   For simplicity, we drop the iteration index k in the proof.

We consider first the case $g^T B g \leq 0$. Here, we have

$$
\begin{aligned}
m(p^c) - m(0) &= m(-\Delta g/\|g\|) - f \\
&= -\frac{\Delta}{\|g\|} \|g\|^2 + \frac{1}{2} \frac{\Delta^2}{\|g\|^2} g^T B g \\
&\leq -\Delta \|g\| \\
&\leq -\|g\| \min\left(\Delta, \frac{\|g\|}{\|B\|} \right),
\end{aligned}
$$

and so (4.21) certainly holds.

For the next case, consider $g^T B g > 0$ and

$$\frac{\|g\|^3}{\Delta g^T B g} \leq 1. \tag{4.22}$$

From (4.12), we have $\tau = \|g\|^3 / \left(\Delta g^T B g \right)$, and so from (4.11) it follows that

$$
\begin{aligned}
m(p^c) - m(0) &= -\frac{\|g\|^4}{g^T B g} + \frac{1}{2} g^T B g \frac{\|g\|^4}{(g^T B g)^2} \\
&= -\frac{1}{2} \frac{\|g\|^4}{g^T B g} \\
&\leq -\frac{1}{2} \frac{\|g\|^4}{\|B\| \|g\|^2} \\
&= -\frac{1}{2} \frac{\|g\|^2}{\|B\|} \\
&\leq -\frac{1}{2} \|g\| \min\left(\Delta, \frac{\|g\|}{\|B\|} \right),
\end{aligned}
$$

so (4.21) holds here too.

In the remaining case, (4.22) does not hold, and therefore

$$g^T B g < \frac{\|g\|^3}{\Delta}. \tag{4.23}$$

From (4.12), we have $\tau = 1$, and using this fact together with (4.23), we obtain

$$
\begin{aligned}
m(p^c) - m(0) &= -\frac{\Delta}{\|g\|}\|g\|^2 + \frac{1}{2}\frac{\Delta^2}{\|g\|^2}g^T B g \\
&\leq -\Delta\|g\| + \frac{1}{2}\frac{\Delta^2}{\|g\|^2}\frac{\|g\|^3}{\Delta} \\
&= -\tfrac{1}{2}\Delta\|g\| \\
&\leq -\tfrac{1}{2}\|g\| \min\left(\Delta, \frac{\|g\|}{\|B\|}\right),
\end{aligned}
$$

yielding the desired result (4.21) once again.   □

To satisfy (4.20), our approximate solution p_k has only to achieve a reduction that is at least some fixed fraction c_2 of the reduction achieved by the Cauchy point. We state the observation formally as a theorem.

Theorem 4.4.

  Let p_k be any vector such that $\|p_k\| \leq \Delta_k$ and $m_k(0) - m_k(p_k) \geq c_2\left(m_k(0) - m_k(p_k^c)\right)$. Then p_k satisfies (4.20) with $c_1 = c_2/2$. In particular, if p_k is the exact solution p_k^* of (4.3), then it satisfies (4.20) with $c_1 = \tfrac{1}{2}$.

PROOF.   Since $\|p_k\| \leq \Delta_k$, we have from Lemma 4.3 that

$$
m_k(0) - m_k(p_k) \geq c_2\left(m_k(0) - m_k(p_k^c)\right) \geq \tfrac{1}{2}c_2\|g_k\| \min\left(\Delta_k, \frac{\|g_k\|}{\|B_k\|}\right),
$$

giving the result.   □

Note that the dogleg and two-dimensional subspace minimization algorithms both satisfy (4.20) with $c_1 = \tfrac{1}{2}$, because they all produce approximate solutions p_k for which $m_k(p_k) \leq m_k(p_k^c)$.

CONVERGENCE TO STATIONARY POINTS

Global convergence results for trust-region methods come in two varieties, depending on whether we set the parameter η in Algorithm 4.1 to zero or to some small positive value. When $\eta = 0$ (that is, the step is taken whenever it produces a lower value of f), we can show that the sequence of gradients $\{g_k\}$ has a limit point at zero. For the more stringent acceptance test with $\eta > 0$, which requires the actual decrease in f to be at least some small fraction of the predicted decrease, we have the stronger result that $g_k \to 0$.

In this section we prove the global convergence results for both cases. We assume throughout that the approximate Hessians B_k are uniformly bounded in norm, and that f

is bounded below on the level set

$$S \overset{\text{def}}{=} \{x \mid f(x) \le f(x_0)\}. \tag{4.24}$$

For later reference, we define an open neighborhood of this set by

$$S(R_0) \overset{\text{def}}{=} \{x \mid \|x - y\| < R_0 \text{ for some } y \in S\},$$

where R_0 is a positive constant.

To allow our results to be applied more generally, we also allow the length of the approximate solution p_k of (4.3) to exceed the trust-region bound, provided that it stays within some fixed multiple of the bound; that is,

$$\|p_k\| \le \gamma \Delta_k, \quad \text{for some constant } \gamma \ge 1. \tag{4.25}$$

The first result deals with the case $\eta = 0$.

Theorem 4.5.

Let $\eta = 0$ in Algorithm 4.1. Suppose that $\|B_k\| \le \beta$ for some constant β, that f is bounded below on the level set S defined by (4.24) and Lipschitz continuously differentiable in the neighborhood $S(R_0)$ for some $R_0 > 0$, and that all approximate solutions of (4.3) satisfy the inequalities (4.20) and (4.25), for some positive constants c_1 and γ. We then have

$$\liminf_{k \to \infty} \|g_k\| = 0. \tag{4.26}$$

PROOF.   By performing some technical manipulation with the ratio ρ_k from (4.4), we obtain

$$|\rho_k - 1| = \left| \frac{(f(x_k) - f(x_k + p_k)) - (m_k(0) - m_k(p_k))}{m_k(0) - m_k(p_k)} \right|$$

$$= \left| \frac{m_k(p_k) - f(x_k + p_k)}{m_k(0) - m_k(p_k)} \right|.$$

Since from Taylor's theorem (Theorem 2.1) we have that

$$f(x_k + p_k) = f(x_k) + g(x_k)^T p_k + \int_0^1 [g(x_k + t p_k) - g(x_k)]^T p_k \, dt,$$

for some $t \in (0, 1)$, it follows from the definition (4.2) of m_k that

$$|m_k(p_k) - f(x_k + p_k)| = \left| \tfrac{1}{2} p_k^T B_k p_k - \int_0^1 [g(x_k + t p_k) - g(x_k)]^T p_k \, dt \right|$$

$$\le (\beta/2)\|p_k\|^2 + \beta_1 \|p_k\|^2, \tag{4.27}$$

where we have used β_1 to denote the Lipschitz constant for g on the set $S(R_0)$, and assumed that $\|p_k\| \le R_0$ to ensure that x_k and $x_k + t p_k$ both lie in the set $S(R_0)$.

Suppose for contradiction that there is $\epsilon > 0$ and a positive index K such that

$$\|g_k\| \ge \epsilon, \qquad \text{for all } k \ge K. \tag{4.28}$$

From (4.20), we have for $k \ge K$ that

$$m_k(0) - m_k(p_k) \ge c_1 \|g_k\| \min\left(\Delta_k, \frac{\|g_k\|}{\|B_k\|}\right) \ge c_1 \epsilon \min\left(\Delta_k, \frac{\epsilon}{\beta}\right). \tag{4.29}$$

Using (4.29), (4.27), and the bound (4.25), we have

$$|\rho_k - 1| \le \frac{\gamma^2 \Delta_k^2 (\beta/2 + \beta_1)}{c_1 \epsilon \min(\Delta_k, \epsilon/\beta)}. \tag{4.30}$$

We now derive a bound on the right-hand-side that holds for all sufficiently small values of Δ_k, that is, for all $\Delta_k \le \bar{\Delta}$, where $\bar{\Delta}$ is defined as follows:

$$\bar{\Delta} = \min\left(\frac{1}{2} \frac{c_1 \epsilon}{\gamma^2 (\beta/2 + \beta_1)}, \frac{R_0}{\gamma}\right). \tag{4.31}$$

The R_0/γ term in this definition ensures that the bound (4.27) is valid (because $\|p_k\| \le \gamma \Delta_k \le \gamma \bar{\Delta} \le R_0$). Note that since $c_1 \le 1$ and $\gamma \ge 1$, we have $\bar{\Delta} \le \epsilon/\beta$. The latter condition implies that for all $\Delta_k \in [0, \bar{\Delta}]$, we have $\min(\Delta_k, \epsilon/\beta) = \Delta_k$, so from (4.30) and (4.31), we have

$$|\rho_k - 1| \le \frac{\gamma^2 \Delta_k^2 (\beta/2 + \beta_1)}{c_1 \epsilon \Delta_k} = \frac{\gamma^2 \Delta_k (\beta/2 + \beta_1)}{c_1 \epsilon} \le \frac{\gamma^2 \bar{\Delta} (\beta/2 + \beta_1)}{c_1 \epsilon} \le \frac{1}{2}.$$

Therefore, $\rho_k > \frac{1}{4}$, and so by the workings of Algorithm 4.1, we have $\Delta_{k+1} \ge \Delta_k$ whenever Δ_k falls below the threshold $\bar{\Delta}$. It follows that reduction of Δ_k (by a factor of $\frac{1}{4}$) can occur in our algorithm only if

$$\Delta_k \ge \bar{\Delta},$$

and therefore we conclude that

$$\Delta_k \ge \min\left(\Delta_K, \bar{\Delta}/4\right) \qquad \text{for all } k \ge K. \tag{4.32}$$

Suppose now that there is an infinite subsequence $\mathcal{K}$ such that $\rho_k \ge \frac{1}{4}$ for $k \in \mathcal{K}$. For

$k \in \mathcal{K}$ and $k \geq K$, we have from (4.29) that

$$
\begin{aligned}
f(x_k) - f(x_{k+1}) = f(x_k) - f(x_k + p_k) \\
\geq \tfrac{1}{4} \left[m_k(0) - m_k(p_k) \right] \\
\geq \tfrac{1}{4} c_1 \epsilon \min(\Delta_k, \epsilon/\beta).
\end{aligned}
$$

Since f is bounded below, it follows from this inequality that

$$
\lim_{k \in \mathcal{K}, \, k \to \infty} \Delta_k = 0,
$$

contradicting (4.32). Hence no such infinite subsequence $\mathcal{K}$ can exist, and we must have $\rho_k < \tfrac{1}{4}$ for all k sufficiently large. In this case, Δ_k will eventually be multiplied by $\tfrac{1}{4}$ at every iteration, and we have $\lim_{k \to \infty} \Delta_k = 0$, which again contradicts (4.32). Hence, our original assertion (4.28) must be false, giving (4.26). □

Our second global convergence result, for the case $\eta > 0$, borrows much of the analysis from the proof above. Our approach here follows that of Schultz, Schnabel, and Byrd [279].

Theorem 4.6.

Let $\eta \in \left(0, \tfrac{1}{4}\right)$ in Algorithm 4.1. Suppose that $\|B_k\| \leq \beta$ for some constant β, that f is bounded below on the level set S (4.24) and Lipschitz continuously differentiable in $S(R_0)$ for some $R_0 > 0$, and that all approximate solutions p_k of (4.3) satisfy the inequalities (4.20) and (4.25) for some positive constants c_1 and γ. We then have

$$
\lim_{k \to \infty} g_k = 0. \tag{4.33}
$$

PROOF.   We consider a particular positive index m with $g_m \neq 0$. Using β_1 again to denote the Lipschitz constant for g on the set $S(R_0)$, we have

$$
\|g(x) - g_m\| \leq \beta_1 \|x - x_m\|,
$$

for all $x \in S(R_0)$. We now define the scalars ϵ and R to satisfy

$$
\epsilon = \tfrac{1}{2} \|g_m\|, \qquad R = \min\left(\frac{\epsilon}{\beta_1}, R_0\right).
$$

Note that the ball

$$
\mathcal{B}(x_m, R) = \{x \mid \|x - x_m\| \leq R\}
$$

is contained in $S(R_0)$, so Lipschitz continuity of g holds inside $\mathcal{B}(x_m, R)$. We have

$$
x \in \mathcal{B}(x_m, R) \implies \|g(x)\| \geq \|g_m\| - \|g(x) - g_m\| \geq \tfrac{1}{2} \|g_m\| = \epsilon.
$$

If the entire sequence $\{x_k\}_{k \geq m}$ stays inside the ball $\mathcal{B}(x_m, R)$, we would have $\|g_k\| \geq \epsilon > 0$

for all $k \geq m$. The reasoning in the proof of Theorem 4.5 can be used to show that this scenario does not occur. Therefore, the sequence $\{x_k\}_{k \geq m}$ eventually leaves $\mathcal{B}(x_m, R)$.

Let the index $l \geq m$ be such that x_{l+1} is the first iterate after x_m outside $\mathcal{B}(x_m, R)$. Since $\|g_k\| \geq \epsilon$ for $k = m, m+1, \ldots, l$, we can use (4.29) to write

$$
\begin{aligned}
f(x_m) - f(x_{l+1}) &= \sum_{k=m}^{l} f(x_k) - f(x_{k+1}) \\
&\geq \sum_{k=m, x_k \neq x_{k+1}}^{l} \eta[m_k(0) - m_k(p_k)] \\
&\geq \sum_{k=m, x_k \neq x_{k+1}}^{l} \eta c_1 \epsilon \min\left(\Delta_k, \frac{\epsilon}{\beta}\right),
\end{aligned}
$$

where we have limited the sum to the iterations k for which $x_k \neq x_{k+1}$, that is, those iterations on which a step was actually taken. If $\Delta_k \leq \epsilon/\beta$ for all $k = m, m+1, \ldots, l$, we have

$$
f(x_m) - f(x_{l+1}) \geq \eta c_1 \epsilon \sum_{k=m, x_k \neq x_{k+1}}^{l} \Delta_k \geq \eta c_1 \epsilon R = \eta c_1 \epsilon \min\left(\frac{\epsilon}{\beta_1}, R_0\right). \tag{4.34}
$$

Otherwise, we have $\Delta_k > \epsilon/\beta$ for some $k = m, m+1, \ldots, l$, and so

$$
f(x_m) - f(x_{l+1}) \geq \eta c_1 \epsilon \frac{\epsilon}{\beta}. \tag{4.35}
$$

Since the sequence $\{f(x_k)\}_{k=0}^{\infty}$ is decreasing and bounded below, we have that

$$
f(x_k) \downarrow f^* \tag{4.36}
$$

for some $f^* > -\infty$. Therefore, using (4.34) and (4.35), we can write

$$
\begin{aligned}
f(x_m) - f^* &\geq f(x_m) - f(x_{l+1}) \\
&\geq \eta c_1 \epsilon \min\left(\frac{\epsilon}{\beta}, \frac{\epsilon}{\beta_1}, R_0\right) \\
&= \frac{1}{2} \eta c_1 \|g_m\| \min\left(\frac{\|g_m\|}{2\beta}, \frac{\|g_m\|}{2\beta_1}, R_0\right) > 0.
\end{aligned}
$$

Since $f(x_m) - f^* \downarrow 0$, we must have $g_m \to 0$, giving the result. $\square$

4.3   ITERATIVE SOLUTION OF THE SUBPROBLEM

In this section, we describe a technique that uses the characterization (4.6) of the subproblem solution, applying Newton's method to find the value of λ which matches the given trust-region radius Δ in (4.5). We also prove the key result Theorem 4.1 concerning the characterization of solutions of (4.3).

The methods of Section 4.1 make no serious attempt to find the exact solution of the subproblem (4.5). They do, however, make some use of the information in the model Hessian B_k, and they have advantages of reasonable implementation cost and nice global convergence properties.

When the problem is relatively small (that is, n is not too large), it may be worthwhile to exploit the model more fully by looking for a closer approximation to the solution of the subproblem. In this section, we describe an approach for finding a good approximation at the cost of a few factorizations of the matrix B (typically three factorization), as compared with a single factorization for the dogleg and two-dimensional subspace minimization methods. This approach is based on the characterization of the exact solution given in Theorem 4.1, together with an ingenious application of Newton's method in one variable. Essentially, the algorithm tries to identify the value of λ for which (4.6) is satisfied by the solution of (4.5).

The characterization of Theorem 4.1 suggests an algorithm for finding the solution p of (4.7). Either $\lambda = 0$ satisfies (4.8a) and (4.8c) with $\|p\| \leq \Delta$, or else we define

$$p(\lambda) = -(B + \lambda I)^{-1} g$$

for λ sufficiently large that $B + \lambda I$ is positive definite and seek a value $\lambda > 0$ such that

$$\|p(\lambda)\| = \Delta. \tag{4.37}$$

This problem is a one-dimensional root-finding problem in the variable λ.

To see that a value of λ with all the desired properties exists, we appeal to the eigendecomposition of B and use it to study the properties of $\|p(\lambda)\|$. Since B is symmetric, there is an orthogonal matrix Q and a diagonal matrix Λ such that $B = Q\Lambda Q^T$, where

$$\Lambda = \mathrm{diag}(\lambda_1, \lambda_2, \ldots, \lambda_n),$$

and $\lambda_1 \leq \lambda_2 \leq \cdots \leq \lambda_n$ are the eigenvalues of B; see (A.16). Clearly, $B + \lambda I = Q(\Lambda + \lambda I)Q^T$, and for $\lambda \neq \lambda_j$, we have

$$p(\lambda) = -Q(\Lambda + \lambda I)^{-1} Q^T g = -\sum_{j=1}^{n} \frac{q_j^T g}{\lambda_j + \lambda} q_j, \tag{4.38}$$

where q_j denotes the jth column of Q. Therefore, by orthonormality of $q_1, q_2, \ldots, q_n$, we have

$$\|p(\lambda)\|^2 = \sum_{j=1}^{n} \frac{\left(q_j^T g\right)^2}{(\lambda_j + \lambda)^2}. \tag{4.39}$$

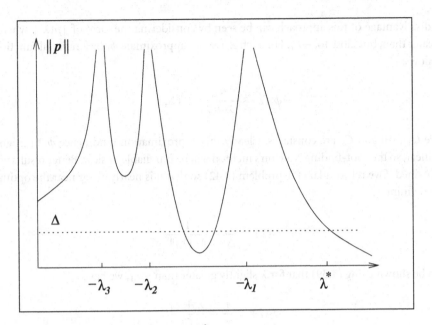

Figure 4.5 $\|p(\lambda)\|$ as a function of λ.

This expression tells us a lot about $\|p(\lambda)\|$. If $\lambda > -\lambda_1$, we have $\lambda_j + \lambda > 0$ for all $j = 1, 2, \ldots, n$, and so $\|p(\lambda)\|$ is a continuous, nonincreasing function of λ on the interval $(-\lambda_1, \infty)$. In fact, we have that

$$\lim_{\lambda \to \infty} \|p(\lambda)\| = 0. \tag{4.40}$$

Moreover, we have when $q_j^T g \neq 0$ that

$$\lim_{\lambda \to -\lambda_j} \|p(\lambda)\| = \infty. \tag{4.41}$$

Figure 4.5 plots $\|p(\lambda)\|$ against λ in a case in whcih $q_1^T g, q_2^T g$, and $q_3^T g$ are all nonzero. Note that the properties (4.40) and (4.41) hold and that $\|p(\lambda)\|$ is a nonincreasing function of λ on $(-\lambda_1, \infty)$. In particular, as is always the case when $q_1^T g \neq 0$, that there is a unique value $\lambda^* \in (-\lambda_1, \infty)$ such that $\|p(\lambda^*)\| = \Delta$. (There may be other, smaller values of λ for which $\|p(\lambda)\| = \Delta$, but these will fail to satisfy (4.8c).)

We now sketch a procedure for identifying the $\lambda^* \in (-\lambda_1, \infty)$ for which $\|p(\lambda^*)\| = \Delta$, which works when $q_1^T g \neq 0$. (We discuss the case of $q_1^T g = 0$ later.) First, note that when B positive definite and $\|B^{-1}g\| \leq \Delta$, the value $\lambda = 0$ satisfies (4.8), so the procedure can be terminated immediately with $\lambda^* = 0$. Otherwise, we could use the root-finding Newton's method (see the Appendix) to find the value of $\lambda > -\lambda_1$ that solves

$$\phi_1(\lambda) = \|p(\lambda)\| - \Delta = 0. \tag{4.42}$$

The disadvantage of this approach can be seen by considering the form of $\|p(\lambda)\|$ when λ is greater than, but close to, $-\lambda_1$. For such λ, we can approximate ϕ_1 by a rational function, as follows:

$$\phi_1(\lambda) \approx \frac{C_1}{\lambda + \lambda_1} + C_2,$$

where $C_1 > 0$ and C_2 are constants. Clearly this approximation (and hence ϕ_1) is highly nonlinear, so the root-finding Newton's method will be unreliable or slow. Better results will be obtained if we reformulate the problem (4.42) so that it is nearly linear near the optimal λ. By defining

$$\phi_2(\lambda) = \frac{1}{\Delta} - \frac{1}{\|p(\lambda)\|},$$

it can be shown using (4.39) that for λ slightly greater than $-\lambda_1$, we have

$$\phi_2(\lambda) \approx \frac{1}{\Delta} - \frac{\lambda + \lambda_1}{C_3}$$

for some $C_3 > 0$. Hence, ϕ_2 is nearly linear near $-\lambda_1$ (see Figure 4.6), and the root-finding

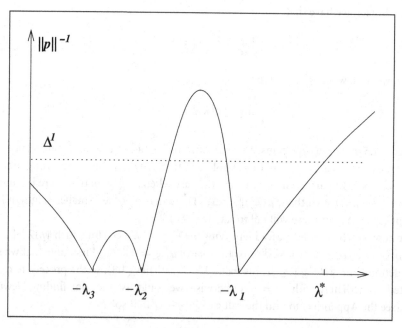

Figure 4.6 $1/\|p(\lambda)\|$ as a function of λ.

Newton's method will perform well, provided that it maintains $\lambda > -\lambda_1$. The root-finding Newton's method applied to ϕ_2 generates a sequence of iterates $\lambda^{(\ell)}$ by setting

$$\lambda^{(\ell+1)} = \lambda^{(\ell)} - \frac{\phi_2\left(\lambda^{(\ell)}\right)}{\phi_2'\left(\lambda^{(\ell)}\right)}. \tag{4.43}$$

After some elementary manipulation, this updating formula can be implemented in the following practical way.

Algorithm 4.3 (Trust Region Subproblem).
  Given $\lambda^{(0)}$, $\Delta > 0$:
  for $\ell = 0, 1, 2, \ldots$
      Factor $B + \lambda^{(\ell)}I = R^T R$;
      Solve $R^T R p_\ell = -g$, $R^T q_\ell = p_\ell$;

      Set

$$\lambda^{(\ell+1)} = \lambda^{(\ell)} + \left(\frac{\|p_\ell\|}{\|q_\ell\|}\right)^2 \left(\frac{\|p_\ell\| - \Delta}{\Delta}\right); \tag{4.44}$$

  end (for).

    Safeguards must be added to this algorithm to make it practical; for instance, when $\lambda^{(\ell)} < -\lambda_1$, the Cholesky factorization $B + \lambda^{(\ell)}I = R^T R$ will not exist. A slightly enhanced version of this algorithm does, however, converge to a solution of (4.37) in most cases.

    The main work in each iteration of this method is, of course, the Cholesky factorization of $B + \lambda^{(\ell)}I$. Practical versions of this algorithm do not iterate until convergence to the optimal λ is obtained with high accuracy, but are content with an approximate solution that can be obtained in two or three iterations.

THE HARD CASE

    Recall that in the discussion above, we assumed that $q_1^T g \neq 0$. In fact, the approach described above can be applied even when the most negative eigenvalue is a multiple eigenvalue (that is, $0 > \lambda_1 = \lambda_2 = \cdots$), provided that $Q_1^T g \neq 0$, where Q_1 is the matrix whose columns span the subspace corresponding to the eigenvalue λ_1. When this condition does not hold, the situation becomes a little complicated, because the limit (4.41) does not hold for $\lambda_j = \lambda_1$ and so there may not be a value $\lambda \in (-\lambda_1, \infty)$ such that $\|p(\lambda)\| = \Delta$ (see Figure 4.7). Moré and Sorensen [214] refer to this case as the *hard case*. At first glance, it is not clear how p and λ can be chosen to satisfy (4.8) in the hard case. Clearly, our root-finding technique will not work, since there is no solution for λ in the open interval $(-\lambda_1, \infty)$. But Theorem 4.1 assures us that the right value of λ lies in the interval $[-\lambda_1, \infty)$, so there is only

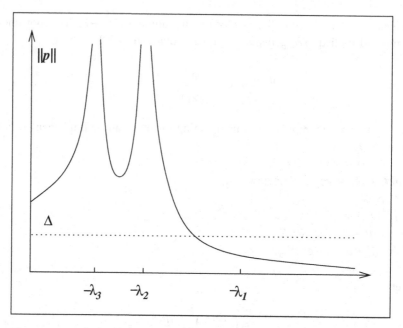

Figure 4.7 The hard case: $\|p(\lambda)\| < \Delta$ for all $\lambda \in (-\lambda_1, \infty)$.

one possibility: $\lambda = -\lambda_1$. To find p, it is not enough to delete the terms for which $\lambda_j = \lambda_1$ from the formula (4.38) and set

$$p = \sum_{j:\lambda_j \neq \lambda_1} \frac{q_j^T g}{\lambda_j + \lambda} q_j.$$

Instead, we note that $(B - \lambda_1 I)$ is singular, so there is a vector z such that $\|z\| = 1$ and $(B - \lambda_1 I)z = 0$. In fact, z is an eigenvector of B corresponding to the eigenvalue λ_1, so by orthogonality of Q we have $q_j^T z = 0$ for $\lambda_j \neq \lambda_1$. It follows from this property that if we set

$$p = \sum_{j:\lambda_j \neq \lambda_1} \frac{q_j^T g}{\lambda_j + \lambda} q_j + \tau z \tag{4.45}$$

for any scalar τ, we have

$$\|p\|^2 = \sum_{j:\lambda_j \neq \lambda_1} \frac{\left(q_j^T g\right)^2}{(\lambda_j + \lambda)^2} + \tau^2,$$

so it is always possible to choose τ to ensure that $\|p\| = \Delta$. It is easy to check that the conditions (4.8) holds for this choice of p and $\lambda = -\lambda_1$.

PROOF OF THEOREM 4.1

We now give a formal proof of Theorem 4.1, the result that characterizes the exact solution of (4.5). The proof relies on the following technical lemma, which deals with the unconstrained minimizers of quadratics and is particularly interesting in the case where the Hessian is positive semidefinite.

Lemma 4.7.

Let m be the quadratic function defined by

$$m(p) = g^T p + \tfrac{1}{2} p^T B p, \tag{4.46}$$

where B is any symmetric matrix. Then the following statements are true.

(i) *m attains a minimum if and only if B is positive semidefinite and g is in the range of B. If B is positive semidefinite, then every p satisfying $Bp = -g$ is a global minimizer of m.*

(ii) *m has a unique minimizer if and only if B is positive definite.*

PROOF.   We prove each of the three claims in turn.

(i) We start by proving the "if" part. Since g is in the range of B, there is a p with $Bp = -g$. For all $w \in R^n$, we have

$$
\begin{aligned}
m(p + w) &= g^T(p + w) + \tfrac{1}{2}(p + w)^T B(p + w) \\
&= (g^T p + \tfrac{1}{2} p^T B p) + g^T w + (Bp)^T w + \tfrac{1}{2} w^T B w \\
&= m(p) + \tfrac{1}{2} w^T B w \\
&\geq m(p), \tag{4.47}
\end{aligned}
$$

since B is positive semidefinite. Hence, p is a minimizer of m.

For the "only if" part, let p be a minimizer of m. Since $\nabla m(p) = Bp + g = 0$, we have that g is in the range of B. Also, we have $\nabla^2 m(p) = B$ positive semidefinite, giving the result.

(ii) For the "if" part, the same argument as in (i) suffices with the additional point that $w^T B w > 0$ whenever $w \neq 0$. For the "only if" part, we proceed as in (i) to deduce that B is positive semidefinite. If B is not positive definite, there is a vector $w \neq 0$ such that $Bw = 0$. Hence, from (4.47), we have $m(p + w) = m(p)$, so the minimizer is not unique, giving a contradiction. $\square$

To illustrate case (i), suppose that

$$
B = \begin{bmatrix} 1 & 0 & 0 \\ 0 & 0 & 0 \\ 0 & 0 & 2 \end{bmatrix},
$$

which has eigenvalues $0, 1, 2$ and is therefore singular. If g is any vector whose second component is zero, then g will be in the range of B, and the quadratic will attain a minimum. But if the second element in g is nonzero, we can decrease $m(\cdot)$ indefinitely by moving along the direction $(0, -g_2, 0)^T$.

We are now in a position to take account of the trust-region bound $\|p\| \leq \Delta$ and hence prove Theorem 4.1.

PROOF. (Theorem 4.1)

Assume first that there is $\lambda \geq 0$ such that the conditions (4.8) are satisfied. Lemma 4.7(i) implies that p^* is a global minimum of the quadratic function

$$\hat{m}(p) = g^T p + \tfrac{1}{2} p^T (B + \lambda I) p = m(p) + \frac{\lambda}{2} p^T p. \tag{4.48}$$

Since $\hat{m}(p) \geq \hat{m}(p^*)$, we have

$$m(p) \geq m(p^*) + \frac{\lambda}{2}((p^*)^T p^* - p^T p). \tag{4.49}$$

Because $\lambda(\Delta - \|p^*\|) = 0$ and therefore $\lambda(\Delta^2 - (p^*)^T p^*) = 0$, we have

$$m(p) \geq m(p^*) + \frac{\lambda}{2}(\Delta^2 - p^T p).$$

Hence, from $\lambda \geq 0$, we have $m(p) \geq m(p^*)$ for all p with $\|p\| \leq \Delta$. Therefore, p^* is a global minimizer of (4.7).

For the converse, we assume that p^* is a global solution of (4.7) and show that there is a $\lambda \geq 0$ that satisfies (4.8).

In the case $\|p^*\| < \Delta$, p^* is an unconstrained minimizer of m, and so

$$\nabla m(p^*) = Bp^* + g = 0, \qquad \nabla^2 m(p^*) = B \text{ positive semidefinite,}$$

and so the properties (4.8) hold for $\lambda = 0$.

Assume for the remainder of the proof that $\|p^*\| = \Delta$. Then (4.8b) is immediately satisfied, and p^* also solves the constrained problem

$$\min m(p) \quad \text{subject to } \|p\| = \Delta.$$

By applying optimality conditions for constrained optimization to this problem (see (12.34)), we find that there is a λ such that the Lagrangian function defined by

$$\mathcal{L}(p, \lambda) = m(p) + \frac{\lambda}{2}(p^T p - \Delta^2)$$

has a stationary point at p^*. By setting $\nabla_p \mathcal{L}(p^*, \lambda)$ to zero, we obtain

$$Bp^* + g + \lambda p^* = 0 \quad \Rightarrow \quad (B + \lambda I)p^* = -g, \tag{4.50}$$

so that (4.8a) holds. Since $m(p) \geq m(p^*)$ for any p with $p^T p = (p^*)^T p^* = \Delta^2$, we have for such vectors p that

$$m(p) \geq m(p^*) + \frac{\lambda}{2}\left((p^*)^T p^* - p^T p\right).$$

If we substitute the expression for g from (4.50) into this expression, we obtain after some rearrangement that

$$\tfrac{1}{2}(p - p^*)^T (B + \lambda I)(p - p^*) \geq 0. \tag{4.51}$$

Since the set of directions

$$\left\{ w : w = \pm \frac{p - p^*}{\|p - p^*\|}, \text{ for some } p \text{ with } \|p\| = \Delta \right\}$$

is dense on the unit sphere, (4.51) suffices to prove (4.8c).

It remains to show that $\lambda \geq 0$. Because (4.8a) and (4.8c) are satisfied by p^*, we have from Lemma 4.7(i) that p^* minimizes $\hat{m}$, so (4.49) holds. Suppose that there are only negative values of λ that satisfy (4.8a) and (4.8c). Then we have from (4.49) that $m(p) \geq m(p^*)$ whenever $\|p\| \geq \|p^*\| = \Delta$. Since we already know that p^* minimizes m for $\|p\| \leq \Delta$, it follows that m is in fact a global, unconstrained minimizer of m. From Lemma 4.7(i) it follows that $Bp = -g$ and B is positive semidefinite. Therefore conditions (4.8a) and (4.8c) are satisfied by $\lambda = 0$, which contradicts our assumption that only negative values of λ can satisfy the conditions. We conclude that $\lambda \geq 0$, completing the proof. $\qquad \square$

CONVERGENCE OF ALGORITHMS BASED ON NEARLY EXACT SOLUTIONS

As we noted in the discussion of Algorithm 4.3, the loop to determine the optimal values of λ and p for the subproblem (4.5) does not iterate until high accuracy is achieved. Instead, it is terminated after two or three iterations with a fairly loose approximation to the true solution. The inexactness in this approximate solution is measured in a different way from the dogleg and subspace minimization algorithms. We can add safeguards to the root-finding Newton method to ensure that the key assumptions of Theorems 4.5 and 4.6 are satisfied by the approximate solution. Specifically, we require that

$$m(0) - m(p) \geq c_1(m(0) - m(p^*)), \tag{4.52a}$$

$$\|p\| \leq \gamma \Delta \tag{4.52b}$$

(where p^* is the exact solution of (4.3)), for some constants $c_1 \in (0, 1]$ and $\gamma > 0$. The condition (4.52a) ensures that the approximate solution achieves a significant fraction of the maximum decrease possible in the model function m. (It is not necessary to know p^*; there are practical termination criteria that imply (4.52a).) One major difference between (4.52) and the earlier criterion (4.20) is that (4.52) makes better use of the second-order part of $m(\cdot)$, that is, the $p^T B p$ term. This difference is illustrated by the case in which $g = 0$ while B has negative eigenvalues, indicating that the current iterate x_k is a saddle point. Here, the right-hand-side of (4.20) is zero (indeed, the algorithms we described earlier would terminate at such a point). The right-hand-side of (4.52) is positive, indicating that decrease in the model function is still possible, so it forces the algorithm to move away from x_k.

The close attention that near-exact algorithms pay to the second-order term is warranted only if this term closely reflects the actual behavior of the function f—in fact, the trust-region Newton method, for which $B = \nabla^2 f(x)$, is the only case that has been treated in the literature. For purposes of global convergence analysis, the use of the exact Hessian allows us to say more about the limit points of the algorithm than merely that they are stationary points. The following result shows that second-order necessary conditions (Theorem 2.3) are satisfied at the limit points.

Theorem 4.8.

Suppose that the assumptions of Theorem 4.6 are satisfied and in addition that f is twice continuously differentiable in the level set S. Suppose that $B_k = \nabla^2 f(x_k)$ for all k, and that the approximate solution p_k of (4.3) at each iteration satisfies (4.52) for some fixed $\gamma > 0$. Then $\lim_{k \to \infty} \|g_k\| = 0$.

If, in addition, the level set S of (4.24) is compact, then either the algorithm terminates at a point x_k at which the second-order necessary conditions (Theorem 2.3) for a local solution hold, or else $\{x_k\}$ has a limit point x^* in S at which the second-order necessary conditions hold.

We omit the proof, which can be found in Moré and Sorensen [214, Section 4].

4.4 LOCAL CONVERGENCE OF TRUST-REGION NEWTON METHODS

Since global convergence of trust-region methods that use exact Hessians $\nabla^2 f(x)$ is established above, we turn our attention now to local convergence issues. The key to attaining the fast rate of convergence usually associated with Newton's method is to show that the trust-region bound eventually does not interfere as we approach a solution. Specifically, we hope that near the solution, the (approximate) solution of the trust-region subproblem is well inside the trust region and becomes closer and closer to the true Newton step. Steps that satisfy the latter property are said to be *asymptotically similar* to Newton steps.

We first prove a general result that applies to any algorithm of the form of Algorithm 4.1 (see Chapter 4) that generates steps that are asymptotically similar to Newton steps whenever the Newton steps easily satisfy the trust-region bound. It shows that the trust-region constraint eventually becomes inactive in algorithms with this property and that superlinear convergence can be attained. The result assumes that the exact Hessian $B_k = \nabla^2 f(x_k)$ is used in (4.3) when x_k is close to a solution x^* that satisfies second-order sufficient conditions (see Theorem 2.4). Moreover, it assumes that the algorithm uses an approximate solution p_k of (4.3) that achieves a similar decrease in the model function m_k as the Cauchy point.

Theorem 4.9.

   Let f be twice Lipschitz continuously differentiable in a neighborhhod of a point x^* at which second-order sufficient conditions (Theorem 2.4) are satisfied. Suppose the sequence $\{x_k\}$ converges to x^* and that for all k sufficiently large, the trust-region algorithm based on (4.3) with $B_k = \nabla^2 f(x_k)$ chooses steps p_k that satisfy the Cauchy-point-based model reduction criterion (4.20) and are asymptotically similar to Newton steps p_k^{N} whenever $\|p_k^{\text{N}}\| \leq \frac{1}{2}\Delta_k$, that is,

$$\|p_k - p_k^{\text{N}}\| = o(\|p_k^{\text{N}}\|). \tag{4.53}$$

Then the trust-region bound Δ_k becomes inactive for all k sufficiently large and the sequence $\{x_k\}$ converges superlinearly to x^*.

PROOF.   We show that $\|p_k^{\text{N}}\| \leq \frac{1}{2}\Delta_k$ and $\|p_k\| \leq \Delta_k$, for all sufficiently large k, so the near-optimal step p_k in (4.53) will eventually always be taken.

   We first seek a lower bound on the predicted reduction $m_k(0) - m_k(p_k)$ for all sufficiently large k. We assume that k is large enough that the $o(\|p_k^{\text{N}}\|)$ term in (4.53) is less than $\|p_k^{\text{N}}\|$. When $\|p_k^{\text{N}}\| \leq \frac{1}{2}\Delta_k$, we then have that $\|p_k\| \leq \|p_k^{\text{N}}\| + o(\|p_k^{\text{N}}\|) \leq 2\|p_k^{\text{N}}\|$, while if $\|p_k^{\text{N}}\| > \frac{1}{2}\Delta_k$, we have $\|p_k\| \leq \Delta_k < 2\|p_k^{\text{N}}\|$. In both cases, then, we have

$$\|p_k\| \leq 2\|p_k^{\text{N}}\| \leq 2 \left\| \nabla^2 f(x_k)^{-1} \right\| \|g_k\|,$$

and so $\|g_k\| \geq \frac{1}{2}\|p_k\| / \left\| \nabla^2 f(x_k)^{-1} \right\|$.

   We have from the relation (4.20) that

$$
\begin{aligned}
m_k(0) &- m_k(p_k) \\
&\geq c_1 \|g_k\| \min \left(\Delta_k, \frac{\|g_k\|}{\left\| \nabla^2 f(x_k) \right\|} \right) \\
&\geq c_1 \frac{\|p_k\|}{2 \left\| \nabla^2 f(x_k)^{-1} \right\|} \min \left(\|p_k\|, \frac{\|p_k\|}{2 \left\| \nabla^2 f(x_k) \right\| \left\| \nabla^2 f(x_k)^{-1} \right\|} \right) \\
&= c_1 \frac{\|p_k\|^2}{4 \left\| \nabla^2 f(x_k)^{-1} \right\|^2 \left\| \nabla^2 f(x_k) \right\|}.
\end{aligned}
$$

Because $x_k \to x^*$, we use continuity of $\nabla^2 f(x)$ and positive definiteness of $\nabla^2 f(x^*)$, to deduce that the following bound holds for all k sufficiently large:

$$\frac{c_1}{4\left\|\nabla^2 f(x_k)^{-1}\right\|^2 \left\|\nabla^2 f(x_k)\right\|} \geq \frac{c_1}{8\left\|\nabla^2 f(x^*)^{-1}\right\|^2 \left\|\nabla^2 f(x^*)\right\|} \overset{\text{def}}{=} c_3,$$

where $c_3 > 0$. Hence, we hae

$$m_k(0) - m_k(p_k) \geq c_3 \|p_k\|^2 \qquad (4.54)$$

for all sufficiently large k. By Lipschitz continuity of $\nabla^2 f(x)$ near x^*, and using Taylor's theorem (Theorem 2.1), we have

$$
\begin{aligned}
&|(f(x_k) - f(x_k + p_k)) - (m_k(0) - m_k(p_k))| \\
&= \left| \tfrac{1}{2} p_k^T \nabla^2 f(x_k) p_k - \tfrac{1}{2} \int_0^1 p_k^T \nabla^2 f(x_k + t p_k) p_k \, dt \right| \\
&\leq \frac{L}{4} \|p_k\|^3,
\end{aligned}
$$

where $L > 0$ is the Lipschitz constant for $\nabla^2 f(\cdot)$. Hence, by definition (4.4) of ρ_k, we have for sufficiently large k that

$$|\rho_k - 1| \leq \frac{\|p_k\|^3 (L/4)}{c_3 \|p_k\|^2} = \frac{L}{4c_3} \|p_k\| \leq \frac{L}{4c_3} \Delta_k. \qquad (4.55)$$

Now, the trust-region radius can be reduced only if $\rho_k < \frac{1}{4}$ (or some other fixed number less than 1), so it is clear from (4.55) that the sequence $\{\Delta_k\}$ is bounded away from zero. Since $x_k \to x^*$, we have $\|p_k^N\| \to 0$ and therefore $\|p_k\| \to 0$ from (4.53). Hence, the trust-region bound is inactive for all k sufficiently large, and the bound $\|p_k^N\| \leq \frac{1}{2}\Delta_k$ is eventually always satisfied.

To prove superlinear convergence, we use the quadratic convergence of Newton's method, proved in Theorem 3.5. In particular, we have from (3.33) that

$$\|x_k + p_k^N - x^*\| = o\left(\|x_k - x^*\|^2\right),$$

which implies that $\|p_k^N\| = O(\|x_k - x^*\|)$. Therefore, using (4.53), we have

$$
\begin{aligned}
&\|x_k + p_k - x^*\| \\
&\leq \|x_k + p_k^N - x^*\| + \|p_k^N - p_k\| = o\left(\|x_k - x^*\|^2\right) + o(\|p_k^N\|) = o\left(\|x_k - x^*\|\right),
\end{aligned}
$$

thus proving superlinear convergence. $\qquad\square$

It is immediate from Theorem 3.5 that if $p_k = p_k^N$ for all k sufficiently large, we have quadratic convergence of $\{x_k\}$ to x^*.

Reasonable implementations of the dogleg, subspace minimization, and nearly-exact algorithm of Section 4.3 with $B_k = \nabla^2 f(x_k)$ eventually use the steps $p_k = p_k^N$ under the conditions of Theorem 4.9, and therefore converge quadratically. In the case of the dogleg and two-dimensional subspace minimization methods, the exact step p_k^N is one of the candidates for p_k—it lies inside the trust region, along the dogleg path, and inside the two-dimensional subspace. Since under the assumptions of Theorem 4.9, p_k^N is the unconstrained minimizer of m_k for k sufficiently large, it is certainly the minimizer in the more restricted domains, so we have $p_k = p_k^N$. For the approach of Section 4.3, if we follow the reasonable strategy of checking whether p_k^N is a solution of (4.3) prior to embarking on Algorithm 4.3, then eventually we will also have $p_k = p_k^N$ also.

4.5 OTHER ENHANCEMENTS

SCALING

As we noted in Chapter 2, optimization problems are often posed with poor scaling—the objective function f is highly sensitive to small changes in certain components of the vector x and relatively insensitive to changes in other components. Topologically, a symptom of poor scaling is that the minimizer x^* lies in a narrow valley, so that the contours of the objective $f(\cdot)$ near x^* tend towards highly eccentric ellipses. Algorithms that fail to compensate for poor scaling can perform badly; see Figure 2.7 for an illustration of the poor performance of the steepest descent approach.

Recalling our definition of a trust region—a region around the current iterate within which the model $m_k(\cdot)$ is an adequate representation of the true objective $f(\cdot)$—it is easy to see that a *spherical* trust region may not be appropriate when f is poorly scaled. Even if the model Hessian B_k is exact, the rapid changes in f along certain directions probably will cause m_k to be a poor approximation to f along these directions. On the other hand, m_k may be a more reliable approximation to f along directions in which f is changing more slowly. Since the shape of our trust region should be such that our confidence in the model is more or less the same at all points on the boundary of the region, we are led naturally to consider *elliptical* trust regions in which the axes are short in the sensitive directions and longer in the less sensitive directions.

Elliptical trust regions can be defined by

$$\|Dp\| \le \Delta, \tag{4.56}$$

where D is a diagonal matrix with positive diagonal elements, yielding the following scaled trust-region subproblem:

$$\min_{p \in \mathbb{R}^n} m_k(p) \stackrel{\text{def}}{=} f_k + g_k^T p + \tfrac{1}{2} p^T B_k p \qquad \text{s.t. } \|Dp\| \le \Delta_k. \tag{4.57}$$

When $f(x)$ is highly sensitive to the value of the ith component x_i, we set the corresponding diagonal element d_{ii} of D to be large, while d_{ii} is smaller for less-sensitive components.

Information to construct the scaling matrix D may be derived from the second derivatives $\partial^2 f / \partial x_i^2$. We can allow D to change from iteration to iteration; most of the theory of this chapter will still apply with minor modifications provided that each d_{ii} stays within some predetermined range $[d_{lo}, d_{hi}]$, where $0 < d_{lo} \leq d_{hi} < \infty$. Of course, we do not need D to be a *precise* reflection of the scaling of the problem, so it is not necessary to devise elaborate heuristics or to perform extensive computations to get it just right.

The following procedure shows how the Cauchy point calculation (Algorithm 4.2) changes when we use a scaled trust region,

Algorithm 4.4 (Generalized Cauchy Point Calculation).
Find the vector p_k^s that solves

$$p_k^s = \arg \min_{p \in \mathbb{R}^n} f_k + g_k^T p \qquad \text{s.t. } \|Dp\| \leq \Delta_k; \tag{4.58}$$

Calculate the scalar $\tau_k > 0$ that minimizes $m_k(\tau p_k^s)$ subject to satisfying the trust-region bound, that is,

$$\tau_k = \arg \min_{\tau > 0} m_k(\tau p_k^s) \qquad \text{s.t. } \|\tau D p_k^s\| \leq \Delta_k; \tag{4.59}$$
$$p_k^c = \tau_k p_k^s.$$

For this scaled version, we find that

$$p_k^s = -\frac{\Delta_k}{\|D^{-1} g_k\|} D^{-2} g_k, \tag{4.60}$$

and that the step length τ_k is obtained from the following modification of (4.12):

$$\tau_k = \begin{cases} 1 & \text{if } g_k^T D^{-2} B_k D^{-2} g_k \leq 0 \\ \min\left(\dfrac{\|D^{-1} g_k\|^3}{\Delta_k g_k^T D^{-2} B_k D^{-2} g_k}, 1 \right) & \text{otherwise.} \end{cases} \tag{4.61}$$

(The details are left as an exercise.)

A simpler alternative for adjusting the definition of the Cauchy point and the various algorithms of this chapter to allow for the elliptical trust region is simply to rescale the variables p in the subproblem (4.57) so that the trust region is spherical in the scaled variables. By defining

$$\tilde{p} \overset{\text{def}}{=} Dp,$$

and by substituting into (4.57), we obtain

$$\min_{\tilde{p}\in\mathbb{R}^n} \tilde{m}_k(\tilde{p}) \overset{\text{def}}{=} f_k + g_k^T D^{-1}\tilde{p} + \tfrac{1}{2}\tilde{p}^T D^{-1} B_k D^{-1}\tilde{p} \qquad \text{s.t. } \|\tilde{p}\| \le \Delta_k.$$

The theory and algorithms can now be derived in the usual way by substituting $\tilde{p}$ for p, $D^{-1}g_k$ for g_k, $D^{-1}B_k D^{-1}$ for B_k, and so on.

TRUST REGIONS IN OTHER NORMS

Trust regions may also be defined in terms of norms other than the Euclidean norm. For instance, we may have

$$\|p\|_1 \le \Delta_k \qquad \text{or} \qquad \|p\|_\infty \le \Delta_k,$$

or their scaled counterparts

$$\|Dp\|_1 \le \Delta_k \qquad \text{or} \qquad \|Dp\|_\infty \le \Delta_k,$$

where D is a positive diagonal matrix as before. Norms such as these offer no obvious advantages for small-medium unconstrained problems, but they may be useful for constrained problems. For instance, for the bound-constrained problem

$$\min_{x\in\mathbb{R}^n} f(x), \qquad \text{subject to } x \ge 0,$$

the trust-region subproblem may take the form

$$\min_{p\in\mathbb{R}^n} m_k(p) = f_k + g_k^T p + \tfrac{1}{2}p^T B_k p \qquad \text{s.t. } x_k + p \ge 0, \|p\| \le \Delta_k. \tag{4.62}$$

When the trust region is defined by a Euclidean norm, the feasible region for (4.62) consists of the intersection of a sphere and the nonnegative orthant—an awkward object, geometrically speaking. When the ∞-norm is used, however, the feasible region is simply the rectangular box defined by

$$x_k + p \ge 0, \qquad p \ge -\Delta_k e, \qquad p \le \Delta_k e,$$

where $e = (1, 1, \dots, 1)^T$, so the solution of the subproblem is easily calculated by using techniques for bound-constrained quadratic programming.

For large problems, in which factorization or formation the model Hessian B_k is not computationally desirable, the use of a trust region defined by $\|\cdot\|_\infty$ will also give rise to a bound-constrained subproblem, which may be more convenient to solve than the standard subproblem (4.3). To our knowledge, there has not been much research on the relative performance of methods that use trust regions of different shapes on large problems.

NOTES AND REFERENCES

One of the earliest works on trust-region methods is Winfield [307]. The influential paper of Powell [244] proves a result like Theorem 4.5 for the case of $\eta = 0$, where the algorithm takes a step whenever it decreases the function value. Powell uses a weaker assumption than ours on the matrices $\|B\|$, but his analysis is more complicated. Moré [211] summarizes developments in algorithms and software before 1982, paying particular attention to the importance of using a scaled trust-region norm.

Byrd, Schnabel, and Schultz [279], [54] provide a general theory for inexact trust-region methods; they introduce the idea of two-dimensional subspace minimization and also focus on proper handling of the case of indefinite B to ensure stronger local convergence results than Theorems 4.5 and 4.6. Dennis and Schnabel [93] survey trust-region methods as part of their overview of unconstrained optimization, providing pointers to many important developments in the literature.

The monograph of Conn, Gould, and Toint [74] is an exhaustive treatment of the state of the art in trust-region methods for both unconstrained and constrained optimization. It includes an comprehensive annotated bibliography of the literature in the area.

✎ EXERCISES

✎   **4.1** Let $f(x) = 10(x_2 - x_1^2)^2 + (1 - x_1)^2$. At $x = (0, -1)$ draw the contour lines of the quadratic model (4.2) assuming that B is the Hessian of f. Draw the family of solutions of (4.3) as the trust region radius varies from $\Delta = 0$ to $\Delta = 2$. Repeat this at $x = (0, 0.5)$.

✎   **4.2** Write a program that implements the dogleg method. Choose B_k to be the exact Hessian. Apply it to solve Rosenbrock's function (2.22). Experiment with the update rule for the trust region by changing the constants in Algorithm 4.1, or by designing your own rules.

✎   **4.3** Program the trust-region method based on Algorithm 7.2. Choose B_k to be the exact Hessian, and use it to minimize the function

$$\min \; f(x) = \sum_{i=1}^{n} \left[(1 - x_{2i-1})^2 + 10(x_{2i} - x_{2i-1}^2)^2 \right]$$

with $n = 10$. Experiment with the starting point and the stopping test for the CG iteration. Repeat the computation with $n = 50$.

Your program should indicate, at every iteration, whether Algorithm 7.2 encountered negative curvature, reached the trust-region boundary, or met the stopping test.

✎ **4.4** Theorem 4.5 shows that the sequence $\{\|g\|\}$ has an accumulation point at zero. Show that if the iterates x stay in a bounded set $\mathcal{B}$, then there is a limit point x_∞ of the sequence $\{x_k\}$ such that $g(x_\infty) = 0$.

✎ **4.5** Show that τ_k defined by (4.12) does indeed identify the minimizer of m_k along the direction $-g_k$.

✎ **4.6** The Cauchy–Schwarz inequality states that for any vectors u and v, we have

$$|u^T v|^2 \leq (u^T u)(v^T v),$$

with equality only when u and v are parallel. When B is positive definite, use this inequality to show that

$$\gamma \overset{\text{def}}{=} \frac{\|g\|^4}{(g^T B g)(g^T B^{-1} g)} \leq 1,$$

with equality only if g and Bg (and $B^{-1}g$) are parallel.

✎ **4.7** When B is positive definite, the *double-dogleg method* constructs a path with three line segments from the origin to the full step. The four points that define the path are

- the origin;

- the unconstrained Cauchy step $p^c = -(g^T g)/(g^T B g)g$;

- a fraction of the full step $\bar{\gamma} p^B = -\bar{\gamma} B^{-1} g$, for some $\bar{\gamma} \in (\gamma, 1]$, where γ is defined in the previous question; and

- the full step $p^B = -B^{-1}g$.

Show that $\|p\|$ increases monotonically along this path.

(Note: The double-dogleg method, as discussed in Dennis and Schnabel [92, Section 6.4.2], was for some time thought to be superior to the standard dogleg method, but later testing has not shown much difference in performance.)

✎ **4.8** Show that (4.43) and (4.44) are equivalent. Hints: Note that

$$\frac{d}{d\lambda}\left(\frac{1}{\|p(\lambda)\|}\right) = \frac{d}{d\lambda}\left(\|p(\lambda)\|^2\right)^{-1/2} = -\frac{1}{2}\left(\|p(\lambda)\|^2\right)^{-3/2}\frac{d}{d\lambda}\|p(\lambda)\|^2,$$

$$\frac{d}{d\lambda}\|p(\lambda)\|^2 = -2\sum_{j=1}^{n}\frac{(q_j^T g)^2}{(\lambda_j + \lambda)^3}$$

(from (4.39)), and

$$\|q\|^2 = \|R^{-T}p\|^2 = p^T(B + \lambda I)^{-1}p = \sum_{j=1}^{n} \frac{(q_j^T g)^2}{(\lambda_j + \lambda)^3}.$$

✎ **4.9** Derive the solution of the two-dimensional subspace minimization problem in the case where B is positive definite.

✎ **4.10** Show that if B is any symmetric matrix, then there exists $\lambda \geq 0$ such that $B + \lambda I$ is positive definite.

✎ **4.11** Verify that the definitions (4.60) for p_k^s and (4.61) for τ_k are valid for the Cauchy point in the case of an elliptical trust region. (Hint: Using the theory of Chapter 12, we can show that the solution of (4.58) satisfies $g_k + \alpha D^2 p_k^s = 0$ for some scalar $\alpha \geq 0$.)

✎ **4.12** The following example shows that the reduction in the model function m achieved by the two-dimensional minimization strategy can be much smaller than that achieved by the exact solution of (4.5).

In (4.5), set

$$g = \left(-\frac{1}{\epsilon}, -1, -\epsilon^2\right)^T,$$

where ϵ is a small positive number. Set

$$B = \text{diag}\left(\frac{1}{\epsilon^3}, 1, \epsilon^3\right), \quad \Delta = 0.5.$$

Show that the solution of (4.5) has components $\left(O(\epsilon), \frac{1}{2} + O(\epsilon), O(\epsilon)\right)^T$ and that the reduction in the model m is $\frac{3}{8} + O(\epsilon)$. For the two-dimensional minimization strategy, show that the solution is a multiple of $B^{-1}g$ and that the reduction in m is $O(\epsilon)$.

CHAPTER **5**

Conjugate Gradient Methods

Our interest in conjugate gradient methods is twofold. First, they are among the most useful techniques for solving large linear systems of equations. Second, they can be adapted to solve nonlinear optimization problems. The remarkable properties of both *linear* and *nonlinear* conjugate gradient methods will be described in this chapter.

The *linear* conjugate gradient method was proposed by Hestenes and Stiefel in the 1950s as an iterative method for solving linear systems with positive definite coefficient matrices. It is an alternative to Gaussian elimination that is well suited for solving large problems. The performance of the linear conjugate gradient method is determined by the distribution of the eigenvalues of the coefficient matrix. By transforming, or *preconditioning*, the linear system, we can make this distribution more favorable and improve the convergence of the method significantly. Preconditioning plays a crucial role in the design of practical conjugate gradient strategies. Our treatment of the linear conjugate gradient method will highlight those properties of the method that are important in optimization.

The first *nonlinear* conjugate gradient method was introduced by Fletcher and Reeves in the 1960s. It is one of the earliest known techniques for solving large-scale nonlinear optimization problems. Over the years, many variants of this original scheme have been proposed, and some are widely used in practice. The key features of these algorithms are that they require no matrix storage and are faster than the steepest descent method.

5.1   THE LINEAR CONJUGATE GRADIENT METHOD

In this section we derive the linear conjugate gradient method and discuss its essential convergence properties. For simplicity, we drop the qualifier "linear" throughout.

The conjugate gradient method is an iterative method for solving a linear system of equations

$$Ax = b, \tag{5.1}$$

where A is an $n \times n$ symmetric positive definite matrix. The problem (5.1) can be stated equivalently as the following minimization problem:

$$\min \phi(x) \stackrel{\text{def}}{=} \tfrac{1}{2}x^T A x - b^T x, \tag{5.2}$$

that is, both (5.1) and (5.2) have the same unique solution. This equivalence will allow us to interpret the conjugate gradient method either as an algorithm for solving linear systems or as a technique for minimizing convex quadratic functions. For future reference, we note that the gradient of ϕ equals the residual of the linear system, that is,

$$\nabla \phi(x) = Ax - b \stackrel{\text{def}}{=} r(x), \tag{5.3}$$

so in particular at $x = x_k$ we have

$$r_k = Ax_k - b. \tag{5.4}$$

CONJUGATE DIRECTION METHODS

One of the remarkable properties of the conjugate gradient method is its ability to generate, in a very economical fashion, a set of vectors with a property known as *conjugacy*. A set of nonzero vectors $\{p_0, p_1, \ldots, p_l\}$ is said to be *conjugate* with respect to the symmetric positive definite matrix A if

$$p_i^T A p_j = 0, \qquad \text{for all } i \neq j. \tag{5.5}$$

It is easy to show that any set of vectors satisfying this property is also linearly independent. (For a geometrical illustration of conjugate directions see Section 9.4.)

The importance of conjugacy lies in the fact that we can minimize $\phi(\cdot)$ in n steps by successively minimizing it along the individual directions in a conjugate set. To verify this claim, we consider the following *conjugate direction* method. (The distinction between the conjugate gradient method and the conjugate direction method will become clear as we proceed.) Given a starting point $x_0 \in \mathbb{R}^n$ and a set of conjugate directions $\{p_0, p_1, \ldots, p_{n-1}\}$, let us generate the sequence $\{x_k\}$ by setting

$$x_{k+1} = x_k + \alpha_k p_k, \tag{5.6}$$

where α_k is the one-dimensional minimizer of the quadratic function $\phi(\cdot)$ along $x_k + \alpha p_k$, given explicitly by

$$\alpha_k = -\frac{r_k^T p_k}{p_k^T A p_k}; \tag{5.7}$$

see (3.55). We have the following result.

Theorem 5.1.

For any $x_0 \in \mathbb{R}^n$ the sequence $\{x_k\}$ generated by the conjugate direction algorithm (5.6), (5.7) converges to the solution x^* of the linear system (5.1) in at most n steps.

PROOF. Since the directions $\{p_i\}$ are linearly independent, they must span the whole space $\mathbb{R}^n$. Hence, we can write the difference between x_0 and the solution x^* in the following way:

$$x^* - x_0 = \sigma_0 p_0 + \sigma_1 p_1 + \cdots + \sigma_{n-1} p_{n-1},$$

for some choice of scalars σ_k. By premultiplying this expression by $p_k^T A$ and using the conjugacy property (5.5), we obtain

$$\sigma_k = \frac{p_k^T A(x^* - x_0)}{p_k^T A p_k}. \tag{5.8}$$

We now establish the result by showing that these coefficients σ_k coincide with the step lengths α_k generated by the formula (5.7).

If x_k is generated by algorithm (5.6), (5.7), then we have

$$x_k = x_0 + \alpha_0 p_0 + \alpha_1 p_1 + \cdots + \alpha_{k-1} p_{k-1}.$$

By premultiplying this expression by $p_k^T A$ and using the conjugacy property, we have that

$$p_k^T A(x_k - x_0) = 0,$$

and therefore

$$p_k^T A(x^* - x_0) = p_k^T A(x^* - x_k) = p_k^T (b - Ax_k) = -p_k^T r_k.$$

By comparing this relation with (5.7) and (5.8), we find that $\sigma_k = \alpha_k$, giving the result.   □

There is a simple interpretation of the properties of conjugate directions. If the matrix A in (5.2) is diagonal, the contours of the function $\phi(\cdot)$ are ellipses whose axes are aligned with the coordinate directions, as illustrated in Figure 5.1. We can find the minimizer of this function by performing one-dimensional minimizations along the coordinate directions

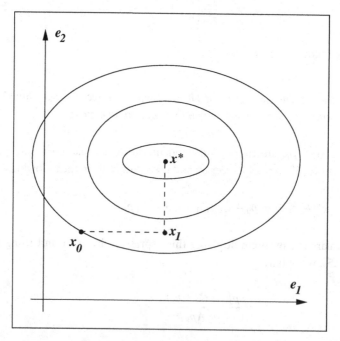

Figure 5.1   Successive minimizations along the coordinate directions find the minimizer of a quadratic with a diagonal Hessian in n iterations.

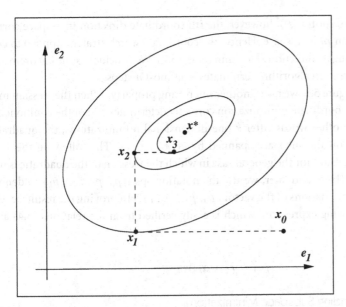

Figure 5.2 Successive minimization along coordinate axes does not find the solution in n iterations, for a general convex quadratic.

$e_1, e_2, \ldots, e_n$ in turn. When A is *not* diagonal, its contours are still elliptical, but they are usually no longer aligned with the coordinate directions. The strategy of successive minimization along these directions in turn no longer leads to the solution in n iterations (or even in a finite number of iterations). This phenomenon is illustrated in the two-dimensional example of Figure 5.2 We can, however, recover the nice behavior of Figure 5.1 if we transform the problem to make A diagonal and then minimize along the coordinate directions. Suppose we transform the problem by defining new variables $\hat{x}$ as

$$\hat{x} = S^{-1}x, \tag{5.9}$$

where S is the $n \times n$ matrix defined by

$$S = [p_0 \; p_1 \; \cdots \; p_{n-1}],$$

where $\{p_0, p_2, \ldots, p_{n-1}\}$ is the set of conjugate directions with respect to A. The quadratic ϕ defined by (5.2) now becomes

$$\hat{\phi}(\hat{x}) \stackrel{\text{def}}{=} \phi(S\hat{x}) = \tfrac{1}{2}\hat{x}^T (S^T A S)\hat{x} - (S^T b)^T \hat{x}.$$

By the conjugacy property (5.5), the matrix $S^T A S$ is diagonal, so we can find the minimizing value of $\hat{\phi}$ by performing n one-dimensional minimizations along the coordinate directions of $\hat{x}$. Because of the relation (5.9), however, the ith coordinate direction in $\hat{x}$-space corresponds to the direction p_i in x-space. Hence, the coordinate search strategy applied to $\hat{\phi}$ is equivalent to the conjugate direction algorithm (5.6), (5.7). We conclude, as in Theorem 5.1, that the conjugate direction algorithm terminates in at most n steps.

Returning to Figure 5.1, we note another interesting property: When the Hessian matrix is diagonal, each coordinate minimization correctly determines one of the components of the solution x^*. In other words, after k one-dimensional minimizations, the quadratic has been minimized on the subspace spanned by $e_1, e_2, \ldots, e_k$. The following theorem proves this important result for the general case in which the Hessian of the quadratic is not necessarily diagonal. (Here and later, we use the notation span$\{p_0, p_1, \ldots, p_k\}$ to denote the set of all linear combinations of the vectors $p_0, p_1, \ldots, p_k$.) In proving the result we will make use of the following expression, which is easily verified from the relations (5.4) and (5.6):

$$r_{k+1} = r_k + \alpha_k A p_k. \tag{5.10}$$

Theorem 5.2 (Expanding Subspace Minimization).

Let $x_0 \in \mathbb{R}^n$ be any starting point and suppose that the sequence $\{x_k\}$ is generated by the conjugate direction algorithm (5.6), (5.7). Then

$$r_k^T p_i = 0, \qquad \text{for } i = 0, 1, \ldots, k - 1, \tag{5.11}$$

and x_k is the minimizer of $\phi(x) = \frac{1}{2}x^T A x - b^T x$ over the set

$$\{x \mid x = x_0 + \text{span}\{p_0, p_1, \ldots, p_{k-1}\}\}. \tag{5.12}$$

PROOF. We begin by showing that a point $\tilde{x}$ minimizes ϕ over the set (5.12) if and only if $r(\tilde{x})^T p_i = 0$, for each $i = 0, 1, \ldots, k - 1$. Let us define $h(\sigma) = \phi(x_0 + \sigma_0 p_0 + \cdots + \sigma_{k-1}p_{k-1})$, where $\sigma = (\sigma_0, \sigma_1, \ldots, \sigma_{k-1})^T$. Since $h(\sigma)$ is a strictly convex quadratic, it has a unique minimizer σ^* that satisfies

$$\frac{\partial h(\sigma^*)}{\partial \sigma_i} = 0, \qquad i = 0, 1, \ldots, k - 1.$$

By the chain rule, this equation implies that

$$\nabla \phi(x_0 + \sigma_0^* p_0 + \cdots + \sigma_{k-1}^* p_{k-1})^T p_i = 0, \qquad i = 0, 1, \ldots, k - 1.$$

By recalling the definition (5.3), we have for the minimizer $\tilde{x} = x_0 + \sigma_0^* p_0 + \sigma_1^* p_2 + \cdots + \sigma_{k-1}^* p_{k-1}$ on the set (5.12) that $r(\tilde{x})^T p_i = 0$, as claimed.

We now use induction to show that x_k satisfies (5.11). For the case $k = 1$, we have from the fact that $x_1 = x_0 + \alpha_0 p_0$ minimizes ϕ along p_0 that $r_1^T p_0 = 0$. Let us now make the induction hypothesis, namely, that $r_{k-1}^T p_i = 0$ for $i = 0, 1, \ldots, k - 2$. By (5.10), we have

$$r_k = r_{k-1} + \alpha_{k-1} A p_{k-1},$$

so that

$$p_{k-1}^T r_k = p_{k-1}^T r_{k-1} + \alpha_{k-1} p_{k-1}^T A p_{k-1} = 0,$$

by the definition (5.7) of α_{k-1}. Meanwhile, for the other vectors $p_i, i = 0, 1, \ldots, k - 2$, we have

$$p_i^T r_k = p_i^T r_{k-1} + \alpha_{k-1} p_i^T A p_{k-1} = 0,$$

where $p_i^T r_{k-1} = 0$ because of the induction hypothesis and $p_i^T A p_{k-1} = 0$ because of conjugacy of the vectors p_i. We have shown that $r_k^T p_i = 0$, for $i = 0, 1, \ldots, k - 1$, so the proof is complete. $\qquad\square$

The fact that the current residual r_k is orthogonal to all previous search directions, as expressed in (5.11), is a property that will be used extensively in this chapter.

The discussion so far has been general, in that it applies to a conjugate direction method (5.6), (5.7) based on *any* choice of the conjugate direction set $\{p_0, p_1, \ldots, p_{n-1}\}$. There are many ways to choose the set of conjugate directions. For instance, the eigenvectors $v_1, v_2, \ldots, v_n$ of A are mutually orthogonal as well as conjugate with respect to A, so these could be used as the vectors $\{p_0, p_1, \ldots, p_{n-1}\}$. For large-scale applications, however, computation of the complete set of eigenvectors requires an excessive amount of computation. An alternative approach is to modify the Gram–Schmidt orthogonalization process to produce a set of conjugate directions rather than a set of orthogonal directions. (This modification is easy to produce, since the properties of conjugacy and orthogonality are closely related in spirit.) However, the Gram–Schmidt approach is also expensive, since it requires us to store the entire direction set.

BASIC PROPERTIES OF THE CONJUGATE GRADIENT METHOD

The conjugate gradient method is a conjugate direction method with a very special property: In generating its set of conjugate vectors, it can compute a new vector p_k by using only the previous vector p_{k-1}. It does *not* need to know all the previous elements $p_0, p_1, \ldots, p_{k-2}$ of the conjugate set; p_k is automatically conjugate to these vectors. This remarkable property implies that the method requires little storage and computation.

In the conjugate gradient method, each direction p_k is chosen to be a linear combination of the negative residual $-r_k$ (which, by (5.3), is the steepest descent direction for the function ϕ) and the previous direction p_{k-1}. We write

$$p_k = -r_k + \beta_k p_{k-1}, \tag{5.13}$$

where the scalar β_k is to be determined by the requirement that p_{k-1} and p_k must be conjugate with respect to A. By premultiplying (5.13) by $p_{k-1}^T A$ and imposing the condition $p_{k-1}^T A p_k = 0$, we find that

$$\beta_k = \frac{r_k^T A p_{k-1}}{p_{k-1}^T A p_{k-1}}.$$

We choose the first search direction p_0 to be the steepest descent direction at the initial point x_0. As in the general conjugate direction method, we perform successive one-dimensional minimizations along each of the search directions. We have thus specified a complete algorithm, which we express formally as follows:

Algorithm 5.1 (CG–Preliminary Version).
   Given x_0;
   Set $r_0 \leftarrow A x_0 - b$, $p_0 \leftarrow -r_0$, $k \leftarrow 0$;
   while $r_k \neq 0$

$$\alpha_k \leftarrow -\frac{r_k^T p_k}{p_k^T A p_k}; \tag{5.14a}$$

$$x_{k+1} \leftarrow x_k + \alpha_k p_k; \tag{5.14b}$$

$$r_{k+1} \leftarrow A x_{k+1} - b; \tag{5.14c}$$

$$\beta_{k+1} \leftarrow \frac{r_{k+1}^T A p_k}{p_k^T A p_k}; \tag{5.14d}$$

$$p_{k+1} \leftarrow -r_{k+1} + \beta_{k+1} p_k; \tag{5.14e}$$

$$k \leftarrow k + 1; \tag{5.14f}$$

   end (while)

This version is useful for studying the essential properties of the conjugate gradient method, but we present a more efficient version later. We show first that the directions $p_0, p_1, \ldots, p_{n-1}$ are indeed conjugate, which by Theorem 5.1 implies termination in n steps. The theorem below establishes this property and two other important properties. First, the residuals r_i are mutually orthogonal. Second, each search direction p_k and residual r_k is contained in the *Krylov subspace of degree k for r_0*, defined as

$$\mathcal{K}(r_0; k) \stackrel{\text{def}}{=} \text{span}\{r_0, A r_0, \ldots, A^k r_0\}. \tag{5.15}$$

Theorem 5.3.

Suppose that the kth iterate generated by the conjugate gradient method is not the solution point x^*. The following four properties hold:

$$r_k^T r_i = 0, \qquad \text{for } i = 0, 1, \ldots, k-1, \tag{5.16}$$

$$\text{span}\{r_0, r_1, \ldots, r_k\} = \text{span}\{r_0, Ar_0, \ldots, A^k r_0\}, \tag{5.17}$$

$$\text{span}\{p_0, p_1, \ldots, p_k\} = \text{span}\{r_0, Ar_0, \ldots, A^k r_0\}, \tag{5.18}$$

$$p_k^T A p_i = 0, \qquad \text{for } i = 0, 1, \ldots, k-1. \tag{5.19}$$

Therefore, the sequence $\{x_k\}$ converges to x^* in at most n steps.

PROOF.   The proof is by induction. The expressions (5.17) and (5.18) hold trivially for $k = 0$, while (5.19) holds by construction for $k = 1$. Assuming now that these three expressions are true for some k (the induction hypothesis), we show that they continue to hold for $k + 1$.

To prove (5.17), we show first that the set on the left-hand side is contained in the set on the right-hand side. Because of the induction hypothesis, we have from (5.17) and (5.18) that

$$r_k \in \text{span}\{r_0, Ar_0, \ldots, A^k r_0\}, \qquad p_k \in \text{span}\{r_0, Ar_0, \ldots, A^k r_0\},$$

while by multiplying the second of these expressions by A, we obtain

$$A p_k \in \text{span}\{Ar_0, \ldots, A^{k+1} r_0\}. \tag{5.20}$$

By applying (5.10), we find that

$$r_{k+1} \in \text{span}\{r_0, Ar_0, \ldots, A^{k+1} r_0\}.$$

By combining this expression with the induction hypothesis for (5.17), we conclude that

$$\text{span}\{r_0, r_1, \ldots, r_k, r_{k+1}\} \subset \text{span}\{r_0, Ar_0, \ldots, A^{k+1} r_0\}.$$

To prove that the reverse inclusion holds as well, we use the induction hypothesis on (5.18) to deduce that

$$A^{k+1} r_0 = A(A^k r_0) \in \text{span}\{A p_0, A p_1, \ldots, A p_k\}.$$

Since by (5.10) we have $A p_i = (r_{i+1} - r_i)/\alpha_i$ for $i = 0, 1, \ldots, k$, it follows that

$$A^{k+1} r_0 \in \text{span}\{r_0, r_1, \ldots, r_{k+1}\}.$$

By combining this expression with the induction hypothesis for (5.17), we find that

$$\text{span}\{r_0, Ar_0, \ldots, A^{k+1}r_0\} \subset \text{span}\{r_0, r_1, \ldots, r_k, r_{k+1}\}.$$

Therefore, the relation (5.17) continues to hold when k is replaced by $k + 1$, as claimed.

We show that (5.18) continues to hold when k is replaced by $k + 1$ by the following argument:

$$
\begin{aligned}
\text{span}\{p_0, p_1, \ldots, &p_k, p_{k+1}\} \\
&= \text{span}\{p_0, p_1, \ldots, p_k, r_{k+1}\} && \text{by (5.14e)} \\
&= \text{span}\{r_0, Ar_0, \ldots, A^k r_0, r_{k+1}\} && \text{by induction hypothesis for (5.18)} \\
&= \text{span}\{r_0, r_1, \ldots, r_k, r_{k+1}\} && \text{by (5.17)} \\
&= \text{span}\{r_0, Ar_0, \ldots, A^{k+1}r_0\} && \text{by (5.17) for } k + 1.
\end{aligned}
$$

Next, we prove the conjugacy condition (5.19) with k replaced by $k+1$. By multiplying (5.14e) by Ap_i, $i = 0, 1, \ldots, k$, we obtain

$$p_{k+1}^T Ap_i = -r_{k+1}^T Ap_i + \beta_{k+1} p_k^T Ap_i. \tag{5.21}$$

By the definition (5.14d) of β_k, the right-hand-side of (5.21) vanishes when $i = k$. For $i \le k - 1$ we need to collect a number of observations. Note first that our induction hypothesis for (5.19) implies that the directions $p_0, p_1, \ldots, p_k$ are conjugate, so we can apply Theorem 5.2 to deduce that

$$r_{k+1}^T p_i = 0, \qquad \text{for } i = 0, 1, \ldots, k. \tag{5.22}$$

Second, by repeatedly applying (5.18), we find that for $i = 0, 1, \ldots, k - 1$, the following inclusion holds:

$$
\begin{aligned}
Ap_i \in A\,\text{span}\{r_0, Ar_0, \ldots, A^i r_0\} &= \text{span}\{Ar_0, A^2 r_0, \ldots, A^{i+1} r_0\} \\
&\subset \text{span}\{p_0, p_1, \ldots, p_{i+1}\}. \tag{5.23}
\end{aligned}
$$

By combining (5.22) and (5.23), we deduce that

$$r_{k+1}^T Ap_i = 0, \qquad \text{for } i = 0, 1, \ldots, k - 1,$$

so the first term in the right-hand-side of (5.21) vanishes for $i = 0, 1, \ldots, k - 1$. Because of the induction hypothesis for (5.19), the second term vanishes as well, and we conclude that $p_{k+1}^T A p_i = 0, i = 0, 1, \ldots, k$. Hence, the induction argument holds for (5.19) also.

It follows that the direction set generated by the conjugate gradient method is indeed a conjugate direction set, so Theorem 5.1 tells us that the algorithm terminates in at most n iterations.

Finally, we prove (5.16) by a noninductive argument. Because the direction set is conjugate, we have from (5.11) that $r_k^T p_i = 0$ for all $i = 0, 1, \ldots, k - 1$ and any $k = 1, 2, \ldots, n - 1$. By rearranging (5.14e), we find that

$$p_i = -r_i + \beta_i p_{i-1},$$

so that $r_i \in \text{span}\{p_i, p_{i-1}\}$ for all $i = 1, \ldots, k - 1$. We conclude that $r_k^T r_i = 0$ for all $i = 1, \ldots, k - 1$. To complete the proof, we note that $r_k^T r_0 = -r_k^T p_0 = 0$, by definition of p_0 in Algorithm 5.1 and by (5.11). $\qquad\square$

The proof of this theorem relies on the fact that the first direction p_0 is the steepest descent direction $-r_0$; in fact, the result does not hold for other choices of p_0. Since the gradients r_k are mutually orthogonal, the term "conjugate gradient method" is actually a misnomer. It is the search directions, not the gradients, that are conjugate with respect to A.

A PRACTICAL FORM OF THE CONJUGATE GRADIENT METHOD

We can derive a slightly more economical form of the conjugate gradient method by using the results of Theorems 5.2 and 5.3. First, we can use (5.14e) and (5.11) to replace the formula (5.14a) for α_k by

$$\alpha_k = \frac{r_k^T r_k}{p_k^T A p_k}.$$

Second, we have from (5.10) that $\alpha_k A p_k = r_{k+1} - r_k$, so by applying (5.14e) and (5.11) once again we can simplify the formula for β_{k+1} to

$$\beta_{k+1} = \frac{r_{k+1}^T r_{k+1}}{r_k^T r_k}.$$

By using these formulae together with (5.10), we obtain the following standard form of the conjugate gradient method.

Algorithm 5.2 (CG).

Given x_0;

Set $r_0 \leftarrow Ax_0 - b$, $p_0 \leftarrow -r_0$, $k \leftarrow 0$;

while $r_k \neq 0$

$$\alpha_k \leftarrow \frac{r_k^T r_k}{p_k^T A p_k}; \tag{5.24a}$$

$$x_{k+1} \leftarrow x_k + \alpha_k p_k; \tag{5.24b}$$

$$r_{k+1} \leftarrow r_k + \alpha_k A p_k; \tag{5.24c}$$

$$\beta_{k+1} \leftarrow \frac{r_{k+1}^T r_{k+1}}{r_k^T r_k}; \tag{5.24d}$$

$$p_{k+1} \leftarrow -r_{k+1} + \beta_{k+1} p_k; \tag{5.24e}$$

$$k \leftarrow k + 1; \tag{5.24f}$$

end (while)

At any given point in Algorithm 5.2 we never need to know the vectors x, r, and p for more than the last two iterations. Accordingly, implementations of this algorithm overwrite old values of these vectors to save on storage. The major computational tasks to be performed at each step are computation of the matrix–vector product Ap_k, calculation of the inner products $p_k^T(Ap_k)$ and $r_{k+1}^T r_{k+1}$, and calculation of three vector sums. The inner product and vector sum operations can be performed in a small multiple of n floating-point operations, while the cost of the matrix–vector product is, of course, dependent on the problem. The CG method is recommended only for large problems; otherwise, Gaussian elimination or other factorization algorithms such as the singular value decomposition are to be preferred, since they are less sensitive to rounding errors. For large problems, the CG method has the advantage that it does not alter the coefficient matrix and (in contrast to factorization techniques) does not produce fill in the arrays holding the matrix. Another key property is that the CG method sometimes approaches the solution quickly, as we discuss next.

RATE OF CONVERGENCE

We have seen that in exact arithmetic the conjugate gradient method will terminate at the solution in at most n iterations. What is more remarkable is that when the distribution of the eigenvalues of A has certain favorable features, the algorithm will identify the solution in many fewer than n iterations. To explain this property, we begin by viewing the expanding subspace minimization property proved in Theorem 5.2 in a slightly different way, using it to show that Algorithm 5.2 is optimal in a certain important sense.

From (5.24b) and (5.18), we have that

$$x_{k+1} = x_0 + \alpha_0 p_0 + \cdots + \alpha_k p_k$$
$$= x_0 + \gamma_0 r_0 + \gamma_1 A r_0 + \cdots + \gamma_k A^k r_0, \qquad (5.25)$$

for some constants γ_i. We now define $P_k^*(\cdot)$ to be a polynomial of degree k with coefficients $\gamma_0, \gamma_1, \ldots, \gamma_k$. Like any polynomial, P_k^* can take either a scalar or a square matrix as its argument. For the matrix argument A, we have

$$P_k^*(A) = \gamma_0 I + \gamma_1 A + \cdots + \gamma_k A^k,$$

which allows us to express (5.25) as follows:

$$x_{k+1} = x_0 + P_k^*(A) r_0. \qquad (5.26)$$

We now show that among all possible methods whose first k steps are restricted to the Krylov subspace $\mathcal{K}(r_0; k)$ given by (5.15), Algorithm 5.2 does the best job of minimizing the distance to the solution after k steps, when this distance is measured by the weighted norm measure $\| \cdot \|_A$ defined by

$$\|z\|_A^2 = z^T A z. \qquad (5.27)$$

(Recall that this norm was used in the analysis of the steepest descent method of Chapter 3.) Using this norm and the definition of ϕ (5.2), and the fact that x^* minimizes ϕ, it is easy to show that

$$\tfrac{1}{2}\|x - x^*\|_A^2 = \tfrac{1}{2}(x - x^*)^T A(x - x^*) = \phi(x) - \phi(x^*). \qquad (5.28)$$

Theorem 5.2 states that x_{k+1} minimizes ϕ, and hence $\|x - x^*\|_A^2$, over the set $x_0 + \mathrm{span}\{p_0, p_1, \ldots, p_k\}$, which by (5.18) is the same as $x_0 + \mathrm{span}\{r_0, A r_0, \ldots, A^k r_0\}$. It follows from (5.26) that the polynomial P_k^* solves the following problem in which the minimum is taken over the space of all possible polynomials of degree k:

$$\min_{P_k} \|x_0 + P_k(A) r_0 - x^*\|_A. \qquad (5.29)$$

We exploit this optimality property repeatedly in the remainder of the section.

Since

$$r_0 = A x_0 - b = A x_0 - A x^* = A(x_0 - x^*),$$

we have that

$$x_{k+1} - x^* = x_0 + P_k^*(A) r_0 - x^* = [I + P_k^*(A) A](x_0 - x^*). \qquad (5.30)$$

Let $0 < \lambda_1 \leq \lambda_2 \leq \cdots \leq \lambda_n$ be the eigenvalues of A, and let $v_1, v_2, \ldots, v_n$ be the corresponding orthonormal eigenvectors, so that

$$A = \sum_{i=1}^{n} \lambda_i v_i v_i^T.$$

Since the eigenvectors span the whole space $\mathbb{R}^n$, we can write

$$x_0 - x^* = \sum_{i=1}^{n} \xi_i v_i, \tag{5.31}$$

for some coefficients ξ_i. It is easy to show that any eigenvector of A is also an eigenvector of $P_k(A)$ for any polynomial P_k. For our particular matrix A and its eigenvalues λ_i and eigenvectors v_i, we have

$$P_k(A)v_i = P_k(\lambda_i)v_i, \qquad i = 1, 2, \ldots, n.$$

By substituting (5.31) into (5.30) we have

$$x_{k+1} - x^* = \sum_{i=1}^{n} [1 + \lambda_i P_k^*(\lambda_i)]\xi_i v_i.$$

By using the fact that $\|z\|_A^2 = z^T A z = \sum_{i=1}^{n} \lambda_i (v_i^T z)^2$, we have

$$\|x_{k+1} - x^*\|_A^2 = \sum_{i=1}^{n} \lambda_i [1 + \lambda_i P_k^*(\lambda_i)]^2 \xi_i^2. \tag{5.32}$$

Since the polynomial P_k^* generated by the CG method is optimal with respect to this norm, we have

$$\|x_{k+1} - x^*\|_A^2 = \min_{P_k} \sum_{i=1}^{n} \lambda_i [1 + \lambda_i P_k(\lambda_i)]^2 \xi_i^2.$$

By extracting the largest of the terms $[1 + \lambda_i P_k(\lambda_i)]^2$ from this expression, we obtain that

$$\|x_{k+1} - x^*\|_A^2 \leq \min_{P_k} \max_{1 \leq i \leq n} [1 + \lambda_i P_k(\lambda_i)]^2 \left(\sum_{j=1}^{n} \lambda_j \xi_j^2 \right)$$

$$= \min_{P_k} \max_{1 \leq i \leq n} [1 + \lambda_i P_k(\lambda_i)]^2 \|x_0 - x^*\|_A^2, \tag{5.33}$$

where we have used the fact that $\|x_0 - x^*\|_A^2 = \sum_{j=1}^{n} \lambda_j \xi_j^2$.

The expression (5.33) allows us to quantify the convergence rate of the CG method by estimating the nonnegative scalar quantity

$$\min_{P_k} \max_{1 \le i \le n} [1 + \lambda_i P_k(\lambda_i)]^2. \tag{5.34}$$

In other words, we search for a polynomial P_k that makes this expression as small as possible. In some practical cases, we can find this polynomial explicitly and draw some interesting conclusions about the properties of the CG method. The following result is an example.

Theorem 5.4.

If A has only r distinct eigenvalues, then the CG iteration will terminate at the solution in at most r iterations.

PROOF.   Suppose that the eigenvalues $\lambda_1, \lambda_2, \ldots, \lambda_n$ take on the r distinct values $\tau_1 < \tau_2 < \cdots < \tau_r$. We define a polynomial $Q_r(\lambda)$ by

$$Q_r(\lambda) = \frac{(-1)^r}{\tau_1 \tau_2 \cdots \tau_r} (\lambda - \tau_1)(\lambda - \tau_2) \cdots (\lambda - \tau_r),$$

and note that $Q_r(\lambda_i) = 0$ for $i = 1, 2, \ldots, n$ and $Q_r(0) = 1$. From the latter observation, we deduce that $Q_r(\lambda) - 1$ is a polynomial of degree r with a root at $\lambda = 0$, so by polynomial division, the function $\bar{P}_{r-1}$ defined by

$$\bar{P}_{r-1}(\lambda) = (Q_r(\lambda) - 1)/\lambda$$

is a polynomial of degree $r - 1$. By setting $k = r - 1$ in (5.34), we have

$$0 \le \min_{P_{r-1}} \max_{1 \le i \le n} [1 + \lambda_i P_{r-1}(\lambda_i)]^2 \le \max_{1 \le i \le n} [1 + \lambda_i \bar{P}_{r-1}(\lambda_i)]^2 = \max_{1 \le i \le n} Q_r^2(\lambda_i) = 0.$$

Hence, the constant in (5.34) is zero for the value $k = r - 1$, so we have by substituting into (5.33) that $\|x_r - x^*\|_A^2 = 0$, and therefore $x_r = x^*$, as claimed.   □

By using similar reasoning, Luenberger [195] establishes the following estimate, which gives a useful characterization of the behavior of the CG method.

Theorem 5.5.

If A has eigenvalues $\lambda_1 \le \lambda_2 \le \cdots \le \lambda_n$, we have that

$$\|x_{k+1} - x^*\|_A^2 \le \left(\frac{\lambda_{n-k} - \lambda_1}{\lambda_{n-k} + \lambda_1} \right)^2 \|x_0 - x^*\|_A^2. \tag{5.35}$$

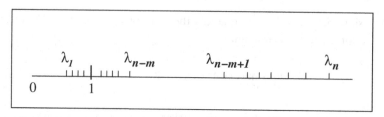

Figure 5.3   Two clusters of eigenvalues.

Without giving details of the proof, we describe how this result is obtained from (5.33). One selects a polynomial $\bar{P}_k$ of degree k such that the polynomial $Q_{k+1}(\lambda) = 1 + \lambda \bar{P}_k(\lambda)$ has roots at the k largest eigenvalues $\lambda_n, \lambda_{n-1}, \dots, \lambda_{n-k+1}$, as well as at the midpoint between λ_1 and λ_{n-k}. It can be shown that the maximum value attained by Q_{k+1} on the remaining eigenvalues $\lambda_1, \lambda_2, \dots, \lambda_{n-k}$ is precisely $(\lambda_{n-k} - \lambda_1)/(\lambda_{n-k} + \lambda_1)$.

We now illustrate how Theorem 5.5 can be used to predict the behavior of the CG method on specific problems. Suppose we have the situation plotted in Figure 5.3, where the eigenvalues of A consist of m large values, with the remaining $n - m$ smaller eigenvalues clustered around 1. If we define $\epsilon = \lambda_{n-m} - \lambda_1$, Theorem 5.5 tells us that after $m + 1$ steps of the conjugate gradient algorithm, we have

$$\|x_{m+1} - x^*\|_A \approx \epsilon \|x_0 - x^*\|_A.$$

For a small value of ϵ, we conclude that the CG iterates will provide a good estimate of the solution after only $m + 1$ steps.

Figure 5.4 shows the behavior of CG on a problem of this type, which has five large eigenvalues with all the smaller eigenvalues clustered between 0.95 and 1.05, and compares this behavior with that of CG on a problem in which the eigenvalues satisfy some random distribution. In both cases, we plot the log of ϕ after each iteration.

For the problem with clustered eigenvalues, Theorem 5.5 predicts a sharp decrease in the error measure at iteration 6. Note, however, that this decrease was achieved one iteration earlier, illustrating the fact that Theorem 5.5 gives only an upper bound, and that the rate of convergence can be faster. By contrast, we observe in Figure 5.4 that for the problem with randomly distributed eigenvalues (dashed line), the convergence rate is slower and more uniform.

Figure 5.4 illustrates another interesting feature: After one more iteration (a total of seven) on the problem with clustered eigenvalues, the error measure drops sharply. An extension of the arguments leading to Theorem 5.4 explains this behavior. It is *almost* true to say that the matrix A has just six distinct eigenvalues: the five large eigenvalues and 1. Then we would expect the error measure to be zero after six iterations. Because the eigenvalues near 1 are slightly spread out, however, the error does not become very small until iteration 7.

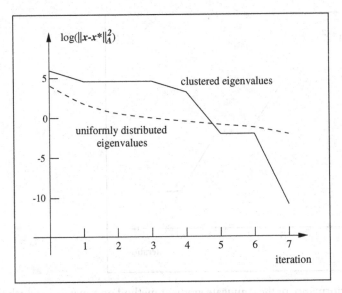

Figure 5.4 Performance of the conjugate gradient method on (a) a problem in which five of the eigenvalues are large and the remainder are clustered near 1, and (b) a matrix with uniformly distributed eigenvalues.

To state this claim more precisely, it is generally true that if the eigenvalues occur in r distinct clusters, the CG iterates will *approximately* solve the problem in about r steps (see [136]). This result can be proved by constructing a polynomial $\bar{P}_{r-1}$ such that $(1+\lambda\bar{P}_{r-1}(\lambda))$ has zeros inside each of the clusters. This polynomial may not vanish at the eigenvalues λ_i, $i = 1, 2, \ldots, n$, but its value will be small at these points, so the constant defined in (5.34) will be small for $k \geq r - 1$. We illustrate this behavior in Figure 5.5, which shows the performance of CG on a matrix of dimension $n = 14$ that has four clusters of eigenvalues: single eigenvalues at 140 and 120, a cluster of 10 eigenvalues very close to 10, with the remaining eigenvalues clustered between 0.95 and 1.05. After four iterations, the error has decreased significantly. After six iterations, the solution is identified to good accuracy.

Another, more approximate, convergence expression for CG is based on the Euclidean condition number of A, which is defined by

$$\kappa(A) = \|A\|_2 \|A^{-1}\|_2 = \lambda_n/\lambda_1.$$

It can be shown that

$$\|x_k - x^*\|_A \leq 2 \left(\frac{\sqrt{\kappa(A)} - 1}{\sqrt{\kappa(A)} + 1} \right)^k \|x_0 - x^*\|_A. \tag{5.36}$$

This bound often gives a large overestimate of the error, but it can be useful in those cases

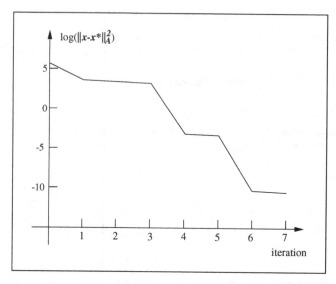

Figure 5.5   Performance of the conjugate gradient method on a matrix in which the eigenvalues occur in four distinct clusters.

where the only information we have about A is estimates of the extreme eigenvalues λ_1 and λ_n. This bound should be compared with that of the steepest descent method given by (3.29), which is identical in form but which depends on the condition number $\kappa(A)$, and not on its square root $\sqrt{\kappa(A)}$.

PRECONDITIONING

We can accelerate the conjugate gradient method by transforming the linear system to improve the eigenvalue distribution of A. The key to this process, which is known as *preconditioning*, is a change of variables from x to $\hat{x}$ via a nonsingular matrix C, that is,

$$\hat{x} = Cx. \tag{5.37}$$

The quadratic ϕ defined by (5.2) is transformed accordingly to

$$\hat{\phi}(\hat{x}) = \tfrac{1}{2}\hat{x}^T(C^{-T}AC^{-1})\hat{x} - (C^{-T}b)^T\hat{x}. \tag{5.38}$$

If we use Algorithm 5.2 to minimize $\hat{\phi}$ or, equivalently, to solve the linear system

$$(C^{-T}AC^{-1})\hat{x} = C^{-T}b,$$

then the convergence rate will depend on the eigenvalues of the matrix $C^{-T}AC^{-1}$ rather than those of A. Therefore, we aim to choose C such that the eigenvalues of $C^{-T}AC^{-1}$

are more favorable for the convergence theory discussed above. We can try to choose C such that the condition number of $C^{-T}AC^{-1}$ is much smaller than the original condition number of A, for instance, so that the constant in (5.36) is smaller. We could also try to choose C such that the eigenvalues of $C^{-T}AC^{-1}$ are clustered, which by the discussion of the previous section ensures that the number of iterates needed to find a good approximate solution is not much larger than the number of clusters.

It is not necessary to carry out the transformation (5.37) explicitly. Rather, we can apply Algorithm 5.2 to the problem (5.38), in terms of the variables $\hat{x}$, and then invert the transformations to reexpress all the equations in terms of x. This process of derivation results in Algorithm 5.3 (Preconditioned Conjugate Gradient), which we now define. It happens that Algorithm 5.3 does not make use of C explicitly, but rather the matrix $M = C^TC$, which is symmetric and positive definite by construction.

Algorithm 5.3 (Preconditioned CG).

Given x_0, preconditioner M;
Set $r_0 \leftarrow Ax_0 - b$;
Solve $My_0 = r_0$ for y_0;
Set $p_0 = -y_0, k \leftarrow 0$;
while $r_k \neq 0$

$$\alpha_k \leftarrow \frac{r_k^T y_k}{p_k^T A p_k}; \qquad (5.39a)$$

$$x_{k+1} \leftarrow x_k + \alpha_k p_k; \qquad (5.39b)$$

$$r_{k+1} \leftarrow r_k + \alpha_k A p_k; \qquad (5.39c)$$

$$\text{Solve } My_{k+1} = r_{k+1}; \qquad (5.39d)$$

$$\beta_{k+1} \leftarrow \frac{r_{k+1}^T y_{k+1}}{r_k^T y_k}; \qquad (5.39e)$$

$$p_{k+1} \leftarrow -y_{k+1} + \beta_{k+1} p_k; \qquad (5.39f)$$

$$k \leftarrow k + 1; \qquad (5.39g)$$

end (while)

If we set $M = I$ in Algorithm 5.3, we recover the standard CG method, Algorithm 5.2. The properties of Algorithm 5.2 generalize to this case in interesting ways. In particular, the orthogonality property (5.16) of the successive residuals becomes

$$r_i^T M^{-1} r_j = 0 \quad \text{for all } i \neq j. \qquad (5.40)$$

In terms of computational effort, the main difference between the preconditioned and unpreconditioned CG methods is the need to solve systems of the form $My = r$ (step (5.39d)).

PRACTICAL PRECONDITIONERS

No single preconditioning strategy is "best" for all conceivable types of matrices: The tradeoff between various objectives—effectiveness of M, inexpensive computation and storage of M, inexpensive solution of $My = r$—varies from problem to problem.

Good preconditioning strategies have been devised for specific types of matrices, in particular, those arising from discretizations of partial differential equations (PDEs). Often, the preconditioner is defined in such a way that the system $My = r$ amounts to a simplified version of the original system $Ax = b$. In the case of a PDE, $My = r$ could represent a coarser discretization of the underlying continuous problem than $Ax = b$. As in many other areas of optimization and numerical analysis, knowledge about the structure and origin of a problem (in this case, knowledge that the system $Ax = b$ is a finite-dimensional representation of a PDE) is the key to devising effective techniques for solving the problem.

General-purpose preconditioners have also been proposed, but their success varies greatly from problem to problem. The most important strategies of this type include symmetric successive overrelaxation (SSOR), incomplete Cholesky, and banded preconditioners. (See [272], [136], and [72] for discussions of these techniques.) *Incomplete Cholesky* is probably the most effective in general. The basic idea is simple: We follow the Cholesky procedure, but instead of computing the exact Cholesky factor L that satisfies $A = LL^T$, we compute an approximate factor $\tilde{L}$ that is sparser than L. (Usually, we require $\tilde{L}$ to be no denser, or not much denser, than the lower triangle of the original matrix A.) We then have $A \approx \tilde{L}\tilde{L}^T$, and by choosing $C = \tilde{L}^T$, we obtain $M = \tilde{L}\tilde{L}^T$ and

$$C^{-T}AC^{-1} = \tilde{L}^{-1}A\tilde{L}^{-T} \approx I,$$

so the eigenvalue distribution of $C^{-T}AC^{-1}$ is favorable. We do not compute M explicitly, but rather store the factor $\tilde{L}$ and solve the system $My = r$ by performing two triangular substitutions with $\tilde{L}$. Because the sparsity of $\tilde{L}$ is similar to that of A, the cost of solving $My = r$ is similar to the cost of computing the matrix–vector product Ap.

There are several possible pitfalls in the incomplete Cholesky approach. One is that the resulting matrix may not be (sufficiently) positive definite, and in this case one may need to increase the values of the diagonal elements to ensure that a value for $\tilde{L}$ can be found. Numerical instability or breakdown can occur during the incomplete factorization because of the sparsity conditions we impose on the factor $\tilde{L}$. This difficulty can be remedied by allowing additional fill-in in $\tilde{L}$, but the denser factor will be more expensive to compute and to apply at each iteration.

5.2 NONLINEAR CONJUGATE GRADIENT METHODS

We have noted that the CG method, Algorithm 5.2, can be viewed as a minimization algorithm for the convex quadratic function ϕ defined by (5.2). It is natural to ask whether we can adapt the approach to minimize general convex functions, or even general nonlinear functions f. In fact, as we show in this section, nonlinear variants of the conjugate gradient are well studied and have proved to be quite successful in practice.

THE FLETCHER–REEVES METHOD

Fletcher and Reeves [107] showed how to extend the conjugate gradient method to nonlinear functions by making two simple changes in Algorithm 5.2. First, in place of the formula (5.24a) for the step length α_k (which minimizes ϕ along the search direction p_k), we need to perform a line search that identifies an approximate minimum of the nonlinear function f along p_k. Second, the residual r, which is simply the gradient of ϕ in Algorithm 5.2 (see (5.3)), must be replaced by the gradient of the nonlinear objective f. These changes give rise to the following algorithm for nonlinear optimization.

Algorithm 5.4 (FR).

Given x_0;
Evaluate $f_0 = f(x_0)$, $\nabla f_0 = \nabla f(x_0)$;
Set $p_0 \leftarrow -\nabla f_0, k \leftarrow 0$;
while $\nabla f_k \neq 0$
    Compute α_k and set $x_{k+1} = x_k + \alpha_k p_k$;

    Evaluate ∇f_{k+1};

$$\beta_{k+1}^{\text{FR}} \leftarrow \frac{\nabla f_{k+1}^T \nabla f_{k+1}}{\nabla f_k^T \nabla f_k}; \tag{5.41a}$$

$$p_{k+1} \leftarrow -\nabla f_{k+1} + \beta_{k+1}^{\text{FR}} p_k; \tag{5.41b}$$

$$k \leftarrow k + 1; \tag{5.41c}$$

end (while)

If we choose f to be a strongly convex quadratic and α_k to be the exact minimizer, this algorithm reduces to the linear conjugate gradient method, Algorithm 5.2. Algorithm 5.4 is appealing for large nonlinear optimization problems because each iteration requires only evaluation of the objective function and its gradient. No matrix operations are required for the step computation, and just a few vectors of storage are required.

To make the specification of Algorithm 5.4 complete, we need to be more precise about the choice of line search parameter α_k. Because of the second term in (5.41b), the search direction p_k may fail to be a descent direction unless α_k satisfies certain conditions.

By taking the inner product of (5.41b) (with k replacing $k + 1$) with the gradient vector ∇f_k, we obtain

$$\nabla f_k^T p_k = -\|\nabla f_k\|^2 + \beta_k^{\text{FR}} \nabla f_k^T p_{k-1}. \tag{5.42}$$

If the line search is exact, so that α_{k-1} is a local minimizer of f along the direction p_{k-1}, we have that $\nabla f_k^T p_{k-1} = 0$. In this case we have from (5.42) that $\nabla f_k^T p_k < 0$, so that p_k is indeed a descent direction. If the line search is not exact, however, the second term in (5.42) may dominate the first term, and we may have $\nabla f_k^T p_k > 0$, implying that p_k is actually a direction of ascent. Fortunately, we can avoid this situation by requiring the step length α_k to satisfy the *strong* Wolfe conditions, which we restate here:

$$f(x_k + \alpha_k p_k) \le f(x_k) + c_1 \alpha_k \nabla f_k^T p_k, \tag{5.43a}$$
$$|\nabla f(x_k + \alpha_k p_k)^T p_k| \le -c_2 \nabla f_k^T p_k, \tag{5.43b}$$

where $0 < c_1 < c_2 < \frac{1}{2}$. (Note that we impose $c_2 < \frac{1}{2}$ here, in place of the looser condition $c_2 < 1$ that was used in the earlier statement (3.7).) By applying Lemma 5.6 below, we can show that condition (5.43b) implies that (5.42) is negative, and we conclude that any line search procedure that yields an α_k satisfying (5.43) will ensure that all directions p_k are descent directions for the function f.

THE POLAK–RIBIÈRE METHOD AND VARIANTS

There are many variants of the Fletcher–Reeves method that differ from each other mainly in the choice of the parameter β_k. An important variant, proposed by Polak and Ribière, defines this parameter as follows:

$$\beta_{k+1}^{\text{PR}} = \frac{\nabla f_{k+1}^T (\nabla f_{k+1} - \nabla f_k)}{\|\nabla f_k\|^2}. \tag{5.44}$$

We refer to the algorithm in which (5.44) replaces (5.41a) as Algorithm PR. It is identical to Algorithm FR when f is a strongly convex quadratic function and the line search is exact, since by (5.16) the gradients are mutually orthogonal, and so $\beta_{k+1}^{\text{PR}} = \beta_{k+1}^{\text{FR}}$. When applied to general nonlinear functions with inexact line searches, however, the behavior of the two algorithms differs markedly. Numerical experience indicates that Algorithm PR tends to be the more robust and efficient of the two.

A surprising fact about Algorithm PR is that the strong Wolfe conditions (5.43) do not guarantee that p_k is always a descent direction. If we define the β parameter as

$$\beta_{k+1}^+ = \max\{\beta_{k+1}^{\text{PR}}, 0\}, \tag{5.45}$$

giving rise to an algorithm we call Algorithm PR+, then a simple adaptation of the strong Wolfe conditions ensures that the descent property holds.

There are many other choices for β_{k+1} that coincide with the Fletcher–Reeves formula β_{k+1}^{FR} in the case where the objective is quadratic and the line search is exact. The Hestenes–Stiefel formula, which defines

$$\beta_{k+1}^{\text{HS}} = \frac{\nabla f_{k+1}^T (\nabla f_{k+1} - \nabla f_k)}{(\nabla f_{k+1} - \nabla f_k)^T p_k}, \tag{5.46}$$

gives rise to an algorithm (called Algorithm HS) that is similar to Algorithm PR, both in terms of its theoretical convergence properties and in its practical performance. Formula (5.46) can be derived by demanding that consecutive search directions be conjugate with respect to the *average Hessian* over the line segment $[x_k, x_{k+1}]$, which is defined as

$$\bar{G}_k \equiv \int_0^1 [\nabla^2 f(x_k + \tau \alpha_k p_k)] d\tau.$$

Recalling from Taylor's theorem (Theorem 2.1) that $\nabla f_{k+1} = \nabla f_k + \alpha_k \bar{G}_k p_k$, we see that for any direction of the form $p_{k+1} = -\nabla f_{k+1} + \beta_{k+1} p_k$, the condition $p_{k+1}^T \bar{G}_k p_k = 0$ requires β_{k+1} to be given by (5.46).

Later, we see that it is possible to guarantee global convergence for any parameter β_k satisfying the bound

$$|\beta_k| \le \beta_k^{\text{FR}}, \tag{5.47}$$

for all $k \ge 2$. This fact suggests the following modification of the PR method, which has performed well on some applications. For all $k \ge 2$ let

$$\beta_k = \begin{cases} -\beta_k^{\text{FR}} & \text{if} & \beta_k^{\text{PR}} < -\beta_k^{\text{FR}} \\ \beta_k^{\text{PR}} & \text{if} & |\beta_k^{\text{PR}}| \le \beta_k^{\text{FR}} \\ \beta_k^{\text{FR}} & \text{if} & \beta_k^{\text{PR}} > \beta_k^{\text{FR}}. \end{cases} \tag{5.48}$$

The algorithm based on this strategy will be denoted by FR-PR.

Other variants of the CG method have recently been proposed. Two choices for β_{k+1} that possess attractive theoretical and computational properties are

$$\beta_{k+1} = \frac{\|\nabla f_{k+1}\|^2}{(\nabla f_{k+1} - \nabla f_k)^T p_k} \tag{5.49}$$

(see [85]) and

$$\beta_{k+1} = \left(\hat{y}_k - 2 p_k \frac{\|\hat{y}_k\|^2}{\hat{y}_k^T p_k} \right)^T \frac{\nabla f_{k+1}}{\hat{y}_k^T p_k}, \qquad \text{with} \qquad \hat{y}_k = \nabla f_{k+1} - \nabla f_k \tag{5.50}$$

(see [161]). These two choices guarantee that p_k is a descent direction, provided the steplength α_k satisfies the Wolfe conditions. The CG algorithms based on (5.49) or (5.50) appear to be competitive with the Polak–Ribière method.

QUADRATIC TERMINATION AND RESTARTS

Implementations of nonlinear conjugate gradient methods usually preserve their close connections with the linear conjugate gradient method. Usually, a quadratic (or cubic) interpolation along the search direction p_k is incorporated into the line search procedure; see Chapter 3. This feature guarantees that when f is a strictly convex quadratic, the step length α_k is chosen to be the exact one-dimensional minimizer, so that the nonlinear conjugate gradient method reduces to the linear method, Algorithm 5.2.

Another modification that is often used in nonlinear conjugate gradient procedures is to *restart* the iteration at every n steps by setting $\beta_k = 0$ in (5.41a), that is, by taking a steepest descent step. Restarting serves to periodically refresh the algorithm, erasing old information that may not be beneficial. We can even prove a strong theoretical result about restarting: It leads to n-step quadratic convergence, that is,

$$\|x_{k+n} - x\| = O\left(\|x_k - x^*\|^2\right). \tag{5.51}$$

After a little thought, this result is not so surprising. Consider a function f that is strongly convex quadratic in a neighborhood of the solution, but is nonquadratic everywhere else. Assuming that the algorithm is converging to the solution in question, the iterates will eventually enter the quadratic region. At some point, the algorithm will be restarted in that region, and from that point onward, its behavior will simply be that of the linear conjugate gradient method, Algorithm 5.2. In particular, finite termination will occur within n steps of the restart. The restart is important, because the finite-termination property and other appealing properties of Algorithm 5.2 hold only when its initial search direction p_0 is equal to the negative gradient.

Even if the function f is not exactly quadratic in the region of a solution, Taylor's theorem (Theorem 2.1) implies that it can still be approximated quite closely by a quadratic, provided that it is smooth. Therefore, while we would not expect termination in n steps after the restart, it is not surprising that substantial progress is made toward the solution, as indicated by the expression (5.51).

Though the result (5.51) is interesting from a theoretical viewpoint, it may not be relevant in a practical context, because nonlinear conjugate gradient methods can be recommended only for solving problems with large n. Restarts may never occur in such problems because an approximate solution may be located in fewer than n steps. Hence, nonlinear CG method are sometimes implemented without restarts, or else they include strategies for restarting that are based on considerations other than iteration counts. The most popular restart strategy makes use of the observation (5.16), which is that the gradients are mutually orthogonal when f is a quadratic function. A restart is performed whenever two consecutive gradients are far from orthogonal, as measured by the test

$$\frac{|\nabla f_k^T \nabla f_{k-1}|}{\|\nabla f_k\|^2} \geq \nu, \tag{5.52}$$

where a typical value for the parameter ν is 0.1.

We could also think of formula (5.45) as a restarting strategy, because p_{k+1} will revert to the steepest descent direction whenever β_k^{PR} is negative. In contrast to (5.52), these restarts are rather infrequent because β_k^{PR} is positive most of the time.

BEHAVIOR OF THE FLETCHER–REEVES METHOD

We now investigate the Fletcher–Reeves algorithm, Algorithm 5.4, a little more closely, proving that it is globally convergent and explaining some of its observed inefficiencies.

The following result gives conditions on the line search under which all search directions are descent directions. It assumes that the level set $\mathcal{L} = \{x : f(x) \leq f(x_0)\}$ is bounded and that f is twice continuously differentiable, so that we have from Lemma 3.1 that there exists a step length α_k satisfying the strong Wolfe conditions.

Lemma 5.6.

   Suppose that Algorithm 5.4 is implemented with a step length α_k that satisfies the strong Wolfe conditions (5.43) with $0 < c_2 < \frac{1}{2}$. Then the method generates descent directions p_k that satisfy the following inequalities:

$$-\frac{1}{1-c_2} \leq \frac{\nabla f_k^T p_k}{\|\nabla f_k\|^2} \leq \frac{2c_2 - 1}{1-c_2}, \qquad \text{for all } k = 0, 1, \ldots . \tag{5.53}$$

PROOF.   Note first that the function $t(\xi) \stackrel{\text{def}}{=} (2\xi - 1)/(1 - \xi)$ is monotonically increasing on the interval $[0, \frac{1}{2}]$ and that $t(0) = -1$ and $t(\frac{1}{2}) = 0$. Hence, because of $c_2 \in (0, \frac{1}{2})$, we have

$$-1 < \frac{2c_2 - 1}{1 - c_2} < 0. \tag{5.54}$$

The descent condition $\nabla f_k^T p_k < 0$ follows immediately once we establish (5.53).

   The proof is by induction. For $k = 0$, the middle term in (5.53) is -1, so by using (5.54), we see that both inequalities in (5.53) are satisfied. Next, assume that (5.53) holds for some $k \geq 1$. From (5.41b) and (5.41a) we have

$$\frac{\nabla f_{k+1}^T p_{k+1}}{\|\nabla f_{k+1}\|^2} = -1 + \beta_{k+1}\frac{\nabla f_{k+1}^T p_k}{\|\nabla f_{k+1}\|^2} = -1 + \frac{\nabla f_{k+1}^T p_k}{\|\nabla f_k\|^2}. \tag{5.55}$$

By using the line search condition (5.43b), we have

$$|\nabla f_{k+1}^T p_k| \leq -c_2 \nabla f_k^T p_k,$$

so by combining with (5.55) and recalling (5.41a), we obtain

$$-1 + c_2 \frac{\nabla f_k^T p_k}{\|\nabla f_k\|^2} \leq \frac{\nabla f_{k+1}^T p_{k+1}}{\|\nabla f_{k+1}\|^2} \leq -1 - c_2 \frac{\nabla f_k^T p_k}{\|\nabla f_k\|^2}.$$

Substituting for the term $\nabla f_k^T p_k / \|\nabla f_k\|^2$ from the left-hand-side of the induction hypothesis (5.53), we obtain

$$-1 - \frac{c_2}{1 - c_2} \leq \frac{\nabla f_{k+1}^T p_{k+1}}{\|\nabla f_{k+1}\|^2} \leq -1 + \frac{c_2}{1 - c_2},$$

which shows that (5.53) holds for $k + 1$ as well.   □

This result used only the second strong Wolfe condition (5.43b); the first Wolfe condition (5.43a) will be needed in the next section to establish global convergence. The bounds on $\nabla f_k^T p_k$ in (5.53) impose a limit on how fast the norms of the steps $\|p_k\|$ can grow, and they will play a crucial role in the convergence analysis given below.

Lemma 5.6 can also be used to explain a weakness of the Fletcher–Reeves method. We will argue that if the method generates a bad direction and a tiny step, then the next direction and next step are also likely to be poor. As in Chapter 3, we let θ_k denote the angle between p_k and the steepest descent direction $-\nabla f_k$, defined by

$$\cos \theta_k = \frac{-\nabla f_k^T p_k}{\|\nabla f_k\| \|p_k\|}. \tag{5.56}$$

Suppose that p_k is a poor search direction, in the sense that it makes an angle of nearly $90°$ with $-\nabla f_k$, that is, $\cos \theta_k \approx 0$. By multiplying both sides of (5.53) by $\|\nabla f_k\|/\|p_k\|$ and using (5.56), we obtain

$$\frac{1 - 2c_2}{1 - c_2} \frac{\|\nabla f_k\|}{\|p_k\|} \leq \cos \theta_k \leq \frac{1}{1 - c_2} \frac{\|\nabla f_k\|}{\|p_k\|}, \qquad \text{for all } k = 0, 1, \ldots. \tag{5.57}$$

From these inequalities, we deduce that $\cos \theta_k \approx 0$ if and only if

$$\|\nabla f_k\| \ll \|p_k\|.$$

Since p_k is almost orthogonal to the gradient, it is likely that the step from x_k to x_{k+1} is tiny, that is, $x_{k+1} \approx x_k$. If so, we have $\nabla f_{k+1} \approx \nabla f_k$, and therefore

$$\beta_{k+1}^{\text{FR}} \approx 1, \tag{5.58}$$

by the definition (5.41a). By using this approximation together with $\|\nabla f_{k+1}\| \approx \|\nabla f_k\| \ll \|p_k\|$ in (5.41b), we conclude that

$$p_{k+1} \approx p_k,$$

so the new search direction will improve little (if at all) on the previous one. It follows that if the condition $\cos\theta_k \approx 0$ holds at some iteration k and if the subsequent step is small, a long sequence of unproductive iterates will follow.

The Polak–Ribière method behaves quite differently in these circumstances. If, as in the previous paragraph, the search direction p_k satisfies $\cos\theta_k \approx 0$ for some k, and if the subsequent step is small, it follows by substituting $\nabla f_k \approx \nabla f_{k+1}$ into (5.44) that $\beta_{k+1}^{\text{PR}} \approx 0$. From the formula (5.41b), we find that the new search direction p_{k+1} will be close to the steepest descent direction $-\nabla f_{k+1}$, and $\cos\theta_{k+1}$ will be close to 1. Therefore, Algorithm PR essentially performs a restart after it encounters a bad direction. The same argument can be applied to Algorithms PR+ and HS. For the FR-PR variant, defined by (5.48), we have noted already that $\beta_{k+1}^{\text{FR}} \approx 1$, and $\beta_{k+1}^{\text{PR}} \approx 0$. The formula (5.48) thus sets $\beta_{k+1} = \beta_{k+1}^{\text{PR}}$, as desired. Thus, the modification (5.48) seems to avoid the inefficiencies of the FR method, while falling back on this method for global convergence.

The undesirable behavior of the Fletcher–Reeves method predicted by the arguments given above can be observed in practice. For example, the paper [123] describes a problem with $n = 100$ in which $\cos\theta_k$ is of order 10^{-2} for hundreds of iterations and the steps $\|x_k - x_{k-1}\|$ are of order 10^{-2}. Algorithm FR requires thousands of iterations to solve this problem, while Algorithm PR requires just 37 iterations. In this example, the Fletcher–Reeves method performs much better if it is periodically restarted along the steepest descent direction, since each restart terminates the cycle of bad steps. In general, Algorithm FR should not be implemented without some kind of restart strategy.

GLOBAL CONVERGENCE

Unlike the linear conjugate gradient method, whose convergence properties are well understood and which is known to be optimal as described above, nonlinear conjugate gradient methods possess surprising, sometimes bizarre, convergence properties. We now present a few of the main results known for the Fletcher–Reeves and Polak–Ribière methods using practical line searches.

For the purposes of this section, we make the following (nonrestrictive) assumptions on the objective function.

Assumptions 5.1.

(i) *The level set $\mathcal{L} := \{x \mid f(x) \le f(x_0)\}$ is bounded;*

(ii) *In some open neighborhood $\mathcal{N}$ of $\mathcal{L}$, the objective function f is Lipschitz continuously differentiable.*

These assumptions imply that there is a constant $\bar{\gamma}$ such that

$$\|\nabla f(x)\| \le \bar{\gamma}, \text{ for all } x \in \mathcal{L}. \tag{5.59}$$

Our main analytical tool in this section is Zoutendijk's theorem—Theorem 3.2 in Chapter 3. It states, that under Assumptions 5.1, any line search iteration of the form $x_{k+1} = x_k + \alpha_k p_k$, where p_k is a descent direction and α_k satisfies the Wolfe conditions (5.43) gives the limit

$$\sum_{k=0}^{\infty} \cos^2 \theta_k \|\nabla f_k\|^2 < \infty. \tag{5.60}$$

We can use this result to prove global convergence for algorithms that are periodically restarted by setting $\beta_k = 0$. If k_1, k_2, and so on denote the iterations on which restarts occur, we have from (5.60) that

$$\sum_{k=k_1,k_2,\dots} \|\nabla f_k\|^2 < \infty. \tag{5.61}$$

If we allow no more than $\bar{n}$ iterations between restarts, the sequence $\{k_j\}_{j=1}^{\infty}$ is infinite, and from (5.61) we have that $\lim_{j\to\infty} \|\nabla f_{k_j}\| = 0$. That is, a subsequence of gradients approaches zero, or equivalently,

$$\liminf_{k\to\infty} \|\nabla f_k\| = 0. \tag{5.62}$$

This result applies equally to restarted versions of all the algorithms discussed in this chapter.

It is more interesting, however, to study the global convergence of *unrestarted* conjugate gradient methods, because for large problems (say $n \ge 1000$) we expect to find a solution in many fewer than n iterations—the first point at which a regular restart would take place. Our study of large sequences of unrestarted conjugate gradient iterations reveals some surprising patterns in their behavior.

We can build on Lemma 5.6 and Zoutendijk's result (5.60) to prove a global convergence result for the Fletcher–Reeves method. While we cannot show that the limit of the sequence of gradients $\{\nabla f_k\}$ is zero, the following result shows that this sequence is not bounded away from zero.

Theorem 5.7 (Al-Baali [3]).

Suppose that Assumptions 5.1 hold, and that Algorithm 5.4 is implemented with a line search that satisfies the strong Wolfe conditions (5.43), with $0 < c_1 < c_2 < \frac{1}{2}$. Then

$$\liminf_{k\to\infty} \|\nabla f_k\| = 0. \tag{5.63}$$

PROOF. The proof is by contradiction. It assumes that the opposite of (5.63) holds, that is, there is a constant $\gamma > 0$ such that

$$\|\nabla f_k\| \geq \gamma, \tag{5.64}$$

for all k sufficiently large. By substituting the left inequality of (5.57) into Zoutendijk's condition (5.60), we obtain

$$\sum_{k=0}^{\infty} \frac{\|\nabla f_k\|^4}{\|p_k\|^2} < \infty. \tag{5.65}$$

By using (5.43b) and (5.53), we obtain that

$$|\nabla f_k^T p_{k-1}| \leq -c_2 \nabla f_{k-1}^T p_{k-1} \leq \frac{c_2}{1 - c_2} \|\nabla f_{k-1}\|^2. \tag{5.66}$$

Thus, from (5.41b) and recalling the definition (5.41a) of β_k^{FR} we obtain

$$\|p_k\|^2 \leq \|\nabla f_k\|^2 + 2\beta_k^{\text{FR}} |\nabla f_k^T p_{k-1}| + (\beta_k^{\text{FR}})^2 \|p_{k-1}\|^2$$

$$\leq \|\nabla f_k\|^2 + \frac{2c_2}{1 - c_2} \beta_k^{\text{FR}} \|\nabla f_{k-1}\|^2 + (\beta_k^{\text{FR}})^2 \|p_{k-1}\|^2$$

$$= \left(\frac{1 + c_2}{1 - c_2} \right) \|\nabla f_k\|^2 + (\beta_k^{\text{FR}})^2 \|p_{k-1}\|^2.$$

Applying this relation repeatedly, and defining $c_3 \stackrel{\text{def}}{=} (1 + c_2)/(1 - c_2) \geq 1$, we have

$$\|p_k\|^2 \leq c_3 \|\nabla f_k\|^2 + (\beta_k^{\text{FR}})^2 (c_3 \|\nabla f_{k-1}\|^2 + (\beta_{k-1}^{\text{FR}})^2 (c_3 \|\nabla f_{k-2}\|^2 +$$

$$\cdots + (\beta_1^{\text{FR}})^2 \|p_0\|^2)) \cdots)$$

$$= c_3 \|\nabla f_k\|^4 \sum_{j=0}^{k} \|\nabla f_j\|^{-2}, \tag{5.67}$$

where we used the facts that

$$(\beta_k^{\text{FR}})^2 (\beta_{k-1}^{\text{FR}})^2 \cdots (\beta_{k-i}^{\text{FR}})^2 = \frac{\|\nabla f_k\|^4}{\|\nabla f_{k-i-1}\|^4}$$

and $p_0 = -\nabla f_0$. By using the bounds (5.59) and (5.64) in (5.67), we obtain

$$\|p_k\|^2 \leq \frac{c_3 \bar{\gamma}^4}{\gamma^2} k, \tag{5.68}$$

which implies that

$$\sum_{k=1}^{\infty} \frac{1}{\|p_k\|^2} \geq \gamma_4 \sum_{k=1}^{\infty} \frac{1}{k}, \tag{5.69}$$

for some positive constant γ_4.

On the other hand, from (5.64) and (5.65), we have that

$$\sum_{k=1}^{\infty} \frac{1}{\|p_k\|^2} < \infty. \tag{5.70}$$

However, if we combine this inequality with (5.69), we obtain that $\sum_{k=1}^{\infty} 1/k < \infty$, which is not true. Hence, (5.64) does not hold, and the claim (5.63) is proved. $\qquad\square$

This global convergence result can be extended to any choice of β_k satisfying (5.47), and in particular to the FR-PR method given by (5.48).

In general, if we can show that there exist constants $c_4, c_5 > 0$ such that

$$\cos\theta_k \geq c_4 \frac{\|\nabla f_k\|}{\|p_k\|}, \qquad \frac{\|\nabla f_k\|}{\|p_k\|} \geq c_5 > 0, \qquad k = 1, 2, \ldots,$$

it follows from (5.60) that

$$\lim_{k\to\infty} \|\nabla f_k\| = 0.$$

In fact, this result can be established for the Polak–Ribière method under the assumption that f is strongly convex and that an exact line search is used.

For general (nonconvex) functions, however, is it not possible to prove a result like Theorem 5.7 for Algorithm PR. This fact is unexpected, since the Polak–Ribière method performs better in practice than the Fletcher–Reeves method. The following surprising result shows that the Polak–Ribière method can cycle infinitely without approaching a solution point, even if an ideal line search is used. (By "ideal" we mean that line search returns a value α_k that is the first positive stationary point for the function $t(\alpha) = f(x_k + \alpha p_k)$.)

Theorem 5.8.

Consider the Polak–Ribière method method (5.44) with an ideal line search. There exists a twice continuously differentiable objective function $f : \mathbb{R}^3 \to \mathbb{R}$ and a starting point $x_0 \in \mathbb{R}^3$ such that the sequence of gradients $\{\|\nabla f_k\|\}$ is bounded away from zero.

The proof of this result, given in [253], is quite complex. It demonstrates the existence of the desired objective function without actually constructing this function explicitly. The result is interesting, since the step length assumed in the proof—the first stationary point—may be accepted by any of the practical line search algorithms currently in use. The proof of Theorem 5.8 requires that some consecutive search directions become almost negatives of each other. In the case of ideal line searches, this happens only if $\beta_k < 0$, so the analysis suggests Algorithm PR+ (see (5.45)), in which we reset β_k to zero whenever it becomes negative. We mentioned earlier that a line search strategy based on a slight modification of the Wolfe conditions guarantees that all search directions generated by Algorithm PR+ are descent directions. Using these facts, it is possible to a prove global convergence result like Theorem 5.7 for Algorithm PR+. An attractive property of the formulae (5.49), (5.50) is that global convergence can be established without introducing any modification to a line search based on the Wolfe conditions.

NUMERICAL PERFORMANCE

Table 5.1 illustrates the performance of Algorithms FR, PR, and PR+ without restarts. For these tests, the parameters in the strong Wolfe conditions (5.43) were chosen to be $c_1 = 10^{-4}$ and $c_2 = 0.1$. The iterations were terminated when

$$\|\nabla f_k\|_\infty < 10^{-5}(1 + |f_k|).$$

If this condition was not satisfied after 10,000 iterations, we declare failure (indicated by a $*$ in the table).

The final column, headed "mod," indicates the number of iterations of Algorithm PR+ for which the adjustment (5.45) was needed to ensure that $\beta_k^{\text{PR}} \geq 0$. Algorithm FR on problem GENROS takes very short steps far from the solution that lead to tiny improvements in the objective function, and convergence was not achieved within the maximum number of iterations.

The Polak–Ribière algorithm, or its variation PR+, are not always more efficient than Algorithm FR, and it has the slight disadvantage of requiring one more vector of storage. Nevertheless, we recommend that users choose Algorithm PR, PR+ or FR-PR, or the methods based on (5.49) and (5.50).

Table 5.1 Iterations and function/gradient evaluations required by three nonlinear conjugate gradient methods on a set of test problems; see [123]

Problem	n	Alg FR it/f-g	Alg PR it/f-g	Alg PR+ it/f-g	mod
CALCVAR3	200	2808/5617	2631/5263	2631/5263	0
GENROS	500	$*$	1068/2151	1067/2149	1
XPOWSING	1000	533/1102	212/473	97/229	3
TRIDIA1	1000	264/531	262/527	262/527	0
MSQRT1	1000	422/849	113/231	113/231	0
XPOWELL	1000	568/1175	212/473	97/229	3
TRIGON	1000	231/467	40/92	40/92	0

NOTES AND REFERENCES

The conjugate gradient method was developed in the 1950s by Hestenes and Stiefel [168] as an alternative to factorization methods for finding solutions of symmetric positive definite systems. It was not until some years later, in one of the most important developments in sparse linear algebra, that this method came to be viewed as an iterative method that could give good approximate solutions to systems in many fewer than n steps. Our presentation of the linear conjugate gradient method follows that of Luenberger [195]. For a history of the development of the conjugate gradient and Lanczos methods see Golub and O'Leary [135].

Interestingly enough, the nonlinear conjugate gradient method of Fletcher and Reeves [107] was proposed after the linear conjugate gradient method had fallen out of favor, but several years before it was rediscovered as an iterative method for linear systems. The Polak–Ribière method was introduced in [237], and the example showing that it may fail to converge on nonconvex problems is given by Powell [253]. Restart procedures are discussed in Powell [248].

Hager and Zhang [161] report some of the best computational results obtained to date with a nonlinear CG method. Their implementation is based on formula (5.50) and uses a high-accuracy line search procedure. The results in Table 5.1 are taken from Gilbert and Nocedal [123]. This paper also describes a line search that guarantees that Algorithm PR+ always generates descent directions and proves global convergence.

Analysis due to Powell [245] provides further evidence of the inefficiency of the Fletcher–Reeves method using exact line searches. He shows that if the iterates enter a region in which the function is the two-dimensional quadratic

$$f(x) = \tfrac{1}{2}x^T x,$$

then the angle between the gradient ∇f_k and the search direction p_k stays constant. Since this angle can be arbitrarily close to $90°$, the Fletcher–Reeves method can be slower than the steepest descent method. The Polak–Ribière method behaves quite differently in these circumstances: If a very small step is generated, the next search direction tends to the steepest descent direction, as argued above. This feature prevents a sequence of tiny steps.

The global convergence of nonlinear conjugate gradient methods has received much attention; see for example Al-Baali [3], Gilbert and Nocedal [123], Dai and Yuan [85], and Hager and Zhang [161]. For recent surveys on CG methods see Gould et al. [147] and Hager and Zhang [162].

Most of the theory on the rate of convergence of conjugate gradient methods assumes that the line search is exact. Crowder and Wolfe [82] show that the rate of convergence is linear, and show by constructing an example that Q-superlinear convergence is not achievable. Powell [245] studies the case in which the conjugate gradient method enters a region where the objective function is quadratic, and shows that either finite termination occurs or the rate of convergence is linear. Cohen [63] and Burmeister [45] prove n-step quadratic convergence (5.51) for general objective functions. Ritter [265] shows that in fact, the rate is *superquadratic*, that is,

$$\|x_{k+n} - x^*\| = o(\|x_k - x^*\|^2).$$

Powell [251] gives a slightly better result and performs numerical tests on small problems to measure the rate observed in practice. He also summarizes rate-of-convergence results for asymptotically exact line searches, such as those obtained by Baptist and Stoer [11] and Stoer [282]. Even faster rates of convergence can be established (see Schuller [278], Ritter [265]), under the assumption that the search directions are uniformly linearly independent, but this assumption is hard to verify and does not often occur in practice.

Nemirovsky and Yudin [225] devote some attention to the *global efficiency* of the Fletcher–Reeves and Polak–Ribière methods with exact line searches. For this purpose they define a measure of "laboriousness" and an "optimal bound" for it among a certain class of iterations. They show that on strongly convex problems not only do the Fletcher–Reeves and Polak–Ribière methods fail to attain the optimal bound, but they may also be slower than the steepest descent method. Subsequently, Nesterov [225] presented an algorithm that attains this optimal bound. It is related to PARTAN, the method of parallel tangents (see, for example, Luenberger [195]). We feel that this approach is unlikely to be effective in practice, but no conclusive investigation has been carried out, to the best of our knowledge.

EXERCISES

5.1 Implement Algorithm 5.2 and use to it solve linear systems in which A is the Hilbert matrix, whose elements are $A_{i,j} = 1/(i + j - 1)$. Set the right-hand-side to $b = (1, 1, \ldots, 1)^T$ and the initial point to $x_0 = 0$. Try dimensions $n = 5, 8, 12, 20$ and report the number of iterations required to reduce the residual below 10^{-6}.

5.2 Show that if the nonzero vectors $p_0, p_1, \ldots, p_l$ satisfy (5.5), where A is symmetric and positive definite, then these vectors are linearly independent. (This result implies that A has at most n conjugate directions.)

5.3 Verify the formula (5.7).

5.4 Show that if $f(x)$ is a strictly convex quadratic, then the function $h(\sigma) \stackrel{\text{def}}{=} f(x_0 + \sigma_0 p_0 + \cdots + \sigma_{k-1} p_{k-1})$ also is a strictly convex quadratic in the variable $\sigma = (\sigma_0, \sigma_1, \ldots, \sigma_{k-1})^T$.

5.5 Verify from the formulae (5.14) that (5.17) and (5.18) hold for $k = 1$.

5.6 Show that (5.24d) is equivalent to (5.14d).

✎   **5.7** Let $\{\lambda_i, v_i\}$ $i = 1, 2, \ldots, n$ be the eigenpairs of the symmetric matrix A. Show that the eigenvalues and eigenvectors of $[I + P_k(A)A]^T A [I + P_k(A)A]$ are $\lambda_i [1 + \lambda_i P_k(\lambda_i)]^2$ and v_i, respectively.

✎   **5.8** Construct matrices with various eigenvalue distributions (clustered and non-clustered) and apply the CG method to them. Comment on whether the behavior can be explained from Theorem 5.5.

✎   **5.9** Derive Algorithm 5.3 by applying the standard CG method in the variables $\hat{x}$ and then transforming back into the original variables.

✎   **5.10** Verify the modified conjugacy condition (5.40).

✎   **5.11** Show that when applied to a quadratic function, with exact line searches, both the Polak–Ribière formula given by (5.44) and the Hestenes–Stiefel formula given by (5.46) reduce to the Fletcher–Reeves formula (5.41a).

✎   **5.12** Prove that Lemma 5.6 holds for any choice of β_k satisfying $|\beta_k| \leq \beta_k^{\text{FR}}$.

CHAPTER 6

Quasi-Newton Methods

In the mid 1950s, W.C. Davidon, a physicist working at Argonne National Laboratory, was using the coordinate descent method (see Section 9.3) to perform a long optimization calculation. At that time computers were not very stable, and to Davidon's frustration, the computer system would always crash before the calculation was finished. So Davidon decided to find a way of accelerating the iteration. The algorithm he developed—the first quasi-Newton algorithm—turned out to be one of the most creative ideas in nonlinear optimization. It was soon demonstrated by Fletcher and Powell that the new algorithm was much faster and more reliable than the other existing methods, and this dramatic advance transformed nonlinear optimization overnight. During the following twenty years, numerous variants were proposed and hundreds of papers were devoted to their study. An interesting historical irony is that Davidon's paper [87] was not accepted for publication; it remained as a technical report for more than thirty years until it appeared in the first issue of the *SIAM Journal on Optimization* in 1991 [88].

Quasi-Newton methods, like steepest descent, require only the gradient of the objective function to be supplied at each iterate. By measuring the changes in gradients, they construct a model of the objective function that is good enough to produce superlinear convergence. The improvement over steepest descent is dramatic, especially on difficult problems. Moreover, since second derivatives are not required, quasi-Newton methods are sometimes more efficient than Newton's method. Today, optimization software libraries contain a variety of quasi-Newton algorithms for solving unconstrained, constrained, and large-scale optimization problems. In this chapter we discuss quasi-Newton methods for small and medium-sized problems, and in Chapter 7 we consider their extension to the large-scale setting.

The development of automatic differentiation techniques has made it possible to use Newton's method without requiring users to supply second derivatives; see Chapter 8. Still, automatic differentiation tools may not be applicable in many situations, and it may be much more costly to work with second derivatives in automatic differentiation software than with the gradient. For these reasons, quasi-Newton methods remain appealing.

6.1 THE BFGS METHOD

The most popular quasi-Newton algorithm is the BFGS method, named for its discoverers Broyden, Fletcher, Goldfarb, and Shanno. In this section we derive this algorithm (and its close relative, the DFP algorithm) and describe its theoretical properties and practical implementation.

We begin the derivation by forming the following quadratic model of the objective function at the current iterate x_k:

$$m_k(p) = f_k + \nabla f_k^T p + \tfrac{1}{2} p^T B_k p. \tag{6.1}$$

Here B_k is an $n \times n$ symmetric positive definite matrix that will be revised or *updated* at every iteration. Note that the function value and gradient of this model at $p = 0$ match f_k and ∇f_k, respectively. The minimizer p_k of this convex quadratic model, which we can write explicitly as

$$p_k = -B_k^{-1} \nabla f_k, \tag{6.2}$$

is used as the search direction, and the new iterate is

$$x_{k+1} = x_k + \alpha_k p_k, \qquad (6.3)$$

where the step length α_k is chosen to satisfy the Wolfe conditions (3.6). This iteration is quite similar to the line search Newton method; the key difference is that the approximate Hessian B_k is used in place of the true Hessian.

Instead of computing B_k afresh at every iteration, Davidon proposed to update it in a simple manner to account for the curvature measured during the most recent step. Suppose that we have generated a new iterate x_{k+1} and wish to construct a new quadratic model, of the form

$$m_{k+1}(p) = f_{k+1} + \nabla f_{k+1}^T p + \tfrac{1}{2} p^T B_{k+1} p.$$

What requirements should we impose on B_{k+1}, based on the knowledge gained during the latest step? One reasonable requirement is that the gradient of m_{k+1} should match the gradient of the objective function f at the latest two iterates x_k and x_{k+1}. Since $\nabla m_{k+1}(0)$ is precisely ∇f_{k+1}, the second of these conditions is satisfied automatically. The first condition can be written mathematically as

$$\nabla m_{k+1}(-\alpha_k p_k) = \nabla f_{k+1} - \alpha_k B_{k+1} p_k = \nabla f_k.$$

By rearranging, we obtain

$$B_{k+1}\alpha_k p_k = \nabla f_{k+1} - \nabla f_k. \qquad (6.4)$$

To simplify the notation it is useful to define the vectors

$$s_k = x_{k+1} - x_k = \alpha_k p_k, \qquad y_k = \nabla f_{k+1} - \nabla f_k, \qquad (6.5)$$

so that (6.4) becomes

$$B_{k+1}s_k = y_k. \qquad (6.6)$$

We refer to this formula as the *secant equation*.

Given the displacement s_k and the change of gradients y_k, the secant equation requires that the symmetric positive definite matrix B_{k+1} map s_k into y_k. This will be possible only if s_k and y_k satisfy the *curvature condition*

$$s_k^T y_k > 0, \qquad (6.7)$$

as is easily seen by premultiplying (6.6) by s_k^T. When f is strongly convex, the inequality (6.7) will be satisfied for any two points x_k and x_{k+1} (see Exercise 6.1). However, this condition will not always hold for nonconvex functions, and in this case we need to enforce (6.7) explicitly, by imposing restrictions on the line search procedure that chooses the step length α. In fact, the condition (6.7) is guaranteed to hold if we impose the Wolfe (3.6) or strong Wolfe conditions (3.7) on the line search. To verify this claim, we note from (6.5) and (3.6b) that $\nabla f_{k+1}^T s_k \geq c_2 \nabla f_k^T s_k$, and therefore

$$y_k^T s_k \geq (c_2 - 1)\alpha_k \nabla f_k^T p_k. \tag{6.8}$$

Since $c_2 < 1$ and since p_k is a descent direction, the term on the right is positive, and the curvature condition (6.7) holds.

When the curvature condition is satisfied, the secant equation (6.6) always has a solution B_{k+1}. In fact, it admits an infinite number of solutions, since the $n(n + 1)/2$ degrees of freedom in a symmetric positive definite matrix exceed the n conditions imposed by the secant equation. The requirement of positive definiteness imposes n additional inequalities—all principal minors must be positive—but these conditions do not absorb the remaining degrees of freedom.

To determine B_{k+1} uniquely, we impose the additional condition that *among all symmetric matrices satisfying the secant equation, B_{k+1} is, in some sense, closest to the current matrix B_k.* In other words, we solve the problem

$$\min_B \|B - B_k\| \tag{6.9a}$$

$$\text{subject to} \quad B = B^T, \quad Bs_k = y_k, \tag{6.9b}$$

where s_k and y_k satisfy (6.7) and B_k is symmetric and positive definite. Different matrix norms can be used in (6.9a), and each norm gives rise to a different quasi-Newton method. A norm that allows easy solution of the minimization problem (6.9) and gives rise to a scale-invariant optimization method is the weighted Frobenius norm

$$\|A\|_W \equiv \left\| W^{1/2} A W^{1/2} \right\|_F, \tag{6.10}$$

where $\| \cdot \|_F$ is defined by $\|C\|_F^2 = \sum_{i=1}^n \sum_{j=1}^n c_{ij}^2$. The weight matrix W can be chosen as *any* matrix satisfying the relation $W y_k = s_k$. For concreteness, the reader can assume that $W = \bar{G}_k^{-1}$ where $\bar{G}_k$ is the *average Hessian* defined by

$$\bar{G}_k = \left[\int_0^1 \nabla^2 f(x_k + \tau\alpha_k p_k)d\tau \right]. \tag{6.11}$$

The property

$$y_k = \bar{G}_k \alpha_k p_k = \bar{G}_k s_k \tag{6.12}$$

follows from Taylor's theorem, Theorem 2.1. With this choice of weighting matrix W, the norm (6.10) is non-dimensional, which is a desirable property, since we do not wish the solution of (6.9) to depend on the units of the problem.

With this weighting matrix and this norm, the unique solution of (6.9) is

$$\text{(DFP)} \qquad B_{k+1} = \left(I - \rho_k y_k s_k^T\right) B_k \left(I - \rho_k s_k y_k^T\right) + \rho_k y_k y_k^T, \qquad (6.13)$$

with

$$\rho_k = \frac{1}{y_k^T s_k}. \qquad (6.14)$$

This formula is called the DFP updating formula, since it is the one originally proposed by Davidon in 1959, and subsequently studied, implemented, and popularized by Fletcher and Powell.

The inverse of B_k, which we denote by

$$H_k = B_k^{-1},$$

is useful in the implementation of the method, since it allows the search direction (6.2) to be calculated by means of a simple matrix–vector multiplication. Using the Sherman–Morrison–Woodbury formula (A.28), we can derive the following expression for the update of the inverse Hessian approximation H_k that corresponds to the DFP update of B_k in (6.13):

$$\text{(DFP)} \qquad H_{k+1} = H_k - \frac{H_k y_k y_k^T H_k}{y_k^T H_k y_k} + \frac{s_k s_k^T}{y_k^T s_k}. \qquad (6.15)$$

Note that the last two terms in the right-hand-side of (6.15) are rank-one matrices, so that H_k undergoes a rank-two modification. It is easy to see that (6.13) is also a rank-two modification of B_k. This is the fundamental idea of quasi-Newton updating: Instead of recomputing the approximate Hessians (or inverse Hessians) from scratch at every iteration, we apply a simple modification that combines the most recently observed information about the objective function with the existing knowledge embedded in our current Hessian approximation.

The DFP updating formula is quite effective, but it was soon superseded by the BFGS formula, which is presently considered to be the most effective of all quasi-Newton updating formulae. BFGS updating can be derived by making a simple change in the argument that led to (6.13). Instead of imposing conditions on the Hessian approximations B_k, we impose similar conditions on their inverses H_k. The updated approximation H_{k+1} must be symmetric and positive definite, and must satisfy the secant equation (6.6), now written as

$$H_{k+1} y_k = s_k.$$

The condition of closeness to H_k is now specified by the following analogue of (6.9):

$$\min_{H} \|H - H_k\| \qquad (6.16a)$$

$$\text{subject to} \quad H = H^T, \qquad H y_k = s_k. \qquad (6.16b)$$

The norm is again the weighted Frobenius norm described above, where the weight matrix W is now any matrix satisfying $Ws_k = y_k$. (For concreteness, we assume again that W is given by the average Hessian $\bar{G}_k$ defined in (6.11).) The unique solution H_{k+1} to (6.16) is given by

$$\text{(BFGS)} \qquad H_{k+1} = (I - \rho_k s_k y_k^T)H_k(I - \rho_k y_k s_k^T) + \rho_k s_k s_k^T, \qquad (6.17)$$

with ρ_k defined by (6.14).

Just one issue has to be resolved before we can define a complete BFGS algorithm: How should we choose the initial approximation H_0? Unfortunately, there is no magic formula that works well in all cases. We can use specific information about the problem, for instance by setting it to the inverse of an approximate Hessian calculated by finite differences at x_0. Otherwise, we can simply set it to be the identity matrix, or a multiple of the identity matrix, where the multiple is chosen to reflect the scaling of the variables.

Algorithm 6.1 (BFGS Method).
   Given starting point x_0, convergence tolerance $\epsilon > 0$,
       inverse Hessian approximation H_0;
   $k \leftarrow 0$;
   while $\|\nabla f_k\| > \epsilon$;
       Compute search direction

$$p_k = -H_k \nabla f_k; \qquad (6.18)$$

       Set $x_{k+1} = x_k + \alpha_k p_k$ where α_k is computed from a line search
          procedure to satisfy the Wolfe conditions (3.6);
      Define $s_k = x_{k+1} - x_k$ and $y_k = \nabla f_{k+1} - \nabla f_k$;
      Compute H_{k+1} by means of (6.17);
      $k \leftarrow k + 1$;
  end (while)

Each iteration can be performed at a cost of $O(n^2)$ arithmetic operations (plus the cost of function and gradient evaluations); there are no $O(n^3)$ operations such as linear system solves or matrix–matrix operations. The algorithm is robust, and its rate of convergence is superlinear, which is fast enough for most practical purposes. Even though Newton's method converges more rapidly (that is, quadratically), its cost per iteration usually is higher, because of its need for second derivatives and solution of a linear system.

We can derive a version of the BFGS algorithm that works with the Hessian approximation B_k rather than H_k. The update formula for B_k is obtained by simply applying the Sherman–Morrison–Woodbury formula (A.28) to (6.17) to obtain

$$\text{(BFGS)} \qquad B_{k+1} = B_k - \frac{B_k s_k s_k^T B_k}{s_k^T B_k s_k} + \frac{y_k y_k^T}{y_k^T s_k}. \qquad (6.19)$$

A naive implementation of this variant is not efficient for unconstrained minimization, because it requires the system $B_k p_k = -\nabla f_k$ to be solved for the step p_k, thereby increasing the cost of the step computation to $O(n^3)$. We discuss later, however, that less expensive implementations of this variant are possible by updating Cholesky factors of B_k.

PROPERTIES OF THE BFGS METHOD

It is usually easy to observe the superlinear rate of convergence of the BFGS method on practical problems. Below, we report the last few iterations of the steepest descent, BFGS, and an inexact Newton method on Rosenbrock's function (2.22). The table gives the value of $\|x_k - x^*\|$. The Wolfe conditions were imposed on the step length in all three methods. From the starting point $(-1.2, 1)$, the steepest descent method required 5264 iterations, whereas BFGS and Newton took only 34 and 21 iterations, respectively to reduce the gradient norm to 10^{-5}.

steepest descent	BFGS	Newton
1.827e-04	1.70e-03	3.48e-02
1.826e-04	1.17e-03	1.44e-02
1.824e-04	1.34e-04	1.82e-04
1.823e-04	1.01e-06	1.17e-08

A few points in the derivation of the BFGS and DFP methods merit further discussion. Note that the minimization problem (6.16) that gives rise to the BFGS update formula does not explicitly require the updated Hessian approximation to be positive definite. It is easy to show, however, that H_{k+1} will be positive definite whenever H_k is positive definite, by using the following argument. First, note from (6.8) that $y_k^T s_k$ is positive, so that the updating formula (6.17), (6.14) is well-defined. For any nonzero vector z, we have

$$z^T H_{k+1} z = w^T H_k w + \rho_k (z^T s_k)^2 \geq 0,$$

where we have defined $w = z - \rho_k y_k (s_k^T z)$. The right hand side can be zero only if $s_k^T z = 0$, but in this case $w = z \neq 0$, which implies that the first term is greater than zero. Therefore, H_{k+1} is positive definite.

To make quasi-Newton updating formulae invariant to transformations in the variables (such as scaling transformations), it is necessary for the objectives (9.9a) and (6.16a) to be invariant under the same transformations. The choice of the weighting matrices W used to define the norms in (9.9a) and (6.16a) ensures that this condition holds. Many other choices of the weighting matrix W are possible, each one of them giving a different update formula. However, despite intensive searches, no formula has been found that is significantly more effective than BFGS.

The BFGS method has many interesting properties when applied to quadratic functions. We discuss these properties later in the more general context of the Broyden family of updating formulae, of which BFGS is a special case.

It is reasonable to ask whether there are situations in which the updating formula such as (6.17) can produce bad results. If at some iteration the matrix H_k becomes a poor approximation to the true inverse Hessian, is there any hope of correcting it? For example, when the inner product $y_k^T s_k$ is tiny (but positive), then it follows from (6.14), (6.17) that H_{k+1} contains very large elements. Is this behavior reasonable? A related question concerns the rounding errors that occur in finite-precision implementation of these methods. Can these errors grow to the point of erasing all useful information in the quasi-Newton approximate Hessian?

These questions have been studied analytically and experimentally, and it is now known that the BFGS formula has very effective self-correcting properties. If the matrix H_k incorrectly estimates the curvature in the objective function, and if this bad estimate slows down the iteration, then the Hessian approximation will tend to correct itself within a few steps. It is also known that the DFP method is less effective in correcting bad Hessian approximations; this property is believed to be the reason for its poorer practical performance. The self-correcting properties of BFGS hold only when an adequate line search is performed. In particular, the Wolfe line search conditions ensure that the gradients are sampled at points that allow the model (6.1) to capture appropriate curvature information.

It is interesting to note that the DFP and BFGS updating formulae are *duals* of each other, in the sense that one can be obtained from the other by the interchanges $s \leftrightarrow y$, $B \leftrightarrow H$. This symmetry is not surprising, given the manner in which we derived these methods above.

IMPLEMENTATION

A few details and enhancements need to be added to Algorithm 6.1 to produce an efficient implementation. The line search, which should satisfy either the Wolfe conditions (3.6) or the strong Wolfe conditions (3.7), should always try the step length $\alpha_k = 1$ first, because this step length will eventually always be accepted (under certain conditions), thereby producing superlinear convergence of the overall algorithm. Computational observations strongly suggest that it is more economical, in terms of function evaluations, to perform a fairly inaccurate line search. The values $c_1 = 10^{-4}$ and $c_2 = 0.9$ are commonly used in (3.6).

As mentioned earlier, the initial matrix H_0 often is set to some multiple βI of the identity, but there is no good general strategy for choosing the multiple β. If β is too large, so that the first step $p_0 = -\beta g_0$ is too long, many function evaluations may be required to find a suitable value for the step length α_0. Some software asks the user to prescribe a value δ for the norm of the first step, and then set $H_0 = \delta \|g_0\|^{-1} I$ to achieve this norm.

A heuristic that is often quite effective is to scale the starting matrix *after* the first step has been computed but before the first BFGS update is performed. We change the provisional value $H_0 = I$ by setting

$$H_0 \leftarrow \frac{y_k^T s_k}{y_k^T y_k} I, \qquad (6.20)$$

before applying the update (6.14), (6.17) to obtain H_1. This formula attempts to make the size of H_0 similar to that of $\nabla^2 f(x_0)^{-1}$, in the following sense. Assuming that the average Hessian defined in (6.11) is positive definite, there exists a square root $\bar{G}_k^{1/2}$ satisfying $\bar{G}_k = \bar{G}_k^{1/2} \bar{G}_k^{1/2}$ (see Exercise 6.6). Therefore, by defining $z_k = \bar{G}_k^{1/2} s_k$ and using the relation (6.12), we have

$$\frac{y_k^T s_k}{y_k^T y_k} = \frac{(\bar{G}_k^{1/2} s_k)^T \bar{G}_k^{1/2} s_k}{(\bar{G}_k^{1/2} s_k)^T \bar{G}_k \bar{G}_k^{1/2} s_k} = \frac{z_k^T z_k}{z_k^T \bar{G}_k z_k}. \qquad (6.21)$$

The *reciprocal* of (6.21) is an approximation to one of the eigenvalues of $\bar{G}_k$, which in turn is close to an eigenvalue of $\nabla^2 f(x_k)$. Hence, the quotient (6.21) itself approximates an eigenvalue of $\nabla^2 f(x_k)^{-1}$. Other scaling factors can be used in (6.20), but the one presented here appears to be the most successful in practice.

In (6.19) we gave an update formula for a BFGS method that works with the Hessian approximation B_k instead of the the inverse Hessian approximation H_k. An efficient implementation of this approach does not store B_k explicitly, but rather the Cholesky factorization $L_k D_k L_k^T$ of this matrix. A formula that updates the factors L_k and D_k *directly* in $O(n^2)$ operations can be derived from (6.19). Since the linear system $B_k p_k = -\nabla f_k$ also can be solved in $O(n^2)$ operations (by performing triangular substitutions with L_k and L_k^T and a diagonal substitution with D_k), the total cost is quite similar to the variant described in Algorithm 6.1. A potential advantage of this alternative strategy is that it gives us the option of modifying diagonal elements in the D_k factor if they are not sufficiently large, to prevent instability when we divide by these elements during the calculation of p_k. However, computational experience suggests no real advantages for this variant, and we prefer the simpler strategy of Algorithm 6.1.

The performance of the BFGS method can degrade if the line search is not based on the Wolfe conditions. For example, some software implements an Armijo backtracking line search (see Section 3.1): The unit step length $\alpha_k = 1$ is tried first and is successively decreased until the sufficient decrease condition (3.6a) is satisfied. For this strategy, there is no guarantee that the curvature condition $y_k^T s_k > 0$ (6.7) will be satisfied by the chosen step, since a step length greater than 1 may be required to satisfy this condition. To cope with this shortcoming, some implementations simply *skip* the BFGS update by setting $H_{k+1} = H_k$ when $y_k^T s_k$ is negative or too close to zero. This approach is not recommended, because the updates may be skipped much too often to allow H_k to capture important curvature information for the objective function f. In Chapter 18 we discuss a *damped* BFGS update that is a more effective strategy for coping with the case where the curvature condition (6.7) is not satisfied.

6.2   THE SR1 METHOD

In the BFGS and DFP updating formulae, the updated matrix B_{k+1} (or H_{k+1}) differs from its predecessor B_k (or H_k) by a rank-2 matrix. In fact, as we now show, there is a simpler rank-1 update that maintains symmetry of the matrix and allows it to satisfy the secant equation. Unlike the rank-two update formulae, this *symmetric-rank-1*, or SR1, update does not guarantee that the updated matrix maintains positive definiteness. Good numerical results have been obtained with algorithms based on SR1, so we derive it here and investigate its properties.

The symmetric rank-1 update has the general form

$$B_{k+1} = B_k + \sigma v v^T,$$

where σ is either $+1$ or -1, and σ and v are chosen so that B_{k+1} satisfies the secant equation (6.6), that is, $y_k = B_{k+1}s_k$. By substituting into this equation, we obtain

$$y_k = B_k s_k + \left[\sigma v^T s_k\right] v. \tag{6.22}$$

Since the term in brackets is a scalar, we deduce that v must be a multiple of $y_k - B_k s_k$, that is, $v = \delta(y_k - B_k s_k)$ for some scalar δ. By substituting this form of v into (6.22), we obtain

$$(y_k - B_k s_k) = \sigma \delta^2 \left[s_k^T(y_k - B_k s_k)\right](y_k - B_k s_k), \tag{6.23}$$

and it is clear that this equation is satisfied if (and only if) we choose the parameters δ and σ to be

$$\sigma = \text{sign}\left[s_k^T(y_k - B_k s_k)\right], \qquad \delta = \pm \left|s_k^T(y_k - B_k s_k)\right|^{-1/2}.$$

Hence, we have shown that the only symmetric rank-1 updating formula that satisfies the secant equation is given by

$$\text{(SR1)} \qquad B_{k+1} = B_k + \frac{(y_k - B_k s_k)(y_k - B_k s_k)^T}{(y_k - B_k s_k)^T s_k}. \tag{6.24}$$

By applying the Sherman–Morrison formula (A.27), we obtain the corresponding update formula for the inverse Hessian approximation H_k:

$$\text{(SR1)} \qquad H_{k+1} = H_k + \frac{(s_k - H_k y_k)(s_k - H_k y_k)^T}{(s_k - H_k y_k)^T y_k}. \tag{6.25}$$

This derivation is so simple that the SR1 formula has been rediscovered a number of times.

It is easy to see that even if B_k is positive definite, B_{k+1} may not have the same property. (The same is, of course, true of H_k.) This observation was considered a major drawback in the early days of nonlinear optimization when only line search iterations were used. However, with the advent of trust-region methods, the SR1 updating formula has proved to be quite useful, and its ability to generate indefinite Hessian approximations can actually be regarded as one of its chief advantages.

The main drawback of SR1 updating is that the denominator in (6.24) or (6.25) can vanish. In fact, even when the objective function is a convex quadratic, there may be steps on which there is no symmetric rank-1 update that satisfies the secant equation. It pays to reexamine the derivation above in the light of this observation.

By reasoning in terms of B_k (similar arguments can be applied to H_k), we see that there are three cases:

1. If $(y_k - B_k s_k)^T s_k \neq 0$, then the arguments above show that there is a unique rank-one updating formula satisfying the secant equation (6.6), and that it is given by (6.24).

2. If $y_k = B_k s_k$, then the only updating formula satisfying the secant equation is simply $B_{k+1} = B_k$.

3. If $y_k \neq B_k s_k$ and $(y_k - B_k s_k)^T s_k = 0$, then (6.23) shows that there is no symmetric rank-one updating formula satisfying the secant equation.

The last case clouds an otherwise simple and elegant derivation, and suggests that numerical instabilities and even breakdown of the method can occur. It suggests that rank-one updating does not provide enough freedom to develop a matrix with all the desired characteristics, and that a rank-two correction is required. This reasoning leads us back to the BFGS method, in which positive definiteness (and thus nonsingularity) of all Hessian approximations is guaranteed.

Nevertheless, we are interested in the SR1 formula for the following reasons.

(i) A simple safeguard seems to adequately prevent the breakdown of the method and the occurrence of numerical instabilities.

(ii) The matrices generated by the SR1 formula tend to be good approximations to the true Hessian matrix—often better than the BFGS approximations.

(iii) In quasi-Newton methods for constrained problems, or in methods for partially separable functions (see Chapters 18 and 7), it may not be possible to impose the curvature condition $y_k^T s_k > 0$, and thus BFGS updating is not recommended. Indeed, in these two settings, indefinite Hessian approximations are desirable insofar as they reflect indefiniteness in the true Hessian.

We now introduce a strategy to prevent the SR1 method from breaking down. It has been observed in practice that SR1 performs well simply by skipping the update if the denominator is small. More specifically, the update (6.24) is applied only if

$$\left| s_k^T (y_k - B_k s_k) \right| \geq r \|s_k\| \, \|y_k - B_k s_k\|, \tag{6.26}$$

where $r \in (0, 1)$ is a small number, say $r = 10^{-8}$. If (6.26) does not hold, we set $B_{k+1} = B_k$. Most implementations of the SR1 method use a skipping rule of this kind.

Why do we advocate skipping of updates for the SR1 method, when in the previous section we discouraged this strategy in the case of BFGS? The two cases are quite different. The condition $s_k^T(y_k - B_k s_k) \approx 0$ occurs infrequently, since it requires certain vectors to be aligned in a specific way. When it does occur, skipping the update appears to have no negative effects on the iteration. This is not surprising, since the skipping condition implies that $s_k^T \bar{G} s_k \approx s_k^T B_k s_k$, where $\bar{G}$ is the average Hessian over the last step—meaning that the curvature of B_k along s_k is already correct. In contrast, the curvature condition $s_k^T y_k \geq 0$ required for BFGS updating may easily fail if the line search does not impose the Wolfe conditions (for example, if the step is not long enough), and therefore skipping the BFGS update can occur often and can degrade the quality of the Hessian approximation.

We now give a formal description of an SR1 method using a trust-region framework, which we prefer over a line search framework because it can accommodate indefinite Hessian approximations more easily.

Algorithm 6.2 (SR1 Trust-Region Method).
    Given starting point x_0, initial Hessian approximation B_0,
        trust-region radius Δ_0, convergence tolerance $\epsilon > 0$,
        parameters $\eta \in (0, 10^{-3})$ and $r \in (0, 1)$;
    $k \leftarrow 0$;
    while $\|\nabla f_k\| > \epsilon$;
        Compute s_k by solving the subproblem

$$\min_s \nabla f_k^T s + \frac{1}{2} s^T B_k s \qquad \text{subject to } \|s\| \leq \Delta_k; \qquad (6.27)$$

        Compute

$$
\begin{aligned}
y_k &= \nabla f(x_k + s_k) - \nabla f_k, \\
\text{ared} &= f_k - f(x_k + s_k) \qquad \text{(actual reduction)} \\
\text{pred} &= -\left(\nabla f_k^T s_k + \frac{1}{2} s_k^T B_k s_k \right) \qquad \text{(predicted reduction)};
\end{aligned}
$$

        if ared/pred $> \eta$
            $x_{k+1} = x_k + s_k$;
        else
            $x_{k+1} = x_k$;
        end (if)

    if ared/pred > 0.75
        if $\|s_k\| \leq 0.8\Delta_k$
            $\Delta_{k+1} = \Delta_k;$
        else
            $\Delta_{k+1} = 2\Delta_k;$
        end (if)
    else if $0.1 \leq$ ared/pred ≤ 0.75
        $\Delta_{k+1} = \Delta_k;$
    else
        $\Delta_{k+1} = 0.5\Delta_k;$
    end (if)
    if (6.26) holds
        Use (6.24) to compute B_{k+1} (even if $x_{k+1} = x_k$);
    else
        $B_{k+1} \leftarrow B_k;$
    end (if)
    $k \leftarrow k + 1;$
end (while)

    This algorithm has the typical form of a trust region method (cf. Algorithm 4.1). For concreteness, we have specified a particular strategy for updating the trust region radius, but other heuristics can be used instead.

    To obtain a fast rate of convergence, it is important for the matrix B_k to be updated even along a failed direction s_k. The fact that the step was poor indicates that B_k is an inadequate approximation of the true Hessian in this direction. Unless the quality of the approximation is improved, steps along similar directions could be generated on later iterations, and repeated rejection of such steps could prevent superlinear convergence.

PROPERTIES OF SR1 UPDATING

    One of the main advantages of SR1 updating is its ability to generate good Hessian approximations. We demonstrate this property by first examining a quadratic function. For functions of this type, the choice of step length does not affect the update, so to examine the effect of the updates, we can assume for simplicity a uniform step length of 1, that is,

$$p_k = -H_k \nabla f_k, \qquad x_{k+1} = x_k + p_k. \tag{6.28}$$

It follows that $p_k = s_k$.

Theorem 6.1.
    Suppose that $f : \mathbb{R}^n \to \mathbb{R}$ is the strongly convex quadratic function $f(x) = b^T x + \frac{1}{2}x^T A x$, where A is symmetric positive definite. Then for any starting point x_0 and any

symmetric starting matrix H_0, the iterates $\{x_k\}$ generated by the SR1 method (6.25), (6.28) converge to the minimizer in at most n steps, provided that $(s_k - H_k y_k)^T y_k \neq 0$ for all k. Moreover, if n steps are performed, and if the search directions p_i are linearly independent, then $H_n = A^{-1}$.

PROOF.   Because of our assumption $(s_k - H_k y_k)^T y_k \neq 0$, the SR1 update is always well-defined. We start by showing inductively that

$$H_k y_j = s_j, \qquad j = 0, 1, \ldots, k - 1. \tag{6.29}$$

In other words, we claim that the secant equation is satisfied not only along the most recent search direction, but along all previous directions.

By definition, the SR1 update satisfies the secant equation, so we have $H_1 y_0 = s_0$. Let us now assume that (6.29) holds for some value $k > 1$ and show that it holds also for $k + 1$. From this assumption, we have from (6.29) that

$$(s_k - H_k y_k)^T y_j = s_k^T y_j - y_k^T (H_k y_j) = s_k^T y_j - y_k^T s_j = 0, \quad \text{all } j < k, \tag{6.30}$$

where the last equality follows because $y_i = As_i$ for the quadratic function we are considering here. By using (6.30) and the induction hypothesis (6.29) in (6.25), we have

$$H_{k+1} y_j = H_k y_j = s_j, \qquad \text{for all } j < k.$$

Since $H_{k+1} y_k = s_k$ by the secant equation, we have shown that (6.29) holds when k is replaced by $k + 1$. By induction, then, this relation holds for all k.

If the algorithm performs n steps and if these steps $\{s_j\}$ are linearly independent, we have

$$s_j = H_n y_j = H_n As_j, \qquad j = 0, 1, \ldots, n - 1.$$

It follows that $H_n A = I$, that is, $H_n = A^{-1}$. Therefore, the step taken at x_n is the Newton step, and so the next iterate x_{n+1} will be the solution, and the algorithm terminates.

Consider now the case in which the steps become linearly dependent. Suppose that s_k is a linear combination of the previous steps, that is,

$$s_k = \xi_0 s_0 + \cdots + \xi_{k-1} s_{k-1}, \tag{6.31}$$

for some scalars ξ_i. From (6.31) and (6.29) we have that

$$H_k y_k = H_k As_k$$
$$= \xi_0 H_k As_0 + \cdots + \xi_{k-1} H_k As_{k-1}$$

$$\begin{aligned}
&= \xi_0 H_k y_0 + \cdots + \xi_{k-1} H_k y_{k-1} \\
&= \xi_0 s_0 + \cdots + \xi_{k-1} s_{k-1} \\
&= s_k.
\end{aligned}$$

Since $y_k = \nabla f_{k+1} - \nabla f_k$ and since $s_k = p_k = -H_k \nabla f_k$ from (6.28), we have that

$$H_k(\nabla f_{k+1} - \nabla f_k) = -H_k \nabla f_k,$$

which, by the nonsingularity of H_k, implies that $\nabla f_{k+1} = 0$. Therefore, x_{k+1} is the solution point. □

The relation (6.29) shows that when f is quadratic, the secant equation is satisfied along all previous search directions, regardless of how the line search is performed. A result like this can be established for BFGS updating only under the restrictive assumption that the line search is exact, as we show in the next section.

For general nonlinear functions, the SR1 update continues to generate good Hessian approximations under certain conditions.

Theorem 6.2.

Suppose that f is twice continuously differentiable, and that its Hessian is bounded and Lipschitz continuous in a neighborhood of a point x^*. Let $\{x_k\}$ be any sequence of iterates such that $x_k \to x^*$ for some $x^* \in \mathbb{R}^n$. Suppose in addition that the inequality (6.26) holds for all k, for some $r \in (0, 1)$, and that the steps s_k are uniformly linearly independent. Then the matrices B_k generated by the SR1 updating formula satisfy

$$\lim_{k \to \infty} \| B_k - \nabla^2 f(x^*) \| = 0.$$

The term "uniformly linearly independent steps" means, roughly speaking, that the steps do not tend to fall in a subspace of dimension less than n. This assumption is usually, but not always, satisfied in practice (see the Notes and References at the end of this chapter).

6.3 THE BROYDEN CLASS

So far, we have described the BFGS, DFP, and SR1 quasi-Newton updating formulae, but there are many others. Of particular interest is the *Broyden class*, a family of updates specified by the following general formula:

$$B_{k+1} = B_k - \frac{B_k s_k s_k^T B_k}{s_k^T B_k s_k} + \frac{y_k y_k^T}{y_k^T s_k} + \phi_k (s_k^T B_k s_k) v_k v_k^T, \tag{6.32}$$

where ϕ_k is a scalar parameter and

$$v_k = \left[\frac{y_k}{y_k^T s_k} - \frac{B_k s_k}{s_k^T B_k s_k}\right]. \tag{6.33}$$

The BFGS and DFP methods are members of the Broyden class—we recover BFGS by setting $\phi_k = 0$ and DFP by setting $\phi_k = 1$ in (6.32). We can therefore rewrite (6.32) as a "linear combination" of these two methods, that is,

$$B_{k+1} = (1 - \phi_k)B_{k+1}^{\text{BFGS}} + \phi_k B_{k+1}^{\text{DFP}}.$$

This relationship indicates that all members of the Broyden class satisfy the secant equation (6.6), since the BGFS and DFP matrices themselves satisfy this equation. Also, since BFGS and DFP updating preserve positive definiteness of the Hessian approximations when $s_k^T y_k > 0$, this relation implies that the same property will hold for the Broyden family if $0 \le \phi_k \le 1$.

Much attention has been given to the so-called *restricted Broyden class*, which is obtained by restricting ϕ_k to the interval $[0, 1]$. It enjoys the following property when applied to quadratic functions. Since the analysis is independent of the step length, we assume for simplicity that each iteration has the form

$$p_k = -B_k^{-1}\nabla f_k, \qquad x_{k+1} = x_k + p_k. \tag{6.34}$$

Theorem 6.3.
Suppose that $f : \mathbb{R}^n \to \mathbb{R}$ is the strongly convex quadratic function $f(x) = b^T x + \frac{1}{2}x^T Ax$, where A is symmetric and positive definite. Let x_0 be any starting point for the iteration (6.34) and B_0 be any symmetric positive definite starting matrix, and suppose that the matrices B_k are updated by the Broyden formula (6.32) with $\phi_k \in [0, 1]$. Define $\lambda_1^k \le \lambda_2^k \le \cdots \le \lambda_n^k$ to be the eigenvalues of the matrix

$$A^{\frac{1}{2}}B_k^{-1}A^{\frac{1}{2}}. \tag{6.35}$$

Then for all k, we have

$$\min\{\lambda_i^k, 1\} \le \lambda_i^{k+1} \le \max\{\lambda_i^k, 1\}, \quad i = 1, 2, \ldots, n. \tag{6.36}$$

Moreover, the property (6.36) does not hold if the Broyden parameter ϕ_k is chosen outside the interval $[0, 1]$.

Let us discuss the significance of this result. If the eigenvalues λ_i^k of the matrix (6.35) are all 1, then the quasi-Newton approximation B_k is identical to the Hessian A of the quadratic objective function. This situation is the ideal one, so we should be hoping for these eigenvalues to be as close to 1 as possible. In fact, relation (6.36) tells us that the eigenvalues $\{\lambda_i^k\}$ converge monotonically (but not strictly monotonically) to 1. Suppose, for example, that at iteration k the smallest eigenvalue is $\lambda_1^k = 0.7$. Then (6.36) tells us that at the next iteration $\lambda_1^{k+1} \in [0.7, 1]$. We cannot be sure that this eigenvalue has actually moved closer to 1, but it is reasonable to expect that it has. In contrast, the first eigenvalue can become smaller than 0.7 if we allow ϕ_k to be outside $[0, 1]$. Significantly, the result of Theorem 6.3 holds even if the line searches are not exact.

Although Theorem 6.3 seems to suggest that the best update formulas belong to the restricted Broyden class, the situation is not at all clear. Some analysis and computational testing suggest that algorithms that allow ϕ_k to be negative (in a strictly controlled manner) may in fact be superior to the BFGS method. The SR1 formula is a case in point: It is a member of the Broyden class, obtained by setting

$$\phi_k = \frac{s_k^T y_k}{s_k^T y_k - s_k^T B_k s_k},$$

but it does not belong to the restricted Broyden class, because this value of ϕ_k may fall outside the interval $[0, 1]$.

In the remaining discussion of this section, we determine more precisely the range of values of ϕ_k that preserve positive definiteness.

The last term in (6.32) is a rank-one correction, which by the interlacing eigenvalue theorem (Theorem A.1) increases the eigenvalues of the matrix when ϕ_k is positive. Therefore B_{k+1} is positive definite for all $\phi_k \geq 0$. On the other hand, by Theorem A.1 the last term in (6.32) decreases the eigenvalues of the matrix when ϕ_k is negative. As we decrease ϕ_k, this matrix eventually becomes singular and then indefinite. A little computation shows that B_{k+1} is singular when ϕ_k has the value

$$\phi_k^c = \frac{1}{1 - \mu_k}, \tag{6.37}$$

where

$$\mu_k = \frac{(y_k^T B_k^{-1} y_k)(s_k^T B_k s_k)}{(y_k^T s_k)^2}. \tag{6.38}$$

By applying the Cauchy–Schwarz inequality (A.5) to (6.38), we see that $\mu_k \geq 1$ and therefore $\phi_k^c \leq 0$. Hence, if the initial Hessian approximation B_0 is symmetric and positive definite, and if $s_k^T y_k > 0$ and $\phi_k > \phi_k^c$ for each k, then all the matrices B_k generated by Broyden's formula (6.32) remain symmetric and positive definite.

When the line search is exact, *all* methods in the Broyden class with $\phi_k \geq \phi_k^c$ generate the same sequence of iterates. This result applies to general nonlinear functions and is based on the observation that when all the line searches are exact, the directions generated by Broyden-class methods differ only in their lengths. The line searches identify the same minima along the chosen search direction, though the values of the step lengths may differ because of the different scaling.

The Broyden class has several remarkable properties when applied with exact line searches to quadratic functions. We state some of these properties in the next theorem, whose proof is omitted.

Theorem 6.4.

Suppose that a method in the Broyden class is applied to the strongly convex quadratic function $f(x) = b^T x + \frac{1}{2} x^T A x$, where x_0 is the starting point and B_0 is any symmetric positive definite matrix. Assume that α_k is the exact step length and that $\phi_k \geq \phi_k^c$ for all k, where ϕ_k^c is defined by (6.37). Then the following statements are true.

   (i) *The iterates are independent of ϕ_k and converge to the solution in at most n iterations.*

  (ii) *The secant equation is satisfied for all previous search directions, that is,*

$$B_k s_j = y_j, \quad j = k-1, k-2, \ldots, 1.$$

(iii) *If the starting matrix is $B_0 = I$, then the iterates are identical to those generated by the conjugate gradient method (see Chapter 5). In particular, the search directions are conjugate, that is,*

$$s_i^T A s_j = 0, \quad \text{for } i \neq j.$$

(iv) *If n iterations are performed, we have $B_n = A$.*

Note that parts (i), (ii), and (iv) of this result echo the statement and proof of Theorem 6.1, where similar results were derived for the SR1 update formula.

We can generalize Theorem 6.4 slightly: It continues to hold if the Hessian approximations remain nonsingular but not necessarily positive definite. (Hence, we could allow ϕ_k to be smaller than ϕ_k^c, provided that the chosen value did not produce a singular updated matrix.) We can also generalize point (iii) as follows. If the starting matrix B_0 is not the identity matrix, then the Broyden-class method is identical to the preconditioned conjugate gradient method that uses B_0 as preconditioner.

We conclude by commenting that results like Theorem 6.4 would appear to be of mainly theoretical interest, since the inexact line searches used in practical implementations of Broyden-class methods (and all other quasi-Newton methods) cause their performance to differ markedly. Nevertheless, it is worth noting that this type of analysis guided much of the development of quasi-Newton methods.

6.4 CONVERGENCE ANALYSIS

In this section we present global and local convergence results for practical implementations of the BFGS and SR1 methods. We give more details for BFGS because its analysis is more general and illuminating than that of SR1. The fact that the Hessian approximations evolve by means of updating formulas makes the analysis of quasi-Newton methods much more complex than that of steepest descent and Newton's method.

Although the BFGS and SR1 methods are known to be remarkably robust in practice, we will not be able to establish truly global convergence results for general nonlinear objective functions. That is, we cannot prove that the iterates of these quasi-Newton methods approach a stationary point of the problem from any starting point and any (suitable) initial Hessian approximation. In fact, it is not yet known if the algorithms enjoy such properties. In our analysis we will either assume that the objective function is convex or that the iterates satisfy certain properties. On the other hand, there are well known local, superlinear convergence results that are true under reasonable assumptions.

Throughout this section we use $\| \cdot \|$ to denote the Euclidean vector or matrix norm, and denote the Hessian matrix $\nabla^2 f(x)$ by $G(x)$.

GLOBAL CONVERGENCE OF THE BFGS METHOD

We study the global convergence of the BFGS method, with a practical line search, when applied to a smooth convex function from an arbitrary starting point x_0 and from any initial Hessian approximation B_0 that is symmetric and positive definite. We state our precise assumptions about the objective function formally, as follows.

Assumption 6.1.

 (i) *The objective function f is twice continuously differentiable.*

 (ii) *The level set $\mathcal{L} = \{x \in \mathbb{R}^n \mid f(x) \leq f(x_0)\}$ is convex, and there exist positive constants m and M such that*

$$m\|z\|^2 \leq z^T G(x) z \leq M\|z\|^2 \tag{6.39}$$

for all $z \in \mathbb{R}^n$ and $x \in \mathcal{L}$.

Part (ii) of this assumption implies that $G(x)$ is positive definite on $\mathcal{L}$ and that f has a unique minimizer x^* in $\mathcal{L}$.

By using (6.12) and (6.39) we obtain

$$\frac{y_k^T s_k}{s_k^T s_k} = \frac{s_k^T \bar{G}_k s_k}{s_k^T s_k} \geq m, \tag{6.40}$$

where $\bar{G}_k$ is the average Hessian defined in (6.11). Assumption 6.1 implies that $\bar{G}_k$ is positive definite, so its square root is well-defined. Therefore, as in (6.21), we have by defining $z_k = \bar{G}_k^{1/2} s_k$ that

$$\frac{y_k^T y_k}{y_k^T s_k} = \frac{s_k^T \bar{G}_k^2 s_k}{s_k^T \bar{G}_k s_k} = \frac{z_k^T \bar{G}_k z_k}{z_k^T z_k} \le M. \tag{6.41}$$

We are now ready to present the global convergence result for the BFGS method. It does not seem to be possible to establish a bound on the condition number of the Hessian approximations B_k, as is done in Section 3.2. Instead, we will introduce two new tools in the analysis, the trace and determinant, to estimate the size of the largest and smallest eigenvalues of the Hessian approximations. The trace of a matrix (denoted by trace($\cdot$)) is the sum of its eigenvalues, while the determinant (denoted by det($\cdot$)) is the product of the eigenvalues; see the Appendix for a brief discussion of their properties.

Theorem 6.5.

 Let B_0 be any symmetric positive definite initial matrix, and let x_0 be a starting point for which Assumption 6.1 is satisfied. Then the sequence $\{x_k\}$ generated by Algorithm 6.1 (with $\epsilon = 0$) converges to the minimizer x^* of f.

PROOF.   We start by defining

$$m_k = \frac{y_k^T s_k}{s_k^T s_k}, \qquad M_k = \frac{y_k^T y_k}{y_k^T s_k}, \tag{6.42}$$

and note from (6.40) and (6.41) that

$$m_k \ge m, \qquad M_k \le M. \tag{6.43}$$

By computing the trace of the BFGS update (6.19), we obtain that

$$\text{trace}(B_{k+1}) = \text{trace}(B_k) - \frac{\|B_k s_k\|^2}{s_k^T B_k s_k} + \frac{\|y_k\|^2}{y_k^T s_k} \tag{6.44}$$

(see Exercise 6.11). We can also show (Exercise 6.10) that

$$\det(B_{k+1}) = \det(B_k) \frac{y_k^T s_k}{s_k^T B_k s_k}. \tag{6.45}$$

We now define

$$\cos \theta_k = \frac{s_k^T B_k s_k}{\|s_k\| \, \|B_k s_k\|}, \qquad q_k = \frac{s_k^T B_k s_k}{s_k^T s_k}, \tag{6.46}$$

so that θ_k is the angle between s_k and $B_k s_k$. We then obtain that

$$\frac{\|B_k s_k\|^2}{s_k^T B_k s_k} = \frac{\|B_k s_k\|^2 \|s_k\|^2}{(s_k^T B_k s_k)^2} \frac{s_k^T B_k s_k}{\|s_k\|^2} = \frac{q_k}{\cos^2 \theta_k}. \tag{6.47}$$

In addition, we have from (6.42) that

$$\det(B_{k+1}) = \det(B_k) \frac{y_k^T s_k}{s_k^T s_k} \frac{s_k^T s_k}{s_k^T B_k s_k} = \det(B_k) \frac{m_k}{q_k}. \tag{6.48}$$

We now combine the trace and determinant by introducing the following function of a positive definite matrix B:

$$\psi(B) = \text{trace}(B) - \ln(\det(B)), \tag{6.49}$$

where $\ln(\cdot)$ denotes the natural logarithm. It is not difficult to show that $\psi(B) > 0$; see Exercise 6.9. By using (6.42) and (6.44)–(6.49), we have that

$$\begin{aligned}
\psi(B_{k+1}) &= \text{trace}(B_k) + M_k - \frac{q_k}{\cos^2 \theta_k} - \ln(\det(B_k)) - \ln m_k + \ln q_k \\
&= \psi(B_k) + (M_k - \ln m_k - 1) \\
&\quad + \left[1 - \frac{q_k}{\cos^2 \theta_k} + \ln \frac{q_k}{\cos^2 \theta_k} \right] + \ln \cos^2 \theta_k.
\end{aligned} \tag{6.50}$$

Now, since the function $h(t) = 1 - t + \ln t$ is nonpositive for all $t > 0$ (see Exercise 6.8), the term inside the square brackets is nonpositive, and thus from (6.43) and (6.50) we have

$$0 < \psi(B_{k+1}) \leq \psi(B_0) + c(k+1) + \sum_{j=0}^{k} \ln \cos^2 \theta_j, \tag{6.51}$$

where we can assume the constant $c = M - \ln m - 1$ to be positive, without loss of generality.

We now relate these expressions to the results given in Section 3.2. Note from the form $s_k = -\alpha_k B_k^{-1} \nabla f_k$ of the quasi-Newton iteration that $\cos \theta_k$ defined by (6.46) is the angle between the steepest descent direction and the search direction, which plays a crucial role in the global convergence theory of Chapter 3. From (3.22), (3.23) we know that the sequence $\|\nabla f_k\|$ generated by the line search algorithm is bounded away from zero only if $\cos \theta_j \to 0$.

Let us then proceed by contradiction and assume that $\cos \theta_j \to 0$. Then there exists $k_1 > 0$ such that for all $j > k_1$, we have

$$\ln \cos^2 \theta_j < -2c,$$

where c is the constant defined above. Using this inequality in (6.51) we find the following relations to be true for all $k > k_1$:

$$0 < \psi(B_0) + c(k+1) + \sum_{j=0}^{k_1} \ln \cos^2 \theta_j + \sum_{j=k_1+1}^{k} (-2c)$$

$$= \psi(B_0) + \sum_{j=0}^{k_1} \ln \cos^2 \theta_j + 2ck_1 + c - ck.$$

However, the right-hand-side is negative for large k, giving a contradiction. Therefore, there exists a subsequence of indices $\{j_k\}_{k=1,2,\dots}$ such that $\cos \theta_{j_k} \geq \delta > 0$. By Zoutendijk's result (3.14) this limit implies that $\liminf \|\nabla f_k\| \to 0$. Since the problem is strongly convex, the latter limit is enough to prove that $x_k \to x^*$. □

Theorem 6.5 has been generalized to the entire restricted Broyden class, *except for the DFP method*. In other words, Theorem 6.5 can be shown to hold for all $\phi_k \in [0, 1)$ in (6.32), but the argument seems to break down as ϕ_k approaches 1 because some of the self-correcting properties of the update are weakened considerably.

An extension of the analysis just given shows that the rate of convergence of the iterates is linear. In particular, we can show that the sequence $\|x_k - x^*\|$ converges to zero rapidly enough that

$$\sum_{k=1}^{\infty} \|x_k - x^*\| < \infty. \tag{6.52}$$

We will not prove this claim, but rather establish that if (6.52) holds, then the rate of convergence is actually superlinear.

SUPERLINEAR CONVERGENCE OF THE BFGS METHOD

The analysis of this section makes use of the Dennis and Moré characterization (3.36) of superlinear convergence. It applies to general nonlinear—not just convex—objective functions. For the results that follow we need to make an additional assumption.

Assumption 6.2.

The Hessian matrix G is Lipschitz continuous at x^*, that is,

$$\|G(x) - G(x^*)\| \leq L\|x - x^*\|,$$

for all x near x^*, where L is a positive constant.

We start by introducing the quantities

$$\tilde{s}_k = G_*^{1/2}s_k, \qquad \tilde{y}_k = G_*^{-1/2}y_k, \qquad \tilde{B}_k = G_*^{-1/2}B_kG_*^{-1/2},$$

where $G_* = G(x^*)$ and x^* is a minimizer of f. Similarly to (6.46), we define

$$\cos\tilde{\theta}_k = \frac{\tilde{s}_k^T\tilde{B}_k\tilde{s}_k}{\|\tilde{s}_k\|\,\|\tilde{B}_k\tilde{s}_k\|}, \qquad \tilde{q}_k = \frac{\tilde{s}_k^T\tilde{B}_k\tilde{s}_k}{\|\tilde{s}_k\|^2},$$

while we echo (6.42) and (6.43) in defining

$$\tilde{M}_k = \frac{\|\tilde{y}_k\|^2}{\tilde{y}_k^T\tilde{s}_k}, \qquad \tilde{m}_k = \frac{\tilde{y}_k^T\tilde{s}_k}{\tilde{s}_k^T\tilde{s}_k}.$$

By pre- and postmultiplying the BFGS update formula (6.19) by $G_*^{-1/2}$ and grouping terms appropriately, we obtain

$$\tilde{B}_{k+1} = \tilde{B}_k - \frac{\tilde{B}_k\tilde{s}_k\tilde{s}_k^T\tilde{B}_k}{\tilde{s}_k^T\tilde{B}_k\tilde{s}_k} + \frac{\tilde{y}_k\tilde{y}_k^T}{\tilde{y}_k^T\tilde{s}_k}.$$

Since this expression has precisely the same form as the BFGS formula (6.19), it follows from the argument leading to (6.50) that

$$\begin{aligned} \psi(\tilde{B}_{k+1}) = {}& \psi(\tilde{B}_k) + (\tilde{M}_k - \ln\tilde{m}_k - 1) \\ &+ \left[1 - \frac{\tilde{q}_k}{\cos^2\tilde{\theta}_k} + \ln\frac{\tilde{q}_k}{\cos^2\tilde{\theta}_k}\right] \\ &+ \ln\cos^2\tilde{\theta}_k. \end{aligned} \qquad (6.53)$$

Recalling (6.12), we have that

$$y_k - G_*s_k = (\bar{G}_k - G_*)s_k,$$

and thus

$$\tilde{y}_k - \tilde{s}_k = G_*^{-1/2}(\bar{G}_k - G_*)G_*^{-1/2}\tilde{s}_k.$$

By Assumption 6.2, and recalling the definition (6.11), we have

$$\|\tilde{y}_k - \tilde{s}_k\| \le \|G_*^{-1/2}\|^2\|\tilde{s}_k\|\,\|\bar{G}_k - G_*\| \le \|G_*^{-1/2}\|^2\|\tilde{s}_k\|L\epsilon_k,$$

where ϵ_k is defined by

$$\epsilon_k = \max\{\|x_{k+1} - x^*\|, \|x_k - x^*\|\}.$$

We have thus shown that

$$\frac{\|\tilde{y}_k - \tilde{s}_k\|}{\|\tilde{s}_k\|} \leq \bar{c}\epsilon_k, \tag{6.54}$$

for some positive constant $\bar{c}$. This inequality and (6.52) play an important role in superlinear convergence, as we now show.

Theorem 6.6.

Suppose that f is twice continuously differentiable and that the iterates generated by the BFGS algorithm converge to a minimizer x^* at which Assumption 6.2 holds. Suppose also that (6.52) holds. Then x_k converges to x^* at a superlinear rate.

PROOF.   From (6.54), we have from the triangle inequality (A.4a) that

$$\|\tilde{y}_k\| - \|\tilde{s}_k\| \leq \bar{c}\epsilon_k\|\tilde{s}_k\|, \qquad \|\tilde{s}_k\| - \|\tilde{y}_k\| \leq \bar{c}\epsilon_k\|\tilde{s}_k\|,$$

so that

$$(1 - \bar{c}\epsilon_k)\|\tilde{s}_k\| \leq \|\tilde{y}_k\| \leq (1 + \bar{c}\epsilon_k)\|\tilde{s}_k\|. \tag{6.55}$$

By squaring (6.54) and using (6.55), we obtain

$$(1 - \bar{c}\epsilon_k)^2\|\tilde{s}_k\|^2 - 2\tilde{y}_k^T\tilde{s}_k + \|\tilde{s}_k\|^2 \leq \|\tilde{y}_k\|^2 - 2\tilde{y}_k^T\tilde{s}_k + \|\tilde{s}_k\|^2 \leq \bar{c}^2\epsilon_k^2\|\tilde{s}_k\|^2,$$

and therefore

$$2\tilde{y}_k^T\tilde{s}_k \geq (1 - 2\bar{c}\epsilon_k + \bar{c}^2\epsilon_k^2 + 1 - \bar{c}^2\epsilon_k^2)\|\tilde{s}_k\|^2 = 2(1 - \bar{c}\epsilon_k)\|\tilde{s}_k\|^2.$$

It follows from the definition of $\tilde{m}_k$ that

$$\tilde{m}_k = \frac{\tilde{y}_k^T\tilde{s}_k}{\|\tilde{s}_k\|^2} \geq 1 - \bar{c}\epsilon_k. \tag{6.56}$$

By combining (6.55) and (6.56), we obtain also that

$$\tilde{M}_k = \frac{\|\tilde{y}_k\|^2}{\tilde{y}_k^T\tilde{s}_k} \leq \frac{1 + \bar{c}\epsilon_k}{1 - \bar{c}\epsilon_k}. \tag{6.57}$$

Since $x_k \to x^*$, we have that $\epsilon_k \to 0$, and thus by (6.57) there exists a positive constant $c > \bar{c}$ such that the following inequalities hold for all sufficiently large k:

$$\tilde{M}_k \leq 1 + \frac{2\bar{c}}{1 - \bar{c}\epsilon_k}\epsilon_k \leq 1 + c\epsilon_k. \tag{6.58}$$

We again make use of the nonpositiveness of the function $h(t) = 1 - t + \ln t$. Therefore, we have

$$\frac{-x}{1-x} - \ln(1-x) = h\left(\frac{1}{1-x}\right) \le 0.$$

Now, for k large enough we can assume that $\bar{c}\epsilon_k < \frac{1}{2}$, and therefore

$$\ln(1 - \bar{c}\epsilon_k) \ge \frac{-\bar{c}\epsilon_k}{1 - \bar{c}\epsilon_k} \ge -2\bar{c}\epsilon_k.$$

This relation and (6.56) imply that for sufficiently large k, we have

$$\ln \tilde{m}_k \ge \ln(1 - \bar{c}\epsilon_k) \ge -2\bar{c}\epsilon_k > -2c\epsilon_k. \tag{6.59}$$

We can now deduce from (6.53), (6.58), and (6.59) that

$$0 < \psi(\tilde{B}_{k+1}) \le \psi(\tilde{B}_k) + 3c\epsilon_k + \ln \cos^2 \tilde{\theta}_k + \left[1 - \frac{\tilde{q}_k}{\cos^2 \tilde{\theta}_k} + \ln \frac{\tilde{q}_k}{\cos^2 \tilde{\theta}_k}\right]. \tag{6.60}$$

By summing this expression and making use of (6.52) we have that

$$\sum_{j=0}^{\infty} \left(\ln \frac{1}{\cos^2 \tilde{\theta}_j} - \left[1 - \frac{\tilde{q}_j}{\cos^2 \tilde{\theta}_j} + \ln \frac{\tilde{q}_j}{\cos^2 \tilde{\theta}_j}\right]\right) \le \psi(\tilde{B}_0) + 3c \sum_{j=0}^{\infty} \epsilon_j < +\infty.$$

Since the term in the square brackets is nonpositive, and since $\ln\left(1/\cos^2 \tilde{\theta}_j\right) \ge 0$ for all j, we obtain the two limits

$$\lim_{j\to\infty} \ln \frac{1}{\cos^2 \tilde{\theta}_j} = 0, \qquad \lim_{j\to\infty} \left(1 - \frac{\tilde{q}_j}{\cos^2 \tilde{\theta}_j} + \ln \frac{\tilde{q}_j}{\cos^2 \tilde{\theta}_j}\right) = 0,$$

which imply that

$$\lim_{j\to\infty} \cos \tilde{\theta}_j = 1, \qquad \lim_{j\to\infty} \tilde{q}_j = 1. \tag{6.61}$$

The essence of the result has now been proven; we need only to interpret these limits in terms of the Dennis–Moré characterization of superlinear convergence.

Recalling (6.47), we have

$$
\frac{\|G_*^{-1/2}(B_k - G_*)s_k\|^2}{\|G_*^{1/2}s_k\|^2} = \frac{\|(\tilde{B}_k - I)\tilde{s}_k\|^2}{\|\tilde{s}_k\|^2}
$$

$$
= \frac{\|\tilde{B}_k\tilde{s}_k\|^2 - 2\tilde{s}_k^T \tilde{B}_k\tilde{s}_k + \tilde{s}_k^T \tilde{s}_k}{\tilde{s}_k^T \tilde{s}_k}
$$

$$
= \frac{\tilde{q}_k^2}{\cos \tilde{\theta}_k^2} - 2\tilde{q}_k + 1.
$$

Since by (6.61) the right-hand-side converges to 0, we conclude that

$$
\lim_{k \to \infty} \frac{\|(B_k - G_*)s_k\|}{\|s_k\|} = 0.
$$

The limit (3.36) and Theorem 3.6 imply that the unit step length $\alpha_k = 1$ will satisfy the Wolfe conditions near the solution, and hence that the rate of convergence is superlinear. □

CONVERGENCE ANALYSIS OF THE SR1 METHOD

The convergence properties of the SR1 method are not as well understood as those of the BFGS method. No global results like Theorem 6.5 or local superlinear results like Theorem 6.6 have been established, except the results for quadratic functions discussed earlier. There is, however, an interesting result for the trust-region SR1 algorithm, Algorithm 6.2. It states that when the objective function has a unique stationary point and the condition (6.26) holds at every step (so that the SR1 update is never skipped) and the Hessian approximations B_k are bounded above, then the iterates converge to x^* at an $(n + 1)$-step superlinear rate. The result does not require exact solution of the trust-region subproblem (6.27).

We state the result formally as follows.

Theorem 6.7.

Suppose that the iterates x_k are generated by Algorithm 6.2. Suppose also that the following conditions hold:

(c1) *The sequence of iterates does not terminate, but remains in a closed, bounded, convex set D, on which the function f is twice continuously differentiable, and in which f has a unique stationary point x^*;*

(c2) *the Hessian $\nabla^2 f(x^*)$ is positive definite, and $\nabla^2 f(x)$ is Lipschitz continuous in a neighborhood of x^*;*

(c3) *the sequence of matrices $\{B_k\}$ is bounded in norm;*

(c4) *condition (6.26) holds at every iteration, where r is some constant in $(0, 1)$.*

Then $\lim_{k\to\infty} x_k = x^*$, *and we have that*

$$\lim_{k\to\infty} \frac{\|x_{k+n+1} - x^*\|}{\|x_k - x^*\|} = 0.$$

Note that the BFGS method does not require the boundedness assumption (c3) to hold. As we have mentioned already, the SR1 update does not necessarily maintain positive definiteness of the Hessian approximations B_k. In practice, B_k may be indefinite at any iteration, which means that the trust region bound may continue to be active for arbitrarily large k. Interestingly, however, it can be shown that the SR1 Hessian approximations tend to be positive definite most of the time. The precise result is that

$$\lim_{k\to\infty} \frac{\text{number of indices } j = 1, 2, \ldots, k \text{ for which } B_j \text{ is positive semidefinite}}{k} = 1,$$

under the assumptions of Theorem 6.7. This result holds regardless of whether the initial Hessian approximation is positive definite or not.

NOTES AND REFERENCES

For a comprehensive treatment of quasi-Newton methods see Dennis and Schnabel [92], Dennis and Moré [91], and Fletcher [101]. A formula for updating the Cholesky factors of the BFGS matrices is given in Dennis and Schnabel [92].

Several safeguards and modifications of the SR1 method have been proposed, but the condition (6.26) is favored in the light of the analysis of Conn, Gould, and Toint [71]. Computational experiments by Conn, Gould, and Toint [70, 73] and Khalfan, Byrd, and Schnabel [181], using both line search and trust-region approaches, indicate that the SR1 method appears to be competitive with the BFGS method. The proof of Theorem 6.7 is given in Byrd, Khalfan, and Schnabel [51].

A study of the convergence of BFGS matrices for nonlinear problems can be found in Ge and Powell [119] and Boggs and Tolle [32]; however, the results are not as satisfactory as for SR1 updating.

The global convergence of the BFGS method was established by Powell [246]. This result was extended to the restricted Broyden class, except for DFP, by Byrd, Nocedal, and Yuan [53]. For a discussion of the self-correcting properties of quasi-Newton methods see Nocedal [229]. Most of the early analysis of quasi-Newton methods was based on the *bounded deterioration* principle. This is a tool for the local analysis that quantifies the worst-case behavior of quasi-Newton updating. Assuming that the starting point is sufficiently close to the solution x^* and that the initial Hessian approximation is sufficiently close to $\nabla^2 f(x^*)$, one can use the bounded deterioration bounds to prove that the iteration cannot stray away from the solution. This property can then be used to show that the quality of the quasi-Newton approximations is good enough to yield superlinear convergence. For details, see Dennis and Moré [91] or Dennis and Schnabel [92].

✎ EXERCISES

✎ **6.1**

(a) Show that if f is strongly convex, then (6.7) holds for any vectors x_k and x_{k+1}.

(b) Give an example of a function of one variable satisfying $g(0) = -1$ and $g(1) = -\frac{1}{4}$ and show that (6.7) does not hold in this case.

✎ **6.2** Show that the second strong Wolfe condition (3.7b) implies the curvature condition (6.7).

✎ **6.3** Verify that (6.19) and (6.17) are inverses of each other.

✎ **6.4** Use the Sherman–Morrison formula (A.27) to show that (6.24) is the inverse of (6.25).

✎ **6.5** Prove the statements (ii) and (iii) given in the paragraph following (6.25).

✎ **6.6** The square root of a matrix A is a matrix $A^{1/2}$ such that $A^{1/2}A^{1/2} = A$. Show that any symmetric positive definite matrix A has a square root, and that this square root is itself symmetric and positive definite. (Hint: Use the factorization $A = UDU^T$ (A.16), where U is orthogonal and D is diagonal with positive diagonal elements.)

✎ **6.7** Use the Cauchy-Schwarz inequality (A.5) to verify that $\mu_k \geq 1$, where μ_k is defined by (6.38).

✎ **6.8** Define $h(t) = 1 - t + \ln t$, and note that $h'(t) = -1 + 1/t$, $h''(t) = -1/t^2 < 0$, $h(1) = 0$, and $h'(1) = 0$. Show that $h(t) \leq 0$ for all $t > 0$.

✎ **6.9** Denote the eigenvalues of the positive definite matrix B by $\lambda_1, \lambda_2, \ldots, \lambda_n$, where $0 < \lambda_1 \leq \lambda_2 \leq \cdots \leq \lambda_n$. Show that the ψ function defined in (6.49) can be written as

$$\psi(B) = \sum_{i=1}^{n}(\lambda_i - \ln \lambda_i).$$

Use this form to show that $\psi(B) > 0$.

✎ **6.10** The object of this exercise is to prove (6.45).

(a) Show that $\det(I + xy^T) = 1 + y^T x$, where x and y are n-vectors. Hint: Assuming that $x \neq 0$, we can find vectors $w_1, w_2, \ldots, w_{n-1}$ such that the matrix Q defined by

$$Q = [x, w_1, w_2, \ldots, w_{n-1}]$$

is nonsingular and $x = Qe_1$, where $e_1 = (1, 0, 0, \ldots, 0)^T$. If we define

$$y^T Q = (z_1, z_2, \ldots, z_n),$$

then

$$z_1 = y^T Qe_1 = y^T Q(Q^{-1}x) = y^T x,$$

and

$$\det(I + xy^T) = \det(Q^{-1}(I + xy^T)Q) = \det(I + e_1 y^T Q).$$

(b) Use a similar technique to prove that

$$\det(I + xy^T + uv^T) = (1 + y^T x)(1 + v^T u) - (x^T v)(y^T u).$$

(c) Use this relation to establish (6.45).

⌀ **6.11** Use the properties of the trace of a symmetric matrix and the formula (6.19) to prove (6.44).

⌀ **6.12** Show that if f satisfies Assumption 6.1 and if the sequence of gradients satisfies $\liminf \|\nabla f_k\| = 0$, then the whole sequence of iterates x converges to the solution x^*.

CHAPTER **7**

Large-Scale Unconstrained Optimization

Many applications give rise to unconstrained optimization problems with thousands or millions of variables. Problems of this size can be solved efficiently only if the storage and computational costs of the optimization algorithm can be kept at a tolerable level. A diverse collection of large-scale optimization methods has been developed to achieve this goal, each being particularly effective for certain problem types. Some of these methods are straightforward adaptations of the methods described in Chapters 3, 4, and 6. Other approaches are modifications of these basic methods that allow approximate steps to be calculated at lower cost in computation and storage. One set of approaches that we have already discussed—the nonlinear conjugate gradient methods of Section 5.2—can be applied to large problems without modification, because of its minimal storage demands and its reliance on only first-order derivative information.

The line search and trust-region Newton algorithms of Chapters 3 and 4 require matrix factorizations of the Hessian matrices $\nabla^2 f_k$. In the large-scale case, these factorizations can be carried out using sparse elimination techniques. Such algorithms have received much attention, and high quality software implementations are available. If the computational cost and memory requirements of these sparse factorization methods are affordable for a given application, and if the Hessian matrix can be formed explicitly, Newton methods based on sparse factorizations constitute an effective approach for solving such problems.

Often, however, the cost of factoring the Hessian is prohibitive, and it is preferable to compute approximations to the Newton step using iterative linear algebra techniques. Section 7.1 discusses inexact Newton methods that use these techniques, in both line search and trust-region frameworks. The resulting algorithms have attractive global convergence properties and may be superlinearly convergent for suitable choices of parameters. They find effective search directions when the Hessian $\nabla^2 f_k$ is indefinite, and may even be implemented in a "Hessian-free" manner, without explicit calculation or storage of the Hessian.

The Hessian approximations generated by the quasi-Newton approaches of Chapter 6 are usually dense, even when the true Hessian is sparse, and the cost of storing and working with these approximations can be excessive for large n. Section 7.2 discusses limited-memory variants of the quasi-Newton approach, which use Hessian approximations that can be stored compactly by using just a few vectors of length n. These methods are fairly robust, inexpensive, and easy to implement, but they do not converge rapidly. Another approach, discussed briefly in Section 7.3, is to define quasi-Newton approximate Hessians B_k that preserve sparsity, for example by mimicking the sparsity pattern of the Hessian.

In Section 7.4, we note that objective functions in large problems often possess a structural property known as *partial separability*, which means they can be decomposed into a sum of simpler functions, each of which depends on only a small subspace of $\mathbb{R}^n$. Effective Newton and quasi-Newton methods that exploit this property have been developed. Such methods usually converge rapidly and are robust, but they require detailed information about the objective function, which can be difficult to obtain in some applications.

We conclude the chapter with a discussion of software for large-scale unconstrained optimization problems.

7.1 INEXACT NEWTON METHODS

Recall from (2.15) that the basic Newton step p_k^{N} is obtained by solving the symmetric $n \times n$ linear system

$$\nabla^2 f_k p_k^{\text{N}} = -\nabla f_k. \tag{7.1}$$

In this section, we describe techniques for obtaining approximations to p_k^{N} that are inexpensive to calculate but are good search directions or steps. These approaches are based on solving (7.1) by using the conjugate gradient (CG) method (see Chapter 5) or the Lanczos method, with modifications to handle negative curvature in the Hessian $\nabla^2 f_k$. Both line search and trust-region approaches are described here. We refer to this family of methods by the general name *inexact Newton methods*.

The use of iterative methods for (7.1) spares us from concerns about the expense of a direct factorization of the Hessian $\nabla^2 f_k$ and the fill-in that may occur during this process. Further, we can customize the solution strategy to ensure that the rapid convergence properties associated with Newton's methods are not lost in the inexact version. In addition, as noted below, we can implement these methods in a Hessian-free manner, so that the Hessian $\nabla^2 f_k$ need not be calculated or stored explicitly at all.

We examine first how the inexactness in the step calculation determines the local convergence properties of inexact Newton methods. We then consider line search and trust-region approaches based on using CG (possibly with preconditioning) to obtain an approximate solution of (7.1). Finally, we discuss the use of the Lanczos method for solving (7.1) approximately.

LOCAL CONVERGENCE OF INEXACT NEWTON METHODS

Most rules for terminating the iterative solver for (7.1) are based on the residual

$$r_k = \nabla^2 f_k p_k + \nabla f_k, \tag{7.2}$$

where p_k is the inexact Newton step. Usually, we terminate the CG iterations when

$$\|r_k\| \leq \eta_k \|\nabla f_k\|, \tag{7.3}$$

where the sequence $\{\eta_k\}$ (with $0 < \eta_k < 1$ for all k) is called the *forcing sequence*.

We now study how the rate of convergence of inexact Newton methods based on (7.1)–(7.3) is affected by the choice of the forcing sequence. The next two theorems apply not just to Newton–CG procedures but to all inexact Newton methods whose steps satisfy (7.2) and (7.3).

Our first result says that local convergence is obtained simply by ensuring that η_k is bounded away from 1.

Theorem 7.1.

Suppose that $\nabla^2 f(x)$ exists and is continuous in a neighborhood of a minimizer x^*, with $\nabla^2 f(x^*)$ is positive definite. Consider the iteration $x_{k+1} = x_k + p_k$ where p_k satisfies (7.3), and assume that $\eta_k \leq \eta$ for some constant $\eta \in [0, 1)$. Then, if the starting point x_0 is sufficiently near x^*, the sequence $\{x_k\}$ converges to x^* and satisfies

$$\|\nabla^2 f(x^*)(x_{k+1} - x^*)\| \leq \hat{\eta} \|\nabla^2 f(x^*)(x_k - x^*)\|, \tag{7.4}$$

for some constant $\hat{\eta}$ with $\eta < \hat{\eta} < 1$.

Rather than giving a rigorous proof of this theorem, we present an informal derivation that contains the essence of the argument and motivates the next result.

Since the Hessian matrix $\nabla^2 f$ is positive definite at x^* and continuous near x^*, there exists a positive constant L such that $\|(\nabla^2 f_k)^{-1}\| \leq L$ for all x_k sufficiently close to x^*. We therefore have from (7.2) that the inexact Newton step satisfies

$$\|p_k\| \leq L(\|\nabla f_k\| + \|r_k\|) \leq 2L\|\nabla f_k\|,$$

where the second inequality follows from (7.3) and $\eta_k < 1$. Using this expression together with Taylor's theorem and the continuity of $\nabla^2 f(x)$, we obtain

$$
\begin{aligned}
\nabla f_{k+1} &= \nabla f_k + \nabla^2 f_k p_k + \int_0^1 [\nabla f(x_k + t p_k) - \nabla f(x_k)] p_k \, dt \\
&= \nabla f_k + \nabla^2 f_k p_k + o(\|p_k\|) \\
&= \nabla f_k - (\nabla f_k - r_k) + o(\|\nabla f_k\|) \\
&= r_k + o(\|\nabla f_k\|).
\end{aligned}
\tag{7.5}
$$

Taking norms and recalling (7.3), we have that

$$\|\nabla f_{k+1}\| \leq \eta_k \|\nabla f_k\| + o(\|\nabla f_k\|) \leq (\eta_k + o(1))) \|\nabla f_k\|. \tag{7.6}$$

When x_k is close enough to x^* that the $o(1)$ term in the last estimate is bounded by $(1-\eta)/2$, we have

$$\|\nabla f_{k+1}\| \leq (\eta_k + (1-\eta)/2)\|\nabla f_k\| \leq \frac{1+\eta}{2}\|\nabla f_k\|, \tag{7.7}$$

so the gradient norm decreases by a factor of $(1+\eta)/2$ at this iteration. By choosing the initial point x_0 sufficiently close to x^*, we can ensure that this rate of decrease occurs at every iteration.

To prove (7.4), we note that under our smoothness assumptions, we have

$$\nabla f_k = \nabla^2 f(x^*)(x_k - x^*) + o(\|x_k - x^*\|).$$

Hence it can be shown that for x_k close to x^*, the gradient ∇f_k differs from the scaled error $\nabla^2 f(x^*)(x_k - x^*)$ by only a relatively small perturbation. A similar estimate holds at x_{k+1}, so (7.4) follows from (7.7).

From (7.6), we have that

$$\frac{\|\nabla f_{k+1}\|}{\|\nabla f_k\|} \leq \eta_k + o(1). \tag{7.8}$$

If $\lim_{k\to\infty} \eta_k = 0$, we have from this expression that

$$\lim_{k\to\infty} \frac{\|\nabla f_{k+1}\|}{\|\nabla f_k\|} = 0,$$

indicating Q-superlinear convergence of the gradient norms $\|\nabla f_k\|$ to zero. Superlinear convergence of the iterates $\{x_k\}$ to x^* can be proved as a consequence.

We can obtain quadratic convergence by making the additional assumption that the Hessian $\nabla^2 f(x)$ is Lipschitz continuous near x^*. In this case, the estimate (7.5) can be tightened to

$$\nabla f_{k+1} = r_k + O\left(\|\nabla f_k\|^2\right).$$

By choosing the forcing sequence so that $\eta_k = O(\|\nabla f_k\|)$, we have from this expression that

$$\|\nabla f_{k+1}\| = O(\|\nabla f_k\|^2),$$

indicating Q-quadratic convergence of the gradient norms to zero (and thus also Q-quadratic convergence of the iterates x_k to x^*). The last two observations are summarized in the following theorem.

Theorem 7.2.

Suppose that the conditions of Theorem 7.1 hold, and assume that the iterates $\{x_k\}$ generated by the inexact Newton method converge to x^*. Then the rate of convergence is superlinear if $\eta_k \to 0$. If in addition, $\nabla^2 f(x)$ is Lipschitz continuous for x near x^* and if $\eta_k = O(\|\nabla f_k\|)$, then the convergence is quadratic.

To obtain superlinear convergence, we can set, for example, $\eta_k = \min\left(0.5, \sqrt{\|\nabla f_k\|}\right)$; the choice $\eta_k = \min(0.5, \|\nabla f_k\|)$ would yield quadratic convergence.

All the results presented in this section, which are proved by Dembo, Eisenstat, and Steihaug [89], are local in nature: They assume that the sequence $\{x_k\}$ eventually enters the near vicinity of the solution x^*. They also assume that the unit step length $\alpha_k = 1$ is taken and hence that globalization strategies do not interfere with rapid convergence. In the following pages we show that inexact Newton strategies can, in fact, be incorporated in practical line search and trust-region implementations of Newton's method, yielding algorithms with good local and global convergence properties. We start with a line search approach.

LINE SEARCH NEWTON–CG METHOD

In the line search Newton–CG method, also known as the *truncated Newton method*, we compute the search direction by applying the CG method to the Newton equations (7.1) and attempt to satisfy a termination test of the form (7.3). However, the CG method is designed to solve positive definite systems, and the Hessian $\nabla^2 f_k$ may have negative eigenvalues when x_k is not close to a solution. Therefore, we terminate the CG iteration as soon as a direction of negative curvature is generated. This adaptation of the CG method produces a search direction p_k that is a descent direction. Moreover, the adaptation guarantees that the fast convergence rate of the pure Newton method is preserved, provided that the step length $\alpha_k = 1$ is used whenever it satisfies the acceptance criteria.

We now describe Algorithm 7.1, a line search algorithm that uses a modification of Algorithm 5.2 as the inner iteration to compute each search direction p_k. For purposes of this algorithm, we write the linear system (7.1) in the form

$$B_k p = -\nabla f_k, \tag{7.9}$$

where B_k represents $\nabla^2 f_k$. For the inner CG iteration, we denote the search directions by d_j and the sequence of iterates that it generates by z_j. When B_k is positive definite, the inner iteration sequence $\{z_j\}$ will converge to the Newton step p_k^N that solves (7.9). At each major iteration, we define a tolerance ϵ_k that specifies the required accuracy of the computed solution. For concreteness, we choose the forcing sequence to be $\eta_k = \min(0.5, \sqrt{\|\nabla f_k\|})$ to obtain a superlinear convergence rate, but other choices are possible.

Algorithm 7.1 (Line Search Newton–CG).

Given initial point x_0;

for $k = 0, 1, 2, \ldots$
    Define tolerance $\epsilon_k = \min(0.5, \sqrt{\|\nabla f_k\|})\|\nabla f_k\|$;
    Set $z_0 = 0, r_0 = \nabla f_k, d_0 = -r_0 = -\nabla f_k$;
    for $j = 0, 1, 2, \ldots$
        if $d_j^T B_k d_j \leq 0$
            if $j = 0$
                return $p_k = -\nabla f_k$;
            else
                return $p_k = z_j$;
        Set $\alpha_j = r_j^T r_j / d_j^T B_k d_j$;
        Set $z_{j+1} = z_j + \alpha_j d_j$;
        Set $r_{j+1} = r_j + \alpha_j B_k d_j$;
        if $\|r_{j+1}\| < \epsilon_k$
            return $p_k = z_{j+1}$;
        Set $\beta_{j+1} = r_{j+1}^T r_{j+1} / r_j^T r_j$;
        Set $d_{j+1} = -r_{j+1} + \beta_{j+1} d_j$;
    end (for)
    Set $x_{k+1} = x_k + \alpha_k p_k$, where α_k satisfies the Wolfe, Goldstein, or
        Armijo backtracking conditions (using $\alpha_k = 1$ if possible);
end

The main differences between the inner loop of Algorithm 7.1 and Algorithm 5.2 are that the specific starting point $z_0 = 0$ is used; the use of a positive tolerance ϵ_k allows the CG iterations to terminate at an inexact solution; and the negative curvature test $d_j^T B_k d_j \leq 0$ ensures that p_k is a descent direction for f at x_k. If negative curvature is detected on the first inner iteration $j = 0$, the returned direction $p_k = -\nabla f_k$ is both a descent direction and a direction of nonpositive curvature for f at x_k.

We can modify the CG iterations in Algorithm 7.1 by introducing preconditioning, in the manner described in Chapter 5.

Algorithm 7.1 is well suited for large problems, but it has a weakness. When the Hessian $\nabla^2 f_k$ is nearly singular, the line search Newton–CG direction can be long and of poor quality, requiring many function evaluations in the line search and giving only a small reduction in the function. To alleviate this difficulty, we can try to normalize the Newton step, but good rules for doing so are difficult to determine. They run the risk of undermining the rapid convergence of Newton's method in the case where the pure Newton step is well scaled. It is preferable to introduce a threshold value into the test $d_j^T B d_j \leq 0$, but good choices of the threshold are difficult to determine. The trust-region Newton–CG method described below deals more effectively with this problematic situation and is therefore preferable, in our opinion.

The line search Newton–CG method does not require explicit knowledge of the Hessian $B_k = \nabla^2 f_k$. Rather, it requires only that we can supply Hessian–vector products of the form $\nabla^2 f_k d$ for any given vector d. When the user cannot easily supply code to calculate second derivatives, or where the Hessian requires too much storage, the techniques of Chapter 8 (automatic differentiation and finite differencing) can be used to calculate these Hessian–vector products. Methods of this type are known as *Hessian-free* Newton methods.

To illustrate the finite-differencing technique briefly, we use the approximation

$$\nabla^2 f_k d \approx \frac{\nabla f(x_k + hd) - \nabla f(x_k)}{h}, \tag{7.10}$$

for some small differencing interval h. It is easy to prove that the accuracy of this approximation is $O(h)$; appropriate choices of h are discussed in Chapter 8. The price we pay for bypassing the computation of the Hessian is one new gradient evaluation per CG iteration.

TRUST-REGION NEWTON–CG METHOD

In Chapter 4, we discussed approaches for finding an approximate solution of the trust-region subproblem (4.3) that produce improvements on the Cauchy point. Here we define a modified CG algorithm for solving the subproblem with these properties. This algorithm, due to Steihaug [281], is specified below as Algorithm 7.2. A complete algorithm for minimizing f is obtained by using Algorithm 7.2 to generate the step p_k required by Algorithm 4.1 of Chapter 4, for some choice of tolerance ϵ_k at each iteration.

We use notation similar to (7.9) to define the trust-region subproblem for which Steihaug's method finds an approximate solution:

$$\min_{p \in R^n} m_k(p) \stackrel{\text{def}}{=} f_k + (\nabla f_k)^T p + \tfrac{1}{2} p^T B_k p \quad \text{subject to } \|p\| \le \Delta_k, \qquad (7.11)$$

where $B_k = \nabla^2 f_k$. As in Algorithm 7.1, we use d_j to denote the search directions of this modified CG iteration and z_j to denote the sequence of iterates that it generates.

Algorithm 7.2 (CG–Steihaug).
Given tolerance $\epsilon_k > 0$;
Set $z_0 = 0$, $r_0 = \nabla f_k$, $d_0 = -r_0 = -\nabla f_k$;
if $\|r_0\| < \epsilon_k$
    return $p_k = z_0 = 0$;
for $j = 0, 1, 2, \dots$
    if $d_j^T B_k d_j \le 0$
        Find τ such that $p_k = z_j + \tau d_j$ minimizes $m_k(p_k)$ in (4.5)
            and satisfies $\|p_k\| = \Delta_k$;
        return p_k;
    Set $\alpha_j = r_j^T r_j / d_j^T B_k d_j$;
    Set $z_{j+1} = z_j + \alpha_j d_j$;
    if $\|z_{j+1}\| \ge \Delta_k$
        Find $\tau \ge 0$ such that $p_k = z_j + \tau d_j$ satisfies $\|p_k\| = \Delta_k$;
        return p_k;
    Set $r_{j+1} = r_j + \alpha_j B_k d_j$;
    if $\|r_{j+1}\| < \epsilon_k$
        return $p_k = z_{j+1}$;
    Set $\beta_{j+1} = r_{j+1}^T r_{j+1} / r_j^T r_j$;
    Set $d_{j+1} = -r_{j+1} + \beta_{j+1} d_j$;
end (for).

The first **if** statement inside the loop stops the method if its current search direction d_j is a direction of nonpositive curvature along B_k, while the second **if** statement inside the loop causes termination if z_{j+1} violates the trust-region bound. In both cases, the method returns the step p_k obtained by intersecting the current search direction with the trust-region boundary.

The choice of the tolerance ϵ_k at each call to Algorithm 7.2 is important in keeping the overall cost of the trust-region Newton–CG method low. Near a well-behaved solution x^*, the trust-region bound becomes inactive, and the method reduces to the inexact Newton method analyzed in Theorems 7.1 and 7.2. Rapid convergence can be obtained in these circumstances by choosing ϵ_k in a similar fashion to Algorithm 7.1.

The essential differences between Algorithm 5.2 and the inner loop of Algorithm 7.2 are that the latter terminates when it violates the trust-region bound $\|p\| \leq \Delta$, when it encounters a direction of negative curvature in $\nabla^2 f_k$, or when it satisfies a convergence tolerance defined by a parameter ϵ_k. In these respects, Algorithm 7.2 is quite similar to the inner loop of Algorithm 7.1.

The initialization of z_0 to zero in Algorithm 7.2 is a crucial feature of the algorithm. Provided $\|\nabla f_k\|_2 \geq \epsilon_k$, Algorithm 7.2 terminates at a point p_k for which $m_k(p_k) \leq m_k(p_k^c)$, that is, when the reduction in model function equals or exceeds that of the Cauchy point. To demonstrate this fact, we consider several cases. First, if $d_0^T B_k d_0 = (\nabla f_k)^T B_k \nabla f_k \leq 0$, then the condition in the first **if** statement is satisfied, and the algorithm returns the Cauchy point $p = -\Delta_k (\nabla f_k)/\|\nabla f_k\|$. Otherwise, Algorithm 7.2 defines z_1 as follows:

$$z_1 = \alpha_0 d_0 = \frac{r_0^T r_0}{d_0^T B_k d_0} d_0 = -\frac{(\nabla f_k)^T \nabla f_k}{(\nabla f_k)^T B_k \nabla f_k} \nabla f_k.$$

If $\|z_1\| < \Delta_k$, then z_1 is exactly the Cauchy point. Subsequent steps of Algorithm 7.2 ensure that the final p_k satisfies $m_k(p_k) \leq m_k(z_1)$. When $\|z_1\| \geq \Delta_k$, on the other hand, the second **if** statement is activated, and Algorithm 7.2 terminates at the Cauchy point, proving our claim. This property is important for global convergence: Since each step is at least as good as the Cauchy point in reducing the model m_k, Algorithm 7.2 is globally convergent.

Another crucial property of the method is that each iterate z_j is larger in norm than its predecessor. This property is another consequence of the initialization $z_0 = 0$. Its main implication is that it is acceptable to stop iterating as soon as the trust-region boundary is reached, because no further iterates giving a lower value of the model function m_k will lie inside the trust region. We state and prove this property formally in the following theorem, which makes use of the expanding subspace property of the conjugate gradient algorithm, described in Theorem 5.2.

Theorem 7.3.

The sequence of vectors $\{z_j\}$ generated by Algorithm 7.2 satisfies

$$0 = \|z_0\|_2 < \cdots < \|z_j\|_2 < \|z_{j+1}\|_2 < \cdots < \|p_k\|_2 \leq \Delta_k.$$

PROOF. We first show that the sequences of vectors generated by Algorithm 7.2 satisfy $z_j^T r_j = 0$ for $j \geq 0$ and $z_j^T d_j > 0$ for $j \geq 1$.

Algorithm 7.2 computes z_{j+1} recursively in terms of z_j; but when all the terms of this recursion are written explicitly, we see that

$$z_j = z_0 + \sum_{i=0}^{j-1} \alpha_i d_i = \sum_{i=0}^{j-1} \alpha_i d_i,$$

since $z_0 = 0$. Multiplying by r_j and applying the expanding subspace property of conjugate gradients (see Theorem 5.2), we obtain

$$z_j^T r_j = \sum_{i=0}^{j-1} \alpha_i d_i^T r_j = 0. \tag{7.12}$$

An induction proof establishes the relation $z_j^T d_j > 0$. By applying the expanding subspace property again, we obtain

$$z_1^T d_1 = (\alpha_0 d_0)^T (-r_1 + \beta_1 d_0) = \alpha_0 \beta_1 d_0^T d_0 > 0.$$

We now make the inductive hypothesis that $z_j^T d_j > 0$ and deduce that $z_{j+1}^T d_{j+1} > 0$. From (7.12), we have $z_{j+1}^T r_{j+1} = 0$, and therefore

$$\begin{aligned} z_{j+1}^T d_{j+1} &= z_{j+1}^T (-r_{j+1} + \beta_{j+1} d_j) \\ &= \beta_{j+1} z_{j+1}^T d_j \\ &= \beta_{j+1} (z_j + \alpha_j d_j)^T d_j \\ &= \beta_{j+1} z_j^T d_j + \alpha_j \beta_{j+1} d_j^T d_j. \end{aligned}$$

Because of the inductive hypothesis and positivity of β_{j+1} and α_j, the last expression is positive.

We now prove the theorem. If Algorithm 7.2 terminates because $d_j^T B_k d_j \leq 0$ or $\|z_{j+1}\|_2 \geq \Delta_k$, then the final point p_k is chosen to make $\|p_k\|_2 = \Delta_k$, which is the largest possible length. To cover all other possibilities in the algorithm, we must show that $\|z_j\|_2 < \|z_{j+1}\|_2$ when $z_{j+1} = z_j + \alpha_j d_j$ and $j \geq 1$. Observe that

$$\|z_{j+1}\|_2^2 = (z_j + \alpha_j d_j)^T (z_j + \alpha_j d_j) = \|z_j\|_2^2 + 2\alpha_j z_j^T d_j + \alpha_j^2 \|d_j\|_2^2.$$

It follows from this expression and our intermediate result that $\|z_j\|_2 < \|z_{j+1}\|_2$, so our proof is complete. $\qquad\square$

From this theorem we see that Algorithm 7.2 sweeps out points z_j that move on some interpolating path from z_1 to the final solution p_k, a path in which every step increases its total distance from the start point. When $B_k = \nabla^2 f_k$ is positive definite, this path may be compared to the path of the dogleg method: Both methods start by minimizing m_k along the negative gradient direction $-\nabla f_k$ and subsequently progress toward p_k^{N}, until the trust-region boundary intervenes. One can show that, when $B_k = \nabla^2 f_k$ is positive definite, Algorithm 7.2 provides a decrease in the model (7.11) that is at least half as good as the optimal decrease [320].

PRECONDITIONING THE TRUST-REGION NEWTON–CG METHOD

As discussed in Chapter 5, preconditioning can be used to accelerate the CG iteration. Preconditioning techniques are based on finding a nonsingular matrix D such that the eigenvalues of $D^{-T} \nabla^2 f_k D^{-1}$ have a more favorable distribution. By generalizing Theorem 7.3, we can show that the iterates z_j generated by a preconditioned variant of Algorithm 7.2 will grow monotonically in the weighted norm $\| D \cdot \|$. To be consistent, we should redefine the trust-region subproblem in terms of the same norm, as follows:

$$\min_{p \in \mathbb{R}^n} m_k(p) \stackrel{\text{def}}{=} f_k + \nabla f_k^T p + \tfrac{1}{2} p^T B_k p \quad \text{subject to} \quad \| Dp \| \le \Delta_k. \tag{7.13}$$

Making the change of variables $\hat{p} = Dp$ and defining

$$\hat{g}_k = D^{-T} \nabla f_k, \qquad \hat{B}_k = D^{-T} (\nabla^2 f_k) D^{-1},$$

we can write (7.13) as

$$\min_{\hat{p} \in \mathbb{R}^n} f_k + \hat{g}_k^T \hat{p} + \tfrac{1}{2} \hat{p}^T \hat{B}_k \hat{p} \quad \text{subject to} \quad \| \hat{p} \| \le \Delta,$$

which has exactly the form of (7.11). We can apply Algorithm 7.2 without any modification to this subproblem, which is equivalent to applying a preconditioned version of Algorithm 7.2 to the problem (7.13).

Many preconditioners can be used within this framework; we discuss some of them in Chapter 5. Of particular interest is *incomplete Cholesky* factorization, which has proved useful in a wide range of optimization problems. The incomplete Cholesky factorization of a positive definite matrix B finds a lower triangular matrix L such that

$$B = LL^T - R,$$

where the amount of fill-in in L is restricted in some way. (For instance, it is constrained to have the same sparsity structure as the lower triangular part of B or is allowed to have a number of nonzero entries similar to that in B.) The matrix R accounts for the inexactness in the approximate factorization. The situation is complicated somewhat by the possible indefiniteness of the Hessian $\nabla^2 f_k$; we must be able to handle this indefiniteness as well as maintain the sparsity. The following algorithm combines incomplete Cholesky and a form of modified Cholesky to define a preconditioner for the trust-region Newton–CG approach.

Algorithm 7.3 (Inexact Modified Cholesky).
  Compute $T = \text{diag}(\| Be_1 \|, \| Be_2 \|, \ldots, \| Be_n \|)$, where e_i is the
    ith coordinate vector;
  Set $\bar{B} \leftarrow T^{-1/2} B T^{-1/2}$; Set $\beta \leftarrow \| \bar{B} \|$;

(compute a shift to ensure positive definiteness)
if $\min_i b_{ii} > 0$
    $\alpha_0 \leftarrow 0$
else
    $\alpha_0 \leftarrow \beta/2$;
for $k = 0, 1, 2, \ldots$
    Attempt to apply incomplete Cholesky algorithm to obtain

$$LL^T = \bar{B} + \alpha_k I;$$

    if the factorization is completed successfully
        stop and return L;
    else
        $\alpha_{k+1} \leftarrow \max(2\alpha_k, \beta/2)$;
end (for)

We can then set the preconditioner to be $D = L^T$, where L is the lower triangular matrix output from Algorithm 7.3. A trust-region Newton–CG method using this preconditioner is implemented in the LANCELOT [72] and TRON [192] codes.

TRUST-REGION NEWTON–LANCZOS METHOD

A limitation of Algorithm 7.2 is that it accepts *any* direction of negative curvature, even when this direction gives an insignificant reduction in the model. Consider, for example, the case where the subproblem (7.11) is

$$\min_p m(p) = 10^{-3} p_1 - 10^{-4} p_1^2 - p_2^2 \qquad \text{subject to } \|p\| \leq 1,$$

where subscripts indicate elements of the vector p. The steepest descent direction at $p = 0$ is $(-10^{-3}, 0)^T$, which is a direction of negative curvature for the model. Algorithm 7.2 would follow this direction to the boundary of the trust region, yielding a reduction in model function m of about 10^{-3}. A step along e_2—also a direction of negative curvature—would yield a much greater reduction of 1.

Several remedies have been proposed. We have seen in Chapter 4 that when the Hessian $\nabla^2 f_k$ contains negative eigenvalues, the search direction should have a significant component along the eigenvector corresponding to the most negative eigenvalue of $\nabla^2 f_k$. This feature would allow the algorithm to move away rapidly from stationary points that are not minimizers. One way to achieve this is to compute a nearly exact solution of the trust-region subproblem (7.11) using the techniques described in Section 4.3. This approach requires the solution of a few linear systems with coefficient matrices of the form

$B_k + \lambda I$. Although this approach is perhaps too expensive in the large-scale case, it generates productive search directions in all cases.

A more practical alternative is to use the *Lanczos method* (see, for example, [136]) rather than the CG method to solve the linear system $B_k p = -\nabla f_k$. The Lanczos method can be seen as a generalization of the CG method that is applicable to indefinite systems, and we can use it to continue the CG process while gathering negative curvature information.

After j steps, the Lanczos method generates an $n \times j$ matrix Q_j with orthogonal columns that span the Krylov subspace (5.15) generated by this method. This matrix has the property that $Q_j^T B Q_j = T_j$, where T_j is an tridiagonal. We can take advantage of this tridiagonal structure and seek to find an approximate solution of the trust-region subproblem in the range of the basis Q_j. To do so, we solve the problem

$$\min_{w \in \mathbb{R}^j} f_k + e_1^T Q_j (\nabla f_k) e_1^T w + \tfrac{1}{2} w^T T_j w \quad \text{subject to } \|w\| \le \Delta_k, \qquad (7.14)$$

where $e_1 = (1, 0, 0, \dots, 0)^T$, and we define the approximate solution of the trust-region subproblem as $p_k = Q_j w$. Since T_j is tridiagonal, problem (7.14) can be solved by factoring the system $T_j + \lambda I$ and following the (nearly) exact approach of Section 4.3.

The Lanczos iteration may be terminated, as in the Newton–CG methods, by a test of the form (7.3). Preconditioning can also be incorporated to accelerate the convergence of the Lanczos iteration. The additional robustness in this trust-region algorithm comes at the cost of a more expensive solution of the subproblem than in the Newton–CG approach. A sophisticated implementation of the Newton–Lanczos approach has been implemented in the GLTR package [145].

7.2 LIMITED-MEMORY QUASI-NEWTON METHODS

Limited-memory quasi-Newton methods are useful for solving large problems whose Hessian matrices cannot be computed at a reasonable cost or are not sparse. These methods maintain simple and compact approximations of Hessian matrices: Instead of storing fully dense $n \times n$ approximations, they save only a few vectors of length n that represent the approximations implicitly. Despite these modest storage requirements, they often yield an acceptable (albeit linear) rate of convergence. Various limited-memory methods have been proposed; we focus mainly on an algorithm known as L-BFGS, which, as its name suggests, is based on the BFGS updating formula. The main idea of this method is to use curvature information from only the most recent iterations to construct the Hessian approximation. Curvature information from earlier iterations, which is less likely to be relevant to the actual behavior of the Hessian at the current iteration, is discarded in the interest of saving storage.

Following our discussion of L-BFGS and its convergence behavior, we discuss its relationship to the nonlinear conjugate gradient methods of Chapter 5. We then discuss implementations of limited-memory schemes that make use of a compact representation of approximate Hessian information. These techniques can be applied not only to L-BFGS but also to limited-memory versions of other quasi-Newton procedures such as SR1. Finally, we discuss quasi-Newton updating schemes that impose a particular sparsity pattern on the approximate Hessian.

LIMITED-MEMORY BFGS

We begin our description of the L-BFGS method by recalling its parent, the BFGS method, which was described in Algorithm 8.1. Each step of the BFGS method has the form

$$x_{k+1} = x_k - \alpha_k H_k \nabla f_k, \qquad (7.15)$$

where α_k is the step length and H_k is updated at every iteration by means of the formula

$$H_{k+1} = V_k^T H_k V_k + \rho_k s_k s_k^T \qquad (7.16)$$

(see (6.17)), where

$$\rho_k = \frac{1}{y_k^T s_k}, \qquad V_k = I - \rho_k y_k s_k^T, \qquad (7.17)$$

and

$$s_k = x_{k+1} - x_k, \qquad y_k = \nabla f_{k+1} - \nabla f_k. \qquad (7.18)$$

Since the inverse Hessian approximation H_k will generally be dense, the cost of storing and manipulating it is prohibitive when the number of variables is large. To circumvent this problem, we store a *modified* version of H_k implicitly, by storing a certain number (say, m) of the vector pairs $\{s_i, y_i\}$ used in the formulas (7.16)–(7.18). The product $H_k \nabla f_k$ can be obtained by performing a sequence of inner products and vector summations involving ∇f_k and the pairs $\{s_i, y_i\}$. After the new iterate is computed, the oldest vector pair in the set of pairs $\{s_i, y_i\}$ is replaced by the new pair $\{s_k, y_k\}$ obtained from the current step (7.18). In this way, the set of vector pairs includes curvature information from the m most recent iterations. Practical experience has shown that modest values of m (between 3 and 20, say) often produce satisfactory results.

We now describe the updating process in a little more detail. At iteration k, the current iterate is x_k and the set of vector pairs is given by $\{s_i, y_i\}$ for $i = k - m, \ldots, k - 1$. We first choose some initial Hessian approximation H_k^0 (in contrast to the standard BFGS iteration, this initial approximation is allowed to vary from iteration to iteration) and find by repeated application of the formula (7.16) that the L-BFGS approximation H_k satisfies the following formula:

$$H_k = \left(V_{k-1}^T \cdots V_{k-m}^T\right) H_k^0 \left(V_{k-m} \cdots V_{k-1}\right)$$
$$+ \rho_{k-m} \left(V_{k-1}^T \cdots V_{k-m+1}^T\right) s_{k-m} s_{k-m}^T \left(V_{k-m+1} \cdots V_{k-1}\right)$$
$$+ \rho_{k-m+1} \left(V_{k-1}^T \cdots V_{k-m+2}^T\right) s_{k-m+1} s_{k-m+1}^T \left(V_{k-m+2} \cdots V_{k-1}\right)$$
$$+ \cdots$$
$$+ \rho_{k-1} s_{k-1} s_{k-1}^T. \tag{7.19}$$

From this expression we can derive a recursive procedure to compute the product $H_k \nabla f_k$ efficiently.

Algorithm 7.4 (L-BFGS two-loop recursion).

$q \leftarrow \nabla f_k$;
for $i = k-1, k-2, \ldots, k-m$
$\qquad \alpha_i \leftarrow \rho_i s_i^T q$;
$\qquad q \leftarrow q - \alpha_i y_i$;
end (for)
$r \leftarrow H_k^0 q$;
for $i = k-m, k-m+1, \ldots, k-1$
$\qquad \beta \leftarrow \rho_i y_i^T r$;
$\qquad r \leftarrow r + s_i(\alpha_i - \beta)$
end (for)
stop with result $H_k \nabla f_k = r$.

Without considering the multiplication $H_k^0 q$, the two-loop recursion scheme requires $4mn$ multiplications; if H_k^0 is diagonal, then n additional multiplications are needed. Apart from being inexpensive, this recursion has the advantage that the multiplication by the initial matrix H_k^0 is isolated from the rest of the computations, allowing this matrix to be chosen freely and to vary between iterations. We may even use an implicit choice of H_k^0 by defining some initial approximation B_k^0 to the Hessian (not its inverse) and obtaining r by solving the system $B_k^0 r = q$.

A method for choosing H_k^0 that has proved effective in practice is to set $H_k^0 = \gamma_k I$, where

$$\gamma_k = \frac{s_{k-1}^T y_{k-1}}{y_{k-1}^T y_{k-1}}. \tag{7.20}$$

As discussed in Chapter 6, γ_k is the scaling factor that attempts to estimate the size of the true Hessian matrix along the most recent search direction (see (6.21)). This choice helps to ensure that the search direction p_k is well scaled, and as a result the step length $\alpha_k = 1$ is accepted in most iterations. As discussed in Chapter 6, it is important that the line search be based on the Wolfe conditions (3.6) or strong Wolfe conditions (3.7), so that BFGS updating is stable.

The limited-memory BFGS algorithm can be stated formally as follows.

Algorithm 7.5 (L-BFGS).

Choose starting point x_0, integer $m > 0$;

$k \leftarrow 0$;

repeat

      Choose H_k^0 (for example, by using (7.20));

      Compute $p_k \leftarrow -H_k \nabla f_k$ from Algorithm 7.4;

      Compute $x_{k+1} \leftarrow x_k + \alpha_k p_k$, where α_k is chosen to

           satisfy the Wolfe conditions;

      if $k > m$

           Discard the vector pair $\{s_{k-m}, y_{k-m}\}$ from storage;

      Compute and save $s_k \leftarrow x_{k+1} - x_k$, $y_k = \nabla f_{k+1} - \nabla f_k$;

      $k \leftarrow k + 1$;

until convergence.

The strategy of keeping the m most recent correction pairs $\{s_i, y_i\}$ works well in practice; indeed no other strategy has yet proved to be consistently better. During its first $m - 1$ iterations, Algorithm 7.5 is equivalent to the BFGS algorithm of Chapter 6 if the initial matrix H_0 is the same in both methods, and if L-BFGS chooses $H_k^0 = H_0$ at each iteration.

Table 7.1 presents results illustrating the behavior of Algorithm 7.5 for various levels of memory m. It gives the number of function and gradient evaluations (nfg) and the total CPU time. The test problems are taken from the CUTE collection [35], the number of variables is indicated by n, and the termination criterion $\|\nabla f_k\| \leq 10^{-5}$ is used. The table shows that the algorithm tends to be less robust when m is small. As the amount of storage increases, the number of function evaluations tends to decrease; but since the cost of each iteration increases with the amount of storage, the best CPU time is often obtained for small values of m. Clearly, the optimal choice of m is problem dependent.

Because some rival algorithms are inefficient, Algorithm 7.5 is often the approach of choice for large problems in which the true Hessian is not sparse. In particular, a Newton

Table 7.1 Performance of Algorithm 7.5.

Problem	n	L-BFGS $m = 3$		L-BFGS $m = 5$		L-BFGS $m = 17$		L-BFGS $m = 29$	
		nfg	time	nfg	time	nfg	time	nfg	time
DIXMAANL	1500	146	16.5	134	17.4	120	28.2	125	44.4
EIGENALS	110	821	21.5	569	15.7	363	16.2	168	12.5
FREUROTH	1000	>999	—	>999	—	69	8.1	38	6.3
TRIDIA	1000	876	46.6	611	41.4	531	84.6	462	127.1

method in which the exact Hessian is computed and factorized is not practical in such circumstances. The L-BFGS approach may also outperform Hessian-free Newton methods such as Newton–CG approaches, in which Hessian–vector products are calculated by finite differences or automatic differentiation. The main weakness of the L-BFGS method is that it converges slowly on ill-conditioned problems—specifically, on problems where the Hessian matrix contains a wide distribution of eigenvalues. On certain applications, the nonlinear conjugate gradient methods discussed in Chapter 5 are competitive with limited-memory quasi-Newton methods.

RELATIONSHIP WITH CONJUGATE GRADIENT METHODS

Limited-memory methods evolved as an attempt to improve nonlinear conjugate gradient methods, and early implementations resembled conjugate gradient methods more than quasi-Newton methods. The relationship between the two classes is the basis of a *memoryless BFGS iteration*, which we now outline.

We start by considering the Hestenes–Stiefel form of the nonlinear conjugate gradient method (5.46). Recalling that $s_k = \alpha_k p_k$, we have that the search direction for this method is given by

$$p_{k+1} = -\nabla f_{k+1} + \frac{\nabla f_{k+1}^T y_k}{y_k^T p_k} p_k = -\left(I - \frac{s_k y_k^T}{y_k^T s_k}\right)\nabla f_{k+1} \equiv -\hat{H}_{k+1}\nabla f_{k+1}. \qquad (7.21)$$

This formula resembles a quasi-Newton iteration, but the matrix $\hat{H}_{k+1}$ is neither symmetric nor positive definite. We could symmetrize it as $\hat{H}_{k+1}^T \hat{H}_{k+1}$, but this matrix does not satisfy the secant equation $\hat{H}_{k+1} y_k = s_k$ and is, in any case, singular. An iteration matrix that is symmetric, positive definite, and satisfies the secant equation is given by

$$H_{k+1} = \left(I - \frac{s_k y_k^T}{y_k^T s_k}\right)\left(I - \frac{y_k s_k^T}{y_k^T s_k}\right) + \frac{s_k s_k^T}{y_k^T s_k}. \qquad (7.22)$$

This matrix is exactly the one obtained by applying a single BFGS update (7.16) to the identity matrix. Hence, an algorithm whose search direction is given by $p_{k+1} = -H_{k+1}\nabla f_{k+1}$, with H_{k+1} defined by (7.22), can be thought of as a "memoryless" BFGS method, in which the previous Hessian approximation is always reset to the identity matrix before updating it and where only the most recent correction pair (s_k, y_k) is kept at every iteration. Alternatively, we can view the method as a variant of Algorithm 7.5 in which $m = 1$ and $H_k^0 = I$ at each iteration.

A more direct connection with conjugate gradient methods can be seen if we consider the memoryless BFGS formula (7.22) in conjunction with an exact line search, for which

$\nabla f_{k+1}^T p_k = 0$ for all k. We then obtain

$$p_{k+1} = -H_{k+1}\nabla f_{k+1} = -\nabla f_{k+1} + \frac{\nabla f_{k+1}^T y_k}{y_k^T p_k} p_k, \qquad (7.23)$$

which is none other than the Hestenes–Stiefel conjugate gradient method. Moreover, it is easy to verify that when $\nabla f_{k+1}^T p_k = 0$, the Hestenes–Stiefel formula reduces to the Polak–Ribière formula (5.44). Even though the assumption of exact line searches is unrealistic, it is intriguing that the BFGS formula is related in this way to the Polak–Ribière and Hestenes–Stiefel methods.

GENERAL LIMITED-MEMORY UPDATING

Limited-memory quasi-Newton approximations are useful in a variety of optimization methods. L-BFGS, Algorithm 7.5, is a line search method for unconstrained optimization that (implicitly) updates an approximation H_k to the inverse of the Hessian matrix. Trust-region methods, on the other hand, require an approximation B_k to the Hessian matrix, not to its inverse. We would also like to develop limited-memory methods based on the SR1 formula, which is an attractive alternative to BFGS; see Chapter 6. In this section we consider limited-memory updating in a general setting and show that by representing quasi-Newton matrices in a compact (or outer product) form, we can derive efficient implementations of *all* popular quasi-Newton update formulas, and their inverses. These compact representations will also be useful in designing limited-memory methods for constrained optimization, where approximations to the Hessian or reduced Hessian of the Lagrangian are needed; see Chapter 18 and Chapter 19.

We will consider only limited-memory methods (such as L-BFGS) that continuously refresh the correction pairs by removing and adding information at each stage. A different approach saves correction pairs until the available storage is exhausted and then discards all correction pairs (except perhaps one) and starts the process anew. Computational experience suggests that this second approach is less effective in practice.

Throughout this chapter we let B_k denote an approximation to a Hessian matrix and H_k the approximation to the inverse. In particular, we always have that $B_k^{-1} = H_k$.

COMPACT REPRESENTATION OF BFGS UPDATING

We now describe an approach to limited-memory updating that is based on representing quasi-Newton matrices in outer-product form. We illustrate it for the case of a BFGS approximation B_k to the Hessian.

Theorem 7.4.

Let B_0 be symmetric and positive definite, and assume that the k vector pairs $\{s_i, y_i\}_{i=0}^{k-1}$ satisfy $s_i^T y_i > 0$. Let B_k be obtained by applying k BFGS updates with these vector pairs to B_0, *using the formula (6.19). We then have that*

$$
B_k = B_0 - \begin{bmatrix} B_0 S_k & Y_k \end{bmatrix} \begin{bmatrix} S_k^T B_0 S_k & L_k \\ L_k^T & -D_k \end{bmatrix}^{-1} \begin{bmatrix} S_k^T B_0 \\ Y_k^T \end{bmatrix}, \tag{7.24}
$$

where S_k and Y_k are the $n \times k$ matrices defined by

$$
S_k = [s_0, \ldots, s_{k-1}], \qquad Y_k = [y_0, \ldots, y_{k-1}], \tag{7.25}
$$

while L_k and D_k are the $k \times k$ matrices

$$
(L_k)_{i,j} = \begin{cases} s_{i-1}^T y_{j-1} & \text{if } i > j, \\ 0 & \text{otherwise,} \end{cases} \tag{7.26}
$$

$$
D_k = \text{diag}\left[s_0^T y_0, \ldots, s_{k-1}^T y_{k-1} \right]. \tag{7.27}
$$

This result can be proved by induction. We note that the conditions $s_i^T y_i > 0$, $i = 0, 1, \ldots, k-1$, ensure that the middle matrix in (7.24) is nonsingular, so that this expression is well defined. The utility of this representation becomes apparent when we consider limited-memory updating.

As in the L-BFGS algorithm, we keep the m most recent correction pairs $\{s_i, y_i\}$ and refresh this set at every iteration by removing the oldest pair and adding a newly generated pair. During the first m iterations, the update procedure described in Theorem 7.4 can be used without modification, except that usually we make the specific choice $B_k^0 = \delta_k I$ for the basic matrix, where $\delta_k = 1/\gamma_k$ and γ_k is defined by (7.20).

At subsequent iterations $k > m$, the update procedure needs to be modified slightly to reflect the changing nature of the set of vector pairs $\{s_i, y_i\}$ for $i = k-m, k-m+1, \ldots, k-1$. Defining the $n \times m$ matrices S_k and Y_k by

$$
S_k = [s_{k-m}, \ldots, s_{k-1}], \qquad Y_k = [y_{k-m}, \ldots, y_{k-1}], \tag{7.28}
$$

we find that the matrix B_k resulting from m updates to the basic matrix $B_0^{(k)} = \delta_k I$ is given by

$$
B_k = \delta_k I - \begin{bmatrix} \delta_k S_k & Y_k \end{bmatrix} \begin{bmatrix} \delta_k S_k^T S_k & L_k \\ L_k^T & -D_k \end{bmatrix}^{-1} \begin{bmatrix} \delta_k S_k^T \\ Y_k^T \end{bmatrix}, \tag{7.29}
$$

where L_k and D_k are now the $m \times m$ matrices defined by

$$
(L_k)_{i,j} = \begin{cases} (s_{k-m-1+i})^T (y_{k-m-1+j}) & \text{if } i > j, \\ 0 & \text{otherwise,} \end{cases}
$$

$$
D_k = \text{diag}\left[s_{k-m}^T y_{k-m}, \ldots, s_{k-1}^T y_{k-1} \right].
$$

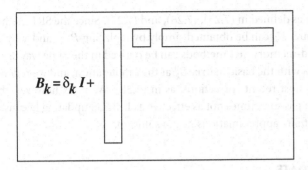

$$B_k = \delta_k I +$$

Figure 7.1
Compact (or outer product) representation of B_k in (7.29).

After the new iterate x_{k+1} is generated, we obtain S_{k+1} by deleting s_{k-m} from S_k and adding the new displacement s_k, and we update Y_{k+1} in a similar fashion. The new matrices L_{k+1} and D_{k+1} are obtained in an analogous way.

Since the middle matrix in (7.29) is small—of dimension $2m$—its factorization requires a negligible amount of computation. The key idea behind the compact representation (7.29) is that the corrections to the basic matrix can be expressed as an outer product of two long and narrow matrices—$[\delta_k S_k \ Y_k]$ and its transpose—with an intervening multiplication by a small $2m \times 2m$ matrix. See Figure 7.1 for a graphical illustration.

The limited-memory updating procedure of B_k requires approximately $2mn + O(m^3)$ operations, and matrix–vector products of the form $B_k v$ can be performed at a cost of $(4m + 1)n + O(m^2)$ multiplications. These operation counts indicate that updating and manipulating the direct limited-memory BFGS matrix B_k is quite economical when m is small.

This approximation B_k can be used in a trust-region method for unconstrained optimization or, more significantly, in methods for bound-constrained and general-constrained optimization. The program L-BFGS-B [322] makes extensive use of compact limited-memory approximations to solve large nonlinear optimization problems with bound constraints. In this situation, projections of B_k into subspaces defined by the constraint gradients must be calculated repeatedly. Several codes for general-constrained optimization, including KNITRO and IPOPT, make use of the compact limited-memory matrix B_k to approximate the Hessian of the Lagrangians; see Section 19.3.

We can derive a formula, similar to (7.24), that provides a compact representation of the inverse BFGS approximation H_k; see [52] for details. An implementation of the unconstrained L-BFGS algorithm based on this expression requires a similar amount of computation as the algorithm described in the previous section.

Compact representations can also be derived for matrices generated by the symmetric rank-one (SR1) formula. If k updates are applied to the symmetric matrix B_0 using the vector pairs $\{s_i, y_i\}_{i=0}^{k-1}$ and the SR1 formula (6.24), the resulting matrix B_k can be expressed as

$$B_k = B_0 + (Y_k - B_0 S_k)(D_k + L_k + L_k^T - S_k^T B_0 S_k)^{-1}(Y_k - B_0 S_k)^T, \qquad (7.30)$$

where S_k, Y_k, D_k, and L_k are as defined in (7.25), (7.26), and (7.27). Since the SR1 method is self-dual, the inverse formula H_k can be obtained simply by replacing B, s, and y by H, y, and s, respectively. Limited-memory SR1 methods can be derived in the same way as the BFGS method. We replace B_0 with the basic matrix B_k^0 at the kth iteration, and we redefine S_k and Y_k to contain the m most recent corrections, as in (7.28). We note, however, that limited-memory SR1 updating is sometimes not as effective as L-BFGS updating because it may not produce positive definite approximations near a solution.

UNROLLING THE UPDATE

The reader may wonder whether limited-memory updating can be implemented in simpler ways. In fact, as we show here, the most obvious implementation of limited-memory BFGS updating is considerably more expensive than the approach based on compact representations discussed in the previous section.

The direct BFGS formula (6.19) can be written as

$$B_{k+1} = B_k - a_k a_k^T + b_k b_k^T, \tag{7.31}$$

where the vectors a_k and b_k are defined by

$$a_k = \frac{B_k s_k}{(s_k^T B_k s_k)^{\frac{1}{2}}}, \qquad b_k = \frac{y_k}{(y_k^T s_k)^{\frac{1}{2}}}. \tag{7.32}$$

We could continue to save the vector pairs $\{s_i, y_i\}$ but use the formula (7.31) to compute matrix–vector products. A limited-memory BFGS method that uses this approach would proceed by defining the basic matrix B_k^0 at each iteration and then updating according to the formula

$$B_k = B_k^0 + \sum_{i=k-m}^{k-1} \left[b_i b_i^T - a_i a_i^T \right]. \tag{7.33}$$

The vector pairs $\{a_i, b_i\}$, $i = k - m, k - m + 1, \ldots, k - 1$, would then be recovered from the stored vector pairs $\{s_i, y_i\}$, $i = k - m, k - m + 1, \ldots, k - 1$, by the following procedure:

Procedure 7.6 (Unrolling the BFGS formula).
   for $i = k - m, k - m + 1, \ldots, k - 1$
      $b_i \leftarrow y_i / (y_i^T s_i)^{1/2}$;
      $a_i \leftarrow B_k^0 s_i + \sum_{j=k-m}^{i-1} \left[(b_j^T s_i) b_j - (a_j^T s_i) a_j \right]$;
      $a_i \leftarrow a_i / (s_i^T a_i)^{1/2}$;
   end (for)

Note that the vectors a_i must be recomputed at each iteration because they all depend on the vector pair $\{s_{k-m}, y_{k-m}\}$, which is removed at the end of iteration k. On the other hand, the vectors b_i and the inner products $b_j^T s_i$ can be saved from the previous iteration, so only the new values b_{k-1} and $b_j^T s_{k-1}$ need to be computed at the current iteration.

By taking all these computations into account, and assuming that $B_k^0 = I$, we find that approximately $\frac{3}{2}m^2n$ operations are needed to determine the limited-memory matrix. The actual computation of the inner product $B_m v$ (for arbitrary $v \in \mathbb{R}^n$) requires $4mn$ multiplications. Overall, therefore, this approach is less efficient than the one based on the compact matrix representation described previously. Indeed, while the product $B_k v$ costs the same in both cases, updating the representation of the limited-memory matrix by using the compact form requires only $2mn$ multiplications, compared to $\frac{3}{2}m^2n$ multiplications needed when the BFGS formula is unrolled.

7.3 SPARSE QUASI-NEWTON UPDATES

We now discuss a quasi-Newton approach to large-scale problems that has intuitive appeal: We demand that the quasi-Newton approximations B_k have the same (or similar) sparsity pattern as the true Hessian. This approach would reduce the storage requirements of the algorithm and perhaps give rise to more accurate Hessian approximations.

Suppose that we know which components of the Hessian may be nonzero at some point in the domain of interest. That is, we know the contents of the set Ω defined by

$$\Omega \overset{\text{def}}{=} \{(i,j) \mid [\nabla^2 f(x)]_{ij} \neq 0 \text{ for some } x \text{ in the domain of } f\}.$$

Suppose also that the current Hessian approximation B_k mirrors the nonzero structure of the exact Hessian, that is, $(B_k)_{ij} = 0$ for $(i,j) \notin \Omega$. In updating B_k to B_{k+1}, then, we could try to find the matrix B_{k+1} that satisfies the secant condition, has the same sparsity pattern, and is as close as possible to B_k. Specifically, we define B_{k+1} to be the solution of the following quadratic program:

$$\min_{B} \|B - B_k\|_F^2 = \sum_{(i,j)\in\Omega} [B_{ij} - (B_k)_{ij}]^2, \tag{7.34a}$$

$$\text{subject to } Bs_k = y_k, \ B = B^T, \text{ and } B_{ij} = 0 \text{ for } (i,j) \notin \Omega. \tag{7.34b}$$

One can show that the solution B_{k+1} of this problem can be obtained by solving an $n \times n$ linear system whose sparsity pattern is Ω, the same as the sparsity of the true Hessian. Once B_{k+1} has been computed, we can use it, within a trust-region method, to obtain the new iterate x_{k+1}. We note that B_{k+1} is not guaranteed to be positive definite.

We omit further details of this approach because it has several drawbacks. The updating process does not possess scale invariance under linear transformations of the variables and, more significantly, its practical performance has been disappointing. The fundamental weakness of this approach is that (7.34a) is an inadequate model and can produce poor Hessian approximations.

An alternative approach is to relax the secant equation, making sure that it is approximately satisfied along the last few steps rather than requiring it to hold strictly on the latest step. To do so, we define S_k and Y_k by (7.28) so that they contain the m most recent difference pairs. We can then define the new Hessian approximation B_{k+1} to be the solution of

$$\min_B \| B S_k - Y_k \|_F^2$$
$$\text{subject to } B = B^T \text{ and } B_{ij} = 0 \text{ for } (i, j) \notin \Omega.$$

This convex optimization problem has a solution, but it is not easy to compute. Moreover, this approach can produce singular or poorly conditioned Hessian approximations. Even though it frequently outperforms methods based on (7.34a), its performance on large problems has not been impressive.

7.4 ALGORITHMS FOR PARTIALLY SEPARABLE FUNCTIONS

In a *separable* unconstrained optimization problem, the objective function can be decomposed into a sum of simpler functions that can be optimized independently. For example, if we have

$$f(x) = f_1(x_1, x_3) + f_2(x_2, x_4, x_6) + f_3(x_5),$$

we can find the optimal value of x by minimizing each function f_i, $i = 1, 2, 3$, independently, since no variable appears in more than one function. The cost of performing m lower-dimensional optimizations is much less in general than the cost of optimizing an n-dimensional function.

In many large problems the objective function $f : \mathbb{R}^n \to \mathbb{R}$ is not separable, but it can still be written as the sum of simpler functions, known as *element functions*. Each element function has the property that it is unaffected when we move along a large number of linearly independent directions. If this property holds, we say that f is *partially separable*. All functions whose Hessians $\nabla^2 f$ are sparse are partially separable, but so are many functions whose Hessian is not sparse. Partial separability allows for economical problem representation, efficient automatic differentiation, and effective quasi-Newton updating.

The simplest form of partial separability arises when the objective function can be written as

$$f(x) = \sum_{i=1}^{ne} f_i(x), \tag{7.35}$$

where each of the element functions f_i depends on only a few components of x. It follows that the gradients ∇f_i and Hessians $\nabla^2 f_i$ of each element function contain just a few nonzeros. By differentiating (7.35), we obtain

$$\nabla f(x) = \sum_{i=1}^{ne} \nabla f_i(x), \qquad \nabla^2 f(x) = \sum_{i=1}^{ne} \nabla^2 f_i(x).$$

A natural question is whether it is more effective to maintain quasi-Newton approximations to each of the element Hessians $\nabla^2 f_i(x)$ separately, rather than approximating the entire Hessian $\nabla^2 f(x)$. We will show that the answer is affirmative, provided that the quasi-Newton approximation fully exploits the structure of each element Hessian.

We introduce the concept by means of a simple example. Consider the objective function

$$f(x) = (x_1 - x_3^2)^2 + (x_2 - x_4^2)^2 + (x_3 - x_2^2)^2 + (x_4 - x_1^2)^2 \qquad (7.36)$$
$$\equiv f_1(x) + f_2(x) + f_3(x) + f_4(x).$$

The Hessians of the element functions f_i are 4×4 sparse, singular matrices with 4 nonzero entries.

Let us focus on f_1; all other element functions have exactly the same form. Even though f_1 is formally a function of all components of x, it depends only on x_1 and x_3, which we call the *element variables* for f_1. We assemble the element variables into a vector that we call $x_{[1]}$, that is,

$$x_{[1]} = \begin{bmatrix} x_1 \\ x_3 \end{bmatrix},$$

and note that

$$x_{[1]} = U_1 x \quad \text{with} \quad U_1 = \begin{bmatrix} 1 & 0 & 0 & 0 \\ 0 & 0 & 1 & 0 \end{bmatrix}.$$

If we define the function ϕ_1 by

$$\phi_1(z_1, z_2) = (z_1 - z_2^2)^2,$$

then we can write $f_1(x) = \phi_1(U_1 x)$. By applying the chain rule to this representation, we obtain

$$\nabla f_1(x) = U_1^T \nabla \phi_1(U_1 x), \qquad \nabla^2 f_1(x) = U_1^T \nabla^2 \phi_1(U_1 x) U_1. \qquad (7.37)$$

In our case, we have

$$\nabla^2 \phi_1(U_1 x) = \begin{bmatrix} 2 & -4x_3 \\ -4x_3 & 12x_3^2 - 4x_1 \end{bmatrix}, \quad \nabla^2 f_1(x) = \begin{bmatrix} 2 & 0 & -4x_3 & 0 \\ 0 & 0 & 0 & 0 \\ -4x_3 & 0 & 12x_3^2 - 4x_1 & 0 \\ 0 & 0 & 0 & 0 \end{bmatrix}.$$

The matrix U_1, known as a *compactifying matrix,* allows us to map the derivative information for the low-dimensional function ϕ_1 into the derivative information for the element function f_1.

Now comes the key idea: Instead of maintaining a quasi-Newton approximation to $\nabla^2 f_1$, we maintain a 2×2 quasi-Newton approximation $B_{[1]}$ of $\nabla^2 \phi_1$ and use the relation (7.37) to transform it into a quasi-Newton approximation to $\nabla^2 f_1$. To update $B_{[1]}$ after a typical step from x to x^+, we record the information

$$s_{[1]} = x_{[1]}^+ - x_{[1]}, \qquad y_{[1]} = \nabla \phi_1(x_{[1]}^+) - \nabla \phi_1(x_{[1]}), \tag{7.38}$$

and use BFGS or SR1 updating to obtain the new approximation $B_{[1]}^+$. We therefore update small, dense quasi-Newton approximations with the property

$$B_{[1]} \approx \nabla^2 \phi_1(U_1 x) = \nabla^2 \phi_1(x_{[1]}). \tag{7.39}$$

To obtain an approximation of the element Hessian $\nabla^2 f_1$, we use the transformation suggested by the relationship (7.37); that is,

$$\nabla^2 f_1(x) \approx U_1^T B_{[1]} U_1.$$

This operation has the effect of mapping the elements of $B_{[1]}$ to the correct positions in the full $n \times n$ Hessian approximation.

The previous discussion concerned only the first element function f_1, but we can treat all other functions f_i in the same way. The full objective function can now be written as

$$f(x) = \sum_{i=1}^{ne} \phi_i(U_i x), \tag{7.40}$$

and we maintain a quasi-Newton approximation $B_{[i]}$ for each of the functions ϕ_i. To obtain a complete approximation to the full Hessian $\nabla^2 f$, we simply sum the element Hessian approximations as follows:

$$B = \sum_{i=1}^{ne} U_i^T B_{[i]} U_i. \tag{7.41}$$

We may use this approximate Hessian in a trust-region algorithm, obtaining an approximate solution p_k of the system

$$B_k p_k = -\nabla f_k. \tag{7.42}$$

We need not assemble B_k explicitly but rather use the conjugate gradient approach to solve (7.42), computing matrix–vector products of the form $B_k v$ by performing operations with the matrices U_i and $B_{[i]}$.

To illustrate the usefulness of this element-by-element updating technique, let us consider a problem of the form (7.36) but this time involving 1000 variables, not just 4. The functions ϕ_i still depend on only two internal variables, so that each Hessian approximation $B_{[i]}$ is a 2×2 matrix. After just a few iterations, we will have sampled enough directions $s_{[i]}$ to make each $B_{[i]}$ an accurate approximation to $\nabla^2 \phi_i$. Hence the full quasi-Newton approximation (7.41) will tend to be a very good approximation to $\nabla^2 f(x)$. By contrast, a quasi-Newton method that ignores the partially separable structure of the objective function will attempt to estimate the total average curvature—the sum of the individual curvatures of the element functions—by approximating the 1000×1000 Hessian matrix. When the number of variables n is large, many iterations will be required before this quasi-Newton approximation is of good quality. Hence an algorithm of this type (for example, standard BFGS or L-BFGS) will require many more iterations than a method based on the partially separable approximate Hessian.

It is not always possible to use the BFGS formula to update the partial Hessian $B_{[i]}$, because there is no guarantee that the curvature condition $s_{[i]}^T y_{[i]} > 0$ will be satisfied. That is, even though the full Hessian $\nabla^2 f(x)$ is at least positive semidefinite at the solution x^*, some of the individual Hessians $\nabla^2 \phi_i(\cdot)$ may be indefinite. One way to overcome this obstacle is to apply the SR1 update to each of the element Hessians. This approach has proved effective in the LANCELOT package [72], which is designed to take full advantage of partial separability.

The main limitations of this quasi-Newton approach are the cost of the step computation (7.42), which is comparable to the cost of a Newton step, and the difficulty of identifying the partially separable structure of a function. The performance of quasi-Newton methods is satisfactory provided that we find the *finest* partially separable decomposition of the problem; see [72]. Furthermore, even when the partially separable structure is known, it may be more efficient to compute a Newton step. For example, the modeling language AMPL automatically detects the partially separable structure of a function f and uses it to compute the Hessian $\nabla^2 f(x)$.

7.5 PERSPECTIVES AND SOFTWARE

Newton–CG methods have been used successfully to solve large problems in a variety of applications. Many of these implementations are developed by engineers and scientists and use problem-specific preconditioners. Freely available packages include TN/TNBC [220] and TNPACK [275]. Software for more general problems, such as LANCELOT [72], KNITRO/CG [50], and TRON [192], employ Newton–CG methods when applied to unconstrained problems. Other packages, such as LOQO [294] implement Newton methods with a sparse factorization modified to ensure positive definiteness. GLTR [145] offers a Newton–Lanczos method. There is insufficient experience to date to say whether the Newton–Lanczos method is significantly better in practice than the Steihaug strategy given in Algorithm 7.2.

Software for computing incomplete Cholesky preconditioners includes the ICFS [193] and MA57 [166] packages. A preconditioner for Newton–CG based on limited-memory BFGS approximations is provided in PREQN [209].

Limited-memory BFGS methods are implemented in LBFGS [194] and M1QN3 [122]; see Gill and Leonard [125] for a variant that requires less storage and appears to be quite efficient. The compact limited-memory representations of Section 7.2 are used in LBFGS-B [322], IPOPT [301], and KNITRO.

The LANCELOT package exploits partial separability. It provides SR1 and BFGS quasi-Newton options as well as a Newton methods. The step computation is obtained by a preconditioned conjugate gradient iteration using trust regions. If f is partially separable, a general affine transformation will not in general preserve the partially separable structure. The quasi-Newton method for partially separable functions described in Section 7.4 is not invariant to affine transformations of the variables, but this is not a drawback because the method is invariant under transformations that preserve separability.

NOTES AND REFERENCES

A complete study of inexact Newton methods is given in [74]. For a discussion of the Newton–Lanczos method see [145]. Other iterative methods for the solution of a trust-region problem have been proposed by Hager [160], and by Rendl and Wolkowicz [263].

For further discussion on the L-BFGS method see Nocedal [228], Liu and Nocedal [194], and Gilbert and Lemaréchal [122]. The last paper also discusses various ways in which the scaling parameter can be chosen. Algorithm 7.4, the two-loop L-BFGS recursion, constitutes an economical procedure for computing the product $H_k \nabla f_k$. It is based, however, on the specific form of the BFGS update formula (7.16), and recursions of this type have not yet been developed (and may not exist) for other members of the Broyden class (for instance, the SR1 and DFP methods). Our discussion of compact representations of limited-memory matrices is based on Byrd, Nocedal, and Schnabel [52].

Sparse quasi-Newton updates have been studied by Toint [288, 289] and Fletcher et al. [102, 104], among others. The concept of partial separability was introduced by Griewank and Toint [156, 155]. For an extensive treatment of the subject see Conn, Gould, and Toint [72].

✎ EXERCISES

✎   **7.1** Code Algorithm 7.5, and test it on the extended Rosenbrock function

$$f(x) = \sum_{i=1}^{n/2} \left[\alpha(x_{2i} - x_{2i-1}^2)^2 + (1 - x_{2i-1})^2 \right],$$

where α is a parameter that you can vary (for example, 1 or 100). The solution is $x^* = (1, 1, \ldots, 1)^T$, $f^* = 0$. Choose the starting point as $(-1, -1, \ldots, -1)^T$. Observe the behavior of your program for various values of the memory parameter m.

✎   **7.2** Show that the matrix $\hat{H}_{k+1}$ in (7.21) is singular.

✎   **7.3** Derive the formula (7.23) under the assumption that line searches are exact.

✎   **7.4** Consider limited-memory SR1 updating based on (7.30). Explain how the storage can be cut in half if the basic matrix B_k^0 is kept fixed for all k. (Hint: Consider the matrix $Q_k = [q_0, \ldots, q_{k-1}] = Y_k - B_0 S_k$.)

✎   **7.5** Write the function defined by

$$f(x) = x_2 x_3 e^{x_1 + x_3 - x_4} + (x_2 x_3)^2 + (x_3 - x_4)$$

in the form (7.40). In particular, give the definition of each of the compactifying transformations U_i.

✎   **7.6** Does the approximation B obtained by the partially separable quasi-Newton updating (7.38), (7.41) satisfy the secant equation $Bs = y$?

✎   **7.7** The minimum surface problem is a classical application of the calculus of variations and can be found in many textbooks. We wish to find the surface of minimum area, defined on the unit square, that interpolates a prescribed continuous function on the boundary of the square. In the standard discretization of this problem, the unknowns are the values of the sought-after function $z(x, y)$ on a $q \times q$ rectangular mesh of points over the unit square.

More specifically, we divide each edge of the square into q intervals of equal length, yielding $(q + 1)^2$ grid points. We label the grid points as

$$x_{(i-1)(q+1)+1}, \ldots, x_{i(q+1)} \quad \text{for } i = 1, 2, \ldots, q + 1,$$

so that each value of i generates a line. With each point we associate a variable z_i that represents the height of the surface at this point. For the $4q$ grid points on the boundary of the unit square, the values of these variables are determined by the given function. The optimization problem is to determine the other $(q + 1)^2 - 4q$ variables z_i so that the total surface area is minimized.

A typical subsquare in this partition looks as follows:

We denote this square by A_j and note that its area is q^2. The desired function is $z(x, y)$, and we wish to compute its surface over A_j. Calculus books show that the area of the surface is given by

$$f_j(x) \equiv \int\int_{(x,y) \in A_j} \sqrt{1 + \left(\frac{\partial z}{\partial x}\right)^2 + \left(\frac{\partial z}{\partial y}\right)^2} \, dx \, dy.$$

Approximate the derivatives by finite differences, and show that f_j has the form

$$f_j(x) = \frac{1}{q^2}\left[1 + \frac{q^2}{2}[(x_j - x_{j+q+1})^2 + (x_{j+1} - x_{j+q})^2]\right]^{\frac{1}{2}}. \tag{7.43}$$

✐ **7.8** Compute the gradient of the element function (7.43) with respect to the full vector x. Show that it contains at most four nonzeros, and that two of these four nonzero components are negatives of the other two. Compute the Hessian of f_j, and show that, among the 16 nonzeros, only three different magnitudes are represented. Also show that this Hessian is singular.

CHAPTER 8

Calculating
Derivatives

Most algorithms for nonlinear optimization and nonlinear equations require knowledge of derivatives. Sometimes the derivatives are easy to calculate by hand, and it is reasonable to expect the user to provide code to compute them. In other cases, the functions are too complicated, so we look for ways to calculate or approximate the derivatives automatically. A number of interesting approaches are available, of which the most important are probably the following.

Finite Differencing. This technique has its roots in Taylor's theorem (see Chapter 2). By observing the change in function values in response to small perturbations of the unknowns near a given point x, we can estimate the response to *infintesimal* perturbations, that is, the derivatives. For instance, the partial derivative of a smooth function $f : \mathbb{R}^n \to \mathbb{R}$ with respect to the ith variable x_i can be approximated by the central-difference formula

$$\frac{\partial f}{\partial x_i} \approx \frac{f(x + \epsilon e_i) - f(x - \epsilon e_i)}{2\epsilon},$$

where ϵ is a small positive scalar and e_i is the ith unit vector, that is, the vector whose elements are all 0 except for a 1 in the ith position.

Automatic Differentiation. This technique takes the view that the computer code for evaluating the function can be broken down into a composition of elementary arithmetic operations, to which the chain rule (one of the basic rules of calculus) can be applied. Some software tools for automatic differentiation (such as ADIFOR [25]) produce new code that calculates both function and derivative values. Other tools (such as ADOL-C [154]) keep a record of the elementary computations that take place while the function evaluation code for a given point x is executing on the computer. This information is processed to produce the derivatives at the same point x.

Symbolic Differentiation. In this technique, the algebraic specification for the function f is manipulated by symbolic manipulation tools to produce new algebraic expressions for each component of the gradient. Commonly used symbolic manipulation tools can be found in the packages Mathematica [311], Maple [304], and Macsyma [197].

In this chapter we discuss the first two approaches: finite differencing and automatic differentiation.

The usefulness of derivatives is not restricted to *algorithms* for optimization. Modelers in areas such as design optimization and economics are often interested in performing post-optimal *sensitivity analysis*, in which they determine the sensitivity of the optimum to small perturbations in the parameter or constraint values. Derivatives are also important in other areas such as nonlinear differential equations and simulation.

8.1 FINITE-DIFFERENCE DERIVATIVE APPROXIMATIONS

Finite differencing is an approach to the calculation of approximate derivatives whose motivation (like that of so many algorithms in optimization) comes from Taylor's theorem. Many software packages perform automatic calculation of finite differences whenever the user is unable or unwilling to supply code to calculate exact derivatives. Although they yield only approximate values for the derivatives, the results are adequate in many situations.

By definition, derivatives are a measure of the sensitivity of the function to infinitesimal changes in the values of the variables. Our approach in this section is to make small, *finite* perturbations in the values of x and examine the resulting *differences* in the function values.

By taking ratios of the function difference to variable difference, we obtain approximations to the derivatives.

APPROXIMATING THE GRADIENT

An approximation to the gradient vector $\nabla f(x)$ can be obtained by evaluating the function f at $(n + 1)$ points and performing some elementary arithmetic. We describe this technique, along with a more accurate variant that requires additional function evaluations.

A popular formula for approximating the partial derivative $\partial f / \partial x_i$ at a given point x is the *forward-difference*, or *one-sided-difference*, approximation, defined as

$$\frac{\partial f}{\partial x_i}(x) \approx \frac{f(x + \epsilon e_i) - f(x)}{\epsilon}. \tag{8.1}$$

The gradient can be built up by simply applying this formula for $i = 1, 2, \dots, n$. This process requires evaluation of f at the point x as well as the n perturbed points $x + \epsilon e_i$, $i = 1, 2, \dots, n$: a total of $(n + 1)$ points.

The basis for the formula (8.1) is Taylor's theorem, Theorem 2.1 in Chapter 2. When f is twice continuously differentiable, we have

$$f(x + p) = f(x) + \nabla f(x)^T p + \tfrac{1}{2} p^T \nabla^2 f(x + tp)p, \quad \text{some } t \in (0, 1) \tag{8.2}$$

(see (2.6)). If we choose L to be a bound on the size of $\|\nabla^2 f(\cdot)\|$ in the region of interest, it follows directly from this formula that the last term in this expression is bounded by $(L/2)\|p\|^2$, so that

$$\left\| f(x + p) - f(x) - \nabla f(x)^T p \right\| \le (L/2)\|p\|^2. \tag{8.3}$$

We now choose the vector p to be ϵe_i, so that it represents a small change in the value of a single component of x (the ith component). For this p, we have that $\nabla f(x)^T p = \nabla f(x)^T e_i = \partial f / \partial x_i$, so by rearranging (8.3), we conclude that

$$\frac{\partial f}{\partial x_i}(x) = \frac{f(x + \epsilon e_i) - f(x)}{\epsilon} + \delta_\epsilon, \quad \text{where } |\delta_\epsilon| \le (L/2)\epsilon. \tag{8.4}$$

We derive the forward-difference formula (8.1) by simply ignoring the error term δ_ϵ in this expression, which becomes smaller and smaller as ϵ approaches zero.

An important issue in implementing the formula (8.1) is the choice of the parameter ϵ. The error expression (8.4) suggests that we should choose ϵ as small as possible. Unfortunately, this expression ignores the roundoff errors that are introduced when the function f is evaluated on a real computer, in floating-point arithmetic. From our discussion in the Appendix (see (A.30) and (A.31)), we know that the quantity **u** known as *unit roundoff* is crucial: It is a bound on the relative error that is introduced whenever an arithmetic operation is performed on two floating-point numbers. ($\mathbf{u}$ is about 1.1×10^{-16} in double-precision IEEE floating-point arithmetic.) The effect of these errors on the final computed value of f depends on the way in which f is computed. It could come from an arithmetic formula, or from a differential equation solver, with or without refinement.

As a rough estimate, let us assume simply that the relative error in the computed f is bounded by $\mathbf{u}$, so that the computed values of $f(x)$ and $f(x + \epsilon e_i)$ are related to the exact values in the following way:

$$|\text{comp}(f(x)) - f(x)| \leq \mathbf{u} L_f,$$
$$|\text{comp}(f(x + \epsilon e_i)) - f(x + \epsilon e_i)| \leq \mathbf{u} L_f,$$

where $\text{comp}(\cdot)$ denotes the computed value, and L_f is a bound on the value of $|f(\cdot)|$ in the region of interest. If we use these computed values of f in place of the exact values in (8.4) and (8.1), we obtain an error that is bounded by

$$(L/2)\epsilon + 2\mathbf{u} L_f/\epsilon. \tag{8.5}$$

Naturally, we would like to choose ϵ to make this error as small as possible; it is easy to see that the minimizing value is

$$\epsilon^2 = \frac{4 L_f \mathbf{u}}{L}.$$

If we assume that the problem is well scaled, then the ratio L_f/L (the ratio of function values to second derivative values) does not exceed a modest size. We can conclude that the following choice of ϵ is fairly close to optimal:

$$\epsilon = \sqrt{\mathbf{u}}. \tag{8.6}$$

(In fact, this value is used in many of the optimization software packages that use finite differencing as an option for estimating derivatives.) For this value of ϵ, we have from (8.5) that the total error in the forward-difference approximation is fairly close to $\sqrt{\mathbf{u}}$.

A more accurate approximation to the derivative can be obtained by using the *central-difference* formula, defined as

$$\frac{\partial f}{\partial x_i}(x) \approx \frac{f(x + \epsilon e_i) - f(x - \epsilon e_i)}{2\epsilon}. \tag{8.7}$$

As we show below, this approximation is more accurate than the forward-difference approximation (8.1). It is also about twice as expensive, since we need to evaluate f at the points x and $x \pm \epsilon e_i$, $i = 1, 2, \ldots, n$: a total of $2n + 1$ points.

The basis for the central difference approximation is again Taylor's theorem. When the second derivatives of f exist and are Lipschitz continuous, we have from (8.2) that

$$f(x + p) = f(x) + \nabla f(x)^T p + \tfrac{1}{2} p^T \nabla^2 f(x + tp)p \quad \text{for some } t \in (0, 1)$$
$$= f(x) + \nabla f(x)^T p + \tfrac{1}{2} p^T \nabla^2 f(x)p + O\left(\|p\|^3\right). \tag{8.8}$$

By setting $p = \epsilon e_i$ and $p = -\epsilon e_i$, respectively, we obtain

$$f(x + \epsilon e_i) = f(x) + \epsilon \frac{\partial f}{\partial x_i} + \frac{1}{2}\epsilon^2 \frac{\partial^2 f}{\partial x_i^2} + O\left(\epsilon^3\right),$$

$$f(x - \epsilon e_i) = f(x) - \epsilon \frac{\partial f}{\partial x_i} + \frac{1}{2}\epsilon^2 \frac{\partial^2 f}{\partial x_i^2} + O\left(\epsilon^3\right).$$

(Note that the final error terms in these two expressions are generally not the same, but they are both bounded by some multiple of ϵ^3.) By subtracting the second equation from the first and dividing by 2ϵ, we obtain the expression

$$\frac{\partial f}{\partial x_i}(x) = \frac{f(x + \epsilon e_i) - f(x - \epsilon e_i)}{2\epsilon} + O\left(\epsilon^2\right).$$

We see from this expression that the error is $O\left(\epsilon^2\right)$, as compared to the $O(\epsilon)$ error in the forward-difference formula (8.1). However, when we take evaluation error in f into account, the accuracy that can be achieved in practice is less impressive; the same assumptions that were used to derive (8.6) lead to an optimal choice of ϵ of about $\mathbf{u}^{1/3}$ and an error of about $\mathbf{u}^{2/3}$. In some situations, the extra few digits of accuracy may improve the performance of the algorithm enough to make the extra expense worthwhile.

APPROXIMATING A SPARSE JACOBIAN

Consider now the case of a vector function $r : \mathbb{R}^n \to \mathbb{R}^m$, such as the residual vector that we consider in Chapter 10 or the system of nonlinear equations from Chapter 11. The matrix $J(x)$ of first derivatives for this function is defined as follows:

$$J(x) = \left[\frac{\partial r_j}{\partial x_i}\right]_{\substack{j=1,2,\dots,m \\ i=1,2,\dots,n}} = \begin{bmatrix} \nabla r_1(x)^T \\ \nabla r_2(x)^T \\ \vdots \\ \nabla r_m(x)^T \end{bmatrix}, \tag{8.9}$$

where $r_j, j = 1, 2, \dots, m$ are the components of r. The techniques described in the previous section can be used to evaluate the full Jacobian $J(x)$ one column at a time. When r is twice continuously differentiable, we can use Taylor's theorem to deduce that

$$\|r(x + p) - r(x) - J(x)p\| \le (L/2)\|p\|^2, \tag{8.10}$$

where L is a Lipschitz constant for J in the region of interest. If we require an approximation to the Jacobian–vector product $J(x)p$ for a given vector p (as is the case with inexact Newton methods for nonlinear systems of equations; see Section 11.1), this expression immediately suggests choosing a small nonzero ϵ and setting

$$J(x)p \approx \frac{r(x + \epsilon p) - r(x)}{\epsilon}, \tag{8.11}$$

an approximation that is accurate to $O(\epsilon)$. A two-sided approximation can be derived from the formula (8.7).

If an approximation to the full Jacobian $J(x)$ is required, we can compute it a column at a time, analogously to (8.1), by setting set $p = \epsilon e_i$ in (8.10) to derive the following estimate of the ith column:

$$\frac{\partial r}{\partial x_i}(x) \approx \frac{r(x + \epsilon e_i) - r(x)}{\epsilon}. \tag{8.12}$$

A full Jacobian estimate can be obtained at a cost of $n + 1$ evaluations of the function r. When the Jacobian is sparse, however, we can often obtain the estimate at a much lower cost, sometimes just three or four evaluations of r. The key is to estimate a number of different columns of the Jacobian simultaneously, by judicious choices of the perturbation vector p in (8.10).

We illustrate the technique with a simple example. Consider the function $r : \mathbb{R}^n \to \mathbb{R}^n$ defined by

$$r(x) = \begin{bmatrix} 2(x_2^3 - x_1^2) \\ 3(x_2^3 - x_1^2) + 2(x_3^3 - x_2^2) \\ 3(x_3^3 - x_2^2) + 2(x_4^3 - x_3^2) \\ \vdots \\ 3(x_n^3 - x_{n-1}^2) \end{bmatrix}. \tag{8.13}$$

Each component of r depends on just two or three components of x, so that each row of the Jacobian contains only two or three nonzero elements. For the case of $n = 6$, the Jacobian has the following structure:

$$
\begin{bmatrix}
\times & \times \\
\times & \times & \times \\
& \times & \times & \times \\
& & \times & \times & \times \\
& & & \times & \times & \times \\
& & & & \times & \times
\end{bmatrix},
\tag{8.14}
$$

where each cross represents a nonzero element, with zeros represented by a blank space.

Staying for the moment with the case $n = 6$, suppose that we wish to compute a finite-difference approximation to the Jacobian. (Of course, it is easy to calculate this particular Jacobian by hand, but there are complicated functions with similar structure for which hand calculation is more difficult.) A perturbation $p = \epsilon e_1$ to the first component of x will affect only the first and second components of r. The remaining components will be unchanged, so that the right-hand-side of formula (8.12) will correctly evaluate to zero in the components 3, 4, 5, 6. It is wasteful, however, to reevaluate these components of r when we know in advance that their values are not affected by the perturbation. Instead, we look for a way to modify the perturbation vector so that it does not have any further effect on components 1 and 2, but *does* produce a change in some of the components 3, 4, 5, 6, which we can then use as the basis of a finite-difference estimate for some *other* column of the Jacobian. It is not hard to see that the additional perturbation ϵe_4 has the desired property: It alters the 3rd, 4th, and 5th elements of r, but leaves the 1st and 2nd elements unchanged. The changes in r as a result of the perturbations ϵe_1 and ϵe_4 do not interfere with each other.

To express this discussion in mathematical terms, we set

$$
p = \epsilon(e_1 + e_4),
$$

and note that

$$
r(x + p)_{1,2} = r(x + \epsilon(e_1 + e_4))_{1,2} = r(x + \epsilon e_1)_{1,2}
\tag{8.15}
$$

(where the notation $[\cdot]_{1,2}$ denotes the subvector consisting of the first and second elements), while

$$
r(x + p)_{3,4,5} = r(x + \epsilon(e_1 + e_4))_{3,4,5} = r(x + \epsilon e_4)_{3,4,5}.
\tag{8.16}
$$

By substituting (8.15) into (8.10), we obtain

$$
r(x + p)_{1,2} = r(x)_{1,2} + \epsilon[J(x)e_1]_{1,2} + O(\epsilon^2).
$$

By rearranging this expression, we obtain the following difference formula for estimating the $(1, 1)$ and $(2, 1)$ elements of the Jacobian matrix:

$$\begin{bmatrix} \dfrac{\partial r_1}{\partial x_1}(x) \\[2ex] \dfrac{\partial r_2}{\partial x_1}(x) \end{bmatrix} = [J(x)e_1]_{1,2} \approx \frac{r(x+p)_{1,2} - r(x)_{1,2}}{\epsilon}. \tag{8.17}$$

A similar argument shows that the nonzero elements of the fourth column of the Jacobian can be estimated by substituting (8.16) into (8.10); we obtain

$$\begin{bmatrix} \dfrac{\partial r_4}{\partial x_3}(x) \\[2ex] \dfrac{\partial r_4}{\partial x_4}(x) \\[2ex] \dfrac{\partial r_4}{\partial x_5}(x) \end{bmatrix} = [J(x)e_4]_{3,4,5} \approx \frac{r(x+p)_{3,4,5} - r(x)_{3,4,5}}{\epsilon}. \tag{8.18}$$

To summarize: We have been able to estimate *two* columns of the Jacobian $J(x)$ by evaluating the function r at the single extra point $x + \epsilon(e_1 + e_4)$.

We can approximate the remainder of $J(x)$ in an economical manner as well. Columns 2 and 5 can be approximated by choosing $p = \epsilon(e_2 + e_5)$, while we can use $p = \epsilon(e_3 + e_6)$ to approximate columns 3 and 6. In total, we need 3 evaluations of the function r (after the initial evaluation at x) to estimate the entire Jacobian matrix.

In fact, for *any* choice of n in (8.13) (no matter how large), three extra evaluations of r are sufficient to approximate the entire Jacobian. The corresponding choices of perturbation vectors p are

$$p = \epsilon(e_1 + e_4 + e_7 + e_{10} + \cdots),$$
$$p = \epsilon(e_2 + e_5 + e_8 + e_{11} + \cdots),$$
$$p = \epsilon(e_3 + e_6 + e_9 + e_{12} + \cdots).$$

In the first of these vectors, the nonzero components are chosen so that no two of the columns $1, 4, 7, \ldots$ have a nonzero element in the same row. The same property holds for the other two vectors and, in fact, points the way to the criterion that we can apply to general problems to decide on a valid set of perturbation vectors.

Algorithms for choosing the perturbation vectors can be expressed conveniently in the language of graphs and graph coloring. For any function $r : \mathbb{R}^n \to \mathbb{R}^m$, we can construct a *column incidence graph* G with n nodes by drawing an arc between nodes i and k if there is some component of r that depends on both x_i and x_k. In other words, the ith and kth columns of the Jacobian $J(x)$ each have a nonzero element in some row j, for some $j = 1, 2, \ldots, m$ and some value of x. (The intersection graph for the function defined in

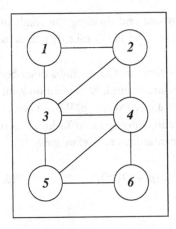

Figure 8.1
Column incidence graph for $r(x)$ defined in (8.13).

(8.13), with $n = 6$, is shown in Figure 8.1.) We now assign each node a "color" according to the following rule: Two nodes can have the same color if there is no arc that connects them. Finally, we choose one perturbation vector corresponding to each color: If nodes $i_1, i_2, \ldots, i_\ell$ have the same color, the corresponding p is $\epsilon(e_{i_1} + e_{i_2} + \cdots + e_{i_\ell})$.

Usually, there are many ways to assign colors to the n nodes in the graph in a way that satisfies the required condition. The simplest way is just to assign each node a different color, but since that scheme produces n perturbation vectors, it is usually not the most efficient approach. It is generally very difficult to find the coloring scheme that uses the fewest possible colors, but there are simple algorithms that do a good job of finding a near-optimal coloring at low cost. Curtis, Powell, and Reid [83] and Coleman and Moré [68] provide descriptions of some methods and performance comparisons. Newsam and Ramsdell [227] show that by considering a more general class of perturbation vectors p, it is possible to evaluate the full Jacobian using no more than n_z evaluations of r (in addition to the evaluation at the point x), where n_z is the maximum number of nonzeros in each row of $J(x)$.

For some functions r with well-studied structures (those that arise from discretizations of differential operators, or those that give rise to banded Jacobians, as in the example above), optimal coloring schemes are known. For the tridiagonal Jacobian of (8.14) and its associated graph in Figure 8.1, the scheme with three colors is optimal.

APPROXIMATING THE HESSIAN

In some situations, the user may be able to provide a routine to calculate the gradient $\nabla f(x)$ but not the Hessian $\nabla^2 f(x)$. We can obtain the Hessian by applying the techniques described above for the vector function r to the gradient ∇f. By using the graph coloring techniques discussed above, sparse Hessians often can be approximated in this manner by using considerably fewer than n perturbation vectors. This approach ignores symmetry of the Hessian, and will usually produce a nonsymmetric approximation. We can recover symmetry by adding the approximation to its transpose and dividing the result by 2. Alternative differencing approaches that take symmetry of $\nabla^2 f(x)$ explicitly into account are discussed below.

Some important algorithms—most notably the Newton–CG methods described in Chapter 7—do not require knowledge of the full Hessian. Instead, each iteration requires us to supply the Hessian–vector product $\nabla^2 f(x)p$, for a given vector p. We can obtain an approximation to this matrix-vector product by appealing once again to Taylor's theorem. When second derivatives of f exist and are Lipschitz continuous near x, we have

$$\nabla f(x + \epsilon p) = \nabla f(x) + \epsilon \nabla^2 f(x) p + O(\epsilon^2), \tag{8.19}$$

so that

$$\nabla^2 f(x) p \approx \frac{\nabla f(x + \epsilon p) - \nabla f(x)}{\epsilon} \tag{8.20}$$

(see also (7.10)). The approximation error is $O(\epsilon)$, and the cost of obtaining the approximation is a single gradient evaluation at the point $x + \epsilon p$. The formula (8.20) corresponds to the forward-difference approximation (8.1). A central-difference formula like (8.7) can be derived by evaluating $\nabla f(x - \epsilon p)$ as well.

For the case in which even gradients are not available, we can use Taylor's theorem once again to derive formulae for approximating the Hessian that use only function values. The main tool is the formula (8.8): By substituting the vectors $p = \epsilon e_i$, $p = \epsilon e_j$, and $p = \epsilon(e_i + e_j)$ into this formula and combining the results appropriately, we obtain

$$\frac{\partial^2 f}{\partial x_i \partial x_j}(x) = \frac{f(x + \epsilon e_i + \epsilon e_j) - f(x + \epsilon e_i) - f(x + \epsilon e_j) + f(x)}{\epsilon^2} + O(\epsilon). \tag{8.21}$$

If we wished to approximate every element of the Hessian with this formula, then we would need to evaluate f at $x + \epsilon(e_i + e_j)$ for all possible i and j (a total of $n(n + 1)/2$ points) as well as at the n points $x + \epsilon e_i$, $i = 1, 2, \ldots, n$. If the Hessian is sparse, we can, of course, reduce this operation count by skipping the evaluation whenever we know the element $\partial^2 f / \partial x_i \partial x_j$ to be zero.

APPROXIMATING A SPARSE HESSIAN

We noted above that a Hessian approximation can be obtained by applying finite-difference Jacobian estimation techniques to the gradient ∇f, treated as a vector function. We now show how symmetry of the Hessian $\nabla^2 f$ can be used to reduce the number of perturbation vectors p needed to obtain a complete approximation, when the Hessian is sparse. The key observation is that, because of symmetry, any estimate of the element $[\nabla^2 f(x)]_{i,j} = \partial^2 f(x)/\partial x_i \partial x_j$ is also an estimate of its symmetric counterpart $[\nabla^2 f(x)]_{j,i}$.

We illustrate the point with the simple function $f : \mathbb{R}^n \to \mathbb{R}$ defined by

$$f(x) = x_1 \sum_{i=1}^{n} i^2 x_i^2. \tag{8.22}$$

It is easy to show that the Hessian $\nabla^2 f$ has the "arrowhead" structure depicted below, for the case of $n = 6$:

$$
\begin{bmatrix}
\times & \times & \times & \times & \times & \times \\
\times & \times & & & & \\
\times & & \times & & & \\
\times & & & \times & & \\
\times & & & & \times & \\
\times & & & & & \times
\end{bmatrix}.
\tag{8.23}
$$

If we were to construct the intersection graph for the function ∇f (analogous to Figure 8.1), we would find that every node is connected to every other node, for the simple reason that row 1 has a nonzero in every column. According to the rule for coloring the graph, then, we would have to assign a different color to every node, which implies that we would need to evaluate ∇f at the $n + 1$ points x and $x + \epsilon e_i$ for $i = 1, 2, \ldots, n$.

We can construct a much more efficient scheme by taking the symmetry into account. Suppose we first use the perturbation vector $p = \epsilon e_1$ to estimate the first column of $\nabla^2 f(x)$. Because of symmetry, the same estimates apply to the first *row* of $\nabla^2 f$. From (8.23), we see that all that remains is to find the diagonal elements $\nabla^2 f(x)_{22}, \nabla^2 f(x)_{33}, \ldots, \nabla^2 f(x)_{66}$. The intersection graph for these remaining elements is completely disconnected, so we can assign them all the same color and choose the corresponding perturbation vector to be

$$p = \epsilon(e_2 + e_3 + \cdots + e_6) = \epsilon(0, 1, 1, 1, 1, 1)^T. \tag{8.24}$$

Note that the second component of ∇f is not affected by the perturbations in components $3, 4, 5, 6$ of the unknown vector, while the third component of ∇f is not affected by perturbations in components $2, 4, 5, 6$ of x, and so on. As in (8.15) and (8.16), we have for each component i that

$$\nabla f(x + p)_i = \nabla f(x + \epsilon(e_2 + e_3 + \cdots + e_6))_i = \nabla f(x + \epsilon e_i)_i.$$

By applying the forward-difference formula (8.1) to each of these individual components, we then obtain

$$\frac{\partial^2 f}{\partial x_i^2}(x) \approx \frac{\nabla f(x + \epsilon e_i)_i - \nabla f(x)_i}{\epsilon} = \frac{\nabla f(x + \epsilon p)_i - \nabla f(x)_i}{\epsilon}, \quad i = 2, 3, \ldots, 6.$$

By exploiting symmetry, we have been able to estimate the entire Hessian by evaluating ∇f only at x and two other points.

Again, graph-coloring techniques can be used to choose the perturbation vectors p economically. We use the *adjacency graph* in place of the intersection graph described earlier. The adjacency graph has n nodes, with arcs connecting nodes i and k whenever $i \neq k$ and $\partial^2 f(x)/(\partial x_i \partial x_k) \neq 0$ for some x. The requirements on the coloring scheme are a little more complicated than before, however. We require not only that connected nodes have different colors, but also that any path of length 3 through the graph contain at least three colors. In other words, if there exist nodes i_1, i_2, i_3, i_4 in the graph that are connected by arcs (i_1, i_2), (i_2, i_3), and (i_3, i_4), then at least three different colors must be used in coloring these four nodes. See Coleman and Moré [69] for an explanation of this rule and for algorithms to compute valid colorings. The perturbation vectors are constructed as before: Whenever the nodes $i_1, i_2, \ldots, i_\ell$ have the same color, we set the corresponding perturbation vector to be $p = \epsilon(e_{i_1} + e_{i_2} + \cdots + e_{i_\ell})$.

8.2   AUTOMATIC DIFFERENTIATION

Automatic differentiation is the generic name for techniques that use the computational representation of a function to produce analytic values for the derivatives. Some techniques produce code for the derivatives at a general point x by manipulating the function code directly. Other techniques keep a record of the computations made during the evaluation of the function at a specific point x and then review this information to produce a set of derivatives at x.

Automatic differentiation techniques are founded on the observation that any function, no matter how complicated, is evaluated by performing a sequence of simple elementary operations involving just one or two arguments at a time. Two-argument operations include addition, multiplication, division, and the power operation a^b. Examples of single-argument operations include the trigonometric, exponential, and logarithmic functions. Another common ingredient of the various automatic differentiation tools is their use of the *chain rule*. This is the well-known rule from elementary calculus that says that if h is a function of the vector $y \in \mathbb{R}^m$, which is in turn a function of the vector $x \in \mathbb{R}^n$, we can write the derivative of h with respect to x as follows:

$$\nabla_x h(y(x)) = \sum_{i=1}^{m} \frac{\partial h}{\partial y_i} \nabla y_i(x). \tag{8.25}$$

See Appendix A for further details.

There are two basic modes of automatic differentiation: the *forward* and *reverse* modes. The difference between them can be illustrated by a simple example. We work through such an example below, and indicate how the techniques can be extended to general functions, including vector functions.

AN EXAMPLE

Consider the following function of 3 variables:

$$f(x) = (x_1 x_2 \sin x_3 + e^{x_1 x_2})/x_3. \tag{8.26}$$

Figure 8.2 shows how the evaluation of this function can be broken down into its elementary operations and also indicates the partial ordering associated with these operations. For instance, the multiplication $x_1 * x_2$ must take place prior to the exponentiation $e^{x_1 x_2}$, or else we would obtain the incorrect result $(e^{x_1})x_2$. This graph introduces the *intermediate variables* $x_4, x_5, \ldots$ that contain the results of intermediate computations; they are distinguished from the *independent variables* x_1, x_2, x_3 that appear at the left of the graph. We can express the evaluation of f in arithmetic terms as follows:

$$\begin{aligned}
x_4 &= x_1 * x_2, \\
x_5 &= \sin x_3, \\
x_6 &= e^{x_4}, \\
x_7 &= x_4 * x_5, \\
x_8 &= x_6 + x_7, \\
x_9 &= x_8/x_3.
\end{aligned} \tag{8.27}$$

The final node x_9 in Figure 8.2 contains the function value $f(x)$. In the terminology of graph theory, node i is the *parent* of node j, and node j the *child* of node i, whenever there is a directed arc from i to j. Any node can be evaluated when the values of all its parents are known, so computation flows through the graph from left to right. Flow of

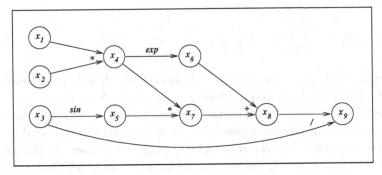

Figure 8.2 Computational graph for $f(x)$ defined in (8.26).

computation in this direction is known as a *forward sweep*. It is important to emphasize that software tools for automatic differentiation do not require the user to break down the code for evaluating the function into its elements, as in (8.27). Identification of intermediate quantities and construction of the computational graph is carried out, explicitly or implicitly, by the software tool itself.

THE FORWARD MODE

In the forward mode of automatic differentiation, we evaluate and carry forward a directional derivative of each intermediate variable x_i in a given direction $p \in \mathbb{R}^n$, simultaneously with the evaluation of x_i itself. For the three-variable example above, we use the following notation for the directional derivative for p associated with each variable:

$$D_p x_i \overset{\text{def}}{=} (\nabla x_i)^T p = \sum_{j=1}^{3} \frac{\partial x_i}{\partial x_j} p_j, \quad i = 1, 2, \ldots, 9, \tag{8.28}$$

where ∇ indicates the gradient with respect to the three independent variables. Our goal is to evaluate $D_p x_9$, which is the same as the directional derivative $\nabla f(x)^T p$. We note immediately that initial values $D_p x_i$ for the independent variables $x_i, i = 1, 2, 3$, are simply the components p_1, p_2, p_3 of p. The direction p is referred to as the *seed vector*.

As soon as the value of x_i at any node is known, we can find the corresponding value of $D_p x_i$ from the chain rule. For instance, suppose we know the values of x_4, $D_p x_4$, x_5, and $D_p x_5$, and we are about to calculate x_7 in Figure 8.2. We have that $x_7 = x_4 x_5$; that is, x_7 is a function of the two variables x_4 and x_5, which in turn are functions of x_1, x_2, x_3. By applying the rule (8.25), we have that

$$\nabla x_7 = \frac{\partial x_7}{\partial x_4} \nabla x_4 + \frac{\partial x_7}{\partial x_5} \nabla x_5 = x_5 \nabla x_4 + x_4 \nabla x_5.$$

By taking the inner product of both sides of this expression with p and applying the definition (8.28), we obtain

$$D_p x_7 = \frac{\partial x_7}{\partial x_4} D_p x_4 + \frac{\partial x_7}{\partial x_5} D_p x_5 = x_5 D_p x_4 + x_4 D_p x_5. \tag{8.29}$$

The directional derivatives $D_p x_i$ are therefore evaluated side by side with the intermediate results x_i, and at the end of the process we obtain $D_p x_9 = D_p f = \nabla f(x)^T p$.

The principle of the forward mode is straightforward enough, but what of its practical implementation and computational requirements? First, we repeat that the user does *not* need to construct the computational graph, break the computation down into elementary operations as in (8.27), or identify intermediate variables. The automatic differentiation software should perform these tasks implicitly and automatically. Nor is it necessary to store the information x_i and $D_p x_i$ for *every* node of the computation graph at once (which is just as well, since this graph can be very large for complicated functions). Once all the children of any node have been evaluated, its associated values x_i and $D_p x_i$ are not needed further and may be overwritten in storage.

The key to practical implementation is the side-by-side evaluation of x_i and $D_p x_i$. The automatic differentiation software associates a scalar $D_p w$ with any scalar w that appears in the evaluation code. Whenever w is used in an arithmetic computation, the software performs an associated operation (based on the chain rule) with the gradient vector $D_p w$. For instance, if w is combined in a division operation with another value y to produce a new value z, that is,

$$z \leftarrow \frac{w}{y},$$

we use w, z, $D_p w$, and $D_p y$ to evaluate the directional derivative $D_p z$ as follows:

$$D_p z \leftarrow \frac{1}{y} D_p w - \frac{w}{y^2} D_p y. \tag{8.30}$$

To obtain the complete gradient vector, we can carry out this procedure simultaneously for the n seed vectors $p = e_1, e_2, \ldots, e_n$. By the definition (8.28), we see that $p = e_j$ implies that $D_p f = \partial f / \partial x_j$, $j = 1, 2, \ldots, n$. We note from the example (8.30) that the additional cost of evaluating f and ∇f (over the cost of evaluating f alone) may be significant. In this example, the single division operation on w and y needed to calculate z gives rise to approximately $2n$ multiplications and n additions in the computation of the gradient elements $D_{e_j} z$, $j = 1, 2, \ldots, n$. It is difficult to obtain an exact bound on the increase in computation, since the costs of retrieving and storing the data should also be taken into account. The storage requirements may also increase by a factor as large as n, since we now have to store n additional scalars $D_{e_j} x_i$, $j = 1, 2, \ldots, n$, alongside each intermediate variable x_i. It is usually possible to make savings by observing that many of these quantities are zero, particularly in the early stages of the computation (that is, toward the left of the computational graph), so sparse data structures can be used to store the vectors $D_{e_j} x_i$, $j = 1, 2, \ldots, n$ (see [27]).

The forward mode of automatic differentiation can be implemented by means of a precompiler, which transforms function evaluation code into extended code that evaluates the derivative vectors as well. An alternative approach is to use the operator-overloading facilities available in languages such as C++ to transparently extend the data structures and operations in the manner described above.

THE REVERSE MODE

The reverse mode of automatic differentiation does not perform function and gradient evaluations concurrently. Instead, after the evaluation of f is complete, it recovers the partial derivatives of f with respect to each variable x_i—independent and intermediate variables alike—by performing a *reverse sweep* of the computational graph. At the conclusion of this process, the gradient vector ∇f can be assembled from the partial derivatives $\partial f/\partial x_i$ with respect to the independent variables $x_i, i = 1, 2, \ldots, n$.

Instead of the gradient vectors $D_p x_i$ used in the forward mode, the reverse mode associates a scalar variable $\bar{x}_i$ with each node in the graph; information about the partial derivative $\partial f/\partial x_i$ is accumulated in $\bar{x}_i$ during the reverse sweep. The $\bar{x}_i$ are sometimes called the *adjoint variables*, and we initialize their values to zero, with the exception of the rightmost node in the graph (node N, say), for which we set $\bar{x}_N = 1$. This choice makes sense because x_N contains the final function value f, so we have $\partial f/\partial x_N = 1$.

The reverse sweep makes use of the following observation, which is again based on the chain rule (8.25): For any node i, the partial derivative $\partial f/\partial x_i$ can be built up from the partial derivatives $\partial f/\partial x_j$ corresponding to its child nodes j according to the following formula:

$$\frac{\partial f}{\partial x_i} = \sum_{j \text{ a child of } i} \frac{\partial f}{\partial x_j} \frac{\partial x_j}{\partial x_i}. \tag{8.31}$$

For each node i, we add the right-hand-side term in (8.31) to $\bar{x}_i$ as soon as it becomes known; that is, we perform the operation

$$\bar{x}_i \mathrel{+}= \frac{\partial f}{\partial x_j} \frac{\partial x_j}{\partial x_i}. \tag{8.32}$$

(In this expression and the ones below, we use the arithmetic notation of the programming language C, in which $x\mathrel{+}=a$ means $x \leftarrow x + a$.) Once contributions have been received from all the child nodes of i, we have $\bar{x}_i = \partial f/\partial x_i$, so we declare node i to be "finalized." At this point, node i is ready to contribute a term to the summation for each of its parent nodes according to the formula (8.31). The process continues in this fashion until all nodes are finalized. Note that for derivative evaluation, the flow of computation in the graph is from children to parents—the opposite direction to the computation flow for function evaluation.

During the reverse sweep, we work with *numerical values*, not with formulae or computer code involving the variables x_i or the partial derivatives $\partial f/\partial x_i$. During the forward sweep—the evaluation of f—we not only calculate the values of each variable x_i, but we also calculate and store the numerical values of each partial derivative $\partial x_j/\partial x_i$. Each of these partial derivatives is associated with a particular arc of the computational graph. The numerical values of $\partial x_j/\partial x_i$ computed during the forward sweep are then used in the formula (8.32) during the reverse sweep.

We illustrate the reverse mode for the example function (8.26). In Figure 8.3 we fill in the graph of Figure 8.2 for a specific evaluation point $x = (1, 2, \pi/2)^T$, indicating the

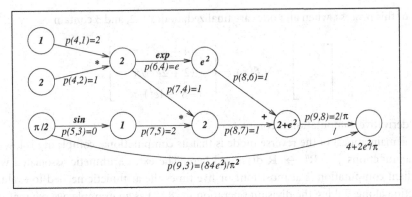

Figure 8.3 Computational graph for $f(x)$ defined in (8.26) showing numerical values of intermediate values and partial derivatives for the point $x = (1, 2, \pi/2)^T$. Notation: $p(j, i) = \partial x_j / \partial x_i$.

numerical values of the intermediate variables $x_4, x_5, \ldots, x_9$ associated with each node and the partial derivatives $\partial x_j / \partial x_i$ associated with each arc.

As mentioned above, we initialize the reverse sweep by setting all the adjoint variables $\bar{x}_i$ to zero, except for the rightmost node, for which we have $\bar{x}_9 = 1$. Since $f(x) = x_9$ and since node 9 has no children, we have $\bar{x}_9 = \partial f / \partial x_9$, and so we can immediately declare node 9 to be finalized.

Node 9 is the child of nodes 3 and 8, so we use formula (8.32) to update the values of $\bar{x}_3$ and $\bar{x}_8$ as follows:

$$\bar{x}_3 += \frac{\partial f}{\partial x_9} \frac{\partial x_9}{\partial x_3} = -\frac{2 + e^2}{(\pi/2)^2} = \frac{-8 - 4e^2}{\pi^2}, \tag{8.33a}$$

$$\bar{x}_8 += \frac{\partial f}{\partial x_9} \frac{\partial x_9}{\partial x_8} = \frac{1}{\pi/2} = \frac{2}{\pi}. \tag{8.33b}$$

Node 3 is not finalized after this operation; it still awaits a contribution from its other child, node 5. On the other hand, node 9 is the only child of node 8, so we can declare node 8 to be finalized with the value $\frac{\partial f}{\partial x_8} = 2/\pi$. We can now update the values of $\bar{x}_i$ at the two parent nodes of node 8 by applying the formula (8.32) once again; that is,

$$\bar{x}_6 += \frac{\partial f}{\partial x_8} \frac{\partial x_8}{\partial x_6} = \frac{2}{\pi};$$

$$\bar{x}_7 += \frac{\partial f}{\partial x_8} \frac{\partial x_8}{\partial x_7} = \frac{2}{\pi}.$$

At this point, nodes 6 and 7 are finalized, so we can use them to update nodes 4 and 5. At the end of this process, when all nodes are finalized, nodes 1, 2, and 3 contain

$$
\begin{bmatrix} \bar{x}_1 \\ \bar{x}_2 \\ \bar{x}_3 \end{bmatrix} = \nabla f(x) = \begin{bmatrix} (4 + 4e^2)/\pi \\ (2 + 2e^2)/\pi \\ (-8 - 4e^2)/\pi^2 \end{bmatrix},
$$

and the derivative computation is complete.

The main appeal of the reverse mode is that its computational complexity is low for the scalar functions $f : \mathbb{R}^n \to \mathbb{R}$ discussed here. The extra arithmetic associated with the gradient computation is at most four or five times the arithmetic needed to evaluate the function alone. Taking the division operation in (8.33) as an example, we see that two multiplications, a division, and an addition are required for (8.33a), while a division and an addition are required for (8.33b). This is about five times as much work as the single division involving these nodes that was performed during the forward sweep.

As we noted above, the forward mode may require up to n times more arithmetic to compute the gradient ∇f than to compute the function f alone, making it appear uncompetitive with the reverse mode. When we consider vector functions $r : \mathbb{R}^n \to \mathbb{R}^m$, the relative costs of the forward and reverse modes become more similar as m increases, as we describe in the next section.

An apparent drawback of the reverse mode is the need to store the entire computational graph, which is needed for the reverse sweep. In principle, storage of this graph is not too difficult to implement. Whenever an elementary operation is performed, we can form and store a new node containing the intermediate result, pointers to the (one or two) parent nodes, and the partial derivatives associated with these arcs. During the reverse sweep, the nodes can be read in the reverse order to that in which they were written, giving a particularly simple access pattern. The process of forming and writing the graph can be implemented as a straightforward extension to the elementary operations via operator overloading (as in ADOL-C [154]). The reverse sweep/gradient evaluation can be invoked as a simple function call.

Unfortunately, the computational graph may require a huge amount of storage. If each node can be stored in 20 bytes, then a function that requires one second of evaluation time on a 100 megaflop computer may produce a graph of up to 2 gigabytes in size. The storage requirements can be reduced, at the cost of some extra arithmetic, by performing partial forward and reverse sweeps on pieces of the computational graph, reevaluating portions of the graph as needed rather than storing the whole structure. Descriptions of this approach, sometimes known as *checkpointing*, can be found in Griewank [150] and Grimm, Pottier, and Rostaing-Schmidt [157]. An implementation of checkpointing in the context of variational data assimilation can be found in Restrepo, Leaf, and Griewank [264].

VECTOR FUNCTIONS AND PARTIAL SEPARABILITY

So far, we have looked at automatic differentiation of general scalar-valued functions $f : \mathbb{R}^n \to \mathbb{R}$. In nonlinear least-squares problems (Chapter 10) and nonlinear equations

(Chapter 11), we have to deal with vector functions $r : \mathbb{R}^n \rightarrow \mathbb{R}^m$ with m components r_j, $j = 1, 2, \ldots, m$. The rightmost column of the computational graph then consists of m nodes, none of which has any children, in place of the single node described above. The forward and reverse modes can be adapted in straightforward ways to find the Jacobian $J(x)$, the $m \times n$ matrix defined in (8.9).

Besides their applications to least-squares and nonlinear-equations problems, automatic differentiation of vector functions is a useful technique for dealing with partially separable functions. We recall that partial separability is commonly observed in large-scale optimization, and we saw in Chapter 7 that there exist efficient quasi-Newton procedures for the minimization of objective functions with this property. Since an automatic procedure for detecting the decomposition of a given function f into its partially separable representation was developed recently by Gay [118], it has become possible to exploit the efficiencies that accrue from this property without asking much information from the user.

In the simplest sense, a function f is partially separable if we can express it in the form

$$f(x) = \sum_{i=1}^{ne} f_i(x), \tag{8.34}$$

where each *element function* $f_i(\cdot)$ depends on just a few components of x. If we construct the vector function r from the partially separable components, that is,

$$r(x) = \begin{bmatrix} f_1(x) \\ f_2(x) \\ \vdots \\ f_{ne}(x) \end{bmatrix},$$

it follows from (8.34) that

$$\nabla f(x) = J(x)^T e, \tag{8.35}$$

where, as usual, $e = (1, 1, \ldots, 1)^T$. Because of the partial separability property, most columns of $J(x)$ contain just a few nonzeros. This structure makes it possible to calculate $J(x)$ efficiently by applying graph-coloring techniques, as we discuss below. The gradient $\nabla f(x)$ can then be recovered from the formula (8.35).

In constrained optimization, it is often beneficial to evaluate the objective function f and the constraint functions c_i, $i \in \mathcal{I} \cup \mathcal{E}$, simultaneously. By doing so, we can take advantage of common expressions (which show up as shared intermediate nodes in the computation graph) and thus can reduce the total workload. In this case, the vector function r can be defined as

$$r(x) = \left[\begin{array}{c} f(x) \\ \left[c_j(x) \right]_{j \in \mathcal{I} \cup \mathcal{E}} \end{array} \right].$$

An example of shared intermediate nodes was seen in Figure 8.2, where x_4 is shared during the computation of x_6 and x_7.

CALCULATING JACOBIANS OF VECTOR FUNCTIONS

The forward mode is the same for vector functions as for scalar functions. Given a seed vector p, we continue to associate quantities $D_p x_i$ with the node that calculates each intermediate variable x_i. At each of the rightmost nodes (containing r_j, $j = 1, 2, \ldots, m$), this variable contains the quantity $D_p r_j = (\nabla r_j)^T p$, $j = 1, 2, \ldots, m$. By assembling these m quantities, we obtain $J(x)p$, the product of the Jacobian and our chosen vector p. As in the case of scalar functions ($m = 1$), we can evaluate the complete Jacobian by setting $p = e_1, e_2, \ldots, e_n$ and evaluating the n quantities $D_{e_j} x_i$ simultaneously. For sparse Jacobians, we can use the coloring techniques outlined above in the context of finite-difference methods to make more intelligent and economical choices of the seed vectors p. The factor of increase in cost of arithmetic, when compared to a single evaluation of r, is about equal to the number of seed vectors used.

The key to applying the reverse mode to a vector function $r(x)$ is to choose seed vectors $q \in \mathbb{R}^m$ and apply the reverse mode to the scalar functions $r(x)^T q$. The result of this process is the vector

$$\nabla [r(x)^T q] = \nabla \left[\sum_{j=1}^{m} q_j r_j(x) \right] = J(x)^T q.$$

Instead of the Jacobian–vector product that we obtain with the forward mode, the reverse mode yields a Jacobian-transpose–vector product. The technique can be implemented by seeding the variables $\bar{x}_i$ in the m dependent nodes that contain $r_1, r_2, \ldots, r_m$, with the components $q_1, q_2, \ldots, q_m$ of the vector q. At the end of the reverse sweep, the node for independent variables $x_1, x_2, \ldots, x_n$ will contain

$$\frac{d}{dx_i} \left[r(x)^T q \right], \quad i = 1, 2, \ldots, n,$$

which are simply the components of $J(x)^T q$.

As usual, we can obtain the full Jacobian by carrying out the process above for the m unit vectors $q = e_1, e_2, \ldots, e_m$. Alternatively, for sparse Jacobians, we can apply the usual coloring techniques to find a smaller number of seed vectors q—the only difference being that the graphs and coloring strategies are defined with reference to the transpose $J(x)^T$ rather than to $J(x)$ itself. The factor of increase in the number of arithmetic operations required, in comparison to an evaluation of r alone, is no more than 5 times the number of seed vectors. (The factor of 5 is the usual overhead from the reverse mode for a scalar function.) The space required for storage of the computational graph is no greater than in the scalar case. As before, we need only store the graph topology information together with the partial derivative associated with each arc.

The forward- and reverse-mode techniques can be combined to cumulatively reveal all the elements of $J(x)$. We can choose a set of seed vectors p for the forward mode to reveal some columns of J, then perform the reverse mode with another set of seed vectors q to reveal the rows that contain the remaining elements.

Finally, we note that for some algorithms, we do not need full knowledge of the Jacobian $J(x)$. For instance, iterative methods such as the inexact Newton method for nonlinear equations (see Section 11.1) require repeated calculation of $J(x)p$ for a succession of vectors p. Each such matrix–vector product can be computed using the forward mode by using a single forward sweep, at a similar cost to evaluation of the function alone.

CALCULATING HESSIANS: FORWARD MODE

So far, we have described how the forward and reverse modes can be applied to obtain first derivatives of scalar and vector functions. We now outline extensions of these techniques to the computation of the Hessian $\nabla^2 f$ of a scalar function f, and evaluation of the Hessian–vector product $\nabla^2 f(x)p$ for a given vector p.

Recall that the forward mode makes use of the quantities $D_p x_i$, each of which stores $(\nabla x_i)^T p$ for each node i in the computational graph and a given vector p. For a given *pair* of seed vectors p and q (both in $\mathbb{R}^n$) we now define another scalar quantity by

$$D_{pq} x_i = p^T (\nabla^2 x_i) q, \tag{8.36}$$

for each node i in the computational graph. We can evaluate these quantities during the forward sweep through the graph, alongside the function values x_i and the first-derivative values $D_p x_i$. The initial values of D_{pq} at the independent variable nodes x_i, $i = 1, 2 \ldots, n$, will be 0, since the second derivatives of x_i are zero at each of these nodes. When the forward sweep is complete, the value of $D_{pq} x_i$ in the rightmost node of the graph will be $p^T \nabla^2 f(x) q$.

The formulae for transformation of the $D_{pq} x_i$ variables during the forward sweep can once again be derived from the chain rule. For instance, if x_i is obtained by adding the values at its two parent nodes, $x_i = x_j + x_k$, the corresponding accumulation operations on $D_p x_i$ and $D_{pq} x_i$ are as follows:

$$D_p x_i = D_p x_j + D_p x_k, \qquad D_{pq} x_i = D_{pq} x_j + D_{pq} x_k. \tag{8.37}$$

The other binary operations $-, \times, /$ are handled similarly. If x_i is obtained by applying the unitary transformation L to x_j, we have

$$x_i = L(x_j), \tag{8.38a}$$

$$D_p x_i = L'(x_j)(D_p x_j), \tag{8.38b}$$

$$D_{pq} x_i = L''(x_j)(D_p x_j)(D_q x_j) + L'(x_j)D_{pq} x_j. \tag{8.38c}$$

We see in (8.38c) that computation of $D_{pq}x_i$ can rely on the first-derivative quantities $D_p x_i$ and $D_q x_i$, so both these quantities must be accumulated during the forward sweep as well.

We could compute a general dense Hessian by choosing the pairs (p, q) to be all possible pairs of unit vectors (e_j, e_k), for $j = 1, 2, \ldots, n$ and $k = 1, 2, \ldots, j$, a total of $n(n + 1)/2$ vector pairs. (Note that we need only evaluate the lower triangle of $\nabla^2 f(x)$, because of symmetry.) When we know the sparsity structure of $\nabla^2 f(x)$, we need evaluate $D_{e_j e_k} x_i$ only for the pairs (e_j, e_k) for which the (j, k) component of $\nabla^2 f(x)$ is possibly nonzero.

The total increase factor for the number of arithmetic operations, compared with the amount of arithmetic to evaluate f alone, is a small multiple of $1 + n + N_z(\nabla^2 f)$, where $N_z(\nabla^2 f)$ is the number of elements of $\nabla^2 f$ that we choose to evaluate. This number reflects the evaluation of the quantities x_i, $D_{e_j} x_i$ $(j = 1, 2, \ldots, n)$, and $D_{e_j e_k} x_i$ for the $N_z(\nabla^2 f)$ vector pairs (e_j, e_k). The "small multiple" results from the fact that the update operations for $D_p x_i$ and $D_{pq} x_i$ may require a few times more operations than the update operation for x_i alone; see, for example, (8.38). One storage location per node of the graph is required for each of the $1 + n + N_z(\nabla^2 f)$ quantities that are accumulated, but recall that storage of node i can be overwritten once all its children have been evaluated.

When we do not need the complete Hessian, but only a matrix–vector product involving the Hessian (as in the Newton–CG algorithm of Chapter 7), the amount of arithmetic is, of course, smaller. Given a vector $q \in \mathbb{R}^n$, we use the techniques above to compute the first-derivative quantities $D_{e_1} x_i, \ldots D_{e_n} x_i$ and $D_q x_i$, as well as the second-derivative quantities $D_{e_1 q} x_i, \ldots, D_{e_n q} x_i$, during the forward sweep. The final node will contain the quantities

$$e_j^T \left(\nabla^2 f(x) \right) q = \left[\nabla^2 f(x) q \right]_j, \quad j = 1, 2, \ldots, n,$$

which are the components of the vector $\nabla^2 f(x)q$. Since $2n + 1$ quantities in addition to x_i are being accumulated during the forward sweep, the increase factor in the number of arithmetic operations increases by a small multiple of $2n$.

An alternative technique for evaluating sparse Hessians is based on the forward-mode propagation of first and second derivatives of *univariate* functions. To motivate this approach, note that the (i, j) element of the Hessian can be expressed as follows:

$$[\nabla^2 f(x)]_{ij} = e_i^T \nabla^2 f(x) e_j$$

$$= \frac{1}{2} \left[(e_i + e_j)^T \nabla^2 f(x)(e_i + e_j) - e_i^T \nabla^2 f(x) e_i - e_j^T \nabla^2 f(x) e_j \right]. \tag{8.39}$$

We can use this interpolation formula to evaluate $[\nabla^2 f(x)]_{ij}$, provided that the second derivatives $D_{pp} x_k$, for $p = e_i$, $p = e_j$, $p = e_i + e_j$, and all nodes x_k, have been evaluated during the forward sweep through the computational graph. In fact, we can evaluate all the nonzero elements of the Hessian, provided that we use the forward mode to evaluate $D_p x_k$ and $D_{pp} x_k$ for a selection of vectors p of the form $e_i + e_j$, where i and j are both indices in $\{1, 2, \ldots, n\}$, possibly with $i = j$.

One advantage of this approach is that it is no longer necessary to propagate "cross terms" of the form $D_{pq} x_k$ for $p \neq q$ (see, for example, (8.37) and (8.38c)). The propagation formulae therefore simplify somewhat. Each $D_{pp} x_k$ is a function of x_ℓ, $D_p x_\ell$, and $D_{pp} x_\ell$ for all parent nodes ℓ of node k.

Note, too, that if we define the univariate function ψ by

$$\psi(t) = f(x + tp), \tag{8.40}$$

then the values of $D_p f$ and $D_{pp} f$, which emerge at the completion of the forward sweep, are simply the first two derivatives of ψ evaluated at $t = 0$; that is,

$$D_p f = p^T \nabla f(x) = \psi'(t)|_{t=0}, \quad D_{pp} f = p^T \nabla^2 f(x) p = \psi''(t)|_{t=0}.$$

Extension of this technique to third, fourth, and higher derivatives is possible. Interpolation formulae analogous to (8.39) can be used in conjunction with higher derivatives of the univariate functions ψ defined in (8.40), again for a suitably chosen set of vectors p, where each p is made up of a sum of unit vectors e_i. For details, see Bischof, Corliss, and Griewank [26].

CALCULATING HESSIANS: REVERSE MODE

We can also devise schemes based on the reverse mode for calculating Hessian–vector products $\nabla^2 f(x) q$, or the full Hessian $\nabla^2 f(x)$. A scheme for obtaining $\nabla^2 f(x) q$ proceeds as follows. We start by using the forward mode to evaluate both f and $\nabla f(x)^T q$, by accumulating the two variables x_i and $D_q x_i$ during the forward sweep in the manner described above. We then apply the reverse mode in the normal fashion to the computed function $\nabla f(x)^T q$. At the end of the reverse sweep, the nodes $i = 1, 2, \ldots, n$ of the computational graph that correspond to the independent variables will contain

$$\frac{\partial}{\partial x_i} (\nabla f(x)^T q) = [\nabla^2 f(x) q]_i, \quad i = 1, 2, \ldots, n.$$

The number of arithmetic operations required to obtain $\nabla^2 f(x)q$ by this procedure increases by only a modest factor, independent of n, over the evaluation of f alone. By the usual analysis for the forward mode, we see that the computation of f and $\nabla f(x)^T q$ jointly requires a small multiple of the operation count for f alone, while the reverse sweep introduces a further factor of at most 5. The total increase factor is approximately 12 over the evaluation of f alone. If the entire Hessian $\nabla^2 f(x)$ is required, we could apply the procedure just described with $q = e_1, e_2, \ldots, e_n$. This approach would introduce an additional factor of n into the operation count, leading to an increase of at most $12n$ over the cost of f alone.

Once again, when the Hessian is sparse with known structure, we may be able to use graph-coloring techniques to evaluate this entire matrix using many fewer than n seed vectors. The choices of q are similar to those used for finite-difference evaluation of the Hessian, described above. The increase in operation count over evaluating f alone is a multiple of up to $12N_c(\nabla^2 f)$, where N_c is the number of seed vectors q used in calculating $\nabla^2 f$.

CURRENT LIMITATIONS

The current generation of automatic differentiation tools has proved its worth through successful application to some large and difficult design optimization problems. However, these tools can run into difficulties with some commonly used programming constructs and some implementations of computer arithmetic. As an example, if the evaluation of $f(x)$ depends on the solution of a partial differential equation (PDE), then the computed value of f may contain truncation error arising from the finite-difference or the finite-element technique that is used to solve the PDE numerically. That is, we have $\hat{f}(x) = f(x) + \tau(x)$, where $\hat{f}(\cdot)$ is the computed value of $f(\cdot)$ and $\tau(\cdot)$ is the truncation error. Though $|\tau(x)|$ is usually small, its derivative $\tau'(x)$ may not be, so the error in the computed derivative $\hat{f}'(x)$ is potentially large. (The finite-difference approximation techniques discussed in Section 8.1 experience the same difficulty.) Similar problems arise when the computer uses piecewise rational functions to approximate trigonometric functions.

Another source of potential difficulty is the presence of branching in the code to improve the speed or accuracy of function evaluation in certain domains. A pathological example is provided by the linear function $f(x) = x - 1$. If we used the following (perverse, but valid) piece of code to evaluate this function,

    `if` $(x = 1.0)$ `then` $f = 0.0$ `else` $f = x - 1.0$,

then by applying automatic differentiation to this procedure we would obtain the derivative value $f'(1) = 0$. For a discussion of such issues and an approach to dealing with them, see Griewank [151, 152].

In conclusion, automatic differentiation should be regarded as a set of increasingly sophisticated techniques that enhances optimization algorithms, allowing them to be applied more widely to practical problems involving complicated functions. By providing sensitivity information, it helps the modeler to extract more information from the results of the computation. Automatic differentiation should not be regarded as a panacea that absolves the user altogether from the responsibility of thinking about derivative calculations.

NOTES AND REFERENCES

A comprehensive and authoritative reference on automatic differentiation is the book of Griewank [152]. The web site www.autodiff.org contains a wealth of current information about theory, software, and applications. A number of edited collections of papers on automatic differentiation have appeared since 1991; see Griewank and Corliss [153], Berz et al. [20], and Bücker et al. [40]. An historical paper of note is Corliss and Rall [78], which includes an extensive bibliography. Software tool development in automatic differentiation makes use not only of forward and reverse modes but also includes "mixed modes" and "cross-country algorithms" that combine the two approaches; see for example Naumann [222].

The field of automatic differentiation grew considerably during the 1990s, and and a number of good software tools appeared. These included ADIFOR [25] and ADIC [28], and ADOL-C [154]. Tools developed in more recent years include TAPENADE, which accepts Fortran code through a web server and returns differentiated code; TAF, a commercial tool that also performs source-to-source automatic differentiation of Fortran codes; OpenAD, which works with Fortran, C, and C++; and TOMLAB/MAD, which works with MATLAB code.

The technique for calculating the gradient of a partially separable function was described by Bischof et al. [24], whereas the computation of the Hessian matrix has been considered by several authors; see, for example, Gay [118].

The work of Coleman and Moré [69] on efficient estimation of Hessians was predated by Powell and Toint [261], who did not use the language of graph coloring but nevertheless devised highly effective schemes. Software for estimating sparse Hessians and Jacobians is described by Coleman, Garbow, and Moré [66, 67]. The recent paper of Gebremedhin, Manne, and Pothen [120] contains a comprehensive discussion of the application of graph coloring to both finite difference and automatic differentiation techniques.

✐ EXERCISES

✐ **8.1** Show that a suitable value for the perturbation ϵ in the central-difference formula is $\epsilon = \mathbf{u}^{1/3}$, and that the accuracy achievable by this formula when the values of f contain roundoff errors of size $\mathbf{u}$ is approximately $\mathbf{u}^{2/3}$. (Use similar assumptions to the ones used to derive the estimate (8.6) for the forward-difference formula.)

✐ **8.2** Derive a central-difference analogue of the Hessian–vector approximation formula (8.20).

8.3 Verify the formula (8.21) for approximating an element of the Hessian using only function values.

8.4 Verify that if the Hessian of a function f has nonzero diagonal elements, then its adjacency graph is a subgraph of the intersection graph for ∇f. In other words, show that any arc in the adjacency graph also belongs to the intersection graph.

8.5 Draw the adjacency graph for the function f defined by (8.22). Show that the coloring scheme in which node 1 has one color while nodes $2, 3, \ldots, n$ have another color is valid. Draw the intersection graph for ∇f.

8.6 Construct the adjacency graph for the function whose Hessian has the nonzero structure

$$
\begin{bmatrix}
\times & \times & \times & \times & & & \\
\times & \times & \times & & \times & & \\
\times & \times & \times & & & \times & \\
\times & & & & \times & & \\
& & \times & & & & \times \\
& & & \times & & & \times
\end{bmatrix},
$$

and find a valid coloring scheme with just four colors.

8.7 Trace the computations performed in the forward mode for the function $f(x)$ in (8.26), expressing the intermediate derivatives ∇x_i, $i = 4, 5, \ldots, 9$ in terms of quantities available at their parent nodes and then in terms of the independent variables x_1, x_2, x_3.

8.8 Formula (8.30) showed the gradient operations associated with scalar division. Derive similar formulae for the following operations:

$$
\begin{aligned}
(s, t) &\to s + t && \text{addition;} \\
t &\to e^t && \text{exponentiation;} \\
t &\to \tan(t) && \text{tangent;} \\
(s, t) &\to s^t.
\end{aligned}
$$

8.9 By calculating the partial derivatives $\partial x_j / \partial x_i$ for the function (8.26) from the expressions (8.27), verify the numerical values for the arcs in Figure 8.3 for the evaluation point $x = (1, 2, \pi/2)^T$. Work through the remaining details of the reverse sweep process, indicating the order in which the nodes become finalized.

✏ **8.10** Using (8.33) as a guide, describe the reverse sweep operations corresponding to the following elementary operations in the forward sweep:

$$x_k \leftarrow x_i x_j \qquad \text{multiplication};$$
$$x_k \leftarrow \cos(x_i) \qquad \text{cosine}.$$

In each case, compare the arithmetic workload in the reverse sweep to the workload required for the forward sweep.

✏ **8.11** Define formulae similar to (8.37) for accumulating the first derivatives $D_p x_i$ and the second derivatives $D_{pq} x_i$ when x_i is obtained from the following three binary operations: $x_i = x_j - x_k$, $x_i = x_j x_k$, and $x_i = x_j / x_k$.

✏ **8.12** By using the definitions (8.28) of $D_p x_i$ and (8.36) of $D_{pq} x_i$, verify the differentiation formulae (8.38) for the unitary operation $x_i = L(x_j)$.

✏ **8.13** Let $a \in \mathbb{R}^n$ be a fixed vector and define f as $f(x) = \frac{1}{2}\left(x^T x + \left(a^T x\right)^2\right)$. Count the number of operations needed to evaluate f, ∇f, $\nabla^2 f$, and the Hessian–vector product $\nabla^2 f(x) p$ for an arbitrary vector p.

CHAPTER **9**

Derivative-Free Optimization

Many practical applications require the optimization of functions whose derivatives are not available. Problems of this kind can be solved, in principle, by approximating the gradient (and possibly the Hessian) using finite differences (see Chapter 8), and using these approximate gradients within the algorithms described in earlier chapters. Even though this finite-difference approach is effective in some applications, it cannot be regarded a general-purpose technique for derivative-free optimization because the number of function evaluations required can be excessive and the approach can be unreliable in the presence of noise. (For the purposes of this chapter we define *noise* to be inaccuracy in the function evaluation.) Because of these shortcomings, various algorithms have been developed that do not attempt to approximate the gradient. Rather, they use the function values at a set of sample points to determine a new iterate by some other means.

Derivative-free optimization (DFO) algorithms differ in the way they use the sampled function values to determine the new iterate. One class of methods constructs a linear or quadratic model of the objective function and defines the next iterate by seeking to minimize this model inside a trust region. We pay particular attention to these model-based approaches because they are related to the unconstrained minimization methods described in earlier chapters. Other widely used DFO methods include the simplex-reflection method of Nelder and Mead, pattern-search methods, conjugate-direction methods, and simulated annealing. In this chapter we briefly discuss these methods, with the exception of simulated annealing, which is a nondeterministic approach and has little in common with the other techniques discussed in this book.

Derivative-free optimization methods are not as well developed as gradient-based methods; current algorithms are effective only for small problems. Although most DFO methods have been adapted to handle simple types of constraints, such as bounds, the efficient treatment of general constraints is still the subject of investigation. Consequently, we limit our discussion to the unconstrained optimization problem

$$\min_{x \in R^n} f(x). \tag{9.1}$$

Problems in which derivatives are not available arise often in practice. The evaluation of $f(x)$ can, for example, be the result of an experimental measurement or a stochastic simulation, with the underlying analytic form of f unknown. Even if the objective function f is known in analytic form, coding its derivatives may be time consuming or impractical. Automatic differentiation tools (Chapter 8) may not be applicable if $f(x)$ is provided only in the form of binary computer code. Even when the source code is available, these tools cannot be applied if the code is written in a combination of languages.

Methods for derivative-free optimization are often used (with mixed success) to minimize problems with nondifferentiable functions or to try to locate the global minimizer of a function. Since we do not treat nonsmooth optimization or global optimization in this book, we will restrict our attention to smooth problems in which f has a continuous derivative. We do, however, discuss the effects of noise in Sections 9.1 and 9.6.

9.1 FINITE DIFFERENCES AND NOISE

As mentioned above, an obvious DFO approach is to estimate the gradient by using finite differences and then employ a gradient-based method. This approach is sometimes successful and should always be considered, but the finite-difference estimates can be inaccurate when the objective function contains noise. We quantify the effect of noise in this section.

Noise can arise in function evaluations for various reasons. If $f(x)$ depends on a stochastic simulation, there will be a random error in the evaluated function because of the finite number of trials in the simulation. When a differential equation solver or some other complex numerical procedure is needed to calculate f, small but nonzero error tolerances that are used during the calculations will produce noise in the value of f.

In many applications, then, the objective function f has the form

$$f(x) = h(x) + \phi(x), \tag{9.2}$$

where h is a smooth function and ϕ represents the noise. Note that we have written ϕ to be a function of x but in practice it need not be. For instance, if the evaluation of f depends on a simulation, the value of ϕ will generally differ at each evaluation, even at the same x. The form (9.2) is, however, useful for illustrating some of the difficulties caused by noise in gradient estimates and for developing algorithms for derivative-free optimization.

Given a difference interval ϵ, recall that the centered finite-difference approximation (8.7) to the gradient of f at x is defined as follows:

$$\nabla_\epsilon f(x) = \left[\frac{f(x + \epsilon e_i) - f(x - \epsilon e_i)}{2\epsilon} \right]_{i=1,2,\ldots,n}, \tag{9.3}$$

where e_i is the ith unit vector (the vector whose only nonzero element is a 1 in the ith position). We wish to relate $\nabla_\epsilon f(x)$ to the gradient of the underlying smooth function $h(x)$, as a function of ϵ and the noise level. For this purpose we define the noise level η to be the largest value of ϕ in a box of edge length 2ϵ centered at x, that is,

$$\eta(x; \epsilon) = \sup_{\|z-x\|_\infty \le \epsilon} |\phi(z)|. \tag{9.4}$$

By applying to the central difference formula (9.3) the argument that led to (8.5), we can establish the following result.

Lemma 9.1.

Suppose that $\nabla^2 h$ is Lipschitz continuous in a neighborhood of the box $\{z \mid \|z - x\|_\infty \le \epsilon\}$ with Lipschitz constant L_h. Then we have

$$\|\nabla_\epsilon f(x) - \nabla h(x)\|_\infty \le L_h \epsilon^2 + \frac{\eta(x; \epsilon)}{\epsilon}. \tag{9.5}$$

Thus the error in the approximation (9.3) comes from both the intrinsic finite difference approximation error (the $O(\epsilon^2)$ term) and the noise (the $\eta(x; \epsilon)/\epsilon$ term). If the noise dominates the difference interval ϵ, we cannot expect any accuracy at all in $\nabla_\epsilon f(x)$, so it will only be pure luck if $-\nabla_\epsilon f(x)$ turns out to be a direction of descent for f.

Instead of computing a tight cluster of function values around the current iterate, as required by a finite-difference approximation to the gradient, it may be preferable to separate these points more widely and use them to construct a model of the objective function. This approach, which we consider in the next section and in Section 9.6, may be more robust to the presence of noise.

9.2 MODEL-BASED METHODS

Some of the most effective algorithms for unconstrained optimization described in the previous chapters compute steps by minimizing a quadratic model of the objective function f. The model is formed by using function and derivative information at the current iterate. When derivatives are not available, we may define the model m_k as the quadratic function that interpolates f at a set of appropriately chosen sample points. Since such a model is usually nonconvex, the model-based methods discussed in this chapter use a trust-region approach to compute the step.

Suppose that at the current iterate x_k we have a set of sample points $Y = \{y^1, y^2, \ldots, y^q\}$, with $y^i \in \mathbb{R}^n$, $i = 1, 2, \ldots, q$. We assume that x_k is an element of this set and that no point in Y has a lower function value than x_k. We wish to construct a quadratic model of the form

$$m_k(x_k + p) = c + g^T p + \tfrac{1}{2} p^T G p. \tag{9.6}$$

We cannot define $g = \nabla f(x_k)$ and $G = \nabla^2 f(x_k)$ because these derivatives are not available. Instead, we determine the scalar c, the vector $g \in R^n$, and the symmetric matrix $G \in R^{n \times n}$ by imposing the interpolation conditions

$$m_k(y^l) = f(y^l), \qquad l = 1, 2, \ldots, q. \tag{9.7}$$

Since there are $\tfrac{1}{2}(n + 1)(n + 2)$ coefficients in the model (9.6) (that is, the components of c, g, and G, taking into account the symmetry of G), the interpolation conditions (9.7) determine m_k uniquely only if

$$q = \tfrac{1}{2}(n + 1)(n + 2). \tag{9.8}$$

In this case, (9.7) can be written as a square linear system of equations in the coefficients of the model. If we choose the interpolation points $y^1, y^2, \ldots, y^q$ so that this linear system is nonsingular, the model m_k will be uniquely determined.

Once m_k has been formed, we compute a step p by approximately solving the trust-region subproblem

$$\min_p m_k(x_k + p), \qquad \text{subject to} \quad \|p\|_2 \le \Delta, \tag{9.9}$$

for some trust-region radius $\Delta > 0$. We can use one of the techniques described in Chapter 4 to solve this subproblem. If $x_k + p$ gives a sufficient reduction in the objective function, the new iterate is defined as $x_{k+1} = x_k + p$, the trust region radius Δ is updated, and a new iteration commences. Otherwise the step is rejected, and the interpolation set Y may be improved or the trust region shrunk.

To reduce the cost of the algorithm, we update the model m_k at every iteration, rather than recomputing it from scratch. In practice, we choose a convenient basis for the space of quadratic polynomials, the most common choices being Lagrange and Newton polynomials. The properties of these bases can be used both to measure appropriateness of the sample set Y and to change this set if necessary. A complete algorithm that treats all these issues effectively is far more complicated than the quasi-Newton methods discussed in Chapter 6. Consequently, we will provide only a broad outline of model-based DFO methods.

As is common in trust-region algorithms, the step-acceptance and trust-region update strategies are based on the ratio between the actual reduction in the function and the reduction predicted by the model, that is,

$$\rho = \frac{f(x_k) - f(x_k^+)}{m_k(x_k) - m_k(x_k^+)}, \tag{9.10}$$

where x_k^+ denotes the trial point. Throughout this section, the integer q is defined by (9.8).

Algorithm 9.1 (Model-Based Derivative-Free Method).
    Choose an interpolation set $Y = \{y^1, y^2, \ldots, y^q\}$ such that the linear system defined by (9.7) is nonsingular, and select x_0 as a point in this set such that $f(x_0) \le f(y^i)$ for all $y^i \in Y$. Choose an initial trust region radius Δ_0, a constant $\eta \in (0, 1)$, and set $k \leftarrow 0$.

    repeat until a convergence test is satisfied:
        Form the quadratic model $m_k(x_k + p)$ that satisfies the interpolation
            conditions (9.7);
        Compute a step p by approximately solving subproblem (9.9);
        Define the trial point as $x_k^+ = x_k + p$;
        Compute the ratio ρ defined by (9.10);
        if $\rho \ge \eta$
            Replace an element of Y by x_k^+;
            Choose $\Delta_{k+1} \ge \Delta_k$;
            Set $x_{k+1} \leftarrow x_k^+$;
            Set $k \leftarrow k + 1$ and go to the next iteration;
        else if the set Y need not be improved
            Choose $\Delta_{k+1} < \Delta_k$;
            Set $x_{k+1} \leftarrow x_k$;
            Set $k \leftarrow k + 1$ and go to the next iteration;
        end (if)

Invoke a geometry-improving procedure to update Y:
      at least one of the points in Y is replaced by some other point,
      with the goal of improving the conditioning of (9.7);
Set $\Delta_{k+1} \leftarrow \Delta_k$;
Choose $\hat{x}$ as an element in Y with lowest function value;
Set $x_k^+ \leftarrow \hat{x}$ and recompute ρ by (9.10);
if $\rho \geq \eta$
      Set $x_{k+1} \leftarrow x_k^+$;
else
      Set $x_{k+1} \leftarrow x_k$;
end (if)
      Set $k \leftarrow k + 1$;
end (repeat)

The case of $\rho \geq \eta$, in which we obtain sufficient reduction in the merit function, is the simplest. In this case we always accept the trial point x_k^+ as the new iterate, include x_k^+ in Y, and remove an element from Y.

When sufficient reduction is not achieved ($\rho < \eta$), we look at two possible causes: inadequacy of the interpolation set Y and a trust region that is too large. The first cause can arise when the iterates become restricted to a low-dimensional surface of $\mathbb{R}^n$ that does not contain the solution. The algorithm could then be converging to a minimizer in this subset. Behavior such as this can be detected by monitoring the conditioning of the linear system defined by the interpolation conditions (9.7). If the condition number is too high, we change Y to improve it, typically by replacing one element of Y with a new element so as to move the interpolation system (9.7) as far away from singularity as possible. If Y seems adequate, we simply decrease the trust region radius Δ, as is done in the methods of Chapter 4.

A good initial choice for Y is given by the vertices and the midpoints of the edges of a simplex in $\mathbb{R}^n$.

The use of quadratic models limits the size of problems that can be solved in practice. Performing $O(n^2)$ function evaluations just to start the algorithm is onerous, even for moderate values of n (say, $n = 50$). In addition, the cost of the iteration is high. Even by updating the model m_k at every iteration, rather than recomputing it from scratch, the number of operations required to construct m_k and compute a step is $O(n^4)$ [257].

To alleviate these drawbacks, we can replace the quadratic model by a linear model in which the matrix G in (9.6) is set to zero. Since such a model contains only $n+1$ parameters, we need to retain only $n+1$ interpolation points in the set Y, and the cost of each iteration is $O(n^3)$. Algorithm 9.1 can be applied with little modification when the model is linear, but it is not rapidly convergent because linear models cannot represent curvature of the problem. Therefore, some model-based algorithms start with $n + 1$ initial points and compute steps using a linear model, but after $q = \frac{1}{2}(n+1)(n+2)$ function values become available, they switch to using quadratic models.

INTERPOLATION AND POLYNOMIAL BASES

We now consider in more detail how to form a model of the objective function using interpolation techniques. We begin by considering a linear model of the form

$$m_k(x_k + p) = f(x_k) + g^T p. \tag{9.11}$$

To determine the vector $g \in \mathbb{R}^n$, we impose the interpolation conditions $m_k(y^l) = f(y^l)$, $l = 1, 2, \ldots, n$, which can be written as

$$(s^l)^T g = f(y^l) - f(x_k), \qquad l = 1, 2, \ldots, n, \tag{9.12}$$

where

$$s^l = y^l - x_k, \qquad l = 1, 2, \ldots, n. \tag{9.13}$$

Conditions (9.12) represent a linear system of equations in which the rows of the coefficient matrix are given by the vectors $(s^l)^T$. It follows that the model (9.11) is determined uniquely by (9.12) if and only if the interpolation points $\{y^1, y^2, \ldots, y^n\}$ are such that the set $\{s^l : l = 1, 2, \ldots, n\}$ is linearly independent. If this condition holds, the simplex formed by the points $x_k, y^1, y^2, \ldots, y^n$ is said to be *nondegenerate*.

Let us now consider how to construct a quadratic model of the form (9.6), with $f = f(x_k)$. We rewrite the model as

$$m_k(x_k + p) = f(x_k) + g^T p + \sum_{i<j} G_{ij} p_i p_j + \frac{1}{2} \sum_i G_{ii} p_i^2 \tag{9.14}$$

$$\stackrel{\text{def}}{=} f(x_k) + \hat{g}^T \hat{p}, \tag{9.15}$$

where we have collected the elements of g and G in the $(q-1)$-vector of unknowns

$$\hat{g} \equiv \left(g^T, \{G_{ij}\}_{i<j}, \left\{ \frac{1}{\sqrt{2}} G_{ii} \right\} \right)^T, \tag{9.16}$$

and where the $(q-1)$-vector $\hat{p}$ is given by

$$\hat{p} \equiv \left(p^T, \{p_i p_j\}_{i<j}, \left\{ \frac{1}{\sqrt{2}} p_i^2 \right\} \right)^T.$$

The model (9.15) has the same form as (9.11), and the determination of the vector of unknown coefficients $\hat{g}$ can be done as in the linear case.

Multivariate quadratic functions can be represented in various ways. The *monomial basis* (9.14) has the advantage that known structure in the Hessian can be imposed easily by setting appropriate elements in G to zero. Other bases are, however, more convenient when one is developing mechanisms for avoiding singularity of the system (9.7).

We denote by $\{\phi_i(\cdot)\}_{i=1}^q$ a basis for the linear space of n-dimensional quadratic functions. The function (9.6) can therefore be expressed as

$$m_k(x) = \sum_{i=1}^{q} \alpha_i \phi_i(x),$$

for some coefficients α_i. The interpolation set $Y = \{y^1, y^2, \ldots, y^q\}$ determines the coefficients α_i uniquely if the determinant

$$\delta(Y) \stackrel{\text{def}}{=} \det \begin{pmatrix} \phi_1(y^1) & \cdots & \phi_1(y^q) \\ \vdots & & \vdots \\ \phi_q(y^1) & \cdots & \phi_q(y^q) \end{pmatrix} \tag{9.17}$$

is nonzero.

As model-based algorithms iterate, the determinant $\delta(Y)$ may approach zero, leading to numerical difficulties or even failure. Several algorithms therefore contain a mechanism for keeping the interpolation points well placed. We now describe one of those mechanisms.

UPDATING THE INTERPOLATION SET

Rather than waiting until the determinant $\delta(Y)$ becomes smaller than a threshold, we may invoke a geometry-improving procedure whenever a trial point does not provide sufficient decrease in f. The goal in this case is to replace one of the interpolation points so that the determinant (9.17) increases in magnitude. To guide us in this exchange, we use the following property of $\delta(Y)$, which we state in terms of Lagrange functions.

For every $y \in Y$, we define the Lagrangian function $L(\cdot, y)$ to be a polynomial of degree at most 2 such that $L(y, y) = 1$ and $L(\hat{y}, y) = 0$ for $\hat{y} \neq y$, $\hat{y} \in Y$. Suppose that the set Y is updated by removing a point y_- and replacing it by some other point y_+, to give the new set Y^+. One can show that (after a suitable normalization and given certain conditions [256])

$$|\delta(Y^+)| \leq |L(y_+, y_-)| \, |\delta(Y)|. \tag{9.18}$$

Algorithm 9.1 can make good use of this inequality to update the interpolation set.

Consider first the case in which trial point x^+ provides sufficient reduction in the objective function ($\rho \geq \eta$). We include x^+ in Y and remove another point y_- from Y.

Motivated by (9.18), we select the outgoing point as follows:

$$y_- = \arg \max_{y \in Y} |L(x^+, y)|.$$

Next, let us consider the case in which the reduction in f is not sufficient ($\rho < \eta$). We first determine whether the set Y should be improved, and for this purpose we use the following rule. We consider Y to be adequate at the current iterate x_k if for all $y^i \in Y$ such that $\|x_k - y^i\| \leq \Delta$ we have that $|\delta(Y)|$ cannot be doubled by replacing one of these interpolation points y^i with any point y inside the trust region. If Y is adequate but the reduction in f was not sufficient, we decrease the trust-region radius and begin a new iteration.

If Y is inadequate, the geometry-improving mechanism is invoked. We choose a point $y_- \in Y$ and replace it by some other point y^+ that is chosen solely with the objective of improving the determinant (9.17). For every point $y^i \in Y$, we define its potential replacement y_r^i as

$$y_r^i = \arg \max_{\|y - x_k\| \leq \Delta} |L(y, y^i)|.$$

The outgoing point y_- is selected as the point for which $|L(y_r^i, y^i)|$ is maximized over all indices $y^i \in Y$.

Implementing these rules efficiently in practice is not simple, and one must also consider several possible difficulties we have not discussed; see [76]. Strategies for improving the position of the interpolation set are the subject of ongoing investigation and new developments are likely in the coming years.

A METHOD BASED ON MINIMUM-CHANGE UPDATING

We now consider a method that be viewed as an extension of the quasi-Newton approach discussed in Chapter 6. The method uses quadratic models but requires only $O(n^3)$ operations per iteration, substantially fewer than the $O(n^4)$ operations required by the methods described above. To achieve this economy, the method retains only $O(n)$ points for the interpolation conditions (9.7) and absorbs the remaining degrees of freedom in the model (9.6) by requiring that the Hessian of the model change as little as possible from one iteration to the next. This least-change property is one of the key ingredients in quasi-Newton methods, the other ingredient being the requirement that the model interpolate the gradient ∇f at the two most recent points. The method we describe now combines the least-change property with interpolation of function values.

At the kth iteration of the algorithm, a new quadratic model m_{k+1} of the form (9.6) is constructed after taking a step from x_k to x_{k+1}. The coefficients $f_{k+1}, g_{k+1}, G_{k+1}$ of the model m_{k+1} are determined as the solution of the problem

$$\min_{f,g,G} \quad \|G - G_k\|_F^2 \tag{9.19a}$$

$$\text{subject to} \quad G \text{ symmetric}$$

$$m(y^l) = f(y^l) \qquad l = 1, 2, \ldots, \hat{q}, \tag{9.19b}$$

where $\| \cdot \|_F$ denotes the Frobenius norm (see (A.9)), G_k is the Hessian of the previous model m_k, and $\hat{q}$ is an integer comparable to n. One can show that the integer $\hat{q}$ must be chosen larger than $n+1$ to guarantee that G_{k+1} is not equal to G_k. An appropriate value in practice is $\hat{q} = 2n + 1$; for this choice the number of interpolation points is roughly twice that used for linear models.

Problem (9.19) is an equality-constrained quadratic program whose KKT conditions can be expressed as a system of equations. Once the model m_{k+1} is determined, we compute a new step by solving a trust-region problem of the form (9.9). In this approach, too, it is necessary to ensure that the geometry of the interpolation set Y is adequate. We therefore impose two minimum requirements. First, the set Y should be such that the equations (9.19b) can be satisfied for any right-hand side. Second, the points y^i should not all lie in a hyperplane. If these two conditions hold, problem (9.19) has a unique solution.

A practical algorithm based on the subproblem (9.19) resembles Algorithm 9.1 in that it contains procedures both for generating new iterates and for improving the geometry of the set Y. The implementation described in [260] contains other features to ensure that the interpolation points are well separated and that steps are not too small. A strength of this method is that it requires only $O(n)$ interpolation points to start producing productive steps. In practice the method often approaches a solution with fewer than $\frac{1}{2}(n+1)(n+2)$ function evaluations. However, since this approach has been developed only recently, there is insufficient numerical experience to assess its full potential.

9.3 COORDINATE AND PATTERN-SEARCH METHODS

Rather than constructing a model of f explicitly based on function values, coordinate search and pattern-search methods look along certain specified directions from the current iterate for a point with a lower function value. If such a point is found, they step to it and repeat the process, possibly modifying the directions of search for the next iteration. If no satisfactory new point is found, the step length along the current search directions may be adjusted, or new search directions may be generated.

We describe first a simple approach of this type that has been used often in practice. We then consider a generalized approach that is potentially more efficient and has stronger theoretical properties.

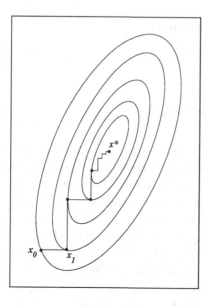

Figure 9.1
Coordinate search method makes slow progress on this function of two variables.

COORDINATE SEARCH METHOD

The coordinate search method (also known as the coordinate descent method or the alternating variables method) cycles through the n coordinate directions $e_1, e_2, \ldots, e_n$, obtaining new iterates by performing a line search along each direction in turn. Specifically, at the first iteration, we fix all components of x except the first one x_1 and find a new value of this component that minimizes (or at least reduces) the objective function. On the next iteration, we repeat the process with the second component x_2, and so on. After n iterations, we return to the first variable and repeat the cycle. Though simple and somewhat intuitive, this method can be quite inefficient in practice, as we illustrate in Figure 9.1 for a quadratic function in two variables. Note that after a few iterations, neither the vertical (x_2) nor the horizontal (x_1) move makes much progress toward the solution at each iteration.

In general, the coordinate search method can iterate infinitely without ever approaching a point where the gradient of the objective function vanishes, even when exact line searches are used. (By contrast, as we showed in Section 3.2, the steepest descent method produces a sequence of iterates $\{x_k\}$ for which $\|\nabla f_k\| \to 0$, under reasonable assumptions.) In fact, a cyclic search along *any* set of linearly independent directions does not guarantee global convergence [243]. Technically speaking, this difficulty arises because the steepest descent search direction $-\nabla f_k$ may become more and more perpendicular to the coordinate search direction. In such circumstances, the Zoutendijk condition (3.14) is satisfied because $\cos \theta_k$ approaches zero rapidly, even when ∇f_k does not approach zero.

When the coordinate search method *does* converge to a solution, it often converges much more slowly than the steepest descent method, and the difference between the two approaches tends to increase with the number of variables. However, coordinate search may still be useful because it does not require calculation of the gradient ∇f_k, and the speed of convergence can be quite acceptable if the variables are loosely coupled in the objective function f.

Many variants of the coordinate search method have been proposed, some of which allow a global convergence property to be proved. One simple variant is a "back-and-forth" approach in which we search along the sequence of directions

$$e_1, e_2, \ldots, e_{n-1}, e_n, e_{n-1}, \ldots, e_2, e_1, e_2, \ldots \quad \text{(repeats)}.$$

Another approach, suggested by Figure 9.1, is first to perform a sequence of coordinate descent steps and then search along the line joining the first and last points in the cycle. Several algorithms, such as that of Hooke and Jeeves, are based on these ideas; see Fletcher [101] and Gill, Murray, and Wright [130].

The pattern-search approach, described next, generalizes coordinate search in that it allows the use of a richer set of search directions at each iteration.

PATTERN-SEARCH METHODS

We consider pattern-search methods that choose a certain set of search directions at each iterate and evaluate f at a given step length along each of these directions. These candidate points form a "frame," or "stencil," around the current iterate. If a point with a significantly lower function value is found, it is adopted as the new iterate, and the center of the frame is shifted to this new point. Whether shifted or not, the frame may then be altered in some way (the set of search directions may be changed, or the step length may grow or shrink), and the process repeats. For certain methods of this type it is possible to prove global convergence results—typically, that there exists a stationary accumulation point.

The presence of noise or other forms of inexactness in the function values may affect the performance of pattern-search algorithms and certainly impacts the convergence theory. Nonsmoothness may also cause undesirable behavior, as can be shown by simple examples, although satisfactory convergence is often observed on nonsmooth problems.

To define pattern-search methods, we introduce some notation. For the current iterate x_k, we define $\mathcal{D}_k$ to be the set of possible search directions and γ_k to be the line search parameter. The frame consists of the points $x_k + \gamma_k p_k$, for all $p_k \in \mathcal{D}_k$. When one of the points in the frame yields a significant decrease in f, we take the step and may also increase γ_k, so as to expand the frame for the next iteration. If none of the points in the frame has a significantly better function value than f_k, we reduce γ_k (contract the frame), set $x_{k+1} = x_k$, and repeat. In either case, we may change the direction set $\mathcal{D}_k$ prior to the next iteration, subject to certain restrictions.

A more precise description of the algorithm follows.

Algorithm 9.2 (Pattern-Search).

Given convergence tolerance γ_{tol}, contraction parameter $\theta_{\max}$,
  sufficient decrease function $\rho : [0, \infty) \to \mathbb{R}$ with $\rho(t)$ an increasing
  function of t and $\rho(t)/t \to 0$ as $t \downarrow 0$;
Choose initial point x_0, initial step length $\gamma_0 > \gamma_{\text{tol}}$, initial direction set $\mathcal{D}_0$;
for $k = 1, 2, \ldots$
     if $\gamma_k \leq \gamma_{\text{tol}}$
        stop;
     if $f(x_k + \gamma_k p_k) < f(x_k) - \rho(\gamma_k)$ for some $p_k \in \mathcal{D}_k$
        Set $x_{k+1} \leftarrow x_k + \gamma_k p_k$ for some such p_k;
        Set $\gamma_{k+1} \leftarrow \phi_k \gamma_k$ for some $\phi_k \geq 1$; (* increase step length *)
     else
        Set $x_{k+1} \leftarrow x_k$;
        Set $\gamma_{k+1} \leftarrow \theta_k \gamma_k$, where $0 < \theta_k \leq \theta_{\max} < 1$;
        end (if)
end (for)

A wise choice of the direction set $\mathcal{D}_k$ is crucial to the practical behavior of this approach and to the theoretical results that can be proved about it. A key condition is that at least one direction in this set should give a direction of descent for f whenever $\nabla f(x_k) \neq 0$ (that is, whenever x_k is not a stationary point). To make this condition specific, we refer to formula (3.12), where we defined the angle between a possible search direction d and the gradient ∇f_k as follows:

$$\cos \theta = \frac{-\nabla f_k^T p}{\|\nabla f_k\| \|p\|}. \tag{9.20}$$

Recall from Theorem 3.2 that global convergence of a line-search method to a stationary point of f could be ensured if the search direction d at each iterate x_k satisfied $\cos \theta \geq \delta$, for some constant $\delta > 0$, and if the line search parameter satisfied certain conditions. In the same spirit, we choose $\mathcal{D}_k$ so that at least one direction $p \in \mathcal{D}_k$ will yield $\cos \theta > \delta$, regardless of the value of ∇f_k. This condition is as follows:

$$\kappa(\mathcal{D}_k) \stackrel{\text{def}}{=} \min_{v \in \mathbb{R}^n} \max_{p \in \mathcal{D}_k} \frac{v^T p}{\|v\| \|p\|} \geq \delta. \tag{9.21}$$

A second condition on $\mathcal{D}_k$ is that the lengths of the vectors in this set are all roughly similar, so that the diameter of the frame formed by this set is captured adequately by the step length parameter γ_k. Thus, we impose the condition

$$\beta_{\min} \leq \|p\| \leq \beta_{\max}, \quad \text{for all } p \in \mathcal{D}_k, \tag{9.22}$$

for some positive constants $\beta_{\min}$ and $\beta_{\max}$ and all k. If the conditions (9.21) and (9.22) hold, we have for any k that

$$-\nabla f_k^T p \geq \kappa(\mathcal{D}_k)\|\nabla f_k\|\|p\| \geq \delta\beta_{\min}\|\nabla f_k\|, \quad \text{for some } p \in \mathcal{D}_k.$$

Examples of sets $\mathcal{D}_k$ that satisfy the properties (9.21) and (9.22) include the coordinate direction set

$$\{e_1, e_2, \ldots, e_n, -e_1, -e_2, \ldots, -e_n\}, \tag{9.23}$$

and the set of $n+1$ vectors defined by

$$p_i = \frac{1}{2n}e - e_i, \quad i = 1, 2, \ldots, n; \quad p_{n+1} = \frac{1}{2n}e, \tag{9.24}$$

where $e = (1, 1, \ldots, 1)^T$. For $n = 3$ these direction sets are sketched in Figure 9.2.

The coordinate descent method described above is similar to the special case of Algorithm 9.2 obtained by setting $\mathcal{D}_k = \{e_i, -e_i\}$ for some $i = 1, 2, \ldots, n$ at each iteration. Note that for this choice of $\mathcal{D}_k$, we have $\kappa(\mathcal{D}_k) = 0$ for all k. Hence, as noted above, $\cos\theta$ can be arbitrarily close to zero at each iteration.

Often, the directions that satisfy the properties (9.21) and (9.22) form only a subset of the direction set $\mathcal{D}_k$, which may contain other directions as well. These additional directions could be chosen heuristically, according to some knowledge of the function f and its scaling, or according to experience on previous iterations. They could also be chosen as linear combinations of the core set of directions (the ones that ensure $\delta > 0$).

Note that Algorithm 9.2 does not require us to choose the point $x_k + \gamma_k p_k$, $p_k \in \mathcal{D}_k$, with the smallest objective value. Indeed, we may save on function evaluations by not evaluating f at all points in the frame, but rather performing the evaluations one at a time and accepting the first candidate point that satisfies the sufficient decrease condition.

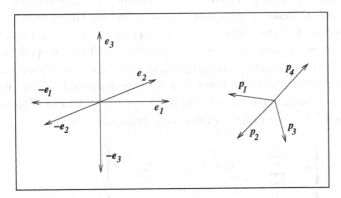

Figure 9.2 Generating search sets in $\mathbb{R}^3$: coordinate direction set (left) and simplex set (right).

Another important detail in the implementation of Algorithm 9.2 is the choice of sufficient decrease function $\rho(t)$. If $\rho(\cdot)$ is chosen to be identically zero, then any candidate point that produces a decrease in f is acceptable as a new iterate. As we have seen in Chapter 3, such a weak condition does not lead to strong global convergence results in general. A more appropriate choice might be $\rho(t) = Mt^{3/2}$, where M is some positive constant.

9.4 A CONJUGATE-DIRECTION METHOD

We have seen in Chapter 5 that the minimizer of a strictly convex quadratic function

$$f(x) = \tfrac{1}{2}x^T A x - b^T x \tag{9.25}$$

can be located by performing one-dimensional minimizations along a set of n conjugate directions. These directions were defined in Chapter 5 as a linear combination of gradients. In this section, we show how to construct conjugate directions using only function values, and we therefore devise an algorithm for minimizing (9.25) that requires only function value calculations. Naturally, we also consider an extension of this approach to the case of a nonlinear objective f.

We use the *parallel subspace property*, which we describe first for the case $n = 2$. Consider two parallel lines $l_1(\alpha) = x_1 + \alpha p$ and $l_2(\alpha) = x_2 + \alpha p$, where x_1, x_2, and p are given vectors in $\mathbb{R}^2$ and α is the scalar parameter that defines the lines. We show below that if x_1^* and x_2^* denote the minimizers of $f(x)$ along l_1 and l_2, respectively, then $x_1^* - x_2^*$ is conjugate to p. Hence, if we perform a one-dimensional minimization along the line joining x_1^* and x_2^*, we will reach the minimizer of f, because we have successively minimized along the two conjugate directions p and $x_2^* - x_1^*$. This process is illustrated in Figure 9.3.

This observation suggests the following algorithm for minimizing a two-dimensional quadratic function f. We choose a set of linearly independent directions, say the coordinate directions e_1 and e_2. From any initial point x_0, we first minimize f along e_2 to obtain the point x_1. We then perform successive minimizations along e_1 and e_2, starting from x_1, to obtain the point z. It follows from the parallel subspace property that $z - x_1$ is conjugate to e_2 because x_1 and z are minimizers along two lines parallel to e_2. Thus, if we perform a one-dimensional search from x_1 along the direction $z - x_1$, we will locate the minimizer of f.

We now state the parallel subspace minimization property in its most general form. Suppose that x_1, x_2 are two distinct points in $\mathbb{R}^n$ and that $\{p_1, p_2, \ldots, p_l\}$ is a set of linearly independent directions in $\mathbb{R}^n$. Let us define the two parallel linear varieties

$$S_1 = \left\{ x_1 + \sum_{i=1}^{l} \alpha_i p_i \mid \alpha_i \in \mathbb{R}, \ i = 1, 2, \ldots, l \right\},$$

$$S_2 = \left\{ x_2 + \sum_{i=1}^{l} \alpha_i p_i \mid \alpha_i \in \mathbb{R}, \ i = 1, 2, \ldots, l \right\}.$$

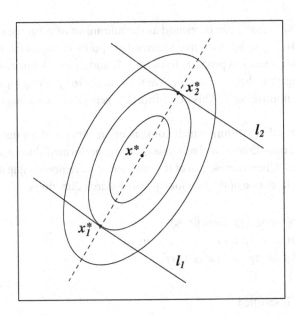

Figure 9.3
Geometric construction of conjugate directions. (The minimizer of f is denoted by x^*.)

If we denote the minimizers of f on S_1 and S_2 by x_1^* and x_2^*, respectively, then $x_2^* - x_1^*$ is conjugate to $p_1, p_2, \ldots, p_l$. It is easy to verify this claim. By the minimization property, we have that

$$\left.\frac{\partial f(x_1^* + \alpha_i p_i)}{\partial \alpha_i}\right|_{\alpha_i=0} = \nabla f(x_1^*)^T p_i = 0, \qquad i = 1, 2, \ldots, l,$$

and similarly for x_2. Therefore we have from (9.25) that

$$\begin{aligned}
0 &= (\nabla f(x_1^*) - \nabla f(x_2^*))^T p_i \\
&= (Ax_1^* - b - Ax_2^* + b)^T p_i \\
&= (x_1^* - x_2^*)^T A p_i, \qquad i = 1, 2, \ldots, l. \qquad (9.26)
\end{aligned}$$

We now consider the case $n = 3$ and show how the parallel subspace property can be used to generate a set of three conjugate directions. We choose a set of linearly independent directions, say e_1, e_2, e_3. From any starting point x_0 we first minimize f along the last direction e_3 to obtain a point x_1. We then perform three successive one-dimensional minimizations, starting from x_1, along the directions e_1, e_2, e_3 and denote the resulting point by z. Next, we minimize f along the direction $p_1 = z - x_1$ to obtain x_2. As noted earlier, $p_1 = z - x_1$ is conjugate to e_3. We note also that x_2 is the minimizer of f on the set $S_1 = \{y + \alpha_1 e_3 + \alpha_2 p_1 \,|\, \alpha_1 \in \mathbb{R}, \ \alpha_2 \in \mathbb{R}\}$, where y is the intermediate point obtained after minimizing along e_1 and e_2.

A new iteration now commences. We discard e_1 and define the new set of search directions as e_2, e_3, p_1. We perform one-dimensional minimizations along e_2, e_3, p_1, starting from x_2, to obtain the point $\hat{z}$. Note that $\hat{z}$ can be viewed as the minimizer of f on the set $S_2 = \{\hat{y} + \alpha_1 e_3 + \alpha_2 p_1 \mid \alpha_1 \in \mathbb{R}, \ \alpha_2 \in \mathbb{R}\}$, for some intermediate point $\hat{y}$. Therefore, by applying the parallel subspace minimization property to the sets S_1 and S_2 just defined, we have that $p_2 = \hat{z} - x_2$ is conjugate to both e_3 and p_1. We then minimize f along p_2 to obtain a point x_3, which is the minimizer of f. This procedure thus generates the conjugate directions e_3, p_1, p_2.

We can now state the general algorithm, which consists of an inner and an outer iteration. In the inner iteration, n one-dimensional minimizations are performed along a set of linearly independent directions. Upon completion of the inner iteration, a new conjugate direction is generated, which replaces one of the previously stored search directions.

Algorithm 9.3 (DFO Method of Conjugate Directions).
Choose an initial point x_0 and set $p_i = e_i$, for $i = 1, 2, \ldots, n$;
Compute x_1 as the minimizer of f along the line $x_0 + \alpha p_n$;
Set $k \leftarrow 1$.

> **repeat** until a convergence test is satisfied
> Set $z_1 \leftarrow x_k$;
> **for** $j = 1, 2, \ldots, n$
> Calculate α_j so that $f(z_j + \alpha_j p_j)$ is minimized;
> $\qquad$ Set $z_{j+1} \leftarrow z_j + \alpha_j p_j$;
> **end (for)**
> $\quad$ Set $p_j \leftarrow p_{j+1}$ for $j = 1, 2, \ldots, n-1$ and $p_n \leftarrow z_{n+1} - z_1$;
> $\quad$ Calculate α_n so that $f(z_{n+1} + \alpha_n p_n)$ is minimized;
> $\quad$ Set $x_{k+1} \leftarrow z_{n+1} + \alpha_n p_n$;
> $\quad$ Set $k \leftarrow k + 1$;
> **end (repeat)**

The line searches can be performed by quadratic interpolation using three function values along each search direction. Since the restriction of (9.25) to a line is a (strictly convex) quadratic, the interpolating quadratic matches it exactly, and the one-dimensional minimizer can easily be computed. Note that at the end of (the outer) iteration k, the directions $p_{n-k}, p_{n-k+1}, \ldots, p_n$ are conjugate by the property mentioned above. Thus the algorithm terminates at the minimizer of (9.25) after $n-1$ iterations, provided none of the conjugate directions is zero. Unfortunately, this possibility cannot be ruled out, and some safeguards described below must be incorporated to improve robustness. In the (usual) case that Algorithm 9.3 terminates after $n-1$ iterations, it will perform $O(n^2)$ function evaluations.

Algorithm 9.3 can be extended to minimize nonquadratic objective functions. The only change is in the line search, which must be performed approximately, using interpolation. Because of the possible nonconvexity, this one-dimensional search must be done with care; see Brent [39] for a treatment of this subject. Numerical experience indicates that this extension of Algorithm 9.3 performs adequately for small-dimensional problems but that sometimes the directions $\{p_i\}$ tend to become linearly dependent. Several modifications of the algorithm have been proposed to guard against this possibility. One such modification measures the degree to which the directions $\{p_i\}$ are conjugate. To do so, we define the scaled directions

$$\hat{p}_i = \frac{p_i}{\sqrt{p_i^T A p_i}}, \qquad i = 1, 2, \ldots, n. \tag{9.27}$$

One can show [239] that the quantity

$$|\det(\hat{p}_1, \hat{p}_2, \ldots, \hat{p}_n)| \tag{9.28}$$

is maximized if and only if the vectors p_i are conjugate with respect to A. This result suggests that we should *not* replace one of the existing search directions in the set $\{p_1, p_2, \ldots, p_n\}$ by the most recently generated conjugate direction if this action causes the quantity (9.28) to decrease.

Procedure 9.4 implements this strategy for the case of the quadratic objective function (9.25). Some algebraic manipulations (which we do not present here) show that we can compute the scaled directions $\hat{p}_i$ without using the Hessian A because the terms $p_i^T A p_i$ are available from the line search along p_i. Further, only comparisons using computed function values are needed to ensure that (9.28) does not increase. The following procedure is invoked immediately after the execution of the inner iteration (or for-loop) of Algorithm 9.3.

Procedure 9.4 (Updating of the Set of Directions).
Find the integer $m \in \{1, 2, \ldots, n\}$ such that $\psi_m = f(x_{m-1}) - f(x_m)$
      is maximized;
Let $f_1 = f(z_1)$, $f_2 = f(z_{n+1})$, and $f_3 = f(2z_{n+1} - z_1)$;
if $f_3 \geq f_1$ or $(f_1 - 2f_2 + f_3)(f_1 - f_2 - \psi_m)^2 \geq \frac{1}{2}\psi_m(f_1 - f_3)^2$
      Keep the set $p_1, p_2, \ldots, p_n$ unchanged and set $x_{k+1} \leftarrow z_{n+1}$;
else
      Set $\hat{p} \leftarrow z_{n+1} - z_1$ and calculate $\hat{\alpha}$ so that $f(z_{n+1} + \hat{\alpha}\hat{p})$ is minimized;
      Set $x_{k+1} \leftarrow z_{n+1} + \alpha\hat{p}$;
      Remove p_m from the set of directions and add $\hat{p}$ to this set;
end (if)

This procedure can be applied to general objective functions by implementing inexact one-dimensional line searches. The resulting conjugate-gradient method has been found to be useful for solving small dimensional problems.

9.5 NELDER–MEAD METHOD

The Nelder–Mead simplex-reflection method has been a popular DFO method since its introduction in 1965 [223]. It takes its name from the fact that at any stage of the algorithm, we keep track of $n + 1$ points of interest in $\mathbb{R}^n$, whose convex hull forms a simplex. (The method has nothing to do with the simplex method for linear programming discussed in Chapter 13.) Given a simplex S with vertices $\{z_1, z_2, \ldots, z_{n+1}\}$, we can define an associated matrix $V(S)$ by taking the n edges along V from one of its vertices (z_1, say), as follows:

$$V(S) = [z_2 - z_1, z_3 - z_1, \ldots, z_{n+1} - z_1].$$

The simplex is said to be *nondegenerate* or *nonsingular* if V is a nonsingular matrix. (For example, a simplex in $\mathbb{R}^3$ is nondegenerate if its four vertices are not coplanar.)

In a single iteration of the Nelder–Mead algorithm, we seek to remove the vertex with the worst function value and replace it with another point with a better value. The new point is obtained by reflecting, expanding, or contracting the simplex along the line joining the worst vertex with the centroid of the remaining vertices. If we cannot find a better point in this manner, we retain only the vertex with the *best* function value, and we shrink the simplex by moving all other vertices toward this value.

We specify a single step of the algorithm after some defining some notation. The $n + 1$ vertices of the current simplex are denoted by $\{x_1, x_2, \ldots, x_{n+1}\}$, where we choose the ordering so that

$$f(x_1) \leq f(x_2) \leq \cdots \leq f(x_{n+1}).$$

The centroid of the best n points is denoted by

$$\bar{x} = \sum_{i=1}^{n} x_i.$$

Points along the line joining $\bar{x}$ and the "worst" vertex x_{n+1} are denoted by

$$\bar{x}(t) = \bar{x} + t(x_{n+1} - \bar{x}).$$

Procedure 9.5 (One Step of Nelder–Mead Simplex).
Compute the reflection point $\bar{x}(-1)$ and evaluate $f_{-1} = f(\bar{x}(-1))$;
if $f(x_1) \leq f_{-1} < f(x_n)$
      (* reflected point is neither best nor worst in the new simplex *)
      replace x_{n+1} by $\bar{x}(-1)$ and go to next iteration;
else if $f_{-1} < f(x_1)$

(* reflected point is better than the current best; try to
      go farther along this direction *)
Compute the expansion point $\bar{x}(-2)$ and evaluate $f_{-2} = f(\bar{x}(-2))$;
if $f_{-2} < f_{-1}$
      replace x_{n+1} by x_{-2} and go to next iteration;
else
      replace x_{n+1} by x_{-1} and go to next iteration;
else if $f_{-1} \geq f(x_n)$
    (* reflected point is still worse than x_n; contract *)
    if $f(x_n) \leq f_{-1} < f(x_{n+1})$
      (* try to perform "outside" contraction *)
      evaluate $f_{-1/2} = \bar{x}(-1/2)$;
      if $f_{-1/2} \leq f_{-1}$
         replace x_{n+1} by $x_{-1/2}$ and go to next iteration;
    else
      (* try to perform "inside" contraction *)
      evaluate $f_{1/2} = \bar{x}(1/2)$;
      if $f_{1/2} < f_{n+1}$
         replace x_{n+1} by $x_{1/2}$ and go to next iteration;
    (* neither outside nor inside contraction was acceptable;
      shrink the simplex toward x_1 *)
replace $x_i \leftarrow (1/2)(x_1 + x_i)$ for $i = 2, 3, \ldots, n + 1$;

Procedure 9.5 is illustrated on a three-dimensional example in Figure 9.4. The worst current vertex is x_3, and the possible replacement points are $\bar{x}(-1), \bar{x}(-2), \bar{x}(-\frac{1}{2}), \bar{x}(\frac{1}{2})$. If none of the replacement points proves to be satisfactory, the simplex is shrunk to the smaller triangle indicated by the dotted line, which retains the best vertex x_1. The scalars t used in defining the candidate points $\bar{x}(t)$ have been assigned the specific (and standard) values $-1, -2, -\frac{1}{2}$, and $\frac{1}{2}$ in our description above. Different choices are also possible, subject to certain restrictions.

Practical performance of the Nelder–Mead algorithm is often reasonable, though stagnation has been observed to occur at nonoptimal points. Restarting can be used when stagnation is detected; see Kelley [178]. Note that unless the final shrinkage step is performed, the average function value

$$\frac{1}{n+1} \sum_{i=1}^{n+1} f(x_i) \tag{9.29}$$

will decrease at each step. When f is convex, even the shrinkage step is guaranteed not to increase the average function value.

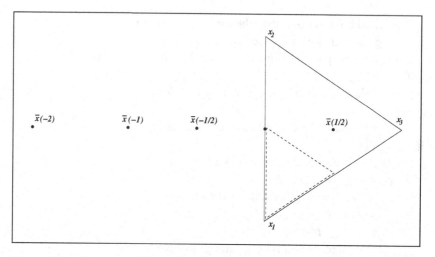

Figure 9.4   One step of the Nelder–Mead simplex method in $\mathbb{R}^3$, showing current simplex (solid triangle with vertices x_1, x_2, x_3), reflection point $\bar{x}(-1)$, expansion point $\bar{x}(-2)$, inside contraction point $\bar{x}(\frac{1}{2})$, outside contraction point $\bar{x}(-\frac{1}{2})$, and shrunken simplex (dotted triangle).

A limited amount of convergence theory has been developed for the Nelder–Mead method in recent years; see, for example, Kelley [179] and Lagarias et al. [186].

9.6   IMPLICIT FILTERING

We now describe an algorithm designed for functions whose evaluations are modeled by (9.2), where h is smooth. This *implicit filtering* approach is, in its simplest form, a variant of the steepest descent algorithm with line search discussed in Chapter 3, in which the gradient ∇f_k is replaced by a finite difference estimate such as (9.3), with a difference parameter ϵ that may not be particularly small.

Implicit filtering works best on functions for which the noise level decreases as the iterates approach a solution. This situation may occur when we have control over the noise level, as is the case when f is obtained by solving a differential equation to a user-specified tolerance, or by running a stochastic simulation for a user-specified number of trials (where an increase in the number of trials usually produces a decrease in the noise). The implicit filtering algorithm decreases ϵ systematically (but, one hopes, not as rapidly as the decay in error) so as to maintain reasonable accuracy in $\nabla_\epsilon f(x)$, given the noise level at the current value of x. For each value of ϵ, it performs an inner loop that is simply an Armijo line search using the search direction $-\nabla_\epsilon f(x)$. If the inner loop is unable to find a satisfactory step length after backtracking at least a_{max} times, we return to the outer loop, choose a smaller value of ϵ, and repeat. A formal specification follows.

Algorithm 9.6 (Implicit Filtering).
Choose a sequence $\{\epsilon_k\} \downarrow 0$, Armijo parameters c and ρ in $(0, 1)$,
    maximum backtracking parameter $a_{\max}$;
Set $k \leftarrow 1$, Choose initial point $x = x_0$;
repeat
    `increment_k` $\leftarrow$ `false`;
    repeat
        Compute $f(x)$ and $\nabla_{\epsilon_k} f(x)$;
        if $\left\| \nabla_{\epsilon_k} f(x) \right\| \leq \epsilon_k$
            `increment_k` $\leftarrow$ `true`;
        else
            Find the smallest integer m between 0 and $a_{\max}$ such that
            $$f\left(x - \rho^m \nabla_{\epsilon_k} f(x)\right) \leq f(x) - c\rho^m \left\| \nabla_{\epsilon_k} f(x) \right\|_2^2;$$
            if no such m exists
                `increment_k` $\leftarrow$ `true`;
            else
                $x \leftarrow x - \rho^m \nabla_\epsilon f(x)$;
    until `increment_k`;
    $x_k \leftarrow x; k \leftarrow k + 1$;
until a termination test is satisfied.

Note that the inner loop in Algorithm 9.6 is essentially the backtracking line search algorithm—Algorithm 3.1 of Chapter 3—with a convergence criterion added to detect whether the minimum appears to have been found to within the accuracy implied by the difference parameter ϵ_k. If the gradient estimate $\nabla_{\epsilon_k} f$ is small, or if the line search fails to find a satisfactory new iterate (indicating that the gradient approximation $\nabla_{\epsilon_k} f(x)$ is insufficiently accurate to produce descent in f), we decrease the difference parameter to ϵ_{k+1} and proceed.

A basic convergence result for Algorithm 9.6 is the following.

Theorem 9.2.
    Suppose that $\nabla^2 h$ is Lipschitz continuous, that Algorithm 9.6 generates an infinite sequence of iterates $\{x_k\}$, and that

$$\lim_{k \to \infty} \epsilon_k^2 + \frac{\eta(x_k; \epsilon_k)}{\epsilon_k} = 0.$$

Suppose, too, that all but a finite number of inner loops in Algorithm 9.6 terminate with $\left\| \nabla_{\epsilon_k} f(x_k) \right\| \leq \epsilon_k$. Then all limit points of the sequence $\{x_k\}$ are stationary.

PROOF. Using $\{\epsilon_k\} \downarrow 0$, we have under our assumptions on inner loop termination that $\nabla_{\epsilon_k} f(x_k) \to 0$. By invoking the error bound (9.5) and noting that the right-hand side of this expression is approaching zero, we conclude that $\nabla h(x_k) \to 0$. Hence all limit points satisfy $\nabla h(x) = 0$, as claimed. $\square$

More sophisticated versions of implicit filtering methods can be derived by using the gradient estimate $\nabla_{\epsilon_k} f$ to construct quasi-Newton approximate Hessians, and thus generating quasi-Newton search directions instead of the negative-approximate-gradient search direction used in Algorithm 9.6.

NOTES AND REFERENCES

A classical reference on derivative-free methods is Brent [39], which focuses primarily on one-dimensional problems and includes discussion of roundoff errors and global minimization. Recent surveys on derivative-free methods include Wright [314], Powell [256], Conn, Scheinberg, and Toint [76], and Kolda, Lewis, and Torczon [183].

The first model-based method for derivative-free optimization was proposed by Winfield [307]. It uses quadratic models, which are determined by the interpolation conditions (9.7), and computes steps by solving a subproblem of the form (9.9). Practical procedures for improving the geometry of the interpolation points were first developed by Powell in the context of model-based methods using linear and quadratic polynomials; see [256] for a review of this work.

Conn, Scheinberg, and Toint [75] propose and analyze model-based methods and study the use of Newton fundamental polynomials. Methods that combine minimum change updating and interpolation are discussed by Powell [258, 260]. Our presentation of model-based methods in Section 9.2 is based on [76, 259, 258].

For a comprehensive discussion of pattern-search methods of the type discussed here, we refer the reader to the review paper of Kolda, Lewis, and Torczon [183], and the references therein.

The method of conjugate directions given in Algorithm 9.3 was proposed by Powell [239]. For a discussion on the rate of convergence of the coordinate descent method and for more references about this method, see Luenberger [195]. For further information on implicit filtering, see Kelley [179] and Choi and Kelley [60] and the references therein.

Software packages that implement model-based methods include COBYLA [258], DFO [75], UOBYQA [257], WEDGE [200], and NEWUOA [260]. The earliest code is COBYLA, which employs linear models. DFO, UOBYQA, and WEDGE use quadratic models, whereas the method based on minimum change updating (9.19) is implemented in NEWUOA. A pattern-search method is implemented in APPS [171], while DIRECT [173] is designed to find a global solution.

✎ EXERCISES

✎ **9.1** Prove Lemma 9.1.

✎ **9.2**

(a) Verify that the number of interpolation conditions to uniquely determine the coefficients in (9.6) are $q = \frac{1}{2}(n+1)(n+2)$.

(b) Verify that the number of vertices and midpoints of the edges of a nondegenerate simplex in R^n add up to $q = \frac{1}{2}(n + 1)(n + 2)$ and can therefore be used as the initial interpolation set in a DFO algorithm.

(c) How many interpolation conditions would be required to determine the coefficients in (9.6) if the matrix G were identically 0? How many if G were diagonal? How many if G were tridiagonal?

✎ **9.3** Describe conditions on the vectors s^l that guarantee that the model (9.14) is uniquely determined.

✎ **9.4** Consider the determination of a quadratic function in two variables.

(a) Show that six points on a line do not determine the quadratic.

(b) Show that six points in a circle in the plane do not uniquely determine the quadratic.

✎ **9.5** Use induction to show that at the end of the outer iteration k of Algorithm 9.3, the directions $p_{n-k}, p_{n-k+1}, \ldots, p_n$ are conjugate. Use this fact to show that if the step lengths α_i in Algorithm 9.3 are never zero, the iteration terminates at the minimizer of (9.25) after at most n outer iterations.

✎ **9.6** Write a program that computes the one-dimensional minimizer of a strictly convex quadratic function f along a direction p using quadratic interpolation. Describe the formulas used in your program.

✎ **9.7** Find the quadratic function

$$m(x_1, x_2) = f + g_1 x_1 + g_2 x_2 + \frac{1}{2} G_{11}^2 x_1^2 + G_{12} x_1 x_2 + \frac{1}{2} G_{22}^2 x_2^2$$

that interpolates the following data: $x_0 = y^1 = (0, 0)^T$, $y^2 = (1, 0)^T$, $y^3 = (2, 0)^T$, $y^4 = (1, 1)^T$, $y^5 = (0, 2)^T$, $y^6 = (0, 1)^T$, and $f(y^1) = 1$, $f(y^2) = 2.0084$, $f(y^3) = 7.0091$, $f(y^4) = 1.0168$, $f(y^5) = -0.9909$, and $f(y^6) = -0.9916$.

✎ **9.8** Find the value of δ for which the coordinate generating set (9.23) satisfies the property (9.21).

✎ **9.9** Show that $\kappa(\mathcal{D}_k) = 0$, where $\kappa(\cdot)$ is defined by (9.21) and $\mathcal{D}_k = \{e_i, -e_i\}$ for any $i = 1, 2, \ldots, n$.

✎ **9.10** (Hard) Prove that the generating set (9.24) satisfies the property (9.21) for a certain value $\delta > 0$, and find this value of δ.

✎ **9.11** Justify the statement that the average function value at the Nelder–Mead simplex points will decrease over one step if any of the points $\bar{x}(-1), \bar{x}(-2), \bar{x}(-\frac{1}{2}), \bar{x}(\frac{1}{2})$ are adopted as a replacement for x_{n+1}.

✏ **9.12** Show that if f is a convex function, the shrinkage step in the Nelder–Mead simplex method will not increase the average value of the function over the simplex vertices defined by (9.29). Show that unless $f(x_1) = f(x_2) = \cdots = f(x_{n+1})$, the average value will in fact decrease.

✏ **9.13** Suppose for the f defined in (9.2), we define the approximate gradient $\nabla_\epsilon f(x)$ by the forward-difference formula

$$\nabla_\epsilon f(x) = \left[\frac{f(x + \epsilon e_i) - f(x)}{\epsilon} \right]_{i=1,2,\dots,n},$$

rather than the central-difference formula (9.3). (This formula requires only half as many function evaluations but is less accurate.) For this definition, prove the following variant of Lemma 9.1: Suppose that $\nabla h(x)$ is Lipschitz continuous in a neighborhood of the box $\{z \mid z \geq x, \ \|z - x\|_\infty \leq \epsilon\}$ with Lipschitz constant L_h. Then we have

$$\|\nabla_\epsilon f(x) - \nabla h(x)\|_\infty \leq L_h \epsilon + \frac{\eta(x; \epsilon)}{\epsilon},$$

where $\eta(x; \epsilon)$ is redefined as follows:

$$\eta(x; \epsilon) = \sup_{z \geq x, \ \|z - x\|_\infty \leq \epsilon} |\phi(z)|.$$

CHAPTER 10

Least-Squares Problems

In least-squares problems, the objective function f has the following special form:

$$f(x) = \tfrac{1}{2} \sum_{j=1}^{m} r_j^2(x), \qquad (10.1)$$

where each r_j is a smooth function from $\mathbb{R}^n$ to $\mathbb{R}$. We refer to each r_j as a *residual*, and we assume throughout this chapter that $m \geq n$.

Least-squares problems arise in many areas of applications, and may in fact be the largest source of unconstrained optimization problems. Many who formulate a parametrized model for a chemical, physical, financial, or economic application use a function of the form (10.1) to measure the discrepancy between the model and the observed behavior of the system (see Example 2.1, for instance). By minimizing this function, they select values for the parameters that best match the model to the data. In this chapter we show how to devise efficient, robust minimization algorithms by exploiting the special structure of the function f and its derivatives.

To see why the special form of f often makes least-squares problems easier to solve than general unconstrained minimization problems, we first assemble the individual components r_j from (10.1) into a *residual vector* $r : \mathbb{R}^n \to \mathbb{R}^m$, as follows

$$r(x) = (r_1(x), r_2(x), \ldots, r_m(x))^T. \tag{10.2}$$

Using this notation, we can rewrite f as $f(x) = \frac{1}{2}\|r(x)\|_2^2$. The derivatives of $f(x)$ can be expressed in terms of the *Jacobian* $J(x)$, which is the $m \times n$ matrix of first partial derivatives of the residuals, defined by

$$J(x) = \left[\frac{\partial r_j}{\partial x_i} \right]_{\substack{j=1,2,\ldots,m \\ i=1,2,\ldots,n}} = \begin{bmatrix} \nabla r_1(x)^T \\ \nabla r_2(x)^T \\ \vdots \\ \nabla r_m(x)^T \end{bmatrix}, \tag{10.3}$$

where each $\nabla r_j(x)$, $j = 1, 2, \ldots, m$ is the gradient of r_j. The gradient and Hessian of f can then be expressed as follows:

$$\nabla f(x) = \sum_{j=1}^{m} r_j(x) \nabla r_j(x) = J(x)^T r(x), \tag{10.4}$$

$$\nabla^2 f(x) = \sum_{j=1}^{m} \nabla r_j(x) \nabla r_j(x)^T + \sum_{j=1}^{m} r_j(x) \nabla^2 r_j(x)$$

$$= J(x)^T J(x) + \sum_{j=1}^{m} r_j(x) \nabla^2 r_j(x). \tag{10.5}$$

In many applications, the first partial derivatives of the residuals and hence the Jacobian matrix $J(x)$ are relatively easy or inexpensive to calculate. We can thus obtain the gradient $\nabla f(x)$ as written in formula (10.4). Using $J(x)$, we also can calculate the first term $J(x)^T J(x)$ in the Hessian $\nabla^2 f(x)$ without evaluating any second derivatives of the functions r_j. This availability of part of $\nabla^2 f(x)$ "for free" is the distinctive feature of least-squares problems. Moreover, this term $J(x)^T J(x)$ is often more important than the second summation term in (10.5), either because the residuals r_j are close to affine near the solution (that is, the $\nabla^2 r_j(x)$ are relatively small) or because of small residuals (that is, the $r_j(x)$ are relatively small). Most algorithms for nonlinear least-squares exploit these structural properties of the Hessian.

The most popular algorithms for minimizing (10.1) fit into the line search and trust-region frameworks described in earlier chapters. They are based on the Newton and quasi-Newton approaches described earlier, with modifications that exploit the particular structure of f.

Section 10.1 contains some background on applications. Section 10.2 discusses linear least-squares problems, which provide important motivation for algorithms for the nonlinear problem. Section 10.3 describes the major algorithms, while Section 10.4 briefly describes a variant of least squares known as orthogonal distance regression.

Throughout this chapter, we use the notation $\|\cdot\|$ to denote the Euclidean norm $\|\cdot\|_2$, unless a subscript indicates that some other norm is intended.

10.1   BACKGROUND

We discuss a simple parametrized model and show how least-squares techniques can be used to choose the parameters that best fit the model to the observed data.

❏ **EXAMPLE 10.1**

We would like to study the effect of a certain medication on a patient. We draw blood samples at certain times after the patient takes a dose, and measure the concentration of the medication in each sample, tabulating the time t_j and concentration y_j for each sample.

Based on our previous experience in such experiments, we find that the following function $\phi(x; t)$ provides a good prediction of the concentration at time t, for appropriate values of the five-dimensional parameter vector $x = (x_1, x_2, x_3, x_4, x_5)$:

$$\phi(x; t) = x_1 + tx_2 + t^2 x_3 + x_4 e^{-x_5 t}. \tag{10.6}$$

We choose the parameter vector x so that this model best agrees with our observation, in some sense. A good way to measure the difference between the predicted model values and the observations is the following least-squares function:

$$\frac{1}{2} \sum_{j=1}^{m} [\phi(x; t_j) - y_j]^2, \tag{10.7}$$

which sums the squares of the discrepancies between predictions and observations at each t_j. This function has precisely the form (10.1) if we define

$$r_j(x) = \phi(x; t_j) - y_j. \tag{10.8}$$

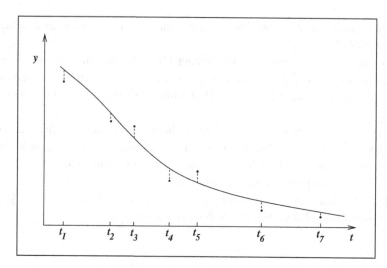

Figure 10.1   Model (10.7) (smooth curve) and the observed measurements, with deviations indicated by vertical dotted lines.

Graphically, each term in (10.7) represents the square of the vertical distance between the curve $\phi(x; t)$ (plotted as a function of t) and the point (t_j, y_j), for a fixed choice of parameter vector x; see Figure 10.1. The minimizer x^* of the least-squares problem is the parameter vector for which the sum of squares of the lengths of the dotted lines in Figure 10.1 is minimized. Having obtained x^*, we use $\phi(x^*; t)$ to estimate the concentration of medication remaining in the patient's bloodstream at any time t.   □

This model is an example of what statisticians call a *fixed-regressor model*. It assumes that the times t_j at which the blood samples are drawn are known to high accuracy, while the observations y_j may contain more or less random errors due to the limitations of the equipment (or the lab technician!)

In general data-fitting problems of the type just described, the ordinate t in the model $\phi(x; t)$ could be a vector instead of a scalar. (In the example above, for instance, t could have two dimensions, with the first dimension representing the time since the drug was admistered and the second dimension representing the weight of the patient. We could then use observations for an entire population of patients, not just a single patient, to obtain the "best" parameters for this model.)

The sum-of-squares function (10.7) is not the only way of measuring the discrepancy between the model and the observations. Other common measures include the maximum absolute value

$$\max_{j=1,2,\ldots,m} |\phi(x; t_j) - y_j| \tag{10.9}$$

and the sum of absolute values

$$\sum_{j=1}^{m} |\phi(x; t_j) - y_j|. \tag{10.10}$$

By using the definitions of the ℓ_∞ and ℓ_1 norms, we can rewrite these two measures as

$$f(x) = \|r(x)\|_\infty, \quad f(x) = \|r(x)\|_1, \tag{10.11}$$

respectively. As we discuss in Chapter 17, the problem of minimizing the functions (10.11) can be reformulated a smooth constrained optimization problem.

In this chapter we focus only on the ℓ_2-norm formulation (10.1). In some situations, there are statistical motivations for choosing the least-squares criterion. Changing the notation slightly, let the discrepancies between model and observation be denoted by ϵ_j, that is,

$$\epsilon_j = \phi(x; t_j) - y_j.$$

It often is reasonable to assume that the ϵ_j's are independent and identically distributed with a certain variance σ^2 and probability density function $g_\sigma(\cdot)$. (This assumption will often be true, for instance, when the model accurately reflects the actual process, and when the errors made in obtaining the measurements y_j do not contain a systematic bias.) Under this assumption, the likelihood of a particular set of observations $y_j, j = 1, 2, \ldots, m$, given that the actual parameter vector is x, is given by the function

$$p(y; x, \sigma) = \prod_{j=1}^{m} g_\sigma(\epsilon_j) = \prod_{j=1}^{m} g_\sigma(\phi(x; t_j) - y_j). \tag{10.12}$$

Given the observations $y_1, y_2, \ldots, y_m$, the "most likely" value of x is obtained by maximizing $p(y; x, \sigma)$ with respect to x. The resulting value of x is called the *maximum likelihood estimate.*

When we assume that the discrepancies follow a *normal* distribution, we have

$$g_\sigma(\epsilon) = \frac{1}{\sqrt{2\pi\sigma^2}} \exp\left(-\frac{\epsilon^2}{2\sigma^2}\right).$$

Substitution in (10.12) yields

$$p(y; x, \sigma) = (2\pi\sigma^2)^{-m/2} \exp\left(-\frac{1}{2\sigma^2} \sum_{j=1}^{m} [\phi(x; t_j) - y_j]^2\right).$$

For any fixed value of the variance σ^2, it is obvious that p is maximized when the sum of squares (10.7) is minimized. To summarize: When the discrepancies are assumed to be independent and identically distributed with a normal distribution function, the maximum likelihood estimate is obtained by minimizing the sum of squares.

The assumptions on ϵ_j in the previous paragraph are common, but they do not describe the only situation for which the minimizer of the sum of squares makes good statistical sense. Seber and Wild [280] describe many instances in which minimization of functions like (10.7), or generalizations of this function such as

$$r(x)^T W r(x), \quad \text{where } W \in \mathbb{R}^{m \times m} \text{ is symmetric,}$$

is the crucial step in obtaining estimates of the parameters x from observed data.

10.2   LINEAR LEAST-SQUARES PROBLEMS

Many models $\phi(x; t)$ in data-fitting problems are linear functions of x. In these cases, the residuals $r_j(x)$ defined by (10.8) also are linear, and the problem of minimizing (10.7) is called a *linear least-squares problem*. We can write the residual vector as $r(x) = Jx - y$ for some matrix J and vector y, both independent of x, so that the objective is

$$f(x) = \tfrac{1}{2} \| Jx - y \|^2, \tag{10.13}$$

where $y = r(0)$. We also have

$$\nabla f(x) = J^T (Jx - y), \qquad \nabla^2 f(x) = J^T J.$$

(Note that the second term in $\nabla^2 f(x)$ (see (10.5)) disappears, because $\nabla^2 r_j = 0$ for all $j = 1, 2, \ldots, m$.) It is easy to see that the $f(x)$ in (10.13) is convex—a property that does not necessarily hold for the nonlinear problem (10.1). Theorem 2.5 tells us that any point x^* for which $\nabla f(x^*) = 0$ is the global minimizer of f. Therefore, x^* must satisfy the following linear system of equations:

$$J^T J x^* = J^T y. \tag{10.14}$$

These are known as the *normal equations* for (10.13).

We outline briefly three major algorithms for the unconstrained linear least-squares problem. We assume in most of our discussion that $m \geq n$ and that J has full column rank.

The first and most obvious algorithm is simply to form and solve the system (10.14) by the following three-step procedure:

- compute the coefficient matrix $J^T J$ and the right-hand-side $J^T y$;

- compute the Cholesky factorization of the symmetric matrix $J^T J$;

- perform two triangular substitutions with the Cholesky factors to recover the solution x^*.

The Cholesky factorization

$$J^T J = \bar{R}^T \bar{R} \tag{10.15}$$

(where $\bar{R}$ is an $n \times n$ upper triangular with positive diagonal elements) is guaranteed to exist when $m \geq n$ and J has rank n. This method is frequently used in practice and is often effective, but it has one significant disadvantage, namely, that the condition number of $J^T J$ is the square of the condition number of J. Since the relative error in the computed solution of a problem is usually proportional to the condition number, the Cholesky-based method may result in less accurate solutions than those obtained from methods that avoid this squaring of the condition number. When J is ill conditioned, the Cholesky factorization process may even break down, since roundoff errors may cause small negative elements to appear on the diagonal during the factorization process.

A second approach is based on a QR factorization of the matrix J. Since the Euclidean norm of any vector is not affected by orthogonal transformations, we have

$$\|Jx - y\| = \|Q^T(Jx - y)\| \tag{10.16}$$

for any $m \times m$ orthogonal matrix Q. Suppose we perform a QR factorization with column pivoting on the matrix J (see (A.24)) to obtain

$$J\Pi = Q \begin{bmatrix} R \\ 0 \end{bmatrix} = [\; Q_1 \quad Q_2 \;] \begin{bmatrix} R \\ 0 \end{bmatrix} = Q_1 R, \tag{10.17}$$

where

Π is an $n \times n$ permutation matrix (hence, orthogonal);
Q is $m \times m$ orthogonal;
Q_1 is the first n columns of Q, while Q_2 contains the last $m - n$ columns;
R is $n \times n$ upper triangular with positive diagonal elements.

By combining (10.16) and (10.17), we obtain

$$\|Jx - y\|_2^2 = \left\| \begin{bmatrix} Q_1^T \\ Q_2^T \end{bmatrix} (J\Pi\Pi^T x - y) \right\|_2^2$$

$$= \left\| \begin{bmatrix} R \\ 0 \end{bmatrix} (\Pi^T x) - \begin{bmatrix} Q_1^T y \\ Q_2^T y \end{bmatrix} \right\|^2$$

$$= \|R(\Pi^T x) - Q_1^T y\|_2^2 + \|Q_2^T y\|^2. \tag{10.18}$$

No choice of x has any effect on the second term of this last expression, but we can minimize $\|Jx - y\|$ by driving the first term to zero, that is, by setting

$$x^* = \Pi R^{-1} Q_1^T y.$$

(In practice, we perform a triangular substitution to solve $Rz = Q_1^T y$, then permute the components of z to obtain $x^* = \Pi z$.)

This QR-based approach does not degrade the conditioning of the problem unnecessarily. The relative error in the final computed solution x^* is usually proportional to the condition number of J, not its square, and this method is usually reliable. Some situations, however, call for greater robustness or more information about the sensitivity of the solution to perturbations in the data (J or y). A third approach, based on the singular-value decomposition (SVD) of J, can be used in these circumstances. Recall from (A.15) that the SVD of J is given by

$$J = U \begin{bmatrix} S \\ 0 \end{bmatrix} V^T = \begin{bmatrix} U_1 & U_2 \end{bmatrix} \begin{bmatrix} S \\ 0 \end{bmatrix} V^T = U_1 S V^T, \tag{10.19}$$

where

    U is $m \times m$ orthogonal;
    U_1 contains the first n columns of U, U_2 the last $m - n$ columns;
    V is $n \times n$ orthogonal;
    S is $n \times n$ diagonal, with diagonal elements $\sigma_1 \geq \sigma_2 \geq \cdots \geq \sigma_n > 0$.

(Note that $J^T J = V S^2 V^T$, so that the columns of V are eigenvectors of $J^T J$ with eigenvalues σ_j^2, $j = 1, 2, \ldots, n$.) By following the same logic that led to (10.18), we obtain

$$\|Jx - y\|^2 = \left\| \begin{bmatrix} S \\ 0 \end{bmatrix} (V^T x) - \begin{bmatrix} U_1^T \\ U_2^T \end{bmatrix} y \right\|^2$$

$$= \|S(V^T x) - U_1^T y\|^2 + \|U_2^T y\|^2. \tag{10.20}$$

Again, the optimum is found by choosing x to make the first term equal to zero; that is,

$$x^* = V S^{-1} U_1^T y.$$

Denoting the ith columns of U and V by $u_i \in \mathbb{R}^m$ and $v_i \in \mathbb{R}^n$, respectively, we have

$$x^* = \sum_{i=1}^n \frac{u_i^T y}{\sigma_i} v_i. \tag{10.21}$$

This formula yields useful information about the sensitivity of x^*. When σ_i is small, x^* is particularly sensitive to perturbations in y that affect $u_i^T y$, and also to perturbations in J that affect this same quantity. Such information is particularly useful when J is nearly rank-deficient, that is, when $\sigma_n/\sigma_1 \ll 1$. It is sometimes worth the extra cost of the SVD algorithm to obtain this sensitivity information.

All three approaches above have their place. The Cholesky-based algorithm is particularly useful when $m \gg n$ and it is practical to store $J^T J$ but not J itself. It can also be less expensive than the alternatives when $m \gg n$ and J is sparse. However, this approach must be modified when J is rank-deficient or ill conditioned to allow pivoting of the diagonal elements of $J^T J$. The QR approach avoids squaring of the condition number and hence may be more numerically robust. While potentially the most expensive, the SVD approach is the most robust and reliable of all. When J is actually rank-deficient, some of the singular values σ_i are exactly zero, and any vector x^* of the form

$$x^* = \sum_{\sigma_i \neq 0} \frac{u_i^T y}{\sigma_i} v_i + \sum_{\sigma_i = 0} \tau_i v_i \tag{10.22}$$

(for arbitrary coefficients τ_i) is a minimizer of (10.20). Frequently, the solution with smallest norm is the most desirable, and we obtain it by setting each $\tau_i = 0$ in (10.22). When J has full rank but is ill conditioned, the last few singular values $\sigma_n, \sigma_{n-1}, \ldots$ are small relative to σ_1. The coefficients $u_i^T y/\sigma_i$ in (10.22) are particularly sensitive to perturbations in $u_i^T y$ when σ_i is small, so an approximate solution that is less sentitive to perturbations than the true solution can be obtained by omitting these terms from the summation.

When the problem is very large, it may be efficient to use iterative techniques, such as the conjugate gradient method, to solve the normal equations (10.14). A direct implementation of conjugate gradients (Algorithm 5.2) requires one matrix vector multiplication with $J^T J$ to be performed at each iteration. This operation can be performed by means of successive multiplications by J and J^T; we need only the ability to perform matrix-vector multiplications with these two matrices to implement this algorithm. Several modifications of the conjugate gradient approach have been proposed that involve a similar amount of work per iteration (one matrix-vector multiplication each with J and J^T) but that have superior numerical properties. Some alternatives are described by Paige and Saunders [234], who propose in particular an algorithm called LSQR which has become the basis of a highly successful code.

10.3 ALGORITHMS FOR NONLINEAR LEAST-SQUARES PROBLEMS

THE GAUSS–NEWTON METHOD

We now describe methods for minimizing the nonlinear objective function (10.1) that exploit the structure in the gradient ∇f (10.4) and Hessian $\nabla^2 f$ (10.5). The simplest of these methods—the Gauss–Newton method—can be viewed as a modified Newton's method with line search. Instead of solving the standard Newton equations $\nabla^2 f(x_k)p = -\nabla f(x_k)$, we solve instead the following system to obtain the search direction p_k^{GN}:

$$J_k^T J_k p_k^{\text{GN}} = -J_k^T r_k. \tag{10.23}$$

This simple modification gives a number of advantages over the plain Newton's method. First, our use of the approximation

$$\nabla^2 f_k \approx J_k^T J_k \tag{10.24}$$

saves us the trouble of computing the individual residual Hessians $\nabla^2 r_j$, $j = 1, 2, \ldots, m$, which are needed in the second term in (10.5). In fact, if we calculated the Jacobian J_k in the course of evaluating the gradient $\nabla f_k = J_k^T r_k$, the approximation (10.24) does not require any additional derivative evaluations, and the savings in computational time can be quite significant in some applications. Second, there are many interesting situations in which the first term $J^T J$ in (10.5) dominates the second term (at least close to the solution x^*), so that $J_k^T J_k$ is a close approximation to $\nabla^2 f_k$ and the convergence rate of Gauss–Newton is similar to that of Newton's method. The first term in (10.5) will be dominant when the norm of each second-order term (that is, $|r_j(x)| \|\nabla^2 r_j(x)\|$) is significantly smaller than the eigenvalues of $J^T J$. As mentioned in the introduction, we tend to see this behavior when either the residuals r_j are small or when they are nearly affine (so that the $\|\nabla^2 r_j\|$ are small). In practice, many least-squares problems have small residuals at the solution, leading to rapid local convergence of Gauss–Newton.

A third advantage of Gauss–Newton is that whenever J_k has full rank and the gradient ∇f_k is nonzero, the direction p_k^{GN} is a descent direction for f, and therefore a suitable direction for a line search. From (10.4) and (10.23) we have

$$(p_k^{\text{GN}})^T \nabla f_k = (p_k^{\text{GN}})^T J_k^T r_k = -(p_k^{\text{GN}})^T J_k^T J_k p_k^{\text{GN}} = -\|J_k p_k^{\text{GN}}\|^2 \leq 0. \tag{10.25}$$

The final inequality is strict unless $J_k p_k^{GN} = 0$, in which case we have by (10.23) and full rank of J_k that $J_k^T r_k = \nabla f_k = 0$; that is, x_k is a stationary point. Finally, the fourth advantage of Gauss–Newton arises from the similarity between the equations (10.23) and the normal equations (10.14) for the linear least-squares problem. This connection tells us that p_k^{GN} is in fact the solution of the linear least-squares problem

$$\min_p \tfrac{1}{2}\|J_k p + r_k\|^2. \tag{10.26}$$

Hence, we can find the search direction by applying linear least-squares algorithms to the subproblem (10.26). In fact, if the QR or SVD-based algorithms are used, there is no need to calculate the Hessian approximation $J_k^T J_k$ in (10.23) explicitly; we can work directly with the Jacobian J_k. The same is true if we use a conjugate-gradient technique to solve (10.26). For this method we need to perform matrix-vector multiplications with $J_k^T J_k$, which can be done by first multiplying by J_k and then by J_k^T.

If the number of residuals m is large while the number of variables n is relatively small, it may be unwise to store the Jacobian J explicitly. A preferable strategy may be to calculate the matrix $J^T J$ and gradient vector $J^T r$ by evaluating r_j and ∇r_j successively for $j = 1, 2, \ldots, m$ and performing the accumulations

$$J^T J = \sum_{i=1}^{m} (\nabla r_j)(\nabla r_j)^T, \quad J^T r = \sum_{i=1}^{m} r_j(\nabla r_j). \tag{10.27}$$

The Gauss–Newton steps can then be computed by solving the system (10.23) of normal equations directly.

The subproblem (10.26) suggests another motivation for the Gauss–Newton search direction. We can view this equation as being obtained from a linear model for the the vector function $r(x_k + p) \approx r_k + J_k p$, substituted into the function $\tfrac{1}{2}\|\cdot\|^2$. In other words, we use the approximation

$$f(x_k + p) = \tfrac{1}{2}\|r(x_k + p)\|^2 \approx \tfrac{1}{2}\|J_k p + r_k\|^2,$$

and choose p_k^{GN} to be the minimizer of this approximation.

Implementations of the Gauss–Newton method usually perform a line search in the direction p_k^{GN}, requiring the step length α_k to satisfy conditions like those discussed in Chapter 3, such as the Armijo and Wolfe conditions; see (3.4) and (3.6).

CONVERGENCE OF THE GAUSS–NEWTON METHOD

The theory of Chapter 3 can applied to study the convergence properties of the Gauss–Newton method. We prove a global convergence result under the assumption that the Jacobians $J(x)$ have their singular values uniformly bounded away from zero in the region of interest; that is, there is a constant $\gamma > 0$ such that

$$\|J(x)z\| \geq \gamma \|z\| \tag{10.28}$$

for all x in a neighborhood $\mathcal{N}$ of the level set

$$\mathcal{L} = \{x \mid f(x) \leq f(x_0)\}, \tag{10.29}$$

where x_0 is the starting point for the algorithm. We assume here and in the rest of the chapter that $\mathcal{L}$ is bounded. Our result is a consequence of Theorem 3.2.

Theorem 10.1.

  Suppose each residual function r_j is Lipschitz continuously differentiable in a neighborhood $\mathcal{N}$ of the bounded level set (10.29), and that the Jacobians $J(x)$ satisfy the uniform full-rank condition (10.28) on $\mathcal{N}$. Then if the iterates x_k are generated by the Gauss–Newton method with step lengths α_k that satisfy (3.6), we have

$$\lim_{k \to \infty} J_k^T r_k = 0.$$

PROOF.   First, we note that the neighborhood $\mathcal{N}$ of the bounded level set $\mathcal{L}$ can be chosen small enough that the following properties are satisfied for some positive constants L and β:

$$|r_j(x)| \leq \beta \quad \text{and} \quad \|\nabla r_j(x)\| \leq \beta,$$
$$|r_j(x) - r_j(\tilde{x})| \leq L\|x - \tilde{x}\| \quad \text{and} \quad \|\nabla r_j(x) - \nabla r_j(\tilde{x})\| \leq L\|x - \tilde{x}\|,$$

for all $x, \tilde{x} \in \mathcal{N}$ and all $j = 1, 2, \ldots, m$. It is easy to deduce that there exists a constant $\bar{\beta} > 0$ such that $\|J(x)^T\| = \|J(x)\| \leq \bar{\beta}$ for all $x \in \mathcal{L}$. In addition, by applying the results concerning Lipschitz continuity of products and sums (see for example (A.43)) to the gradient $\nabla f(x) = \sum_{j=1}^m r_j(x) \nabla r_j(x)$, we can show that ∇f is Lipschitz continuous. Hence, the assumptions of Theorem 3.2 are satisfied.

  We check next that the angle θ_k between the search direction p_k^{GN} and the negative gradient $-\nabla f_k$ is uniformly bounded away from $\pi/2$. From (3.12), (10.25), and (10.28), we have for $x = x_k \in \mathcal{L}$ and $p^{\text{GN}} = p_k^{\text{GN}}$ that

$$\cos \theta_k = -\frac{(\nabla f)^T p^{\text{GN}}}{\|p^{\text{GN}}\| \|\nabla f\|} = \frac{\|J p^{\text{GN}}\|^2}{\|p^{\text{GN}}\| \|J^T J p^{\text{GN}}\|} \geq \frac{\gamma^2 \|p^{\text{GN}}\|^2}{\bar{\beta}^2 \|p^{\text{GN}}\|^2} = \frac{\gamma^2}{\bar{\beta}^2} > 0.$$

It follows from (3.14) in Theorem 3.2 that $\nabla f(x_k) \to 0$, giving the result.   □

  If J_k is rank-deficient for some k (so that a condition like (10.28) is not satisfied), the coefficient matrix in (10.23) is singular. The system (10.23) still has a solution, however, because of the equivalence between this linear system and the minimization problem (10.26).

In fact, there are infinitely many solutions for p_k^{GN} in this case; each of them has the form of (10.22). However, there is no longer an assurance that $\cos \theta_k$ is uniformly bounded away from zero, so we cannot prove a result like Theorem 10.1.

The convergence of Gauss–Newton to a solution x^* can be rapid if the leading term $J_k^T J_k$ dominates the second-order term in the Hessian (10.5). Suppose that x_k is close to x^* and that assumption (10.28) is satisfied. Then, applying an argument like the Newton's method analysis (3.31), (3.32), (3.33) in Chapter 3, we have for a unit step in the Gauss–Newton direction that

$$
\begin{aligned}
x_k + p_k^{\text{GN}} - x^* &= x_k - x^* - [J^T J(x_k)]^{-1} \nabla f(x_k) \\
&= [J^T J(x_k)]^{-1} \big[J^T J(x_k)(x_k - x^*) + \nabla f(x^*) - \nabla f(x_k) \big],
\end{aligned}
$$

where $J^T J(x)$ is shorthand notation for $J(x)^T J(x)$. Using $H(x)$ to denote the second-order term in (10.5), we have from (A.57) that

$$
\begin{aligned}
\nabla f(x_k) - \nabla f(x^*) &= \int_0^1 J^T J(x^* + t(x_k - x^*))(x_k - x^*) \, dt \\
&\quad + \int_0^1 H(x^* + t(x_k - x^*))(x_k - x^*) \, dt.
\end{aligned}
$$

A similar argument as in (3.32), (3.33), assuming Lipschitz continuity of $H(\cdot)$ near x^*, shows that

$$
\begin{aligned}
\|x_k &+ p_k^{\text{GN}} - x^*\| \\
&\leq \int_0^1 \|[J^T J(x_k)]^{-1} H(x^* + t(x_k - x^*))\| \|x_k - x^*\| \, dt + O(\|x_k - x^*\|^2) \\
&\approx \|[J^T J(x^*)]^{-1} H(x^*)\| \|x_k - x^*\| + O(\|x_k - x^*\|^2).
\end{aligned} \tag{10.30}
$$

Hence, if $\|[J^T J(x^*)]^{-1} H(x^*)\| \ll 1$, we can expect a unit step of Gauss–Newton to move us much closer to the solution x^*, giving rapid local convergence. When $H(x^*) = 0$, the convergence is actually quadratic.

When n and m are both large and the Jacobian $J(x)$ is sparse, the cost of computing steps exactly by factoring either J_k or $J_k^T J_k$ at each iteration may become quite expensive relative to the cost of function and gradient evaluations. In this case, we can design *inexact* variants of the Gauss–Newton algorithm that are analogous to the inexact Newton algorithms discussed in Chapter 7. We simply replace the Hessian $\nabla^2 f(x_k)$ in these methods by its approximation $J_k^T J_k$. The positive semidefiniteness of this approximation simplifies the resulting algorithms in several places.

THE LEVENBERG–MARQUARDT METHOD

Recall that the Gauss–Newton method is like Newton's method with line search, except that we use the convenient and often effective approximation (10.24) for the Hessian. The Levenberg–Marquardt method can be obtained by using the same Hessian approximation, but replacing the line search with a trust-region strategy. The use of a trust region avoids one of the weaknesses of Gauss–Newton, namely, its behavior when the Jacobian $J(x)$ is rank-deficient, or nearly so. Since the same Hessian approximations are used in each case, the local convergence properties of the two methods are similar.

The Levenberg–Marquardt method can be described and analyzed using the trust-region framework of Chapter 4. (In fact, the Levenberg–Marquardt method is sometimes considered to be the progenitor of the trust-region approach for general unconstrained optimization discussed in Chapter 4.) For a spherical trust region, the subproblem to be solved at each iteration is

$$\min_p \tfrac{1}{2}\|J_k p + r_k\|^2, \quad \text{subject to } \|p\| \le \Delta_k, \tag{10.31}$$

where $\Delta_k > 0$ is the trust-region radius. In effect, we are choosing the model function $m_k(\cdot)$ in (4.3) to be

$$m_k(p) = \tfrac{1}{2}\|r_k\|^2 + p^T J_k^T r_k + \tfrac{1}{2} p^T J_k^T J_k p. \tag{10.32}$$

We drop the iteration counter k during the rest of this section and concern ourselves with the subproblem (10.31). The results of Chapter 4 allow us to characterize the solution of (10.31) in the following way: When the solution p^{GN} of the Gauss–Newton equations (10.23) lies strictly inside the trust region (that is, $\|p^{\text{GN}}\| < \Delta$), then this step p^{GN} also solves the subproblem (10.31). Otherwise, there is a $\lambda > 0$ such that the solution $p = p^{\text{LM}}$ of (10.31) satisfies $\|p\| = \Delta$ and

$$\left(J^T J + \lambda I\right) p = -J^T r. \tag{10.33}$$

This claim is verified in the following lemma, which is a straightforward consequence of Theorem 4.1 from Chapter 4.

Lemma 10.2.

The vector p^{LM} is a solution of the trust-region subproblem

$$\min_p \|J p + r\|^2, \quad \text{subject to } \|p\| \le \Delta,$$

if and only if p^{LM} is feasible and there is a scalar $\lambda \ge 0$ such that

$$(J^T J + \lambda I)p^{\text{LM}} = -J^T r, \tag{10.34a}$$
$$\lambda(\Delta - \|p^{\text{LM}}\|) = 0. \tag{10.34b}$$

Proof.   In Theorem 4.1, the semidefiniteness condition (4.8c) is satisfied automatically, since $J^T J$ is positive semidefinite and $\lambda \geq 0$. The two conditions (10.34a) and (10.34b) follow from (4.8a) and (4.8b), respectively. $\square$

Note that the equations (10.33) are just the normal equations for the following linear least-squares problem:

$$\min_p \frac{1}{2} \left\| \begin{bmatrix} J \\ \sqrt{\lambda} I \end{bmatrix} p + \begin{bmatrix} r \\ 0 \end{bmatrix} \right\|^2 . \tag{10.35}$$

Just as in the Gauss–Newton case, the equivalence between (10.33) and (10.35) gives us a way of solving the subproblem without computing the matrix–matrix product $J^T J$ and its Cholesky factorization.

IMPLEMENTATION OF THE LEVENBERG–MARQUARDT METHOD

To find a value of λ that approximately matches the given Δ in Lemma 10.2, we can use the rootfinding algorithm described in Chapter 4. It is easy to safeguard this procedure: The Cholesky factor R is guaranteed to exist whenever the current estimate $\lambda^{(\ell)}$ is positive, since the approximate Hessian $B = J^T J$ is already positive semidefinite. Because of the special structure of B, we do not need to compute the Cholesky factorization of $B + \lambda I$ from scratch in each iteration of Algorithm 4.1. Rather, we present an efficient technique for finding the following QR factorization of the coefficient matrix in (10.35):

$$\begin{bmatrix} R_\lambda \\ 0 \end{bmatrix} = Q_\lambda^T \begin{bmatrix} J \\ \sqrt{\lambda} I \end{bmatrix} \tag{10.36}$$

(Q_λ orthogonal, R_λ upper triangular). The upper triangular factor R_λ satisfies $R_\lambda^T R_\lambda = (J^T J + \lambda I)$.

We can save computer time in the calculation of the factorization (10.36) by using a combination of Householder and Givens transformations. Suppose we use Householder transformations to calculate the QR factorization of J alone as

$$J = Q \begin{bmatrix} R \\ 0 \end{bmatrix} . \tag{10.37}$$

We then have

$$\begin{bmatrix} R \\ 0 \\ \sqrt{\lambda} I \end{bmatrix} = \begin{bmatrix} Q^T & \\ & I \end{bmatrix} \begin{bmatrix} J \\ \sqrt{\lambda} I \end{bmatrix} . \tag{10.38}$$

The leftmost matrix in this formula is upper triangular except for the n nonzero terms of the matrix λI. These can be eliminated by a sequence of $n(n + 1)/2$ Givens rotations, in which the diagonal elements of the upper triangular part are used to eliminate the nonzeros of λI and the fill-in terms that arise in the process. The first few steps of this process are as follows:

rotate row n of R with row n of $\sqrt{\lambda}I$, to eliminate the (n, n) element of $\sqrt{\lambda}I$;

rotate row $n - 1$ of R with row $n - 1$ of $\sqrt{\lambda}I$ to eliminate the $(n - 1, n - 1)$ element of the latter matrix. This step introduces fill-in in position $(n - 1, n)$ of $\sqrt{\lambda}I$, which is eliminated by rotating row n of R with row $n - 1$ of $\sqrt{\lambda}I$, to eliminate the fill-in element at position $(n - 1, n)$;

rotate row $n - 2$ of R with row $n - 2$ of $\sqrt{\lambda}I$, to eliminate the $(n - 2)$ diagonal in the latter matrix. This step introduces fill-in in the $(n - 2, n - 1)$ and $(n - 2, n)$ positions, which we eliminate by $\cdots$

and so on. If we gather all the Givens rotations into a matrix $\bar{Q}_\lambda$, we obtain from (10.38) that

$$
\bar{Q}_\lambda^T \begin{bmatrix} R \\ 0 \\ \sqrt{\lambda}I \end{bmatrix} = \begin{bmatrix} R_\lambda \\ 0 \\ 0 \end{bmatrix},
$$

and hence (10.36) holds with

$$
Q_\lambda = \begin{bmatrix} Q & \\ & I \end{bmatrix} \bar{Q}_\lambda.
$$

The advantage of this combined approach is that when the value of λ is changed in the rootfinding algorithm, we need only recalculate $\bar{Q}_\lambda$ and not the Householder part of the factorization (10.38). This feature can save a lot of computation in the case of $m \gg n$, since just $O(n^3)$ operations are required to recalculate $\bar{Q}_\lambda$ and R_λ for each value of λ, after the initial cost of $O(mn^2)$ operations needed to calculate Q in (10.37).

Least-squares problems are often poorly scaled. Some of the variables could have values of about 10^4, while other variables could be of order 10^{-6}. If such wide variations are ignored, the algorithms above may encounter numerical difficulties or produce solutions of poor quality. One way to reduce the effects of poor scaling is to use an *ellipsoidal* trust region in place of the spherical trust region defined above. The step is confined to an ellipse in which the lengths of the principal axes are related to the typical values of the corresponding variables. Analytically, the trust-region subproblem becomes

$$
\min_p \tfrac{1}{2} \|J_k p + r_k\|^2, \quad \text{subject to } \|D_k p\| \le \Delta_k, \tag{10.39}
$$

where D_k is a diagonal matrix with positive diagonal entries (cf. (7.13)). Instead of (10.33), the solution of (10.39) satisfies an equation of the form

$$\left(J_k^T J_k + \lambda D_k^2\right) p_k^{\text{LM}} = -J_k^T r_k, \tag{10.40}$$

and, equivalently, solves the linear least-squares problem

$$\min_p \left\| \begin{bmatrix} J_k \\ \sqrt{\lambda} D_k \end{bmatrix} p + \begin{bmatrix} r_k \\ 0 \end{bmatrix} \right\|^2. \tag{10.41}$$

The diagonals of the scaling matrix D_k can change from iteration to iteration, as we gather information about the typical range of values for each component of x. If the variation in these elements is kept within certain bounds, then the convergence theory for the spherical case continues to hold, with minor modifications. Moreover, the technique described above for calculating R_λ needs no modification. Seber and Wild [280] suggest choosing the diagonals of D_k^2 to match those of $J_k^T J_k$, to make the algorithm invariant under diagonal scaling of the components of x. This approach is analogous to the technique of scaling by diagonal elements of the Hessian, which was described in Section 4.5 in the context of trust-region algorithms for unconstrained optimization.

For problems in which m and n are large and $J(x)$ is sparse, we may prefer to solve (10.31) or (10.39) approximately using the CG-Steihaug algorithm, Algorithm 7.2 from Chapter 7, with $J_k^T J_k$ replacing the exact Hessian $\nabla^2 f_k$. Positive semidefiniteness of the matrix $J_k^T J_k$ makes for some simplification of this algorithm, because negative curvature cannot arise. It is not necessary to calculate $J_k^T J_k$ explicitly to implement Algorithm 7.2; the matrix-vector products required by the algorithm can be found by forming matrix-vector products with J_k and J_k^T separately.

CONVERGENCE OF THE LEVENBERG–MARQUARDT METHOD

It is not necessary to solve the trust-region problem (10.31) exactly in order for the Levenberg–Marquardt method to enjoy global convergence properties. The following convergence result can be obtained as a direct consequence of Theorem 4.6.

Theorem 10.3.

Let $\eta \in \left(0, \frac{1}{4}\right)$ in Algorithm 4.1 of Chapter 4, and suppose that the level set $\mathcal{L}$ defined in (10.29) is bounded and that the residual functions $r_j(\cdot)$, $j = 1, 2, \ldots, m$ are Lipschitz continuously differentiable in a neighborhood $\mathcal{N}$ of $\mathcal{L}$. Assume that for each k, the approximate solution p_k of (10.31) satisfies the inequality

$$m_k(0) - m_k(p_k) \geq c_1 \|J_k^T r_k\| \min\left(\Delta_k, \frac{\|J_k^T r_k\|}{\|J_k^T J_k\|}\right), \tag{10.42}$$

for some constant $c_1 > 0$, and in addition $\|p_k\| \leq \gamma \Delta_k$ for some constant $\gamma \geq 1$. We then have that

$$\lim_{k \to \infty} \nabla f_k = \lim_{k \to \infty} J_k^T r_k = 0.$$

PROOF.   The smoothness assumption on $r_j(\cdot)$ implies that we can choose a constant $M > 0$ such that $\|J_k^T J_k\| \leq M$ for all iterates k. Note too that the objective f is bounded below (by zero). Hence, the assumptions of Theorem 4.6 are satisfied, and the result follows immediately.   □

As in Chapter 4, there is no need to calculate the right-hand-side in the inequality (10.42) or to check it explicitly. Instead, we can simply require the decrease given by our approximate solution p_k of (10.31) to at least match the decrease given by the Cauchy point, which can be calculated inexpensively in the same way as in Chapter 4. If we use the iterative CG-Steihaug approach, Algorithm 7.2, the condition (10.42) is satisfied automatically for $c_1 = 1/2$, since the Cauchy point is the first estimate of p_k computed by this approach, while subsequent estimates give smaller values for the model function.

The local convergence behavior of Levenberg–Marquardt is similar to the Gauss–Newton method. Near a solution x^* at which the first term of the Hessian $\nabla^2 f(x^*)$ (10.5) dominates the second term, the model function in (10.31), the trust region becomes inactive and the algorithm takes Gauss–Newton steps, giving the rapid local convergence expression (10.30).

METHODS FOR LARGE-RESIDUAL PROBLEMS

In large-residual problems, the quadratic model in (10.31) is an inadequate representation of the function f because the second-order part of the Hessian $\nabla^2 f(x)$ is too significant to be ignored. In data-fitting problems, the presence of large residuals may indicate that the model is inadequate or that errors have been made in monitoring the observations. Still, the practitioner may need to solve the least-squares problem with the current model and data, to indicate where improvements are needed in the weighting of observations, modeling, or data collection.

On large-residual problems, the asymptotic convergence rate of Gauss–Newton and Levenberg–Marquardt algorithms is only linear—slower than the superlinear convergence rate attained by algorithms for general unconstrained problems, such as Newton or quasi-Newton. If the individual Hessians $\nabla^2 r_j$ are easy to calculate, it may be better to ignore the structure of the least-squares objective and apply Newton's method with trust region or line search to the problem of minimizing f. Quasi-Newton methods, which attain a superlinear convergence rate without requiring calculation of $\nabla^2 r_j$, are another option. However, the behavior of both Newton and quasi-Newton on early iterations (before reaching a neighborhood of the solution) may be inferior to Gauss–Newton and Levenberg–Marquardt.

Of course, we often do not know beforehand whether a problem will turn out to have small or large residuals at the solution. It seems reasonable, therefore, to consider *hybrid algorithms*, which would behave like Gauss–Newton or Levenberg–Marquardt if the residuals turn out to be small (and hence take advantage of the cost savings associated with these methods) but switch to Newton or quasi-Newton steps if the residuals at the solution appear to be large.

There are a couple of ways to construct hybrid algorithms. One approach, due to Fletcher and Xu (see Fletcher [101]), maintains a sequence of positive definite Hessian approximations B_k. If the Gauss–Newton step from x_k reduces the function f by a certain fixed amount (say, a factor of 5), then this step is taken and B_k is overwritten by $J_k^T J_k$. Otherwise, a direction is computed using B_k, and the new point x_{k+1} is obtained by performing a line search. In either case, a BFGS-like update is applied to B_k to obtain a new approximation B_{k+1}. In the zero-residual case, the method eventually always takes Gauss–Newton steps (giving quadratic convergence), while it eventually reduces to BFGS in the nonzero-residual case (giving superlinear convergence). Numerical results in Fletcher [101, Tables 6.1.2, 6.1.3] show good results for this approach on small-, large-, and zero-residual problems.

A second way to combine Gauss–Newton and quasi-Newton ideas is to maintain approximations to just the second-order part of the Hessian. That is, we maintain a sequence of matrices S_k that approximate the summation term $\sum_{j=1}^{m} r_j(x_k)\nabla^2 r_j(x_k)$ in (10.5), and then use the overall Hessian approximation

$$B_k = J_k^T J_k + S_k$$

in a trust-region or line search model for calculating the step p_k. Updates to S_k are devised so that the approximate Hessian B_k, or its constituent parts, mimics the behavior of the corresponding exact quantities over the step just taken. The update formula is based on a *secant equation*, which arises also in the context of unconstrained minimization (6.6) and nonlinear equations (11.27). In the present instance, there are a number of different ways to define the secant equation and to specify the other conditions needed for a complete update formula for S_k. We describe the algorithm of Dennis, Gay, and Welsch [90], which is probably the best-known algorithm in this class because of its implementation in the well-known NL2SOL package.

In [90], the secant equation is motivated in the following way. Ideally, S_{k+1} should be a close approximation to the exact second-order term at $x = x_{k+1}$; that is,

$$S_{k+1} \approx \sum_{j=1}^{m} r_j(x_{k+1})\nabla^2 r_j(x_{k+1}).$$

Since we do not want to calculate the individual Hessians $\nabla^2 r_j$ in this formula, we could replace each of them with an approximation $(B_j)_{k+1}$ and impose the condition that $(B_j)_{k+1}$

should mimic the behavior of its exact counterpart $\nabla^2 r_j$ over the step just taken; that is,

$$(B_j)_{k+1}(x_{k+1} - x_k) = \nabla r_j(x_{k+1}) - \nabla r_j(x_k)$$

$$= (\text{row } j \text{ of } J(x_{k+1}))^T - (\text{row } j \text{ of } J(x_k))^T.$$

This condition leads to a secant equation on S_{k+1}, namely,

$$S_{k+1}(x_{k+1} - x_k) = \sum_{j=1}^{m} r_j(x_{k+1})(B_j)_{k+1}(x_{k+1} - x_k)$$

$$= \sum_{j=1}^{m} r_j(x_{k+1}) \left[(\text{row } j \text{ of } J(x_{k+1}))^T - (\text{row } j \text{ of } J(x_k))^T \right]$$

$$= J_{k+1}^T r_{k+1} - J_k^T r_{k+1}.$$

As usual, this condition does not completely specify the new approximation S_{k+1}. Dennis, Gay, and Welsch add requirements that S_{k+1} be symmetric and that the difference $S_{k+1} - S_k$ from the previous estimate S_k be minimized in a certain sense, and derive the following update formula:

$$S_{k+1} = S_k + \frac{(y^\sharp - S_k s)y^T + y(y^\sharp - S_k s)^T}{y^T s} - \frac{(y^\sharp - S_k s)^T s}{(y^T s)^2} yy^T, \qquad (10.43)$$

where

$$s = x_{k+1} - x_k,$$
$$y = J_{k+1}^T r_{k+1} - J_k^T r_k,$$
$$y^\sharp = J_{k+1}^T r_{k+1} - J_k^T r_{k+1}.$$

Note that (10.43) is a slight variant on the DFP update for unconstrained minimization. It would be identical if $y^\sharp$ and y were the same.

Dennis, Gay, and Welsch use their approximate Hessian $J_k^T J_k + S_k$ in conjunction with a trust-region strategy, but a few more features are needed to enhance its performance. One deficiency of its basic update strategy for S_k is that this matrix is not guaranteed to vanish as the iterates approach a zero-residual solution, so it can interfere with superlinear convergence. This problem is avoided by scaling S_k prior to its update; we replace S_k by $\tau_k S_k$ on the right-hand-side of (10.43), where

$$\tau_k = \min \left(1, \frac{|s^T y^\sharp|}{|s^T S_k s|} \right).$$

A final modification in the overall algorithm is that the S_k term is omitted from the Hessian approximation when the resulting Gauss–Newton model produces a sufficiently good step.

10.4 ORTHOGONAL DISTANCE REGRESSION

In Example 10.1 we assumed that no errors were made in noting the time at which the blood samples were drawn, so that the differences between the model $\phi(x; t_j)$ and the observation y_j were due to inadequacy in the model or measurement errors in y_j. We assumed that any errors in the ordinates—the times t_j—are tiny by comparison with the errors in the observations. This assumption often is reasonable, but there are cases where the answer can be seriously distorted if we fail to take possible errors in the ordinates into account. Models that take these errors into account are known in the statistics literature as *errors-in-variables models* [280, Chapter 10], and the resulting optimization problems are referred to as *total least squares* in the case of a linear model (see Golub and Van Loan [136, Chapter 5]) or as *orthogonal distance regression* in the nonlinear case (see Boggs, Byrd, and Schnabel [30]).

We formulate this problem mathematically by introducing perturbations δ_j for the ordinates t_j, as well as perturbations ϵ_j for y_j, and seeking the values of these $2m$ perturbations that minimize the discrepancy between the model and the observations, as measured by a weighted least-squares objective function. To be precise, we relate the quantities t_j, y_j, δ_j, and ϵ_j by

$$y_j = \phi(x; t_j + \delta_j) + \epsilon_j, \qquad j = 1, 2, \ldots, m, \tag{10.44}$$

and define the minimization problem as

$$\min_{x, \delta_j, \epsilon_j} \tfrac{1}{2} \sum_{j=1}^{m} w_j^2 \epsilon_j^2 + d_j^2 \delta_j^2, \qquad \text{subject to (10.44).} \tag{10.45}$$

The quantities w_i and d_i are weights, selected either by the modeler or by some automatic estimate of the relative significance of the error terms.

It is easy to see how the term "orthogonal distance regression" originates when we graph this problem; see Figure 10.2. If all the weights w_i and d_i are equal, then each term in the summation (10.45) is simply the shortest distance between the point (t_j, y_j) and the curve $\phi(x; t)$ (plotted as a function of t). The shortest path between each point and the curve is orthogonal to the curve at the point of intersection.

Using the constraints (10.44) to eliminate the variables ϵ_j from (10.45), we obtain the unconstrained least-squares problem

$$\min_{x, \delta} F(x, \delta) = \tfrac{1}{2} \sum_{j=1}^{m} w_j^2 [y_j - \phi(x; t_j + \delta_j)]^2 + d_j^2 \delta_j^2 = \tfrac{1}{2} \sum_{j=1}^{2m} r_j^2(x, \delta), \tag{10.46}$$

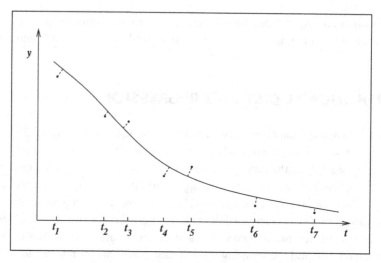

Figure 10.2 Orthogonal distance regression minimizes the sum of squares of the distance from each point to the curve.

where $\delta = (\delta_1, \delta_2, \ldots, \delta_m)^T$ and we have defined

$$
r_j(x, \delta) = \begin{cases} w_j[\phi(x; t_j + \delta_j) - y_j], & j = 1, 2, \ldots, m, \\ d_{j-m}\delta_{j-m}, & j = m+1, \ldots, 2m. \end{cases} \tag{10.47}
$$

Note that (10.46) is now a standard least-squares problem with $2m$ residuals and $m + n$ unknowns, which we can solve by using the techniques in this chapter. A naive implementation of this strategy may, however, be quite expensive, since the number of parameters $(2n)$ and the number of observations $(m + n)$ may both be much larger than for the original problem.

Fortunately, the Jacobian matrix for (10.46) has a special structure that can be exploited in implementing the Gauss–Newton or Levenberg–Marquardt methods. Many of its components are zero; for instance, we have

$$
\frac{\partial r_j}{\partial \delta_i} = \frac{\partial [\phi(t_j + \delta_j; x) - y_j]}{\partial \delta_i} = 0, \quad i, j = 1, 2, \ldots, m, \ i \neq j,
$$

and

$$
\frac{\partial r_j}{\partial x_i} = 0, \quad j = m+1, \ldots, 2m, \ i = 1, 2, \ldots, n.
$$

Additionally, we have for $j = 1, 2, \ldots, m$ and $i = 1, 2, \ldots, m$ that

$$\frac{\partial r_{m+j}}{\partial \delta_i} = \begin{cases} d_j & \text{if } i = j, \\ 0 & \text{otherwise.} \end{cases}$$

Hence, we can partition the Jacobian of the residual function r defined by (10.47) into blocks and write

$$J(x, \delta) = \begin{bmatrix} \hat{J} & V \\ 0 & D \end{bmatrix}, \tag{10.48}$$

where V and D are $m \times m$ diagonal matrices and $\hat{J}$ is the $m \times n$ matrix of partial derivatives of the functions $w_j \phi(t_j + \delta_j; x)$ with respect to x. Boggs, Byrd, and Schnabel [30]) apply the Levenberg–Marquardt algorithm to (10.46) and note that block elimination can be used to solve the subproblems (10.33), (10.35) efficiently. Given the partitioning (10.48), we can partition the step vector p and the residual vector r accordingly as

$$p = \begin{bmatrix} p_x \\ p_\delta \end{bmatrix}, \qquad r = \begin{bmatrix} \hat{r}_1 \\ \hat{r}_2 \end{bmatrix},$$

and write the normal equations (10.33) in the partitioned form

$$\begin{bmatrix} \hat{J}^T \hat{J} + \lambda I & \hat{J}^T V \\ V \hat{J} & V^2 + D^2 + \lambda I \end{bmatrix} \begin{bmatrix} p_x \\ p_\delta \end{bmatrix} = - \begin{bmatrix} \hat{J}^T \hat{r}_1 \\ V \hat{r}_1 + D \hat{r}_2 \end{bmatrix}. \tag{10.49}$$

Since the lower right submatrix $V^2 + D^2 + \lambda I$ is diagonal, it is easy to eliminate p_δ from this system and obtain a smaller $n \times n$ system to be solved for p_x alone. The total cost of finding a step is only marginally greater than for the $m \times n$ problem arising from the standard least-squares model.

NOTES AND REFERENCES

Algorithms for linear least squares are discussed comprehensively by Björck [29], who includes detailed error analyses of the different algorithms and software listings. He considers not just the basic problem (10.13) but also the situation in which there are bounds (for example, $x \geq 0$) or linear constraints (for example, $Ax \geq b$) on the variables. Golub and Van Loan [136, Chapter 5] survey the state of the art, including discussion of the suitability of the different approaches (for example, normal equations vs. QR factorization) for different problem types. A classical reference on linear least-squares is Lawson and Hanson [188].

Very large nonlinear least-squares problems arise in numerous areas of application, such as medical imaging, geophysics, economics, and engineering design. In many instances, both the number of variables n and the number of residuals m is large, but it is also quite common that only m is large.

The original description of the Levenberg–Marquardt algorithm [190, 203] did not make the connection with the trust-region concept. Rather, it adjusted the value of λ in (10.33) directly, increasing or decreasing it by a certain factor according to whether or not the previous trial step was effective in decreasing $f(\cdot)$. (The heuristics for adjusting λ were analogous to those used for adjusting the trust-region radius Δ_k in Algorithm 4.1.) Similar convergence results to Theorem 10.3 can be proved for algorithms that use this approach (see, for instance, Osborne [231]), independently of trust-region analysis. The connection with trust regions was firmly established by Moré [210].

Wright and Holt [318] present an inexact Levenberg–Marquardt approach for large-scale nonlinear least squares that manipulates the parameter λ directly rather than making use of the connection to trust-region algorithms. This method takes steps $\bar{p}_k$ that, analogously to (7.2) and (7.3) in Chapter 7, satisfy the system

$$\left\| \left(J_k^T J_k + \lambda_k I \right) \bar{p}_k + J_k^T r_k \right\| \leq \eta_k \| J_k^T r_k \|, \qquad \text{for some } \eta_k \in [0, \eta],$$

where $\eta \in (0, 1)$ is a constant and $\{\eta_k\}$ is a forcing sequence. A ratio of actual to predicted decrease is used to decide whether the step $\bar{p}_k$ should be taken, and convergence to stationary points can be proved under certain assumptions. The method can be implemented efficiently by using Algorithm LSQR of Paige and Saunders [234] to calculate the approximate solution of (10.35) since, for a small marginal cost, this algorithm can compute approximate solutions for a number of different values of λ_k simultaneously. Hence, we can compute values of $\bar{p}_k$ corresponding to a range of values of λ_k, and choose the actual step to be the one corresponding to the smallest λ_k for which the actual-predicted decrease ratio is satisfactory.

Nonlinear least squares software is fairly prevalent because of the high demand for it. Major numerical software libraries such as IMSL, HSL, NAG, and SAS, as well as programming environments such as Mathematica and Matlab, contain robust nonlinear least-squares implementations. Other high quality implmentations include DFNLP, MINPACK, NL2SOL, and NLSSOL; see Moré and Wright [217, Chapter 3]. The nonlinear programming packages LANCELOT, KNITRO, and SNOPT provide large-scale implementations of the Gauss–Newton and Levenberg–Marquardt methods. The orthogonal distance regression algorithm is implemented by ORDPACK [31].

All these routines (which can be accessed through the web) give the user the option of either supplying Jacobians explicitly or else allowing the code to compute them by finite differencing. (In the latter case, the user need only write code to compute the residual vector $r(x)$; see Chapter 8.) Seber and Wild [280, Chapter 15] describe some of the important practical issues in selecting software for statistical applications.

✐ EXERCISES

✐ **10.1** Let J be an $m \times n$ matrix with $m \geq n$, and let $y \in \mathbb{R}^m$ be a vector.

(a) Show that J has full column rank if and only if $J^T J$ is nonsingular.

(b) Show that J has full column rank if and only if $J^T J$ is positive definite.

✐ **10.2** Show that the function $f(x)$ in (10.13) is convex.

✐ **10.3** Show that

(a) if Q is an orthogonal matrix, then $\|Qx\| = \|x\|$ for any vector x;

(b) the matrices $\bar{R}$ in (10.15) and R in (10.17) are identical if $\Pi = I$, provided that J has full column rank n.

✐ **10.4**

(a) Show that x^* defined in (10.22) is a minimizer of (10.13).

(b) Find $\|x^*\|$ and conclude that this norm is minimized when $\tau_i = 0$ for all i with $\sigma_i = 0$.

✐ **10.5** Suppose that each residual function r_j and its gradient are Lipschitz continuous with Lipschitz constant L, that is,

$$\|r_j(x) - r_j(\tilde{x})\| \leq L\|x - \tilde{x}\|, \quad \|\nabla r_j(x) - \nabla r_j(\tilde{x})\| \leq L\|x - \tilde{x}\|$$

for all $j = 1, 2, \ldots, m$ and all $x, \tilde{x} \in \mathcal{D}$, where $\mathcal{D}$ is a compact subset of $\mathbb{R}^n$. Assume also that the r_j are bounded on $\mathcal{D}$, that is, there exists $M > 0$ such that $|r_j(x)| \leq M$ for all $j = 1, 2, \ldots, m$ and all $x \in \mathcal{D}$. Find Lipschitz constants for the Jacobian J (10.3) and the gradient ∇f (10.4) over $\mathcal{D}$.

✐ **10.6** Express the solution p of (10.33) in terms of the singular-value decomposition of $J(x)$ and the scalar λ. Express its squared-norm $\|p\|^2$ in these same terms, and show that

$$\lim_{\lambda \to 0} p = \sum_{\sigma_i \neq 0} \frac{u_i^T r}{\sigma_i} v_i.$$

CHAPTER 11

Nonlinear Equations

In many applications we do not need to optimize an objective function explicitly, but rather to find values of the variables in a model that satisfy a number of given relationships. When these relationships take the form of n equalities—the same number of equality conditions as variables in the model—the problem is one of solving a system of *nonlinear equations*. We write this problem mathematically as

$$r(x) = 0, \tag{11.1}$$

where $r : \mathbb{R}^n \to \mathbb{R}^n$ is a vector function, that is,

$$r(x) = \begin{bmatrix} r_1(x) \\ r_2(x) \\ \vdots \\ r_n(x) \end{bmatrix}.$$

In this chapter, we assume that each function $r_i : \mathbb{R}^n \to \mathbb{R}$, $i = 1, 2, \ldots, n$, is smooth. A vector x^* for which (11.1) is satisfied is called a *solution* or *root* of the nonlinear equations. A simple example is the system

$$r(x) = \begin{bmatrix} x_2^2 - 1 \\ \sin x_1 - x_2 \end{bmatrix} = 0,$$

which is a system of $n = 2$ equations with infinitely many solutions, two of which are $x^* = (3\pi/2, -1)^T$ and $x^* = (\pi/2, 1)^T$. In general, the system (11.1) may have no solutions, a unique solution, or many solutions.

The techniques for solving nonlinear equations overlap in their motivation, analysis, and implementation with optimization techniques discussed in earlier chapters. In both optimization and nonlinear equations, Newton's method lies at the heart of many important algorithms. Features such as line searches, trust regions, and inexact solution of the linear algebra subproblems at each iteration are important in both areas, as are other issues such as derivative evaluation and global convergence.

Because some important algorithms for nonlinear equations proceed by minimizing a sum of squares of the equations, that is,

$$\min_x \sum_{i=1}^{n} r_i^2(x),$$

there are particularly close connections with the nonlinear least-squares problem discussed in Chapter 10. The differences are that in nonlinear equations, the number of equations *equals* the number of variables (instead of exceeding the number of variables, as is typically the case in Chapter 10), and that we expect all equations to be satisfied at the solution, rather than just minimizing the sum of squares. This point is important because the nonlinear equations may represent physical or economic constraints such as conservation laws or consistency principles, which must hold exactly in order for the solution to be meaningful.

Many applications require us to solve a sequence of closely related nonlinear systems, as in the following example.

❏ **Example 11.1**   (Rheinboldt; see [212])

An interesting problem in control is to analyze the stability of an aircraft in response to the commands of the pilot. The following is a simplified model based on force-balance equations, in which gravity terms have been neglected.

The equilibrium equations for a particular aircraft are given by a system of 5 equations in 8 unknowns of the form

$$F(x) \equiv Ax + \phi(x) = 0, \tag{11.2}$$

where $F : \mathbb{R}^8 \to \mathbb{R}^5$, the matrix A is given by

$$A = \begin{bmatrix} -3.933 & 0.107 & 0.126 & 0 & -9.99 & 0 & -45.83 & -7.64 \\ 0 & -0.987 & 0 & -22.95 & 0 & -28.37 & 0 & 0 \\ 0.002 & 0 & -0.235 & 0 & 5.67 & 0 & -0.921 & -6.51 \\ 0 & 1.0 & 0 & -1.0 & 0 & -0.168 & 0 & 0 \\ 0 & 0 & -1.0 & 0 & -0.196 & 0 & -0.0071 & 0 \end{bmatrix},$$

and the nonlinear part is defined by

$$\phi(x) = \begin{bmatrix} -0.727x_2x_3 + 8.39x_3x_4 - 684.4x_4x_5 + 63.5x_4x_2 \\ 0.949x_1x_3 + 0.173x_1x_5 \\ -0.716x_1x_2 - 1.578x_1x_4 + 1.132x_4x_2 \\ -x_1x_5 \\ x_1x_4 \end{bmatrix}.$$

The first three variables x_1, x_2, x_3, represent the rates of roll, pitch, and yaw, respectively, while x_4 is the incremental angle of attack and x_5 the sideslip angle. The last three variables x_6, x_7, x_8 are the controls; they represent the deflections of the elevator, aileron, and rudder, respectively.

For a given choice of the control variables x_6, x_7, x_8 we obtain a system of 5 equations and 5 unknowns. If we wish to study the behavior of the aircraft as the controls are changed, we need to solve a system of nonlinear equations with unknowns $x_1, x_2, \ldots, x_5$ for each setting of the controls.

❏

Despite the many similarities between nonlinear equations and unconstrained and least-squares optimization algorithms, there are also some important differences. To obtain quadratic convergence in optimization we require second derivatives of the objective function, whereas knowledge of the first derivatives is sufficient in nonlinear equations.

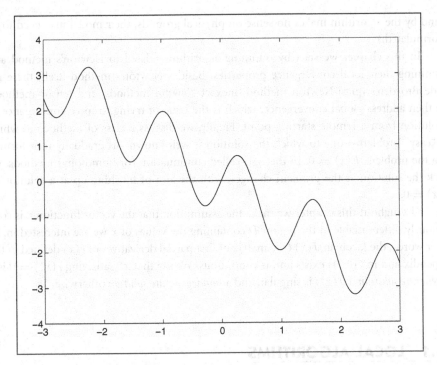

Figure 11.1   The function $r(x) = \sin(5x) - x$ has three roots.

Quasi-Newton methods are perhaps less useful in nonlinear equations than in optimization. In unconstrained optimization, the objective function is the natural choice of merit function that gauges progress towards the solution, but in nonlinear equations various merit functions can be used, all of which have some drawbacks. Line search and trust-region techniques play an equally important role in optimization, but one can argue that trust-region algorithms have certain theoretical advantages in solving nonlinear equations.

Some of the difficulties that arise in trying to solve nonlinear equations can be illustrated by a simple scalar example ($n = 1$). Suppose we have

$$r(x) = \sin(5x) - x, \tag{11.3}$$

as plotted in Figure 11.1. From this figure we see that there are three solutions of the problem $r(x) = 0$, also known as *roots of* r, located at zero and approximately ± 0.519148. This situation of multiple solutions is similar to optimization problems where, for example, a function may have more than one local minimum. It is not *quite* the same, however: In the case of optimization, one of the local minima may have a lower function value than the others (making it a "better" solution), while in nonlinear equations all solutions are equally good from a mathematical viewpoint. (If the modeler decides that the solution found by the algorithm makes no sense on physical grounds, their model may need to be reformulated.)

In this chapter we start by outlining algorithms related to Newton's method and examining their local convergence properties. Besides Newton's method itself, these include Broyden's quasi-Newton method, inexact Newton methods, and tensor methods. We then address global convergence, which is the issue of trying to force convergence to a solution from a remote starting point. Finally, we discuss a class of methods in which an "easy" problem—one to which the solution is well known—is gradually transformed into the problem $F(x) = 0$. In these so-called continuation (or homotopy) methods, we track the solution as the problem changes, with the aim of finishing up at a solution of $F(x) = 0$.

Throughout this chapter we make the assumption that the vector function r is continuously differentiable in the region $\mathcal{D}$ containing the values of x we are interested in. In other words, the Jacobian $J(x)$ (the matrix of first partial derivatives of $r(x)$ defined in the Appendix and in (10.3)) exists and is continuous. We say that x^* satisfying $r(x^*) = 0$ is a *degenerate solution* if $J(x^*)$ is singular, and a *nondegenerate solution* otherwise.

11.1 LOCAL ALGORITHMS

NEWTON'S METHOD FOR NONLINEAR EQUATIONS

Recall from Theorem 2.1 that Newton's method for minimizing $f : \mathbb{R}^n \to \mathbb{R}$ forms a quadratic model function by taking the first three terms of the Taylor series approximation of f around the current iterate x_k. The Newton step is the vector that minimizes this model. In the case of nonlinear equations, Newton's method is derived in a similar way, but with a *linear* model, one that involves function values and first derivatives of the functions $r_i(x)$, $i = 1, 2, \ldots, m$ at the current iterate x_k. We justify this strategy by referring to the following multidimensional variant of Taylor's theorem.

Theorem 11.1.

Suppose that $r : \mathbb{R}^n \to \mathbb{R}^n$ is continuously differentiable in some convex open set $\mathcal{D}$ and that x and $x + p$ are vectors in $\mathcal{D}$. We then have that

$$r(x + p) = r(x) + \int_0^1 J(x + tp)p \, dt. \tag{11.4}$$

We can define a linear model $M_k(p)$ of $r(x_k + p)$ by approximating the second term on the right-hand-side of (11.4) by $J(x)p$, and writing

$$M_k(p) \stackrel{\text{def}}{=} r(x_k) + J(x_k)p. \tag{11.5}$$

Newton's method, in its pure form, chooses the step p_k to be the vector for which $M_k(p_k) = 0$, that is, $p_k = -J(x_k)^{-1}r(x_k)$. We define it formally as follows.

Algorithm 11.1 (Newton's Method for Nonlinear Equations).

Choose x_0;
for $k = 0, 1, 2, \ldots$
     Calculate a solution p_k to the Newton equations

$$J(x_k)p_k = -r(x_k); \tag{11.6}$$

    $x_{k+1} \leftarrow x_k + p_k$;
end (for)

We use a linear model to derive the Newton step, rather than a quadratic model as in unconstrained optimization, because the linear model normally has a solution and yields an algorithm with rapid convergence properties. In fact, Newton's method for unconstrained optimization (see (2.15)) can be derived by applying Algorithm 11.1 to the nonlinear equations $\nabla f(x) = 0$. We see also in Chapter 18 that sequential quadratic programming for equality-constrained optimization can be derived by applying Algorithm 11.1 to the nonlinear equations formed by the first-order optimality conditions (18.3) for this problem. Another connection is with the Gauss–Newton method for nonlinear least squares; the formula (11.6) is equivalent to (10.23) in the usual case in which $J(x_k)$ is nonsingular.

When the iterate x_k is close to a nondegenerate root x^*, Newton's method converges superlinearly, as we show in Theorem 11.2 below. Potential shortcomings of the method include the following.

- When the starting point is remote from a solution, Algorithm 11.1 can behave erratically. When $J(x_k)$ is singular, the Newton step may not even be defined.

- First-derivative information (the Jacobian matrix J) may be difficult to obtain.

- It may be too expensive to find and calculate the Newton step p_k exactly when n is large.

- The root x^* in question may be degenerate, that is, $J(x^*)$ may be singular.

An example of a degenerate problem is the scalar function $r(x) = x^2$, which has a single degenerate root at $x^* = 0$. Algorithm 11.1, when started from any nonzero x_0, generates the sequence of iterates

$$x_k = \frac{1}{2^k}x_0,$$

which converges to the solution 0, but only at a linear rate.

As we show later in this chapter, Newton's method can be modified and enhanced in various ways to get around most of these problems. The variants we describe form the basis of much of the available software for solving nonlinear equations.

We summarize the local convergence properties of Algorithm 11.1 in the following theorem. For part of this result, we make use of a Lipschitz continuity assumption on the Jacobian, by which we mean that there is a constant β_L such that

$$\|J(x_0) - J(x_1)\| \le \beta_L \|x_0 - x_1\|, \tag{11.7}$$

for all x_0 and x_1 in the domain in question.

Theorem 11.2.

Suppose that r is continuously differentiable in a convex open set $\mathcal{D} \subset \mathbf{R}^n$. Let $x^* \in \mathcal{D}$ be a nondegenerate solution of $r(x) = 0$, and let $\{x_k\}$ be the sequence of iterates generated by Algorithm 11.1. Then when $x_k \in \mathcal{D}$ is sufficiently close to x^*, we have

$$x_{k+1} - x^* = o(\|x_k - x^*\|), \tag{11.8}$$

indicating local Q-superlinear convergence. When r is Lipschitz continuously differentiable near x^*, we have for all x_k sufficiently close to x^* that

$$x_{k+1} - x^* = O(\|x_k - x^*\|^2), \tag{11.9}$$

indicating local Q-quadratic convergence.

PROOF.   Since $r(x^*) = 0$, we have from Theorem 11.1 that

$$r(x_k) = r(x_k) - r(x^*) = J(x_k)(x_k - x^*) + w(x_k, x^*), \tag{11.10}$$

where

$$w(x_k, x^*) = \int_0^1 \left[J(x_k + t(x^* - x_k)) - J(x_k)\right](x_k - x^*). \tag{11.11}$$

From (A.12) and continuity of J, we have

$$\begin{aligned}
\|w(x_k, x^*)\| &= \left\| \int_0^1 [J(x^* + t(x^* - x_k)) - J(x_k)](x_k - x^*)\,dt \right\| \\
&\le \int_0^1 \|J(x^* + t(x^* - x_k)) - J(x_k)\| \, \|x_k - x^*\| \, dt \\
&= o(\|x_k - x^*\|).
\end{aligned} \tag{11.12}$$

Since $J(x^*)$ is nonsingular, there is a radius $\delta > 0$ and a positive constant β^* such that for all x in the ball $\mathcal{B}(x^*, \delta)$ defined by

$$\mathcal{B}(x^*, \delta) = \{x \mid \|x - x^*\| \le \delta\}, \tag{11.13}$$

we have that

$$\|J(x)^{-1}\| \leq \beta^* \quad \text{and} \quad x \in \mathcal{D}. \tag{11.14}$$

Assuming that $x_k \in \mathcal{B}(x^*, \delta)$, and recalling the definition (11.6), we multiply both sides of (11.10) by $J(x_k)^{-1}$ to obtain

$$-p_k = (x_k - x^*) + \|J(x_k)^{-1}\| o(\|x_k - x^*\|),$$
$$\Rightarrow \quad x_k + p_k - x^* = o(\|x_k - x^*\|),$$
$$\Rightarrow \quad x_{k+1} - x^* = o(\|x_k - x^*\|), \tag{11.15}$$

which yields (11.8).

When the Lipschitz continuity assumption (11.7) is satisfied, we can obtain a sharper estimate for the remainder term $w(x_k, x^*)$ defined in (11.11). By using (11.7) in (11.12), we obtain

$$\|w(x_k, x^*)\| = O(\|x_k - x^*\|^2). \tag{11.16}$$

By multiplying (11.10) by $J(x_k)^{-1}$ as above, we obtain

$$-p_k - (x_k - x^*) = J(x_k)^{-1} w(x_k, x^*),$$

so the estimate (11.9) follows as in (11.15). $\qquad\square$

INEXACT NEWTON METHODS

Instead of solving (11.6) exactly, inexact Newton methods use search directions p_k that satisfy the condition

$$\|r_k + J_k p_k\| \leq \eta_k \|r_k\|, \quad \text{for some } \eta_k \in [0, \eta], \tag{11.17}$$

where $\eta \in [0, 1)$ is a constant. As in Chapter 7, we refer to $\{\eta_k\}$ as the *forcing sequence*. Different methods make different choices of the forcing sequence, and they use different algorithms for finding the approximate solutions p_k. The general framework for this class of methods can be stated as follows.

Framework 11.2 (Inexact Newton for Nonlinear Equations).
  Given $\eta \in [0, 1)$;
  Choose x_0;
  for $k = 0, 1, 2, \ldots$
      Choose forcing parameter $\eta_k \in [0, \eta]$;
      Find a vector p_k that satisfies (11.17);
      $x_{k+1} \leftarrow x_k + p_k$;
  end (for)

The convergence theory for these methods depends only on the condition (11.17) and not on the particular technique used to calculate p_k. The most important methods in this class, however, make use of iterative techniques for solving linear systems of the form $Jp = -r$, such as GMRES (Saad and Schultz [273], Walker [302]) or other Krylov-space methods. Like the conjugate-gradient algorithm of Chapter 5 (which is not directly applicable here, since the coefficient matrix J is not symmetric positive definite), these methods typically require us to perform a matrix–vector multiplication of the form Jd for some d at each iteration, and to store a number of work vectors of length n. GMRES requires an additional vector to be stored at each iteration, so must be restarted periodically (often every 10 or 20 iterations) to keep memory requirements at a reasonable level.

The matrix–vector products Jd can be computed without explicit knowledge of the Jacobian J. A finite-difference approximation to Jd that requires one evaluation of $r(\cdot)$ is given by the formula (8.11). Calculation of Jd exactly (at least, to within the limits of finite-precision arithmetic) can be performed by using the forward mode of automatic differentiation, at a cost of at most a small multiple of an evaluation of $r(\cdot)$. Details of this procedure are given in Section 8.2.

We do not discuss the iterative methods for sparse linear systems here, but refer the interested reader to Kelley [177] and Saad [272] for comprehensive descriptions and implementations of the most interesting techniques. We prove a local convergence theorem for the method, similar to Theorem 11.2.

Theorem 11.3.

Suppose that r is continuously differentiable in a convex open set $\mathcal{D} \subset \mathbb{R}^n$. Let $x^* \in \mathcal{D}$ be a nondegenerate solution of $r(x) = 0$, and let $\{x_k\}$ be the sequence of iterates generated by the Framework 11.2. Then when $x_k \in \mathcal{D}$ is sufficiently close to x^*, the following are true:

(i) *If η in (11.17) is sufficiently small, the convergence of $\{x_k\}$ to x^* is Q-linear.*

(ii) *If $\eta_k \to 0$, the convergence is Q-superlinear.*

(iii) *If, in addition, $J(\cdot)$ is Lipschitz continuous in a neighborhood of x^* and $\eta_k = O(\|r_k\|)$, the convergence is Q-quadratic.*

PROOF.   We first rewrite (11.17) as

$$J(x_k)p_k + r(x_k) = v_k, \quad \text{where } \|v_k\| \le \eta_k \|r(x_k)\|. \tag{11.18}$$

Since x^* is a nondegenerate root, we have as in (11.14) that there is a radius $\delta > 0$ such that $\|J(x)^{-1}\| \le \beta^*$ for some constant β^* and all $x \in \mathcal{B}(x^*, \delta)$. By multiplying both sides of (11.18) by $J(x_k)^{-1}$ and rearranging, we find that

$$\left\| p_k + J(x_k)^{-1}r(x_k) \right\| = \left\| J(x_k)^{-1}v_k \right\| \le \beta^* \eta_k \|r(x_k)\|. \tag{11.19}$$

As in (11.10), we have that

$$r(x) = J(x)(x - x^*) + w(x, x^*),$$ (11.20)

where $\rho(x) \stackrel{\text{def}}{=} \|w(x, x^*)\|/\|x - x^*\| \to 0$ as $x \to x^*$. By reducing δ if necessary, we have from this expression that the following bound holds for all $x \in \mathcal{B}(x^*, \delta)$:

$$\|r(x)\| \le 2\|J(x^*)\| \, \|x - x^*\| + o(\|x - x^*\|) \le 4\|J(x^*)\| \, \|x - x^*\|.$$ (11.21)

We now set $x = x_k$ in (11.20), and use (11.19) and (11.21) to obtain

$$
\begin{aligned}
\|x_k + p_k - x^*\| = \left\| p_k + J(x_k)^{-1}(r(x_k) - w(x_k, x^*)) \right\| \\
\le \beta^* \eta_k \|r(x_k)\| + \|J(x_k)^{-1}\| \|w(x_k, x^*)\| \\
\le \left[4\|J(x^*)\|\beta^* \eta_k + \beta^* \rho(x_k) \right] \|x_k - x^*\|.
\end{aligned}
$$ (11.22)

By choosing x_k close enough to x^* that $\rho(x_k) \le 1/(4\beta^*)$, and choosing $\eta = 1/(8\|J(x^*)\|\beta^*)$, we have that the term in square brackets in (11.22) is at most $1/2$. Hence, since $x_{k+1} = x_k + p_k$, this formula indicates Q-linear convergence of $\{x_k\}$ to x^*, proving part (i).

Part (ii) follows immediately from the fact that the term in brackets in (11.22) goes to zero as $x_k \to x^*$ and $\eta_k \to 0$. For part (iii), we combine the techniques above with the logic of the second part of the proof of Theorem 11.2. Details are left as an exercise. $\square$

BROYDEN'S METHOD

Secant methods, also known as quasi-Newton methods, do not require calculation of the Jacobian $J(x)$. Instead, they construct their own approximation to this matrix, updating it at each iteration so that it mimics the behavior of the true Jacobian J over the step just taken. The approximate Jacobian, which we denote at iteration k by B_k, is then used to construct a linear model analogous to (11.5), namely

$$M_k(p) = r(x_k) + B_k p.$$ (11.23)

We obtain the step by setting this model to zero. When B_k is nonsingular, we have the following explicit formula (cf. (11.6)):

$$p_k = -B_k^{-1} r(x_k).$$ (11.24)

The requirement that the approximate Jacobian should mimic the behavior of the true Jacobian can be specified as follows. Let s_k denote the step from x_k to x_{k+1}, and let y_k

be the corresponding change in r, that is,

$$s_k = x_{k+1} - x_k, \qquad y_k = r(x_{k+1}) - r(x_k). \tag{11.25}$$

From Theorem 11.1, we have that s_k and y_k are related by the expression

$$y_k = \int_0^1 J(x_k + ts_k)s_k \, dt \approx J(x_{k+1})s_k + o(\|s_k\|). \tag{11.26}$$

We require the updated Jacobian approximation B_{k+1} to satisfy the following equation, which is known as the *secant equation*,

$$y_k = B_{k+1}s_k, \tag{11.27}$$

which ensures that B_{k+1} and $J(x_{k+1})$ have similar behavior along the direction s_k. (Note the similarity with the secant equation (6.6) in quasi-Newton methods for unconstrained optimization; the motivation is the same in both cases.) The secant equation does not say anything about how B_{k+1} should behave along directions orthogonal to s_k. In fact, we can view (11.27) as a system of n linear equations in n^2 unknowns, where the unknowns are the components of B_{k+1}, so for $n > 1$ the equation (11.27) does not determine all the components of B_{k+1} uniquely. (The scalar case of $n = 1$ gives rise to the scalar secant method; see (A.60).)

The most successful practical algorithm is Broyden's method, for which the update formula is

$$B_{k+1} = B_k + \frac{(y_k - B_k s_k)s_k^T}{s_k^T s_k}. \tag{11.28}$$

The Broyden update makes the smallest possible change to the Jacobian (as measured by the Euclidean norm $\|B_k - B_{k+1}\|_2$) that is consistent with (11.27), as we show in the following Lemma.

Lemma 11.4 (Dennis and Schnabel [92, Lemma 8.1.1]).

  Among all matrices B satisfying $Bs_k = y_k$, the matrix B_{k+1} defined by (11.28) minimizes the difference $\|B - B_k\|$.

PROOF. Let B be any matrix that satisfies $Bs_k = y_k$. By the properties of the Euclidean norm (see (A.10)) and the fact that $\|ss^T/s^Ts\| = 1$ for any vector s (see Exercise 11.1), we have

$$\|B_{k+1} - B_k\| = \left\| \frac{(y_k - B_k s_k)s_k^T}{s_k^T s_k} \right\|$$

$$= \left\| \frac{(B - B_k)s_k s_k^T}{s_k^T s_k} \right\| \le \|B - B_k\| \left\| \frac{s_k s_k^T}{s_k^T s_k} \right\| = \|B - B_k\|.$$

Hence, we have that

$$B_{k+1} \in \arg \min_{B \, : \, y_k = B s_k} \|B - B_k\|,$$

and the result is proved. □

In the specification of the algorithm below, we allow a line search to be performed along the search direction p_k, so that $s_k = \alpha p_k$ for some $\alpha > 0$ in the formula (11.25). (See below for details about line-search methods.)

Algorithm 11.3 (Broyden).

Choose x_0 and a nonsingular initial Jacobian approximation B_0;
for $k = 0, 1, 2, \ldots$

Calculate a solution p_k to the linear equations

$$B_k p_k = -r(x_k); \tag{11.29}$$

Choose α_k by performing a line search along p_k;
$x_{k+1} \leftarrow x_k + \alpha_k p_k$;
$s_k \leftarrow x_{k+1} - x_k$;
$y_k \leftarrow r(x_{k+1}) - r(x_k)$;
Obtain B_{k+1} from the formula (11.28);
end (for)

Under certain assumptions, Broyden's method converges *superlinearly*, that is,

$$\|x_{k+1} - x^*\| = o(\|x_k - x^*\|). \tag{11.30}$$

This local convergence rate is fast enough for most practical purposes, though not as fast as the Q-quadratic convergence of Newton's method.

We illustrate the difference between the convergence rates of Newton's and Broyden's method with a small example. The function $r : \mathbb{R}^2 \to \mathbb{R}^2$ defined by

$$r(x) = \begin{bmatrix} (x_1 + 3)(x_2^3 - 7) + 18 \\ \sin(x_2 e^{x_1} - 1) \end{bmatrix} \tag{11.31}$$

has a nondegenerate root at $x^* = (0, 1)^T$. We start both methods from the point $x_0 = (-0.5, 1.4)^T$, and use the exact Jacobian $J(x_0)$ at this point as the initial Jacobian approximation B_0. Results are shown in Table 11.1.

Newton's method clearly exhibits Q-quadratic convergence, which is characterized by doubling of the exponent of the error at each iteration. Broyden's method takes twice as

Table 11.1 Convergence of Iterates in Broyden and Newton Methods

Iteration k	$\|x_k - x^*\|_2$	
	Broyden	Newton
0	0.64×10^0	0.64×10^0
1	0.62×10^{-1}	0.62×10^{-1}
2	0.52×10^{-3}	0.21×10^{-3}
3	0.25×10^{-3}	0.18×10^{-7}
4	0.43×10^{-4}	0.12×10^{-15}
5	0.14×10^{-6}	
6	0.57×10^{-9}	
7	0.18×10^{-11}	
8	0.87×10^{-15}	

Table 11.2 Convergence of Function Norms in Broyden and Newton Methods

Iteration k	$\|r(x_k)\|_2$	
	Broyden	Newton
0	0.74×10^1	0.74×10^1
1	0.59×10^0	0.59×10^0
2	0.20×10^{-2}	0.23×10^{-2}
3	0.21×10^{-2}	0.16×10^{-6}
4	0.37×10^{-3}	0.22×10^{-15}
5	0.12×10^{-5}	
6	0.49×10^{-8}	
7	0.15×10^{-10}	
8	0.11×10^{-18}	

many iterations as Newton's, and reduces the error at a rate that accelerates slightly towards the end. The function norms $\|r(x_k)\|$ approach zero at a similar rate to the iteration errors $\|x_k - x^*\|$. As in (11.10), we have that

$$r(x_k) = r(x_k) - r(x^*) \approx J(x^*)(x_k - x^*),$$

so by nonsingularity of $J(x^*)$, the norms of $r(x_k)$ and $(x_k - x^*)$ are bounded above and below by multiples of each other. For our example problem (11.31), convergence of the sequence of function norms in the two methods is shown in Table 11.2.

The convergence analysis of Broyden's method is more complicated than that of Newton's method. We state the following result without proof.

Theorem 11.5.

Suppose the assumptions of Theorem 11.2 hold. Then there are positive constants ϵ and δ such that if the starting point x_0 and the starting approximate Jacobian B_0 satisfy

$$\|x_0 - x^*\| \leq \delta, \quad \|B_0 - J(x^*)\| \leq \epsilon, \tag{11.32}$$

the sequence $\{x_k\}$ generated by Broyden's method (11.24), (11.28) is well-defined and converges Q-superlinearly to x^*.

The second condition in (11.32)—that the initial Jacobian approximation B_0 must be close to the true Jacobian at the solution $J(x^*)$—is difficult to guarantee in practice. In contrast to the case of unconstrained minimization, a good choice of B_0 can be critical to the performance of the algorithm. Some implementations of Broyden's method recommend choosing B_0 to be $J(x_0)$, or some finite-difference approximation to this matrix.

The Broyden matrix B_k will be dense in general, even if the true Jacobian J is sparse. Therefore, when n is large, an implementation of Broyden's method that stores B_k as a full $n \times n$ matrix may be inefficient. Instead, we can use limited-memory methods in which B_k is stored implicitly in the form of a number of vectors of length n, while the system (11.29) is solved by a technique based on application of the Sherman–Morrison–Woodbury formula (A.28). These methods are similar to the ones described in Chapter 7 for large-scale unconstrained optimization.

TENSOR METHODS

In tensor methods, the linear model $M_k(p)$ used by Newton's method (11.5) is augmented with an extra term that aims to capture some of the nonlinear, higher-order, behavior of r. By doing so, it achieves more rapid and reliable convergence to degenerate roots, in particular, to roots x^* for which the Jacobian $J(x^*)$ has rank $n-1$ or $n-2$. We give a broad outline of the method here, and refer to Schnabel and Frank [277] for details.

We use $\hat{M}_k(p)$ to denote the model function on which tensor methods are based; this function has the form

$$\hat{M}_k(p) = r(x_k) + J(x_k)p + \tfrac{1}{2}T_k pp, \tag{11.33}$$

where T_k is a tensor defined by n^3 elements $(T_k)_{ijl}$ whose action on a pair of arbitrary vectors u and v in $\mathbb{R}^n$ is defined by

$$(T_k uv)_i = \sum_{j=1}^{n}\sum_{l=1}^{n}(T_k)_{ijl}u_j v_l.$$

If we followed the reasoning behind Newton's method, we could consider building T_k from the *second* derivatives of r at the point x_k, that is,

$$(T_k)_{ijl} = [\nabla^2 r_i(x_k)]_{jl}.$$

For instance, in the example (11.31), we have that

$$(T(x)uv)_1 = u^T \nabla^2 r_1(x)v = u^T \begin{bmatrix} 0 & 3x_2^2 \\ 3x_2^2 & 6x_2(x_1 + 3) \end{bmatrix} v$$

$$= 3x_2^2(u_1 v_2 + u_2 v_1) + 6x_2(x_1 + 3)u_2 v_2.$$

However, use of the exact second derivatives is not practical in most instances. If we were to store this information explicitly, about $n^3/2$ memory locations would be needed, about n times the requirements of Newton's method. Moreover, there may be no vector p for which $\hat{M}_k(p) = 0$, so the step may not even be defined.

Instead, the approach described in [277] defines T_k in a way that requires little additional storage, but which gives $\hat{M}_k$ some potentially appealing properties. Specifically, T_k is chosen so that $\hat{M}_k(p)$ interpolates the function $r(x_k + p)$ at some previous iterates visited by the algorithm. That is, we require that

$$\hat{M}_k(x_{k-j} - x_k) = r(x_{k-j}), \quad \text{for } j = 1, 2, \dots, q, \tag{11.34}$$

for some integer $q > 0$. By substituting from (11.33), we see that T_k must satisfy the condition

$$\tfrac{1}{2} T_k s_{jk} s_{jk} = r(x_{k-j}) - r(x_k) - J(x_k)s_{jk},$$

where

$$s_{jk} \overset{\text{def}}{=} x_{k-j} - x_k, \ j = 1, 2, \dots, q.$$

In [277] it is shown that this condition can be ensured by choosing T_k so that its action on arbitrary vectors u and v is

$$T_k uv = \sum_{j=1}^{q} a_j (s_{jk}^T u)(s_{jk}^T v),$$

where a_j, $j = 1, 2, \dots, q$, are vectors of length n. The number of interpolating points q is typically chosen to be quite modest, usually less than $\sqrt{n}$. This T_k can be stored in $2nq$ locations, which contain the vectors a_j and s_{jk} for $j = 1, 2, \dots, q$. Note the connection between this idea and Broyden's method, which also chooses information in the model (albeit in the *first-order* part of the model) to interpolate the function value at the previous iterate.

This technique can be refined in various ways. The points of interpolation can be chosen to make the collection of directions s_{jk} more linearly independent. There may still not be a vector p for which $\hat{M}_k(p) = 0$, but we can instead take the step to be the vector that minimizes $\|\hat{M}_k(p)\|_2^2$, which can be found by using a specialized least-squares technique. There is no assurance that the step obtained in this way is a descent direction for the merit function $\frac{1}{2}\|r(x)\|^2$ (which is discussed in the next section), and in this case it can be replaced by the standard Newton direction $-J_k^{-1}r_k$.

11.2 PRACTICAL METHODS

We now consider practical variants of the Newton-like methods discussed above, in which line-search and trust-region modifications to the steps are made in order to ensure better global convergence behavior.

MERIT FUNCTIONS

As mentioned above, neither Newton's method (11.6) nor Broyden's method (11.24), (11.28) with unit step lengths can be guaranteed to converge to a solution of $r(x) = 0$ unless they are started close to that solution. Sometimes, components of the unknown or function vector or the Jacobian will blow up. Another, more exotic, kind of behavior is *cycling*, where the iterates move between distinct regions of the parameter space without approaching a root. An example is the scalar function

$$r(x) = -x^5 + x^3 + 4x,$$

which has five nondegenerate roots. When started from the point $x_0 = 1$, Newton's method produces a sequence of iterates that oscillates between 1 and -1 (see Exercise 11.3) without converging to any of the roots.

The Newton and Broyden methods can be made more robust by using line-search and trust-region techniques similar to those described in Chapters 3 and 4. Before describing these techniques, we need to define a *merit function*, which is a scalar-valued function of x that indicates whether a new iterate is better or worse than the current iterate, in the sense of making progress toward a root of r. In unconstrained optimization, the objective function f is itself a natural merit function; most algorithms for minimizing f require a decrease in f at each iteration. In nonlinear equations, the merit function is obtained by combining the n components of the vector r in some way.

The most widely used merit function is the sum of squares, defined by

$$f(x) = \tfrac{1}{2}\|r(x)\|^2 = \tfrac{1}{2}\sum_{i=1}^{n} r_i^2(x). \tag{11.35}$$

(The factor 1/2 is introduced for convenience.) Any root x^* of r obviously has $f(x^*) = 0$, and since $f(x) \geq 0$ for all x, each root is a minimizer of f. However, local minimizers of f are not roots of r if f is strictly positive at the point in question. Still, the merit function

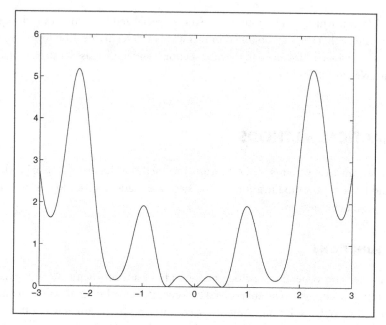

Figure 11.2 Plot of $\frac{1}{2}[\sin(5x) - x]^2$, showing its many local minima.

(11.35) has been used successfully in many applications and is implemented in a number of software packages.

The merit function for the example (11.3) is plotted in Figure 11.2. It shows three local minima corresponding to the three roots, but there are many other local minima (for example, those at around ± 1.53053). Local minima like these that are not roots of f satisfy an interesting property. Since

$$\nabla f(x^*) = J(x^*)^T r(x^*) = 0, \qquad (11.36)$$

we can have $r(x^*) \neq 0$ only if $J(x^*)$ is singular.

Since local minima for the sum-of-squares merit function may be points of attraction for the algorithms described in this section, global convergence results for the algorithms discussed here are less satisfactory than for similar algorithms applied to unconstrained optimization.

Other merit functions are also used in practice. One such is the ℓ_1 norm merit function defined by

$$f_1(x) = \|r(x)\|_1 = \sum_{i=1}^{m} |r_i(x)|.$$

This function is studied in Chapters 17 and 18 in the context of algorithms for constrained optimization.

LINE SEARCH METHODS

We can obtain algorithms with global convergence properties by applying the line-search approach of Chapter 3 to the sum-of-squares merit function $f(x) = \frac{1}{2}\|r(x)\|^2$. When it is well defined, the Newton step

$$J(x_k)p_k = -r(x_k) \tag{11.37}$$

is a descent direction for $f(\cdot)$ whenever $r_k \neq 0$, since

$$p_k^T \nabla f(x_k) = -p_k^T J_k^T r_k = -\|r_k\|^2 < 0. \tag{11.38}$$

Step lengths α_k are chosen by one of the procedures of Chapter 3, and the iterates are defined by the formula

$$x_{k+1} = x_k + \alpha_k p_k, \qquad k = 0, 1, 2, \ldots. \tag{11.39}$$

For the case of line searches that choose α_k to satisfy the Wolfe conditions (3.6), we have the following convergence result, which follows directly from Theorem 3.2.

Theorem 11.6.

Suppose that $J(x)$ is Lipschitz continuous in a neighborhood $\mathcal{D}$ of the level set $\mathcal{L} = \{x : f(x) \leq f(x_0)\}$, and that $\|J(x)\|$ and $\|r(x)\|$ are bounded above on $\mathcal{D}$. Suppose that a line-search algorithm (11.39) is applied to f, where the search directions p_k satisfy $p_k^T \nabla f_k < 0$ while the step lengths α_k satisfy the Wolfe conditions (3.6). Then we have that the Zoutendijk condition holds, that is,

$$\sum_{k \geq 0} \cos^2 \theta_k \|J_k^T r_k\|^2 < \infty,$$

where

$$\cos \theta_k = \frac{-p_k^T \nabla f(x_k)}{\|p_k\| \|\nabla f(x_k)\|}. \tag{11.40}$$

We omit the proof, which verifies that ∇f is Lipschitz continuous on $\mathcal{D}$ and that f is bounded below (by 0) on $\mathcal{D}$, and then applies Theorem 3.2.

Provided that the sequence of iterates satisfies

$$\cos \theta_k \geq \delta, \quad \text{for some } \delta \in (0, 1) \text{ and all } k \text{ sufficiently large}, \tag{11.41}$$

Theorem 11.6 guarantees that $J_k^T r_k \to 0$, meaning that the iterates approach stationarity of the merit function f. Moreover, if we know that $\|J(x_k)^{-1}\|$ is bounded then we must have $r_k \to 0$.

We now investigate the values of $\cos\theta_k$ for the directions generated by the Newton and inexact Newton methods. From (11.40) and (11.38), we have for the exact Newton step (11.6) that

$$\cos\theta_k = -\frac{p_k^T\nabla f(x_k)}{\|p_k\|\|\nabla f(x_k)\|} = \frac{\|r_k\|^2}{\|J_k^{-1}r_k\|\|J_k^T r_k\|} \geq \frac{1}{\|J_k^T\|\|J_k^{-1}\|} = \frac{1}{\kappa(J_k)}. \tag{11.42}$$

When p_k is an inexact Newton direction—that is, one that satisfies the condition (11.17)—we have that

$$\|r_k + J_k p_k\|^2 \leq \eta_k^2\|r_k\|^2 \Rightarrow 2p_k^T J_k^T r_k + \|r_k\|^2 + \|J_k p_k\|^2 \leq \eta^2\|r_k\|^2$$
$$\Rightarrow p_k^T\nabla f_k = p_k^T J_k^T r_k \leq [(\eta^2 - 1)/2]\|r_k\|^2.$$

Meanwhile,

$$\|p_k\| \leq \|J_k^{-1}\|\,[\|r_k + J_k p_k\| + \|r_k\|] \leq \|J_k^{-1}\|(\eta + 1)\|r_k\|,$$

and

$$\|\nabla f_k\| = \|J_k^T r_k\| \leq \|J_k\|\|r_k\|.$$

By combining these estimates, we obtain

$$\cos\theta_k = -\frac{p_k^T\nabla f_k}{\|p_k\|\|\nabla f_k\|} \geq \frac{1 - \eta^2}{2\|J_k\|\|J_k^{-1}\|(1 + \eta)} \geq \frac{1 - \eta}{2\kappa(J_k)}.$$

We conclude that a bound of the form (11.41) is satisfied both for the exact and inexact Newton methods, provided that the condition number $\kappa(J_k)$ is bounded.

When $\kappa(J_k)$ is large, however, this lower bound is close to zero, and use of the Newton direction may cause poor performance of the algorithm. In fact, the following example shows that condition $\cos\theta_k$ can converge to zero, causing the algorithm to fail. This example highlights a fundamental weakness of the line-search approach.

❏ **EXAMPLE 11.2**   (POWELL [241])

Consider the problem of finding a solution of the nonlinear system

$$r(x) = \begin{bmatrix} x_1 \\ \dfrac{10x_1}{(x_1 + 0.1)} + 2x_2^2 \end{bmatrix}, \tag{11.43}$$

with unique solution $x^* = 0$. We try to solve this problem using the Newton iteration (11.37), (11.39) where α_k is chosen to minimize f along p_k. It is proved in [241] that, starting from the point $(3, 1)^T$, the iterates converge to $(1.8016, 0)^T$ (to four digits of accuracy). However, this point is not a solution of (11.43). In fact, it is not even a stationary point of f, and a step from this point in the direction $-\nabla f$ will produce a decrease in both components of r. To verify these claims, note that the Jacobian of r, which is

$$J(x) = \begin{bmatrix} 1 & 0 \\ \dfrac{1}{(x_1 + 0.1)^2} & 4x_2 \end{bmatrix},$$

is singular at all x for which $x_2 = 0$. For such points, we have

$$\nabla f(x) = \begin{bmatrix} x_1 + \dfrac{10x_1}{(x_1 + 0.1)^3} \\ 0 \end{bmatrix},$$

so that the gradient points in the direction of the positive x_1 axis whenever $x_1 > 0$. The point $(1.8016, 0)^T$ is therefore not a stationary point of f.

For this example, a calculation shows that the Newton step generated from an iterate that is close to (but not quite on) the x_1 axis tends to be parallel to the x_2 axis, making it nearly orthogonal to the gradient $\nabla f(x)$. That is, $\cos \theta_k$ for the Newton direction may be arbitrarily close to zero. □

In this example, a Newton method with exact line searches is attracted to a point of no interest at which the Jacobian is singular. Since systems of nonlinear equations often contain singular points, this behavior gives cause for concern.

To prevent this undesirable behavior and ensure that (11.41) holds, we may have to modify the Newton direction. One possibility is to add some multiple $\lambda_k I$ of the identity to $J_k^T J_k$, and define the step p_k to be

$$p_k = -(J_k^T J_k + \lambda_k I)^{-1} J_k^T r_k. \tag{11.44}$$

For any $\lambda_k > 0$ the matrix in parentheses is nonsingular, and if λ_k is bounded away from zero, a condition of the form (11.41) is satisfied. Therefore, some practical algorithms choose λ_k adaptively to ensure that the matrix in (11.44) does not approach singularity. This approach is analogous to the classical Levenberg-Marquardt algorithm discussed in Chapter 10. To implement it without forming $J_k^T J_k$ explicitly and performing trial Cholesky factorizations of the matrices $(J_k^T J_k + \lambda I)$, we can use the technique (10.36) illustrated earlier for the least-squares case. This technique uses the fact that the Cholesky factor of $(J_k^T J_k + \lambda I)$ is identical to R^T, where R is the upper triangular factor from the QR factorization of the matrix

$$\begin{bmatrix} J_k \\ \sqrt{\lambda} I \end{bmatrix}. \tag{11.45}$$

A combination of Householder and Givens transformations can be used, as for (10.36), and the savings noted in the discussion following (10.36) continue to hold if we need to perform this calculation for several candidate values of λ_k.

The drawback of this Levenberg-Marquardt approach is that it is difficult to choose λ_k. If too large, we can destroy the fast rate of convergence of Newton's method. (Note that p_k approaches a multiple of $-J_k^T r_k$ as $\lambda_k \uparrow \infty$, so the step becomes small and tends to point in the steepest-descent direction for f.) If λ_k is too small, the algorithm can be inefficient in the presence of Jacobian singularities. A more satisfactory approach is to follow the trust-region approach described below, which chooses λ_k indirectly.

We conclude by specifying an algorithm based on Newton-like steps and line searches that regularizes the step calculations where necessary. Several details are deliberately left vague; we refer the reader to the papers cited above for details.

Algorithm 11.4 (Line Search Newton-like Method).
  Given c_1, c_2 with $0 < c_1 < c_2 < \frac{1}{2}$;
  Choose x_0;
  for $k = 0, 1, 2, \ldots$
        Calculate a Newton-like step from (11.6) (regularizing with (11.44)
              if J_k appears to be near-singular), or (11.17) or (11.24);
        if $\alpha = 1$ satisfies the Wolfe conditions (3.6)
              Set $\alpha_k = 1$;
        else
              Perform a line search to find $\alpha_k > 0$ that satisfies (3.6);
        end (if)
        $x_{k+1} \leftarrow x_k + \alpha_k p_k$;
  end (for)

TRUST-REGION METHODS

The most widely used trust-region methods for nonlinear equations simply apply Algorithm 4.1 from Chapter 4 to the merit function $f(x) = \frac{1}{2} \|r(x)\|_2^2$, using $B_k = J(x_k)^T J(x_k)$ as the approximate Hessian in the model function $m_k(p)$, which is defined as follows:

$$m_k(p) = \frac{1}{2} \|r_k + J_k p\|_2^2 = f_k + p^T J_k^T r_k + \frac{1}{2} p^T J_k^T J_k p_k.$$

The step p_k is generated by finding an approximate solution of the subproblem

$$\min_{p} m_k(p), \qquad \text{subject to } \|p\| \le \Delta_k, \tag{11.46}$$

where Δ_k is the radius of the trust region. The ratio ρ_k of actual to predicted reduction (see (4.4)), which plays a critical role in many trust-region algorithms, is therefore

$$\rho_k = \frac{\|r(x_k)\|^2 - \|r(x_k + p_k)\|^2}{\|r(x_k)\|^2 - \|r(x_k) + J(x_k)p_k\|^2}. \tag{11.47}$$

We can state the trust-region framework that results from this model as follows.

Algorithm 11.5 (Trust-Region Method for Nonlinear Equations).
  Given $\bar{\Delta} > 0$, $\Delta_0 \in (0, \bar{\Delta})$, and $\eta \in \left[0, \frac{1}{4}\right)$:
  for $k = 0, 1, 2, \ldots$
        Calculate p_k as an (approximate) solution of (11.46);
        Evaluate ρ_k from (11.47);
        if $\rho_k < \frac{1}{4}$
            $\Delta_{k+1} = \frac{1}{4}\|p_k\|$;
        else
            if $\rho_k > \frac{3}{4}$ and $\|p_k\| = \Delta_k$
                $\Delta_{k+1} = \min(2\Delta_k, \bar{\Delta})$;
            else
                $\Delta_{k+1} = \Delta_k$;
            end (if)
        end (if)
        if $\rho_k > \eta$
            $x_{k+1} = x_k + p_k$;
        else
            $x_{k+1} = x_k$;
        end (if)
  end (for).

    The dogleg method is a special case of the trust-region algorithm, Algorithm 4.1, that constructs an approximate solution to (11.46) based on the Cauchy point p_k^c and the unconstrained minimizer of m_k. The Cauchy point is

$$p_k^c = -\tau_k(\Delta_k/\|J_k^T r_k\|)J_k^T r_k, \tag{11.48}$$

where

$$\tau_k = \min\left\{1, \|J_k^T r_k\|^3/(\Delta_k r_k^T J_k(J_k^T J_k)J_k^T r_k)\right\}; \tag{11.49}$$

By comparing with the general definition (4.11), (4.12) we see that it is not necessary to consider the case of an indefinite Hessian approximation in $m_k(p)$, since the model Hessian $J_k^T J_k$ that we use is positive semidefinite. The unconstrained minimizer of $m_k(p)$ is unique when J_k is nonsingular. In this case, we denote it by p_k^J and write

$$p_k^J = -(J_k^T J_k)^{-1}(J_k^T r_k) = -J_k^{-1} r_k.$$

The selection of p_k in the dogleg method proceeds as follows.

Procedure 11.6 (Dogleg).
   Calculate p_k^c;
   if $\|p_k^c\| = \Delta_k$
      $p_k \leftarrow p_k^c$;
   else
      Calculate p_k^J;
      $p_k \leftarrow p_k^c + \tau(p_k^J - p_k^c)$, where τ is the largest value in $[0, 1]$
         such that $\|p_k\| \leq \Delta_k$;
   end (if).

Lemma 4.2 shows that when J_k is nonsingular, the vector p_k chosen above is the minimizer of m_k along the piecewise linear path that leads from the origin to the Cauchy point and then to the unconstrained minimizer p_k^J. Hence, the reduction in model function at least matches the reduction obtained by the Cauchy point, which can be estimated by specializing the bound (4.20) to the least-squares case by writing

$$m_k(0) - m_k(p_k) \geq c_1 \|J_k^T r_k\| \min\left(\Delta_k, \frac{\|J_k^T r_k\|}{\|J_k^T J_k\|}\right), \tag{11.50}$$

where c_1 is some positive constant.

From Theorem 4.1, we know that the exact solution of (11.46) has the form

$$p_k = -(J_k^T J_k + \lambda_k I)^{-1} J_k^T r_k, \tag{11.51}$$

for some $\lambda_k \geq 0$, and that $\lambda_k = 0$ if the unconstrained solution p_k^J satisfies $\|p_k^J\| \leq \Delta_k$. (Note that (11.51) is identical to the formula (10.34a) from Chapter 10. In fact, the Levenberg–Marquardt approach for nonlinear equations is a special case of the same algorithm for nonlinear least-squares problems.) The Levenberg–Marquardt algorithm uses the techniques of Section 4.3 to search for the value of λ_k that satisfies (11.51). The procedure described in the "exact" trust-region algorithm, Algorithm 4.3, is based on Cholesky factorizations, but as in Chapter 10, we can replace these by specialized algorithms to compute the QR factorization of the matrix (11.45). Even if the exact λ_k corresponding to the solution of (11.46) is not found, the p_k calculated from (11.51) will still yield global convergence if it satisfies the condition (11.50) for some value of c_1, together with

$$\|p_k\| \leq \gamma \Delta_k, \quad \text{for some constant } \gamma \geq 1. \tag{11.52}$$

The dogleg method requires just one linear system to be solved per iteration, whereas methods that search for the exact solution of (11.46) require several such systems to be solved. As in Chapter 4, there is a tradeoff to be made between the amount of effort to spend on each iteration and the total number of function and derivative evaluations required.

We can also consider alternative trust-region approaches that are based on different merit functions and different definitions of the trust region. An algorithm based on the ℓ_1 merit function with an ℓ_∞-norm trust region gives rise to subproblems of the form

$$\min_p \|J_k p + r_k\|_1 \quad \text{subject to } \|p\|_\infty \leq \Delta, \tag{11.53}$$

which can be formulated and solved using linear programming techniques. This approach is closely related to the $S\ell_1QP$ and SLQP approaches for nonlinear programming discussed in Section 18.5.

Global convergence results of Algorithm 11.5 when the steps p_k satisfy (11.50) and (11.52) are given in the following theorem, which can be proved by referring directly to Theorems 4.5 and 4.6. The first result is for $\eta = 0$, in which the algorithm accepts all steps that produce a decrease in the merit function f_k, while the second (stronger) result requires a strictly positive choice of η.

Theorem 11.7.

Suppose that $J(x)$ is Lipschitz continuous and that $\|J(x)\|$ is bounded above in a neighborhood $\mathcal{D}$ of the level set $\mathcal{L} = \{x : f(x) \leq f(x_0)\}$. Suppose in addition that all approximate solutions of (11.46) satisfy the bounds (11.50) and (11.52). Then if $\eta = 0$ in Algorithm 11.5, we have that

$$\liminf_{k \to \infty} \|J_k^T r_k\| = 0,$$

while if $\eta \in \left(0, \frac{1}{4}\right)$, we have

$$\lim_{k \to \infty} \|J_k^T r_k\| = 0.$$

We turn now to local convergence of the trust-region algorithm for the case in which the subproblem (11.46) is solved exactly. We assume that the sequence $\{x_k\}$ converges to a nondegenerate solution x^* of the nonlinear equations $r(x) = 0$. The significance of this result is that the algorithmic enhancements needed for global convergence do not, in well-designed algorithms, interfere with the fast local convergence properties described in Section 11.1.

Theorem 11.8.

   Suppose that the sequence $\{x_k\}$ generated by Algorithm 11.5 converges to a nondegenerate solution x^* of the problem $r(x) = 0$. Suppose also that $J(x)$ is Lipschitz continuous in an open neighborhood $\mathcal{D}$ of x^* and that the trust-region subproblem (11.46) is solved exactly for all sufficiently large k. Then the sequence $\{x_k\}$ converges quadratically to x^*.

PROOF.   We prove this result by showing that there is an index K such that the trust-region radius is not reduced further after iteration K; that is, $\Delta_k \geq \Delta_K$ for all $k \geq K$. We then show that the algorithm eventually takes the pure Newton step at every iteration, so that quadratic convergence follows from Theorem 11.2.

   Let p_k denote the exact solution of (11.46). Note first that p_k will simply be the unconstrained Newton step $-J_k^{-1} r_k$ whenever this step satisfies the trust-region bound. Otherwise, we have $\|J_k^{-1} r_k\| > \Delta_k$, while the solution p_k satisfies $\|p_k\| \leq \Delta_k$. In either case, we have

$$\|p_k\| \leq \|J_k^{-1} r_k\|. \tag{11.54}$$

   We consider the ratio ρ_k of actual to predicted reduction defined by (11.47). We have directly from the definition that

$$|1 - \rho_k| \leq \frac{\big|\|r_k + J_k p_k\|^2 - \|r(x_k + p_k)\|^2\big|}{\|r(x_k)\|^2 - \|r(x_k) + J(x_k)p_k\|^2}. \tag{11.55}$$

From Theorem 11.1, we have for the second term in the numerator that

$$\|r(x_k + p_k)\|^2 = \|[r(x_k) + J(x_k)p_k] + w(x_k, x_k + p_k)\|^2, \tag{11.56}$$

where $w(\cdot, \cdot)$ is defined as in (11.11). Because of Lipschitz continuity of J with Lipschitz constant β_L (11.7), we have

$$\|w(x_k, x_k + p_k)\| \leq \int_0^1 \|J(x_k + tp_k) - J(x_k)\| \|p_k\| \, dt$$
$$\leq \int_0^1 \beta_L \|p_k\|^2 \, dt = (\beta_L/2)\|p_k\|^2,$$

so that using (11.56) and the fact that $\|r_k + J_k p_k\| \leq \|r_k\| = f(x_k)^{1/2}$ (since p_k is the solution of (11.46)), we can bound the numerator as follows:

$$\big|\|r_k + J_k p_k\|^2 - \|r(x_k + p_k)\|^2\big|$$
$$\leq 2\|r_k + J_k p_k\| \|w(x_k, x_k + p_k)\| + \|w(x_k, x_k + p_k)\|^2$$
$$\leq f(x_k)^{1/2} \beta_L \|p_k\|^2 + (\beta_L/2)^2 \|p_k\|^4$$
$$\leq \epsilon(x_k) \|p_k\|^2, \tag{11.57}$$

where we define

$$\epsilon(x_k) = f(x_k)^{1/2}\beta_L + (\beta_L/2)^2\|p_k\|^2.$$

Since $x_k \to x^*$ by assumption, it follows that $f(x_k) \to 0$ and $\|r_k\| \to 0$. Because x^* is a nondegenerate root, we have as in (11.14) that $\|J(x_k)^{-1}\| \le \beta^*$ for all k sufficiently large, so from (11.54), we have

$$\|p_k\| \le \|J_k^{-1}r_k\| \le \beta^*\|r_k\| \to 0. \tag{11.58}$$

Hence, $\epsilon(x_k) \to 0$.

Turning now to the denominator of (11.55), we define $\bar{p}_k$ to be a step of the same length as the solution p_k in the Newton direction $-J_k^{-1}r_k$, that is,

$$\bar{p}_k = -\frac{\|p_k\|}{\|J_k^{-1}r_k\|}J_k^{-1}r_k.$$

Since $\bar{p}_k$ is feasible for (11.46), and since p_k is optimal for this subproblem, we have

$$\begin{aligned}
\|r_k\|^2 - \|r_k + J_k p_k\|^2 &\ge \|r_k\|^2 - \left\|r_k - \frac{\|p_k\|}{\|J_k^{-1}r_k\|}r_k\right\|^2 \\
&= 2\frac{\|p_k\|}{\|J_k^{-1}r_k\|}\|r_k\|^2 - \frac{\|p_k\|^2}{\|J_k^{-1}r_k\|^2}\|r_k\|^2 \\
&\ge \frac{\|p_k\|}{\|J_k^{-1}r_k\|}\|r_k\|^2,
\end{aligned}$$

where for the last inequality we have used (11.54). By using (11.58) again, we have from this bound that

$$\|r_k\|^2 - \|r_k + J_k p_k\|^2 \ge \frac{\|p_k\|}{\|J_k^{-1}r_k\|}\|r_k\|^2 \ge \frac{1}{\beta^*}\|p_k\|\|r_k\|. \tag{11.59}$$

By substituting (11.57) and (11.59) into (11.55), and then applying (11.58) again, we have

$$|1 - \rho_k| \le \frac{\beta^*\epsilon(x_k)\|p_k\|^2}{\|p_k\|\|r_k\|} \le (\beta^*)^2\epsilon(x_k) \to 0. \tag{11.60}$$

Therefore, for all k sufficiently large, we have $\rho_k > \frac{1}{4}$, and so the trust region radius Δ_k will not be increased beyond this point. As claimed, there is an index K such that

$$\Delta_k \ge \Delta_K, \quad \text{for all } k \ge K.$$

Since $\|J_k^{-1}r_k\| \leq \beta^*\|r_k\| \to 0$, the Newton step $-J_k^{-1}r_k$ will eventually be smaller than Δ_K (and hence Δ_k), so it will eventually always be accepted as the solution of (11.46). The result now follows from Theorem 11.2. $\qquad\qquad\Box$

We can replace the assumption that $x_k \to x^*$ with an assumption that the nondegenerate solution x^* is just one of the limit points of the sequence. (In fact, this condition implies that $x_k \to x^*$; see Exercise 11.9.)

11.3 CONTINUATION/HOMOTOPY METHODS

MOTIVATION

We mentioned above that Newton-based methods all suffer from one shortcoming: Unless $J(x)$ is nonsingular in the region of interest—a condition that often cannot be guaranteed—they are in danger of converging to a local minimum of the merit function rather that is not a solution of the nonlinear system. Continuation methods, which we outline in this section, are more likely to converge to a solution of $r(x) = 0$ in difficult cases. Their underlying motivation is simple to describe: Rather than dealing with the original problem $r(x) = 0$ directly, we set up an "easy" system of equations for which the solution is obvious. We then gradually transform the easy system into the original system $r(x)$, and follow the solution as it moves from the solution of the easy problem to the solution of the original problem.

One simple way to define the so-called *homotopy map* $H(x, \lambda)$ is as follows:

$$H(x, \lambda) = \lambda r(x) + (1 - \lambda)(x - a), \qquad (11.61)$$

where λ is a scalar parameter and $a \in \mathbb{R}^n$ is a fixed vector. When $\lambda = 0$, (11.61) defines the artificial, easy problem $H(x, 0) = x - a$, whose solution is obviously $x = a$. When $\lambda = 1$, we have $H(x, 1) = r(x)$, the original system of equations.

To solve $r(x) = 0$, consider the following algorithm: First, set $\lambda = 0$ in (11.61) and set $x = a$. Then, increase λ from 0 to 1 in small increments, and for each value of λ, calculate the solution of the system $H(x, \lambda) = 0$. The final value of x corresponding to $\lambda = 1$ will solve the original problem $r(x) = 0$.

This naive approach sounds plausible, and Figure 11.3 illustrates a situation in which it would be successful. In this figure, there is a unique solution x of the system $H(x, \lambda) = 0$ for each value of λ in the range $[0, 1]$. The trajectory of points (x, λ) for which $H(x, \lambda) = 0$ is called the *zero path*.

Unfortunately, however, the approach often fails, as illustrated in Figure 11.4. Here, the algorithm follows the lower branch of the curve from $\lambda = 0$ to $\lambda = \lambda_T$, but it then loses the trail unless it is lucky enough to jump to the top branch of the path. The value λ_T is

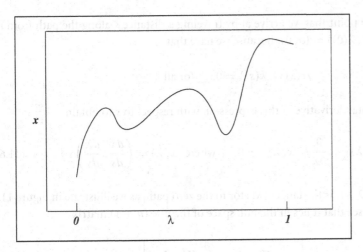

Figure 11.3 Plot of a zero path: Trajectory of points (x, λ) with $H(x, \lambda) = 0$.

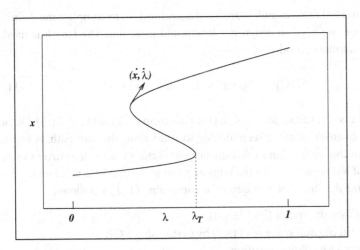

Figure 11.4 Zero path with turning points. The path joining $(a, 0)$ to $(x^*, 1)$ cannot be followed by increasing λ monotonically from 0 to 1.

known as a *turning point*, since at this point we can follow the path smoothly only if we no longer insist on increasing λ at every step. In fact, practical continuation methods work by doing exactly as Figure 11.4 suggests, that is, they follow the zero path explicitly, even if this means allowing λ to decrease from time to time.

PRACTICAL CONTINUATION METHODS

In one practical technique, we model the zero path by allowing both x and λ to be functions of an independent variable s that represents arc length along the path. That is,

$(x(s), \lambda(s))$ is the point that we arrive at by traveling a distance s along the path from the initial point $(x(0), \lambda(0)) = (a, 0)$. Because we have that

$$H(x(s), \lambda(s)) = 0, \quad \text{for all } s \geq 0,$$

we can take the total derivative of this expression with respect to s to obtain

$$\frac{\partial}{\partial x} H(x, \lambda) \dot{x} + \frac{\partial}{\partial \lambda} H(x, \lambda) \dot{\lambda} = 0, \quad \text{where } (\dot{x}, \dot{\lambda}) = \left(\frac{dx}{ds}, \frac{d\lambda}{ds} \right). \tag{11.62}$$

The vector $(\dot{x}(s), \dot{\lambda}(s))$ is the tangent vector to the zero path, as we illustrate in Figure 11.4. From (11.62), we see that it lies in the null space of the $n \times (n+1)$ matrix

$$\left[\begin{array}{cc} \dfrac{\partial}{\partial x} H(x, \lambda) & \dfrac{\partial}{\partial \lambda} H(x, \lambda) \end{array} \right]. \tag{11.63}$$

When this matrix has full rank, its null space has dimension 1, so to complete the definition of $(\dot{x}, \dot{\lambda})$ in this case, we need to assign it a length and direction. The length is fixed by imposing the normalization condition

$$\|\dot{x}(s)\|^2 + |\dot{\lambda}(s)|^2 = 1, \quad \text{for all } s, \tag{11.64}$$

which ensures that s is the true arc length along the path from $(0, a)$ to $(x(s), \lambda(s))$. We need to choose the sign to ensure that we keep moving forward along the zero path. A heuristic that works well is to choose the sign so that the tangent vector $(\dot{x}, \dot{\lambda})$ at the current value of s makes an angle of less than $\pi/2$ with the tangent point at the previous value of s.

We can outline the complete procedure for computing $(\dot{x}, \dot{\lambda})$ as follows:

Procedure 11.7 (Tangent Vector Calculation).

Compute a vector in the null space of (11.63) by performing a QR factorization with column pivoting,

$$Q^T \left[\begin{array}{cc} \dfrac{\partial}{\partial x} H(x, \lambda) & \dfrac{\partial}{\partial \lambda} H(x, \lambda) \end{array} \right] \Pi = \left[\begin{array}{cc} R & w \end{array} \right],$$

where Q is $n \times n$ orthogonal, R is $n \times n$ upper triangular, Π is an $(n+1) \times (n+1)$ permutation matrix, and $w \in \mathbb{R}^n$.

Set

$$v = \Pi \left[\begin{array}{c} R^{-1} w \\ -1 \end{array} \right];$$

Set $(\dot{x}, \dot{\lambda}) = \pm v/\|v\|_2$, where the sign is chosen to satisfy the angle
criterion mentioned above.

Details of the QR factorization procedure are given in the Appendix.

Since we can obtain the tangent at any given point (x, λ) and since we know the initial
point $(x(0), \lambda(0)) = (a, 0)$, we can trace the zero path by calling a standard initial-value
first-order ordinary differential equation solver, terminating the algorithm when it finds a
value of s for which $\lambda(s) = 1$.

A second approach for following the zero path is quite similar to the one just described,
except that it takes an algebraic viewpoint instead of a differential-equations viewpoint.
Given a current point (x, λ), we compute the tangent vector $(\dot{x}, \dot{\lambda})$ as above, and take a
small step (of length ϵ, say) along this direction to produce a "predictor" point (x^P, λ^P);
that is,

$$(x^P, \lambda^P) = (x, \lambda) + \epsilon(\dot{x}, \dot{\lambda}).$$

Usually, this new point will not lie exactly on the zero path, so we apply some "corrector"
iterations to bring it back to the path, thereby identifying a new iterate (x^+, λ^+) that satisfies
$H(x^+, \lambda^+) = 0$. (This process is illustrated in Figure 11.5.) During the corrections, we
choose a component of the predictor step (x^P, λ^P)—one of the components that has been
changing most rapidly during the past few steps—and hold this component fixed during
the correction process. If the index of this component is i, and if we use a pure Newton
corrector process (often adequate, since (x^P, λ^P) is usually quite close to the target point

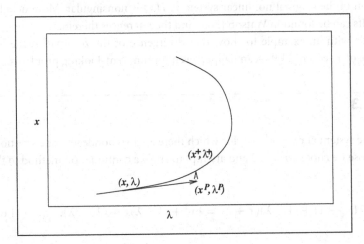

Figure 11.5 The algebraic predictor–corrector procedure, using λ as the fixed
variable in the correction process.

$(x^+, \lambda^+))$, the steps will have the form

$$
\left[
\begin{array}{c|c}
\dfrac{\partial H}{\partial x} & \dfrac{\partial H}{\partial \lambda} \\
\hline
\multicolumn{2}{c}{e_i}
\end{array}
\right]
\left[
\begin{array}{c}
\delta x \\
\delta \lambda
\end{array}
\right]
=
\left[
\begin{array}{c}
-H \\
0
\end{array}
\right],
$$

where the quantities $\partial H/\partial x$, $\partial H/\partial \lambda$, and H are evaluated at the latest point of the corrector process. The last row of this system serves to fix the ith component of $(\delta x, \delta \lambda)$ at zero; the vector $e_i \in \mathbb{R}^{n+1}$ is a vector with $n+1$ components containing all zeros, except for a 1 in the location i that corresponds to the fixed component. Note that in Figure 11.5 the λ component is chosen to be fixed on the current iteration. On the following iteration, it may be more appropriate to choose x as the fixed component, as we reach the turning point in λ.

The two variants on path-following described above are able to follow curves like those depicted in Figure 11.4 to a solution of the nonlinear system. They rely, however, on the $n \times (n+1)$ matrix in (11.63) having full rank for all (x, λ) along the path, so that the tangent vector is well-defined. The following result shows that full rank is guaranteed under certain assumptions.

Theorem 11.9 (Watson [305]).

Suppose that r is twice continuously differentiable. Then for almost all vectors $a \in \mathbb{R}^n$, there is a zero path emanating from $(0, a)$ along which the $n \times (n+1)$ matrix (11.63) has full rank. If this path is bounded for $\lambda \in [0, 1)$, then it has an accumulation point $(\bar{x}, 1)$ such that $r(\bar{x}) = 0$. Furthermore, if the Jacobian $J(\bar{x})$ is nonsingular, the zero path between $(a, 0)$ and $(\bar{x}, 1)$ has finite arc length.

The theorem assures us that unless we are unfortunate in the choice of a, the algorithms described above can be applied to obtain a path that either diverges or else leads to a point $\bar{x}$ that is a solution of the original nonlinear system if $J(\bar{x})$ is nonsingular. More detailed convergence results can be found in Watson [305] and the references therein.

We conclude with an example to show that divergence of the zero path—the less desirable outcome of Theorem 11.9—can happen even for innocent-looking problems.

❏ EXAMPLE 11.3

Consider the system $r(x) = x^2 - 1$, for which there are two nondegenerate solutions $+1$ and -1. Suppose we choose $a = -2$ and attempt to apply a continuation method to the function

$$
H(x, \lambda) = \lambda(x^2 - 1) + (1 - \lambda)(x + 2) = \lambda x^2 + (1 - \lambda)x + (2 - 3\lambda), \qquad (11.65)
$$

obtained by substituting into (11.61). The zero paths for this function are plotted in Figure 11.6. As can be seen from that diagram, there is no zero path that joins $(-2, 0)$

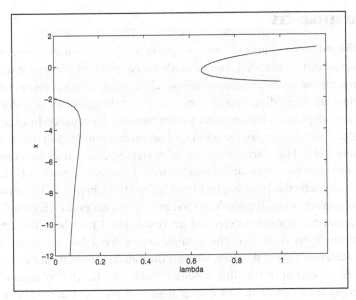

Figure 11.6 Zero paths for the example in which $H(x, \lambda) = \lambda(x^2-1)+(1-\lambda)(x+2)$. There is no continuous zero path from $\lambda = 0$ to $\lambda = 1$.

to either $(1, 1)$ or $(-1, 1)$, so the continuation methods fail on this example. We can find the values of λ for which no solution exists by using the formula for a quadratic root to obtain

$$x = \frac{-(1-\lambda) \pm \sqrt{(1-\lambda)^2 - 4\lambda(2-3\lambda)}}{2\lambda}.$$

Now, when the term in the square root is negative, the corresponding values of x are complex, that is, there are no real roots x. It is easy to verify that such is the case when

$$\lambda \in \left(\frac{5-2\sqrt{3}}{13}, \frac{5+2\sqrt{3}}{13} \right) \approx (0.118, 0.651).$$

Note that the zero path starting from $(-2, 0)$ becomes unbounded, which is one of the possible outcomes of Theorem 11.9. ❐

This example indicates that continuation methods may fail to produce a solution even to a fairly simple system of nonlinear equations. However, it is generally true that they are more reliable than the merit-function methods described earlier in the chapter. The extra robustness comes at a price, since continuation methods typically require significantly more computational effort than the merit-function methods.

NOTES AND REFERENCES

Nonlinear differential equations and integral equations are a rich source of nonlinear equations. When formulated as finite-dimensional nonlinear equations, the unknown vector x is a discrete approximation to the (infinite-dimensional) solution. In other applications, the vector x is intrinsically finite-dimensional; it may represent the quantities of materials to be transported between pairs of cities in a distribution network, for instance. In all cases, the equations r_i enforce consistency, conservation, and optimality principles in the model. Moré [212] and Averick et al. [10] discuss a number of interesting practical applications.

For analysis of the convergence of Broyden's method, including proofs of Theorem 11.5, see Dennis and Schnabel [92, Chapter 8] and Kelley [177, Chapter 6]. Details on a limited-memory implementation of Broyden's method are given by Kelley [177, Section 7.3].

Example 11.2 and the algorithm described by Powell [241] have been influential beyond the field of nonlinear equations. The example shows that a line-search method may not be able to achieve sufficient decrease, whereas the Cauchy step in the trust-region approach is designed to guarantee that this condition holds and hence that reasonable convergence properties are guaranteed. The dogleg algorithm proposed in [241] can be viewed as one of the first modern trust-region methods.

✎ EXERCISES

✎ **11.1** Show that for any vector $s \in \mathbb{R}^n$, we have

$$\left\| \frac{s s^T}{s^T s} \right\| = 1,$$

where $\| \cdot \|$ denotes the Euclidean matrix norm.

✎ **11.2** Consider the function $r : \mathbb{R} \rightarrow \mathbb{R}$ defined by $r(x) = x^q$, where q is an integer greater than 2. Note that $x^* = 0$ is the sole root of this function and that it is degenerate. Show that Newton's method converges Q-linearly, and find the value of the convergence ratio r in (A.34).

✎ **11.3** Show that Newton's method applied to the function $r(x) = -x^5 + x^3 + 4x$ starting from $x_0 = 1$ produces the cyclic behavior described in the text. Find the roots of this function, and check that they are nondegenerate.

✎ **11.4** For the scalar function $r(x) = \sin(5x) - x$, show that the sum-of-squares merit function has infinitely many local minima, and find a general formula for such points.

✎ **11.5** When $r : \mathbb{R}^n \rightarrow \mathbb{R}^n$, show that the function

$$\phi(\lambda) = \left\| (J^T J + \lambda I)^{-1} J^T r \right\|$$

is monotonically decreasing in λ unless $J^T r = 0$. (Hint: Use the singular-value decomposition of J.)

✏ **11.6** Prove part (iii) of Theorem 11.3.

✏ **11.7** Consider a line-search Newton method in which the step length α_k is chosen to be the exact minimizer of the merit function $f(\cdot)$; that is,

$$\alpha_k = \arg\min_\alpha f(x_k - \alpha J_k^{-1} r_k).$$

Show that if $J(x)$ is nonsingular at the solution x^*, then $\alpha_k \to 1$ as $x_k \to x^*$.

✏ **11.8** Let $J \in \mathbb{R}^{n \times m}$ and $r \in \mathbb{R}^n$ and suppose that $J J^T r = 0$. Show that $J^T r = 0$. (Hint: This doesn't even take one line!)

✏ **11.9** Suppose we replace the assumption of $x_k \to x^*$ in Theorem 11.8 by an assumption that the nondegenerate solution x^* is a limit point of x^*. By adding some logic to the proof of this result, show that in fact x^* is the only possible limit point of the sequence. (Hint: Show that $\| J_{k+1}^{-1} r_{k+1} \| \le \frac{1}{2} \| J_k^{-1} r_k \|$ for all k sufficiently large, and hence that for any constant $\epsilon > 0$, the sequence $\{x_k\}$ satisfies $\| x_k - x^* \| \le \epsilon$ for all k sufficiently large.)

✏ **11.10** Consider the following modification of our example of failure of continuation methods:

$$r(x) = x^2 - 1, \quad a = \tfrac{1}{2}.$$

Show that for this example there is a zero path for $H(x, \lambda) = \lambda(x^2 - 1) + (1 - \lambda)(x - a)$ that connects $(\tfrac{1}{2}, 0)$ to $(1, 1)$, so that continuation methods should work for this choice of starting point.

Theory of Constrained Optimization

The second part of this book is about minimizing functions subject to constraints on the variables. A general formulation for these problems is

$$\min_{x \in \mathbb{R}^n} f(x) \quad \text{subject to} \quad \begin{cases} c_i(x) = 0, & i \in \mathcal{E}, \\ c_i(x) \geq 0, & i \in \mathcal{I}, \end{cases} \tag{12.1}$$

where f and the functions c_i are all smooth, real-valued functions on a subset of $\mathbb{R}^n$, and $\mathcal{I}$ and $\mathcal{E}$ are two finite sets of indices. As before, we call f the *objective* function, while c_i,

$i \in \mathcal{E}$ are the *equality constraints* and c_i, $i \in \mathcal{I}$ are the *inequality constraints*. We define the *feasible set* Ω to be the set of points x that satisfy the constraints; that is,

$$\Omega = \{x \mid c_i(x) = 0, \ i \in \mathcal{E}; \ c_i(x) \geq 0, \ i \in \mathcal{I}\}, \qquad (12.2)$$

so that we can rewrite (12.1) more compactly as

$$\min_{x \in \Omega} f(x). \qquad (12.3)$$

In this chapter we derive mathematical characterizations of the solutions of (12.3). As in the unconstrained case, we discuss optimality conditions of two types. *Necessary* conditions are conditions that must be satisfied by any solution point (under certain assumptions). *Sufficient* conditions are those that, if satisfied at a certain point x^*, guarantee that x^* is in fact a solution.

For the unconstrained optimization problem of Chapter 2, the optimality conditions were as follows:

Necessary conditions: Local unconstrained minimizers have $\nabla f(x^*) = 0$ and $\nabla^2 f(x^*)$ positive semidefinite.

Sufficient conditions: Any point x^* at which $\nabla f(x^*) = 0$ and $\nabla^2 f(x^*)$ is positive definite is a strong local minimizer of f.

In this chapter, we derive analogous conditions to characterize the solutions of constrained optimization problems.

LOCAL AND GLOBAL SOLUTIONS

We have seen already that global solutions are difficult to find even when there are no constraints. The situation may be improved when we add constraints, since the feasible set might exclude many of the local minima and it may be comparatively easy to pick the global minimum from those that remain. However, constraints can also make things more difficult. As an example, consider the problem

$$\min (x_2 + 100)^2 + 0.01 x_1^2, \quad \text{subject to } x_2 - \cos x_1 \geq 0, \qquad (12.4)$$

illustrated in Figure 12.1. Without the constraint, the problem has the unique solution $(0, -100)^T$. With the constraint, there are local solutions near the points

$$x^{(k)} = (k\pi, -1)^T, \quad \text{for } k = \pm 1, \pm 3, \pm 5, \ldots.$$

Definitions of the different types of local solutions are simple extensions of the corresponding definitions for the unconstrained case, except that now we restrict consideration to the *feasible* points in the neighborhood of x^*. We have the following definition.

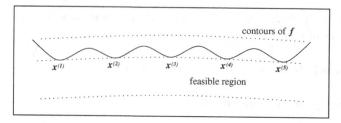

Figure 12.1 Constrained problem with many isolated local solutions.

A vector x^* is a *local solution* of the problem (12.3) if $x^* \in \Omega$ and there is a neighborhood $\mathcal{N}$ of x^* such that $f(x) \geq f(x^*)$ for $x \in \mathcal{N} \cap \Omega$.

Similarly, we can make the following definitions:

A vector x^* is a *strict local solution* (also called a *strong local solution*) if $x^* \in \Omega$ and there is a neighborhood $\mathcal{N}$ of x^* such that $f(x) > f(x^*)$ for all $x \in \mathcal{N} \cap \Omega$ with $x \neq x^*$.

A point x^* is an *isolated local solution* if $x^* \in \Omega$ and there is a neighborhood $\mathcal{N}$ of x^* such that x^* is the only local solution in $\mathcal{N} \cap \Omega$.

Note that isolated local solutions are strict, but that the reverse is not true (see Exercise 12.2).

SMOOTHNESS

Smoothness of objective functions and constraints is an important issue in characterizing solutions, just as in the unconstrained case. It ensures that the objective function and the constraints all behave in a reasonably predictable way and therefore allows algorithms to make good choices for search directions.

We saw in Chapter 2 that graphs of nonsmooth functions contain "kinks" or "jumps" where the smoothness breaks down. If we plot the feasible region for any given constrained optimization problem, we usually observe many kinks and sharp edges. Does this mean that the constraint functions that describe these regions are nonsmooth? The answer is often no, because the nonsmooth boundaries can often be described by a collection of smooth constraint functions. Figure 12.2 shows a diamond-shaped feasible region in $\mathbb{R}^2$ that could be described by the single nonsmooth constraint

$$\|x\|_1 = |x_1| + |x_2| \leq 1. \tag{12.5}$$

It can also be described by the following set of smooth (in fact, linear) constraints:

$$x_1 + x_2 \leq 1, \quad x_1 - x_2 \leq 1, \quad -x_1 + x_2 \leq 1, \quad -x_1 - x_2 \leq 1. \tag{12.6}$$

Each of the four constraints represents one edge of the feasible polytope. In general, the constraint functions are chosen so that each one represents a smooth piece of the boundary of Ω.

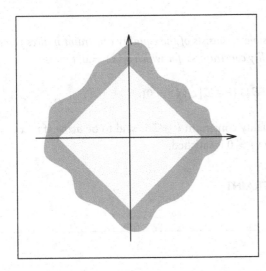

Figure 12.2
A feasible region with a
nonsmooth boundary can be
described by smooth constraints.

Nonsmooth, unconstrained optimization problems can sometimes be reformulated as smooth constrained problems. An example is the unconstrained minimization of a function

$$f(x) = \max(x^2, x), \tag{12.7}$$

which has kinks at $x = 0$ and $x = 1$, and the solution at $x^* = 0$. We obtain a smooth, constrained formulation of this problem by adding an artificial variable t and writing

$$\min t \quad \text{s.t.} \quad t \geq x, \quad t \geq x^2. \tag{12.8}$$

Reformulation techniques such as (12.6) and (12.8) are used often in cases where f is a maximum of a collection of functions or when f is a 1-norm or ∞-norm of a vector function.

In the examples above we expressed inequality constraints in a slightly different way from the form $c_i(x) \geq 0$ that appears in the definition (12.1). However, any collection of inequality constraints with $\geq$ and $\leq$ and nonzero right-hand-sides can be expressed in the form $c_i(x) \geq 0$ by simple rearrangement of the inequality.

12.1 EXAMPLES

To introduce the basic principles behind the characterization of solutions of constrained optimization problems, we work through three simple examples. The discussion here is informal; the ideas introduced will be made rigorous in the sections that follow.

We start by noting one important item of terminology that recurs throughout the rest of the book.

Definition 12.1.

   The active set $\mathcal{A}(x)$ *at any feasible* x *consists of the equality constraint indices from* $\mathcal{E}$ *together with the indices of the inequality constraints* i *for which* $c_i(x) = 0$; *that is,*

$$\mathcal{A}(x) = \mathcal{E} \cup \{i \in \mathcal{I} \mid c_i(x) = 0\}.$$

   At a feasible point x, the inequality constraint $i \in \mathcal{I}$ is said to be *active* if $c_i(x) = 0$ and *inactive* if the strict inequality $c_i(x) > 0$ is satisfied.

A SINGLE EQUALITY CONSTRAINT

❏ EXAMPLE 12.1

   Our first example is a two-variable problem with a single equality constraint:

$$\min x_1 + x_2 \quad \text{s.t.} \quad x_1^2 + x_2^2 - 2 = 0 \tag{12.9}$$

(see Figure 12.3). In the language of (12.1), we have $f(x) = x_1 + x_2$, $\mathcal{I} = \emptyset$, $\mathcal{E} = \{1\}$, and $c_1(x) = x_1^2 + x_2^2 - 2$. We can see by inspection that the feasible set for this problem is the circle of radius $\sqrt{2}$ centered at the origin—just the boundary of this circle, not its interior. The solution x^* is obviously $(-1, -1)^T$. From any other point on the circle, it is easy to find a way to move that *stays feasible* (that is, remains on the circle) while *decreasing* f. For instance, from the point $x = (\sqrt{2}, 0)^T$ any move in the clockwise direction around the circle has the desired effect.

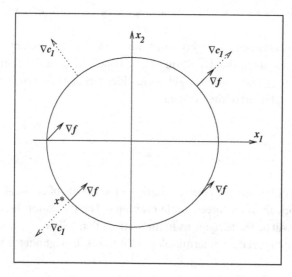

Figure 12.3
Problem (12.9), showing constraint and function gradients at various feasible points.

We also see from Figure 12.3 that at the solution x^*, the *constraint normal* $\nabla c_1(x^*)$ is parallel to $\nabla f(x^*)$. That is, there is a scalar λ_1^* (in this case $\lambda_1^* = -1/2$) such that

$$\nabla f(x^*) = \lambda_1^* \nabla c_1(x^*). \tag{12.10}$$

$\square$

We can derive (12.10) by examining first-order Taylor series approximations to the objective and constraint functions. To retain feasibility with respect to the function $c_1(x) = 0$, we require any small (but nonzero) step s to satisfy that $c_1(x + s) = 0$; that is,

$$0 = c_1(x + s) \approx c_1(x) + \nabla c_1(x)^T s = \nabla c_1(x)^T s. \tag{12.11}$$

Hence, the step s retains feasibility with respect to c_1, to first order, when it satisfies

$$\nabla c_1(x)^T s = 0. \tag{12.12}$$

Similarly, if we want s to produce a decrease in f, we would have so that

$$0 > f(x + s) - f(x) \approx \nabla f(x)^T s,$$

or, to first order,

$$\nabla f(x)^T s < 0. \tag{12.13}$$

Existence of a small step s that satisfies both (12.12) and (12.13) strongly suggests existence of a direction d (where the size of d is *not* small; we could have $d \approx s/\|s\|$ to ensure that the norm of d is close to 1) with the same properties, namely

$$\nabla c_1(x)^T d = 0 \quad \text{and} \quad \nabla f(x)^T d < 0. \tag{12.14}$$

If, on the other hand, there is *no* direction d with the properties (12.14), then is it likely that we cannot find a small step s with the properties (12.12) and (12.13). In this case, x^* would appear to be a local minimizer.

By drawing a picture, the reader can check that the only way that a d satisfying (12.14) does *not* exist is if $\nabla f(x)$ and $\nabla c_1(x)$ are parallel, that is, if the condition $\nabla f(x) = \lambda_1 \nabla c_1(x)$ holds at x, for some scalar λ_1. If in fact $\nabla f(x)$ and $\nabla c_1(x)$ are *not* parallel, we can set

$$\bar{d} = -\left(I - \frac{\nabla c_1(x) \nabla c_1(x)^T}{\|\nabla c_1(x)\|^2}\right) \nabla f(x); \qquad d = \frac{\bar{d}}{\|\bar{d}\|}. \tag{12.15}$$

It is easy to verify that this d satisfies (12.14).

By introducing the *Lagrangian function*

$$\mathcal{L}(x, \lambda_1) = f(x) - \lambda_1 c_1(x), \qquad (12.16)$$

and noting that $\nabla_x \mathcal{L}(x, \lambda_1) = \nabla f(x) - \lambda_1 \nabla c_1(x)$, we can state the condition (12.10) equivalently as follows: At the solution x^*, there is a scalar λ_1^* such that

$$\nabla_x \mathcal{L}(x^*, \lambda_1^*) = 0. \qquad (12.17)$$

This observation suggests that we can search for solutions of the equality-constrained problem (12.9) by seeking stationary points of the Lagrangian function. The scalar quantity λ_1 in (12.16) is called a *Lagrange multiplier* for the constraint $c_1(x) = 0$.

Though the condition (12.10) (equivalently, (12.17)) appears to be *necessary* for an optimal solution of the problem (12.9), it is clearly not *sufficient*. For instance, in Example 12.1, condition (12.10) is satisfied at the point $x = (1, 1)^T$ (with $\lambda_1 = \frac{1}{2}$), but this point is obviously not a solution—in fact, it *maximizes* the function f on the circle. Moreover, in the case of equality-constrained problems, we cannot turn the condition (12.10) into a sufficient condition simply by placing some restriction on the sign of λ_1. To see this, consider replacing the constraint $x_1^2 + x_2^2 - 2 = 0$ by its negative $2 - x_1^2 - x_2^2 = 0$ in Example 12.1. The solution of the problem is not affected, but the value of λ_1^* that satisfies the condition (12.10) changes from $\lambda_1^* = -\frac{1}{2}$ to $\lambda_1^* = \frac{1}{2}$.

A SINGLE INEQUALITY CONSTRAINT

❏ **EXAMPLE 12.2**

This is a slight modification of Example 12.1, in which the equality constraint is replaced by an inequality. Consider

$$\min x_1 + x_2 \qquad \text{s.t.} \qquad 2 - x_1^2 - x_2^2 \geq 0, \qquad (12.18)$$

for which the feasible region consists of the circle of problem (12.9) and its interior (see Figure 12.4). Note that the constraint normal ∇c_1 points toward the interior of the feasible region at each point on the boundary of the circle. By inspection, we see that the solution is still $(-1, -1)^T$ and that the condition (12.10) holds for the value $\lambda_1^* = \frac{1}{2}$. However, this inequality-constrained problem differs from the equality-constrained problem (12.9) of Example 12.1 in that the sign of the Lagrange multiplier plays a significant role, as we now argue.

❏

As before, we conjecture that a given feasible point x is *not* optimal if we can find a small step s that both retains feasibility and decreases the objective function f to first order. The main difference between problems (12.9) and (12.18) comes in the handling of the feasibility condition. As in (12.13), the step s improves the objective function, to first order, if $\nabla f(x)^T s < 0$. Meanwhile, s retains feasibility if

$$0 \leq c_1(x + s) \approx c_1(x) + \nabla c_1(x)^T s,$$

so, to first order, feasibility is retained if

$$c_1(x) + \nabla c_1(x)^T s \geq 0. \tag{12.19}$$

In determining whether a step s exists that satisfies both (12.13) and (12.19), we consider the following two cases, which are illustrated in Figure 12.4.

Case I: Consider first the case in which x lies *strictly inside* the circle, so that the strict inequality $c_1(x) > 0$ holds. In this case, *any* step vector s satisfies the condition (12.19), provided only that its length is sufficiently small. In fact, whenever $\nabla f(x) \neq 0$, we can obtain a step s that satisfies both (12.13) and (12.19) by setting

$$s = -\alpha \nabla f(x),$$

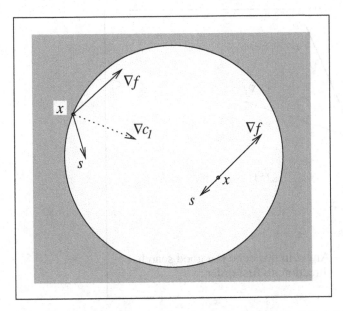

Figure 12.4   Improvement directions s from two feasible points x for the problem (12.18) at which the constraint is active and inactive, respectively.

for any positive scalar α sufficiently small. However, this definition does not give a step s with the required properties when

$$\nabla f(x) = 0, \tag{12.20}$$

Case II: Consider now the case in which x lies on the boundary of the circle, so that $c_1(x) = 0$. The conditions (12.13) and (12.19) therefore become

$$\nabla f(x)^T s < 0, \qquad \nabla c_1(x)^T s \geq 0.$$

The first of these conditions defines an open half-space, while the second defines a closed half-space, as illustrated in Figure 12.5. It is clear from this figure that the intersection of these two regions is empty only when $\nabla f(x)$ and $\nabla c_1(x)$ point in the same direction, that is, when

$$\nabla f(x) = \lambda_1 \nabla c_1(x), \qquad \text{for some } \lambda_1 \geq 0. \tag{12.21}$$

Note that the sign of the multiplier is significant here. If (12.10) were satisfied with a *negative* value of λ_1, then $\nabla f(x)$ and $\nabla c_1(x)$ would point in opposite directions, and we see from Figure 12.5 that the set of directions that satisfy both (12.13) and (12.19) would make up an entire open half-plane.

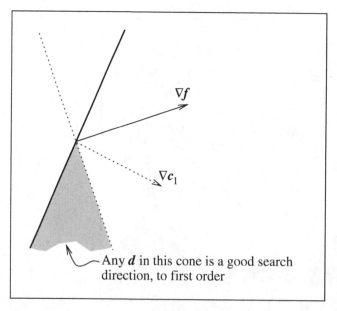

Any d in this cone is a good search direction, to first order

Figure 12.5 A direction d that satisfies both (12.13) and (12.19) lies in the intersection of a closed half-plane and an open half-plane.

The optimality conditions for both cases I and II can again be summarized neatly with reference to the Lagrangian function $\mathcal{L}$ defined in (12.16). When no first-order feasible descent direction exists at some point x^*, we have that

$$\nabla_x \mathcal{L}(x^*, \lambda_1^*) = 0, \quad \text{for some } \lambda_1^* \geq 0, \tag{12.22}$$

where we also require that

$$\lambda_1^* c_1(x^*) = 0. \tag{12.23}$$

Condition (12.23) is known as a *complementarity condition*; it implies that the Lagrange multiplier λ_1 can be strictly positive *only when the corresponding constraint* c_1 *is active*. Conditions of this type play a central role in constrained optimization, as we see in the sections that follow. In case I, we have that $c_1(x^*) > 0$, so (12.23) requires that $\lambda_1^* = 0$. Hence, (12.22) reduces to $\nabla f(x^*) = 0$, as required by (12.20). In case II, (12.23) allows λ_1^* to take on a nonnegative value, so (12.22) becomes equivalent to (12.21).

TWO INEQUALITY CONSTRAINTS

❏ EXAMPLE 12.3

Suppose we add an extra constraint to the problem (12.18) to obtain

$$\min x_1 + x_2 \quad \text{s.t.} \quad 2 - x_1^2 - x_2^2 \geq 0, \quad x_2 \geq 0, \tag{12.24}$$

for which the feasible region is the half-disk illustrated in Figure 12.6. It is easy to see that the solution lies at $(-\sqrt{2}, 0)^T$, a point at which both constraints are active. By repeating the arguments for the previous examples, we would expect a direction d of first-order feasible descent to satisfy

$$\nabla c_i(x)^T d \geq 0, \quad i \in \mathcal{I} = \{1, 2\}, \quad \nabla f(x)^T d < 0. \tag{12.25}$$

However, it is clear from Figure 12.6 that no such direction can exist when $x = (-\sqrt{2}, 0)^T$. The conditions $\nabla c_i(x)^T d \geq 0$, $i = 1, 2$, are both satisfied only if d lies in the quadrant defined by $\nabla c_1(x)$ and $\nabla c_2(x)$, but it is clear by inspection that all vectors d in this quadrant satisfy $\nabla f(x)^T d \geq 0$.

Let us see how the Lagrangian and its derivatives behave for the problem (12.24) and the solution point $(-\sqrt{2}, 0)^T$. First, we include an additional term $\lambda_i c_i(x)$ in the Lagrangian for each additional constraint, so the definition of $\mathcal{L}$ becomes

$$\mathcal{L}(x, \lambda) = f(x) - \lambda_1 c_1(x) - \lambda_2 c_2(x),$$

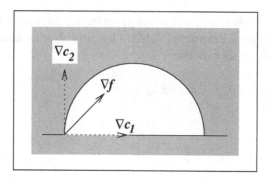

Figure 12.6
Problem (12.24), illustrating the gradients of the active constraints and objective at the solution.

where $\lambda = (\lambda_1, \lambda_2)^T$ is the vector of Lagrange multipliers. The extension of condition (12.22) to this case is

$$\nabla_x \mathcal{L}(x^*, \lambda^*) = 0, \quad \text{for some } \lambda^* \geq 0, \tag{12.26}$$

where the inequality $\lambda^* \geq 0$ means that all components of λ^* are required to be nonnegative. By applying the complementarity condition (12.23) to both inequality constraints, we obtain

$$\lambda_1^* c_1(x^*) = 0, \quad \lambda_2^* c_2(x^*) = 0. \tag{12.27}$$

When $x^* = (-\sqrt{2}, 0)^T$, we have

$$\nabla f(x^*) = \begin{bmatrix} 1 \\ 1 \end{bmatrix}, \quad \nabla c_1(x^*) = \begin{bmatrix} 2\sqrt{2} \\ 0 \end{bmatrix}, \quad \nabla c_2(x^*) = \begin{bmatrix} 0 \\ 1 \end{bmatrix},$$

so that it is easy to verify that $\nabla_x \mathcal{L}(x^*, \lambda^*) = 0$ when we select λ^* as follows:

$$\lambda^* = \begin{bmatrix} 1/(2\sqrt{2}) \\ 1 \end{bmatrix}.$$

Note that both components of λ^* are positive, so that (12.26) is satisfied.

We consider now some other feasible points that are *not* solutions of (12.24), and examine the properties of the Lagrangian and its gradient at these points.

For the point $x = (\sqrt{2}, 0)^T$, we again have that both constraints are active (see Figure 12.7). However, it s easy to identify vectors d that satisfies (12.25): $d = (-1, 0)^T$ is one such vector (there are many others). For this value of x it is easy to verify that the condition $\nabla_x \mathcal{L}(x, \lambda) = 0$ is satisfied only when $\lambda = (-1/(2\sqrt{2}), 1)^T$. Note that the first component λ_1 is negative, so that the conditions (12.26) are not satisfied at this point.

Finally, we consider the point $x = (1, 0)^T$, at which only the second constraint c_2 is active. Since any small step s away from this point will continue to satisfy $c_1(x + s) > 0$, we need to consider only the behavior of c_2 and f in determining whether s is indeed a feasible

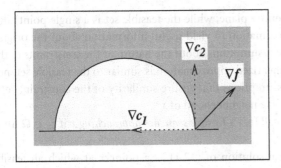

Figure 12.7
Problem (12.24), illustrating the gradients of the active constraints and objective at a nonoptimal point.

descent step. Using the same reasoning as in the earlier examples, we find that the direction of feasible descent d must satisfy

$$\nabla c_2(x)^T d \geq 0, \qquad \nabla f(x)^T d < 0. \tag{12.28}$$

By noting that

$$\nabla f(x) = \begin{bmatrix} 1 \\ 1 \end{bmatrix}, \qquad \nabla c_2(x) = \begin{bmatrix} 0 \\ 1 \end{bmatrix},$$

it is easy to verify that the vector $d = \left(-\frac{1}{2}, \frac{1}{4}\right)^T$ satisfies (12.28) and is therefore a descent direction.

To show that optimality conditions (12.26) and (12.27) fail, we note first from (12.27) that since $c_1(x) > 0$, we must have $\lambda_1 = 0$. Therefore, in trying to satisfy $\nabla_x \mathcal{L}(x, \lambda) = 0$, we are left to search for a value λ_2 such that $\nabla f(x) - \lambda_2 \nabla c_2(x) = 0$. No such λ_2 exists, and thus this point fails to satisfy the optimality conditions. ❑

12.2 TANGENT CONE AND CONSTRAINT QUALIFICATIONS

In this section we define the tangent cone $T_\Omega(x^*)$ to the closed convex set Ω at a point $x^* \in \Omega$, and also the set $\mathcal{F}(x^*)$ of first-order feasible directions at x^*. We also discuss *constraint qualifications*. In the previous section, we determined whether or not it was possible to take a feasible descent step away from a given feasible point x by examining the first derivatives of f and the constraint functions c_i. We used the first-order Taylor series expansion of these functions about x to form an approximate problem in which both objective and constraints are linear. This approach makes sense, however, only when the linearized approximation captures the essential geometric features of the feasible set near the point x in question. If, near x, the linearization is fundamentally different from the feasible set (for instance, it is an entire plane, while the feasible set is a single point) then we cannot expect the linear approximation to yield useful information about the original problem. Hence, we need to make assumptions about the nature of the constraints c_i that are active at x to ensure that the linearized approximation is similar to the feasible set, near x. Constraint qualifications are assumptions that ensure similarity of the constraint set Ω and its linearized approximation, in a neighborhood of x^*.

Given a feasible point x, we call $\{z_k\}$ a *feasible sequence approaching x* if $z_k \in \Omega$ for all k sufficiently large and $z_k \to x$.

Later, we characterize a local solution of (12.1) as a point x at which all feasible sequences approaching x have the property that $f(z_k) \geq f(x)$ for all k sufficiently large, and we will derive practical, verifiable conditions under which this property holds. We lay the groundwork in this section by characterizing the directions in which we can step away from x while remaining feasible.

A *tangent* is a limiting direction of a feasible sequence.

Definition 12.2.

The vector d is said to be a tangent *(or tangent vector) to Ω at a point x if there are a feasible sequence $\{z_k\}$ approaching x and a sequence of positive scalars $\{t_k\}$ with $t_k \to 0$ such that*

$$\lim_{k \to \infty} \frac{z_k - x}{t_k} = d. \tag{12.29}$$

The set of all tangents to Ω at x^* is called the tangent cone *and is denoted by $T_\Omega(x^*)$.*

It is easy to see that the tangent cone is indeed a cone, according to the definition (A.36). If d is a tangent vector with corresponding sequences $\{z_k\}$ and $\{t_k\}$, then by replacing each t_k by $\alpha^{-1} t_k$, for any $\alpha > 0$, we find that $\alpha d \in T_\Omega(x^*)$ also. We obtain that $0 \in T_\Omega(x)$ by setting $z_k \equiv x$ in the definition of feasible sequence.

We turn now to the linearized feasible direction set, which we define as follows.

Definition 12.3.

Given a feasible point x and the active constraint set $\mathcal{A}(x)$ of Definition 12.1, the set of linearized feasible directions $\mathcal{F}(x)$ is

$$\mathcal{F}(x) = \left\{ d \mid \begin{array}{ll} d^T \nabla c_i(x) = 0, & \text{for all } i \in \mathcal{E}, \\ d^T \nabla c_i(x) \geq 0, & \text{for all } i \in \mathcal{A}(x) \cap \mathcal{I} \end{array} \right\}.$$

As with the tangent cone, it is easy to verify that $\mathcal{F}(x)$ is a cone, according to the definition (A.36).

It is important to note that the definition of tangent cone does not rely on the algebraic specification of the set Ω, only on its geometry. The linearized feasible direction set does, however, depend on the definition of the constraint functions c_i, $i \in \mathcal{E} \cup \mathcal{I}$.

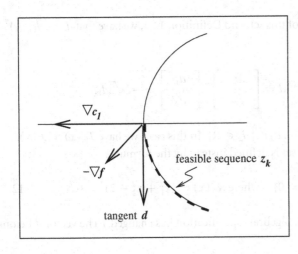

Figure 12.8
Constraint normal, objective gradient, and feasible sequence for problem (12.9).

We illustrate the tangent cone and the linearized feasible direction set by revisiting Examples 12.1 and 12.2.

❏ **EXAMPLE 12.4**   (EXAMPLE 12.1, REVISITED)

Figure 12.8 shows the problem (12.9), the equality-constrained problem in which the feasible set is a circle of radius $\sqrt{2}$, near the nonoptimal point $x = (-\sqrt{2}, 0)^T$. The figure also shows a feasible sequence approaching x. This sequence could be defined analytically by the formula

$$z_k = \begin{bmatrix} -\sqrt{2 - 1/k^2} \\ -1/k \end{bmatrix}. \tag{12.30}$$

By choosing $t_k = \|z_k - x\|$, we find that $d = (0, -1)^T$ is a tangent. Note that the objective function $f(x) = x_1 + x_2$ increases as we move along the sequence (12.30); in fact, we have $f(z_{k+1}) > f(z_k)$ for all $k = 2, 3, \ldots$. It follows that $f(z_k) < f(x)$ for $k = 2, 3, \ldots$, so x cannot be a solution of (12.9).

Another feasible sequence is one that approaches $x = (-\sqrt{2}, 0)^T$ from the opposite direction. Its elements are defined by

$$z_k = \begin{bmatrix} -\sqrt{2 - 1/k^2} \\ 1/k \end{bmatrix}.$$

It is easy to show that f *decreases* along this sequence and that the tangents corresponding to this sequence are $d = (0, \alpha)^T$. In summary, the tangent cone at $x = (-\sqrt{2}, 0)^T$ is $\{(0, d_2)^T \mid d_2 \in \mathbb{R}\}$.

For the definition (12.9) of this set, and Definition 12.3, we have that $d = (d_1, d_2)^T \in \mathcal{F}(x)$ if

$$0 = \nabla c_1(x)^T d = \begin{bmatrix} 2x_1 \\ 2x_2 \end{bmatrix}^T \begin{bmatrix} d_1 \\ d_2 \end{bmatrix} = -2\sqrt{2}d_1.$$

Therefore, we obtain $\mathcal{F}(x) = \{(0, d_2)^T \mid d_2 \in \mathbb{R}\}$. In this case, we have $T_\Omega(x) = \mathcal{F}(x)$.

Suppose that the feasible set is defined instead by the formula

$$\Omega = \{x \mid c_1(x) = 0\}, \quad \text{where} \quad c_1(x) = (x_1^2 + x_2^2 - 2)^2 = 0. \tag{12.31}$$

(Note that Ω is the same, but its algebraic specification has changed.) The vector d belongs to the linearized feasible set if

$$0 = \nabla c_1(x)^T d = \begin{bmatrix} 4(x_1^2 + x_2^2 - 2)x_1 \\ 4(x_1^2 + x_2^2 - 2)x_2 \end{bmatrix}^T \begin{bmatrix} d_1 \\ d_2 \end{bmatrix} = \begin{bmatrix} 0 \\ 0 \end{bmatrix}^T \begin{bmatrix} d_1 \\ d_2 \end{bmatrix},$$

which is true for all $(d_1, d_2)^T$. Hence, we have $\mathcal{F}(x) = \mathbb{R}^2$, so for this algebraic specification of Ω, the tangent cone and linearized feasible sets differ.

⧉

❑ **EXAMPLE 12.5**   (EXAMPLE 12.2, REVISITED)

We now reconsider problem (12.18) in Example 12.2. The solution $x = (-1, -1)^T$ is the same as in the equality-constrained case, but there is a much more extensive collection of feasible sequences that converge to any given feasible point (see Figure 12.9).

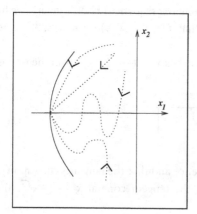

Figure 12.9
Feasible sequences converging to a particular feasible point for the region defined by $x_1^2 + x_2^2 \leq 2$.

From the point $x = (-\sqrt{2}, 0)^T$, the various feasible sequences defined above for the equality-constrained problem are still feasible for (12.18). There are also infinitely many feasible sequences that converge to $x = (-\sqrt{2}, 0)^T$ along a straight line from the interior of the circle. These sequences have the form

$$z_k = (-\sqrt{2}, 0)^T + (1/k)w,$$

where w is any vector whose first component is positive ($w_1 > 0$). The point z_k is feasible provided that $\|z_k\| \le \sqrt{2}$, that is,

$$(-\sqrt{2} + w_1/k)^2 + (w_2/k)^2 \le 2,$$

which is true when $k \ge (w_1^2 + w_2^2)/(2\sqrt{2}w_1)$. In addition to these straight-line feasible sequences, we can also define an infinite variety of sequences that approach $(-\sqrt{2}, 0)^T$ along a curve from the interior of the circle. To summarize, the tangent cone to this set at $(-\sqrt{2}, 0)^T$ is $\{(w_1, w_2)^T \mid w_1 \ge 0\}$.

For the definition (12.18) of this feasible set, we have from Definition 12.3 that $d \in \mathcal{F}(x)$ if

$$0 \le \nabla c_1(x)^T d = \begin{bmatrix} -2x_1 \\ -2x_2 \end{bmatrix}^T \begin{bmatrix} d_1 \\ d_2 \end{bmatrix} = 2\sqrt{2}d_1.$$

Hence, we obtain $\mathcal{F}(x) = T_\Omega(x)$ for this particular algebraic specification of the feasible set. ∎

Constraint qualifications are conditions under which the linearized feasible set $\mathcal{F}(x)$ is similar to the tangent cone $T_\Omega(x)$. In fact, most constraint qualifications ensure that these two sets are identical. As mentioned earlier, these conditions ensure that the $\mathcal{F}(x)$, which is constructed by linearizing the algebraic description of the set Ω at x, captures the essential geometric features of the set Ω in the vicinity of x, as represented by $T_\Omega(x)$.

Revisiting Example 12.4, we see that both $T_\Omega(x)$ and $\mathcal{F}(x)$ consist of the vertical axis, which is qualitatively similar to the set $\Omega - \{x\}$ in the neighborhood of x. As a further example, consider the constraints

$$c_1(x) = 1 - x_1^2 - (x_2 - 1)^2 \ge 0, \qquad c_2(x) = -x_2 \ge 0, \tag{12.32}$$

for which the feasible set is the single point $\Omega = \{(0, 0)^T\}$ (see Figure 12.10). For this point $x = (0, 0)^T$, it is obvious that that tangent cone is $T_\Omega(x) = \{(0, 0)^T\}$, since all feasible sequences approaching x must have $z_k = x = (0, 0)^T$ for all k sufficiently large. Moreover, it is easy to show that linearized approximation to the feasible set $\mathcal{F}(x)$ is

$$\mathcal{F}(x^*) = \{(d_1, 0)^T \mid d_1 \in \mathbb{R}\},$$

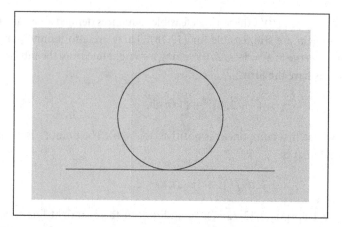

Figure 12.10 Problem (12.32), for which the feasible set is the single point of intersection between circle and line.

that is, the entire horizontal axis. In this case, the linearized feasible direction set does not capture the geometry of the feasible set, so constraint qualifications are not satisfied.

The constraint qualification most often used in the design of algorithms is the subject of the next definition.

Definition 12.4 (LICQ).

Given the point x and the active set $\mathcal{A}(x)$ defined in Definition 12.1, we say that the linear independence constraint qualification (LICQ) *holds if the set of active constraint gradients $\{\nabla c_i(x), i \in \mathcal{A}(x)\}$ is linearly independent.*

Note that this condition is *not* satisfied for the examples (12.32) and (12.31). In general, if LICQ holds, none of the active constraint gradients can be zero. We mention other constraint qualifications in Section 12.6.

12.3 FIRST-ORDER OPTIMALITY CONDITIONS

In this section, we state first-order necessary conditions for x^* to be a local minimizer and show how these conditions are satisfied on a small example. The proof of the result is presented in subsequent sections.

As a preliminary to stating the necessary conditions, we define the Lagrangian function for the general problem (12.1).

$$\mathcal{L}(x, \lambda) = f(x) - \sum_{i \in \mathcal{E} \cup \mathcal{I}} \lambda_i c_i(x). \qquad (12.33)$$

(We had previously defined special cases of this function for the examples of Section 12.1.)

The necessary conditions defined in the following theorem are called *first-order conditions* because they are concerned with properties of the gradients (first-derivative vectors) of the objective and constraint functions. These conditions are the foundation for many of the algorithms described in the remaining chapters of the book.

Theorem 12.1 (First-Order Necessary Conditions).

Suppose that x^* is a local solution of (12.1), that the functions f and c_i in (12.1) are continuously differentiable, and that the LICQ holds at x^*. Then there is a Lagrange multiplier vector λ^*, with components λ_i^*, $i \in \mathcal{E} \cup \mathcal{I}$, such that the following conditions are satisfied at (x^*, λ^*)

$$\nabla_x \mathcal{L}(x^*, \lambda^*) = 0, \tag{12.34a}$$
$$c_i(x^*) = 0, \quad \text{for all } i \in \mathcal{E}, \tag{12.34b}$$
$$c_i(x^*) \geq 0, \quad \text{for all } i \in \mathcal{I}, \tag{12.34c}$$
$$\lambda_i^* \geq 0, \quad \text{for all } i \in \mathcal{I}, \tag{12.34d}$$
$$\lambda_i^* c_i(x^*) = 0, \quad \text{for all } i \in \mathcal{E} \cup \mathcal{I}. \tag{12.34e}$$

The conditions (12.34) are often known as the *Karush–Kuhn–Tucker conditions*, or *KKT conditions* for short. The conditions (12.34e) are *complementarity conditions*; they imply that either constraint i is active or $\lambda_i^* = 0$, or possibly both. In particular, the Lagrange multipliers corresponding to inactive inequality constraints are zero, we can omit the terms for indices $i \notin \mathcal{A}(x^*)$ from (12.34a) and rewrite this condition as

$$0 = \nabla_x \mathcal{L}(x^*, \lambda^*) = \nabla f(x^*) - \sum_{i \in \mathcal{A}(x^*)} \lambda_i^* \nabla c_i(x^*). \tag{12.35}$$

A special case of complementarity is important and deserves its own definition.

Definition 12.5 (Strict Complementarity).

Given a local solution x^* of (12.1) and a vector λ^* satisfying (12.34), we say that the strict complementarity condition holds if exactly one of λ_i^* and $c_i(x^*)$ is zero for each index $i \in \mathcal{I}$. In other words, we have that $\lambda_i^* > 0$ for each $i \in \mathcal{I} \cap \mathcal{A}(x^*)$.

Satisfaction of the strict complementarity property usually makes it easier for algorithms to determine the active set $\mathcal{A}(x^*)$ and converge rapidly to the solution x^*.

For a given problem (12.1) and solution point x^*, there may be many vectors λ^* for which the conditions (12.34) are satisfied. When the LICQ holds, however, the optimal λ^* is unique (see Exercise 12.17).

The proof of Theorem 12.1 is quite complex, but it is important to our understanding of constrained optimization, so we present it in the next section. First, we illustrate the KKT conditions with another example.

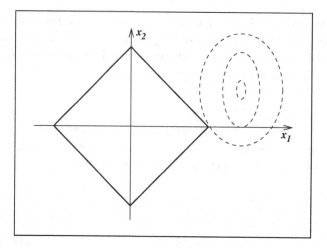

Figure 12.11 Inequality-constrained problem (12.36) with solution at $(1, 0)^T$.

❏ EXAMPLE 12.6

Consider the feasible region illustrated in Figure 12.2 and described by the four constraints (12.6). By restating the constraints in the standard form of (12.1) and including an objective function, the problem becomes

$$\min_x \left(x_1 - \frac{3}{2}\right)^2 + \left(x_2 - \frac{1}{2}\right)^4 \quad \text{s.t.} \quad \begin{bmatrix} 1 - x_1 - x_2 \\ 1 - x_1 + x_2 \\ 1 + x_1 - x_2 \\ 1 + x_1 + x_2 \end{bmatrix} \geq 0. \quad (12.36)$$

It is fairly clear from Figure 12.11 that the solution is $x^* = (1, 0)^T$. The first and second constraints in (12.36) are active at this point. Denoting them by c_1 and c_2 (and the inactive constraints by c_3 and c_4), we have

$$\nabla f(x^*) = \begin{bmatrix} -1 \\ -\frac{1}{2} \end{bmatrix}, \quad \nabla c_1(x^*) = \begin{bmatrix} -1 \\ -1 \end{bmatrix}, \quad \nabla c_2(x^*) = \begin{bmatrix} -1 \\ 1 \end{bmatrix}.$$

Therefore, the KKT conditions (12.34a)–(12.34e) are satisfied when we set

$$\lambda^* = \left(\tfrac{3}{4}, \tfrac{1}{4}, 0, 0\right)^T.$$

❏

12.4   FIRST-ORDER OPTIMALITY CONDITIONS: PROOF

We now develop a proof of Theorem 12.1. A number of key subsidiary results are required, so the development is quite long. However, a complete treatment is worthwhile, since these results are so fundamental to the field of optimization.

RELATING THE TANGENT CONE AND THE FIRST-ORDER FEASIBLE DIRECTION SET

The following key result uses a constraint qualification (LICQ) to relate the tangent cone of Definition 12.2 to the set $\mathcal{F}$ of first-order feasible directions of Definition 12.3. In the proof below and in later results, we use the notation $A(x^*)$ to represent the matrix whose rows are the active constraint gradients at the optimal point, that is,

$$A(x^*)^T = [\nabla c_i(x^*)]_{i \in \mathcal{A}(x^*)}, \tag{12.37}$$

where the active set $\mathcal{A}(x^*)$ is defined as in Definition 12.1.

Lemma 12.2.

Let x^* be a feasible point. The following two statements are true.

(i) $T_\Omega(x^*) \subset \mathcal{F}(x^*)$.

(ii) *If the LICQ condition is satisfied at x^*, then $\mathcal{F}(x^*) = T_\Omega(x^*)$.*

PROOF.   Without loss of generality, let us assume that all the constraints $c_i(\cdot)$, $i = 1, 2, \ldots, m$, are active at x^*. (We can arrive at this convenient ordering by simply dropping all inactive constraints—which are irrelevant in some neighborhood of x^*—and renumbering the active constraints that remain.)

To prove (i), let $\{z_k\}$ and $\{t_k\}$ be the sequences for which (12.29) is satisfied, that is,

$$\lim_{k \to \infty} \frac{z_k - x^*}{t_k} = d.$$

(Note in particular that $t_k > 0$ for all k.) From this definition, we have that

$$z_k = x^* + t_k d + o(t_k). \tag{12.38}$$

By taking $i \in \mathcal{E}$ and using Taylor's theorem, we have that

$$
\begin{aligned}
0 &= \frac{1}{t_k} c_i(z_k) \\
&= \frac{1}{t_k} \left[c_i(x^*) + t_k \nabla c_i(x^*)^T d + o(t_k) \right] \\
&= \nabla c_i(x^*)^T d + \frac{o(t_k)}{t_k}.
\end{aligned}
$$

By taking the limit as $k \to \infty$, the last term in this expression vanishes, and we have $\nabla c_i(x^*)^T d = 0$, as required. For the active inequality constraints $i \in \mathcal{A}(x^*) \cap \mathcal{I}$, we have similarly that

$$
\begin{aligned}
0 &\leq \frac{1}{t_k} c_i(z_k) \\
&= \frac{1}{t_k} \left[c_i(x^*) + t_k \nabla c_i(x^*)^T d + o(t_k) \right] \\
&= \nabla c_i(x^*)^T d + \frac{o(t_k)}{t_k}.
\end{aligned}
$$

Hence, by a similar limiting argument, we have that $\nabla c_i(x^*)^T d \geq 0$, as required.

For (ii), we use the implicit function theorem (see the Appendix or Lang [187, p. 131] for a statement of this result). First, since the LICQ holds, we have from Definition 12.4 that the $m \times n$ matrix $A(x^*)$ of active constraint gradients has full row rank m. Let Z be a matrix whose columns are a basis for the null space of $A(x^*)$; that is,

$$
Z \in \mathbb{R}^{n \times (n-m)}, \qquad Z \text{ has full column rank}, \qquad A(x^*)Z = 0. \tag{12.39}
$$

(See the related discussion in Chapter 16.) Choose $d \in \mathcal{F}(x^*)$ arbitrarily, and suppose that $\{t_k\}_{k=0}^{\infty}$ is any sequence of positive scalars such $\lim_{k \to \infty} t_k = 0$. Define the parametrized system of equations $R : \mathbb{R}^n \times \mathbb{R} \to \mathbb{R}^n$ by

$$
R(z,t) = \begin{bmatrix} c(z) - t A(x^*)d \\ Z^T(z - x^* - td) \end{bmatrix} = \begin{bmatrix} 0 \\ 0 \end{bmatrix}. \tag{12.40}
$$

We claim that the solutions $z = z_k$ of this system for small $t = t_k > 0$ give a feasible sequence that approaches x^* and satisfies the definition (12.29).

At $t = 0$, $z = x^*$, and the Jacobian of R at this point is

$$
\nabla_z R(x^*, 0) = \begin{bmatrix} A(x^*) \\ Z^T \end{bmatrix}, \tag{12.41}
$$

which is nonsingular by construction of Z. Hence, according to the implicit function theorem, the system (12.40) has a unique solution z_k for all values of t_k sufficiently small. Moreover, we have from (12.40) and Definition 12.3 that

$$
i \in \mathcal{E} \Rightarrow c_i(z_k) = t_k \nabla c_i(x^*)^T d = 0, \tag{12.42a}
$$
$$
i \in \mathcal{A}(x^*) \cap \mathcal{I} \Rightarrow c_i(z_k) = t_k \nabla c_i(x^*)^T d \geq 0, \tag{12.42b}
$$

so that z_k is indeed feasible.

It remains to verify that (12.29) holds for this choice of $\{z_k\}$. Using the fact that $R(z_k, t_k) = 0$ for all k together with Taylor's theorem, we find that

$$
0 = R(z_k, t_k) = \left[\begin{array}{c} c(z_k) - t_k A(x^*)d \\ Z^T(z_k - x^* - t_k d) \end{array} \right]
$$

$$
= \left[\begin{array}{c} A(x^*)(z_k - x^*) + o(\|z_k - x^*\|) - t_k A(x^*)d \\ Z^T(z_k - x^* - t_k d) \end{array} \right]
$$

$$
= \left[\begin{array}{c} A(x^*) \\ Z^T \end{array} \right] (z_k - x^* - t_k d) + o(\|z_k - x^*\|).
$$

By dividing this expression by t_k and using nonsingularity of the coefficient matrix in the first term, we obtain

$$
\frac{z_k - x^*}{t_k} = d + o\left(\frac{\|z_k - x^*\|}{t_k} \right),
$$

from which it follows that (12.29) is satisfied (for $x = x^*$). Hence, $d \in T_\Omega(x^*)$ for an arbitrary $d \in \mathcal{F}(x^*)$, so the proof of (ii) is complete. $\qquad\square$

A FUNDAMENTAL NECESSARY CONDITION

As mentioned above, a local solution of (12.1) is a point x at which all feasible sequences have the property that $f(z_k) \geq f(x)$ for all k sufficiently large. The following result shows that if such a sequence exists, then its limiting directions must make a nonnegative inner product with the objective function gradient.

Theorem 12.3.

   If x^* is a local solution of (12.1), then we have

$$
\nabla f(x^*)^T d \geq 0, \quad \text{for all } d \in T_\Omega(x^*). \tag{12.43}
$$

PROOF. Suppose for contradiction that there is a tangent d for which $\nabla f(x^*)^T d < 0$. Let $\{z_k\}$ and $\{t_k\}$ be the sequences satisfying Definition 12.2 for this d. We have that

$$
f(z_k) = f(x^*) + (z_k - x^*)^T \nabla f(x^*) + o(\|z_k - x^*\|)
$$

$$
= f(x^*) + t_k d^T \nabla f(x^*) + o(t_k),
$$

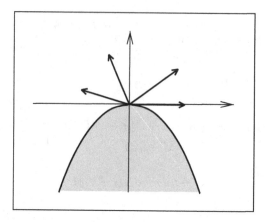

Figure 12.12
Problem (12.44), showing various limiting directions of feasible sequences at the point $(0, 0)^T$.

where the second line follows from (12.38). Since $d^T \nabla f(x^*) < 0$, the remainder term is eventually dominated by the first-order term, that is,

$$f(z_k) < f(x^*) + \tfrac{1}{2} t_k d^T \nabla f(x^*), \quad \text{for all } k \text{ sufficiently large.}$$

Hence, given any open neighborhood of x^*, we can choose k sufficiently large that z_k lies within this neighborhood and has a lower value of the objective f. Therefore, x^* is not a local solution. □

The converse of this result is not necessarily true. That is, we may have $\nabla f(x^*)^T d \geq 0$ for all $d \in T_\Omega(x^*)$, yet x^* is not a local minimizer. An example is the following problem in two unknowns, illustrated in Figure 12.12

$$\min x_2 \quad \text{subject to } x_2 \geq -x_1^2. \tag{12.44}$$

This problem is actually unbounded, but let us examine its behavior at $x^* = (0, 0)^T$. It is not difficult to show that all limiting directions d of feasible sequences must have $d_2 \geq 0$, so that $\nabla f(x^*)^T d = d_2 \geq 0$. However, x^* is clearly not a local minimizer; the point $(\alpha, -\alpha^2)^T$ for $\alpha > 0$ has a smaller function value than x^*, and can be brought arbitrarily close to x^* by setting α sufficiently small.

FARKAS' LEMMA

The most important step in proving Theorem 12.1 is a classical theorem of the alternative known as *Farkas' Lemma*. This lemma considers a cone K defined as follows:

$$K = \{By + Cw \mid y \geq 0\}, \tag{12.45}$$

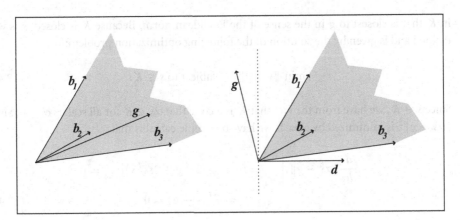

Figure 12.13 Farkas' Lemma: Either $g \in K$ (left) or there is a separating hyperplane (right).

where B and C are matrices of dimension $n \times m$ and $n \times p$, respectively, and y and w are vectors of appropriate dimensions. Given a vector $g \in \mathbb{R}^n$, Farkas' Lemma states that one (and only one) of two alternatives is true. Either $g \in K$, or else there is a vector $d \in \mathbb{R}^n$ such that

$$g^T d < 0, \quad B^T d \geq 0, \quad C^T d = 0. \tag{12.46}$$

The two cases are illustrated in Figure 12.13 for the case of B with three columns, C null, and $n = 2$. Note that in the second case, the vector d defines a *separating hyperplane*, which is a plane in $\mathbb{R}^n$ that separates the vector g from the cone K.

Lemma 12.4 (Farkas).
  Let the cone K be defined as in (12.45). Given any vector $g \in \mathbb{R}^n$, we have either that $g \in K$ or that there exists $d \in \mathbb{R}^n$ satisfying (12.46), but not both.

PROOF.   We show first that the two alternatives cannot hold simultaneously. If $g \in K$, there exist vectors $y \geq 0$ and w such that $g = By + Cw$. If there also exists a d with the property (12.46), we have by taking inner products that

$$0 > d^T g = d^T By + d^T Cw = (B^T d)^T y + (C^T d)^T w \geq 0,$$

where the final inequality follows from $C^T d = 0$, $B^T d \geq 0$, and $y \geq 0$. Hence, we cannot have both alternatives holding at once.

  We now show that *one* of the alternatives holds. To be precise, we show how to construct d with the properties (12.46) in the case that $g \notin K$. For this part of the proof, we need to use the property that K is a *closed* set—a fact that is intuitively obvious but not trivial to prove (see Lemma 12.15 in the Notes and References below). Let $\hat{s}$ be the vector in K that is closest to g in the sense of the Euclidean norm. Because K is closed, $\hat{s}$ is well defined and is given by the solution of the following optimization problem:

$$\min \|s - g\|_2^2 \quad \text{subject to } s \in K. \tag{12.47}$$

Since $\hat{s} \in K$, we have from the fact that K is a cone that $\alpha\hat{s} \in K$ for all scalars $\alpha \geq 0$. Since $\|\alpha\hat{s} - g\|_2^2$ is minimized by $\alpha = 1$, we have by simple calculus that

$$\frac{d}{d\alpha} \|\alpha\hat{s} - g\|_2^2 \Big|_{\alpha=1} = 0 \Rightarrow \left(-2\hat{s}^T g + 2t\hat{s}^T\hat{s}\right)\Big|_{\alpha=1} = 0$$

$$\Rightarrow \hat{s}^T(\hat{s} - g) = 0. \tag{12.48}$$

Now, let s be any other vector in K. Since K is convex, we have by the minimizing property of $\hat{s}$ that

$$\|\hat{s} + \theta(s - \hat{s}) - g\|_2^2 \geq \|\hat{s} - g\|_2^2 \quad \text{for all } \theta \in [0, 1],$$

and hence

$$2\theta(s - \hat{s})^T(\hat{s} - g) + \theta^2\|s - \hat{s}\|_2^2 \geq 0.$$

By dividing this expression by θ and taking the limit as $\theta \downarrow 0$, we have $(s - \hat{s})^T(\hat{s} - g) \geq 0$. Therefore, because of (12.48),

$$s^T(\hat{s} - g) \geq 0, \quad \text{for all } s \in K. \tag{12.49}$$

We claim now that the vector

$$d = \hat{s} - g$$

satisfies the conditions (12.46). Note that $d \neq 0$ because $g \notin K$. We have from (12.48) that

$$d^T g = d^T(\hat{s} - d) = (\hat{s} - g)^T\hat{s} - d^T d = -\|d\|_2^2 < 0,$$

so that d satisfies the first property in (12.46).

From (12.49), we have that $d^T s \geq 0$ for all $s \in K$, so that

$$d^T(By + Cw) \geq 0 \quad \text{for all } y \geq 0 \text{ and all } w.$$

By fixing $y = 0$ we have that $(C^T d)^T w \geq 0$ for all w, which is true only if $C^T d = 0$. By fixing $w = 0$, we have that $(B^T d)^T y \geq 0$ for all $y \geq 0$, which is true only if $B^T d \geq 0$. Hence, d also satisfies the second and third properties in (12.46) and our proof is complete. $\square$

By applying Lemma 12.4 to the cone N defined by

$$N = \left\{ \sum_{i \in \mathcal{A}(x^*)} \lambda_i \nabla c_i(x^*), \quad \lambda_i \geq 0 \text{ for } i \in \mathcal{A}(x^*) \cap \mathcal{I} \right\}, \qquad (12.50)$$

and setting $g = \nabla f(x^*)$, we have that either

$$\nabla f(x^*) = \sum_{i \in \mathcal{A}(x^*)} \lambda_i \nabla c_i(x^*) = A(x^*)^T \lambda^*, \quad \lambda_i \geq 0 \text{ for } i \in \mathcal{A}(x^*) \cap \mathcal{I}, \qquad (12.51)$$

or else there is a direction d such that $d^T \nabla f(x^*) < 0$ and $d \in \mathcal{F}(x^*)$.

PROOF OF THEOREM 12.1

Lemmas 12.2 and 12.4 can be combined to give the KKT conditions described in Theorem 12.1. We work through the final steps of the proof here. Suppose that $x^* \in \mathbb{R}^n$ is a feasible point at which the LICQ holds. The theorem claims that if x^* is a local solution for (12.1), then there is a vector $\lambda^* \in \mathbb{R}^m$ that satisfies the conditions (12.34).

We show first that there are multipliers λ_i, $i \in \mathcal{A}(x^*)$, such that (12.51) is satisfied. Theorem 12.3 tells us that $d^T \nabla f(x^*) \geq 0$ for all tangent vectors $d \in T_\Omega(x^*)$. From Lemma 12.2, since LICQ holds, we have that $T_\Omega(x^*) = \mathcal{F}(x^*)$. By putting these two statements together, we find that $d^T \nabla f(x^*) \geq 0$ for all $d \in \mathcal{F}(x^*)$. Hence, from Lemma 12.4, there is a vector λ for which (12.51) holds, as claimed.

We now define the vector λ^* by

$$\lambda_i^* = \begin{cases} \lambda_i, & i \in \mathcal{A}(x^*), \\ 0, & i \in \mathcal{I} \backslash \mathcal{A}(x^*), \end{cases} \qquad (12.52)$$

and show that this choice of λ^*, together with our local solution x^*, satisfies the conditions (12.34). We check these conditions in turn.

- The condition (12.34a) follows immediately from (12.51) and the definitions (12.33) of the Lagrangian function and (12.52) of λ^*.

- Since x^* is feasible, the conditions (12.34b) and (12.34c) are satisfied.

- We have from (12.51) that $\lambda_i^* \geq 0$ for $i \in \mathcal{A}(x^*) \cap \mathcal{I}$, while from (12.52), $\lambda_i^* = 0$ for $i \in \mathcal{I} \backslash \mathcal{A}(x^*)$. Hence, $\lambda_i^* \geq 0$ for $i \in \mathcal{I}$, so that (12.34d) holds.

- We have for $i \in \mathcal{A}(x^*) \cap \mathcal{I}$ that $c_i(x^*) = 0$, while for $i \in \mathcal{I} \backslash \mathcal{A}(x^*)$, we have $\lambda_i^* = 0$. Hence $\lambda_i^* c_i(x^*) = 0$ for $i \in \mathcal{I}$, so that (12.34e) is satisfied as well.

The proof is complete.

12.5 SECOND-ORDER CONDITIONS

So far, we have described first-order conditions—the KKT conditions—which tell us how the first derivatives of f and the active constraints c_i are related to each other at a solution x^*. When these conditions are satisfied, a move along any vector w from $\mathcal{F}(x^*)$ either increases the first-order approximation to the objective function (that is, $w^T \nabla f(x^*) > 0$), or else keeps this value the same (that is, $w^T \nabla f(x^*) = 0$).

What role do the *second* derivatives of f and the constraints c_i play in optimality conditions? We see in this section that second derivatives play a "tiebreaking" role. For the directions $w \in \mathcal{F}(x^*)$ for which $w^T \nabla f(x^*) = 0$, we cannot determine from first derivative information alone whether a move along this direction will increase or decrease the objective function f. Second-order conditions examine the second derivative terms in the Taylor series expansions of f and c_i, to see whether this extra information resolves the issue of increase or decrease in f. Essentially, the second-order conditions concern the curvature of the Lagrangian function in the "undecided" directions—the directions $w \in \mathcal{F}(x^*)$ for which $w^T \nabla f(x^*) = 0$.

Since we are discussing second derivatives, stronger smoothness assumptions are needed here than in the previous sections. For the purpose of this section, f and c_i, $i \in \mathcal{E} \cup \mathcal{I}$, are all assumed to be twice continuously differentiable.

Given $\mathcal{F}(x^*)$ from Definition 12.3 and some Lagrange multiplier vector λ^* satisfying the KKT conditions (12.34), we define the *critical cone* $\mathcal{C}(x^*, \lambda^*)$ as follows:

$$\mathcal{C}(x^*, \lambda^*) = \{w \in \mathcal{F}(x^*) \mid \nabla c_i(x^*)^T w = 0, \text{ all } i \in \mathcal{A}(x^*) \cap \mathcal{I} \text{ with } \lambda_i^* > 0\}.$$

Equivalently,

$$w \in \mathcal{C}(x^*, \lambda^*) \iff \begin{cases} \nabla c_i(x^*)^T w = 0, & \text{for all } i \in \mathcal{E}, \\ \nabla c_i(x^*)^T w = 0, & \text{for all } i \in \mathcal{A}(x^*) \cap \mathcal{I} \text{ with } \lambda_i^* > 0, \\ \nabla c_i(x^*)^T w \geq 0, & \text{for all } i \in \mathcal{A}(x^*) \cap \mathcal{I} \text{ with } \lambda_i^* = 0. \end{cases} \quad (12.53)$$

The critical cone contains those directions w that would tend to "adhere" to the active inequality constraints even when we were to make small changes to the objective (those indices $i \in \mathcal{I}$ for which the Lagrange multiplier component λ_i^* is positive), as well as to the equality constraints. From the definition (12.53) and the fact that $\lambda_i^* = 0$ for all inactive components $i \in \mathcal{I} \backslash \mathcal{A}(x^*)$, it follows immediately that

$$w \in \mathcal{C}(x^*, \lambda^*) \implies \lambda_i^* \nabla c_i(x^*)^T w = 0 \text{ for all } i \in \mathcal{E} \cup \mathcal{I}. \quad (12.54)$$

Hence, from the first KKT condition (12.34a) and the definition (12.33) of the Lagrangian function, we have that

$$w \in \mathcal{C}(x^*, \lambda^*) \implies w^T \nabla f(x^*) = \sum_{i \in \mathcal{E} \cup \mathcal{I}} \lambda_i^* w^T \nabla c_i(x^*) = 0. \quad (12.55)$$

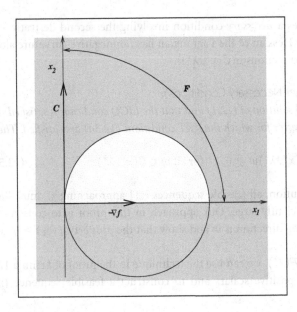

Figure 12.14
Problem (12.56), showing
$\mathcal{F}(x^*)$ and $\mathcal{C}(x^*, \lambda^*)$.

Hence the critical cone $\mathcal{C}(x^*, \lambda^*)$ contains directions from $\mathcal{F}(x^*)$ for which it is not clear from first derivative information alone whether f will increase or decrease.

❑ **EXAMPLE 12.7**

Consider the problem

$$\min x_1 \quad \text{subject to } x_2 \geq 0, \ 1 - (x_1 - 1)^2 - x_2^2 \geq 0, \tag{12.56}$$

illustrated in Figure 12.14. It is not difficult to see that the solution is $x^* = (0, 0)^T$, with active set $\mathcal{A}(x^*) = \{1, 2\}$ and a unique optimal Lagrange multiplier $\lambda^* = (0, 0.5)^T$. Since the gradients of the active constraints at x^* are $(0, 1)^T$ and $(2, 0)^T$, respectively, the LICQ holds, so the optimal multiplier is unique. The linearized feasible set is then

$$\mathcal{F}(x^*) = \{d \mid d \geq 0\},$$

while the critical cone is

$$\mathcal{C}(x^*, \lambda^*) = \{(0, w_2)^T \mid w_2 \geq 0\}.$$

❑

The first theorem defines a *necessary* condition involving the second derivatives: If x^* is a local solution, then the Hessian of the Lagrangian has nonnegative curvature along critical directions (that is, the directions in $C(x^*, \lambda^*)$).

Theorem 12.5 (Second-Order Necessary Conditions).

Suppose that x^* is a local solution of (12.1) and that the LICQ condition is satisfied. Let λ^* be the Lagrange multiplier vector for which the KKT conditions (12.34) are satisfied. Then

$$w^T \nabla_{xx}^2 \mathcal{L}(x^*, \lambda^*) w \geq 0, \quad \text{for all } w \in C(x^*, \lambda^*). \tag{12.57}$$

PROOF.   Since x^* is a local solution, all feasible sequences $\{z_k\}$ approaching x^* must have $f(z_k) \geq f(x^*)$ for all k sufficiently large. Our approach in this proof is to construct a feasible sequence whose limiting direction is w and show that the property $f(z_k) \geq f(x^*)$ implies that (12.57) holds.

Since $w \in C(x^*, \lambda^*) \subset \mathcal{F}(x^*)$, we can use the technique in the proof of Lemma 12.2 to choose a sequence $\{t_k\}$ of positive scalars and to construct a feasible sequence $\{z_k\}$ approaching x^* such that

$$\lim_{k \to \infty} \frac{z_k - x^*}{t_k} = w, \tag{12.58}$$

which we can write also as (12.58) that

$$z_k - x^* = t_k w + o(t_k). \tag{12.59}$$

Because of the construction technique for $\{z_k\}$, we have from formula (12.42) that

$$c_i(z_k) = t_k \nabla c_i(x^*)^T w, \quad \text{for all } i \in \mathcal{A}(x^*) \tag{12.60}$$

From (12.33), (12.60), and (12.54), we have

$$
\begin{aligned}
\mathcal{L}(z_k, \lambda^*) &= f(z_k) - \sum_{i \in \mathcal{E} \cup \mathcal{I}} \lambda_i^* c_i(z_k) \\
&= f(z_k) - t_k \sum_{i \in \mathcal{A}(x^*)} \lambda_i^* \nabla c_i(x^*)^T w \\
&= f(z_k),
\end{aligned}
\tag{12.61}
$$

On the other hand, we can perform a Taylor series expansion to obtain an estimate of $\mathcal{L}(z_k, \lambda^*)$ near x^*. By using Taylor's theorem expression (2.6) and continuity of the Hessians $\nabla^2 f$ and $\nabla^2 c_i$, $i \in \mathcal{E} \cup \mathcal{I}$, we obtain

$$
\begin{aligned}
\mathcal{L}(z_k, \lambda^*) &= \mathcal{L}(x^*, \lambda^*) + (z_k - x^*)^T \nabla_x \mathcal{L}(x^*, \lambda^*) \\
&\quad + \tfrac{1}{2}(z_k - x^*)^T \nabla_{xx}^2 \mathcal{L}(x^*, \lambda^*)(z_k - x^*) + o(\|z_k - x^*\|^2).
\end{aligned}
\tag{12.62}
$$

By the complementarity conditions (12.34e), we have $\mathcal{L}(x^*, \lambda^*) = f(x^*)$. From (12.34a), the second term on the right-hand side is zero. Hence, using (12.59), we can rewrite (12.62) as

$$\mathcal{L}(z_k, \lambda^*) = f(x^*) + \tfrac{1}{2}t_k^2 w^T \nabla_{xx}^2 \mathcal{L}(x^*, \lambda^*) + o(t_k^2). \tag{12.63}$$

By substituting into (12.63), we obtain

$$f(z_k) = f(x^*) + \tfrac{1}{2}t_k^2 w^T \nabla_{xx}^2 \mathcal{L}(x^*, \lambda^*)w + o(t_k^2). \tag{12.64}$$

If $w^T \nabla_{xx}^2 \mathcal{L}(x^*, \lambda^*)w < 0$, then (12.64) would imply that $f(z_k) < f(x^*)$ for all k sufficiently large, contradicting the fact that x^* is a local solution. Hence, the condition (12.57) must hold, as claimed. $\qquad\square$

Sufficient conditions are conditions on f and $c_i, i \in \mathcal{E} \cup \mathcal{I}$, that ensure that x^* is a local solution of the problem (12.1). (They take the opposite tack to necessary conditions, which assume that x^* is a local solution and deduce properties of f and c_i, for the active indices i.) The second-order sufficient condition stated in the next theorem looks very much like the necessary condition just discussed, but it differs in that the constraint qualification is not required, and the inequality in (12.57) is replaced by a strict inequality.

Theorem 12.6 (Second-Order Sufficient Conditions).
Suppose that for some feasible point $x^* \in \mathbb{R}^n$ there is a Lagrange multiplier vector λ^* such that the KKT conditions (12.34) are satisfied. Suppose also that

$$w^T \nabla_{xx}^2 \mathcal{L}(x^*, \lambda^*)w > 0, \quad \text{for all } w \in \mathcal{C}(x^*, \lambda^*), w \neq 0. \tag{12.65}$$

Then x^* is a strict local solution for (12.1).

PROOF.   First, note that the set $\bar{\mathcal{C}} = \{d \in \mathcal{C}(x^*, \lambda^*) \mid \|d\| = 1\}$ is a compact subset of $\mathcal{C}(x^*, \lambda^*)$, so by (12.65), the minimizer of $d^T \nabla_{xx}^2 \mathcal{L}(x^*, \lambda^*)d$ over this set is a strictly positive number, say σ. Since $\mathcal{C}(x^*, \lambda^*)$ is a cone, we have that $(w/\|w\|) \in \bar{\mathcal{C}}$ if and only if $w \in \mathcal{C}(x^*, \lambda^*), w \neq 0$. Therefore, condition (12.65) by

$$w^T \nabla_{xx}^2 \mathcal{L}(x^*, \lambda^*)w \geq \sigma \|w\|^2, \quad \text{for all } w \in \mathcal{C}(x^*, \lambda^*), \tag{12.66}$$

for $\sigma > 0$ defined as above. (Note that this inequality holds trivially for $w = 0$.)

We prove the result by showing that every feasible sequence $\{z_k\}$ approaching x^* has $f(z_k) \geq f(x^*) + (\sigma/4)\|z_k - x^*\|^2$, for all k sufficiently large. Suppose for contradiction that this is *not* the case, and that there is a sequence $\{z_k\}$ approaching x^* with

$$f(z_k) < f(x^*) + (\sigma/4)\|z_k - x^*\|^2, \quad \text{for all } k \text{ sufficiently large.} \tag{12.67}$$

By taking a subsequence if necessary, we can identify a limiting direction d such that

$$\lim_{k \to \infty} \frac{z_k - x^*}{\|z_k - x^*\|} = d. \tag{12.68}$$

We have from Lemma 12.2(i) and Definition 12.3 that $d \in \mathcal{F}(x^*)$. From (12.33) and the facts that $\lambda_i^* \geq 0$ and $c_i(z_k) \geq 0$ for $i \in \mathcal{I}$ and $c_i(z_k) = 0$ for $i \in \mathcal{E}$, we have that

$$\mathcal{L}(z_k, \lambda^*) = f(z_k) - \sum_{i \in \mathcal{A}(x^*)} \lambda_i^* c_i(z_k) \leq f(z_k), \tag{12.69}$$

while the Taylor series approximation (12.63) from the proof of Theorem 12.5 continues to hold.

If d were *not* in $\mathcal{C}(x^*, \lambda^*)$, we could identify some index $j \in \mathcal{A}(x^*) \cap \mathcal{I}$ such that the strict positivity condition

$$\lambda_j^* \nabla c_j(x^*)^T d > 0 \tag{12.70}$$

is satisfied, while for the remaining indices $i \in \mathcal{A}(x^*)$, we have

$$\lambda_i^* \nabla c_i(x^*)^T d \geq 0.$$

From Taylor's theorem and (12.68), we have for this particular value of j that

$$\begin{aligned}
\lambda_j^* c_j(z_k) &= \lambda_j^* c_j(x^*) + \lambda_j^* \nabla c_j(x^*)^T (z_k - x^*) + o(\|z_k - x^*\|) \\
&= \|z_k - x^*\| \lambda_j^* \nabla c_j(x^*)^T d + o(\|z_k - x^*\|).
\end{aligned}$$

Hence, from (12.69), we have that

$$\begin{aligned}
\mathcal{L}(z_k, \lambda^*) &= f(z_k) - \sum_{i \in \mathcal{A}(x^*)} \lambda_i^* c_i(z_k) \\
&\leq f(z_k) - \lambda_j^* c_j(z_k) \\
&\leq f(z_k) - \|z_k - x^*\| \lambda_j^* \nabla c_j(x^*)^T d + o(\|z_k - x^*\|). \tag{12.71}
\end{aligned}$$

From the Taylor series estimate (12.63), we have meanwhile that

$$\mathcal{L}(z_k, \lambda^*) = f(x^*) + O(\|z_k - x^*\|^2),$$

and by combining with (12.71), we obtain

$$f(z_k) \geq f(x^*) + \|z_k - x^*\| \lambda_j^* \nabla c_j(x^*)^T d + o(\|z_k - x^*\|).$$

Because of (12.70), this inequality is incompatible with (12.67). We conclude that $d \in C(x^*, \lambda^*)$, and hence $d^T \nabla_{xx}^2 \mathcal{L}(x^*, \lambda^*) d \geq \sigma$.

By combining the Taylor series estimate (12.63) with (12.69) and using (12.68), we obtain

$$
\begin{aligned}
f(z_k) &\geq f(x^*) + \tfrac{1}{2}(z_k - x^*)^T \nabla_{xx}^2 \mathcal{L}(x^*, \lambda^*)(z_k - x^*) + o(\|z_k - x^*\|^2) \\
&= f(x^*) + \tfrac{1}{2} d^T \nabla_{xx}^2 \mathcal{L}(x^*, \lambda^*) d \|z_k - x^*\|^2 + o(\|z_k - x^*\|^2) \\
&\geq f(x^*) + (\sigma/2)\|z_k - x^*\|^2 + o(\|z_k - x^*\|^2).
\end{aligned}
$$

This inequality yields the contradiction to (12.67). We conclude that every feasible sequence $\{z_k\}$ approaching x^* must satisfy $f(z_k) \geq f(x^*) + (\sigma/4)\|z_k - x^*\|^2$, for all k sufficiently large, so x^* is a strict local solution. □

❏ **EXAMPLE 12.8**   (EXAMPLE 12.2, ONE MORE TIME)

We now return to Example 12.2 to check the second-order conditions for problem (12.18). In this problem we have $f(x) = x_1 + x_2$, $c_1(x) = 2 - x_1^2 - x_2^2$, $\mathcal{E} = \emptyset$, and $\mathcal{I} = \{1\}$. The Lagrangian is

$$
\mathcal{L}(x, \lambda) = (x_1 + x_2) - \lambda_1(2 - x_1^2 - x_2^2),
$$

and it is easy to show that the KKT conditions (12.34) are satisfied by $x^* = (-1, -1)^T$, with $\lambda_1^* = \tfrac{1}{2}$. The Lagrangian Hessian at this point is

$$
\nabla_{xx}^2 \mathcal{L}(x^*, \lambda^*) = \begin{bmatrix} 2\lambda_1^* & 0 \\ 0 & 2\lambda_1^* \end{bmatrix} = \begin{bmatrix} 1 & 0 \\ 0 & 1 \end{bmatrix}.
$$

This matrix is positive definite, so it certainly satisfies the conditions of Theorem 12.6. We conclude that $x^* = (-1, -1)^T$ is a strict local solution for (12.18). (In fact, it is the global solution of this problem, since, as we note later, this problem is a convex programming problem.) ❏

❏ **EXAMPLE 12.9**

For a more complex example, consider the problem

$$
\min \; -0.1(x_1 - 4)^2 + x_2^2 \quad \text{s.t.} \quad x_1^2 + x_2^2 - 1 \geq 0, \tag{12.72}
$$

in which we seek to minimize a nonconvex function over the *exterior* of the unit circle. Obviously, the objective function is not bounded below on the feasible region, since we can take the feasible sequence

$$\begin{bmatrix} 10 \\ 0 \end{bmatrix}, \quad \begin{bmatrix} 20 \\ 0 \end{bmatrix}, \quad \begin{bmatrix} 30 \\ 0 \end{bmatrix}, \quad \begin{bmatrix} 40 \\ 0 \end{bmatrix},$$

and note that $f(x)$ approaches $-\infty$ along this sequence. Therefore, no global solution exists, but it may still be possible to identify a strict local solution on the boundary of the constraint. We search for such a solution by using the KKT conditions (12.34) and the second-order conditions of Theorem 12.6.

By defining the Lagrangian for (12.72) in the usual way, it is easy to verify that

$$\nabla_x \mathcal{L}(x, \lambda) = \begin{bmatrix} -0.2(x_1 - 4) - 2\lambda_1 x_1 \\ 2x_2 - 2\lambda_1 x_2 \end{bmatrix}, \tag{12.73a}$$

$$\nabla_{xx}^2 \mathcal{L}(x, \lambda) = \begin{bmatrix} -0.2 - 2\lambda_1 & 0 \\ 0 & 2 - 2\lambda_1 \end{bmatrix}. \tag{12.73b}$$

The point $x^* = (1, 0)^T$ satisfies the KKT conditions with $\lambda_1^* = 0.3$ and the active set $\mathcal{A}(x^*) = \{1\}$. To check that the second-order sufficient conditions are satisfied at this point, we note that

$$\nabla c_1(x^*) = \begin{bmatrix} 2 \\ 0 \end{bmatrix},$$

so that the set $\mathcal{C}$ defined in (12.53) is simply

$$\mathcal{C}(x^*, \lambda^*) = \{(0, w_2)^T \mid w_2 \in \mathbb{R}\}.$$

Now, by substituting x^* and λ^* into (12.73b), we have for any $w \in \mathcal{C}(x^*, \lambda^*)$ with $w \neq 0$ that $w_2 \neq 0$ and thus

$$w^T \nabla_{xx}^2 \mathcal{L}(x^*, \lambda^*) w = \begin{bmatrix} 0 \\ w_2 \end{bmatrix}^T \begin{bmatrix} -0.4 & 0 \\ 0 & 1.4 \end{bmatrix} \begin{bmatrix} 0 \\ w_2 \end{bmatrix} = 1.4 w_2^2 > 0.$$

Hence, the second-order sufficient conditions are satisfied, and we conclude from Theorem 12.6 that $(1, 0)^T$ is a strict local solution for (12.72).

❑

SECOND-ORDER CONDITIONS AND PROJECTED HESSIANS

The second-order conditions are sometimes stated in a form that is slightly weaker but easier to verify than (12.57) and (12.65). This form uses a two-sided projection of the Lagrangian Hessian $\nabla^2_{xx}\mathcal{L}(x^*, \lambda^*)$ onto subspaces that are related to $\mathcal{C}(x^*, \lambda^*)$.

The simplest case is obtained when the multiplier λ^* that satisfies the KKT conditions (12.34) is unique (as happens, for example, when the LICQ condition holds) *and* strict complementarity holds. In this case, the definition (12.53) of $\mathcal{C}(x^*, \lambda^*)$ reduces to

$$\mathcal{C}(x^*, \lambda^*) = \text{Null} \left[\nabla c_i(x^*)^T \right]_{i \in \mathcal{A}(x^*)} = \text{Null } A(x^*),$$

where $A(x^*)$ is defined as in (12.37). In other words, $\mathcal{C}(x^*, \lambda^*)$ is the null space of the matrix whose rows are the active constraint gradients at x^*. As in (12.39), we can define the matrix Z with full column rank whose columns span the space $\mathcal{C}(x^*, \lambda^*)$; that is,

$$\mathcal{C}(x^*, \lambda^*) = \{ Zu \mid u \in \mathbb{R}^{|\mathcal{A}(x^*)|} \}.$$

Hence, the condition (12.57) in Theorem 12.5 can be restated as

$$u^T Z^T \nabla^2_{xx}\mathcal{L}(x^*, \lambda^*)Zu \geq 0 \quad \text{for all } u,$$

or, more succinctly,

$$Z^T \nabla^2_{xx}\mathcal{L}(x^*, \lambda^*)Z \text{ is positive semidefinite.}$$

Similarly, the condition (12.65) in Theorem 12.6 can be restated as

$$Z^T \nabla^2_{xx}\mathcal{L}(x^*, \lambda^*)Z \text{ is positive definite.}$$

As we show next, Z can be computed numerically, so that the positive (semi)definiteness conditions can actually be checked by forming these matrices and finding their eigenvalues.

One way to compute the matrix Z is to apply a QR factorization to the matrix of active constraint gradients whose null space we seek. In the simplest case above (in which the multiplier λ^* is unique and strictly complementarity holds), we define $A(x^*)$ as in (12.37) and write the QR factorization of its transpose as

$$A(x^*)^T = Q \begin{bmatrix} R \\ 0 \end{bmatrix} = \begin{bmatrix} Q_1 & Q_2 \end{bmatrix} \begin{bmatrix} R \\ 0 \end{bmatrix} = Q_1 R, \qquad (12.74)$$

where R is a square upper triangular matrix and Q is $n \times n$ orthogonal. If R is nonsingular, we can set $Z = Q_2$. If R is singular (indicating that the active constraint gradients are linearly dependent), a slight enhancement of this procedure that makes use of column pivoting during the QR procedure can be used to identify Z.

12.6   OTHER CONSTRAINT QUALIFICATIONS

We now reconsider constraint qualifications, the conditions discussed in Sections 12.2 and 12.4 that ensure that the linearized approximation to the feasible set Ω captures the essential shape of Ω in a neighborhood of x^*.

One situation in which the linearized feasible direction set $\mathcal{F}(x^*)$ is obviously an adequate representation of the actual feasible set occurs when all the active constraints are already linear; that is,

$$c_i(x) = a_i^T x + b_i, \tag{12.75}$$

for some $a_i \in \mathbb{R}^n$ and $b_i \in \mathbb{R}$. It is not difficult to prove a version of Lemma 12.2 for this situation.

Lemma 12.7.

  Suppose that at some $x^* \in \Omega$, all active constraints $c_i(\cdot), i \in \mathcal{A}(x^*)$, are linear functions. Then $\mathcal{F}(x^*) = T_\Omega(x^*)$.

PROOF.   We have from Lemma 12.2 (i) that $T_\Omega(x^*) \subset \mathcal{F}(x^*)$. To prove that $\mathcal{F}(x^*) \subset T_\Omega(x^*)$, we choose an arbitrary $w \in \mathcal{F}(x^*)$ and show that $w \in T_\Omega(x^*)$. By Definition 12.3 and the form (12.75) of the constraints, we have

$$\mathcal{F}(x^*) = \left\{ d \;\middle|\; \begin{array}{ll} a_i^T d = 0, & \text{for all } i \in \mathcal{E}, \\ a_i^T d \geq 0, & \text{for all } i \in \mathcal{A}(x) \cap \mathcal{I} \end{array} \right\}.$$

First, note that there is a positive scalar $\bar{t}$ such that the inactive constraint remain inactive at $x^* + tw$, for all $t \in [0, \bar{t}]$, that is,

$$c_i(x^* + tw) > 0, \quad \text{for all } i \in \mathcal{I} \backslash \mathcal{A}(x^*) \text{ and all } t \in [0, \bar{t}].$$

Now define the sequence z_k by

$$z_k = x^* + (\bar{t}/k)w, \quad k = 1, 2, \dots.$$

Since $a_i^T w \geq 0$ for all $i \in \mathcal{I} \cap \mathcal{A}(x^*)$, we have

$$c_i(z_k) = c_i(z_k) - c_i(x^*) = a_i^T (z_k - x^*) = \frac{\bar{t}}{k} a_i^T w \geq 0, \quad \text{for all } i \in \mathcal{I} \cap \mathcal{A}(x^*),$$

so that z_k is feasible with respect to the active inequality constraints $c_i, i \in \mathcal{I} \cap \mathcal{A}(x^*)$. By the choice of $\bar{t}$, we find that z_k is also feasible with respect to the inactive inequality constraints

$i \in \mathcal{I} \backslash \mathcal{A}(x^*)$, and it is easy to show that $c_i(z_k) = 0$ for the equality constraints $i \in \mathcal{E}$. Hence, z_k is feasible for each $k = 1, 2, \ldots$. In addition, we have that

$$\frac{z_k - x^*}{(\bar{t}/k)} = \frac{(\bar{t}/k)w}{(\bar{t}/k)} = w,$$

so that indeed w is the limiting direction of $\{z_k\}$. Hence, $w \in T_\Omega(x^*)$, and the proof is complete.   □

We conclude from this result that the condition that all active constraints be linear is another possible constraint qualification. It is neither weaker nor stronger than the LICQ condition, that is, there are situations in which one condition is satisfied but not the other (see Exercise 12.12).

Another useful generalization of the LICQ is the Mangasarian–Fromovitz constraint qualification (MFCQ).

Definition 12.6 (MFCQ).

We say that the Mangasarian–Fromovitz constraint qualification (MFCQ) holds if there exists a vector $w \in \mathbb{R}^n$ such that

$$\begin{aligned} \nabla c_i(x^*)^T w > 0, && \text{for all } i \in \mathcal{A}(x^*) \cap \mathcal{I}, \\ \nabla c_i(x^*)^T w = 0, && \text{for all } i \in \mathcal{E}, \end{aligned}$$

and the set of equality constraint gradients $\{\nabla c_i(x^*), i \in \mathcal{E}\}$ is linearly independent.

Note the *strict* inequality involving the active inequality constraints.

The MFCQ is a weaker condition than LICQ. If LICQ is satisfied, then the system of equalities defined by

$$\begin{aligned} \nabla c_i(x^*)^T w = 1, && \text{for all } i \in \mathcal{A}(x^*) \cap \mathcal{I}, \\ \nabla c_i(x^*)^T w = 0, && \text{for all } i \in \mathcal{E}, \end{aligned}$$

has a solution w, by full rank of the active constraint gradients. Hence, we can choose the w of Definition 12.6 to be precisely this vector. On the other hand, it is easy to construct examples in which the MFCQ is satisfied but the LICQ is not; see Exercise 12.13.

It is possible to prove a version of the first-order necessary condition result (Theorem 12.1) in which MFCQ replaces LICQ in the assumptions. MFCQ gives rise to the nice property that it is equivalent to boundedness of the set of Lagrange multiplier vectors λ^* for which the KKT conditions (12.34) are satisfied. (In the case of LICQ, this set consists of a unique vector λ^*, and so is trivially bounded.)

Note that constraint qualifications are *sufficient* conditions for the linear approximation to be adequate, not necessary conditions. For instance, consider the set defined by $x_2 \geq -x_1^2$ and $x_2 \leq x_1^2$ and the feasible point $x^* = (0, 0)^T$. None of the constraint qualifications we have discussed are satisfied, but the linear approximation $\mathcal{F}(x^*) = \{(w_1, 0)^T \mid w_1 \in \mathbb{R}\}$ accurately reflects the geometry of the feasible set near x^*.

12.7   A GEOMETRIC VIEWPOINT

Finally, we mention an alternative first-order optimality condition that depends only on the geometry of the feasible set Ω and not on its particular algebraic description in terms of the constraint functions $c_i, i \in \mathcal{E} \cup \mathcal{I}$. In geometric terms, our problem (12.1) can be stated as

$$\min \ f(x) \quad \text{subject to } x \in \Omega, \tag{12.76}$$

where Ω is the feasible set.

To prove a "geometric" first-order condition, we need to define the normal cone to the set Ω at a feasible point x.

Definition 12.7.

The normal cone *to the set Ω at the point $x \in \Omega$ is defined as*

$$N_\Omega(x) = \{v \mid v^T w \leq 0 \text{ for all } w \in T_\Omega(x)\}, \tag{12.77}$$

where $T_\Omega(x)$ is the tangent cone of Definition 12.2. Each vector $v \in N_\Omega(x)$ is said to be a normal vector.

Geometrically, each normal vector v makes an angle of at least $\pi/2$ with every tangent vector.

The first-order necessary condition for (12.76) is delightfully simple.

Theorem 12.8.

Suppose that x^* is a local minimizer of f in Ω. Then

$$-\nabla f(x^*) \in N_\Omega(x^*). \tag{12.78}$$

PROOF.   Given any $d \in T_\Omega(x^*)$, we have for the sequences $\{t_k\}$ and $\{z_k\}$ in Definition 12.2 that

$$z_k \in \Omega, \quad z_k = x^* + t_k d + o(t_k), \quad \text{for all } k. \tag{12.79}$$

Since x^* is a local solution, we must have

$$f(z_k) \geq f(x^*)$$

for all k sufficiently large. Hence, since f is continuously differentiable, we have from Taylor's theorem (2.4) that

$$f(z_k) - f(x^*) = t_k \nabla f(x^*)^T d + o(t_k) \geq 0.$$

By dividing by t_k and taking limits as $k \to \infty$, we have

$$\nabla f(x^*)^T d \geq 0.$$

Recall that d was an arbitrary member of $T_\Omega(x^*)$, so we have $-\nabla f(x^*)^T d \leq 0$ for all $d \in T_\Omega(x^*)$. We conclude from Definition 12.7 that $-\nabla f(x^*) \in N_\Omega(x^*)$. ☐

This result suggests a close relationship between $N_\Omega(x^*)$ and the conic combination of active constraint gradients given by (12.50). When the linear independence constraint qualification holds, identical (to within a change of sign).

Lemma 12.9.

Suppose that the LICQ assumption (Definition 12.4) holds at x^*. Then t the normal cone $N_\Omega(x^*)$ is simply $-N$, where N is the set defined in (12.50).

PROOF. The proof follows from Farkas' Lemma (Lemma 12.4) and Definition 12.7 of $N_\Omega(x^*)$. From Lemma 12.4, we have that

$$g \in N \implies g^T d \geq 0 \text{ for all } d \in \mathcal{F}(x^*).$$

Since we have $\mathcal{F}(x^*) = T_\Omega(x^*)$ from Lemma 12.2, it follows by switching the sign of this expression that

$$g \in -N \implies g^T d \leq 0 \text{ for all } d \in T_\Omega(x^*).$$

We conclude from Definition 12.7 that $N_\Omega(x^*) = -N$, as claimed. ☐

12.8 LAGRANGE MULTIPLIERS AND SENSITIVITY

The importance of Lagrange multipliers in optimality theory should be clear, but what of their intuitive significance? We show in this section that each Lagrange multiplier λ_i^* tells us something about the *sensitivity* of the optimal objective value $f(x^*)$ to the presence of the constraint c_i. To put it another way, λ_i^* indicates how hard f is "pushing" or "pulling" the solution x^* against the particular constraint c_i.

We illustrate this point with some informal analysis. When we choose an inactive constraint $i \notin \mathcal{A}(x^*)$ such that $c_i(x^*) > 0$, the solution x^* and function value $f(x^*)$ are indifferent to whether this constraint is present or not. If we perturb c_i by a tiny amount, it will still be inactive and x^* will still be a local solution of the optimization problem. Since $\lambda_i^* = 0$ from (12.34e), the Lagrange multiplier indicates accurately that constraint i is not significant.

Suppose instead that constraint i is active, and let us perturb the right-hand-side of this constraint a little, requiring, say, that $c_i(x) \geq -\epsilon \|\nabla c_i(x^*)\|$ instead of $c_i(x) \geq 0$. Suppose that ϵ is sufficiently small that the perturbed solution $x^*(\epsilon)$ still has the same set of active constraints, and that the Lagrange multipliers are not much affected by the perturbation. (These conditions can be made more rigorous with the help of strict complementarity and second-order conditions.) We then find that

$$-\epsilon \|\nabla c_i(x^*)\| = c_i(x^*(\epsilon)) - c_i(x^*) \approx (x^*(\epsilon) - x^*)^T \nabla c_i(x^*),$$
$$0 = c_j(x^*(\epsilon)) - c_j(x^*) \approx (x^*(\epsilon) - x^*)^T \nabla c_j(x^*),$$

$$\text{for all } j \in \mathcal{A}(x^*) \text{ with } j \neq i.$$

The value of $f(x^*(\epsilon))$, meanwhile, can be estimated with the help of (12.34a). We have

$$f(x^*(\epsilon)) - f(x^*) \approx (x^*(\epsilon) - x^*)^T \nabla f(x^*)$$
$$= \sum_{j \in \mathcal{A}(x^*)} \lambda_j^* (x^*(\epsilon) - x^*)^T \nabla c_j(x^*)$$
$$\approx -\epsilon \|\nabla c_i(x^*)\| \lambda_i^*.$$

By taking limits, we see that the family of solutions $x^*(\epsilon)$ satisfies

$$\frac{df(x^*(\epsilon))}{d\epsilon} = -\lambda_i^* \|\nabla c_i(x^*)\|. \tag{12.80}$$

A sensitivity analysis of this problem would conclude that if $\lambda_i^* \|\nabla c_i(x^*)\|$ is large, then the optimal value is sensitive to the placement of the ith constraint, while if this quantity is small, the dependence is not too strong. If λ_i^* is exactly zero for some active constraint, small perturbations to c_i in some directions will hardly affect the optimal objective value at all; the change is zero, to first order.

This discussion motivates the definition below, which classifies constraints according to whether or not their corresponding Lagrange multiplier is zero.

Definition 12.8.

Let x^* be a solution of the problem (12.1), and suppose that the KKT conditions (12.34) are satisfied. We say that an inequality constraint c_i is strongly active or binding if $i \in \mathcal{A}(x^*)$ and $\lambda_i^* > 0$ for some Lagrange multiplier λ^* satisfying (12.34). We say that c_i is weakly active if $i \in \mathcal{A}(x^*)$ and $\lambda_i^* = 0$ for all λ^* satisfying (12.34).

Note that the analysis above is independent of scaling of the individual constraints. For instance, we might change the formulation of the problem by replacing some active constraint c_i by $10c_i$. The new problem will actually be equivalent (that is, it has the same feasible set and same solution), but the optimal multiplier λ_i^* corresponding to c_i will be replaced by $\lambda_i^*/10$. However, since $\|\nabla c_i(x^*)\|$ is replaced by $10\|\nabla c_i(x^*)\|$, the product $\lambda_i^*\|\nabla c_i(x^*)\|$ does not change. If, on the other hand, we replace the objective function f by $10f$, the multipliers λ_i^* in (12.34) all will need to be replaced by $10\lambda_i^*$. Hence in (12.80) we see that the sensitivity of f to perturbations has increased by a factor of 10, which is exactly what we would expect.

12.9 DUALITY

In this section we present some elements of the duality theory for nonlinear programming. This theory is used to motivate and develop some important algorithms, including the augmented Lagrangian algorithms of Chapter 17. In its full generality, duality theory ranges beyond nonlinear programming to provide important insight into the fields of convex nonsmooth optimization and even discrete optimization. Its specialization to linear programming proved central to the development of that area; see Chapter 13. (We note that the discussion of linear programming duality in Section 13.1 can be read without consulting this section first.)

Duality theory shows how we can construct an alternative problem from the functions and data that define the original optimization problem. This alternative "dual" problem is related to the original problem (which is sometimes referred to in this context as the "primal" for purposes of contrast) in fascinating ways. In some cases, the dual problem is easier to solve computationally than the original problem. In other cases, the dual can be used to obtain easily a lower bound on the optimal value of the objective for the primal problem. As remarked above, the dual has also been used to design algorithms for solving the primal problem.

Our results in this section are mostly restricted to the special case of (12.1) in which there are no equality constraints and the objective f and the negatives of the inequality constraints $-c_i$ are all convex functions. For simplicity we assume that there are m inequality constraints labelled $1, 2, \ldots, m$ and rewrite (12.1) as follows:

$$\min_{x \in R^n} f(x) \quad \text{subject to} \quad c_i(x) \geq 0, \quad i = 1, 2, \ldots, m.$$

If we assemble the constraints into a vector function

$$c(x) \stackrel{\text{def}}{=} (c_1(x), c_2(x), \ldots, c_m(x))^T,$$

we can write the problem as

$$\min_{x \in R^n} f(x) \quad \text{subject to} \quad c(x) \geq 0, \tag{12.81}$$

for which the Lagrangian function (12.16) with Lagrange multiplier vector $\lambda \in \mathbb{R}^m$ is

$$\mathcal{L}(x, \lambda) = f(x) - \lambda^T c(x).$$

We define the dual objective function $q : \mathbb{R}^n \rightarrow \mathbb{R}$ as follows:

$$q(\lambda) \stackrel{\text{def}}{=} \inf_x \mathcal{L}(x, \lambda). \tag{12.82}$$

In many problems, this infimum is $-\infty$ for some values of λ. We define the domain of q as the set of λ values for which q is finite, that is,

$$\mathcal{D} \stackrel{\text{def}}{=} \{\lambda \mid q(\lambda) > -\infty\}. \tag{12.83}$$

Note that calculation of the infimum in (12.82) requires finding the *global* minimizer of the function $\mathcal{L}(\cdot, \lambda)$ for the given λ which, as we have noted in Chapter 2, may be extremely difficult in practice. However, when f and $-c_i$ are convex functions and $\lambda \geq 0$ (the case in which we are most interested), the function $\mathcal{L}(\cdot, \lambda)$ is also convex. In this situation, all local minimizers are global minimizers (as we verify in Exercise 12.4), so computation of $q(\lambda)$ becomes a more practical proposition.

The dual problem to (12.81) is defined as follows:

$$\max_{\lambda \in \mathbb{R}^n} q(\lambda) \qquad \text{subject to } \lambda \geq 0. \tag{12.84}$$

❑ EXAMPLE 12.10

Consider the problem

$$\min_{(x_1, x_2)} 0.5(x_1^2 + x_2^2) \quad \text{subject to } x_1 - 1 \geq 0. \tag{12.85}$$

The Lagrangian is

$$\mathcal{L}(x_1, x_2, \lambda_1) = 0.5(x_1^2 + x_2^2) - \lambda_1(x_1 - 1).$$

If we hold λ_1 fixed, this is a convex function of $(x_1, x_2)^T$. Therefore, the infimum with respect to $(x_1, x_2)^T$ is achieved when the partial derivatives with respect to x_1 and x_2 are zero, that is,

$$x_1 - \lambda_1 = 0, \quad x_2 = 0.$$

By substituting these infimal values into $\mathcal{L}(x_1, x_2, \lambda_1)$ we obtain the dual objective (12.82):

$$q(\lambda_1) = 0.5(\lambda_1^2 + 0) - \lambda_1(\lambda_1 - 1) = -0.5\lambda_1^2 + \lambda_1.$$

Hence, the dual problem (12.84) is

$$\max_{\lambda_1 \geq 0} -0.5\lambda_1^2 + \lambda_1, \tag{12.86}$$

which clearly has the solution $\lambda_1 = 1$. $\square$

In the remainder of this section, we show how the dual problem is related to (12.81). Our first result concerns concavity of q.

Theorem 12.10.
The function q defined by (12.82) is concave and its domain $\mathcal{D}$ is convex.

PROOF.   For any λ^0 and λ^1 in $\mathbb{R}^m$, any $x \in \mathbb{R}^n$, and any $\alpha \in [0, 1]$, we have

$$\mathcal{L}(x, (1 - \alpha)\lambda^0 + \alpha\lambda^1) = (1 - \alpha)\mathcal{L}(x, \lambda^0) + \alpha\mathcal{L}(x, \lambda^1).$$

By taking the infimum of both sides in this expression, using the definition (12.82), and using the results that the infimum of a sum is greater than or equal to the sum of infimums, we obtain

$$q((1 - \alpha)\lambda^0 + \alpha\lambda^1) \geq (1 - \alpha)q(\lambda^0) + \alpha q(\lambda^1),$$

confirming concavity of q. If both λ^0 and λ^1 belong to $\mathcal{D}$, this inequality implies that $q((1 - \alpha)\lambda^0 + \alpha\lambda^1) \geq -\infty$ also, and therefore $(1 - \alpha)\lambda^0 + \alpha\lambda^1 \in \mathcal{D}$, verifying convexity of $\mathcal{D}$. $\square$

The optimal value of the dual problem (12.84) gives a lower bound on the optimal objective value for the primal problem (12.81). This observation is a consequence of the following *weak duality* result.

Theorem 12.11 (Weak Duality).
For any $\bar{x}$ feasible for (12.81) and any $\bar{\lambda} \geq 0$, we have $q(\bar{\lambda}) \leq f(\bar{x})$.

PROOF.

$$q(\bar{\lambda}) = \inf_x f(x) - \bar{\lambda}^T c(x) \leq f(\bar{x}) - \bar{\lambda}^T c(\bar{x}) \leq f(\bar{x}),$$

where the final inequality follows from $\bar{\lambda} \geq 0$ and $c(\bar{x}) \geq 0$. $\square$

For the remaining results, we note that the KKT conditions (12.34) specialized to (12.81) are as follows:

$$\nabla f(\bar{x}) - \nabla c(\bar{x})\bar{\lambda} = 0, \tag{12.87a}$$

$$c(\bar{x}) \geq 0, \tag{12.87b}$$

$$\bar{\lambda} \geq 0, \tag{12.87c}$$

$$\bar{\lambda}_i c_i(\bar{x}) = 0, \quad i = 1, 2, \ldots, m, \tag{12.87d}$$

where $\nabla c(x)$ is the $n \times m$ matrix defined by $\nabla c(x) = [\nabla c_1(x), \nabla c_2(x), \ldots, \nabla c_m(x)]$.

The next result shows that optimal Lagrange multipliers for (12.81) are solutions of the dual problem (12.84) under certain conditions. It is essentially due to Wolfe [309].

Theorem 12.12.

Suppose that $\bar{x}$ is a solution of (12.81) and that f and $-c_i, i = 1, 2, \ldots, m$ are convex functions on $\mathbb{R}^n$ that are differentiable at $\bar{x}$. Then any $\bar{\lambda}$ for which $(\bar{x}, \bar{\lambda})$ satisfies the KKT conditions (12.87) is a solution of (12.84).

PROOF.   Suppose that $(\bar{x}, \bar{\lambda})$ satisfies (12.87). We have from $\bar{\lambda} \geq 0$ that $\mathcal{L}(\cdot, \bar{\lambda})$ is a convex and differentiable function. Hence, for any x, we have

$$\mathcal{L}(x, \bar{\lambda}) \geq \mathcal{L}(\bar{x}, \bar{\lambda}) + \nabla_x \mathcal{L}(\bar{x}, \bar{\lambda})^T (x - \bar{x}) = \mathcal{L}(\bar{x}, \bar{\lambda}),$$

where the last equality follows from (12.87a). Therefore, we have

$$q(\bar{\lambda}) = \inf_x \mathcal{L}(x, \bar{\lambda}) = \mathcal{L}(\bar{x}, \bar{\lambda}) = f(\bar{x}) - \bar{\lambda}^T c(\bar{x}) = f(\bar{x}),$$

where the last equality follows from (12.87d). Since from Theorem 12.11, we have $q(\lambda) \leq f(\bar{x})$ for all $\lambda \geq 0$ it follows immediately from $q(\bar{\lambda}) = f(\bar{x})$ that $\bar{\lambda}$ is a solution of (12.84). $\qquad \square$

Note that if the functions are continuously differentiable and a constraint qualification such as LICQ holds at $\bar{x}$, then an optimal Lagrange multiplier is guaranteed to exist, by Theorem 12.1.

In Example 12.10, we see that $\lambda_1 = 1$ is both an optimal Lagrange multiplier for the problem (12.85) and a solution of (12.86). Note too that the optimal objective for both problems is 0.5.

We prove a partial converse of Theorem 12.12, which shows that solutions to the dual problem (12.84) can sometimes be used to derive solutions to the original problem (12.81). The essential condition is strict convexity of the function $\mathcal{L}(\cdot, \hat{\lambda})$ for a certain value $\hat{\lambda}$. We note that this condition holds if either f is strictly convex (as is the case in Example 12.10) or if c_i is strictly convex for some $i = 1, 2, \ldots, m$ with $\hat{\lambda}_i > 0$.

Theorem 12.13.

    Suppose that f and $-c_i$, $i = 1, 2, \ldots, m$ are convex and continuously differentiable on $\mathbb{R}^n$. Suppose that $\bar{x}$ is a solution of (12.81) at which LICQ holds. Suppose that $\hat{\lambda}$ solves (12.84) and that the infimum in $\inf_x \mathcal{L}(x, \hat{\lambda})$ is attained at $\hat{x}$. Assume further than $\mathcal{L}(\cdot, \hat{\lambda})$ is a strictly convex function. Then $\bar{x} = \hat{x}$ (that is, $\hat{x}$ is the unique solution of (12.81)), and $f(\bar{x}) = \mathcal{L}(\hat{x}, \hat{\lambda})$.

PROOF.   Assume for contradiction that $\bar{x} \neq \hat{x}$. From Theorem 12.1, because of the LICQ assumption, there exists $\bar{\lambda}$ satisfying (12.87). Hence, from Theorem 12.12, we have that $\bar{\lambda}$ also solves (12.84), so that

$$\mathcal{L}(\bar{x}, \bar{\lambda}) = q(\bar{\lambda}) = q(\hat{\lambda}) = \mathcal{L}(\hat{x}, \hat{\lambda}).$$

Because $\hat{x} = \arg\min_x \mathcal{L}(x, \hat{\lambda})$, we have from Theorem 2.2 that $\nabla_x \mathcal{L}(\hat{x}, \hat{\lambda}) = 0$. Moreover, by strict convexity of $\mathcal{L}(\cdot, \hat{\lambda})$, it follows that

$$\mathcal{L}(\bar{x}, \hat{\lambda}) - \mathcal{L}(\hat{x}, \hat{\lambda}) > \nabla_x \mathcal{L}(\hat{x}, \hat{\lambda})^T (\bar{x} - \hat{x}) = 0.$$

Hence, we have

$$\mathcal{L}(\bar{x}, \hat{\lambda}) > \mathcal{L}(\hat{x}, \hat{\lambda}) = \mathcal{L}(\bar{x}, \bar{\lambda}),$$

so in particular we have

$$-\hat{\lambda}^T c(\bar{x}) > -\bar{\lambda}^T c(\bar{x}) = 0,$$

where the final equality follows from (12.87d). Since $\hat{\lambda} \geq 0$ and $c(\bar{x}) \geq 0$, this yields the contradiction, and we conclude that $\hat{x} = \bar{x}$, as claimed. $\square$

    In Example 12.10, at the dual solution $\lambda_1 = 1$, the infimum of $\mathcal{L}(x_1, x_2, \lambda_1)$ is achieved at $(x_1, x_2) = (1, 0)^T$, which is the solution of the original problem (12.85).

    An slightly different form of duality that is convenient for computations, known as the *Wolfe dual* [309], can be stated as follows:

$$\max_{x,\lambda} \mathcal{L}(x, \lambda) \tag{12.88a}$$

$$\text{subject to } \nabla_x \mathcal{L}(x, \lambda) = 0, \quad \lambda \geq 0. \tag{12.88b}$$

The following results explains the relationship of the Wolfe dual to (12.81).

Theorem 12.14.

    Suppose that f and $-c_i$, $i = 1, 2, \ldots, m$ are convex and continuously differentiable on $\mathbb{R}^n$. Suppose that $(\bar{x}, \bar{\lambda})$ is a solution pair of (12.81) at which LICQ holds. Then $(\bar{x}, \bar{\lambda})$ solves the problem (12.88).

PROOF.   From the KKT conditions (12.87) we have that $(\bar{x}, \bar{\lambda})$ satisfies (12.88b), and that $\mathcal{L}(\bar{x}, \bar{\lambda}) = f(\bar{x})$. Therefore for any pair (x, λ) that satisfies (12.88b) we have that

$$
\begin{aligned}
\mathcal{L}(\bar{x}, \bar{\lambda}) &= f(\bar{x}) \\
&\geq f(\bar{x}) - \lambda^T c(\bar{x}) \\
&= \mathcal{L}(\bar{x}, \lambda) \\
&\geq \mathcal{L}(x, \lambda) + \nabla_x \mathcal{L}(x, \lambda)^T (\bar{x} - x) \\
&= \mathcal{L}(x, \lambda),
\end{aligned}
$$

where the second inequality follows from the convexity of $\mathcal{L}(\cdot, \lambda)$. We have therefore shown that $(\bar{x}, \bar{\lambda})$ maximizes $\mathcal{L}$ over the constraints (12.88b), and hence solves (12.88).   □

❏ **EXAMPLE 12.11**   (LINEAR PROGRAMMING)

An important special case of (12.81) is the linear programming problem

$$
\min c^T x \quad \text{subject to} \quad Ax - b \geq 0, \tag{12.89}
$$

for which the dual objective is

$$
q(\lambda) = \inf_x \left[c^T x - \lambda^T (Ax - b) \right] = \inf_x \left[(c - A^T \lambda)^T x + b^T \lambda \right].
$$

If $c - A^T \lambda \neq 0$, the infimum is clearly $-\infty$ (we can set x to be a large negative multiple of $-(c - A^T \lambda)$ to make q arbitrarily large and negative). When $c - A^T \lambda = 0$, on the other hand, the dual objective is simply $b^T \lambda$. In maximizing q, we can exclude λ for which $c - A^T \lambda \neq 0$ from consideration (the maximum obviously cannot be attained at a point λ for which $q(\lambda) = -\infty$). Hence, we can write the dual problem (12.84) as follows:

$$
\max_\lambda b^T \lambda \quad \text{subject to } A^T \lambda = c, \ \lambda \geq 0. \tag{12.90}
$$

The Wolfe dual of (12.89) can be written as

$$
\max_\lambda c^T x - \lambda^T (Ax - b) \quad \text{subject to } A^T \lambda = c, \ \lambda \geq 0,
$$

and by substituting the constraint $A^T \lambda - c = 0$ into the objective we obtain (12.90) again.

For some matrices A, the dual problem (12.90) may be computationally easier to solve than the original problem (12.89). We discuss the possibilities further in Chapter 13.   ❏

☐ **EXAMPLE 12.12** (CONVEX QUADRATIC PROGRAMMING)

Consider

$$\min \frac{1}{2}x^T G x + c^T x \quad \text{subject to} \quad Ax - b \geq 0, \tag{12.91}$$

where G is a symmetric positive definite matrix. The dual objective for this problem is

$$q(\lambda) = \inf_x \mathcal{L}(x, \lambda) = \inf_x \frac{1}{2}x^T G x + c^T x - \lambda^T (Ax - b). \tag{12.92}$$

Since G is positive definite, since $\mathcal{L}(\cdot, \lambda)$ is a strictly convex quadratic function, the infimum is achieved when $\nabla_x \mathcal{L}(x, \lambda) = 0$, that is,

$$Gx + c - A^T \lambda = 0. \tag{12.93}$$

Hence, we can substitute for x in the infimum expression and write the dual objective explicitly as follows:

$$q(\lambda) = -\frac{1}{2}(A^T \lambda - c)^T G^{-1} (A^T \lambda - c)^T + b^T \lambda.$$

Alternatively, we can write the Wolfe dual form (12.88) by retaining x as a variable and including the constraint (12.93) explicitly in the dual problem, to obtain

$$\max_{(\lambda, x)} \quad \frac{1}{2}x^T G x + c^T x - \lambda^T (Ax - b) \tag{12.94}$$

$$\text{subject to} \quad Gx + c - A^T \lambda = 0, \quad \lambda \geq 0.$$

To make it clearer that the objective is concave, we can use the constraint to substitute $(c - A^T \lambda)^T x = -x^T Gx$ in the objective, and rewrite the dual formulation as follows:

$$\max_{(\lambda, x)} -\frac{1}{2}x^T G x + \lambda^T b, \quad \text{subject to} \quad Gx + c - A^T \lambda = 0, \lambda \geq 0. \tag{12.95}$$

☐

Note that the Wolfe dual form requires only positive *semidefiniteness* of G.

NOTES AND REFERENCES

The theory of constrained optimization is discussed in many books on numerical optimization. The discussion in Fletcher [101, Chapter 9] is similar to ours, though a little terser, and includes additional material on duality. Bertsekas [19, Chapter 3] emphasizes the role of duality and discusses sensitivity of the solution with respect to the active constraints in some detail. The classic treatment of Mangasarian [198] is particularly notable for its thorough description of constraint qualifications. It also has an extensive discussion of theorems of the alternative [198, Chapter 2], placing Farkas' Lemma firmly in the context of other related results.

The KKT conditions were described in a 1951 paper of Kuhn and Tucker [185], though they were derived earlier (and independently) in an unpublished 1939 master's thesis of W. Karush. Lagrange multipliers and optimality conditions for general problems (including nonsmooth problems) are described in the deep and wide-ranging article of Rockafellar [270].

Duality theory for nonlinear programming is described in the books of Rockafellar [198] and Bertsekas [19]; the latter treatment is particularly extensive and general. The material in Section 12.9 is adapted from these sources.

We return to our claim that the set N defined by

$$N = \{By + Ct \mid y \geq 0\},$$

(where B and C are matrices of dimension $n \times m$ and $n \times p$, respectively, and y and t are vectors of appropriate dimensions; see (12.45)) is a closed set. This fact is needed in the proof of Lemma 12.4 to ensure that the solution of the projection subproblem (12.47) is well-defined. The following technical result is well known; the proof given below is due to R. Byrd.

Lemma 12.15.

   The set N is closed.

PROOF. By splitting t into positive and negative parts, it is easy to see that

$$N = \left\{ \begin{bmatrix} B & C & -C \end{bmatrix} \begin{bmatrix} y \\ t^+ \\ t^- \end{bmatrix} \middle| \begin{bmatrix} y \\ t^+ \\ t^- \end{bmatrix} \geq 0 \right\}.$$

Hence, we can assume without loss of generality that N has the form

$$N = \{By \mid y \geq 0\}.$$

Suppose that B has dimensions $n \times m$.

First, we show that for any $s \in N$, we can write $s = B_I y_I$ with $y_I \geq 0$, where $I \subset \{1, 2, \ldots, m\}$, B_I is the column submatrix of B indexed by I with full column rank, and I has minimum cardinality. To prove this claim, we assume for contradiction that $K \subset \{1, 2, \ldots, m\}$ is an index set with minimal cardinality such that $s = B_K y_K$, $y_K \geq 0$, yet the columns of B_K are linearly dependent. Since K is minimal, y_K has no zero components. We then have a nonzero vector w such that $B_K w = 0$. Since $s = B_K(y_K + \tau w)$ for any τ, we can increase or decrease τ from 0 until one or more components of $y_K + \tau w$ become zero, while the other components remain positive. We define $\bar{K}$ by removing the indices from K that correspond to zero components of $y_K + \tau w$, and define $\bar{y}_{\bar{K}}$ to be the vector of strictly positive components of $y_K + \tau w$. We then have that $s = B_{\bar{K}} \bar{y}_{\bar{K}}$ and $\bar{y}_{\bar{K}} \geq 0$, contradicting our assumption that K was the set of minimal cardinality with this property.

Now let $\{s^k\}$ be a sequence with $s^k \in N$ for all k and $s^k \to s$. We prove the lemma by showing that $s \in N$. By the claim of the previous paragraph, for all k we can write $s^k = B_{I_k} y^k_{I_k}$ with $y^k_{I_k} \geq 0$, I_k is minimal, and the columns of B_{I_k} are linearly independent. Since there only finitely many possible choices of index set I_k, at least one index set occurs infinitely often in the sequence. By choosing such an index set I, we can take a subsequence if necessary and assume without loss of generality that $I_k \equiv I$ for all k. We then have that $s^k = A_I y^k_I$ with $y^k_I \geq 0$ and A_I has full column rank. Because of the latter property, we have that $A_I^T A_I$ is invertible, so that y^k_I is defined uniquely as follows:

$$y^k_I = (A_I^T A_I)^{-1} A_I^T s^k, \qquad k = 0, 1, 2, \ldots.$$

By taking limits and using $s^k \to s$, we have that

$$y^k_I \to y_I \overset{\text{def}}{=} (A_I^T A_I)^{-1} A_I^T s,$$

and moreover $y_I \geq 0$, since $y^k_I \geq 0$ for all k. Hence we can write $s = B_I y_I$ with $y_I \geq 0$, and therefore $s \in N$. $\qquad\square$

✎ EXERCISES

✎ **12.1** The following example from [268] with a single variable $x \in \mathbb{R}$ and a single equality constraint shows that strict local solutions are not necessarily isolated. Consider

$$\min_x x^2 \qquad \text{subject to } c(x) = 0, \text{ where } \quad c(x) = \begin{cases} x^6 \sin(1/x) = 0 & \text{if } x \neq 0 \\ 0 & \text{if } x = 0. \end{cases}$$

$$(12.96)$$

(a) Show that the constraint function is twice continuously differentiable at all x (including at $x = 0$) and that the feasible points are $x = 0$ and $x = 1/(k\pi)$ for all nonzero integers k.

(b) Verify that each feasible point except $x = 0$ is an isolated local solution by showing that there is a neighborhood $\mathcal{N}$ around each such point within which it is the only feasible point.

(c) Verify that $x = 0$ is a global solution and a strict local solution, but not an isolated local solution

✐ **12.2** Is an isolated local solution necessarily a strict local solution? Explain.

✐ **12.3** Does problem (12.4) have a finite or infinite number of local solutions? Use the first-order optimality conditions (12.34) to justify your answer.

✐ **12.4** If f is convex and the feasible region Ω is convex, show that local solutions of the problem (12.3) are also global solutions. Show that the set of global solutions is convex. (Hint: See Theorem 2.5.)

✐ **12.5** Let $v : \mathbb{R}^n \to \mathbb{R}^m$ be a smooth vector function and consider the unconstrained optimization problems of minimizing $f(x)$ where

$$f(x) = \|v(x)\|_\infty, \qquad f(x) = \max_{i=1,2,\ldots,m} v_i(x).$$

Reformulate these (generally nonsmooth) problems as smooth constrained optimization problems.

✐ **12.6** Can you perform a smooth reformulation as in the previous question when f is defined by

$$f(x) = \min_{i=1,2,\ldots,m} f_i(x)?$$

(N.B. "min" not "max.") Why or why not?

✐ **12.7** Show that the vector defined by (12.15) satisfies (12.14) when the first-order optimality condition (12.10) is not satisfied.

✐ **12.8** Verify that for the sequence $\{z_k\}$ defined by (12.30), the function $f(x) = x_1 + x_2$ satisfies $f(z_{k+1}) > f(z_k)$ for $k = 2, 3, \ldots$. (Hint: Consider the trajectory $z(s) = (-\sqrt{2 - 1/s^2}, -1/s)^T$ and show that the function $h(s) \stackrel{\text{def}}{=} f(z(s))$ has $h'(s) > 0$ for all $s \geq 2$.)

✐ **12.9** Consider the problem (12.9). Specify *two* feasible sequences that approach the maximizing point $(1, 1)^T$, and show that neither sequence is a decreasing sequence for f.

✐ **12.10** Verify that neither the LICQ nor the MFCQ holds for the constraint set defined by (12.32) at $x^* = (0, 0)^T$.

✐ **12.11** Consider the feasible set Ω in $\mathbb{R}^2$ defined by $x_2 \geq 0, x_2 \leq x_1^2$.

(a) For $x^* = (0, 0)^T$, write down $T_\Omega(x^*)$ and $\mathcal{F}(x^*)$.

(b) Is LICQ satisfied at x^*? Is MFCQ satisfied?

(c) If the objective function is $f(x) = -x_2$, verify that that KKT conditions (12.34) are satisfied at x^*.

(d) Find a feasible sequence $\{z_k\}$ approaching x^* with $f(z_k) < f(x^*)$ for all k.

✐   **12.12** It is trivial to construct an example of a feasible set and a feasible point x^* at which the LICQ is satisfied but the constraints are nonlinear. Give an example of the reverse situation, that is, where the active constraints are linear but the LICQ is not satisfied.

✐   **12.13** Show that for the feasible region defined by

$$(x_1 - 1)^2 + (x_2 - 1)^2 \leq 2,$$
$$(x_1 - 1)^2 + (x_2 + 1)^2 \leq 2,$$
$$x_1 \geq 0,$$

the MFCQ is satisfied at $x^* = (0, 0)^T$ but the LICQ is not satisfied.

✐   **12.14** Consider the half space defined by $H = \{x \in \mathbb{R}^n \mid a^T x + \alpha \geq 0\}$ where $a \in \mathbb{R}^n$ and $\alpha \in \mathbb{R}$ are given. Formulate and solve the optimization problem for finding the point x in H that has the smallest Euclidean norm.

✐   **12.15** Consider the following modification of (12.36), where t is a parameter to be fixed prior to solving the problem:

$$\min_x \; \left(x_1 - \frac{3}{2}\right)^2 + (x_2 - t)^4 \qquad \text{s.t.} \qquad \begin{bmatrix} 1 - x_1 - x_2 \\ 1 - x_1 + x_2 \\ 1 + x_1 - x_2 \\ 1 + x_1 + x_2 \end{bmatrix} \geq 0. \qquad (12.97)$$

(a) For what values of t does the point $x^* = (1, 0)^T$ satisfy the KKT conditions?

(b) Show that when $t = 1$, only the first constraint is active at the solution, and find the solution.

✐   **12.16** (Fletcher [101]) Solve the problem

$$\min_x \; x_1 + x_2 \text{ subject to } x_1^2 + x_2^2 = 1$$

by eliminating the variable x_2. Show that the choice of sign for a square root operation during the elimination process is critical; the "wrong" choice leads to an incorrect answer.

✐   **12.17** Prove that when the KKT conditions (12.34) and the LICQ are satisfied at a point x^*, the Lagrange multiplier λ^* in (12.34) is unique.

✐ **12.18** Consider the problem of finding the point on the parabola $y = \frac{1}{5}(x - 1)^2$ that is closest to $(x, y) = (1, 2)$, in the Euclidean norm sense. We can formulate this problem as

$$\min \ f(x, y) = (x - 1)^2 + (y - 2)^2 \quad \text{subject to } (x - 1)^2 = 5y.$$

(a) Find all the KKT points for this problem. Is the LICQ satisfied?

(b) Which of these points are solutions?

(c) By directly substituting the constraint into the objective function and eliminating the variable x, we obtain an unconstrained optimization problem. Show that the solutions of this problem cannot be solutions of the original problem.

✐ **12.19** Consider the problem

$$\min_{x \in \mathbb{R}^2} f(x) = -2x_1 + x_2 \quad \text{subject to } \begin{cases} (1 - x_1)^3 - x_2 & \geq \quad 0 \\ x_2 + 0.25x_1^2 - 1 & \geq \quad 0. \end{cases}$$

The optimal solution is $x^* = (0, 1)^T$, where both constraints are active.

(a) Do the LICQ hold at this point?

(b) Are the KKT conditions satisfied?

(c) Write down the sets $\mathcal{F}(x^*)$ and $\mathcal{C}(x^*, \lambda^*)$.

(d) Are the second-order necessary conditions satisfied? Are the second-order sufficient conditions satisfied?

✐ **12.20** Find the minima of the function $f(x) = x_1 x_2$ on the unit circle $x_1^2 + x_2^2 = 1$. Illustrate this problem geometrically.

✐ **12.21** Find the *maxima* of $f(x) = x_1 x_2$ over the unit disk defined by the inequality constraint $1 - x_1^2 - x_2^2 \geq 0$.

✐ **12.22** Show that for (12.1), the feasible set Ω is convex if $c_i, i \in \mathcal{E}$ are linear functions and $-c_i, i \in \mathcal{I}$ are convex functions.

CHAPTER *13*

Linear Programming: The Simplex Method

Dantzig's development of the simplex method in the late 1940s marks the start of the modern era in optimization. This method made it possible for economists to formulate large models and analyze them in a systematic and efficient way. Dantzig's discovery coincided with the development of the first electronic computers, and the simplex method became one of the earliest important applications of this new and revolutionary technology. From those days to the present, computer implementations of the simplex method have been continually improved and refined. They have benefited particularly from interactions with numerical analysis, a branch of mathematics that also came into its own with the appearance of electronic computers, and have now reached a high level of sophistication.

Today, linear programming and the simplex method continue to hold sway as the most widely used of all optimization tools. Since 1950, generations of workers in management, economics, finance, and engineering have been trained in the techniques of formulating linear models and solving them with simplex-based software. Often, the situations they model are actually nonlinear, but linear programming is appealing because of the advanced state of the software, guaranteed convergence to a global minimum, and the fact that uncertainty in the model makes a linear model more appropriate than an overly complex nonlinear model. Nonlinear programming may replace linear programming as the method of choice in some applications as the nonlinear software improves, and a new class of methods known as interior-point methods (see Chapter 14) has proved to be faster for some linear programming problems, but the continued importance of the simplex method is assured for the foreseeable future.

LINEAR PROGRAMMING

Linear programs have a linear objective function and linear constraints, which may include both equalities and inequalities. The feasible set is a polytope, a convex, connected set with flat, polygonal faces. The contours of the objective function are planar. Figure 13.1 depicts a linear program in two-dimensional space, in which the contours of the objective function are indicated by dotted lines. The solution in this case is unique—a single vertex. A simple reorientation of the polytope or the objective gradient c could however make the solution non-unique; the optimal value $c^T x$ could take on the same value over an entire edge. In higher dimensions, the set of optimal points can be a single vertex, an edge or face, or even the entire feasible set. The problem has no solution if the feasible set is empty (the *infeasible* case) or if the objective function is unbounded below on the feasible region (the *unbounded* case).

Linear programs are usually stated and analyzed in the following *standard form*:

$$\min c^T x, \quad \text{subject to } Ax = b, x \geq 0, \tag{13.1}$$

where c and x are vectors in $\mathbb{R}^n$, b is a vector in $\mathbb{R}^m$, and A is an $m \times n$ matrix. Simple devices can be used to transform any linear program to this form. For instance, given the problem

$$\min c^T x, \quad \text{subject to } Ax \leq b$$

(without any bounds on x), we can convert the inequality constraints to equalities by introducing a vector of *slack variables* z and writing

$$\min c^T x, \quad \text{subject to } Ax + z = b, z \geq 0. \tag{13.2}$$

This form is still not quite standard, since not all the variables are constrained to be

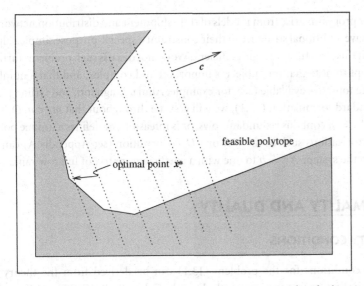

Figure 13.1   A linear program in two dimensions with solution at x^*.

nonnegative. We deal with this by *splitting* x into its nonnegative and nonpositive parts, $x = x^+ - x^-$, where $x^+ = \max(x, 0) \geq 0$ and $x^- = \max(-x, 0) \geq 0$. The problem (13.2) can now be written as

$$
\min \begin{bmatrix} c \\ -c \\ 0 \end{bmatrix}^T \begin{bmatrix} x^+ \\ x^- \\ z \end{bmatrix}, \text{ s.t. } \begin{bmatrix} A & -A & I \end{bmatrix} \begin{bmatrix} x^+ \\ x^- \\ z \end{bmatrix} = b, \begin{bmatrix} x^+ \\ x^- \\ z \end{bmatrix} \geq 0,
$$

which clearly has the same form as (13.1).

Inequality constraints of the form $x \leq u$ or $Ax \geq b$ always can be converted to equality constraints by adding or subtracting *slack variables* to make up the difference between the left- and right-hand sides. Hence,

$$
x \leq u \Leftrightarrow x + w = u, \ w \geq 0,
$$
$$
Ax \geq b \Leftrightarrow Ax - y = b, \ y \geq 0.
$$

(When we subtract the variables from the left hand side, as in the second case, they are sometimes known as *surplus variables*.) We can also convert a "maximize" objective max $c^T x$ into the "minimize" form of (13.1) by simply negating c to obtain: min $(-c)^T x$.

We say that the linear program (13.1) is *infeasible* if the feasible set is empty. We say that the problem (13.1) is *unbounded* if the objective function is unbounded below on the feasible region, that is, there is a sequence of points x^k feasible for (13.1) such that $c^T x^k \downarrow -\infty$. (Of course, unbounded problems have no solution).

Many linear programs arise from models of transshipment and distribution networks. These problems have additional structure in their constraints; special-purpose simplex algorithms that exploit this structure are highly efficient. We do not discuss such problems further in this book, except to note that the subject is important and complex, and that a number of fine texts on the topic are available (see, for example, Ahuja, Magnanti, and Orlin [1]).

For the standard formulation (13.1), we will assume throughout that $m < n$. Otherwise, the system $Ax = b$ contains redundant rows, or is infeasible, or defines a unique point. When $m \geq n$, factorizations such as the QR or LU factorization (see Appendix A) can be used to transform the system $Ax = b$ to one with a coefficient matrix of full row rank.

13.1 OPTIMALITY AND DUALITY

OPTIMALITY CONDITIONS

Optimality conditions for the problem (13.1) can be derived from the theory of Chapter 12. Only the first-order conditions—the Karush–Kuhn–Tucker (KKT) conditions—are needed. Convexity of the problem ensures that these conditions are sufficient for a global minimum. We do not need to refer to the second-order conditions from Chapter 12, which are not informative in any case because the Hessian of the Lagrangian for (13.1) is zero.

The theory we developed in Chapter 12 make derivation of optimality and duality results for linear programming much easier than in other treatments, where this theory is developed more or less from scratch.

The KKT conditions follow from Theorem 12.1. As stated in Chapter 12, this theorem requires linear independence of the active constraint gradients (LICQ). However, as we noted in Section 12.6, the result continues to hold for *dependent* constraints provided they are linear, as is the case here.

We partition the Lagrange multipliers for the problem (13.1) into two vectors λ and s, where $\lambda \in \mathbb{R}^m$ is the multiplier vector for the equality constraints $Ax = b$, while $s \in \mathbb{R}^n$ is the multiplier vector for the bound constraints $x \geq 0$. Using the definition (12.33), we can write the Lagrangian function for (13.1) as

$$\mathcal{L}(x, \lambda, s) = c^T x - \lambda^T (Ax - b) - s^T x. \tag{13.3}$$

Applying Theorem 12.1, we find that the first-order necessary conditions for x^* to be a solution of (13.1) are that there exist vectors λ and s such that

$$A^T \lambda + s = c, \tag{13.4a}$$
$$Ax = b, \tag{13.4b}$$
$$x \geq 0, \tag{13.4c}$$
$$s \geq 0, \tag{13.4d}$$
$$x_i s_i = 0, \quad i = 1, 2, \ldots, n. \tag{13.4e}$$

The complementarity condition (13.4e), which essentially says that at least one of the components x_i and s_i must be zero for each $i = 1, 2, \ldots, n$, is often written in the alternative form $x^T s = 0$. Because of the nonnegativity conditions (13.4c), (13.4d), the two forms are identical.

Let (x^*, λ^*, s^*) denote a vector triple that satisfies (13.4). By combining the three equalities (13.4a), (13.4d), and (13.4e), we find that

$$c^T x^* = (A^T \lambda^* + s^*)^T x^* = (Ax^*)^T \lambda^* = b^T \lambda^*. \tag{13.5}$$

As we shall see in a moment, $b^T \lambda$ is the objective function for the dual problem to (13.1), so (13.5) indicates that the primal and dual objectives are equal for vector triples (x, λ, s) that satisfy (13.4).

It is easy to show directly that the conditions (13.4) are *sufficient* for x^* to be a global solution of (13.1). Let $\bar{x}$ be any other feasible point, so that $A\bar{x} = b$ and $\bar{x} \geq 0$. Then

$$c^T \bar{x} = (A\lambda^* + s^*)^T \bar{x} = b^T \lambda^* + \bar{x}^T s^* \geq b^T \lambda^* = c^T x^*. \tag{13.6}$$

We have used (13.4) and (13.5) here; the inequality relation follows trivially from $\bar{x} \geq 0$ and $s^* \geq 0$. The inequality (13.6) tells us that no other feasible point can have a lower objective value than $c^T x^*$. We can say more: The feasible point $\bar{x}$ is optimal if and only if

$$\bar{x}^T s^* = 0,$$

since otherwise the inequality in (13.6) is strict. In other words, when $s_i^* > 0$, then we must have $\bar{x}_i = 0$ for *all* solutions $\bar{x}$ of (13.1).

THE DUAL PROBLEM

Given the data c, b, and A, which defines the problem (13.1), we can define another, closely related, problem as follows:

$$\max b^T \lambda, \quad \text{subject to } A^T \lambda \leq c. \tag{13.7}$$

This problem is called the *dual problem* for (13.1). In contrast, (13.1) is often referred to as the *primal*. We can restate (13.7) in a slightly different form by introducing a vector of *dual slack variables* s, and writing

$$\max b^T \lambda, \quad \text{subject to } A^T \lambda + s = c, \quad s \geq 0. \tag{13.8}$$

The variables (λ, s) in this problem are sometimes jointly referred to collectively as *dual variables*.

The primal and dual problems present two different viewpoints on the same data. Their close relationship becomes evident when we write down the KKT conditions for (13.7). Let us first restate (13.7) in the form

$$\min -b^T \lambda \text{ subject to } c - A^T \lambda \geq 0,$$

to fit the formulation (12.1) from Chapter 12. By using $x \in \mathbb{R}^n$ to denote the Lagrange multipliers for the constraints $A^T \lambda \leq c$, we see that the Lagrangian function is

$$\bar{\mathcal{L}}(\lambda, x) = -b^T \lambda - x^T (c - A^T \lambda).$$

Using Theorem 12.1 again, we find the first-order necessary conditions for λ to be optimal for (13.7) to be that there exists x such that

$$Ax = b, \tag{13.9a}$$
$$A^T \lambda \leq c, \tag{13.9b}$$
$$x \geq 0, \tag{13.9c}$$
$$x_i (c - A^T \lambda)_i = 0, \quad i = 1, 2, \ldots, n. \tag{13.9d}$$

Defining $s = c - A^T \lambda$ (as in (13.8)), we find that the conditions (13.9) and (13.4) are identical! The optimal Lagrange multipliers λ in the primal problem are the optimal variables in the dual problem, while the optimal Lagrange multipliers x in the dual problem are the optimal variables in the primal problem.

Analogously to (13.6), we can show that (13.9) are in fact *sufficient* conditions for a solution of the dual problem (13.7). Given x^* and λ^* satisfying these conditions (so that the triple $(x, \lambda, s) = (x^*, \lambda^*, c - A^T \lambda^*)$ satisfies (13.4)), we have for any other dual feasible point $\bar{\lambda}$ (with $A^T \bar{\lambda} \leq c$) that

$$
\begin{aligned}
b^T \bar{\lambda} &= (x^*)^T A^T \bar{\lambda} \\
&= (x^*)^T (A^T \bar{\lambda} - c) + c^T x^* \\
&\leq c^T x^* \qquad &&\text{because } A^T \bar{\lambda} - c \leq 0 \text{ and } x^* \geq 0 \\
&= b^T \lambda^* \qquad &&\text{from (13.5).}
\end{aligned}
$$

Hence λ^* achieves the maximum of the dual objective $b^T \lambda$ over the dual feasible region $A^T \lambda \leq c$, so it solves the dual problem (13.7).

The primal–dual relationship is symmetric; by taking the dual of the dual problem (13.7), we recover the primal problem (13.1). We leave the proof of this claim as an exercise.

Given a feasible vector x for the primal (satisfying $Ax = b$ and $x \geq 0$) and a feasible point (λ, s) for the dual (satisfying $A^T \lambda + s = c, s \geq 0$), we have as in (13.6) that

$$c^T x - b^T \lambda = (c - A^T \lambda)^T x = s^T x \geq 0. \tag{13.10}$$

Therefore we have $c^T x \geq b^T \lambda$ (that is, the dual objective is a lower bound on the primal objective) when both the primal and dual variables are feasible—a result known as *weak duality*.

The following *strong duality* result is fundamental to the theory of linear programming.

Theorem 13.1 (Strong Duality).

 (i) *If either problem (13.1) or (13.7) has a (finite) solution, then so does the other, and the objective values are equal.*

 (ii) *If either problem (13.1) or (13.7) is unbounded, then the other problem is infeasible.*

PROOF.   For (i), suppose that (13.1) has a finite optimal solution x^*. It follows from Theorem 12.1 that there are vectors λ^* and s^* such that (x^*, λ^*, s^*) satisfies (13.4). We noted above that (13.4) and (13.9) are equivalent, and that (13.9) are sufficient conditions for λ^* to be a solution of the dual problem (13.7). Moreover, it follows from (13.5) that $c^T x^* = b^T \lambda^*$, as claimed.

A symmetric argument holds if we start by assuming that the dual problem (13.7) has a solution.

To prove (ii), suppose that the primal is unbounded, that is, there is a sequence of points $x^k, k = 1, 2, 3, \ldots$ such that

$$c^T x^k \downarrow -\infty, \qquad A x^k = b, \qquad x^k \geq 0.$$

Suppose too that the dual (13.7) is feasible, that is, there exists a vector $\bar{\lambda}$ such that $A^T \bar{\lambda} \leq c$. From the latter inequality together with $x^k \geq 0$, we have that $\bar{\lambda}^T A x^k \leq c^T x^k$, and therefore

$$\bar{\lambda}^T b = \bar{\lambda}^T A x^k \leq c^T x^k \downarrow -\infty,$$

yielding a contradiction. Hence, the dual must be infeasible.

A similar argument can be used to show that unboundedness of the dual implies infeasibility of the primal.   □

As we showed in the discussion following Theorem 12.1, the multiplier values λ and s for (13.1) indicate the sensitivity of the optimal objective value to perturbations in the constraints. In fact, the process of finding (λ, s) for a given optimal x is often called *sensitivity analysis*. Considering the case of perturbations to the vector b (the right-hand side in (13.1) and objective gradient in (13.7)), we can make an informal argument to illustrate the sensitivity. Suppose that this small change produces small perturbations in the primal and dual solutions, and that the vectors Δs and Δx have zeros in the same locations as s and x, respectively. Since x and s are complementary (see (13.4e)) it follows that

$$0 = x^T s = x^T \Delta s = (\Delta x)^T s = (\Delta x)^T \Delta s.$$

We have from Theorem 13.1 that the optimal objectives of the primal and dual problems are equal, for both the original and perturbed problems, so

$$c^T x = b^T \lambda, \qquad c^T (x + \Delta x) = (b + \Delta b)^T (\lambda + \Delta \lambda).$$

Moreover, by feasibility of the perturbed solutions in the perturbed problems, we have

$$A(x + \Delta x) = b + \Delta b, \qquad A^T \Delta \lambda = -\Delta s.$$

Hence, the change in optimal objective due to the perturbation is as follows:

$$
\begin{aligned}
c^T \Delta x &= (b + \Delta b)^T (\lambda + \Delta \lambda) - b^T \lambda \\
&= (b + \Delta b)^T \Delta \lambda + (\Delta b)^T \lambda \\
&= (x + \Delta x)^T A^T \Delta \lambda + (\Delta b)^T \lambda \\
&= (x + \Delta x)^T \Delta s + (\Delta b)^T \lambda \\
&= (\Delta b)^T \lambda.
\end{aligned}
$$

In particular, if $\Delta b = \epsilon e_j$, where e_j is the jth unit vector in $\mathbb{R}^m$, we have for all ϵ sufficiently small that

$$c^T \Delta x = \epsilon \lambda_j. \tag{13.11}$$

That is, the change in optimal objective is λ_j times the size of the perturbation to b_j, if the perturbation is small.

13.2   GEOMETRY OF THE FEASIBLE SET

BASES AND BASIC FEASIBLE POINTS

We assume for the remainder of the chapter that

$$\text{The matrix } A \text{ in (13.1) has full row rank.} \tag{13.12}$$

In practice, a preprocessing phase is applied to the user-supplied data to remove some redundancies from the given constraints and eliminate some of the variables. Reformulation by adding slack, surplus, and artificial variables can also result in A satisfying the property (13.12).

Each iterate generated by the simplex method is a *basic feasible point* of (13.1). A vector x is a basic feasible point if it is feasible and if there exists a subset $\mathcal{B}$ of the index set $\{1, 2, \ldots, n\}$ such that

- B contains exactly m indices;

- $i \notin B \implies x_i = 0$ (that is, the bound $x_i \geq 0$ can be inactive only if $i \in B$);

- The $m \times m$ matrix B defined by

$$B = [A_i]_{i \in B} \tag{13.13}$$

is nonsingular, where A_i is the ith column of A.

A set B satisfying these properties is called a *basis* for the problem (13.1). The corresponding matrix B is called the *basis matrix*.

The simplex method's strategy of examining only basic feasible points will converge to a solution of (13.1) only if

(a) the problem *has* basic feasible points; and

(b) at least one such point is a *basic optimal point*, that is, a solution of (13.1) that is also a basic feasible point.

Happily, both (a) and (b) are true under reasonable assumptions, as the following result (sometimes known as the fundamental theorem of linear programming) shows.

Theorem 13.2.

(i) *If (13.1) has a nonempty feasible region, then there is at least one basic feasible point;*

(ii) *If (13.1) has solutions, then at least one such solution is a basic optimal point.*

(iii) *If (13.1) is feasible and bounded, then it has an optimal solution.*

PROOF.   Among all feasible vectors x, choose one with the minimal number of nonzero components, and denote this number by p. Without loss of generality, assume that the nonzeros are $x_1, x_2, \ldots, x_p$, so we have

$$\sum_{i=1}^{p} A_i x_i = b.$$

Suppose first that the columns $A_1, A_2, \ldots, A_p$ are linearly dependent. Then we can express one of them (A_p, say) in terms of the others, and write

$$A_p = \sum_{i=1}^{p-1} A_i z_i, \tag{13.14}$$

for some scalars $z_1, z_2, \ldots, z_{p-1}$. It is easy to check that the vector

$$x(\epsilon) = x + \epsilon(z_1, z_2, \ldots, z_{p-1}, -1, 0, 0, \ldots, 0)^T = x + \epsilon z \qquad (13.15)$$

satisfies $Ax(\epsilon) = b$ for any scalar ϵ. In addition, since $x_i > 0$ for $i = 1, 2, \ldots, p$, we also have $x_i(\epsilon) > 0$ for the same indices $i = 1, 2, \ldots, p$ and all ϵ sufficiently small in magnitude. However, there is a value $\bar{\epsilon} \in (0, x_p]$ such that $x_i(\bar{\epsilon}) = 0$ for some $i = 1, 2, \ldots, p$. Hence, $x(\bar{\epsilon})$ is feasible and has at most $p - 1$ nonzero components, contradicting our choice of p as the minimal number of nonzeros.

Therefore, columns $A_1, A_2, \ldots, A_p$ must be linearly independent, and so $p \leq m$. If $p = m$, we are done, since then x is a basic feasible point and $\mathcal{B}$ is simply $\{1, 2, \ldots, m\}$. Otherwise $p < m$ and, because A has full row rank, we can choose $m - p$ columns from among $A_{p+1}, A_{p+2}, \ldots, A_n$ to build up a set of m linearly independent vectors. We construct $\mathcal{B}$ by adding the corresponding indices to $\{1, 2, \ldots, p\}$. The proof of (i) is complete.

The proof of (ii) is quite similar. Let x^* be a solution with a minimal number of nonzero components p, and assume again that $x_1^*, x_2^*, \ldots, x_p^*$ are the nonzeros. If the columns $A_1, A_2, \ldots, A_p$ are linearly dependent, we define

$$x^*(\epsilon) = x^* + \epsilon z,$$

where z is chosen exactly as in (13.14), (13.15). It is easy to check that $x^*(\epsilon)$ will be feasible for all ϵ sufficiently small, both positive and negative. Hence, since x^* is optimal, we must have

$$c^T(x^* + \epsilon z) \geq c^T x^* \quad \Rightarrow \quad \epsilon c^T z \geq 0$$

for all ϵ sufficiently small (positive and negative). Therefore, $c^T z = 0$ and so $c^T x^*(\epsilon) = c^T x^*$ for all ϵ. The same logic as in the proof of (i) can be applied to find $\bar{\epsilon} > 0$ such that $x^*(\bar{\epsilon})$ is feasible and optimal, with at most $p - 1$ nonzero components. This contradicts our choice of p as the minimal number of nonzeros, so the columns $A_1, A_2, \ldots, A_p$ must be linearly independent. We can now apply the same reasoning as above to conclude that x^* is already a basic feasible point and therefore a basic optimal point.

The final statement (iii) is a consequence of finite termination of the simplex method. We comment on the latter property in the next section. $\qquad \square$

The terminology we use here is not quite standard, as the following table shows:

our terminology	terminology used elsewhere
basic feasible point	basic feasible solution
basic optimal point	optimal basic feasible solution

The standard terms arose because "solution" and "feasible solution" were originally used as synonyms for "feasible point." However, as the discipline of optimization developed,

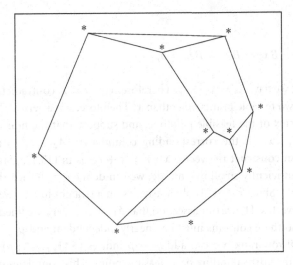

Figure 13.2
Vertices of a
three-dimensional polytope
(indicated by ∗).

the word "solution" took on a more specific and intuitive meaning (as in "solution to the problem"). We maintain consistency with the rest of the book by following this more modern usage.

VERTICES OF THE FEASIBLE POLYTOPE

The feasible set defined by the linear constraints is a polytope, and the *vertices* of this polytope are the points that do not lie on a straight line between two other points in the set. Geometrically, they are easily recognizable; see Figure 13.2. Algebraically, the vertices are exactly the basic feasible points defined above. We therefore have an important relationship between the algebraic and geometric viewpoints and a useful aid to understanding how the simplex method works.

Theorem 13.3.

All basic feasible points for (13.1) are vertices of the feasible polytope $\{x \mid Ax = b, x \geq 0\}$, and vice versa.

PROOF. Let x be a basic feasible point and assume without loss of generality that $\mathcal{B} = \{1, 2, \ldots, m\}$. The matrix $B = [A_i]_{i=1,2,\ldots,m}$ is therefore nonsingular, and

$$x_{m+1} = x_{m+2} = \cdots = x_n = 0. \tag{13.16}$$

Suppose that x lies on a straight line between two other feasible points y and z. Then we can find $\alpha \in (0, 1)$ such that $x = \alpha y + (1 - \alpha)z$. Because of (13.16) and the fact that α and $1 - \alpha$ are both positive, we must have $y_i = z_i = 0$ for $i = m + 1, m + 2, \ldots, n$. Writing $x_{\text{B}} = (x_1, x_2, \ldots, x_m)^T$ and defining y_{B} and z_{B} likewise, we have from $Ax = Ay = Az = b$

that

$$Bx_B = By_B = Bz_B = b,$$

and so, by nonsingularity of B, we have $x_B = y_B = z_B$. Therefore, $x = y = z$, contradicting our assertion that y and z are two feasible points *other* than x. Therefore, x is a vertex.

Conversely, let x be a vertex of the feasible polytope, and suppose that the nonzero components of x are $x_1, x_2, \ldots, x_p$. If the corresponding columns $A_1, A_2, \ldots, A_p$ are linearly dependent, then we can construct the vector $x(\epsilon) = x + \epsilon z$ as in (13.15). Since $x(\epsilon)$ is feasible for all ϵ with sufficiently small magnitude, we can define $\hat{\epsilon} > 0$ such that $x(\hat{\epsilon})$ and $x(-\hat{\epsilon})$ are both feasible. Since $x = x(0)$ obviously lies on a straight line between these two points, it cannot be a vertex. Hence our assertion that $A_1, A_2, \ldots, A_p$ are linearly dependent must be incorrect, so these columns must be linearly independent and $p \leq m$. If $p < m$, and since A has full row rank, we can add $m - p$ indices to $\{1, 2, \ldots, p\}$ to form a basis $\mathcal{B}$, for which x is the corresponding basic feasible point. This completes our proof. $\qquad\square$

We conclude this discussion of the geometry of the feasible set with a definition of *degeneracy*. This term has a variety of meanings in optimization, as we discuss in Chapter 16. For the purposes of this chapter, we use the following definition.

Definition 13.1 (Degeneracy).

A basis $\mathcal{B}$ is said to be degenerate if $x_i = 0$ for some $i \in \mathcal{B}$, where x is the basic feasible solution corresponding to $\mathcal{B}$. A linear program (13.1) is said to be degenerate if it has at least one degenerate basis.

13.3   THE SIMPLEX METHOD

OUTLINE

In this section we give a detailed description of the simplex method for (13.1). There are actually a number of variants the simplex method; the one described here is sometimes known as the *revised simplex method*. (We will describe an alternative known as the *dual simplex method* in Section 13.6.)

As we described above, all iterates of the simplex method are basic feasible points for (13.1) and therefore vertices of the feasible polytope. Most steps consist of a move from one vertex to an adjacent one for which the basis $\mathcal{B}$ differs in exactly one component. On most steps (but not all), the value of the primal objective function $c^T x$ is decreased. Another type of step occurs when the problem is unbounded: The step is an edge along which the objective function is reduced, and along which we can move infinitely far without ever reaching a vertex.

The major issue at each simplex iteration is to decide which index to remove from the basis $\mathcal{B}$. Unless the step is a direction of unboundedness, a single index must be removed from $\mathcal{B}$ and replaced by another from outside $\mathcal{B}$. We can gain some insight into how this decision is made by looking again at the KKT conditions (13.4).

From $\mathcal{B}$ and (13.4), we can derive values for not just the primal variable x but also the dual variables (λ, s), as we now show. First, define the nonbasic index set $\mathcal{N}$ as the complement of $\mathcal{B}$, that is,

$$\mathcal{N} = \{1, 2, \ldots, n\}\backslash\mathcal{B}. \tag{13.17}$$

Just as B is the basic matrix, whose columns are A_i for $i \in \mathcal{B}$, we use N to denote the nonbasic matrix $N = [A_i]_{i\in\mathcal{N}}$. We also partition the n-element vectors x, s, and c according to the index sets $\mathcal{B}$ and $\mathcal{N}$, using the notation

$$x_{\text{B}} = [x_i]_{i\in\mathcal{B}}, \quad x_{\text{N}} = [x_i]_{i\in\mathcal{N}},$$
$$s_{\text{B}} = [s_i]_{i\in\mathcal{B}}, \quad s_{\text{N}} = [s_i]_{i\in\mathcal{N}},$$
$$c_{\text{B}} = [c_i]_{i\in\mathcal{B}}, \quad c_{\text{N}} = [c_i]_{i\in\mathcal{N}}.$$

From the KKT condition (13.4b), we have that

$$Ax = Bx_{\text{B}} + Nx_{\text{N}} = b.$$

The primal variable x for this simplex iterate is defined as

$$x_{\text{B}} = B^{-1}b, \quad x_{\text{N}} = 0. \tag{13.18}$$

Since we are dealing only with basic feasible points, we know that B is nonsingular and that $x_{\text{B}} \geq 0$, so this choice of x satisfies two of the KKT conditions: the equality constraints (13.4b) and the nonnegativity condition (13.4c).

We choose s to satisfy the complementarity condition (13.4e) by setting $s_{\text{B}} = 0$. The remaining components λ and s_{N} can be found by partitioning this condition into c_{B} and c_{N} components and using $s_{\text{B}} = 0$ to obtain

$$B^T\lambda = c_{\text{B}}, \quad N^T\lambda + s_{\text{N}} = c_{\text{N}}. \tag{13.19}$$

Since B is square and nonsingular, the first equation uniquely defines λ as

$$\lambda = B^{-T}c_{\text{B}}. \tag{13.20}$$

The second equation in (13.19) implies a value for s_{N}:

$$s_{\text{N}} = c_{\text{N}} - N^T\lambda = c_{\text{N}} - (B^{-1}N)^T c_{\text{B}}. \tag{13.21}$$

Computation of the vector s_N is often referred to as *pricing*. The components of s_N are often called the *reduced costs* of the nonbasic variables x_N.

The only KKT condition that we have not enforced explicitly is the nonnegativity condition $s \geq 0$. The basic components s_B certainly satisfy this condition, by our choice $s_B = 0$. If the vector s_N defined by (13.21) also satisfies $s_N \geq 0$, we have found an optimal vector triple (x, λ, s), so the algorithm can terminate and declare success. Usually, however, one or more of the components of s_N are negative. The new index to enter the basis $\mathcal{B}$—the *entering index*—is chosen to be *one of the indices $q \in \mathcal{N}$ for which $s_q < 0$*. As we show below, the objective $c^T x$ will decrease when we allow x_q to become positive if and only if (i) $s_q < 0$ and (ii) it is possible to increase x_q away from zero while maintaining feasibility of x. Our procedure for altering $\mathcal{B}$ and changing x and s can be described accordingly as follows:

- allow x_q to increase from zero during the next step;

- fix all other components of x_N at zero, and figure out the effect of increasing x_q on the current basic vector x_B, given that we want to stay feasible with respect to the equality constraints $Ax = b$;

- keep increasing x_q until one of the components of x_B (x_p, say) is driven to zero, or determining that no such component exists (the unbounded case);

- remove index p (known as the *leaving index*) from $\mathcal{B}$ and replace it with the entering index q.

This process of selecting entering and leaving indices, and performing the algebraic operations necessary to keep track of the values of the variables x, λ, and s, is sometimes known as *pivoting*.

We now formalize the pivoting procedure in algebraic terms. Since both the new iterate x^+ and the current iterate x should satisfy $Ax = b$, and since $x_N = 0$ and $x_i^+ = 0$ for $i \in \mathcal{N} \backslash \{q\}$, we have

$$Ax^+ = Bx_B^+ + A_q x_q^+ = Bx_B = Ax.$$

By multiplying this expression by B^{-1} and rearranging, we obtain

$$x_B^+ = x_B - B^{-1} A_q x_q^+. \tag{13.22}$$

Geometrically speaking, (13.22) is usually a move along an edge of the feasible polytope that decreases $c^T x$. We continue to move along this edge until a new vertex is encountered. At this vertex, a new constraint $x_p \geq 0$ must have become active, that is, one of the components x_p, $p \in \mathcal{B}$, has decreased to zero. We then remove this index p from the basis $\mathcal{B}$ and replace it by q.

We now show how the step defined by (13.22) affects the value of $c^T x$. From (13.22), we have

$$c^T x^+ = c_B^T x_B^+ + c_q x_q^+ = c_B^T x_B - c_B^T B^{-1} A_q x_q^+ + c_q x_q^+. \qquad (13.23)$$

From (13.20) we have $c_B^T B^{-1} = \lambda^T$, while from the second equation in (13.19), since $q \in \mathcal{N}$, we have $A_q^T \lambda = c_q - s_q$. Therefore,

$$c_B^T B^{-1} A_q x_q^+ = \lambda^T A_q x_q^+ = (c_q - s_q) x_q^+,$$

so by substituting in (13.23) we obtain

$$c^T x^+ = c_B^T x_B - (c_q - s_q) x_q^+ + c_q x_q^+ = c^T x + s_q x_q^+. \qquad (13.24)$$

Since q was chosen to have $s_q < 0$, it follows that the step (13.22) produces a decrease in the primal objective function $c^T x$ whenever $x_q^+ > 0$.

It is possible that we can increase x_q^+ to ∞ without ever encountering a new vertex. In other words, the constraint $x_B^+ = x_B - B^{-1} A_q x_q^+ \geq 0$ holds for all positive values of x_q^+. When this happens, the linear program is *unbounded*; the simplex method has identified a ray that lies entirely within the feasible polytope along which the objective $c^T x$ decreases to $-\infty$.

Figure 13.3 shows a path traversed by the simplex method for a problem in $\mathbb{R}^2$. In this example, the optimal vertex x^* is found in three steps.

If the basis $\mathcal{B}$ is *nondegenerate* (see Definition 13.1), then we are guaranteed that $x_q^+ > 0$, so we can be assured of a strict decrease in the objective function $c^T x$ at this step. If

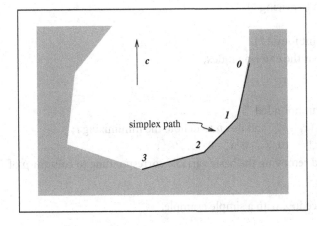

Figure 13.3
Simplex iterates for
a two-dimensional
problem.

the *problem* (13.1) is nondegenerate, we can ensure a decrease in $c^T x$ at *every* step, and can therefore prove the following result concerning termination of the simplex method.

Theorem 13.4.

Provided that the linear program (13.1) is nondegenerate and bounded, the simplex method terminates at a basic optimal point.

PROOF.   The simplex method cannot visit the same basic feasible point x at two different iterations, because it attains a strict decrease at each iteration. Since the number of possible bases B is finite (there are only a finite number of ways to choose a subset of m indices from $\{1, 2, \ldots, n\}$), and since each basis defines a single basic feasible point, there are only a finite number of basic feasible points. Hence, the number of iterations is finite. Moreover, since the method is always able to take a step away from a nonoptimal basic feasible point, and since the problem is not unbounded, the method must terminate at a basic optimal point.   □

This result gives us a proof of Theorem 13.2 (iii) in the case in which the linear program is nondegenerate. The proof of finite termination is considerably more complex when nondegeneracy of (13.1) is not assumed, as we discuss at the end of Section 13.5.

A SINGLE STEP OF THE METHOD

We have covered most of the mechanics of taking a single step of the simplex method. To make subsequent discussions easier to follow, we summarize our description.

Procedure 13.1 (One Step of Simplex).
    Given $B, \mathcal{N}, x_B = B^{-1}b \geq 0, x_N = 0$;
    Solve $B^T \lambda = c_B$ for λ,
    Compute $s_N = c_N - N^T \lambda$; (* pricing *)
    if $s_N \geq 0$
            stop; (* optimal point found *)
    Select $q \in \mathcal{N}$ with $s_q < 0$ as the entering index;
    Solve $Bd = A_q$ for d;
    if $d \leq 0$
            stop; (* problem is unbounded *)
    Calculate $x_q^+ = \min_{i \mid d_i > 0} (x_B)_i / d_i$, and use p to denote the minimizing i;
    Update $x_B^+ = x_B - dx_q^+, x_N^+ = (0, \ldots, 0, x_q^+, 0, \ldots, 0)^T$;
    Change B by adding q and removing the basic variable corresponding to column p of B.

We illustrate this procedure with a simple example.

❑ EXAMPLE 13.1

Consider the problem

$$\min -4x_1 - 2x_2 \quad \text{subject to}$$
$$x_1 + x_2 + x_3 = 5,$$
$$2x_1 + (1/2)x_2 + x_4 = 8,$$
$$x \geq 0.$$

Suppose we start with the basis $\mathcal{B} = \{3, 4\}$, for which we have

$$x_B = \begin{bmatrix} x_3 \\ x_4 \end{bmatrix} = \begin{bmatrix} 5 \\ 8 \end{bmatrix}, \quad \lambda = \begin{bmatrix} 0 \\ 0 \end{bmatrix}, \quad s_N = \begin{bmatrix} s_1 \\ s_2 \end{bmatrix} = \begin{bmatrix} -3 \\ -2 \end{bmatrix},$$

and an objective value of $c^T x = 0$. Since both elements of s_N are negative, we could choose either 1 or 2 to be the entering variable. Suppose we choose $q = 1$. We obtain $d = (1, 2)^T$, so we cannot (yet) conclude that the problem is unbounded. By performing the ratio calculation, we find that $p = 2$ (corresponding to the index 4) and $x_1^+ = 4$. We update the basic and nonbasic index sets to $\mathcal{B} = \{3, 1\}$ and $\mathcal{N} = \{4, 2\}$, and move to the next iteration.

At the second iteration, we have

$$x_B = \begin{bmatrix} x_3 \\ x_1 \end{bmatrix} = \begin{bmatrix} 1 \\ 4 \end{bmatrix}, \quad \lambda = \begin{bmatrix} 0 \\ -3/2 \end{bmatrix}, \quad s_N = \begin{bmatrix} s_4 \\ s_2 \end{bmatrix} = \begin{bmatrix} 3/2 \\ -5/4 \end{bmatrix},$$

with an objective value of -12. We see that s_N has one negative component, corresponding to the index $q = 2$, so we select this index to enter the basis. We obtain $d = (3/2, -1/2)^T$, so again we do not detect unboundedness. Continuing, we find that the maximum value of x_2^+ is $4/3$, and that $p = 1$, which indicates that index 3 will leave the basis $\mathcal{B}$. We update the index sets to $\mathcal{B} = \{2, 1\}$ and $\mathcal{N} = \{4, 3\}$ and continue.

At the start of the third iteration, we have

$$x_B = \begin{bmatrix} x_2 \\ x_1 \end{bmatrix} = \begin{bmatrix} 4/3 \\ 11/3 \end{bmatrix}, \quad \lambda = \begin{bmatrix} -5/3 \\ -2/3 \end{bmatrix}, \quad s_N = \begin{bmatrix} s_4 \\ s_3 \end{bmatrix} = \begin{bmatrix} 7/3 \\ 5/3 \end{bmatrix},$$

with an objective value of $c^T x = -41/3$. We see that $s_N \geq 0$, so the optimality test is satisfied, and we terminate. ❑

We need to flesh out Procedure 13.1 with specifics of three important aspects of the implementation:

- Linear algebra issues—maintaining an LU factorization of B that can be used to solve for λ and d.

- Selection of the entering index q from among the negative components of s_N. (In general, there are many such components.)

- Handling of degenerate bases and degenerate steps, in which it is not possible to choose a positive value of x_q^+ without violating feasibility.

Proper handling of these issues is crucial to the efficiency of a simplex implementation. We give some details in the next three sections.

13.4 LINEAR ALGEBRA IN THE SIMPLEX METHOD

We have to solve two linear systems involving the matrix B at each step; namely,

$$B^T \lambda = c_B, \qquad Bd = A_q. \tag{13.25}$$

We never calculate the inverse basis matrix B^{-1} explicitly just to solve these systems. Instead, we calculate or maintain some factorization of B—usually an LU factorization—and use triangular substitutions with the factors to recover λ and d. It is less expensive to update the factorization than to calculate it afresh at each iteration because the basis matrix B changes by just a single column between iterations.

The standard factorization/updating procedures start with an LU factorization of B at the first iteration of the simplex algorithm. Since in practical applications B is large and sparse, its rows and columns are rearranged during the factorization to maintain both numerical stability and sparsity of the L and U factors. One successful pivot strategy that trades off between these two aims was proposed by Markowitz in 1957 [202]; it is still used as the basis of many practical sparse LU algorithms. Other considerations may also enter into our choice of row and column reordering of B. For example, it may help to improve the efficiency of the updating procedure if as many as possible of the leading columns of U contain just a single nonzero, on the diagonal. Many heuristics have been devised for choosing row and column permutations that produce this and other desirable structural features.

Let us assume for simplicity that row and column permutations are already incorporated in B, so that we write the initial LU factorization as

$$LU = B, \tag{13.26}$$

(L is unit lower triangular, U is upper triangular). The system $Bd = A_q$ can then be solved by the following two-step procedure:

$$L\bar{d} = A_q, \qquad Ud = \bar{d}. \tag{13.27}$$

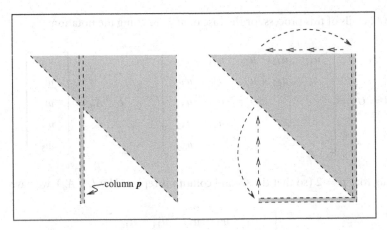

Figure 13.4   Left: $L^{-1}B^+$, which is upper triangular except for the column occupied by A_p. Right: After cyclic row and column permutation P_1, the non–upper triangular part of $P_1 L^{-1} B^+ P_1^T$ appears in the last row.

Similarly, the system $B^T \lambda = c_B$ is solved by performing the following two triangular substitutions:

$$U^T \bar{\lambda} = c_B, \qquad L^T \lambda = \bar{\lambda}.$$

We now discuss a procedure for updating the factors L and U after one step of the simplex method, when the index p is removed from the basis $\mathcal{B}$ and replaced by the index q. The corresponding change to the basis matrix B is that the column B_p is removed from B and replaced by A_q. We call the resulting matrix B^+ and note that if we rewrite (13.26) as $U = L^{-1}B$, the modified matrix $L^{-1}B^+$ will be upper triangular except in column p. That is, $L^{-1}B^+$ has the form shown on the left in Figure 13.4.

We now perform a cyclic permutation that moves column p to the last column position m and moves columns $p+1, p+2, \ldots, m$ one position to the left to make room for it. If we apply the same permutation to rows p through m, the net effect is to move the non-upper triangular part to the last row of the matrix, as shown in Figure 13.4. If we denote the permutation matrix by P_1, the matrix illustrated at right in Figure 13.4 is $P_1 L^{-1} B^+ P_1^T$.

Finally, we perform sparse Gaussian elimination on the matrix $P_1 L^{-1} B^+ P_1^T$ to restore upper triangular form. That is, we find L_1 and U_1 (lower and upper triangular, respectively) such that

$$P_1 L^{-1} B^+ P_1^T = L_1 U_1. \tag{13.28}$$

It is easy to show that L_1 and U_1 have a simple form. The lower triangular matrix L_1 differs from the identity only in the last row, while U_1 is identical to $P_1 L^{-1} B^+ P_1^T$ except that the (m, m) element is changed and the off-diagonal elements in the last row are eliminated.

We give details of this process for the case of $m = 5$. Using the notation

$$L^{-1}B = U = \begin{bmatrix} u_{11} & u_{12} & u_{13} & u_{14} & u_{15} \\ & u_{22} & u_{23} & u_{24} & u_{25} \\ & & u_{33} & u_{34} & u_{35} \\ & & & u_{44} & u_{45} \\ & & & & u_{55} \end{bmatrix}, \qquad L^{-1}A_q = \begin{bmatrix} w_1 \\ w_2 \\ w_3 \\ w_4 \\ w_5 \end{bmatrix},$$

and supposing that $p = 2$ (so that the second column is replaced by $L^{-1}A_q$), we have

$$L^{-1}B^+ = \begin{bmatrix} u_{11} & w_1 & u_{13} & u_{14} & u_{15} \\ & w_2 & u_{23} & u_{24} & u_{25} \\ & w_3 & u_{33} & u_{34} & u_{35} \\ & w_4 & & u_{44} & u_{45} \\ & w_5 & & & u_{55} \end{bmatrix}.$$

After the cyclic permutation P_1, we have

$$P_1 L^{-1} B^+ P_1^T = \begin{bmatrix} u_{11} & u_{13} & u_{14} & u_{15} & w_1 \\ & u_{33} & u_{34} & u_{35} & w_3 \\ & & u_{44} & u_{45} & w_4 \\ & & & u_{55} & w_5 \\ & u_{23} & u_{24} & u_{25} & w_2 \end{bmatrix}. \tag{13.29}$$

The factors L_1 and U_1 are now as follows:

$$L_1 = \begin{bmatrix} 1 & & & & \\ & 1 & & & \\ & & 1 & & \\ & & & 1 & \\ 0 & l_{52} & l_{53} & l_{54} & 1 \end{bmatrix}, \qquad U_1 = \begin{bmatrix} u_{11} & u_{13} & u_{14} & u_{15} & w_1 \\ & u_{33} & u_{34} & u_{35} & w_3 \\ & & u_{44} & u_{45} & w_4 \\ & & & u_{55} & w_5 \\ & & & & \hat{w}_2 \end{bmatrix},$$

$$\tag{13.30}$$

for certain values of l_{52}, l_{53}, l_{54}, and $\hat{w}_2$ (see Exercise 13.10).

The result of this updating process is the factorization (13.28), which we can rewrite as follows:

$$B^+ = L^+ U^+, \quad \text{where } L^+ = L P_1^T L_{1'}, \ U^+ = U_1 P_1. \tag{13.31}$$

There is no need to calculate L^+ and U^+ explicitly. Rather, the nonzero elements in L_1 and the last column of U_1, and the permutation information in P_1, can be stored in compact form, so that triangular substitutions involving L^+ and U^+ can be performed by applying a number of permutations and sparse triangular substitutions involving these factors. The factorization updates from subsequent simplex steps are stored and applied in a similar fashion.

The procedure we have just outlined is due to Forrest and Tomlin [110]. It is quite efficient, because it requires the storage of little data at each update and does not require much movement of data in memory. Its major disadvantage is possible numerical instability. Large elements in the factors of a matrix are a sure indicator of instability, and the multipliers in the L_1 factor (l_{52} in (13.30), for example) may be very large. An earlier scheme of Bartels and Golub [12] allowed swapping of rows to avoid these problems. For instance, if $|u_{33}| < |u_{23}|$ in (13.29), we could swap rows 2 and 5 to ensure that the subsequent multiplier l_{52} in the L_1 factor does not exceed 1 in magnitude. This improved stability comes at a price: The lower right corner of the upper triangular factor may become more dense during each update.

Although the update information for each iteration (the permutation matrices and the sparse triangular factors) can often be stored in a highly compact form, the total amount of space may build up to unreasonable levels after many such updates have been performed. As the number of updates builds up, so does the time needed to solve for the vectors d and λ in Procedure 13.1. If an unstable updating procedure is used, numerical errors may also come into play, blocking further progress by the simplex algorithm. For all these reasons, most simplex implementations periodically calculate a fresh LU factorization of the current basis matrix B and discard the accumulated updates. The new factorization uses the same permutation strategies that we apply to the very first factorization, which balance the requirements of stability, sparsity, and structure.

13.5   OTHER IMPORTANT DETAILS

PRICING AND SELECTION OF THE ENTERING INDEX

There are usually many negative components of s_N at each step. How do we choose one of these to become the index that enters the basis? Ideally, we would like to choose the sequence of entering indices q that gets us to the solution x^* in the fewest possible steps, but we rarely have the global perspective needed to implement this strategy. Instead, we use more shortsighted but practical strategies that obtain a significant decrease in $c^T x$ on just the present iteration. There is usually a tradeoff between the effort spent on finding a good entering index and the amount of decrease in $c^T x$ resulting from this choice. Different pivot strategies resolve this tradeoff in different ways.

Dantzig's original selection rule is one of the simplest. It chooses q such that s_q is the most negative component of $s_N = N^T \lambda$. This rule, which is motivated by (13.24), gives the maximum improvement in $c^T x$ per unit increase in the entering variable x_q. A large reduction in $c^T x$ is not guaranteed, however. It could be that we can increase x_q^+ only a tiny amount from zero (or not at all) before reaching the next vertex.

Calculation of the entire vector s_N from (13.21) requires a multiplication by N^T, which can be expensive when the matrix N is very large. *Partial pricing* strategies calculate only a subvector of s_N and make the choice of entering variable from among the negative entries in this subvector. To give all the indices in $\mathcal{N}$ a chance to enter the basis, these strategies cycle through the nonbasic elements, periodically changing the subvector of s_N they evaluate so that no nonbasic index is ignored for too long.

Neither of these strategies guarantees that we can make a substantial move along the chosen edge before reaching a new vertex. *Multiple pricing* strategies are more thorough: For a small subset of indices $q \in \mathcal{N}$, they evaluate s_q and, if $s_q < 0$, the maximum value of x_q^+ that maintains feasibility of x^+ and the consequent change $s_q x_q^+$ in the objective function (see (13.24)). Calculation of x_q^+ requires evaluation of $d = B^{-1} A_q$ as in Procedure 13.1, which is not cheap. Subsequent iterations deal with this same index subset until we reach an iteration at which all s_q are nonnegative for q in the subset. At this point, the full vector s_N is computed, a new subset of nonbasic indices is chosen, and the cycle begins again. This approach has the advantage that the columns of the matrix N outside the current subset of priced components need not be accessed at all, so memory access in the implementation is quite localized.

Naturally, it is possible to devise heuristics that combine partial and multiple pricing in various imaginative ways.

A sophisticated rule known as *steepest edge* chooses the "most downhill" direction from among all the candidates—the one that produces the largest decrease in $c^T x$ per unit distance moved along the edge. (By contrast, Dantzig's rule maximizes the decrease in $c^T x$ per unit change in x_q^+, which is not the same thing, as a small change in x_q^+ can correspond to a large distance moved along the edge.) During the pivoting step, the overall change in x is

$$x^+ = \begin{bmatrix} x_B^+ \\ x_N^+ \end{bmatrix} = \begin{bmatrix} x_B \\ x_N \end{bmatrix} + \begin{bmatrix} -B^{-1}A_q \\ e_q \end{bmatrix} x_q^+ = x + \eta_q x_q^+, \qquad (13.32)$$

where e_q is the unit vector with a 1 in the position corresponding to the index $q \in \mathcal{N}$ and zeros elsewhere, and the vector η_q is defined as

$$\eta_q = \begin{bmatrix} -B^{-1}A_q \\ e_q \end{bmatrix} = \begin{bmatrix} -d \\ e_q \end{bmatrix}; \qquad (13.33)$$

see (13.25). The change in $c^T x$ per unit step along η_q is given by

$$\frac{c^T \eta_q}{\|\eta_q\|}. \qquad (13.34)$$

The steepest-edge rule chooses $q \in \mathcal{N}$ to minimize this quantity.

If we had to compute each η_i by solving $Bd = A_i$ for each $i \in \mathcal{N}$, the steepest-edge strategy would be prohibitively expensive. Goldfarb and Reid [134] showed that the measure (13.34) of edge steepness for all indices $i \in \mathcal{N}$ can, in fact, be updated quite economically at each iteration. We outline their steepest-edge procedure by showing how each $c^T \eta_i$ and $\|\eta_i\|$ can be updated at the current iteration.

First, note that we already know the numerator $c^T \eta_i$ in (13.34) without calculating η_i, because by taking the inner product of (13.32) with c and using (13.24), we have that $c^T \eta_i = s_i$. To investigate the change in denominator $\|\eta_i\|$ at this step, we define $\gamma_i = \|\eta_i\|^2$, where this quantity is defined before and after the update as follows:

$$\gamma_i = \|\eta_i\|^2 = \|B^{-1} A_i\|^2 + 1, \tag{13.35a}$$
$$\gamma_i^+ = \|\eta_i^+\|^2 = \|(B^+)^{-1} A_i\|^2 + 1. \tag{13.35b}$$

Assume without loss of generality that the entering column A_q replaces the first column of the basis matrix B (that is, $p = 1$), and that this column corresponds to the index t. We can then express the update to B as follows:

$$B^+ = B + (A_q - A_t)e_1^T = B + (A_q - Be_1)e_1^T, \tag{13.36}$$

where $e_1 = (1, 0, 0, \dots, 0)^T$. By applying the Sherman–Morrison formula (A.27) to the rank-one update formula in (13.36), we obtain

$$(B^+)^{-1} = B^{-1} - \frac{(B^{-1} A_q - e_1)e_1^T B^{-1}}{1 + e_1^T (B^{-1} A_q - e_1)} = B^{-1} - \frac{(d - e_1)e_1^T B^{-1}}{e_1^T d},$$

where again we have used the fact that $d = B^{-1} A_q$ (see (13.25)). Therefore, we have that

$$(B^+)^{-1} A_i = B^{-1} A_i - \frac{e_1^T B^{-1} A_i}{e_1^T d}(d - e_1).$$

By substituting for $(B^+)^{-1} A_i$ in (13.35) and performing some simple manipulation, we obtain

$$\gamma_i^+ = \gamma_i - 2 \left(\frac{e_1^T B^{-1} A_i}{e_1^T d} \right) A_i^T B^{-T} d + \left(\frac{e_1^T B^{-1} A_i}{e_1^T d} \right)^2 \gamma_q. \tag{13.37}$$

Once we solve the following two linear systems to obtain $\hat{d}$ and r:

$$B^T \hat{d} = d, \qquad B^T r = e_1. \tag{13.38}$$

The formula (13.37) then becomes

$$\gamma_i^+ = \gamma_i - 2 \left(\frac{r^T A_i}{r^T A_q} \right) \hat{d}^T A_i + \left(\frac{r^T A_i}{r^T A_q} \right)^2 \gamma_q. \tag{13.39}$$

Hence, the entire set of γ_i values, for $i \in \mathcal{N}$ with $i \neq q$, can be calculated by solving the two systems (13.38) and then evaluating the inner products $r^T A_i$ and $\hat{d}^T A_i$, for each i.

The steepest-edge strategy does not guarantee that we can take a long step before reaching another vertex, but it has proved to be highly effective in practice.

STARTING THE SIMPLEX METHOD

The simplex method requires a basic feasible starting point x and a corresponding initial basis $\mathcal{B} \subset \{1, 2, \ldots, n\}$ with $|\mathcal{B}| = m$ such that the basis matrix B defined by (13.13) is nonsingular and $x_{\text{B}} = B^{-1}b \geq 0$ and $x_{\text{N}} = 0$. The problem of finding this initial point and basis may itself be nontrivial—in fact, its difficulty is equivalent to that of actually solving a linear program. We describe here the *two-phase approach* that is commonly used to deal with this difficulty in practical implementations.

In Phase I of this approach we set up an auxiliary linear program based on the data of (13.1), and solve it with the simplex method. The Phase-I problem is designed so that an initial basis and initial basic feasible point is trivial to find, and so that its solution gives a basic feasible initial point for the second phase. In Phase II, a second linear program similar to the original problem (13.1) is solved, with the Phase-I solution as a starting point. The solution of the original problem (13.1) can be extracted easily from the solution of the Phase-II problem.

In Phase I we introduce artificial variables z into (13.1) and redefine the objective function to be the sum of these artificial variables, as follows:

$$\min e^T z, \quad \text{subject to } Ax + Ez = b, (x, z) \geq 0, \tag{13.40}$$

where $z \in \mathbb{R}^m$, $e = (1, 1, \ldots, 1)^T$, and E is a diagonal matrix whose diagonal elements are

$$E_{jj} = +1 \ \text{if } b_j \geq 0, \quad E_{jj} = -1 \ \text{if } b_j < 0.$$

It is easy to see that the point (x, z) defined by

$$x = 0, \quad z_j = |b_j|, \quad j = 1, 2, \ldots, m, \tag{13.41}$$

is a basic feasible point for (13.40). Obviously, this point satisfies the constraints in (13.40), while the initial basis matrix B is simply the diagonal matrix E, which is clearly nonsingular.

At any feasible point for (13.40), the artificial variables z represent the amounts by which the constraints $Ax = b$ are violated by the x component. The objective function is simply the sum of these violations, so by minimizing this sum we are forcing x to become feasible for the original problem (13.1). It is not difficult to see that the Phase-I problem (13.40) has an optimal objective value of zero if and only if the original problem (13.1) is feasible, by using the following argument: If there exists a vector $(\tilde{x}, \tilde{z})$ that is feasible for (13.40) such that $e^T \tilde{z} = 0$, we must have $\tilde{z} = 0$, and therefore $A\tilde{x} = b$ and $\tilde{x} \geq 0$, so $\tilde{x}$ is feasible for (13.1). Conversely, if $\tilde{x}$ is feasible for (13.1), then the point $(\tilde{x}, 0)$ is feasible for (13.40) with an objective value of 0. Since the objective in (13.40) is obviously nonnegative at all feasible points, then $(\tilde{x}, 0)$ must be optimal for (13.40), verifying our claim.

In Phase I, we apply the simplex method to (13.40) from the initial point (13.41). This linear program cannot be unbounded, because its objective function is bounded below by 0, so the simplex method will terminate at an optimal point (assuming that it does not cycle; see below). If the objective $e^T z$ is positive at this solution, we conclude by the argument above that the original problem (13.1) is infeasible. Otherwise, the simplex method identifies a point $(\tilde{x}, \tilde{z})$ with $e^T \tilde{z} = 0$, which is also a basic feasible point for the following *Phase-II problem:*

$$\min c^T x \text{ subject to } Ax + z = b, \quad x \geq 0, \quad 0 \geq z \geq 0. \qquad (13.42)$$

Note that this problem differs from (13.40) in that the objective function is replaced by the original objective $c^T x$, while upper bounds of 0 have been imposed on z. In fact, (13.42) is equivalent to (13.1), because any solution (and indeed any feasible point) must have $z = 0$. We need to retain the artificial variables z in Phase II, however, since some components of z may still be present in the optimal basis from Phase I that we are using as the initial basis for (13.42), though of course the values $\tilde{z}_j$ of these components must be zero. In fact, we can modify (13.42) to include *only* those components of z that are present in the optimal basis for (13.40).

The problem (13.42) is not quite in standard form because of the two-sided bounds on z. However, it is easy to modify the simplex method described above to handle upper and lower bounds on the variables (we omit the details). We can customize the simplex algorithm slightly by deleting each component of z from the problem (13.42) as soon as it is swapped out of the basis. This strategy ensures that components of z do not repeatedly enter and leave the basis, thereby avoiding unnecessary simplex iterations.

If (x^*, z^*) is a basic solution of (13.42), it must have $z^* = 0$, and so x^* is a solution of (13.1). In fact, x^* is a basic feasible point for (13.1), though this claim is not completely obvious because the final basis $\mathcal{B}$ for the Phase-II problem may still contain components of z^*, making it unsuitable as an optimal basis for (13.1). Since A has full row rank, however, we can construct an optimal basis for (13.1) in a postprocessing phase: Extract from $\mathcal{B}$ any components of z that are present, and replace them with nonbasic components of x in a way that maintains nonsingularity of the submatrix B defined by (13.13).

A final point to note is that in many problems we do not need to add a complete set of m artificial variables to form the Phase-I problem. This observation is particularly relevant when slack and surplus variables have already been added to the problem formulation, as in (13.2), to obtain a linear program with inequality constraints in standard form (13.1). Some of these slack/surplus variables can play the roles of artificial variables, making it unnecessary to include such variables explicitly.

   We illustrate this point with the following example.

❑ **EXAMPLE 13.2**

Consider the inequality-constrained linear program defined by

$$\min 3x_1 + x_2 + x_3 \quad \text{subject to}$$
$$2x_1 + x_2 + x_3 \leq 2,$$
$$x_1 - x_2 - x_3 \leq -1,$$
$$x \geq 0.$$

By adding slack variables to both inequality constraints, we obtain the following equivalent problem in standard form:

$$\min 3x_1 + x_2 + x_3 \quad \text{subject to}$$
$$2x_1 + x_2 + x_3 + x_4 = 2,$$
$$x_1 - x_2 - x_3 + x_5 = -1,$$
$$x \geq 0.$$

By inspection, it is easy to see that the vector $x = (0, 0, 0, 2, 0)$ is feasible with respect to the first linear constraint and the lower bound $x \geq 0$, though it does not satisfy the second constraint. Hence, in forming the Phase-I problem, we add just a single artificial variable z_2 to the second constraint and obtain

$$\min z_2 \quad \text{subject to} \tag{13.43}$$
$$2x_1 + x_2 + x_3 + x_4 = 2, \tag{13.44}$$
$$x_1 - x_2 - x_3 + x_5 - z_2 = -1, \tag{13.45}$$
$$(x, z_2) \geq 0. \tag{13.46}$$

It is easy to see that the vector $(x, z_2) = ((0, 0, 0, 2, 0), 1)$ is feasible with respect to (13.43). In fact, it is a basic feasible point, since the corresponding basis matrix B is

$$B = \begin{bmatrix} 1 & 0 \\ 0 & -1 \end{bmatrix},$$

which is clearly nonsingular. In this example, the variable x_4 plays the role of artificial variable for the first constraint. There was no need to add an explicit artificial variable z_1.   ❑

DEGENERATE STEPS AND CYCLING

As noted above, the simplex method may encounter situations in which for the entering index q, we cannot set x_q^+ any greater than zero in (13.22) without violating the nonnegativity condition $x^+ \geq 0$. By referring to Procedure 13.1, we see that these situations arise when there is i with $(x_B)_i = 0$ and $d_i < 0$, where d is defined by (13.25). Steps of this type are called *degenerate steps*. On such steps, the components of x do not change and, therefore, the objective function $c^T x$ does not decrease. However, the steps may still be useful because they change the basis $\mathcal{B}$ (by replacing one index), and the updated B may be closer to the optimal basis. In other words, the degenerate step may be laying the groundwork for reductions in $c^T x$ on later steps.

Sometimes, however, a phenomenon known as *cycling* can occur. After a number of successive degenerate steps, we may return to an earlier basis $\mathcal{B}$. If we continue to apply the algorithm from this point using the same rules for selecting entering and leaving indices, we will repeat the same cycle ad infinitum, never converging.

Cycling was once thought to be a rare phenomenon, but in recent times it has been observed frequently in the large linear programs that arise as relaxations of integer programming problems. Since integer programs are an important source of linear programs, practical simplex codes usually incorporate a cycling avoidance strategy.

In the remainder of this section, we describe a *perturbation strategy* and its close relative, the *lexicographic strategy*.

Suppose that a degenerate basis is encountered at some simplex iteration, at which the basis is $\hat{\mathcal{B}}$ and the basis matrix is $\hat{B}$, say. We consider a modified linear program in which we add a small perturbation to the right-hand side of the constraints in (13.1), as follows:

$$b(\epsilon) = b + \hat{B} \begin{bmatrix} \epsilon \\ \epsilon^2 \\ \vdots \\ \epsilon^m \end{bmatrix},$$

where ϵ is a very small positive number. This perturbation in b induces a perturbation in the components of the basic solution vector; we have

$$x_{\hat{B}}(\epsilon) = x_{\hat{B}} + \begin{bmatrix} \epsilon \\ \epsilon^2 \\ \vdots \\ \epsilon^m \end{bmatrix}. \tag{13.47}$$

Retaining the perturbation for subsequent iterations, we see that subsequent basic solutions have the form

$$
x_B(\epsilon) = x_B + B^{-1}\hat{B}
\begin{bmatrix}
\epsilon \\
\epsilon^2 \\
\vdots \\
\epsilon^m
\end{bmatrix}
= x_B + \sum_{k=1}^{m}(B^{-1}\hat{B})_{\cdot k}\epsilon^k,
\qquad (13.48)
$$

where $(B^{-1}\hat{B})_{\cdot k}$ denotes the kth column of $B^{-1}\hat{B}$ and x_B represents the basic solution for the unperturbed right-hand side b.

From (13.47), we have that for all ϵ sufficiently small (but positive), $(x_{\hat{B}}(\epsilon))_i > 0$ for all i. Hence, the basis is nondegenerate for the perturbed problem, and we can perform a step of the simplex method that produces a nonzero (but tiny) decrease in the objective.

Indeed, if we retain the perturbation over all subsequent iterations, and provided that the initial choice of ϵ was small enough, we claim that *all* subsequent bases visited by the algorithm are nondegenerate. We prove this claim by contradiction, by assuming that there is some basis matrix B such that $(x_B(\epsilon))_i = 0$ for some i and all ϵ sufficiently small. From (13.48), we see that this can happen only when $(x_B)_i = 0$ and $(B^{-1}\bar{B})_{ik} = 0$ for $k = 1, 2, \ldots, m$. The latter relation implies that the ith row of $B^{-1}\bar{B}$ is zero, which cannot occur, because both B and $\bar{B}$ are nonsingular.

We conclude that, provided the initial choice of ϵ is sufficiently small to ensure nondegeneracy of all subsequent bases, no basis is visited more than once by the simplex method and therefore, by the same logic as in the proof of Theorem 13.4, the method terminates finitely at a solution of the perturbed problem. The perturbation can be removed in a postprocessing phase, by resetting $x_B = B^{-1}b$ for the final basis B and the original right-hand side b.

The question remains of how to choose ϵ small enough at the point at which the original degenerate basis $\hat{B}$ is encountered. The *lexicographic strategy* finesses this issue by not making an explicit choice of ϵ, but rather keeping track of the dependence of each basic variable on each power of ϵ. When it comes to selecting the leaving variable, it chooses the index p that minimizes $(x_B(\epsilon))_i/d_i$ over all variables in the basis, for all sufficiently small ϵ. (The choice of p is uniquely defined by this procedure, as we can show by an argument similar to the one above concerning nondenegeracy of each basis.) We can extend the pivot procedure slightly to update the dependence of each basic variable on the powers of ϵ at each iteration, including the variable x_q that has just entered the basis.

13.6   THE DUAL SIMPLEX METHOD

Here we describe another variant of the simplex method that is useful in a variety of situations and is often faster on many practical problems than the variant described above. This *dual simplex method* uses many of the same concepts and methodology described above, such as the splitting of the matrix A into column submatrices B and N and the generation of iterates (x, λ, s) that satisfy the complementarity condition $x^T s = 0$. The method of Section 13.3 starts with a feasible x (with $x_B \geq 0$ and $x_N = 0$) and a corresponding dual iterate (λ, s) for which $s_B = 0$ but s_N is not necessarily nonnegative. After making systematic column interchanges between B and N, it finally reaches a feasible dual point (λ, s) at which $s_N \geq 0$, thus yielding a solution of both the primal problem (13.1) and the dual (13.8). By contrast, the dual simplex method starts with a point (λ, s) feasible for (13.8), at which $s_N \geq 0$ and $s_B = 0$, and a corresponding primal feasible point x for which $x_N = 0$ but x_B is not necessarily nonnegative. By making systematic column interchanges between B and N, it finally reaches a feasible primal point x for which $x_B \geq 0$, signifying optimality. Note that although the matrix B used in this algorithm is a nonsingular column submatrix of A, it is no longer correct to refer to it as a basis matrix, since it does not satisfy the feasibility condition $x_B = B^{-1}b \geq 0$.

We now describe a single step of this method in a similar fashion to Section 13.3, though the details are a little more complicated here. As mentioned above, we commence each step with submatrices B and N of A, and corresponding sets $\mathcal{B}$ and $\mathcal{N}$. The primal and dual variables corresponding to these sets are defined as follows (cf. (13.18), (13.20), and (13.21)):

$$x_B = B^{-1}b, \qquad x_N = 0, \tag{13.49a}$$

$$\lambda = B^{-T}c_B, \tag{13.49b}$$

$$s_B = c_B - B^T\lambda = 0, \qquad s_N = c_N - N^T\lambda \geq 0, \tag{13.49c}$$

If $x_B \geq 0$, the current point (x, λ, s) satisfies the optimality conditions (13.4), and we are done. Otherwise, we select a *leaving index* $q \in \mathcal{B}$ such that $x_q < 0$. Our aim is to move x_q to zero (thereby ensuring that nonnegativity holds for this component), while allowing s_q to increase away from zero. We will also identify an *entering index* $r \in \mathcal{N}$, such that s_r becomes zero on this step while x_r increases away from zero. Hence, the index q will move from $\mathcal{B}$ to $\mathcal{N}$, while r will move from $\mathcal{N}$ to $\mathcal{B}$. How do we choose r, and how are x, λ, and s changed on this step? The description below provides the answer. We use (x^+, λ^+, s^+) to denote the updated values of our variables, after this step is taken.

First, let e_q the vector of length m that contains all zeros except for a 1 in the position occupied by index q in the set $\mathcal{B}$. Since we increase s_q away from zero while fixing the remaining components of s_B at zero, the updated value s_B^+ will have the form

$$s_B^+ = s_B + \alpha e_q \tag{13.50}$$

for some positive scalar α to be determined. We write the corresponding update to λ as

$$\lambda^+ = \lambda + \alpha v, \tag{13.51}$$

for some vector v. In fact, since s_B^+ and λ^+ must satisfy the first equation in (13.49c), we must have

$$s_B^+ = c_B - B^T \lambda^+$$
$$\Rightarrow s_B + \alpha e_q = c_B - B^T (\lambda + \alpha v)$$
$$\Rightarrow e_q = -B^T v, \tag{13.52}$$

which is a system of equations that we can solve to obtain v.

To see how the dual objective value $b^T \lambda$ changes as a result of this step, we use (13.52) and the fact that $x_q = x_B^T e_q$ to obtain

$$
\begin{aligned}
b^T \lambda^+ &= b^T \lambda + \alpha b^T v \\
&= b^T \lambda - \alpha b^T B^{-T} e_q \quad &\text{from (13.52)} \\
&= b^T \lambda - \alpha x_B^T e_q \quad &\text{from (13.49a)} \\
&= b^T \lambda - \alpha x_q \quad &\text{by definition of } e_q.
\end{aligned}
$$

Since $x_q < 0$ and since our aim is to maximize the dual objective, we would like to choose α as large as possible. The upper bound on α is provided by the constraint $s_N^+ \geq 0$. Similarly to (13.49c), we have

$$s_N^+ = c_N - N^T \lambda^+ = s_N - \alpha N^T v = s_N - \alpha w,$$

where we have defined

$$w = N^T v = -N^T B^{-T} e_q.$$

The largest α for which $s_N^+ \geq 0$ is given by the formula

$$\alpha = \min_{j \in \mathcal{N}, w_j > 0} \frac{s_j}{w_j}.$$

We define the entering index r to be the index at which the minimum in this expression is achieved. Note that

$$s_r^+ = 0 \quad \text{and} \quad w_r = A_r^T v > 0, \tag{13.53}$$

where, as usual, A_r denotes the rth column of A.

Having now identified how λ and s are updated on this step, we need to figure out how x changes. For the leaving index q, we need to set $x_q^+ = 0$, while for the entering index r we can allow x_r^+ to be nonzero. We denote the direction of change for x_B to be the vector

d, defined by the following linear system:

$$Bd = \sum_{i \in B} A_i d_i = A_r. \qquad (13.54)$$

Since from (13.49a), we have

$$\sum_{i \in B} A_i x_i = b,$$

we have that

$$\sum_{i \in B} A_i (x_i - \gamma d_i) + A_r \gamma = b, \qquad (13.55)$$

for any scalar γ. To ensure that $x_q^+ = 0$, we set

$$\gamma = \frac{x_q}{d_q}, \qquad (13.56)$$

which is well defined only if d_q is nonzero. In fact, we have that $d_q < 0$, since

$$d_q = d^T e_q = A_r^T B^{-T} e_q = -A_r^T v = -w_r < 0,$$

where we have used the definition of e_q along with (13.54), (13.52), and (13.53) to derive these relationships. Since $x_q < 0$, it follows from (13.56) that $\gamma > 0$. Following (13.55) we can define the updated vector x^+ as follows:

$$x_i^+ = \begin{cases} x_i - \gamma d_i, & \text{for } i \in B \text{ with } i \neq q, \\ 0, & \text{for } i = q, \\ 0, & \text{for } i \in \mathcal{N} \text{ with } i \neq r, \\ \gamma, & \text{for } i = r. \end{cases}$$

13.7 PRESOLVING

Presolving (also known as preprocessing) is carried out in practical linear programming codes to reduce the size of the user-defined linear programming problem before passing it to the solver. A variety of techniques—some obvious, some ingenious—are used to eliminate certain variables, constraints, and bounds from the problem. Often the reduction in problem size is quite dramatic, and the linear programming algorithm takes much less time when applied to the presolved problem than when applied to the original problem. Presolving is beneficial regardless of what algorithm is used to solve the linear program; it is used both in simplex and interior-point codes. Infeasibility may also be detected by the presolver, eliminating the need to call the linear programming algorithm at all.

We mention just a few of the more straightforward preprocessing techniques here, referring the interested reader to Andersen and Andersen [4] for a more comprehensive list. For the purpose of this discussion, we assume that the linear program is formulated with both lower and upper bounds on x, that is,

$$\min c^T x, \quad \text{subject to } Ax = b, l \leq x \leq u, \tag{13.57}$$

where some components l_i of the lower bound vector may be $-\infty$ and some upper bounds u_i may be $+\infty$.

Consider first a *row singleton*, which happens when one of the equality constraints involves just one of the variables. Specifically, if constraint k involves only variable j (that is, $A_{kj} \neq 0$, but $A_{ki} = 0$ for all $i \neq j$), we can immediately set $x_j = b_k/A_{kj}$ and eliminate x_j from the problem. Note that if this value of x_j violates its bounds (that is, $x_j < l_j$ or $x_j > u_j$), we can declare the problem to be infeasible, and terminate.

Another obvious technique is the *free column singleton*, in which there is a variable x_j that occurs in only one of the equality constraints, and is free (that is, its lower bound is $-\infty$ and its upper bound is $+\infty$). In this case, we have for some k that $A_{kj} \neq 0$ while $A_{lj} = 0$ for all $l \neq k$. Here we can simply use constraint k to eliminate x_j from the problem, setting

$$x_j = \frac{b_k - \sum_{p \neq j} A_{kp} x_p}{A_{kj}}.$$

Once the values of x_p for $p \neq j$ have been obtained by solving a reduced linear program, we can substitute into this formula to recover x_j prior to returning the result to the user. This substitution does not require us to modify any other constraints, but it will change the cost vector c in general, whenever $c_j \neq 0$. We will need to make the replacement

$$c_p \leftarrow c_p - c_j A_{kp}/A_{kj}, \quad \text{for all } p \neq j.$$

In this case, we can also determine the dual variable associated with constraint k. Since x_j is a free variable, there is no dual slack associated with it, so the jth dual constraint becomes

$$\sum_{l=1}^{m} A_{lj} \lambda_l = c_j \quad \Rightarrow \quad A_{kj} \lambda_k = c_j,$$

from which we deduce that $\lambda_k = c_j/A_{kj}$.

Perhaps the simplest preprocessing check is for the presence of *zero rows and columns* in A. If $A_{ki} = 0$ for all $i = 1, 2, \ldots, n$, then provided that the right-hand side is also zero ($b_k = 0$), we can simply delete this row from the problem and set the corresponding

Lagrange multiplier λ_k to an arbitrary value. For a zero column—say, $A_{kj} = 0$ for all $k = 1, 2, \ldots, m$—we can determing the optimal value of x_j by inspecting its cost coefficient c_j and its bounds l_j and u_j. If $c_j < 0$, we set $x_j = u_j$ to minimize the product $c_j x_j$. (We are free to do so because x_j is not restricted by any of the equality constraints.) If $c_j < 0$ and $u_j = +\infty$, then the problem is unbounded. Similarly, if $c_j > 0$, we set $x_j = l_j$, or else declare unboundedness if $l_j = -\infty$.

A somewhat more subtle presolving technique is to check for *forcing or dominated constraints*. Rather than give a general specification, we illustrate this case with a simple example. Suppose that one of the equality constraints is as follows:

$$5x_1 - x_4 + 2x_5 = 10,$$

where the variables in question have the following bounds:

$$0 \le x_1 \le 1, \quad -1 \le x_4 \le 5, \quad 0 \le x_5 \le 2.$$

It is not hard to see that the equality constraint can only be satisfied if x_1 and x_5 are at their upper bounds and x_4 is at its lower bound. Any other feasible values of these variables would result in the left-hand side of the equality constraint being strictly less than 10. Hence, we can set $x_1 = 1$, $x_4 = -1$, $x_5 = 2$ and eliminate these variables, and the equality constraint, from the problem.

We use a similar example to illustrate dominated constraints. Suppose that we have the following constraint involving three variables:

$$2x_2 + x_6 - 3x_7 = 8,$$

where the variables in question have the following bounds:

$$-10 \le x_2 \le 10, \quad 0 \le x_6 \le 1, \quad 0 \le x_7 \le 2.$$

By rearranging the constraint and using the bounds on x_6 and x_7, we find that

$$x_2 = 4 - (1/2)x_6 + (3/2)x_7 \le 4 - 0 + (3/2)2 = 7.$$

and similarly, using the opposite bounds on x_6 and x_7 we obtain $x_2 \ge 7/2$. We conclude that the stated bounds of -10 and 10 on x_2 are redundant, since x_2 is implicitly confined to an even smaller interval by the combination of the equality constraint and the bounds on x_6 and x_7. Hence, we can drop the bounds on x_2 from the formulation and treat it as a free variable.

Presolving techniques are applied recursively, because the elimination of certain variables or constraints may create situations that allow further eliminations. As a trivial example, suppose that the following two equality constraints are present in the problem:

$$3x_2 = 6, \quad x_2 + 4x_5 = 10.$$

The first of these constraints is a row singleton, which we can use to set $x_2 = 2$ and eliminate this variable and constraint. After substitution, the second constraint becomes $4x_5 = 10 - x_2 = 8$, which is again a row singleton. We can therefore set $x_5 = 2$ and eliminate this variable and constraint as well.

Relatively little information about presolving techniques has appeared in the literature, in part because they have commercial value as an important component of linear programming software.

13.8 WHERE DOES THE SIMPLEX METHOD FIT?

In linear programming, as in all optimization problems in which inequality constraints are present, the fundamental task of the algorithm is to determine which of these constraints are active at the solution (see Definition 12.1 and which are inactive. The simplex method belongs to a general class of algorithms for constrained optimization known as *active set methods*, which explicitly maintain estimates of the active and inactive index sets that are updated at each step of the algorithm. (At each iteration, the basis B is our current estimate of the inactive set, that is, the set of indices i for which we suspect that $x_i > 0$ at the solution of the linear program.) Like most active set methods, the simplex method makes only modest changes to these index sets at each step; a single index is exchanged between B into $\mathcal{N}$.

Active set algorithms for quadratic programming, bound-constrained optimization, and nonlinear programming use the same basic strategy as simplex of making an explicit estimate of the active set and taking a step toward the solution of a reduced problem in which the constraints in this estimated active set are satisfied as equalities. When nonlinearity enters the problem, many of the features that make the simplex method so effective no longer apply. For example, it is no longer true in general that at least $n - m$ of the bounds $x \geq 0$ are active at the solution, and the specialized linear algebra techniques described in Section 13.5 no longer apply. Nevertheless, the simplex method is rightly viewed as the antecedent of the active set class of methods for constrained optimization.

One undesirable feature of the simplex method attracted attention from its earliest days. Though highly efficient on almost all practical problems (the method generally requires at most $2m$ to $3m$ iterations, where m is the row dimension of the constraint matrix in (13.1)), there are pathological problems on which the algorithm performs very poorly. Klee and Minty [182] presented an n-dimensional problem whose feasible polytope has 2^n vertices, for which the simplex method visits every single vertex before reaching the optimal point! This example verified that the complexity of the simplex method is *exponential;*

roughly speaking, its running time may be an exponential function of the dimension of the problem. For many years, theoreticians searched for a linear programming algorithm that has *polynomial complexity*, that is, an algorithm in which the running time is bounded by a polynomial function of the amount of storage required to define the problem. In the late 1970s, Khachiyan [180] described an *ellipsoid method* that indeed has polynomial complexity but turned out to be impractical. In the mid-1980s, Karmarkar [175] described a polynomial algorithm that approaches the solution through the interior of the feasible polytope rather than working its way around the boundary as the simplex method does. Karmarkar's announcement marked the start of intense research in the field of *interior-point methods*, which are the subject of the next chapter.

NOTES AND REFERENCES

The standard reference for the simplex method is Dantzig's book [86]. Later excellent texts include Chvátal [61] and Vanderbei [293].

Further information on steepest-edge pivoting can be found in Goldfarb and Reid [134] and Goldfarb and Forrest [133].

An alternative procedure for performing the Phase-I calculation of an initial basis was described by Wolfe [310]. This technique does not require artificial variables to be introduced in the problem formulation, but rather starts at any point x that satisfies $Ax = b$ with at most m nonzero components in x. (Note that we do not require the basic part x_B to consist of all positive components.) Phase I then consists in solving the problem

$$\min_x \sum_{x_i < 0} -x_i \quad \text{subject to } Ax = b,$$

and terminating when an objective value of 0 is attained. This problem is *not* a linear program—its objective is only piecewise linear—but it can be solved by the simplex method nonetheless. The key is to *redefine* the cost vector f at each iteration x such that $f_i = -1$ for $x_i < 0$ and $f_i = 0$ otherwise.

✐ EXERCISES

✐   **13.1** Convert the following linear program to standard form:

$$\max_{x,y} c^T x + d^T y \quad \text{subject to } A_1 x = b_1, \ A_2 x + B_2 y \le b_2, \ l \le y \le u,$$

where there are no explicit bounds on x.

✐   **13.2** Verify that the dual of (13.8) is the original primal problem (13.1).

✐ **13.3** Complete the proof of Theorem 13.1 by showing that if the dual (13.7) is unbounded above, the primal (13.1) must be infeasible.

✐ **13.4** Theorem 13.1 does not exclude the possibility that both primal and dual are infeasible. Give a simple linear program for which such is the case.

✐ **13.5** Show that the dual of the linear program

$$\min c^T x \text{ subject to } Ax \geq b, \ x \geq 0,$$

is

$$\max b^T \lambda \text{ subject to } A^T \lambda \leq c, \ \lambda \geq 0.$$

✐ **13.6** Show that when $m \leq n$ and the rows of A are linearly dependent in (13.1), then the matrix B in (13.13) is singular, and therefore there are no basic feasible points.

✐ **13.7** Consider the overdetermined linear system $Ax = b$ with m rows and n columns ($m > n$). When we apply Gaussian elimination with complete pivoting to A, we obtain

$$PAQ = L \begin{bmatrix} U_{11} & U_{12} \\ 0 & 0 \end{bmatrix},$$

where P and Q are permutation matrices, L is $m \times m$ lower triangular, U_{11} is $\bar{m} \times \bar{m}$ upper triangular and nonsingular, U_{12} is $\bar{m} \times (n - \bar{m})$, and $\bar{m} \leq n$ is the rank of A.

(a) Show that the system $Ax = b$ is feasible if the last $m - \bar{m}$ components of $L^{-1} Pb$ are zero, and infeasible otherwise.

(b) When $\bar{m} = n$, find the unique solution of $Ax = b$.

(c) Show that the reduced system formed from the first $\bar{m}$ rows of PA and the first $\bar{m}$ components of Pb is equivalent to $Ax = b$ (i.e., a solution of one system also solves the other).

✐ **13.8** Verify formula (13.37).

✐ **13.9** Consider the following linear program:

$$\begin{aligned} \min -5x_1 - x_2 \quad & \text{subject to} \\ x_1 + x_2 &\leq 5, \\ 2x_1 + (1/2)x_2 &\leq 8, \\ x &\geq 0. \end{aligned}$$

(a) Add slack variables x_3 and x_4 to convert this problem to standard form.

(b) Using Procedure 13.1, solve this problem using the simplex method, showing at each step the basis and the vectors λ, s_N, and x_B, and the value of the objective function. (The initial choice of B for which $x_B \geq 0$ should be obvious once you have added the slacks in part (a).)

✎   **13.10** Calculate the values of l_{52}, l_{53}, l_{54}, and $\hat{w}_2$ in (13.30), by equating the last row of $L_1 U_1$ to the last row of the matrix in (13.29).

✎   **13.11** By extending the procedure (13.27) appropriately, show how the factorization (13.31) can be used to solve linear systems with coefficient matrix B^+ efficiently.

CHAPTER *14*

Linear Programming: Interior-Point Methods

In the 1980s it was discovered that many large linear programs could be solved efficiently by using formulations and algorithms from nonlinear programming and nonlinear equations. One characteristic of these methods was that they required all iterates to satisfy the inequality constraints in the problem *strictly*, so they became known as interior-point methods. By the early 1990s, a subclass of interior-point methods known as *primal-dual methods* had distinguished themselves as the most efficient practical approaches, and proved to be strong competitors to the simplex method on large problems. These methods are the focus of this chapter.

Interior-point methods arose from the search for algorithms with better theoretical properties than the simplex method. As we mentioned in Chapter 13, the simplex method can be inefficient on certain pathological problems. Roughly speaking, the time required to solve a linear program may be exponential in the size of the problem, as measured by the number of unknowns and the amount of storage needed for the problem data. For almost all practical problems, the simplex method is much more efficient than this bound would suggest, but its poor worst-case complexity motivated the development of new algorithms with better guaranteed performance. The first such method was the *ellipsoid method*, proposed by Khachiyan [180], which finds a solution in time that is at worst polynomial in the problem size. Unfortunately, this method approaches its worst-case bound on *all* problems and is not competitive with the simplex method in practice.

Karmarkar's *projective algorithm* [175], announced in 1984, also has the polynomial complexity property, but it came with the added attraction of good practical behavior. The initial claims of excellent performance on large linear programs were never fully borne out, but the announcement prompted a great deal of research activity which gave rise to many new methods. All are related to Karmarkar's original algorithm, and to the log-barrier approach described in Chapter 19, but many of the approaches can be motivated and analyzed independently of the earlier methods.

Interior-point methods share common features that distinguish them from the simplex method. Each interior-point iteration is expensive to compute and can make significant progress towards the solution, while the simplex method usually requires a larger number of inexpensive iterations. Geometrically speaking, the simplex method works its way around the boundary of the feasible polytope, testing a sequence of vertices in turn until it finds the optimal one. Interior-point methods approach the boundary of the feasible set only in the limit. They may approach the solution either from the interior or the exterior of the feasible region, but they never actually lie on the boundary of this region.

In this chapter, we outline some of the basic ideas behind primal-dual interior-point methods, including the relationship to Newton's method and homotopy methods and the concept of the central path. We sketch the important methods in this class, and give a comprehensive convergence analysis of a particular interior-point method known as a *long-step path-following* method. We describe in some detail a practical predictor-corrector algorithm proposed by Mehrotra, which is the basis of much of the current generation of software.

14.1 PRIMAL-DUAL METHODS

OUTLINE

We consider the linear programming problem in standard form; that is,

$$\min c^T x, \quad \text{subject to } Ax = b, x \geq 0, \tag{14.1}$$

where c and x are vectors in $\mathbb{R}^n$, b is a vector in $\mathbb{R}^m$, and A is an $m \times n$ matrix with full row rank. (As in Chapter 13, we can preprocess the problem to remove dependent rows from A if necessary.) The dual problem for (14.1) is

$$\max b^T \lambda, \quad \text{subject to } A^T \lambda + s = c, s \geq 0, \tag{14.2}$$

where λ is a vector in $\mathbb{R}^m$ and s is a vector in $\mathbb{R}^n$. As shown in Chapter 13, solutions of (14.1),(14.2) are characterized by the Karush–Kuhn–Tucker conditions (13.4), which we restate here as follows:

$$A^T \lambda + s = c, \tag{14.3a}$$

$$Ax = b, \tag{14.3b}$$

$$x_i s_i = 0, \quad i = 1, 2, \ldots, n, \tag{14.3c}$$

$$(x, s) \geq 0. \tag{14.3d}$$

Primal-dual methods find solutions (x^*, λ^*, s^*) of this system by applying variants of Newton's method to the three equalities in (14.3) and modifying the search directions and step lengths so that the inequalities $(x, s) \geq 0$ are satisfied *strictly* at every iteration. The equations (14.3a), (14.3b), (14.3c) are linear or only mildly nonlinear and so are not difficult to solve by themselves. However, the problem becomes much more difficult when we add the nonnegativity requirement (14.3d), which gives rise to all the complications in the design and analysis of interior-point methods.

To derive primal-dual interior-point methods we restate the optimality conditions (14.3) in a slightly different form by means of a mapping F from $\mathbb{R}^{2n+m}$ to $\mathbb{R}^{2n+m}$:

$$F(x, \lambda, s) = \begin{bmatrix} A^T \lambda + s - c \\ Ax - b \\ XSe \end{bmatrix} = 0, \tag{14.4a}$$

$$(x, s) \geq 0, \tag{14.4b}$$

where

$$X = \text{diag}(x_1, x_2, \ldots, x_n), \qquad S = \text{diag}(s_1, s_2, \ldots, s_n), \tag{14.5}$$

and $e = (1, 1, \ldots, 1)^T$. Primal-dual methods generate iterates (x^k, λ^k, s^k) that satisfy the bounds (14.4b) strictly, that is, $x^k > 0$ and $s^k > 0$. This property is the origin of the term *interior-point*. By respecting these bounds, the methods avoid spurious solutions, that is, points that satisfy $F(x, \lambda, s) = 0$ but not $(x, s) \geq 0$. Spurious solutions abound, and do not provide useful information about solutions of (14.1) or (14.2), so it makes sense to exclude them altogether from the region of search.

Like most iterative algorithms in optimization, primal-dual interior-point methods have two basic ingredients: a procedure for determining the step and a measure of the desirability of each point in the search space. An important component of the measure of desirability is the average value of the pairwise products $x_i s_i$, $i = 1, 2, \ldots, n$, which are all positive when $x > 0$ and $s > 0$. This quantity is known as the *duality measure* and is defined as follows:

$$\mu = \frac{1}{n} \sum_{i=1}^{n} x_i s_i = \frac{x^T s}{n}. \tag{14.6}$$

The procedure for determining the search direction has its origins in Newton's method for the nonlinear equations (14.4a). Newton's method forms a linear model for F around the current point and obtains the search direction $(\Delta x, \Delta \lambda, \Delta s)$ by solving the following system of linear equations:

$$J(x, \lambda, s) \begin{bmatrix} \Delta x \\ \Delta \lambda \\ \Delta s \end{bmatrix} = -F(x, \lambda, s),$$

where J is the Jacobian of F. (See Chapter 11 for a detailed discussion of Newton's method for nonlinear systems.) If we use the notation r_c and r_b for the first two block rows in F, that is,

$$r_b = Ax - b, \qquad r_c = A^T \lambda + s - c, \tag{14.7}$$

we can write the Newton equations as follows:

$$\begin{bmatrix} 0 & A^T & I \\ A & 0 & 0 \\ S & 0 & X \end{bmatrix} \begin{bmatrix} \Delta x \\ \Delta \lambda \\ \Delta s \end{bmatrix} = \begin{bmatrix} -r_c \\ -r_b \\ -XSe \end{bmatrix}. \tag{14.8}$$

Usually, a full step along this direction would violate the bound $(x, s) \geq 0$, so we perform a line search along the Newton direction and define the new iterate as

$$(x, \lambda, s) + \alpha(\Delta x, \Delta \lambda, \Delta s),$$

for some line search parameter $\alpha \in (0, 1]$. We often can take only a small step along this direction ($\alpha \ll 1$) before violating the condition $(x, s) > 0$. Hence, the pure Newton direction (14.8), sometimes known as the *affine scaling direction*, often does not allow us to make much progress toward a solution.

Most primal-dual methods use a less aggressive Newton direction, one that does not aim directly for a solution of (14.3a), (14.3b), (14.3c) but rather for a point whose pairwise products $x_i s_i$ are reduced to a lower average value—not all the way to zero. Specifically, we take a Newton step toward the a point for which $x_i s_i = \sigma \mu$, where μ is the current duality measure and $\sigma \in [0, 1]$ is the reduction factor that we wish to achieve in the duality measure on this step. The modified step equation is then

$$
\begin{bmatrix} 0 & A^T & I \\ A & 0 & 0 \\ S & 0 & X \end{bmatrix} \begin{bmatrix} \Delta x \\ \Delta \lambda \\ \Delta s \end{bmatrix} = \begin{bmatrix} -r_c \\ -r_b \\ -XSe + \sigma \mu e \end{bmatrix}. \tag{14.9}
$$

We call σ the *centering parameter,* for reasons to be discussed below. When $\sigma > 0$, it usually is possible to take a longer step α along the direction defined by (14.16) before violating the bounds $(x, s) \geq 0$.

At this point, we have specified most of the elements of a path-following primal-dual interior-point method. The general framework for such methods is as follows.

Framework 14.1 (Primal-Dual Path-Following).
  Given (x^0, λ^0, s^0) with $(x^0, s^0) > 0$;
  for $k = 0, 1, 2, \ldots$
      Choose $\sigma_k \in [0, 1]$ and solve

$$
\begin{bmatrix} 0 & A^T & I \\ A & 0 & 0 \\ S^k & 0 & X^k \end{bmatrix} \begin{bmatrix} \Delta x^k \\ \Delta \lambda^k \\ \Delta s^k \end{bmatrix} = \begin{bmatrix} -r_c^k \\ -r_b^k \\ -X^k S^k e + \sigma_k \mu_k e \end{bmatrix}, \tag{14.10}
$$

      where $\mu_k = (x^k)^T s^k / n$;
  Set

$$
(x^{k+1}, \lambda^{k+1}, s^{k+1}) = (x^k, \lambda^k, s^k) + \alpha_k (\Delta x^k, \Delta \lambda^k, \Delta s^k), \tag{14.11}
$$

      choosing α_k so that $(x^{k+1}, s^{k+1}) > 0$.
  end (for).

The choices of centering parameter σ_k and step length α_k are crucial to the performance of the method. Techniques for controlling these parameters, directly and indirectly, give rise to a wide variety of methods with diverse properties.

Although software for implementing interior-point methods does not usually start from a point (x^0, λ^0, s^0) that is feasible with respect to the linear equations (14.3a) and (14.3b), most of the historical development of theory and algorithms assumed that these conditions are satisfied. In the remainder of this section, we discuss this feasible case, showing that a comprehensive convergence analysis can be presented in just a few pages, using only basic mathematical tools and concepts. Analysis of the infeasible case follows the same principles, but is considerably more complicated in the details, so we do not present it here. In Section 14.2, however, we describe a complete practical algorithm that does not require starting from a feasible initial point.

To begin our discussion and analysis of feasible interior-point methods, we introduce the concept of the *central path*, and then describe neighborhoods of this path.

THE CENTRAL PATH

The primal-dual *feasible set* $\mathcal{F}$ and *strictly feasible set* $\mathcal{F}^o$ are defined as follows:

$$\mathcal{F} = \{(x, \lambda, s) \mid Ax = b, A^T\lambda + s = c, (x, s) \geq 0\}, \tag{14.12a}$$
$$\mathcal{F}^o = \{(x, \lambda, s) \mid Ax = b, A^T\lambda + s = c, (x, s) > 0\}. \tag{14.12b}$$

The central path $\mathcal{C}$ is an arc of strictly feasible points that plays a vital role in primal-dual algorithms. It is parametrized by a scalar $\tau > 0$, and each point $(x_\tau, \lambda_\tau, s_\tau) \in \mathcal{C}$ satisfies the following equations:

$$A^T\lambda + s = c, \tag{14.13a}$$
$$Ax = b, \tag{14.13b}$$
$$x_i s_i = \tau, \quad i = 1, 2, \dots, n, \tag{14.13c}$$
$$(x, s) > 0. \tag{14.13d}$$

These conditions differ from the KKT conditions only in the term τ on the right-hand side of (14.13c). Instead of the complementarity condition (14.3c), we require that the pairwise products $x_i s_i$ have the same (positive) value for all indices i. From (14.13), we can define the central path as

$$\mathcal{C} = \{(x_\tau, \lambda_\tau, s_\tau) \mid \tau > 0\}.$$

It can be shown that $(x_\tau, \lambda_\tau, s_\tau)$ is defined uniquely for each $\tau > 0$ if and only if $\mathcal{F}^o$ is nonempty.

The conditions (14.13) are also the optimality conditions for a logarithmic-barrier formulation of the problem (14.1). By introducing log-barrier terms for the nonnegativity constraints, with barrier parameter $\tau > 0$, we obtain

$$\min c^T x - \tau \sum_{i=1}^{n} \ln x_i, \quad \text{subject to } Ax = b. \tag{14.14}$$

The KKT conditions (12.34) for this problem, with Lagrange multiplier λ for the equality constraint, are as follows:

$$c_i - \frac{\tau}{x_i} - A_{\cdot i}^T \lambda, \quad i = 1, 2, \ldots, n, \quad Ax = b.$$

Since the objective in (14.14) is strictly convex, these conditions are sufficient as well as necessary for optimality. We recover (14.13) by defining $s_i = \tau/x_i, i = 1, 2, \ldots, n$.

Another way of defining $\mathcal{C}$ is to use the mapping F defined in (14.4) and write

$$F(x_\tau, \lambda_\tau, s_\tau) = \begin{bmatrix} 0 \\ 0 \\ \tau e \end{bmatrix}, \qquad (x_\tau, s_\tau) > 0. \tag{14.15}$$

The equations (14.13) approximate (14.3) more and more closely as τ goes to zero. If $\mathcal{C}$ converges to anything as $\tau \downarrow 0$, it must converge to a primal-dual solution of the linear program. The central path thus guides us to a solution along a route that maintains positivity of the x and s components and decreases the pairwise products $x_i s_i, i = 1, 2, \ldots, n$ to zero at the same rate.

Most primal-dual algorithms take Newton steps toward points on $\mathcal{C}$ for which $\tau > 0$, rather than pure Newton steps for F. Since these steps are biased toward the interior of the nonnegative orthant defined by $(x, s) \geq 0$, it usually is possible to take longer steps along them than along the pure Newton (affine scaling) steps, before violating the positivity condition.

In the feasible case of $(x, \lambda, s) \in \mathcal{F}$, we have $r_b = 0$ and $r_c = 0$, so the search direction satisfies a special case of (14.8), that is,

$$\begin{bmatrix} 0 & A^T & I \\ A & 0 & 0 \\ S & 0 & X \end{bmatrix} \begin{bmatrix} \Delta x \\ \Delta \lambda \\ \Delta s \end{bmatrix} = \begin{bmatrix} 0 \\ 0 \\ -XSe + \sigma \mu e \end{bmatrix}, \tag{14.16}$$

where μ is the duality measure defined by (14.6) and $\sigma \in [0, 1]$ is the centering parameter. When $\sigma = 1$, the equations (14.16) define a *centering direction*, a Newton step toward the point $(x_\mu, \lambda_\mu, s_\mu) \in \mathcal{C}$, at which all the pairwise products $x_i s_i$ are identical to the current average value of μ. Centering directions are usually biased strongly toward the interior of the nonnegative orthant and make little, if any, progress in reducing the duality measure μ. However, by moving closer to $\mathcal{C}$, they set the scene for a substantial reduction in μ on the next iteration. At the other extreme, the value $\sigma = 0$ gives the standard Newton (affine scaling) step. Many algorithms use intermediate values of σ from the open interval $(0, 1)$ to trade off between the twin goals of reducing μ and improving centrality.

CENTRAL PATH NEIGHBORHOODS AND PATH-FOLLOWING METHODS

Path-following algorithms explicitly restrict the iterates to a neighborhood of the central path $\mathcal{C}$ and follow $\mathcal{C}$ to a solution of the linear program. By preventing the iterates from coming too close to the boundary of the nonnegative orthant, they ensure that it is possible to take a nontrivial step along each search direction. Mopreover, by forcing the duality measure μ_k to zero as $k \to \infty$, we ensure that the iterates (x^k, λ^k, s^k) come closer and closer to satisfying the KKT conditions (14.3).

The two most interesting neighborhoods of $\mathcal{C}$ are

$$\mathcal{N}_2(\theta) = \{(x, \lambda, s) \in \mathcal{F}^o \mid \|XSe - \mu e\|_2 \leq \theta\mu\}, \qquad (14.17)$$

for some $\theta \in [0, 1)$, and

$$\mathcal{N}_{-\infty}(\gamma) = \{(x, \lambda, s) \in \mathcal{F}^o \mid x_i s_i \geq \gamma\mu \quad \text{all } i = 1, 2, \ldots, n\}, \qquad (14.18)$$

for some $\gamma \in (0, 1]$. (Typical values of the parameters are $\theta = 0.5$ and $\gamma = 10^{-3}$.) If a point lies in $\mathcal{N}_{-\infty}(\gamma)$, each pairwise product $x_i s_i$ must be at least some small multiple γ of their average value μ. This requirement is actually quite modest, and we can make $\mathcal{N}_{-\infty}(\gamma)$ encompass most of the feasible region $\mathcal{F}$ by choosing γ close to zero. The $\mathcal{N}_2(\theta)$ neighborhood is more restrictive, since certain points in $\mathcal{F}^o$ do not belong to $\mathcal{N}_2(\theta)$ no matter how close θ is chosen to its upper bound of 1.

By keeping all iterates inside one or other of these neighborhoods, path-following methods reduce all the pairwise products $x_i s_i$ to zero at more or less the same rate. Figure 14.1 shows the projection of the central path $\mathcal{C}$ onto the primal variables for a typical problem, along with a typical neighborhood $\mathcal{N}$.

Path-following methods are akin to homotopy methods for general nonlinear equations, which also define a path to be followed to the solution. Traditional homotopy methods stay in a tight tubular neighborhood of their path, making incremental changes to the parameter and chasing the homotopy path all the way to a solution. For primal-dual methods, this neighborhood is horn-shaped rather than tubular, and it tends to be broad and loose for larger values of the duality measure μ. It narrows as $\mu \to 0$, however, because of the positivity requirement $(x, s) > 0$.

The algorithm we specify below, a special case of Framework 14.1, is known as a *long-step path-following* algorithm. This algorithm can make rapid progress because of its use of the wide neighborhood $\mathcal{N}_{-\infty}(\gamma)$, for γ close to zero. It depends on two parameters σ_{min} and σ_{max}, which are lower and upper bounds on the centering parameter σ_k. The search direction is, as usual, obtained by solving (14.10), and we choose the step length α_k to be as large as possible, subject to the requirement that we stay inside $\mathcal{N}_{-\infty}(\gamma)$.

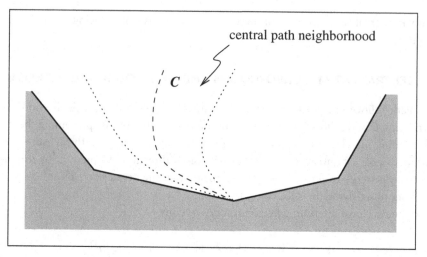

Figure 14.1 Central path, projected into space of primal variables x, showing a typical neighborhood $\mathcal{N}$.

Here and in later analysis, we use the notation

$$(x^k(\alpha), \lambda^k(\alpha), s^k(\alpha)) \stackrel{\text{def}}{=} (x^k, \lambda^k, s^k) + \alpha(\Delta x^k, \Delta \lambda^k, \Delta s^k), \qquad (14.19a)$$

$$\mu_k(\alpha) \stackrel{\text{def}}{=} x^k(\alpha)^T s^k(\alpha)/n. \qquad (14.19b)$$

Algorithm 14.2 (Long-Step Path-Following).

Given γ, $\sigma_{\min}$, $\sigma_{\max}$ with $\gamma \in (0, 1)$, $0 < \sigma_{\min} \le \sigma_{\max} < 1$,
    and $(x^0, \lambda^0, s^0) \in \mathcal{N}_{-\infty}(\gamma)$;
for $k = 0, 1, 2, \ldots$
    Choose $\sigma_k \in [\sigma_{\min}, \sigma_{\max}]$;
    Solve (14.10) to obtain $(\Delta x^k, \Delta \lambda^k, \Delta s^k)$;
    Choose α_k as the largest value of α in $[0, 1]$ such that

$$(x^k(\alpha), \lambda^k(\alpha), s^k(\alpha)) \in \mathcal{N}_{-\infty}(\gamma); \qquad (14.20)$$

    Set $(x^{k+1}, \lambda^{k+1}, s^{k+1}) = (x^k(\alpha_k), \lambda^k(\alpha_k), s^k(\alpha_k))$;
end (for).

Typical behavior of the algorithm is illustrated in Figure 14.2 for the case of $n = 2$. The horizontal and vertical axes in this figure represent the pairwise products $x_1 s_1$ and $x_2 s_2$, so the central path $\mathcal{C}$ is the line emanating from the origin at an angle of 45°. (A point at the origin of this illustration is a primal-dual solution if it also satisfies the feasibility conditions

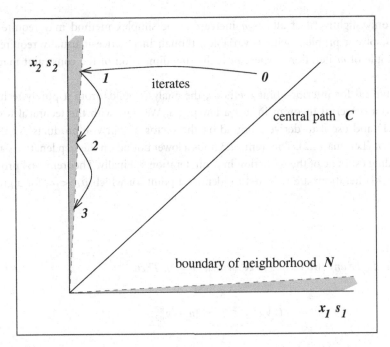

Figure 14.2   Iterates of Algorithm 14.2, plotted in (xs) space.

(14.3a), (14.3b), and (14.3d).) In the unusual geometry of Figure 14.2, the search directions $(\Delta x^k, \Delta \lambda^k, \Delta s^k)$ transform to curves rather than straight lines.

As Figure 14.2 shows (and the analysis confirms), the lower bound σ_{min} on the centering parameter ensures that each search direction starts out by moving away from the boundary of $\mathcal{N}_{-\infty}(\gamma)$ and into the relative interior of this neighborhood. That is, small steps along the search direction improve the centrality. Larger values of α take us outside the neighborhood again, since the error in approximating the nonlinear system (14.15) by the linear step equations (14.16) becomes more pronounced as α increases. Still, we are guaranteed that a certain minimum step can be taken before we reach the boundary of $\mathcal{N}_{-\infty}(\gamma)$, as we show in the analysis below.

The analysis of Algorithm 14.2 appears in the next few pages. With judicious choices of σ_k, this algorithm is fairly efficient in practice. With a few more modifications, it becomes the basis of a truly competitive method, as we discuss in Section 14.2.

Our aim in the analysis below is to show that given some small tolerance $\epsilon > 0$, the algorithm requires $O(n|\log \epsilon|)$ iterations to reduce the duality measure by a factor of ϵ, that is, to identify a point (x^k, λ^k, s^k) for which $\mu_k \le \epsilon \mu_0$. For small ϵ, the point (x^k, λ^k, s^k) satisfies the primal-dual optimality conditions except for perturbations of about ϵ in the right-hand side of (14.3c), so it is usually very close to a primal-dual solution of the original linear program. The $O(n|\log \epsilon|)$ estimate is a worst-case bound on the number of iterations required; on practical problems, the number of iterations required appears to increase only slightly (if at all) as n increases. The simplex method may require 2^n iterations to solve a problem with n variables, though in practice it usually requires a modest multiple of m iterations, where m is the row dimension of the constraint matrix A in (14.1).

As is typical for interior-point methods, the analysis builds from a purely techni-cal lemma to a powerful theorem in just a few pages. We start with the technical result (Lemma 14.1) and use it to derive a bound on the vector of pairwise products $\Delta x_i \Delta s_i$, $i = 1, 2, \ldots, n$ (Lemma 14.2). Theorem 14.3 finds a lower bound on the step length α_k and a corresponding estimate of the reduction in μ on iteration k. Finally, Theorem 14.4 proves that $O(n|\log \epsilon|)$ iterations are required to identify a point for which $\mu_k < \epsilon$, for a given tolerance $\epsilon \in (0, 1)$.

Lemma 14.1.

Let u and v be any two vectors in $\mathbb{R}^n$ with $u^T v \geq 0$. Then

$$\|UVe\|_2 \leq 2^{-3/2} \|u + v\|_2^2,$$

where

$$U = \text{diag}(u_1, u_2, \ldots, u_n), \qquad V = \text{diag}(v_1, v_2, \ldots, v_n).$$

PROOF.   (When the subscript is omitted from $\| \cdot \|$, we mean $\| \cdot \|_2$, as is our convention throughout the book.) First, note that for any two scalars α and β with $\alpha\beta \geq 0$, we have from the algebraic-geometric mean inequality that

$$\sqrt{|\alpha\beta|} \leq \frac{1}{2}|\alpha + \beta|. \tag{14.21}$$

Since $u^T v \geq 0$, we have

$$0 \leq u^T v = \sum_{u_i v_i \geq 0} u_i v_i + \sum_{u_i v_i < 0} u_i v_i = \sum_{i \in \mathcal{P}} |u_i v_i| - \sum_{i \in \mathcal{M}} |u_i v_i|, \tag{14.22}$$

where we partitioned the index set $\{1, 2, \ldots, n\}$ as

$$\mathcal{P} = \{i \mid u_i v_i \geq 0\}, \qquad \mathcal{M} = \{i \mid u_i v_i < 0\}.$$

Now,

$$
\begin{aligned}
\|UVe\| &= \left(\|[u_i v_i]_{i\in P}\|^2 + \|[u_i v_i]_{i\in M}\|^2\right)^{1/2} \\
&\leq \left(\|[u_i v_i]_{i\in P}\|_1^2 + \|[u_i v_i]_{i\in M}\|_1^2\right)^{1/2} && \text{since } \|\cdot\|_2 \leq \|\cdot\|_1 \\
&\leq \left(2\,\|[u_i v_i]_{i\in P}\|_1^2\right)^{1/2} && \text{from (14.22)} \\
&\leq \sqrt{2}\,\left\|\left[\frac{1}{4}(u_i + v_i)^2\right]_{i\in P}\right\|_1 && \text{from (14.21)} \\
&= 2^{-3/2}\sum_{i\in P}(u_i + v_i)^2 \\
&\leq 2^{-3/2}\sum_{i=1}^{n}(u_i + v_i)^2 \\
&\leq 2^{-3/2}\|u + v\|^2,
\end{aligned}
$$

completing the proof. $\square$

For the next result, we omit the iteration counter k from (14.10), and define the diagonal matrices ΔX and ΔS similarly to (14.5), as follows:

$$
\Delta X = \mathrm{diag}(\Delta x_1, \Delta x_2, \ldots, \Delta x_n), \qquad \Delta S = \mathrm{diag}(\Delta s_1, \Delta s_2, \ldots, \Delta s_n)..
$$

Lemma 14.2.
   If $(x, \lambda, s) \in \mathcal{N}_{-\infty}(\gamma)$, then

$$
\|\Delta X \Delta Se\| \leq 2^{-3/2}(1 + 1/\gamma)n\mu.
$$

PROOF.   It is easy to show using (14.10) that

$$
\Delta x^T \Delta s = 0. \tag{14.23}
$$

By multiplying the last block row in (14.10) by $(XS)^{-1/2}$ and using the definition $D = X^{1/2}S^{-1/2}$, we obtain

$$
D^{-1}\Delta x + D\Delta s = (XS)^{-1/2}(-XSe + \sigma\mu e). \tag{14.24}
$$

Because $(D^{-1}\Delta x)^T(D\Delta s) = \Delta x^T \Delta s = 0$, we can apply Lemma 14.1 with $u = D^{-1}\Delta x$ and $v = D\Delta s$ to obtain

$$
\begin{aligned}
\|\Delta X \Delta Se\| &= \|(D^{-1}\Delta X)(D\Delta S)e\| \\
&\leq 2^{-3/2}\|D^{-1}\Delta x + D\Delta s\|^2 && \text{from Lemma 14.1} \\
&= 2^{-3/2}\|(XS)^{-1/2}(-XSe + \sigma\mu e)\|^2 && \text{from (14.24).}
\end{aligned}
$$

Expanding the squared Euclidean norm and using such relationships as $x^T s = n\mu$ and

$e^T e = n$, we obtain

$$
\begin{aligned}
\|\Delta X \Delta S e\| &\leq 2^{-3/2} \left[x^T s - 2\sigma \mu e^T e + \sigma^2 \mu^2 \sum_{i=1}^{n} \frac{1}{x_i s_i} \right] \\
&\leq 2^{-3/2} \left[x^T s - 2\sigma \mu e^T e + \sigma^2 \mu^2 \frac{n}{\gamma \mu} \right] \qquad \text{since } x_i s_i \geq \gamma \mu \\
&\leq 2^{-3/2} \left[1 - 2\sigma + \frac{\sigma^2}{\gamma} \right] n\mu \\
&\leq 2^{-3/2}(1 + 1/\gamma)n\mu,
\end{aligned}
$$

as claimed. $\qquad\qquad\qquad\qquad\qquad\qquad\qquad\qquad\qquad\qquad\qquad\qquad\qquad\qquad\qquad$ □

Theorem 14.3.

Given the parameters γ, $\sigma_{\min}$, and $\sigma_{\max}$ in Algorithm 14.2, there is a constant δ independent of n such that

$$
\mu_{k+1} \leq \left(1 - \frac{\delta}{n}\right) \mu_k, \tag{14.25}
$$

for all $k \geq 0$.

PROOF. We start by proving that

$$
\left(x^k(\alpha), \lambda^k(\alpha), s^k(\alpha)\right) \in \mathcal{N}_{-\infty}(\gamma) \text{ for all } \alpha \in \left[0, 2^{3/2}\gamma \frac{1-\gamma}{1+\gamma} \frac{\sigma_k}{n}\right], \tag{14.26}
$$

where $\left(x^k(\alpha), \lambda^k(\alpha), s^k(\alpha)\right)$ is defined as in (14.19). It follows that the step length α_k is at least as long as the upper bound of this interval, that is,

$$
\alpha_k \geq 2^{3/2} \frac{\sigma_k}{n} \gamma \frac{1-\gamma}{1+\gamma}. \tag{14.27}
$$

For any $i = 1, 2, \ldots, n$, we have from Lemma 14.2 that

$$
|\Delta x_i^k \Delta s_i^k| \leq \|\Delta X^k \Delta S^k e\|_2 \leq 2^{-3/2}(1 + 1/\gamma)n\mu_k. \tag{14.28}
$$

Using (14.10), we have from $x_i^k s_i^k \geq \gamma \mu_k$ and (14.28) that

$$
\begin{aligned}
x_i^k(\alpha) s_i^k(\alpha) &= \left(x_i^k + \alpha \Delta x_i^k\right)\left(s_i^k + \alpha \Delta s_i^k\right) \\
&= x_i^k s_i^k + \alpha \left(x_i^k \Delta s_i^k + s_i^k \Delta x_i^k\right) + \alpha^2 \Delta x_i^k \Delta s_i^k \\
&\geq x_i^k s_i^k(1 - \alpha) + \alpha \sigma_k \mu_k - \alpha^2 |\Delta x_i^k \Delta s_i^k| \\
&\geq \gamma(1 - \alpha)\mu_k + \alpha \sigma_k \mu_k - \alpha^2 2^{-3/2}(1 + 1/\gamma)n\mu_k.
\end{aligned}
$$

By summing the n components of the equation $S^k \Delta x^k + X^k \Delta s^k = -X^k S^k e + \sigma_k \mu_k e$ (the third block row from (14.10)), and using (14.23) and the definition of μ_k and $\mu_k(\alpha)$ (see (14.19)), we obtain

$$\mu_k(\alpha) = (1 - \alpha(1 - \sigma_k))\mu_k.$$

From these last two formulas, we can see that the proximity condition

$$x_i^k(\alpha)s_i^k(\alpha) \geq \gamma \mu_k(\alpha)$$

is satisfied, provided that

$$\gamma(1 - \alpha)\mu_k + \alpha\sigma_k\mu_k - \alpha^2 2^{-3/2}(1 + 1/\gamma)n\mu_k \geq \gamma(1 - \alpha + \alpha\sigma_k)\mu_k.$$

Rearranging this expression, we obtain

$$\alpha\sigma_k\mu_k(1 - \gamma) \geq \alpha^2 2^{-3/2}n\mu_k(1 + 1/\gamma),$$

which is true if

$$\alpha \leq \frac{2^{3/2}}{n}\sigma_k\gamma\frac{1 - \gamma}{1 + \gamma}.$$

We have proved that $(x^k(\alpha), \lambda^k(\alpha), s^k(\alpha))$ satisfies the proximity condition for $\mathcal{N}_{-\infty}(\gamma)$ when α lies in the range stated in (14.26). It is not difficult to show that $(x^k(\alpha), \lambda^k(\alpha), s^k(\alpha)) \in \mathcal{F}^o$ for all α in the given range. Hence, we have proved (14.26) and therefore (14.27).

We complete the proof of the theorem by estimating the reduction in μ on the kth step. Because of (14.23), (14.27), and the last block row of (14.16), we have

$$\begin{aligned}
\mu_{k+1} &= x^k(\alpha_k)^T s^k(\alpha_k)/n \\
&= \left[(x^k)^T s^k + \alpha_k\left((x^k)^T \Delta s^k + (s^k)^T \Delta x^k\right) + \alpha_k^2 (\Delta x^k)^T \Delta s^k\right]/n \\
&= \mu_k + \alpha_k\left(-(x^k)^T s^k/n + \sigma_k\mu_k\right) \\
&= (1 - \alpha_k(1 - \sigma_k))\mu_k \\
&\leq \left(1 - \frac{2^{3/2}}{n}\gamma\frac{1 - \gamma}{1 + \gamma}\sigma_k(1 - \sigma_k)\right)\mu_k.
\end{aligned} \tag{14.29}$$

Now, the function $\sigma(1 - \sigma)$ is a concave quadratic function of σ, so on any given interval it attains its minimum value at one of the endpoints. Hence, we have

$$\sigma_k(1 - \sigma_k) \geq \min\{\sigma_{\min}(1 - \sigma_{\min}), \sigma_{\max}(1 - \sigma_{\max})\}, \quad \text{for all } \sigma_k \in [\sigma_{\min}, \sigma_{\max}].$$

The proof is completed by substituting this estimate into (14.29) and setting

$$\delta = 2^{3/2} \gamma \frac{1 - \gamma}{1 + \gamma} \min \{\sigma_{\min}(1 - \sigma_{\min}), \sigma_{\max}(1 - \sigma_{\max})\}. \qquad \square$$

We conclude with a result that a reduction of a factor of ϵ in the duality measure μ can be obtained in $O(n \log 1/\epsilon)$ iterations.

Theorem 14.4.

Given $\epsilon \in (0, 1)$ and $\gamma \in (0, 1)$, suppose the starting point in Algorithm 14.2 satisfies $(x^0, \lambda^0, s^0) \in \mathcal{N}_{-\infty}(\gamma)$. Then there is an index K with $K = O(n \log 1/\epsilon)$ such that

$$\mu_k \leq \epsilon \mu_0, \qquad \text{for all } k \geq K.$$

PROOF. By taking logarithms of both sides in (14.25), we obtain

$$\log \mu_{k+1} \leq \log \left(1 - \frac{\delta}{n}\right) + \log \mu_k.$$

By applying this formula repeatedly, we have

$$\log \mu_k \leq k \log \left(1 - \frac{\delta}{n}\right) + \log \mu_0.$$

The following well-known estimate for the log function,

$$\log(1 + \beta) \leq \beta, \qquad \text{for all } \beta > -1,$$

implies that

$$\log(\mu_k/\mu_0) \leq k \left(-\frac{\delta}{n}\right).$$

Therefore, the condition $\mu_k/\mu_0 \leq \epsilon$ is satisfied if we have

$$k \left(-\frac{\delta}{n}\right) \leq \log \epsilon.$$

This inequality holds for all k that satisfy

$$k \geq K \stackrel{\text{def}}{=} \frac{n}{\delta} \log \frac{1}{\epsilon} = \frac{n}{\delta} |\log \epsilon|,$$

$\square$

so the proof is complete.

14.2 PRACTICAL PRIMAL-DUAL ALGORITHMS

Practical implementations of interior-point algorithms follow the spirit of the previous section, in that strict positivity of x^k and s^k is maintained throughout and each step is a Newton-like step involving a centering component. However, most implementations work with an infeasible starting point and infeasible iterations. Several aspects of "theoretical" algorithms are typically ignored, while several enhancements are added that have a significant effect on practical performance. In this section, we describe the algorithmic enhancements that are found in a typical implementation of an infeasible-interior-point method, and present the resulting method as Algorithm 14.3. Many of the techniques of this section are described in the paper of Mehrotra [207], which can be consulted for further details.

CORRECTOR AND CENTERING STEPS

A key feature of practical algorithms is their use of corrector steps that compensate for the linearization error made by the Newton (affine-scaling) step in modeling the equation $x_i s_i = 0, i = 1, 2, \ldots, n$ (see (14.3c)). Consider the affine-scaling direction $(\Delta x, \Delta \lambda, \Delta s)$ defined by

$$
\begin{bmatrix} 0 & A^T & I \\ A & 0 & 0 \\ S & 0 & X \end{bmatrix} \begin{bmatrix} \Delta x^{\text{aff}} \\ \Delta \lambda^{\text{aff}} \\ \Delta s^{\text{aff}} \end{bmatrix} = \begin{bmatrix} -r_c \\ -r_b \\ -XSe \end{bmatrix}, \tag{14.30}
$$

(where r_b and r_c are defined in (14.7)). If we take a full step in this direction, we obtain

$$
(x_i + \Delta x_i^{\text{aff}})(s_i + \Delta s_i^{\text{aff}})
$$
$$
= x_i s_i + x_i \Delta s_i^{\text{aff}} + s_i \Delta x_i^{\text{aff}} + \Delta x_i^{\text{aff}} \Delta s_i^{\text{aff}} = \Delta x_i^{\text{aff}} \Delta s_i^{\text{aff}}.
$$

That is, the updated value of $x_i s_i$ is $\Delta x_i^{\text{aff}} \Delta s_i^{\text{aff}}$ rather than the ideal value 0. We can solve the following system to obtain a step $(\Delta x^{\text{cor}}, \Delta \lambda^{\text{cor}}, \Delta s^{\text{cor}})$ that attempts to correct for this deviation from the ideal:

$$
\begin{bmatrix} 0 & A^T & I \\ A & 0 & 0 \\ S & 0 & X \end{bmatrix} \begin{bmatrix} \Delta x^{\text{cor}} \\ \Delta \lambda^{\text{cor}} \\ \Delta s^{\text{cor}} \end{bmatrix} = \begin{bmatrix} 0 \\ 0 \\ -\Delta X^{\text{aff}} \Delta S^{\text{aff}} e \end{bmatrix}. \tag{14.31}
$$

In many cases, the combined step $(\Delta x^{\text{aff}}, \Delta \lambda^{\text{aff}}, \Delta s^{\text{aff}}) + (\Delta x^{\text{cor}}, \Delta \lambda^{\text{cor}}, \Delta s^{\text{cor}})$ does a better job of reducing the duality measure than does the affine-scaling step alone.

Like theoretical algorithms such as the one analysed in Section 14.1, practical algorithms make use of centering steps, with an adaptive choice of the centering parameter σ_k. The affine-scaling step can be used as the basis of a successful heuristic for choosing σ_k.

Roughly speaking, if the affine-scaling step (multiplied by a steplength to maintain non-negativity of x and s) reduces the duality measure significantly, there is not much need for centering, so a smaller value of σ_k is appropriate. Conversely, if not much progress can be made along this direction before reaching the boundary of the nonnegative orthant, a larger value of σ_k will ensure that the next iterate is more centered, so a longer step will be possible from this next point. Specifically, this scheme calculates the maximum allowable steplengths along the affine-scaling direction (14.30) as follows:

$$\alpha_{\text{aff}}^{\text{pri}} \stackrel{\text{def}}{=} \min\left(1, \min_{i:\Delta x_i^{\text{aff}}<0} -\frac{x_i}{\Delta x_i^{\text{aff}}}\right), \tag{14.32a}$$

$$\alpha_{\text{aff}}^{\text{dual}} \stackrel{\text{def}}{=} \min\left(1, \min_{i:\Delta s_i^{\text{aff}}<0} -\frac{s_i}{\Delta s_i^{\text{aff}}}\right), \tag{14.32b}$$

and then defines μ_{aff} to be the value of μ that would be obtained by using these steplengths, that is,

$$\mu_{\text{aff}} = (x + \alpha_{\text{aff}}^{\text{pri}}\Delta x^{\text{aff}})^T (s + \alpha_{\text{aff}}^{\text{dual}}\Delta s^{\text{aff}})/n. \tag{14.33}$$

The centering parameter σ is chosen according to the following heuristic (which does not have a solid analytical justification, but appears to work well in practice):

$$\sigma = \left(\frac{\mu_{\text{aff}}}{\mu}\right)^3. \tag{14.34}$$

To summarize, computation of the search direction requires the solution of two linear systems. First, the system (14.30) is solved to obtain the affine-scaling direction, also known as the *predictor step*. This step is used to define the right-hand side for the corrector step (see (14.31)) and to calculate the centering parameter from (14.33), (14.34). Second, the search direction is calculated by solving

$$\begin{bmatrix} 0 & A^T & I \\ A & 0 & 0 \\ S & 0 & X \end{bmatrix} \begin{bmatrix} \Delta x \\ \Delta \lambda \\ \Delta s \end{bmatrix} = \begin{bmatrix} -r_c \\ -r_b \\ -XSe - \Delta X^{\text{aff}}\Delta S^{\text{aff}}e + \sigma\mu e \end{bmatrix}. \tag{14.35}$$

Note that the predictor, corrector, and centering contributions have been aggregated on the right-hand side of this system. The coefficient matrix in both linear systems (14.30) and (14.35) is the same. Thus, the factorization of the matrix needs to be computed only once, and the marginal cost of solving the second system is relatively small.

STEP LENGTHS

Practical implementations typically do not enforce membership of the central path neighborhoods $\mathcal{N}_2$ and $\mathcal{N}_{-\infty}$ defined in the previous section. Rather, they calculate the maximum steplengths that can be taken in the x and s variables (separately) without violating nonnegativity, then take a steplength of slightly less than this maximum (but no greater than 1). Given an iterate (x^k, λ^k, s^k) with $(x^k, s^k) > 0$, and a step $(\Delta x^k, \Delta \lambda^k, \Delta s^k)$, it is easy to show that the quantities $\alpha_{k,\max}^{\text{pri}}$ and $\alpha_{k,\max}^{\text{dual}}$ defined as follows:

$$\alpha_{k,\max}^{\text{pri}} \overset{\text{def}}{=} \min_{i:\Delta x_i^k < 0} -\frac{x_i^k}{\Delta x_i^k}, \qquad \alpha_{k,\max}^{\text{dual}} \overset{\text{def}}{=} \min_{i:\Delta s_i^k < 0} -\frac{s_i^k}{\Delta s_i^k}, \qquad (14.36)$$

are the largest values of α for which $x^k + \alpha \Delta x^k \geq 0$ and $s^k + \alpha \Delta s^k \geq 0$, respectively. (Note that these formulae are similar to the ratio test used in the simplex method to determine the index that enters the basis.) Practical algorithms then choose the steplengths to lie in the *open* intervals defined by these maxima, that is,

$$\alpha_k^{\text{pri}} \in (0, \alpha_{k,\max}^{\text{pri}}), \qquad \alpha_k^{\text{dual}} \in (0, \alpha_{k,\max}^{\text{dual}}),$$

and then obtain a new iterate by setting

$$x^{k+1} = x^k + \alpha_k^{\text{pri}} \Delta x^k, \qquad (\lambda^{k+1}, s^{k+1}) = (\lambda^k, s^k) + \alpha_k^{\text{dual}} (\Delta \lambda^k, \Delta s^k).$$

If the step $(\Delta x^k, \Delta \lambda^k, \Delta s^k)$ rectifies the infeasibility in the KKT conditions (14.3a) and (14.3b), that is,

$$A\Delta x^k = -r_b^k = -(Ax^k - b), \qquad A^T \Delta \lambda^k + \Delta s^k = -r_c^k = -(A^T \lambda^k + s^k - c),$$

it is easy to show that the infeasibilities at the new iterate satisfy

$$r_b^{k+1} = \left(1 - \alpha_k^{\text{pri}}\right) r_b^k, \qquad r_c^{k+1} = \left(1 - \alpha_k^{\text{dual}}\right) r_c^k. \qquad (14.37)$$

The following formula is used to calculate steplengths in many practical implementations

$$\alpha_k^{\text{pri}} = \min(1, \eta_k \alpha_{k,\max}^{\text{pri}}), \qquad \alpha_k^{\text{dual}} = \min(1, \eta_k \alpha_{k,\max}^{\text{dual}}), \qquad (14.38)$$

where $\eta_k \in [0.9, 1.0)$ is chosen so that $\eta_k \to 1$ as the iterates approach the primal-dual solution, to accelerate the asymptotic convergence.

STARTING POINT

Choice of starting point is an important practical issue with a significant effect on the robustness of the algorithm. A poor choice (x^0, λ^0, s^0) satisfying only the minimal conditions $x^0 > 0$ and $s^0 > 0$ often leads to failure of convergence. We describe here a heuristic that finds a starting point that satisfies the equality constraints in the primal and dual problems reasonably well, while maintaining positivity of the x and s components and avoiding excessively large values of these components.

First, we find a vector $\tilde{x}$ of minimum norm satisfying the primal constraint $Ax = b$, and a vector $(\tilde{\lambda}, \tilde{s})$ satisfy the dual constraint $A^T\lambda + s = c$ such that $\tilde{s}$ has minimum norm. That is, we solve the problems

$$\min_{x} \tfrac{1}{2}x^Tx \quad \text{subject to } Ax = b, \tag{14.39a}$$

$$\min_{(\lambda,s)} \tfrac{1}{2}s^Ts \quad \text{subject to } A^T\lambda + s = c. \tag{14.39b}$$

It is not difficult to show that $\tilde{x}$ and $(\tilde{\lambda}, \tilde{s})$ can be written explicitly as follows:

$$\tilde{x} = A^T(AA^T)^{-1}b, \quad \tilde{\lambda} = (AA^T)^{-1}Ac, \quad \tilde{s} = c - A^T\tilde{\lambda}. \tag{14.40}$$

In general, $\tilde{x}$ and $\tilde{s}$ will have nonpositive components, so are not suitable for use as a starting point. We define

$$\delta_x = \max(-(3/2)\min_i \tilde{x}_i, 0), \quad \delta_s = \max(-(3/2)\min_i \tilde{s}_i, 0),$$

and adjust the x and s vectors as follows:

$$\hat{x} = \tilde{x} + \delta_x e, \quad \hat{s} = \tilde{s} + \delta_s e,$$

where, as usual, $e = (1, 1, \ldots, 1)^T$. Clearly, we have $\hat{x} \geq 0$ and $\hat{s} \geq 0$. To ensure that the components of x^0 and s^0 are not too close to zero and not too dissimilar, we add two more scalars defined as follows:

$$\hat{\delta}_x = \frac{1}{2}\frac{\hat{x}^T\hat{s}}{e^T\hat{s}}, \quad \hat{\delta}_s = \frac{1}{2}\frac{\hat{x}^T\hat{s}}{e^T\hat{x}}$$

(Note that $\hat{\delta}_x$ is the average size of the components of $\hat{x}$, weighted by the corresponding components of $\hat{s}$; similarly for $\hat{\delta}_s$.) Finally, we define the starting point as follows:

$$x^0 = \hat{x} + \hat{\delta}_x e, \quad \lambda^0 = \tilde{\lambda}, \quad s^0 = \hat{s} + \hat{\delta}_s e.$$

The computational cost of finding (x^0, λ^0, s^0) by this scheme is about the same as one step of the primal-dual method.

In some cases, we have prior knowledge about the solution, possibly in the form of a solution to a similar linear program. The use of such "warm-start" information in constructing a starting point is discussed in Section 14.4.

A PRACTICAL ALGORITHM

We now give a formal specification of a practical algorithm.

Algorithm 14.3 (Predictor-Corrector Algorithm (Mehrotra [207])).
Calculate (x^0, λ^0, s^0) as described above;

for $k = 0, 1, 2, \ldots$
    Set $(x, \lambda, s) = (x^k, \lambda^k, s^k)$ and solve (14.30) for $(\Delta x^{\text{aff}}, \Delta \lambda^{\text{aff}}, \Delta s^{\text{aff}})$;
    Calculate $\alpha_{\text{aff}}^{\text{pri}}, \alpha_{\text{aff}}^{\text{dual}}$, and μ_{aff} as in (14.32) and (14.33);
    Set centering parameter to $\sigma = (\mu_{\text{aff}}/\mu)^3$;
    Solve (14.35) for $(\Delta x, \Delta \lambda, \Delta s)$;
    Calculate α_k^{pri} and α_k^{dual} from (14.38);
    Set

$$x^{k+1} = x^k + \alpha_k^{\text{pri}} \Delta x,$$
$$(\lambda^{k+1}, s^{k+1}) = (\lambda^k, s^k) + \alpha_k^{\text{dual}} (\Delta \lambda, \Delta s);$$

end (for).

No convergence theory is available for Mehrotra's algorithm, at least in the form in which it is described above. In fact, there are examples for which the algorithm diverges. Simple safeguards could be incorporated into the method to force it into the convergence framework of existing methods or to improve its robustness, but many practical codes do not implement these safeguards, because failures are rare.

When presented with a linear program that is infeasible or unbounded, the algorithm above typically diverges, with the infeasibilities r_b^k and r_c^k and/or the duality measure μ_k going to ∞. Since the symptoms of infeasibility and unboundedness are fairly easy to recognize, interior-point codes contain heuristics to detect and report these conditions. More rigorous approaches for detecting infeasibility and unboundedness make use of the homogeneous self-dual formulation; see Wright [316, Chapter 9] and the references therein for a discussion. A more recent approach that applies directly to infeasible-interior-point methods is described by Todd [286].

SOLVING THE LINEAR SYSTEMS

Most of the computational effort in primal-dual methods is taken up in solving linear systems such as (14.9), (14.30), and (14.35). The coefficient matrix in these systems is usually large and sparse, since the constraint matrix A is itself large and sparse in most applications.

The special structure in the step equations allows us to reformulate them as systems with more compact symmetric coefficient matrices, which are easier and cheaper to factor than the original sparse form.

We apply the reformulation procedures to the following general form of the linear system:

$$
\begin{bmatrix} 0 & A^T & I \\ A & 0 & 0 \\ S & 0 & X \end{bmatrix} \begin{bmatrix} \Delta x \\ \Delta \lambda \\ \Delta s \end{bmatrix} = \begin{bmatrix} -r_c \\ -r_b \\ -r_{xs} \end{bmatrix}.
\tag{14.41}
$$

Since x and s are strictly positive, the diagonal matrices X and S are nonsingular. We can eliminate Δs and add $-X^{-1}$ times the third equation in this system to the first equation to obtain

$$
\begin{bmatrix} -D^{-2} & A^T \\ A & 0 \end{bmatrix} \begin{bmatrix} \Delta x \\ \Delta \lambda \end{bmatrix} = \begin{bmatrix} -r_c + X^{-1}r_{xs} \\ -r_b \end{bmatrix},
\tag{14.42a}
$$

$$
\Delta s = -X^{-1}r_{xs} - X^{-1}S\Delta x,
\tag{14.42b}
$$

where we have introduced the notation

$$
D = S^{-1/2} X^{1/2}.
\tag{14.43}
$$

This form of the step equations usually is known as the *augmented system*. We can go further and eliminate Δx and add AD^2 times the first equation to the second equation in (14.42a) to obtain

$$
AD^2 A^T \Delta \lambda = -r_b - AXS^{-1}r_c + AS^{-1}r_{xs}
\tag{14.44a}
$$

$$
\Delta s = -r_c - A^T \Delta \lambda,
\tag{14.44b}
$$

$$
\Delta x = -S^{-1}r_{xs} - XS^{-1}\Delta s,
\tag{14.44c}
$$

where the expressions for Δs and Δs are obtained from the original system (14.41). The form (14.44a) often is called the *normal-equations* form, because the system (14.44a) can be viewed as the normal equations (10.14) for a certain linear least-squares problem with coefficient matrix DA^T.

Most implementations of primal-dual methods are based on formulations like (14.44). They use direct sparse Cholesky algorithms to factor the matrix $AD^2 A^T$, and then perform triangular solves with the resulting sparse factors to obtain the step $\Delta \lambda$ from (14.44a). The steps Δs and Δx are recovered from (14.44b) and (14.44c). General-purpose sparse Cholesky software can be applied to $AD^2 A^T$, but modifications are needed because $AD^2 A^T$ may be ill-conditioned or singular. (Ill conditioning of this system is often observed during the final stages of a primal-dual algorithm, when the elements of the diagonal weighting matrix D^2 take on both huge and tiny values.) The Cholesky technique may encounter diagonal elements that are very small, zero or (because of roundoff error) slightly negative. One approach for handling this eventuality is to skip a step of the factorization, setting the component of $\Delta\lambda$ that corresponds to the faulty diagonal element to zero. We refer to Wright [317] for details of this and other approaches.

A disadvantage of the normal-equations formulation is that if A contains any dense columns, the entire matrix AD^2A^T is also dense. Hence, practical software identifies dense and nearly-dense columns, excludes them from the matrix product AD^2A^T, and performs the Cholesky factorization of the resulting sparse matrix. Then, a device such as a Sherman-Morrison-Woodbury update is applied to account for the excluded columns. We refer the reader to Wright [316, Chapter 11] for further details.

The formulation (14.42) has received less attention than (14.44), mainly because algorithms and software for factoring sparse symmetric indefinite matrices are more complicated, slower, and less prevalent than sparse Cholesky algorithms. Nevertheless, the formulation (14.42) is cleaner and more flexible than (14.44) in a number of respects. It normally avoids the fill-in caused by dense columns in A in the matrix product AD^2A^T. Moreover, it allows free variables (components of x with no explicit lower or upper bounds) to be handled directly in the formulation. (The normal equations form must resort to various artificial devices to express such variables, otherwise it is not possible to perform the block elimination that leads to the system (14.44a).)

14.3   OTHER PRIMAL-DUAL ALGORITHMS AND EXTENSIONS

OTHER PATH-FOLLOWING METHODS

Framework 14.1 is the basis of a number of other algorithms of the path-following variety. They are less important from a practical viewpoint, but we mention them here because of their elegance and their strong theoretical properties.

Some path-following methods choose conservative values for the centering parameter σ (that is, σ only slightly less than 1) so that unit steps (that is, a steplength of $\alpha = 1$) can be taken along the resulting direction from (14.16) without leaving the chosen neighborhood. These methods, which are known as *short-step* path-following methods, make only slow progress toward the solution because they require the iterates to stay inside a restrictive $\mathcal{N}_2$ neighborhood (14.17). From a theoretical point of view, however, they have the advantage of better complexity. (A result similar to Theorem 14.4 holds with n replaced by $n^{1/2}$ in the complexity estimate.)

Better results are obtained with the *predictor-corrector* method, due to Mizuno, Todd, and Ye [208], which uses two $\mathcal{N}_2$ neighborhoods, nested one inside the other. (Despite the similar terminology, this algorithm is quite distinct from Algorithm 14.3 of Section 14.2.) Every second step of this method is a predictor step, which starts in the inner neighborhood and moves along the affine-scaling direction (computed by setting $\sigma = 0$ in (14.16)) to the boundary of the outer neighborhood. The gap between neighborhood boundaries is wide enough to allow this step to make significant progress in reducing μ. Alternating with the predictor steps are corrector steps (computed with $\sigma = 1$ and $\alpha = 1$), which take the next iterate back inside the inner neighborhood in preparation for the next predictor step. The predictor-corrector algorithm produces a sequence of duality measures μ_k that converge superlinearly to zero, in contrast to the linear convergence that characterizes most methods.

POTENTIAL-REDUCTION METHODS

Potential-reduction methods take steps of the same form as path-following methods, but they do not explicitly follow the central path C and can be motivated independently of it. They use a logarithmic potential function to measure the worth of each point in $\mathcal{F}^o$ and aim to achieve a certain fixed reduction in this function at each iteration. The primal-dual potential function, which we denote generically by Φ, usually has two important properties:

$$\Phi \to \infty \text{ if } x_i s_i \to 0 \text{ for some } i, \text{ while } \mu = x^T s/n \not\to 0, \tag{14.45a}$$

$$\Phi \to -\infty \text{ if and only if } (x, \lambda, s) \to \Omega. \tag{14.45b}$$

The first property (14.45a) prevents any one of the pairwise products $x_i s_i$ from approaching zero independently of the others, and therefore keeps the iterates away from the boundary of the nonnegative orthant. The second property (14.45b) relates Φ to the solution set Ω. If our algorithm forces Φ to $-\infty$, then (14.45b) ensures that the sequence approaches the solution set.

An interesting primal-dual potential function is defined by

$$\Phi_\rho(x, s) = \rho \log x^T s - \sum_{i=1}^{n} \log x_i s_i, \tag{14.46}$$

for some parameter $\rho > n$ (see Tanabe [283] and Todd and Ye [287]). Like all algorithms based on Framework 14.1, potential-reduction algorithms obtain their search directions by solving (14.10), for some $\sigma_k \in (0, 1)$, and they take steps of length α_k along these directions. For instance, the step length α_k may be chosen to approximately minimize Φ_ρ along the computed direction. By fixing $\sigma_k = n/(n + \sqrt{n})$ for all k, one can guarantee constant reduction in Φ_ρ at every iteration. Hence, Φ_ρ will approach $-\infty$, forcing convergence. Adaptive and heuristic choices of σ_k and α_k are also covered by the theory, provided that they at least match the reduction in Φ_ρ obtained from the conservative theoretical values of these parameters.

EXTENSIONS

Primal-dual methods for linear programming can be extended to wider classes of problems. There are simple extensions of the algorithm to the monotone linear complementarity problem (LCP) and convex quadratic programming problems for which the convergence and polynomial complexity properties of the linear programming algorithms are retained. The monotone LCP is the problem of finding vectors x and s in $\mathbb{R}^n$ that satisfy the following conditions:

$$ s = Mx + q, \qquad (x, s) \geq 0, \qquad x^T s = 0, \tag{14.47} $$

where M is a positive semidefinite $n \times n$ matrix and $q \in \mathbb{R}^n$. The similarity between (14.47) and the KKT conditions (14.3) is obvious: The last two conditions in (14.47) correspond to (14.3d) and (14.3c), respectively, while the condition $s = Mx + q$ is similar to the equations (14.3a) and (14.3b). For practical instances of the problem (14.47), see Cottle, Pang, and Stone [80]. Interior-point methods for monotone LCP have a close correspondence to algorithms for linear programming. The duality measure (14.6) is redefined to be the *complementarity measure* (with the same definition $\mu = x^T s/n$), and the conditions that must be satisfied by the solution can be stated similarly to (14.4) as follows:

$$ \begin{bmatrix} Mx + q - s \\ X S e \end{bmatrix} = 0, \quad (x, s) \geq 0. $$

The general formula for a path-following step is defined analogously to (14.9) as follows:

$$ \begin{bmatrix} M & -I \\ S & X \end{bmatrix} \begin{bmatrix} \Delta x \\ \Delta s \end{bmatrix} = \begin{bmatrix} -(Mx + q - s) \\ -X S e + \sigma \mu e \end{bmatrix}, $$

where $\sigma \in [0, 1]$. Using these and similar adaptations, an extension of the practical method of Section 14.2 can also be derived.

Extensions to convex quadratic programs are discussed in Section 16.6. Their adaptation to nonlinear programming problems is the subject of Chapter 19.

Interior-point methods are highly effective in solving *semidefinite programming* problems, a class of problems involving symmetric matrix variables that are constrained to be positive semidefinite. Semidefinite programming, which has been the topic of concentrated research since the early 1990s, has applications in many areas, including control theory and combinatorial optimization. Further information on this increasingly important topic can be found in the survey papers of Todd [285] and Vandenberghe and Boyd [292] and the books of Nesterov and Nemirovskii [226], Boyd et al. [37], and Boyd and Vandenberghe [38].

14.4 PERSPECTIVES AND SOFTWARE

The appearance of interior-point methods in the 1980s presented the first serious challenge to the dominance of the simplex method as a practical means of solving linear programming problems. By about 1990, interior-point codes had emerged that incorporated the techniques described in Section 14.2 and that were superior on many large problems to the simplex codes available at that time. The years that followed saw significant improvements in simplex software, evidenced by the appearance of packages such as CPLEX and XPRESS-MP. These improvements were due to algorthmic advances such as steepest-edge pivoting (see Goldfarb and Forrest [133]) and improved pricing heuristics, and also to close attention to the nuts and bolts of efficient implementation. The efficiency of interior-point codes also continued to improve, through improvements in the linear algebra for solving the step equations and through the use of higher-order correctors in the step calculation (see Gondzio [138]). During this period, a number of good interior-point codes became freely available (such as PCx [84], HOPDM [137], BPMPD, and LIPSOL [321]) and found their way into many applications.

In general, simplex codes are faster on problems of small-medium dimensions, while interior-point codes are competitive and often faster on large problems. However, this rule is certainly not hard-and-fast; it depends strongly on the structure of the particular application. Interior-point methods are generally not able to take full advantage of prior knowledge about the solution, such as an estimate of the solution itself or an estimate of the optimal basis. Hence, interior-point methods are less useful than simplex approaches in situations in which "warm-start" information is readily available. One situation of this type involves branch-and-bound algorithms for solving integer programs, where each node in the branch-and-bound tree requires the solution of a linear program that differs only slightly from one already solved in the parent node. In other situations, we may wish to solve a sequence of linear programs in which the data is perturbed slightly to investigate sensitivity of the solutions to various perturbations, or in which we approximate a non-linear optimization problem by a sequence of linear programs. Yıldırım and Wright [319] describe how a given point (such as an approximate solution) can be modified to obtain a starting point that is theoretically valid, in that it allows complexity results to be proved that depend on the quality of the given point. In practice, however, these techniques can be expected to provide only a modest improvement in algorithmic performance (perhaps a factor of between 2 and 5) over a "cold" starting point such as the one described in Section 14.2.

Interior-point software has the advantage that it is easy to program, relative to the simplex method. The most complex operation is the solution of the large linear systems at each iteration to compute the step; software to perform this linear algebra operation is readily available. The interior-point code LIPSOL [321] is written entirely in the Matlab language, apart from a small amount of FORTRAN code that interfaces to the linear algebra software. The code PCx [84] is written in C, but also is easy for the interested user to comprehend and modify. It is even possible for a non-expert in optimization to write an efficient interior-point implementation from scratch that is customized to their particular application.

NOTES AND REFERENCES

For more details on the material of this chapter, see the book by Wright [316].

As noted in the text, Karmarkar's method arose from a search for linear programming algorithms with better worst-case behavior than the simplex method. The first algorithm with polynomial complexity, Khachiyan's ellipsoid algorithm [180], was a computational disappointment. In contrast, the execution times required by Karmarkar's method were not too much greater than simplex codes at the time of its introduction, particularly for large linear programs. Karmarkar's is a *primal* algorithm; that is, it is described, motivated, and implemented purely in terms of the primal problem (14.1) without reference to the dual. At each iteration, Karmarkar's algorithm performs a projective transformation on the primal feasible set that maps the current iterate x^k to the center of the set and takes a step in the feasible steepest descent direction for the transformed space. Progress toward optimality is measured by a logarithmic potential function. Descriptions of the algorithm can be found in Karmarkar's original paper [175] and in Fletcher [101, Section 8.7].

Karmarkar's method falls outside the scope of this chapter, and in any case, its practical performance does not appear to match the most efficient primal-dual methods. The algorithms we discussed in this chapter have polynomial complexity, like Karmarkar's method.

Many of the algorithmic ideas that have been examined since 1984 actually had their genesis in three works that preceded Karmarkar's paper. The first of these is the book of Fiacco and McCormick [98] on logarithmic barrier functions (originally proposed by Frisch [115]), which proves existence of the central path, among many other results. Further analysis of the central path was carried out by McLinden [205], in the context of nonlinear complementarity problems. Finally, there is Dikin's paper [94], in which an interior-point method known as primal affine-scaling was originally proposed. The outburst of research on primal-dual methods, which culminated in the efficient software packages available today, dates to the seminal paper of Megiddo [206].

Todd gives an excellent survey of potential reduction methods in [284]. He relates the primal-dual potential reduction method mentioned above to pure primal potential reduction methods, including Karmarkar's original algorithm, and discusses extensions to special classes of nonlinear problems.

For an introduction to complexity theory and its relationship to optimization, see the book by Vavasis [297].

Andersen et al. [6] cover many of the practical issues relating to implementation of interior-point methods. In particular, they describe an alternative scheme for choosing the initial point, for the case in which upper bounds are also present on the variables.

✐ **EXERCISES**

✐   **14.1** This exercise illustrates the fact that the bounds $(x, s) \geq 0$ are essential in relating solutions of the system (14.4a) to solutions of the linear program (14.1) and its dual. Consider the following linear program in $\mathbb{R}^2$:

$$\min x_1, \quad \text{subject to } x_1 + x_2 = 1, \quad (x_1, x_2) \geq 0.$$

Show that the primal-dual solution is

$$x^* = \begin{pmatrix} 0 \\ 1 \end{pmatrix}, \quad \lambda^* = 0, \quad s^* = \begin{pmatrix} 1 \\ 0 \end{pmatrix}.$$

Also verify that the system $F(x, \lambda, s) = 0$ has the spurious solution

$$x = \begin{pmatrix} 1 \\ 0 \end{pmatrix}, \quad \lambda = 1, \quad s = \begin{pmatrix} 0 \\ -1 \end{pmatrix},$$

which has no relation to the solution of the linear program.

✐   **14.2**

(i) Show that $\mathcal{N}_2(\theta_1) \subset \mathcal{N}_2(\theta_2)$ when $0 \leq \theta_1 < \theta_2 < 1$ and that $\mathcal{N}_{-\infty}(\gamma_1) \subset \mathcal{N}_{-\infty}(\gamma_2)$ for $0 < \gamma_2 \leq \gamma_1 \leq 1$.

(ii) Show that $\mathcal{N}_2(\theta) \subset \mathcal{N}_{-\infty}(\gamma)$ if $\gamma \leq 1 - \theta$.

✐   **14.3** Given an arbitrary point $(x, \lambda, s) \in \mathcal{F}^o$, find the range of γ values for which $(x, \lambda, s) \in \mathcal{N}_{-\infty}(\gamma)$. (The range depends on x and s.)

✐   **14.4** For $n = 2$, find a point $(x, s) > 0$ for which the condition

$$\| X S e - \mu e \|_2 \leq \theta \mu$$

is *not* satisfied for any $\theta \in [0, 1)$.

✐   **14.5** Prove that the neighborhoods $\mathcal{N}_{-\infty}(1)$ (see (14.18)) and $\mathcal{N}_2(0)$ (see (14.17)) coincide with the central path $\mathcal{C}$.

✐   **14.6** In the long-step path-following method (Algorithm 14.2), give a procedure for calculating the maximum value of α such that (14.20) is satisfied.

✐   **14.7** Show that Φ_ρ defined by (14.46) has the property (14.45a).

✐   **14.8** Prove that the coefficient matrix in (14.16) is nonsingular if and only if A has full row rank.

📎 **14.9** Given $(\Delta x, \Delta \lambda, \Delta s)$ satisfying (14.10), prove (14.23).

📎 **14.10** Given an iterate (x^k, λ^k, s^k) with $(x^k, s^k) > 0$, show that the quantities $\alpha_{\max}^{\text{pri}}$ and $\alpha_{\max}^{\text{dual}}$ defined by (14.36) are the largest values of α such that $x^k + \alpha \Delta x^k \geq 0$ and $s^k + \alpha \Delta s^k \geq 0$, respectively.

📎 **14.11** Verify (14.37).

📎 **14.12** Given that X and S are diagonal with positive diagonal elements, show that the coefficient matrix in (14.44a) is symmetric and positive definite if and only if A has full row rank. Does this result continue to hold if we replace D by a diagonal matrix in which exactly m of the diagonal elements are positive and the remainder are zero? (Here m is the number of rows of A.)

📎 **14.13** Given a point (x, λ, s) with $(x, s) > 0$, consider the trajectory $\mathcal{H}$ defined by

$$
F\left(\hat{x}(\tau), \hat{\lambda}(\tau), \hat{s}(\tau)\right) =
\begin{bmatrix}
(1 - \tau)(A^T\lambda + s - c) \\
(1 - \tau)(Ax - b) \\
(1 - \tau)XSe
\end{bmatrix},
\qquad (\hat{x}(\tau), \hat{s}(\tau)) \geq 0,
$$

for $\tau \in [0, 1]$, and note that $\left(\hat{x}(0), \hat{\lambda}(0), \hat{s}(0)\right) = (x, \lambda, s)$, while the limit of $\left(\hat{x}(\tau), \hat{\lambda}(\tau), \hat{s}(\tau)\right)$ as $\tau \uparrow 1$ will lie in the primal-dual solution set of the linear program. Find equations for the first, second, and third derivatives of $\mathcal{H}$ with respect to τ at $\tau = 0$. Hence, write down a Taylor series approximation to $\mathcal{H}$ near the point (x, λ, s).

📎 **14.14** Consider the following linear program, which contains "free variables" denoted by y:

$$
\min c^T x + d^T y, \quad \text{subject to } A_1 x + A_2 y = b, x \geq 0.
$$

By introducing Lagrange multipliers λ for the equality constraints and s for the bounds $x \geq 0$, write down optimality conditions for this problem in an analogous fashion to (14.3). Following (14.4) and (14.16), use these conditions to derive the general step equations for a primal-dual interior-point method. Express these equations in augmented system form analogously to (14.42) and explain why it is not possible to reduce further to a formulation like (14.44) in which the coefficient matrix is symmetric positive definite.

📎 **14.15** Program Algorithm 14.3 in Matlab. Choose $\eta = 0.99$ uniformly in (14.38). Test your code on a linear programming problem (14.1) generated by choosing A randomly, and then setting x, s, b, and c as follows:

$$
x_i = \begin{cases} \text{random positive number} & i = 1, 2, \ldots, m, \\ 0 & i = m+1, m+2, \ldots, n, \end{cases}
$$

$$
s_i = \begin{cases} \text{random positive number} & i = m+1, m+2, \ldots, n \\ 0 & i = 1, 2, \ldots, m, \end{cases}
$$

$$
\lambda = \text{random vector},
$$

$$
c = A^T \lambda + s,
$$

$$
b = Ax.
$$

Choose the starting point (x^0, λ^0, s^0) with the components of x^0 and s^0 set to large positive values.

14.17 Show that the solutions of the problems (14.39) are given explicitly by (14.40).

CHAPTER **15**

Fundamentals of Algorithms for Nonlinear Constrained Optimization

In this chapter, we begin our discussion of algorithms for solving the general constrained optimization problem

$$\min_{x \in \mathbb{R}^n} f(x) \quad \text{subject to} \quad \begin{aligned} c_i(x) &= 0, \quad i \in \mathcal{E}, \\ c_i(x) &\geq 0, \quad i \in \mathcal{I}, \end{aligned} \tag{15.1}$$

where the objective function f and the constraint functions c_i are all smooth, real-valued functions on a subset of $\mathbb{R}^n$, and $\mathcal{I}$ and $\mathcal{E}$ are finite index sets of inequality and equality constraints, respectively. In Chapter 12, we used this general statement of the problem to derive optimality conditions that characterize its solutions. This theory is useful for motivating the various algorithms discussed in the remainder of the book, which differ from each other in fundamental ways but are all iterative in nature. They generate a sequence of estimates of the solution x^* that, we hope, tend toward a solution. In some cases, they also generate a sequence of guesses for the Lagrange multipliers associated with the constraints. As in the chapters on unconstrained optimization, we study only algorithms for finding local solutions of (15.1); the problem of finding a global solution is outside the scope of this book.

We note that this chapter is not concerned with individual algorithms themselves, but rather with fundamental concepts and building blocks that are common to more than one algorithm. After reading Sections 15.1 and 15.2, the reader may wish to glance at the material in Sections 15.3, 15.4, 15.5, and 15.6, and return to these sections as needed during study of subsequent chapters.

15.1 CATEGORIZING OPTIMIZATION ALGORITHMS

We now catalog the algorithmic approaches presented in the rest of the book. No standard taxonomy exists for nonlinear optimization algorithms; in the remaining chapters we have grouped the various approaches as follows.

I. In Chapter 16 we study algorithms for solving *quadratic programming* problems. We consider this category separately because of its intrinsic importance, because its particular characteristics can be exploited by efficient algorithms, and because quadratic programming subproblems need to be solved by sequential quadratic programming methods and certain interior-point methods for nonlinear programming. We discuss active set, interior-point, and gradient projection methods.

II. In Chapter 17 we discuss *penalty* and *augmented Lagrangian* methods. By combining the objective function and constraints into a *penalty function*, we can attack problem (15.1) by solving a sequence of unconstrained problems. For example, if only equality constraints are present in (15.1), we can define the quadratic penalty function as

$$f(x) + \frac{\mu}{2} \sum_{i \in \mathcal{E}} c_i^2(x), \qquad (15.2)$$

where $\mu > 0$ is referred to as a *penalty parameter*. We minimize this unconstrained function, for a series of increasing values of μ, until the solution of the constrained optimization problem is identified to sufficient accuracy.

If we use an *exact* penalty function, it may be possible to find a local solution of (15.1) by solving a single unconstrained optimization problem. For the equality-constrained problem, the function defined by

$$f(x) + \mu \sum_{i \in \mathcal{E}} |c_i(x)|,$$

is usually an exact penalty function, for a sufficiently large value of $\mu > 0$. Although they often are nondifferentiable, exact penalty functions can be minimized by solving a sequence of smooth subproblems.

In *augmented Lagrangian methods*, we define a function that combines the properties of the Lagrangian function (12.33) and the quadratic penalty function (15.2). This so-called augmented Lagrangian function has the following form for equality-constrained problems:

$$\mathcal{L}_A(x, \lambda; \mu) = f(x) - \sum_{i \in \mathcal{E}} \lambda_i c_i(x) + \frac{\mu}{2} \sum_{i \in \mathcal{E}} c_i^2(x).$$

Methods based on this function fix λ to some estimate of the optimal Lagrange multiplier vector and fix μ to some positive value, then find a value of x that approximately minimizes $\mathcal{L}_A(\cdot, \lambda; \mu)$. At this new x-iterate, λ and μ may be updated; then the process is repeated. This approach avoids certain drawbacks associated with the minimization of the quadratic penalty function (15.2).

III. In Chapter 18 we describe *sequential quadratic programming* (SQP) methods, which model (15.1) by a quadratic programming subproblem at each iterate and define the search direction to be the solution of this subproblem. In the basic SQP method, we define the search direction p_k at the iterate (x_k, λ_k) to be the solution of

$$\min_p \quad \tfrac{1}{2} p^T \nabla_{xx}^2 \mathcal{L}(x_k, \lambda_k) p + \nabla f(x_k)^T p \qquad (15.3a)$$

$$\text{subject to} \quad \nabla c_i(x_k)^T p + c_i(x_k) = 0, \quad i \in \mathcal{E}, \qquad (15.3b)$$

$$\nabla c_i(x_k)^T p + c_i(x_k) \geq 0, \quad i \in \mathcal{I}, \qquad (15.3c)$$

where $\mathcal{L}$ is the Lagrangian function defined in (12.33). The objective in this subproblem is an approximation to the change in the Lagrangian function in moving from x_k to $x_k + p$, while the constraints are linearizations of the constraints in (15.1). A trust-region constraint may be added to (15.3) to control the length and quality of the step, and quasi-Newton approximate Hessians can be used in place of $\nabla_{xx}^2 \mathcal{L}(x_k, \lambda_k)$. In a variant called *sequential linear-quadratic programming*, the step p_k is computed in two stages. First, we solve a linear program that is defined by omitting the first (quadratic) term from the objective (15.3a) and adding a trust-region constraint to (15.3). Next, we obtain the step p_k by solving an equality-constrained subproblem in which the constraints active at the solution of the linear program are imposed as equalities, while all other constraints are ignored.

IV. In Chapter 19 we study *interior-point methods for nonlinear programming*. These methods can be viewed as extensions of the primal-dual interior-point methods for linear programming discussed in Chapter 14. We can also view them as *barrier methods* that generate steps by solving the problem

$$\min_{x,s} \quad f(x) - \mu \sum_{i=1}^{m} \log s_i \tag{15.4a}$$

$$\text{subject to} \quad c_i(x) = 0, \quad i \in \mathcal{E}, \tag{15.4b}$$

$$c_i(x) - s_i = 0, \quad i \in \mathcal{I}, \tag{15.4c}$$

for some positive value of the barrier parameter μ, where the variables $s_i > 0$ are slacks. Interior-point methods constitute the newest class of methods for nonlinear programming and have already proved to be formidable competitors of sequential quadratic programming methods.

The algorithms in categories I, III, and IV make use of elimination techniques, in which the constraints are used to eliminate some of the degrees of freedom in the problem. As a background to those algorithms, we discuss elimination in Section 15.3. In later sections we discuss merit functions and filters, which are important mechanisms for promoting convergence of nonlinear programming algorithms from remote starting points.

15.2   THE COMBINATORIAL DIFFICULTY OF INEQUALITY-CONSTRAINED PROBLEMS

One of the main challenges in solving nonlinear programming problems lies in dealing with inequality constraints—in particular, in deciding which of these constraints are active at the solution and which are not. One approach, which is the essence of active-set methods, starts by making a guess of the optimal active set $\mathcal{A}^*$, that is, the set of constraints that are satisfied as equalities at a solution. We call our guess the *working set* and denote it by $\mathcal{W}$. We then solve a problem in which the constraints in the working set are imposed as equalities and the constraints not in $\mathcal{W}$ are ignored. We then check to see if there is a choice of Lagrange multipliers such that the solution x^* obtained for this $\mathcal{W}$ satisfies the KKT conditions (12.34). If so, we accept x^* as a local solution of (15.1). Otherwise, we make a different choice of $\mathcal{W}$ and repeat the process. This approach is based on the observation that, in general, it is much simpler to solve equality-constrained problems than to solve nonlinear programs.

The number of choices for working set $\mathcal{W}$ may be very large—up to $2^{|\mathcal{I}|}$, where $|\mathcal{I}|$ is the number of inequality constraints. We arrive at this estimate by observing that we can make one of two choices for each $i \in \mathcal{I}$: to include it in $\mathcal{W}$ or leave it out. Since the number of possible working sets grows exponentially with the number of inequalities—a phenomenon which we refer to as the *combinatorial difficulty* of nonlinear programming—we cannot hope to design a practical algorithm by considering all possible choices for $\mathcal{W}$.

The following example suggests that even for a small number of inequality constraints, determination of the optimal active set is not a simple task.

❑ **EXAMPLE 15.1**

Consider the problem

$$\min_{x,y} \ f(x,y) \stackrel{\text{def}}{=} \tfrac{1}{2}(x-2)^2 + \tfrac{1}{2}(y-\tfrac{1}{2})^2 \tag{15.5}$$

$$\text{subject to} \quad \begin{aligned} (x+1)^{-1} - y - \tfrac{1}{4} &\geq 0, \\ x &\geq 0, \\ y &\geq 0. \end{aligned}$$

We label the constraints, in order, with the indices 1 through 3. Figure 15.1 illustrates the contours of the objective function (dashed circles). The feasible region is the region enclosed by the curve and the two axes. We see that only the first constraint is active at the solution, which is $(x^*, y^*)^T = (1.953, 0.089)^T$.

Let us now apply the working-set approach described above to (15.5), considering all $2^3 = 8$ possible choices of $\mathcal{W}$.

We consider first the possibility that no constraints are active at the solution, that is, $\mathcal{W} = \emptyset$. Since $\nabla f = (x-2, y-1/2)^T$, we see that the unconstrained minimum of f lies outside the feasible region. Hence, the optimal active set cannot be empty.

There are seven further possibilities. First, all three constraints could be active (that is, $\mathcal{W} = \{1, 2, 3\}$). A glance at Figure 15.1 shows that this does not happen for our problem; the three constraints do not share a common point of intersection. Three further possibilities are obtained by making a single constraint active (that is, $\mathcal{W} = \{1\}$, $\mathcal{W} = \{2\}$, and $\mathcal{W} = \{3\}$),

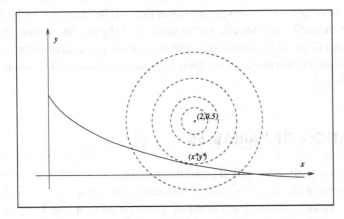

Figure 15.1 Graphical illustration of problem (15.5).

while the final three possibilities are obtained by making exactly two constraints active (that is, $\mathcal{W} = \{1, 2\}$, $\mathcal{W} = \{1, 3\}$, and $\mathcal{W} = \{2, 3\}$). We consider three of these cases in detail.

- $\mathcal{W} = \{2\}$; that is, only the constraint $x = 0$ is active. If we minimize f enforcing only this constraint, we obtain the point $(0, 1/2)^T$. A check of the KKT conditions (12.34) shows that no matter how we choose the Lagrange multipliers, we cannot satisfy all these conditions at $(0, 1/2)^T$. (We must have $\lambda_1 = \lambda_3 = 0$ to satisfy (12.34e), which implies that we must set $\lambda_2 = -2$ to satisfy (12.34a); but this value of λ_2 violates the condition (12.34d).)

- $\mathcal{W} = \{1, 3\}$, which yields the single feasible point $(3, 0)^T$. Since constraint 2 is inactive at this point, we have $\lambda_2 = 0$, so by solving (12.34a) for the other Lagrange multipliers, we obtain $\lambda_1 = -16$ and $\lambda_3 = -16.5$. These values are negative, so they violate (12.34d), and $x = (3, 0)^T$ cannot be a solution of (15.1).

- $\mathcal{W} = \{1\}$. Solving the equality-constrained problem in which the first constraint is active, we obtain $(x, y)^T = (1.953, 0.089)^T$ with Lagrange multiplier $\lambda_1 = 0.411$. It is easy to see that by setting $\lambda_2 = \lambda_3 = 0$, the remaining KKT conditions (12.34) are satisfied, so we conclude that this is a KKT point. Furthermore, it is easy to show that the second-order sufficient conditions are satisfied, as the Hessian of the Lagrangian is positive definite.

⊓⊔

Even for this small example, we see that it is exhausting to consider all possible choices for $\mathcal{W}$. Figure 15.1 suggests, however, that some choices of $\mathcal{W}$ can be eliminated from consideration if we make use of knowledge of the functions that define the problem, and their derivatives. In fact, the active set methods described in Chapter 16 use this kind of information to make a series of educated guesses for the working set, avoiding choices of $\mathcal{W}$ that obviously will not lead to a solution of (15.1).

A different approach is followed by interior-point (or barrier) methods discussed in Chapter 19. These methods generate iterates that stay away from the boundary of the feasible region defined by the inequality constraints. As the solution of the nonlinear program is approached, the barrier effects are weakened to permit an increasingly accurate estimate of the solution. In this manner, interior-point methods avoid the combinatorial difficulty of nonlinear programming.

15.3   ELIMINATION OF VARIABLES

When dealing with constrained optimization problems, it is natural to try to use the constraints to eliminate some of the variables from the problem, to obtain a simpler problem with fewer degrees of freedom. Elimination techniques must be used with care, however, as they may alter the problem or introduce ill conditioning.

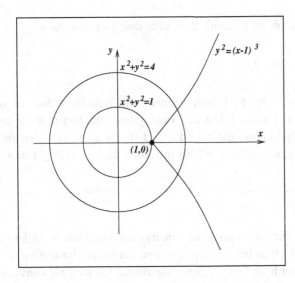

Figure 15.2
The danger of nonlinear elimination.

We begin with an example in which it is safe and convenient to eliminate variables. In the problem

$$\min f(x) = f(x_1, x_2, x_3, x_4) \quad \text{subject to} \quad x_1 + x_3^2 - x_4 x_3 = 0,$$
$$-x_2 + x_4 + x_3^2 = 0,$$

there is no risk in setting

$$x_1 = x_4 x_3 - x_3^2, \qquad x_2 = x_4 + x_3^2,$$

to obtain a function of two variables

$$h(x_3, x_4) = f(x_4 x_3 - x_3^2, x_4 + x_3^2, x_3, x_4),$$

which we can minimize using the unconstrained optimization techniques described in earlier chapters.

The dangers of nonlinear elimination are illustrated in the following example.

❑ **EXAMPLE 15.2**   (FLETCHER [101])

Consider the problem

$$\min x^2 + y^2 \qquad \text{subject to } (x - 1)^3 = y^2.$$

The contours of the objective function and the constraints are illustrated in Figure 15.2, which shows that the solution is $(x, y) = (1, 0)$.

We attempt to solve this problem by eliminating y. By doing so, we obtain

$$h(x) = x^2 + (x - 1)^3.$$

Clearly, $h(x) \to -\infty$ as $x \to -\infty$. By blindly applying this transformation we may conclude that the problem is unbounded, but this view ignores the fact that the constraint $(x - 1)^3 = y^2$ implicitly imposes the bound $x \geq 1$ that is active at the solution. Hence, if we wish to eliminate y, we should explicitly introduce the bound $x \geq 1$ into the problem.

❒

This example shows that the use of nonlinear equations to eliminate variables may result in errors that can be difficult to trace. For this reason, nonlinear elimination is not used by most optimization algorithms. Instead, many algorithms linearize the constraints and apply elimination techniques to the simplified problem. We now describe systematic procedures for performing variable elimination using linear constraints.

SIMPLE ELIMINATION USING LINEAR CONSTRAINTS

We consider the minimization of a nonlinear function subject to a set of linear equality constraints,

$$\min \ f(x) \quad \text{subject to } Ax = b, \tag{15.6}$$

where A is an $m \times n$ matrix with $m \leq n$. Suppose for simplicity that A has full row rank. (If such is not the case, we find either that the problem is inconsistent or that some of the constraints are redundant and can be deleted without affecting the solution of the problem.) Under this assumption, we can find a subset of m columns of A that is linearly independent. If we gather these columns into an $m \times m$ matrix B and define an $n \times n$ permutation matrix P that swaps these columns to the first m column positions in A, we can write

$$AP = [B \,|\, N], \tag{15.7}$$

where N denotes the $n - m$ remaining columns of A. (The notation here is consistent with that of Chapter 13, where we discussed similar concepts in the context of linear programming.) We define the subvectors $x_B \in \mathbb{R}^m$ and $x_N \in \mathbb{R}^{n-m}$ as follows:

$$\begin{bmatrix} x_B \\ x_N \end{bmatrix} = P^T x, \tag{15.8}$$

and call x_B the *basic variables* and B the *basis matrix*. Noting that $PP^T = I$, we can rewrite the constraint $Ax = b$ as

$$b = Ax = AP(P^T x) = Bx_B + Nx_N.$$

By rearranging this formula, we deduce that the basic variables can be expressed as follows:

$$x_B = B^{-1}b - B^{-1}Nx_N. \tag{15.9}$$

We can therefore compute a feasible point for the constraints $Ax = b$ by choosing *any* value of x_N and then setting x_B according to the formula (15.9). The problem (15.6) is therefore equivalent to the unconstrained problem

$$\min_{x_N} h(x_N) \overset{\text{def}}{=} f\left(P\begin{bmatrix} B^{-1}b - B^{-1}Nx_N \\ x_N \end{bmatrix}\right). \tag{15.10}$$

We refer to the substitution in (15.9) as *simple elimination of variables*.

This discussion shows that a nonlinear optimization problem with linear equality constraints is, from a mathematical point of view, the same as an unconstrained problem.

❑ **EXAMPLE 15.3**

Consider the problem

$$\min \; \sin(x_1 + x_2) + x_3^2 + \frac{1}{3}(x_4 + x_5^4 + x_6/2) \tag{15.11a}$$

$$\text{subject to} \quad \begin{aligned} 8x_1 - 6x_2 + x_3 + 9x_4 + 4x_5 &= 6 \\ 3x_1 + 2x_2 - x_4 + 6x_5 + 4x_6 &= -4. \end{aligned} \tag{15.11b}$$

By defining the permutation matrix P so as to reorder the components of x as $x^T = (x_3, x_6, x_1, x_2, x_4, x_5)^T$, we find that the coefficient matrix AP is

$$AP = \begin{bmatrix} 1 & 0 & 8 & -6 & 9 & 4 \\ 0 & 4 & 3 & 2 & -1 & 6 \end{bmatrix}.$$

The basis matrix B is diagonal and therefore easy to invert. We obtain from (15.9) that

$$\begin{bmatrix} x_3 \\ x_6 \end{bmatrix} = -\begin{bmatrix} 8 & -6 & 9 & 4 \\ \frac{3}{4} & \frac{1}{2} & -\frac{1}{4} & \frac{3}{2} \end{bmatrix} \begin{bmatrix} x_1 \\ x_2 \\ x_4 \\ x_5 \end{bmatrix} + \begin{bmatrix} 6 \\ -1 \end{bmatrix}. \tag{15.12}$$

By substituting for x_3 and x_6 in (15.11a), the problem becomes

$$\min_{x_1, x_2, x_4, x_5} \sin(x_1 + x_2) + (8x_1 - 6x_2 + 9x_4 + 4x_5 - 6)^2 \tag{15.13}$$

$$+ \frac{1}{3}(x_4 + x_5^4 - [(1/2) + (3/8)x_1 + (1/4)x_2 - (1/8)x_4 + (3/4)x_5]).$$

We could have chosen two other columns of the coefficient matrix A (that is, two variables other than x_3 and x_6) as the basis for elimination in the system (15.11b), but the matrix $B^{-1}N$ would not have been so simple. ❑

A set of m independent columns can be selected, in general, by means of Gaussian elimination. In the parlance of linear algebra, we can compute the row echelon form of the matrix and choose the pivot columns as the columns of the basis B. Ideally, we would like B to be easy to factor and well conditioned. A technique that suits these purposes is a sparse Gaussian elimination approach that attempts to preserve sparsity while keeping rounding errors under control. A well-known implementation of this algorithm is MA48 from the HSL library [96]. As we discuss below, however, there is no guarantee that the Gaussian elimination process will identify the best choice of basis matrix.

There is an interesting interpretation of the simple elimination-of-variables approach that we have just described. To simplify the notation, we will assume from now on that the coefficient matrix is already given to us so that the basic columns appear in the first m positions, that is, $P = I$.

From (15.8) and (15.9) we see that any feasible point x for the linear constraints in (15.6) can be written as

$$\begin{bmatrix} x_B \\ x_N \end{bmatrix} = x = Yb + Zx_N, \tag{15.14}$$

where

$$Y = \begin{bmatrix} B^{-1} \\ 0 \end{bmatrix}, \quad Z = \begin{bmatrix} -B^{-1}N \\ I \end{bmatrix}. \tag{15.15}$$

Note that Z has $n - m$ linearly independent columns (because of the presence of the identity matrix in the lower block) and that it satisfies $AZ = 0$. Therefore, Z is a *basis for the null space* of A. In addition, the columns of Y and the columns of Z form a linearly independent set. We note also from (15.15), (15.7) that Yb is a particular solution of the linear constraints $Ax = b$.

In other words, the simple elimination technique expresses feasible points as the sum of a particular solution of $Ax = b$ (the first term in (15.14)) plus a displacement along the

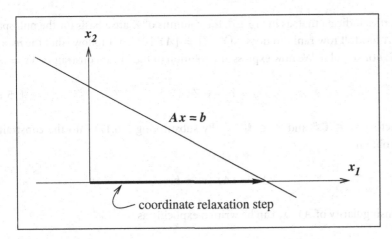

Figure 15.3   Simple elimination, showing the coordinate relaxation step obtained by choosing the basis to be the first column of A.

null space of the constraints (the second term in (15.14)). The relations (15.14), (15.15) indicate that the particular Yb solution is obtained by holding $n-m$ components of x at zero while relaxing the other m components (the ones in x_{B}) until they reach the constraints. The particular solution Yb is sometimes known as the coordinate relaxation step. In Figure 15.3, we see the coordinate relaxation step Yb obtained by choosing the basis matrix B to be the first column of A. If we were to choose B to be the second column of A, the coordinate relaxation step would lie along the x_2 axis.

Simple elimination is inexpensive but can give rise to numerical instabilities. If the feasible set in Figure 15.3 consisted of a line that was almost parallel to the x_1 axis, the coordinate relaxation along this axis would be very large in magnitude. We would then be computing x as the difference of very large vectors, giving rise to numerical cancellation. In that situation it would be preferable to choose a particular solution along the x_2 axis, that is, to select a different basis. Selection of the best basis is, therefore, not a straightforward task in general. To overcome the dangers of an excessively large coordinate relaxation step, we could define the particular solution Yb as the minimum-norm step to the constraints. This approach is a special case of more general elimination strategies, which we now describe.

GENERAL REDUCTION STRATEGIES FOR LINEAR CONSTRAINTS

To generalize (15.14) and (15.15), we choose matrices $Y \in \mathbb{R}^{n \times m}$ and $Z \in \mathbb{R}^{n \times (n-m)}$ with the following properties:

$$[Y \mid Z] \in \mathbb{R}^{n \times n} \text{ is nonsingular,} \quad AZ = 0. \tag{15.16}$$

These properties indicate that, as in (15.15), the columns of Z are a basis for the null space of A. Since A has full row rank, so does $A[Y \mid Z] = [AY \mid 0]$, so it follows that the $m \times m$ matrix AY is nonsingular. We now express any solution of the linear constraints $Ax = b$ as

$$x = Y x_Y + Z x_Z, \tag{15.17}$$

for some vectors $x_Y \in \mathbb{R}^m$ and $x_Z \in \mathbb{R}^{n-m}$. By substituting (15.17) into the constraints $Ax = b$, we obtain

$$Ax = (AY) x_Y = b;$$

hence by nonsingularity of AY, x_Y can be written explicitly as

$$x_Y = (AY)^{-1} b. \tag{15.18}$$

By substituting this expression into (15.17), we conclude that any vector x of the form

$$x = Y(AY)^{-1} b + Z x_Z \tag{15.19}$$

satisfies the constraints $Ax = b$ for any choice of $x_Z \in \mathbb{R}^{n-m}$. Therefore, the problem (15.6) can be restated equivalently as the following unconstrained problem

$$\min_{x_Z} \ f(Y(AY)^{-1} b + Z x_Z). \tag{15.20}$$

Ideally, we would like to choose Y in such a way that the matrix AY is as well conditioned as possible, since it needs to be factorized to give the particular solution $Y(AY)^{-1} b$. We can do this by computing Y and Z by means of a QR factorization of A^T, which has the form

$$A^T \Pi = \begin{bmatrix} Q_1 & Q_2 \end{bmatrix} \begin{bmatrix} R \\ 0 \end{bmatrix}, \tag{15.21}$$

where $\begin{bmatrix} Q_1 & Q_2 \end{bmatrix}$ is orthogonal. The submatrices Q_1 and Q_2 have orthonormal columns and are of dimension $n \times m$ and $n \times (n-m)$, while R is $m \times m$ upper triangular and nonsingular and Π is an $m \times m$ permutation matrix. (See the discussion following (A.24) in the Appendix for further details.) We now define

$$Y = Q_1, \quad Z = Q_2, \tag{15.22}$$

so that the columns of Y and Z form an *orthonormal basis* of $\mathbb{R}^n$. If we expand (15.21) and do a little rearrangement, we obtain

$$AY = \Pi R^T, \quad AZ = 0.$$

Therefore, Y and Z have the desired properties, and the condition number of AY is the same as that of R, which in turn is the same as that of A itself. From (15.19) we see that any solution of $Ax = b$ can be expressed as

$$x = Q_1 R^{-T} \Pi^T b + Q_2 x_z,$$

for some vector x_z. The computation $R^{-T} \Pi^T b$ can be carried out inexpensively, at the cost of a single triangular substitution.

A simple computation shows that the particular solution $Q_1 R^{-T} \Pi^T b$ can also be written as $A^T (A A^T)^{-1} b$. This vector is the solution of the following problem:

$$\min \|x\|^2 \qquad \text{subject to } Ax = b;$$

that is, it is the minimum-norm solution of $Ax = b$. See Figure 15.5 for an illustration of this step.

Elimination via the orthogonal basis (15.22) is ideal from the point of view of numerical stability. The main cost associated with this reduction strategy is in computing the QR factorization (15.21). Unfortunately, for problems in which A is large and sparse, a sparse QR factorization can be much more costly to compute than the sparse Gaussian elimination strategy used in simple elimination. Therefore, other elimination strategies have been developed that seek a compromise between these two techniques; see Exercise 15.7.

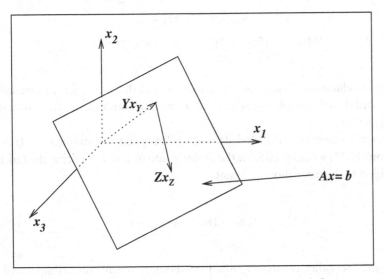

Figure 15.4   General elimination: Case in which $A \in \mathbb{R}^{1\times3}$, showing the particular solution and a step in the null space of A.

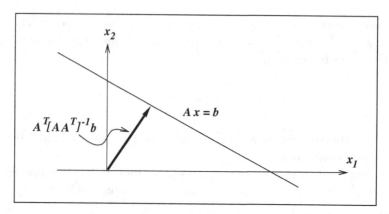

Figure 15.5 The minimum-norm step.

EFFECT OF INEQUALITY CONSTRAINTS

Elimination of variables is not always beneficial if inequality constraints are present alongside the equalities. For instance, if problem (15.11) had the additional constraint $x \geq 0$, then after eliminating the variables x_3 and x_6, we would be left with the problem of minimizing the function in (15.13) subject to the constraints

$$(x_1, x_2, x_4, x_5) \geq 0,$$
$$8x_1 - 6x_2 + 9x_4 + 4x_5 \leq 6,$$
$$(3/4)x_1 + (1/2)x_2 - (1/4)x_4 + (3/2)x_5 \leq -1.$$

Hence, the cost of eliminating the equality constraints (15.11b) is to make the inequalities more complicated than the simple bounds $x \geq 0$. For many algorithms, this transformation will not yield any benefit.

If, however, problem (15.11) included the general inequality constraint $3x_1 + 2x_3 \geq 1$, the elimination (15.12) would transform the problem into one of minimizing the function in (15.13) subject to the inequality constraint

$$- 13x_1 + 12x_2 - 18x_4 - 8x_5 \geq -11. \tag{15.23}$$

In this case, the inequality constraint would not become much more complicated after elimination of the equality constraints, so it is probably worthwhile to perform the elimination.

15.4 MERIT FUNCTIONS AND FILTERS

Suppose that an algorithm for solving the nonlinear programming problem (15.1) generates a step that reduces the objective function but increases the violation of the constraints. Should we accept this step?

This question is not easy to answer. We must look for a way to balance the twin (often competing) goals of reducing the objective function and satisfying the constraints. Merit functions and filters are two approaches for achieving this balance. In a typical constrained optimization algorithm, a step p will be accepted only if it leads to a sufficient reduction in the merit function ϕ or if it is acceptable to the filter. These concepts are explained in the rest of the section.

MERIT FUNCTIONS

In unconstrained optimization, the objective function f is the natural choice for the merit function. All the unconstrained optimization methods described in this book require that f be decreased at each step (or at least within a certain number of iterations). In feasible methods for constrained optimization in which the starting point and all subsequent iterates satisfy all the constraints in the problem, the objective function is still an appropriate merit function. On the other hand, algorithms that allow iterates to violate the constraints require some means to assess the quality of the steps and iterates. The merit function in this case combines the objective with measures of constraint violation.

A popular choice of merit function for the nonlinear programming problem (15.1) is the ℓ_1 penalty function defined by

$$\phi_1(x; \mu) = f(x) + \mu \sum_{i \in \mathcal{E}} |c_i(x)| + \mu \sum_{i \in \mathcal{I}} [c_i(x)]^-, \qquad (15.24)$$

where we use the notation $[z]^- = \max\{0, -z\}$. The positive scalar μ is the *penalty parameter*, which determines the weight that we assign to constraint satisfaction relative to minimization of the objective. The ℓ_1 merit function ϕ_1 is not differentiable because of the presence of the absolute value and $[\cdot]^-$ functions, but it has the important property of being *exact*.

Definition 15.1 (Exact Merit Function).
A merit function $\phi(x; \mu)$ is exact if there is a positive scalar μ^* such that for any $\mu > \mu^*$, any local solution of the nonlinear programming problem (15.1) is a local minimizer of $\phi(x; \mu)$.

We show in Theorem 17.3 that, under certain assumptions, the ℓ_1 merit function $\phi_1(x; \mu)$ is exact and that the threshold value μ^* is given by

$$\mu^* = \max\{|\lambda_i^*|, \ i \in \mathcal{E} \cup \mathcal{I}\},$$

where the λ_i^* denote the Lagrange multipliers associated with an optimal solution x^*. Since the optimal Lagrange multipliers are, however, not known in advance, algorithms based on the ℓ_1 merit function contain rules for adjusting the penalty parameter whenever there is reason to believe that it is not large enough (or is excessively large). These rules depend on the choice of optimization algorithm and are discussed in the next chapters.

Another useful merit function is the exact ℓ_2 function, which for equality-constrained problems takes the form

$$\phi_2(x; \mu) = f(x) + \mu\|c(x)\|_2. \tag{15.25}$$

This function is nondifferentiable because the 2-norm term is not squared; its derivative is not defined at x for which $c(x) = 0$.

Some merit functions are both smooth and exact. To ensure that both properties hold, we must include additional terms in the merit function. For equality-constrained problems, *Fletcher's augmented Lagrangian* is given by

$$\phi_F(x; \mu) = f(x) - \lambda(x)^T c(x) + \tfrac{1}{2}\mu \sum_{i \in \mathcal{E}} c_i(x)^2, \tag{15.26}$$

where $\mu > 0$ is the penalty parameter and

$$\lambda(x) = [A(x)A(x)^T]^{-1}A(x)\nabla f(x). \tag{15.27}$$

(Here $A(x)$ denotes the Jacobian of $c(x)$.) Although this merit function has some interesting theoretical properties, it has practical limitations, including the expense of solving for $\lambda(x)$ in (15.27).

A quite different merit function is the (standard) augmented Lagrangian in x and λ, which for equality-constrained problems has the form

$$\mathcal{L}_A(x, \lambda; \mu) = f(x) - \lambda^T c(x) + \tfrac{1}{2}\mu\|c(x)\|_2^2. \tag{15.28}$$

We assess the acceptability of a trial point (x^+, λ^+) by comparing the value of $\mathcal{L}_A(x^+, \lambda^+; \mu)$ with the value at the current iterate, (x, λ). Strictly speaking, $\mathcal{L}_A$ is not a merit function in the sense that a solution (x^*, λ^*) of the nonlinear programming problem is not in general a minimizer of $\mathcal{L}_A(x, \lambda; \mu)$ but only a stationary point. Although some sequential quadratic programming methods use $\mathcal{L}_A$ successfully as a merit function by adaptively modifying μ and λ, we will not consider its use as a merit function further. Instead, we will focus primarily on the nonsmooth exact penalty functions ϕ_1 and ϕ_2.

A trial step $x^+ = x + \alpha p$ generated by a line search algorithm will be accepted if it produces a *sufficient decrease* in the merit function $\phi(x; \mu)$. One way to define this concept is analogous to the condition (3.4) used in unconstrained optimization, where the amount of decrease is not too small relative to the predicted change in the function over the step. The ℓ_1 and ℓ_2 merit functions are not differentiable, but they have a directional derivative. (See (A.51) for background on directional derivatives.) We write the directional derivative of $\phi(x; \mu)$ in the direction p as

$$D(\phi(x; \mu); p).$$

In a line search method, the sufficient decrease condition requires the steplength parameter $\alpha > 0$ to be small enough that the inequality

$$\phi(x + \alpha p; \mu) \leq \phi(x; \mu) + \eta \alpha D(\phi(x; \mu); p), \qquad (15.29)$$

is satisfied for some $\eta \in (0, 1)$.

Trust-region methods typically use a quadratic model $q(p)$ to estimate the value of the merit function ϕ after a step p; see Section 18.5. The sufficient decrease condition can be stated in terms of a decrease in this model, as follows

$$\phi(x + p; \mu) \leq \phi(x; \mu) - \eta(q(0) - q(p)), \qquad (15.30)$$

for some $\eta \in (0, 1)$. (The final term in (15.30) is positive, because the step p is computed to decrease the model q.)

FILTERS

Filter techniques are step acceptance mechanisms based on ideas from multiobjective optimization. Our derivation starts with the observation that nonlinear programming has two goals: minimization of the objective function and the satisfaction of the constraints. If we define a measure of infeasibility as

$$h(x) = \sum_{i \in \mathcal{E}} |c_i(x)| + \sum_{i \in \mathcal{I}} [c_i(x)]^-, \qquad (15.31)$$

we can write these two goals as

$$\min_x f(x) \quad \text{and} \quad \min_x h(x). \qquad (15.32)$$

Unlike merit functions, which combine both problems into a single minimization problem, filter methods keep the two goals in (15.32) separate. Filter methods accept a trial step x^+ as a new iterate if the pair $(f(x^+), h(x^+))$ is not *dominated* by a previous pair $(f_l, h_l) = (f(x_l), h(x_l))$ generated by the algorithm. These concepts are defined as follows.

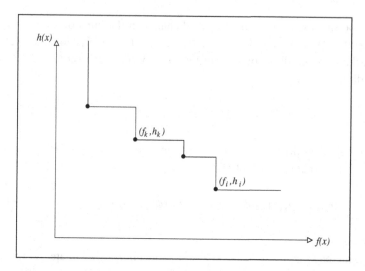

Figure 15.6   Graphical illustration of a filter with four pairs.

Definition 15.2.

(a) *A pair* (f_k, h_k) *is said to dominate another pair* (f_l, h_l) *if both* $f_k \leq f_l$ *and* $h_k \leq h_l$.

(b) *A filter is a list of pairs* (f_l, h_l) *such that no pair dominates any other.*

(c) *An iterate* x_k *is said to be acceptable to the filter if* (f_k, h_k) *is not dominated by any pair in the filter.*

When an iterate x_k is acceptable to the filter, we (normally) add (f_k, h_k) to the filter and remove any pairs that are dominated by (f_k, h_k). Figure 15.6 shows a filter where each pair (f_l, h_l) in the filter is represented as a black dot. Every point in the filter creates an (infinite) rectangular region, and their union defines the set of pairs not acceptable to the filter. More specifically, a trial point x^+ is acceptable to the filter if (f^+, h^+) lies below or to the left of the solid line in Figure 15.6.

To compare the filter and merit function approaches, we plot in Figure 15.7 the contour line of the set of pairs (f, h) such that $f + \mu h = f_k + \mu h_k$, where x_k is the current iterate. The region to the left of this line corresponds to the set of pairs that reduce the merit function $\phi(x; \mu) = f(x) + \mu h(x)$; clearly this set is quite different from the set of points acceptable to the filter.

If a trial step $x^+ = x_k + \alpha_k p_k$ generated by a line search method gives a pair (f^+, h^+) that is acceptable to the filter, we set $x_{k+1} = x^+$; otherwise, a backtracking line search is performed. In a trust-region method, if the step is not acceptable to the filter, the trust region is reduced, and a new step is computed.

Several enhancements to this filter technique are needed to obtain global convergence and good practical performance. We need to ensure, first of all, that we do not accept a point whose (f, h) pair is very close to the current pair (f_k, h_k) or to another pair in the filter. We

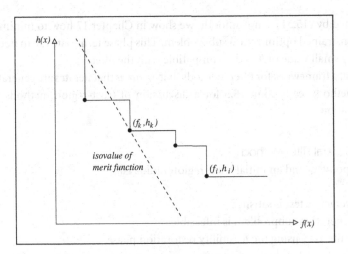

Figure 15.7   Comparing the filter and merit function techniques.

do so by modifying the acceptability criterion and imposing a sufficient decrease condition. A trial iterate x^+ is acceptable to the filter if, for all pairs (f_j, h_j) in the filter, we have that

$$f(x^+) \le f_j - \beta h_j \qquad \text{or} \qquad h(x^+) \le h_j - \beta h_j, \tag{15.33}$$

for $\beta \in (0, 1)$. Although this condition is effective in practice using, say $\beta = 10^{-5}$, for purposes of analysis it may be advantageous to replace the first inequality by

$$f(x^+) \le f_j - \beta h^+.$$

A second enhancement addresses some problematic aspects of the filter mechanism. Under certain circumstances, the search directions generated by line search methods may require arbitrarily small steplengths α_k to be acceptable to the filter. This phenomenon can cause the algorithm to stall and fail. To guard against this situation, if the backtracking line search generates a steplength that is smaller than a given threshold α_{min}, the algorithm switches to a *feasibility restoration phase*, which we describe below. Similarly, in a trust-region method, if a sequence of trial steps is rejected by the filter, the trust-region radius may be decreased so much that the trust-region subproblem becomes infeasible (see Section 18.5). In this case, too, the feasibility restoration phase is invoked. (Other mechanisms could be employed to handle this situation, but as we discuss below, the feasibility restoration phase can help the algorithm achieve other useful goals.)

The feasibility restoration phase aims exclusively to reduce the constraint violation, that is, to find an approximate solution to the problem

$$\min_x \; h(x).$$

Although $h(x)$ defined by (15.31) is not smooth, we show in Chapter 17 how to minimize it using a smooth constrained optimization subproblem. This phase terminates at an iterate that has a sufficiently small value of h and is compatible with the filter.

We now present a framework for filter methods that assumes that iterates are generated by a trust-region method; see Section 18.5 for a discussion of trust-region methods for constrained optimization.

Algorithm 15.1 (General Filter Method).
    Choose a starting point x_0 and an initial trust-region radius Δ_0;
    Set $k \leftarrow 0$;
    repeat until a convergence test is satisfied
        if the step-generation subproblem is infeasible
            Compute x_{k+1} using the feasibility restoration phase;
        else
            Compute a trial iterate $x^+ = x_k + p_k$;
            if (f^+, h^+) is acceptable to the filter
                Set $x_{k+1} = x^+$ and add (f_{k+1}, h_{k+1}) to the filter;
                Choose Δ_{k+1} such that $\Delta_{k+1} \geq \Delta_k$;
                Remove all pairs from the filter that are dominated
                      by (f_{k+1}, h_{k+1});
            else
                Reject the step, set $x_{k+1} = x_k$;
                Choose $\Delta_{k+1} < \Delta_k$;
            end if
        end if
        $k \leftarrow k + 1$;
    end repeat

Other enhancements of this simple filter framework are used in practice; they depend on the choice of algorithm and will be discussed in subsequent chapters.

15.5 THE MARATOS EFFECT

Some algorithms based on merit functions or filters may fail to converge rapidly because they reject steps that make good progress toward a solution. This undesirable phenomenon is often called the *Maratos effect*, because it was first observed by Maratos [199]. It is illustrated by the following example, in which steps p_k, which would yield quadratic convergence if accepted, cause an increase both in the objective function value and the constraint violation.

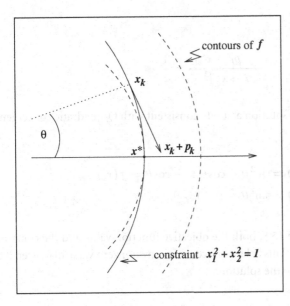

contours of f

x_k

θ

x^* $x_k + p_k$

constraint $x_1^2 + x_2^2 = 1$

Figure 15.8
Maratos Effect: Example 15.4.
Note that the constraint is no
longer satisfied after the step
from x_k to $x_k + p_k$, and the
objective value has increased.

☐ **EXAMPLE 15.4**   (POWELL [255])

Consider the problem

$$\min \ f(x_1, x_2) = 2(x_1^2 + x_2^2 - 1) - x_1, \quad \text{subject to} \quad x_1^2 + x_2^2 - 1 = 0. \quad (15.34)$$

One can verify (see Figure 15.8) that the optimal solution is $x^* = (1, 0)^T$, that the
corresponding Lagrange multiplier is $\lambda^* = \frac{3}{2}$, and that $\nabla_{xx}^2 \mathcal{L}(x^*, \lambda^*) = I$.
   Let us consider an iterate x_k of the form $x_k = (\cos\theta, \sin\theta)^T$, which is feasible for any
value of θ. Suppose that our algorithm computes the following step:

$$p_k = \begin{pmatrix} \sin^2\theta \\ -\sin\theta\cos\theta \end{pmatrix}, \quad (15.35)$$

which yields a trial point

$$x_k + p_k = \begin{pmatrix} \cos\theta + \sin^2\theta \\ \sin\theta(1 - \cos\theta) \end{pmatrix}.$$

By using elementary trigonometric identities, we have that

$$\|x_k + p_k - x^*\|_2 = 2\sin^2(\theta/2), \quad \|x_k - x^*\|_2 = 2|\sin(\theta/2)|,$$

and therefore

$$\frac{\|x_k + p_k - x^*\|_2}{\|x_k - x^*\|_2^2} = \frac{1}{2}.$$

Hence, this step approaches the solution at a rate consistent with Q-quadratic convergence. However, we have that

$$f(x_k + p_k) = \sin^2\theta - \cos\theta > -\cos\theta = f(x_k),$$
$$c(x_k + p_k) = \sin^2\theta > c(x_k) = 0,$$

so that, as can be seen in Figure 15.8, both the objective function value and the constraint violation increase over this step. This behavior occurs for any nonzero value of θ, even if the initial point is arbitrarily close to the solution. ∎

On the example above, any algorithm that requires reduction of a merit function of the form

$$\phi(x; \mu) = f(x) + \mu h(c(x)),$$

where $h(\cdot)$ is a nonnegative function satisfying $h(0) = 0$, will reject the good step (15.35). (Examples of such merit functions include the ϕ_1 and ϕ_2 penalty functions.) The step (15.35) will also be rejected by the filter mechanism described above because the pair $(f(x_k + p_k), h(x_k + p_k))$ is dominated by (f_k, h_k). Therefore, all these approaches will suffer from the Maratos effect.

If no remedial measures are taken, the Maratos effect can slow optimization methods by interfering with good steps away from the solution and by preventing superlinear convergence. Strategies for avoiding the Maratos effect include the following.

1. We can use a merit function that does not suffer from the Maratos effect. An example is Fletcher's augmented Lagrangian function (15.26).

2. We can use a second-order correction in which we add to p_k a step $\hat{p}_k$, which is computed at $c(x_k + p_k)$ and which decreases the constraint violation.

3. We can allow the merit function ϕ to increase on certain iterations; that is, we can use a nonmonotone strategy.

We discuss the last two approaches in the next section.

15.6 SECOND-ORDER CORRECTION AND NONMONOTONE TECHNIQUES

By adding a correction term that decreases the constraint violation, various algorithms are able to overcome the difficulties associated with the Maratos effect. We describe this technique with respect to the equality-constrained problem, in which the constraint is $c(x) = 0$, where $c : \mathbb{R}^n \to \mathbb{R}^{|\mathcal{E}|}$.

Given a step p_k, the second-order correction step $\hat{p}_k$ is defined to be

$$\hat{p}_k = -A_k^T (A_k A_k^T)^{-1} c(x_k + p_k), \tag{15.36}$$

where $A_k = A(x_k)$ is the Jacobian of c at x_k. Note that $\hat{p}_k$ has the property that it satisfies a linearization of the constraint c at the point $x_k + p_k$, that is,

$$A_k \hat{p}_k + c(x_k + p_k) = 0.$$

In fact, $\hat{p}_k$ is the minimum-norm solution of this equation. (A different interpretation of the second-order correction is given in Section 18.3.)

The effect of the correction step $\hat{p}_k$ is to decrease the quantity $\|c(x)\|$ to the order of $\|x_k - x^*\|^3$, provided the primary step p_k satisfies $A_k p_k + c(x_k) = 0$. This estimate indicates that the step from from x_k to $x_k + p_k + \hat{p}_k$ will decrease the merit function, at least near the solution. The cost of this enhancement includes the additional evaluation of the constraint function c at $x_k + p_k$ and the linear algebra required to calculate the step $\hat{p}_k$ from (15.36).

We now describe an algorithm that uses a merit function together with a line-search strategy and a second-order correction step. We assume that the search direction p_k and the penalty parameter μ_k are computed so that p_k is a descent direction for the merit function, that is, $D(\phi(x_k; \mu); p_k) < 0$. In Chapters 18 and 19, we discuss how to accomplish these goals. The key feature of the algorithm is that, if the full step $\alpha_k = 1$ does not produce satisfactory descent in the merit function, we try the second-order correction step *before* backtracking along the original direction p_k.

Algorithm 15.2 (Generic Algorithm with Second-Order Correction).
    Choose parameters $\eta \in (0, 0.5)$ and τ_1, τ_2 with $0 < \tau_1 < \tau_2 < 1$;
    Choose initial point x_0; set $k \leftarrow 0$;
    repeat until a convergence test is satisfied:
        Compute a search direction p_k;
        Set $\alpha_k \leftarrow 1$, newpoint $\leftarrow$ false;
        while newpoint = false
            if $\phi(x_k + \alpha_k p_k; \mu) \leq \phi(x_k; \mu) + \eta \alpha_k D(\phi(x_k; \mu); p_k)$
                Set $x_{k+1} \leftarrow x_k + \alpha_k p_k$;
                Set newpoint $\leftarrow$ true;

> **else if** $\alpha_k = 1$
>> Compute $\hat{p}_k$ from (15.36);
>> **if** $\phi(x_k + p_k + \hat{p}_k; \mu) \leq \phi(x_k; \mu) + \eta D(\phi(x_k; \mu); p_k)$
>>> Set $x_{k+1} \leftarrow x_k + p_k + \hat{p}_k$;
>>> Set newpoint $\leftarrow$ true;
>> **else**
>>> Choose new α_k in $[\tau_1\alpha_k, \tau_2\alpha_k]$;
>> **end**
> **else**
>> Choose new α_k in $[\tau_1\alpha_k, \tau_2\alpha_k]$;
> **end**
> **end while**
end repeat

In this algorithm, the full second-order correction step $\hat{p}_k$ is discarded if does not produce a reduction in the merit function. We do not backtrack along the direction $p_k + \hat{p}_k$ because it is not guaranteed to be a descent direction for the merit function. A variation of this algorithm applies the second-order correction step only if the sufficient decrease condition (15.29) is violated as a result of an increase in the norm of the constraints.

The second-order correction strategy is effective in practice. The cost of performing the extra constraint function evaluation and an additional backsolve in (15.36) is outweighed by added robustness and efficiency.

NONMONOTONE (WATCHDOG) STRATEGY

The inefficiencies caused by the Maratos effect can also be avoided by occasionally accepting steps that increase the merit function; such steps are called *relaxed steps*. There is a limit to our tolerance, however. If a sufficient reduction of the merit function has not been obtained within a certain number of iterates of the relaxed step ($\hat{t}$ iterates, say), then we return to the iterate before the relaxed step and perform a normal iteration, using a line search or some other technique to force a reduction in the merit function.

In contrast with the second-order correction, which aims only to improve satisfaction of the constraints, this nonmonotone strategy always takes regular steps p_k of the algorithm that aim both for improved feasibility and optimality. The hope is that any increase in the merit function over a single step will be temporary, and that subsequent steps will more than compensate for it.

We now describe a particular instance of the nonmonotone approach called the *watchdog strategy*. We set $\hat{t} = 1$, so that we allow the merit function to increase on just a single step before insisting on a sufficient decrease in the merit function. As above, we focus our discussion on a line search algorithm that uses a nonsmooth merit function ϕ. We assume that the penalty parameter μ is not changed until a successful cycle has been completed. To simplify the notation, we omit the dependence of ϕ on μ and write the merit function as $\phi(x)$ and the directional derivative as $D(\phi(x); p_k)$.

Algorithm 15.3 (Watchdog).

  Choose a constant $\eta \in (0, 0.5)$ and an initial point x_0;

  Set $k \leftarrow 0, S \leftarrow \{0\}$;

  repeat until a termination test is satisfied

        Compute a step p_k;

        Set $x_{k+1} \leftarrow x_k + p_k$;

        if $\phi(x_{k+1}) \leq \phi(x_k) + \eta D(\phi(x_k); p_k)$

           $k \leftarrow k + 1, S \leftarrow S \cup \{k\}$;

      else

           Compute a search direction p_{k+1} from x_{k+1};

           Find α_{k+1} such that

$$\phi(x_{k+2}) \leq \phi(x_{k+1}) + \eta \alpha_{k+1} D(\phi(x_{k+1}); p_{k+1});$$

           Set $x_{k+2} \leftarrow x_{k+1} + \alpha_{k+1} p_{k+1}$;

           if $\phi(x_{k+1}) \leq \phi(x_k)$ **or** $\phi(x_{k+2}) \leq \phi(x_k) + \eta D(\phi(x_k); p_k)$

               $k \leftarrow k + 2, S \leftarrow S \cup \{k\}$;

           else if $\phi(x_{k+2}) > \phi(x_k)$

               (* return to x_k and search along p_k *)

               Find α_k such that $\phi(x_{k+3}) \leq \phi(x_k) + \eta \alpha_k D(\phi(x_k); p_k)$;

               Compute $x_{k+3} = x_k + \alpha_k p_k$;

               $k \leftarrow k + 3, S \leftarrow S \cup \{k\}$;

           else

               Compute a direction p_{k+2} from x_{k+2};

               Find α_{k+2} such that

$$\phi(x_{k+3}) \leq \phi(x_{k+2}) + \eta \alpha_{k+2} D(\phi(x_{k+2}); p_{k+2});$$

               Set $x_{k+3} \leftarrow x_{k+2} + \alpha_{k+2} p_{k+2}$;

               $k \leftarrow k + 3, S \leftarrow S \cup \{k\}$;

           end

      end

  end (repeat)

The set S is not required by the algorithm and is introduced only to identify the iterates for which a sufficient merit function reduction was obtained. Note that at least a third of the iterates have their indices in S. By using this fact, one can show that various constrained optimization methods that use the watchdog technique are globally convergent. One can also show that for all sufficiently large k, the step length is $\alpha_k = 1$ and the convergence rate is superlinear.

In practice, it may be advantageous to allow increases in the merit function for more than one iteration. Values of $\hat{t}$ such as 5 or 8 are typical. As this discussion indicates, careful implementations of the watchdog technique have a certain degree of complexity, but the added complexity is worthwhile because the approach has good practical performance. A potential advantage of the watchdog technique over the second-order correction strategy is that it may require fewer evaluations of the constraint functions. In the best case, most of the steps will be full steps, and there will rarely be a need to return to an earlier point.

NOTES AND REFERENCES

Techniques for eliminating linear constraints are described, for example, in Fletcher [101] and Gill, Murray, and Wright [131]. For a thorough discussion of merit functions see Boggs and Tolle [33] and Conn, Gould, and Toint [74]. Some of the earliest references on nonmonotone methods include Grippo, Lampariello and Lucidi [158], and Chamberlain et al [57]; see [74] for a review of nonmonotone techniques and an extensive list of references. The concept of a filter was introduced by Fletcher and Leyffer [105]; our discussion of filters is based on that paper. Second-order correction steps are motivated and discussed in Fletcher [101].

✎ EXERCISES

✎   **15.1** In Example 15.1, consider these three choices of the working set: $\mathcal{W} = \{3\}$, $\mathcal{W} = \{1, 2\}$, $\mathcal{W} = \{2, 3\}$. Show that none of these working sets are the optimal active set for (15.5).

✎   **15.2** For the problem in Example 15.3, perform simple elimination of the variables x_2 and x_5 to obtain an unconstrained problem in the remaining variables x_1, x_3, x_4, and x_6. Similarly to (15.12), express the eliminated variables explicitly in terms of the retained variables.

✎   **15.3** Do the following problems have solutions? Explain.

$$\min x_1 + x_2 \quad \text{subject to } x_1^2 + x_2^2 = 2,\ 0 \le x_1 \le 1,\ 0 \le x_2 \le 1;$$
$$\min x_1 + x_2 \quad \text{subject to } x_1^2 + x_2^2 \le 1,\ x_1 + x_2 = 3;$$
$$\min x_1 x_2 \quad \text{subject to } x_1 + x_2 = 2.$$

✎   **15.4** Show that if in Example 15.2 we eliminate x in terms of y, then the correct solution of the problem is obtained by performing unconstrained minimization.

✎   **15.5** Show that the basis matrices (15.15) are linearly independent.

✎   **15.6** Show that the particular solution $Q_1 R^{-T} \Pi^T b$ of $Ax = b$ is identical to $A^T (AA^T)^{-1} b$.

15.7 In this exercise we compute basis matrices that attempt to compromise between the orthonormal basis (15.22) and simple elimination (15.15). We assume that the basis matrix is given by the first m columns of A, so that $P = I$ in (15.7), and define

$$Y = \begin{bmatrix} I \\ (B^{-1}N)^T \end{bmatrix}, \quad Z = \begin{bmatrix} -B^{-1}N \\ I \end{bmatrix}. \tag{15.37}$$

(a) Show that the columns of Y and Z are no longer of norm 1 and that the relations $AZ = 0$ and $Y^T Z = 0$ hold. Therefore, the columns of Y and Z form a linearly independent set, showing that (15.37) is a valid choice of the basis matrices.

(b) Show that the particular solution $Y(AY)^{-1}b$ defined by this choice of Y is, as in the orthogonal factorization approach, the minimum-norm solution of $Ax = b$. More specifically, show that

$$Y(AY)^{-1} = A^T(AA^T)^{-1}.$$

It follows that the matrix $Y(AY)^{-1}$ is independent of the choice of basis matrix B in (15.7), and its conditioning is determined by that of A alone. (Note, however, that the matrix Z still depends explicitly on B, so a careful choice of B is needed to ensure well conditioning in this part of the computation.)

15.8 Verify that by adding the inequality constraint $3x_1 + 2x_3 \geq 1$ to the problem (15.11), the elimination (15.12) transforms the problem into one of minimizing the function (15.13) subject to the inequality constraint (15.23).

CHAPTER *16*

Quadratic Programming

An optimization problem with a quadratic objective function and linear constraints is called a quadratic program. Problems of this type are important in their own right, and they also arise as subproblems in methods for general constrained optimization, such as sequential quadratic programming (Chapter 18), augmented Lagrangian methods (Chapter 17), and interior-point methods (Chapter 19).

The general quadratic program (QP) can be stated as

$$\min_{x} \quad q(x) = \tfrac{1}{2}x^{T}Gx + x^{T}c \tag{16.1a}$$

$$\text{subject to} \quad a_{i}^{T}x = b_{i}, \qquad i \in \mathcal{E}, \tag{16.1b}$$

$$a_{i}^{T}x \geq b_{i}, \qquad i \in \mathcal{I}, \tag{16.1c}$$

where G is a symmetric $n \times n$ matrix, $\mathcal{E}$ and $\mathcal{I}$ are finite sets of indices, and c, x, and $\{a_{i}\}, i \in \mathcal{E} \cup \mathcal{I}$, are vectors in $\mathbb{R}^{n}$. Quadratic programs can always be solved (or shown to be infeasible) in a finite amount of computation, but the effort required to find a solution depends strongly on the characteristics of the objective function and the number of inequality constraints. If the Hessian matrix G is positive semidefinite, we say that (16.1) is a *convex QP*, and in this case the problem is often similar in difficulty to a linear program. (*Strictly convex QPs* are those in which G is positive definite.) *Nonconvex QPs*, in which G is an indefinite matrix, can be more challenging because they can have several stationary points and local minima.

In this chapter we focus primarily on convex quadratic programs. We start by considering an interesting application of quadratic programming.

❑ **EXAMPLE 16.1**   (PORTFOLIO OPTIMIZATION)

Every investor knows that there is a tradeoff between risk and return: To increase the expected return on investment, an investor must be willing to tolerate greater risks. Portfolio theory studies how to model this tradeoff given a collection of n possible investments with returns $r_{i}, i = 1, 2, \ldots, n$. The returns r_{i} are usually not known in advance and are often assumed to be random variables that follow a normal distribution. We can characterize these variables by their expected value $\mu_{i} = E[r_{i}]$ and their variance $\sigma_{i}^{2} = E[(r_{i} - \mu_{i})^{2}]$. The variance measures the fluctuations of the variable r_{i} about its mean, so that larger values of σ_{i} indicate riskier investments. The returns are not in general independent, and we can define correlations between pairs of returns as follows:

$$\rho_{ij} = \frac{E[(r_{i} - \mu_{i})(r_{j} - \mu_{j})]}{\sigma_{i}\sigma_{j}}, \qquad \text{for } i, j = 1, 2, \ldots, n.$$

The correlation measures the tendency of the return on investments i and j to move in the same direction. Two investments whose returns tend to rise and fall together have a positive correlation; the nearer ρ_{ij} is to 1, the more closely the two investments track each other. Investments whose returns tend to move in opposite directions have a negative correlation.

An investor constructs a portfolio by putting a fraction x_{i} of the available funds into investment i, for $i = 1, 2, \ldots, n$. Assuming that all available funds are invested and that short-selling is not allowed, the constraints are $\sum_{i=1}^{n} x_{i} = 1$ and $x \geq 0$. The return on the portfolio is given by

$$R = \sum_{i=1}^{n} x_i r_i. \tag{16.2}$$

To measure the desirability of the portfolio, we need to obtain measures of its expected return and variance. The expected return is simply

$$E[R] = E\left[\sum_{i=1}^{n} x_i r_i\right] = \sum_{i=1}^{n} x_i E[r_i] = x^T \mu,$$

while the variance is given by

$$\mathrm{Var}[R] = E[(R - E[R])^2] = \sum_{i=1}^{n}\sum_{j=1}^{n} x_i x_j \sigma_i \sigma_j \rho_{ij} = x^T G x,$$

where the $n \times n$ symmetric positive semidefinite matrix G defined by

$$G_{ij} = \rho_{ij}\sigma_i\sigma_j$$

is called the *covariance matrix*.

Ideally, we would like to find a portfolio for which the expected return $x^T \mu$ is large while the variance $x^T G x$ is small. In the model proposed by Markowitz [201], we combine these two aims into a single objective function with the aid of a "risk tolerance parameter" denoted by κ, and we solve the following problem to find the optimal portfolio:

$$\max x^T \mu - \kappa x^T G x, \quad \text{subject to} \quad \sum_{i=1}^{n} x_i = 1, \ x \geq 0.$$

The value chosen for the nonnegative parameter κ depends on the preferences of the individual investor. Conservative investors, who place more emphasis on minimizing risk in their portfolio, would choose a large value of κ to increase the weight of the variance measure in the objective function. More daring investors, who are prepared to take on more risk in the hope of a higher expected return, would choose a smaller value of κ.

The difficulty in applying this portfolio optimization technique to real-life investing lies in defining the expected returns, variances, and correlations for the investments in question. Financial professionals often combine historical data with their own insights and expectations to produce values of these quantities.

□

16.1 EQUALITY-CONSTRAINED QUADRATIC PROGRAMS

We begin our discussion of algorithms for quadratic programming by considering the case in which only equality constraints are present. Techniques for this special case are applicable also to problems with inequality constraints since, as we see later in this chapter, some algorithms for general QP require the solution of an equality-constrained QP at each iteration.

PROPERTIES OF EQUALITY-CONSTRAINED QPs

For simplicity, we write the equality constraints in matrix form and state the equality-constrained QP as follows:

$$\min_x \quad q(x) \stackrel{\text{def}}{=} \tfrac{1}{2}x^T G x + x^T c \tag{16.3a}$$

$$\text{subject to} \quad Ax = b, \tag{16.3b}$$

where A is the $m \times n$ Jacobian of constraints (with $m \leq n$) whose rows are a_i^T, $i \in \mathcal{E}$ and b is the vector in $\mathbb{R}^m$ whose components are b_i, $i \in \mathcal{E}$. For the present, we assume that A has full row rank (rank m) so that the constraints (16.3b) are consistent. (In Section 16.8 we discuss the case in which A is rank deficient.)

The first-order necessary conditions for x^* to be a solution of (16.3) state that there is a vector λ^* such that the following system of equations is satisfied:

$$\begin{bmatrix} G & -A^T \\ A & 0 \end{bmatrix} \begin{bmatrix} x^* \\ \lambda^* \end{bmatrix} = \begin{bmatrix} -c \\ b \end{bmatrix}. \tag{16.4}$$

These conditions are a consequence of the general result for first-order optimality conditions, Theorem 12.1. As in Chapter 12, we call λ^* the vector of Lagrange multipliers. The system (16.4) can be rewritten in a form that is useful for computation by expressing x^* as $x^* = x + p$, where x is some estimate of the solution and p is the desired step. By introducing this notation and rearranging the equations, we obtain

$$\begin{bmatrix} G & A^T \\ A & 0 \end{bmatrix} \begin{bmatrix} -p \\ \lambda^* \end{bmatrix} = \begin{bmatrix} g \\ h \end{bmatrix}, \tag{16.5}$$

where

$$h = Ax - b, \qquad g = c + Gx, \qquad p = x^* - x. \tag{16.6}$$

The matrix in (16.5) is called the Karush–Kuhn–Tucker (KKT) matrix, and the following result gives conditions under which it is nonsingular. As in Chapter 15, we use Z to denote the $n \times (n - m)$ matrix whose columns are a basis for the null space of A. That is, Z has full rank and satisfies $AZ = 0$.

Lemma 16.1.

  Let A have full row rank, and assume that the reduced-Hessian matrix $Z^T G Z$ is positive definite. Then the KKT matrix

$$K = \begin{bmatrix} G & A^T \\ A & 0 \end{bmatrix} \tag{16.7}$$

is nonsingular, and hence there is a unique vector pair (x^*, λ^*) satisfying (16.4).

PROOF.   Suppose there are vectors w and v such that

$$\begin{bmatrix} G & A^T \\ A & 0 \end{bmatrix} \begin{bmatrix} w \\ v \end{bmatrix} = 0. \tag{16.8}$$

Since $Aw = 0$, we have from (16.8) that

$$0 = \begin{bmatrix} w \\ v \end{bmatrix}^T \begin{bmatrix} G & A^T \\ A & 0 \end{bmatrix} \begin{bmatrix} w \\ v \end{bmatrix} = w^T G w.$$

Since w lies in the null space of A, it can be written as $w = Zu$ for some vector $u \in \mathbf{R}^{n-m}$. Therefore, we have

$$0 = w^T G w = u^T Z^T G Z u,$$

which by positive definiteness of $Z^T G Z$ implies that $u = 0$. Therefore, $w = 0$, and by (16.8), $A^T v = 0$. Full row rank of A then implies that $v = 0$. We conclude that equation (16.8) is satisfied only if $w = 0$ and $v = 0$, so the matrix is nonsingular, as claimed.   □

❑ **EXAMPLE 16.2**

  Consider the quadratic programming problem

$$\min q(x) = 3x_1^2 + 2x_1x_2 + x_1x_3 + 2.5x_2^2 + 2x_2x_3 + 2x_3^2 - 8x_1 - 3x_2 - 3x_3,$$
$$\text{subject to} \quad x_1 + x_3 = 3, \quad x_2 + x_3 = 0. \tag{16.9}$$

We can write this problem in the form (16.3) by defining

$$
G = \begin{bmatrix} 6 & 2 & 1 \\ 2 & 5 & 2 \\ 1 & 2 & 4 \end{bmatrix}, \quad
c = \begin{bmatrix} -8 \\ -3 \\ -3 \end{bmatrix}, \quad
A = \begin{bmatrix} 1 & 0 & 1 \\ 0 & 1 & 1 \end{bmatrix}, \quad
b = \begin{bmatrix} 3 \\ 0 \end{bmatrix}.
$$

The solution x^* and optimal Lagrange multiplier vector λ^* are given by

$$
x^* = (2, -1, 1)^T, \qquad \lambda^* = (3, -2)^T.
$$

In this example, the matrix G is positive definite, and the null-space basis matrix can be defined as in (15.15), giving

$$
Z = (-1, -1, 1)^T. \tag{16.10}
$$

❑

We have seen that when the conditions of Lemma 16.1 are satisfied, there is a unique vector pair (x^*, λ^*) that satisfies the first-order necessary conditions for (16.3). In fact, the second-order sufficient conditions (see Theorem 12.6) are also satisfied at (x^*, λ^*), so x^* is a strict local minimizer of (16.3). In fact, we can use a direct argument to show that x^* is a *global* solution of (16.3).

Theorem 16.2.
Let A have full row rank and assume that the reduced-Hessian matrix $Z^T G Z$ is positive definite. Then the vector x^* satisfying (16.4) is the unique global solution of (16.3).

PROOF.   Let x be any other feasible point (satisfying $Ax = b$), and as before, let p denote the difference $x^* - x$. Since $Ax^* = Ax = b$, we have that $Ap = 0$. By substituting into the objective function (16.3a), we obtain

$$
\begin{aligned}
q(x) &= \tfrac{1}{2}(x^* - p)^T G(x^* - p) + c^T(x^* - p) \\
&= \tfrac{1}{2} p^T G p - p^T G x^* - c^T p + q(x^*).
\end{aligned} \tag{16.11}
$$

From (16.4) we have that $Gx^* = -c + A^T \lambda^*$, so from $Ap = 0$ we have that

$$
p^T G x^* = p^T(-c + A^T \lambda^*) = -p^T c.
$$

By substituting this relation into (16.11), we obtain

$$
q(x) = \tfrac{1}{2} p^T G p + q(x^*).
$$

Since p lies in the null space of A, we can write $p = Zu$ for some vector $u \in \mathbb{R}^{n-m}$, so that

$$q(x) = \tfrac{1}{2} u^T Z^T G Z u + q(x^*).$$

By positive definiteness of $Z^T G Z$, we conclude that $q(x) > q(x^*)$ except when $u = 0$, that is, when $x = x^*$. Therefore, x^* is the unique global solution of (16.3). $\qquad \square$

When the reduced Hessian matrix $Z^T G Z$ is positive semidefinite with zero eigenvalues, the vector x^* satisfying (16.4) is a local minimizer but not a strict local minimizer. If the reduced Hessian has negative eigenvalues, then x^* is only a stationary point, not a local minimizer.

16.2 DIRECT SOLUTION OF THE KKT SYSTEM

In this section we discuss efficient methods for solving the KKT system (16.5). The first important observation is that if $m \geq 1$, the KKT matrix is always indefinite. We define the inertia of a symmetric matrix K to be the scalar triple that indicates the numbers n_+, n_-, and n_0 of positive, negative, and zero eigenvalues, respectively, that is,

$$\text{inertia}(K) = (n_+, n_-, n_0).$$

The following result characterizes the inertia of the KKT matrix.

Theorem 16.3.

Let K be defined by (16.7), and suppose that A has rank m. Then

$$\text{inertia}(K) = \text{inertia}(Z^T G Z) + (m, m, 0).$$

Therefore, if $Z^T G Z$ is positive definite, $\text{inertia}(K) = (n, m, 0)$.

The proof of this result is given in [111], for example. Note that the assumptions of this theorem are satisfied by Example 16.2. Hence, if we construct the 5×5 matrix K using the data of this example, we obtain $\text{inertia}(K) = (3, 2, 0)$.

Knowing that the KKT system is indefinite, we now describe the main direct techniques used to solve (16.5).

FACTORING THE FULL KKT SYSTEM

One option for solving (16.5) is to perform a triangular factorization on the full KKT matrix and then perform backward and forward substitution with the triangular factors. Because of indefiniteness, we cannot use the Cholesky factorization. We could use Gaussian elimination with partial pivoting (or a sparse variant thereof) to obtain the L and U factors, but this approach has the disadvantage that it ignores the symmetry.

The most effective strategy in this case is to use a *symmetric indefinite factorization*, which we have discussed in Chapter 3 and the Appendix. For a general symmetric matrix K, this factorization has the form

$$P^T K P = L B L^T, \tag{16.12}$$

where P is a permutation matrix, L is unit lower triangular, and B is block-diagonal with either 1×1 or 2×2 blocks. The symmetric permutations defined by the matrix P are introduced for numerical stability of the computation and, in the case of large sparse K, for maintaining sparsity. The computational cost of the symmetric indefinite factorization (16.12) is typically about half the cost of sparse Gaussian elimination.

To solve (16.5), we first compute the factorization (16.12) of the coefficient matrix. We then perform the following sequence of operations to arrive at the solution:

$$
\begin{aligned}
\text{solve } Lz = P^T \begin{bmatrix} g \\ h \end{bmatrix} \qquad & \text{to obtain } z; \\
\text{solve } B\hat{z} = z \qquad & \text{to obtain } \hat{y}; \\
\text{solve } L^T \bar{z} = \hat{z} \qquad & \text{to obtain } \bar{z}; \\
\text{set } \begin{bmatrix} -p \\ \lambda^* \end{bmatrix} = P\bar{z}. &
\end{aligned}
$$

Since multiplications with the permutation matrices P and P^T can be performed by simply rearranging vector components, they are inexpensive. Solution of the system $B\hat{z} = z$ entails solving a number of small 1×1 and 2×2 systems, so the number of operations is a small multiple of the system dimension $(m + n)$, again inexpensive. Triangular substitutions with L and L^T are more costly. Their precise cost depends on the amount of sparsity, but is usually significantly less than the cost of performing the factorization (16.12).

This approach of factoring the full $(n + m) \times (n + m)$ KKT matrix (16.7) is quite effective on many problems. It may be expensive, however, when the heuristics for choosing the permutation matrix P are not able to maintain sparsity in the L factor, so that L becomes much more dense than the original coefficient matrix.

SCHUR-COMPLEMENT METHOD

Assuming that G is positive definite, we can multiply the first equation in (16.5) by AG^{-1} and then subtract the second equation to obtain a linear system in the vector λ^* alone:

$$(AG^{-1}A^T)\lambda^* = (AG^{-1}g - h). \tag{16.13}$$

We solve this symmetric positive definite system for λ^* and then recover p from the first equation in (16.5) by solving

$$Gp = A^T\lambda^* - g. \tag{16.14}$$

This approach requires us to perform operations with G^{-1}, as well as to compute the factorization of the $m \times m$ matrix $AG^{-1}A^T$. Therefore, it is most useful when:

- G is well conditioned and easy to invert (for instance, when G is diagonal or block-diagonal); or

- G^{-1} is known explicitly through a quasi-Newton updating formula; or

- the number of equality constraints m is small, so that the number of backsolves needed to form the matrix $AG^{-1}A^T$ is not too large.

The name "Schur-Complement method" derives from the fact that, by applying block Gaussian elimination to (16.7) using G as the pivot, we obtain the block upper triangular system

$$\begin{bmatrix} G & A^T \\ 0 & -AG^{-1}A^T \end{bmatrix}. \tag{16.15}$$

In linear algebra terminology, the matrix $AG^{-1}A^T$ is the Schur complement of G in the matrix K of (16.7). By applying this block elimination technique to the system (16.5), and performing a block backsolve, we obtain (16.13), (16.14).

We can use an approach like the Schur-complement method to derive an explicit inverse formula for the KKT matrix in (16.5). This formula is

$$\begin{bmatrix} G & A^T \\ A & 0 \end{bmatrix}^{-1} = \begin{bmatrix} C & E \\ E^T & F \end{bmatrix}, \tag{16.16}$$

with

$$C = G^{-1} - G^{-1}A^T(AG^{-1}A^T)^{-1}AG^{-1},$$
$$E = G^{-1}A^T(AG^{-1}A^T)^{-1},$$
$$F = -(AG^{-1}A^T)^{-1}.$$

The solution of (16.5) can be obtained by multiplying its right-hand side by this inverse matrix. If we take advantage of common expressions, and group the terms appropriately, we recover the approach (16.13), (16.14).

NULL-SPACE METHOD

The null-space method does not require nonsingularity of G and therefore has wider applicability than the Schur-complement method. It assumes only that the conditions of Lemma 16.1 hold, namely, that A has full row rank and that $Z^T G Z$ is positive definite. However, it requires knowledge of the null-space basis matrix Z. Like the Schur-complement method, it exploits the block structure in the KKT system to decouple (16.5) into two smaller systems.

Suppose that we partition the vector p in (16.5) into two components, as follows:

$$p = Y p_Y + Z p_Z, \qquad (16.17)$$

where Z is the $n \times (n - m)$ null-space matrix, Y is any $n \times m$ matrix such that $[Y \mid Z]$ is nonsingular, p_Y is an m-vector, and p_Z is an $(n - m)$-vector. The matrices Y and Z were discussed in Section 15.3, where Figure 15.4 shows that $Y x_Y$ is a particular solution of $Ax = b$, while $Z x_Z$ is a displacement along these constraints.

By substituting p into the second equation of (16.5) and recalling that $AZ = 0$, we obtain

$$(AY) p_Y = -h. \qquad (16.18)$$

Since A has rank m and $[Y \mid Z]$ is $n \times n$ nonsingular, the product $A[Y \mid Z] = [AY \mid 0]$ has rank m. Therefore, AY is a nonsingular $m \times m$ matrix, and p_Y is well determined by the equations (16.18). Meanwhile, we can substitute (16.17) into the first equation of (16.5) to obtain

$$-G Y p_Y - G Z p_Z + A^T \lambda^* = g$$

and multiply by Z^T to obtain

$$(Z^T G Z) p_Z = -Z^T G Y p_Y - Z^T g. \qquad (16.19)$$

This system can be solved by performing a Cholesky factorization of the reduced-Hessian matrix $Z^T G Z$ to determine p_Z. We therefore can compute the total step $p = Y p_Y + Z p_Z$. To obtain the Lagrange multiplier, we multiply the first block row in (16.5) by Y^T to obtain the linear system

$$(AY)^T \lambda^* = Y^T (g + Gp), \qquad (16.20)$$

which can be solved for λ^*.

❏ **EXAMPLE 16.3**

Consider the problem (16.9) given in Example 16.2. We can choose

$$Y = \begin{bmatrix} 2/3 & -1/3 \\ -1/3 & 2/3 \\ 1/3 & 1/3 \end{bmatrix}$$

and set Z as in (16.10). Note that $AY = I$.

Suppose we have $x = (0, 0, 0)^T$ in (16.6). Then

$$h = Ax - b = -b, \qquad g = c + Gx = c = \begin{bmatrix} -8 \\ -3 \\ -3 \end{bmatrix}.$$

Simple calculation shows that

$$p_Y = \begin{bmatrix} 3 \\ 0 \end{bmatrix}, \qquad p_Z = \begin{bmatrix} 0 \end{bmatrix},$$

so that

$$p = x^* - x = Yp_Y + Zp_Z = \begin{bmatrix} 2 \\ -1 \\ 1 \end{bmatrix}.$$

After recovering λ^* from (16.20), we conclude that

$$x^* = \begin{bmatrix} 2 \\ -1 \\ 1 \end{bmatrix}, \qquad \lambda^* = \begin{bmatrix} 3 \\ -2 \end{bmatrix}.$$

❏

The null-space approach can be very effective when the number of degrees of freedom $n - m$ is small. Its main limitation lies in the need for the null-space matrix Z which, as we have seen in Chapter 15, can be expensive to compute in some large problems. The matrix Z is not uniquely defined and, if it is poorly chosen, the reduced system (16.19) may become ill conditioned. If we choose Z to have orthonormal columns, as is normally done in software for small and medium-sized problems, then the conditioning of $Z^T G Z$ is at least as good as that of G itself. When A is large and sparse, however, an orthonormal Z is expensive to compute, so for practical reasons we are often forced to use one of the less reliable choices of Z described in Chapter 15.

It is difficult to give hard and fast rules about the relative effectiveness of null-space and Schur-complement methods, because factors such as fill-in during computation of Z vary significantly even among problems of the same dimension. In general, we can recommend the Schur-complement method if G is positive definite and $AG^{-1}A^T$ can be computed relatively cheaply (because G is easy to invert or because m is small relative to n). Otherwise, the null-space method is often preferable, in particular when it is much more expensive to compute factors of G than to compute the null-space matrix Z and the factors of $Z^T GZ$.

16.3 ITERATIVE SOLUTION OF THE KKT SYSTEM

An alternative to the direct factorization techniques discussed in the previous section is to use an iterative method to solve the KKT system (16.5). Iterative methods are suitable for solving very large systems and often lend themselves well to parallelization. The conjugate gradient (CG) method is not recommended for solving the full system (16.5) as written, because it can be unstable on systems that are not positive definite. Better options are Krylov methods for general linear or symmetric indefinite systems. Candidates include the GMRES, QMR, and LSQR methods; see the Notes and References at the end of the chapter. Other iterative methods can be derived from the null-space approach by applying the conjugate gradient method to the reduced system (16.19). Methods of this type are key to the algorithms of Chapters 18 and 19, and are discussed in the remainder of this section. We assume throughout that $Z^T GZ$ is positive definite.

CG APPLIED TO THE REDUCED SYSTEM

We begin our discussion of iterative null-space methods by deriving the underlying equations in the notation of the equality-constrained QP (16.3). Expressing the solution of the quadratic program (16.3) as

$$x^* = Yx_Y + Zx_Z,\qquad(16.21)$$

for some vectors $x_Z \in \mathbb{R}^{n-m}$, $x_Y \in \mathbb{R}^m$, the constraints $Ax = b$ yield

$$AYx_Y = b,\qquad(16.22)$$

which determines the vector x_Y. In Chapter 15, various practical choices of Y are described, some of which allow (16.22) to be solved economically. Substituting (16.21) into (16.3), we see that x_Z solves the unconstrained reduced problem

$$\min_{x_Z} \tfrac{1}{2}x_Z{}^T Z^T GZx_Z + x_Z{}^T c_Z,$$

where

$$c_z = Z^T GY x_Y + Z^T c. \tag{16.23}$$

The solution x_Z satisfies the linear system

$$Z^T GZ x_Z = -c_z. \tag{16.24}$$

Since $Z^T GZ$ is positive definite, we can apply the CG method to this linear system and substitute x_Z into (16.21) to obtain a solution of (16.3).

As discussed in Chapter 5, preconditioning can improve the rate of convergence of the CG iteration, so we assume that a preconditioner W_{ZZ} is given. The preconditioned CG method (Algorithm 5.3) applied to the $(n - m)$-dimensional reduced system (16.24) is as follows. (We denote the steps produced by the CG iteration by d_Z.)

Algorithm 16.1 (Preconditioned CG for Reduced Systems).
Choose an initial point x_Z;
Compute $r_Z = Z^T GZ x_Z + c_z$, $g_Z = W_{ZZ}^{-1} r_Z$, and $d_Z = -g_Z$;
repeat

$$\alpha \leftarrow r_Z^T g_Z / d_Z^T Z^T GZ d_Z; \tag{16.25a}$$

$$x_Z \leftarrow x_Z + \alpha d_Z; \tag{16.25b}$$

$$r_Z^+ \leftarrow r_Z + \alpha Z^T GZ d_Z; \tag{16.25c}$$

$$g_Z^+ \leftarrow W_{ZZ}^{-1} r_Z^+; \tag{16.25d}$$

$$\beta \leftarrow (r_Z^+)^T g_Z^+ / r_Z^T g_Z; \tag{16.25e}$$

$$d_Z \leftarrow -g_Z^+ + \beta d_Z; \tag{16.25f}$$

$$g_Z \leftarrow g_Z^+; \quad r_Z \leftarrow r_Z^+; \tag{16.25g}$$

until a termination test is satisfied.

This iteration may be terminated when, for example, $r_Z^T W_{ZZ}^{-1} r_Z$ is sufficiently small.

In this approach, it is not necessary to form the reduced Hessian $Z^T GZ$ explicitly because the CG method requires only that we compute matrix-vector products involving this matrix. In fact, it is not even necessary to form Z explicitly as long as we are able to compute products of Z and Z^T with arbitrary vectors. For some choices of Z, these products are much cheaper to compute than Z itself, as we have seen in Chapter 15.

The preconditioner W_{ZZ} is a symmetric, positive definite matrix of dimension $n - m$, which might be chosen to cluster the eigenvalues of $W_{ZZ}^{-1/2}(Z^T GZ)W_{ZZ}^{-1/2}$ and to reduce the span between the smallest and largest eigenvalues. An ideal choice of preconditioner is one for which $W_{ZZ}^{-1/2}(Z^T GZ)W_{ZZ}^{-1/2} = I$, that is, $W_{ZZ} = Z^T GZ$. Motivated by this ideal, we consider preconditioners of the form

$$W_{ZZ} = Z^T HZ, \tag{16.26}$$

where H is a symmetric matrix such that $Z^T H Z$ is positive definite. Some choices of H are discussed below. Preconditioners of the form (16.26) allow us to apply the CG method in n-dimensional space, as we discuss next.

THE PROJECTED CG METHOD

It is possible to design a modification of the Algorithm 16.1 that avoids operating with the null-space basis Z, provided we use a preconditioner of the form (16.26) and a particular solution of the equation $Ax = b$. This approach works implicitly with an orthogonal matrix Z and is not affected by ill conditioning in A or by a poor choice of Z.

After the solution x_z of (16.24) has been computed by using Algorithm 16.1, it must be multiplied by Z and substituted in (16.21) to give the solution of the quadratic program (16.3). Alternatively, we may rewrite Algorithm 16.1 to work directly with the vector $x = Zx_z + Yx_Y$, where the Yx_Y term is fixed at the start and the x_z term is updated (implicitly) within each iteration. To specify this form of the CG algorithm, we introduce the n-vectors x, r, g, and d, which satisfy $x = Zx_z + Yx_Y$, $Z^T r = r_z$, $g = Zg_z$, and $d = Zd_z$, respectively. We also define the scaled $n \times n$ projection matrix P as follows:

$$P = Z(Z^T H Z)^{-1} Z^T, \tag{16.27}$$

where H is the preconditioning matrix from (16.26). The CG iteration in n-dimensional space can be specified as follows.

Algorithm 16.2 (Projected CG Method).
  Choose an initial point x satisfying $Ax = b$;
  Compute $r = Gx + c$, $g = Pr$, and $d = -g$;
  repeat

$$\alpha \leftarrow r^T g / d^T G d; \tag{16.28a}$$
$$x \leftarrow x + \alpha d; \tag{16.28b}$$
$$r^+ \leftarrow r + \alpha G d; \tag{16.28c}$$
$$g^+ \leftarrow P r^+; \tag{16.28d}$$
$$\beta \leftarrow (r^+)^T g^+ / r^T g; \tag{16.28e}$$
$$d \leftarrow -g^+ + \beta d; \tag{16.28f}$$
$$g \leftarrow g^+; \qquad r \leftarrow r^+; \tag{16.28g}$$

until a convergence test is satisfied.

A practical stop test is to terminate when $r^T g = r^T Pr$ is smaller than a prescribed tolerance.

Note that the vector g^+, which we call the *preconditioned residual,* has been defined to be in the null space of A. As a result, in exact arithmetic, all the search directions d generated by Algorithm 16.2 also lie in the null space of A, and thus the iterates x all satisfy $Ax = b$. It is not difficult to verify (see Exercise 16.14) that the iteration is well defined if $Z^T GZ$ and $Z^T HZ$ are positive definite. The reader can also verify that the iterates x generated by Algorithm 16.2 are related to the iterates x_z of Algorithm 16.1 via (16.21).

Two simple choices of the preconditioning matrix H are $H = \text{diag}(|G_{ii}|)$ and $H = I$. In some applications, it is effective to define H as a block diagonal submatrix of G.

Algorithm 16.2 makes use of the null-space basis Z only through the operator (16.27). It is possible, however, to compute Pr without knowing a representation of the null-space basis Z. For simplicity, we first consider the case in which $H = I$, so that P is the orthogonal projection operator onto the null space of A. We use P_I to denote this special case of P, that is,

$$P_I = Z(Z^T Z)^{-1} Z^T. \tag{16.29}$$

The computation of the preconditioned residual $g^+ = P_I r^+$ in (16.28d) can be performed in two ways. The first is to express P_I by the equivalent formula

$$P_I = I - A^T (AA^T)^{-1} A \tag{16.30}$$

and thus compute $g^+ = P_I r^+$. We can then write $g^+ = r^+ - A^T v^+$, where v^+ is the solution of the system

$$AA^T v^+ = Ar^+. \tag{16.31}$$

This approach for computing the projection $g^+ = P_I r^+$ is called the *normal equations approach;* the system (16.31) can be solved by using a Cholesky factorization of AA^T.

The second approach is to express the projection (16.28d) as the solution of the augmented system

$$\begin{bmatrix} I & A^T \\ A & 0 \end{bmatrix} \begin{bmatrix} g^+ \\ v^+ \end{bmatrix} = \begin{bmatrix} r^+ \\ 0 \end{bmatrix}, \tag{16.32}$$

which can be solved by means of a symmetric indefinite factorization, as discussed earlier. We call this approach the *augmented system approach.*

We suppose now that the preconditioning has the general form of (16.27) and (16.28d). When H is nonsingular, we can compute g^+ as follows:

$$g^+ = Pr^+, \quad \text{where} \quad P = H^{-1}\left(I - A^T(AH^{-1}A^T)^{-1}AH^{-1}\right). \qquad (16.33)$$

Otherwise, when $z^T H z \neq 0$ for all nonzero z with $Az = 0$, we can find g^+ as the solution of the system

$$\begin{bmatrix} H & A^T \\ A & 0 \end{bmatrix} \begin{bmatrix} g^+ \\ v^+ \end{bmatrix} = \begin{bmatrix} r^+ \\ 0 \end{bmatrix}. \qquad (16.34)$$

While (16.33) is unappealing when H^{-1} does not have a simple form, (16.34) is a useful generalization of (16.32). A "perfect" preconditioner is obtained by taking $H = G$, but other choices for H are also possible, provided that $Z^T H Z$ is positive definite. The matrix in (16.34) is often called a *constraint preconditioner*.

None of these procedures for computing the projection makes use of a null-space basis Z; only the factorization of matrices involving A is required. Significantly, all these forms allow us to compute an initial point satisfying $Ax = b$. The operator $g^+ = P_I r^+$ relies on a factorization of AA^T from which we can compute $x = A^T(AA^T)^{-1}b$, while factorizations of the system matrices in (16.32) and (16.34) allow us to find a suitable x by solving

$$\begin{bmatrix} I & A^T \\ A & 0 \end{bmatrix} \begin{bmatrix} x \\ y \end{bmatrix} = \begin{bmatrix} 0 \\ b \end{bmatrix} \quad \text{or} \quad \begin{bmatrix} H & A^T \\ A & 0 \end{bmatrix} \begin{bmatrix} x \\ y \end{bmatrix} = \begin{bmatrix} 0 \\ b \end{bmatrix}.$$

Therefore we can compute an initial point for Algorithm 16.2 at the cost of one backsolve, using the factorization of the system needed to perform the projection operators.

We point out that these approaches for computing g^+ can give rise to significant round-off errors, so the use of iterative refinement is recommended to improve accuracy.

16.4 INEQUALITY-CONSTRAINED PROBLEMS

In the remainder of the chapter we discuss several classes of algorithms for solving convex quadratic programs that contain both inequality and equality constraints. *Active-set methods* have been widely used since the 1970s and are effective for small- and medium-sized problems. They allow for efficient detection of unboundedness and infeasibility and typically return an accurate estimate of the optimal active set. *Interior-point methods* are more recent, having become popular in the 1990s. They are well suited for large problems but may not be the most effective when a series of related QPs must be solved. We also study a special type of active-set methods called a *gradient projection method,* which is most effective when the only constraints in the problem are bounds on the variables.

OPTIMALITY CONDITIONS FOR INEQUALITY-CONSTRAINED PROBLEMS

We begin our discussion with a brief review of the optimality conditions for inequality-constrained quadratic programming, then discuss some of the less obvious properties of the solutions.

Theorem 12.1 can be applied to (16.1) by noting that the Lagrangian for this problem is

$$\mathcal{L}(x, \lambda) = \tfrac{1}{2} x^T G x + x^T c - \sum_{i \in \mathcal{I} \cup \mathcal{E}} \lambda_i (a_i^T x - b_i). \tag{16.35}$$

As in Definition 12.1, the active set $\mathcal{A}(x^*)$ consists of the indices of the constraints for which equality holds at x^*:

$$\mathcal{A}(x^*) = \left\{ i \in \mathcal{E} \cup \mathcal{I} \mid a_i^T x^* = b_i \right\}. \tag{16.36}$$

By specializing the KKT conditions (12.34) to this problem, we find that any solution x^* of (16.1) satisfies the following first-order conditions, for some Lagrange multipliers λ_i^*, $i \in \mathcal{A}(x^*)$:

$$Gx^* + c - \sum_{i \in \mathcal{A}(x^*)} \lambda_i^* a_i = 0, \tag{16.37a}$$

$$a_i^T x^* = b_i, \qquad \text{for all } i \in \mathcal{A}(x^*), \tag{16.37b}$$

$$a_i^T x^* \geq b_i, \qquad \text{for all } i \in \mathcal{I} \backslash \mathcal{A}(x^*), \tag{16.37c}$$

$$\lambda_i^* \geq 0, \qquad \text{for all } i \in \mathcal{I} \cap \mathcal{A}(x^*). \tag{16.37d}$$

A technical point: In Theorem 12.1 we assumed that the linear independence constraint qualification (LICQ) was satisfied. As mentioned in Section 12.6, this theorem still holds if we replace LICQ by other constraint qualifications, such as linearity of the constraints, which is certainly satisfied for quadratic programming. Hence, in the optimality conditions for quadratic programming given above, we need not assume that the active constraints are linearly independent at the solution.

For convex QP, when G is positive semidefinite, the conditions (16.37) are in fact sufficient for x^* to be a global solution, as we now prove.

Theorem 16.4.

If x^* satisfies the conditions (16.37) for some $\lambda_i^*, i \in \mathcal{A}(x^*)$, and G is positive semidefinite, then x^* is a global solution of (16.1).

PROOF. If x is any other feasible point for (16.1), we have that $a_i^T x = b_i$ for all $i \in \mathcal{E}$ and $a_i^T x \geq b_i$ for all $i \in \mathcal{A}(x^*) \cap \mathcal{I}$. Hence, $a_i^T (x - x^*) = 0$ for all $i \in \mathcal{E}$ and $a_i^T (x - x^*) \geq 0$ for all $i \in \mathcal{A}(x^*) \cap \mathcal{I}$. Using these relationships, together with (16.37a) and (16.37d), we have that

$$(x - x^*)^T (Gx^* + c) = \sum_{i \in \mathcal{E}} \lambda_i^* a_i^T (x - x^*) + \sum_{i \in \mathcal{A}(x^*) \cap \mathcal{I}} \lambda_i^* a_i^T (x - x^*) \geq 0. \qquad (16.38)$$

By elementary manipulation, we find that

$$
\begin{aligned}
q(x) &= q(x^*) + (x - x^*)^T (Gx^* + c) + \tfrac{1}{2}(x - x^*)^T G(x - x^*) \\
&\geq q(x^*) + \tfrac{1}{2}(x - x^*)^T G(x - x^*) \\
&\geq q(x^*),
\end{aligned}
$$

where the first inequality follows from (16.38) and the second inequality follows from positive semidefiniteness of G. We have shown that $q(x) \geq q(x^*)$ for any feasible x, so x^* is a global solution. $\qquad \square$

By a trivial modification of this proof, we see that x^* is actually the unique global solution when G is positive definite.

We can also apply the theory from Section 12.5 to derive second-order optimality conditions for (16.1). Second-order sufficient conditions for x^* to be a local minimizer are satisfied if $Z^T G Z$ is positive definite, where Z is defined to be a null-space basis matrix for the active constraint Jacobian matrix, which is the matrix whose rows are a_i^T for all $i \in \mathcal{A}(x^*)$. In this case, x^* is a strict local solution, according to Theorem 12.6.

When G is not positive definite, the general problem (16.1) may have more than one strict local solution. As mentioned above, such problems are called "nonconvex QPs" or "indefinite QPs," and they cause some complications for algorithms. Examples of indefinite QPs are illustrated in Figure 16.1. On the left we have plotted the feasible region and the contours of a quadratic objective $q(x)$ in which G has one positive and one negative eigenvalue. We have indicated by $+$ or $-$ that the function tends toward plus or minus infinity in that direction. Note that x^{**} is a local maximizer, x^* a local minimizer, and the center of the box is a stationary point. The picture on the right in Figure 16.1, in which both eigenvalues of G are negative, shows a global maximizer at $\tilde{x}$ and local minimizers at x_* and x_{**}.

DEGENERACY

A second property that causes difficulties for some algorithms is *degeneracy*. Confusingly, this term has been given a variety of meanings. It refers to situations in which

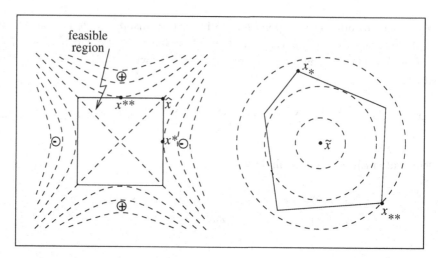

Figure 16.1 Nonconvex quadratic programs.

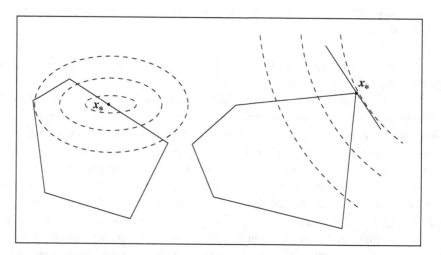

Figure 16.2 Degenerate solutions of quadratic programs.

(a) the active constraint gradients a_i, $i \in \mathcal{A}(x^*)$, are linearly dependent at the solution x^*, and/or

(b) the strict complementarity condition of Definition 12.5 fails to hold, that is, there is some index $i \in \mathcal{A}(x^*)$ such that all Lagrange multipliers satisfying (16.37) have $\lambda_i^* = 0$. (Such constraints are *weakly active* according to Definition 12.8.)

Two examples of degeneracy are shown in Figure 16.2. In the left-hand picture, there is a single active constraint at the solution x_*, which is also an unconstrained minimizer of the objective function. In the notation of (16.37a), we have that $Gx_* + c = 0$, so that the lone Lagrange multiplier must be zero. In the right-hand picture, three constraints are active at the solution x_*. Since each of the three constraint gradients is a vector in $\mathbb{R}^2$, they must be linearly dependent.

Lack of strict complementarity is also illustrated by the problem

$$\min\ x_1^2 + (x_2 + 1)^2 \qquad \text{subject to } x \geq 0,$$

which has a solution at $x^* = 0$ at which both constraints are active. Strict complementarity does not hold at x^* because the Lagrange multiplier associated with the active constraint $x_1 \geq 0$ is zero.

Degeneracy can cause problems for algorithms for two main reasons. First, linear dependence of the active constraint gradients can cause numerical difficulties in the step computation because certain matrices that we need to factor become rank deficient. Second, when the problem contains weakly active constraints, it is difficult for the algorithm to determine whether these constraints are active at the solution. In the case of active-set methods and gradient projection methods (described below), this indecisiveness can cause the algorithm to zigzag as the iterates move on and off the weakly active constraints on successive iterations. Safeguards must be used to prevent such behavior.

16.5 ACTIVE-SET METHODS FOR CONVEX QPs

We now describe active-set methods for solving quadratic programs of the form (16.1) containing equality and inequality constraints. We consider only the convex case, in which the matrix G in (16.1a) is positive semidefinite. The case in which G is an indefinite matrix raises complications in the algorithms and is outside the scope of this book. We refer to Gould [147] for a discussion of nonconvex QPs.

If the contents of the *optimal* active set (16.36) were known in advance, we could find the solution x^* by applying one of the techniques for equality-constrained QP of Sections 16.2 and 16.3 to the problem

$$\min_x\ q(x) = \tfrac{1}{2}x^T G x + x^T c \quad \text{subject to} \quad a_i^T x = b_i, \ \ i \in \mathcal{A}(x^*).$$

Of course, we usually do not have prior knowledge of $\mathcal{A}(x^*)$ and, as we now see, determination of this set is the main challenge facing algorithms for inequality-constrained QP.

We have already encountered an active-set approach for linear programming in Chapter 13, namely, the simplex method. In essence, the simplex method starts by making a guess of the optimal active set, then repeatedly uses gradient and Lagrange multiplier information to drop one index from the current estimate of $\mathcal{A}(x^*)$ and add a new index, until optimality is detected. Active-set methods for QP differ from the simplex method in that the iterates (and the solution x^*) are not necessarily vertices of the feasible region.

Active-set methods for QP come in three varieties: *primal, dual,* and *primal-dual.* We restrict our discussion to primal methods, which generate iterates that remain feasible with respect to the primal problem (16.1) while steadily decreasing the objective function $q(x)$.

Primal active-set methods find a step from one iterate to the next by solving a quadratic subproblem in which some of the inequality constraints (16.1c), and all the equality constraints (16.1b), are imposed as equalities. This subset is referred to as the *working set* and is denoted at the kth iterate x_k by $\mathcal{W}_k$. An important requirement we impose on $\mathcal{W}_k$ is that the gradients a_i of the constraints in the working set be linearly independent, even when the full set of active constraints at that point has linearly dependent gradients.

Given an iterate x_k and the working set $\mathcal{W}_k$, we first check whether x_k minimizes the quadratic q in the subspace defined by the working set. If not, we compute a step p by solving an equality-constrained QP subproblem in which the constraints corresponding to the working set $\mathcal{W}_k$ are regarded as equalities and all other constraints are temporarily disregarded. To express this subproblem in terms of the step p, we define

$$p = x - x_k, \qquad g_k = Gx_k + c.$$

By substituting for x into the objective function (16.1a), we find that

$$q(x) = q(x_k + p) = \tfrac{1}{2}p^T Gp + g_k^T p + \rho_k,$$

where $\rho_k = \tfrac{1}{2}x_k^T Gx_k + c^T x_k$ is independent of p. Since we can drop ρ_k from the objective without changing the solution of the problem, we can write the QP subproblem to be solved at the kth iteration as follows:

$$\min_{p} \quad \tfrac{1}{2}p^T Gp + g_k^T p \tag{16.39a}$$

$$\text{subject to} \quad a_i^T p = 0, \quad i \in \mathcal{W}_k. \tag{16.39b}$$

We denote the solution of this subproblem by p_k. Note that for each $i \in \mathcal{W}_k$, the value of $a_i^T x$ does not change as we move along p_k, since we have $a_i^T(x_k + \alpha p_k) = a_i^T x_k = b_i$ for all α. Since the constraints in $\mathcal{W}_k$ were satisfied at x_k, they are also satisfied at $x_k + \alpha p_k$, for any value of α. Since G is positive definite, the solution of (16.39) can be computed by any of the techniques described in Section 16.2.

Supposing for the moment that the optimal p_k from (16.39) is nonzero, we need to decide how far to move along this direction. If $x_k + p_k$ is feasible with respect to all the constraints, we set $x_{k+1} = x_k + p_k$. Otherwise, we set

$$x_{k+1} = x_k + \alpha_k p_k, \tag{16.40}$$

where the step-length parameter α_k is chosen to be the largest value in the range $[0, 1]$ for which all constraints are satisfied. We can derive an explicit definition of α_k by considering what happens to the constraints $i \notin \mathcal{W}_k$, since the constraints $i \in \mathcal{W}_k$ will certainly be satisfied regardless of the choice of α_k. If $a_i^T p_k \geq 0$ for some $i \notin \mathcal{W}_k$, then for all $\alpha_k \geq 0$ we have $a_i^T(x_k + \alpha_k p_k) \geq a_i^T x_k \geq b_i$. Hence, constraint i will be satisfied for all nonnegative choices of the step-length parameter. Whenever $a_i^T p_k < 0$ for some $i \notin \mathcal{W}_k$, however, we have that $a_i^T(x_k + \alpha_k p_k) \geq b_i$ only if

$$\alpha_k \leq \frac{b_i - a_i^T x_k}{a_i^T p_k}.$$

To maximize the decrease in q, we want α_k to be as large as possible in $[0, 1]$ subject to retaining feasibility, so we obtain the following definition:

$$\alpha_k \overset{\text{def}}{=} \min \left(1, \min_{i \notin \mathcal{W}_k, \, a_i^T p_k < 0} \frac{b_i - a_i^T x_k}{a_i^T p_k} \right). \tag{16.41}$$

We call the constraints i for which the minimum in (16.41) is achieved the *blocking constraints*. (If $\alpha_k = 1$ and no new constraints are active at $x_k + \alpha_k p_k$, then there are no blocking constraints on this iteration.) Note that it is quite possible for α_k to be zero, because we could have $a_i^T p_k < 0$ for some constraint i that is active at x_k but not a member of the current working set $\mathcal{W}_k$.

If $\alpha_k < 1$, that is, the step along p_k was blocked by some constraint not in $\mathcal{W}_k$, a new working set $\mathcal{W}_{k+1}$ is constructed by adding one of the blocking constraints to $\mathcal{W}_k$.

We continue to iterate in this manner, adding constraints to the working set until we reach a point $\hat{x}$ that minimizes the quadratic objective function over its current working set $\hat{\mathcal{W}}$. It is easy to recognize such a point because the subproblem (16.39) has solution $p = 0$. Since $p = 0$ satisfies the optimality conditions (16.5) for (16.39), we have that

$$\sum_{i \in \hat{\mathcal{W}}} a_i \hat{\lambda}_i = g = G\hat{x} + c, \tag{16.42}$$

for some Lagrange multipliers $\hat{\lambda}_i$, $i \in \hat{\mathcal{W}}$. It follows that $\hat{x}$ and $\hat{\lambda}$ satisfy the first KKT condition (16.37a), if we define the multipliers corresponding to the inequality constraints that are not in the working set to be zero. Because of the control imposed on the step length, $\hat{x}$ is also feasible with respect to all the constraints, so the second and third KKT conditions (16.37b) and (16.37c) are satisfied at this point.

We now examine the signs of the multipliers corresponding to the inequality constraints in the working set, that is, the indices $i \in \hat{\mathcal{W}} \cap \mathcal{I}$. If these multipliers are all nonnegative, the fourth KKT condition (16.37d) is also satisfied, so we conclude that $\hat{x}$ is a KKT point for the original problem (16.1). In fact, since G is positive semidefinite, we have

470 CHAPTER 16. QUADRATIC PROGRAMMING

from Theorem 16.4 that $\hat{x}$ is a global solution of (16.1). (As noted after Theorem 16.4, $\hat{x}$ is a strict local minimizer and the unique global solution if G is positive definite.)

If, on the other hand, one or more of the multipliers $\hat{\lambda}_j$, $j \in \hat{\mathcal{W}} \cap \mathcal{I}$, is negative, the condition (16.37d) is not satisfied and the objective function $q(\cdot)$ may be decreased by dropping one of these constraints, as shown in Section 12.3. Thus, we remove an index j corresponding to one of the negative multipliers from the working set and solve a new subproblem (16.39) for the new step. We show in the following theorem that this strategy produces a direction p at the next iteration that is feasible with respect to the dropped constraint. We continue to assume that the constraint gradients a_i for i in the working set are linearly independent. After the algorithm has been fully stated, we discuss how this property can be maintained.

Theorem 16.5.

Suppose that the point $\hat{x}$ satisfies first-order conditions for the equality-constrained subproblem with working set $\hat{\mathcal{W}}$; that is, equation (16.42) is satisfied along with $a_i^T \hat{x} = b_i$ for all $i \in \hat{\mathcal{W}}$. Suppose, too, that the constraint gradients a_i, $i \in \hat{\mathcal{W}}$, are linearly independent and that there is an index $j \in \hat{\mathcal{W}}$ such that $\hat{\lambda}_j < 0$. Let p be the solution obtained by dropping the constraint j and solving the following subproblem:

$$\min_p \tfrac{1}{2} p^T G p + (G\hat{x} + c)^T p, \tag{16.43a}$$

$$\text{subject to } a_i^T p = 0, \text{ for all } i \in \hat{\mathcal{W}} \text{ with } i \neq j. \tag{16.43b}$$

Then p is a feasible direction for constraint j, that is, $a_j^T p \geq 0$. Moreover, if p satisfies second-order sufficient conditions for (16.43), then we have that $a_j^T p > 0$, and that p is a descent direction for $q(\cdot)$.

PROOF. Since p solves (16.43), we have from the results of Section 16.1 that there are multipliers $\tilde{\lambda}_i$, for all $i \in \hat{\mathcal{W}}$ with $i \neq j$, such that

$$\sum_{i \in \hat{\mathcal{W}}, i \neq j} \tilde{\lambda}_i a_i = Gp + (G\hat{x} + c). \tag{16.44}$$

In addition, we have by second-order necessary conditions that if Z is a null-space basis vector for the matrix

$$\left[a_i^T \right]_{i \in \hat{\mathcal{W}}, i \neq j},$$

then $Z^T G Z$ is positive semidefinite. Clearly, p has the form $p = Zp_z$ for some vector p_z, so it follows that $p^T G p \geq 0$.

We have made the assumption that $\hat{x}$ and $\hat{W}$ satisfy the relation (16.42). By subtracting (16.42) from (16.44), we obtain

$$\sum_{i \in \hat{W}, i \neq j} (\tilde{\lambda}_i - \hat{\lambda}_i) a_i - \hat{\lambda}_j a_j = Gp. \qquad (16.45)$$

By taking inner products of both sides with p and using the fact that $a_i^T p = 0$ for all $i \in \hat{W}$ with $i \neq j$, we have that

$$-\hat{\lambda}_j a_j^T p = p^T Gp. \qquad (16.46)$$

Since $p^T Gp \geq 0$ and $\hat{\lambda}_j < 0$ by assumption, it follows that $a_j^T p \geq 0$.

If the second-order sufficient conditions of Section 12.5 are satisfied, we have that $Z^T G Z$ defined above is positive definite. From (16.46), we can have $a_j^T p = 0$ only if $p^T Gp = p_z^T Z^T G Z p_z = 0$, which happens only if $p_z = 0$ and $p = 0$. But if $p = 0$, then by substituting into (16.45) and using linear independence of a_i for $i \in \hat{W}$, we must have that $\hat{\lambda}_j = 0$, which contradicts our choice of j. We conclude that $p^T Gp > 0$ in (16.46), and therefore $a_j^T p > 0$ whenever p satisfies the second-order sufficient conditions for (16.43).

The claim that p is a descent direction for $q(\cdot)$ is proved in Theorem 16.6 below.   □

While any index j for which $\hat{\lambda}_j < 0$ usually will yield a direction p along which the algorithm can make progress, the most negative multiplier is often chosen in practice (and in the algorithm specified below). This choice is motivated by the sensitivity analysis given in Chapter 12, which shows that the rate of decrease in the objective function when one constraint is removed is proportional to the magnitude of the Lagrange multiplier for that constraint. As in linear programming, however, the step along the resulting direction may be short (as when it is blocked by a new constraint), so the amount of decrease in q is not guaranteed to be greater than for other possible choices of j.

We conclude with a result that shows that whenever p_k obtained from (16.39) is nonzero and satisfies second-order sufficient optimality conditions for the current working set, it is a direction of strict descent for $q(\cdot)$.

Theorem 16.6.

Suppose that the solution p_k of (16.39) is nonzero and satisfies the second-order sufficient conditions for optimality for that problem. Then the function $q(\cdot)$ is strictly decreasing along the direction p_k.

PROOF.   Since p_k satisfies the second-order conditions, that is, $Z^T G Z$ is positive definite for the matrix Z whose columns are a basis of the null space of the constraints (16.39b), we have by applying Theorem 16.2 to (16.39) that p_k is the unique global solution of (16.39). Since $p = 0$ is also a feasible point for (16.39), its objective value in (16.39a) must be larger than that of p_k, so we have

$$\tfrac{1}{2} p_k^T G p_k + g_k^T p_k < 0.$$

Since $p_k^T G p_k \geq 0$ by convexity, this inequality implies that $g_k^T p_k < 0$. Therefore, we have

$$q(x_k + \alpha_k p_k) = q(x_k) + \alpha g_k^T p_k + \tfrac{1}{2} \alpha^2 p_k^T G p_k < q(x_k),$$

for all $\alpha > 0$ sufficiently small. $\qquad\square$

When G is positive definite—the *strictly* convex case—the second-order sufficient conditions are satisfied for *all* feasible subproblems of the form (16.39). Hence, it follows from the result above that we obtain a strict decrease in $q(\cdot)$ whenever $p_k \neq 0$. This fact is significant when we discuss finite termination of the algorithm.

SPECIFICATION OF THE ACTIVE-SET METHOD FOR CONVEX QP

Having described the active-set algorithm for convex QP, we now present the following formal specification. We assume that the objective function q is bounded in the feasible set (16.1b), (16.1c).

Algorithm 16.3 (Active-Set Method for Convex QP).
Compute a feasible starting point x_0;
Set $\mathcal{W}_0$ to be a subset of the active constraints at x_0;
for $k = 0, 1, 2, \ldots$
    Solve (16.39) to find p_k;
    if $p_k = 0$
        Compute Lagrange multipliers $\hat{\lambda}_i$ that satisfy (16.42),
            with $\hat{\mathcal{W}} = \mathcal{W}_k$;
        if $\hat{\lambda}_i \geq 0$ for all $i \in \mathcal{W}_k \cap \mathcal{I}$
            stop with solution $x^* = x_k$;
        else
            $j \leftarrow \arg\min_{j \in \mathcal{W}_k \cap \mathcal{I}} \hat{\lambda}_j$;
            $x_{k+1} \leftarrow x_k$; $\mathcal{W}_{k+1} \leftarrow \mathcal{W}_k \backslash \{j\}$;
    else (* $p_k \neq 0$ *)
        Compute α_k from (16.41);
        $x_{k+1} \leftarrow x_k + \alpha_k p_k$;
        if there are blocking constraints
            Obtain $\mathcal{W}_{k+1}$ by adding one of the blocking
                constraints to $\mathcal{W}_k$;
        else
            $\mathcal{W}_{k+1} \leftarrow \mathcal{W}_k$;
end (**for**)

Various techniques can be used to determine an initial feasible point. One such is to use the "Phase I" approach for linear programming described in Chapter 13. Though no significant modifications are needed to generalize this method from linear programming to quadratic programming, we describe a variant here that allows the user to supply an initial estimate $\tilde{x}$ of the vector x. This estimate need not be feasible, but a good choice based on knowledge of the QP may reduce the work needed in the Phase I step.

Given $\tilde{x}$, we define the following feasibility linear program:

$$\min_{(x,z)} e^T z$$

$$\text{subject to } a_i^T x + \gamma_i z_i = b_i, \qquad i \in \mathcal{E},$$
$$a_i^T x + \gamma_i z_i \geq b_i, \qquad i \in \mathcal{I},$$
$$z \geq 0,$$

where $e = (1, 1, \ldots, 1)^T$, $\gamma_i = -\text{sign}(a_i^T \tilde{x} - b_i)$ for $i \in \mathcal{E}$, and $\gamma_i = 1$ for $i \in \mathcal{I}$. A feasible initial point for this problem is then

$$x = \tilde{x}, \qquad z_i = |a_i^T \tilde{x} - b_i| \ (i \in \mathcal{E}), \qquad z_i = \max(b_i - a_i^T \tilde{x}, 0) \ (i \in \mathcal{I}).$$

It is easy to verify that if $\tilde{x}$ is feasible for the original problem (16.1), then $(\tilde{x}, 0)$ is optimal for the feasibility subproblem. In general, if the original problem has feasible points, then the optimal objective value in the subproblem is zero, and any solution of the subproblem yields a feasible point for the original problem. The initial working set $\mathcal{W}_0$ for Algorithm 16.3 can be found by taking a linearly independent subset of the active constraints at the solution of the feasibility problem.

An alternative approach is a penalty (or "big M") method, which does away with the "Phase I" and instead includes a measure of infeasibility in the objective that is guaranteed to be zero at the solution. That is, we introduce a scalar artificial variable η into (16.1) to measure the constraint violation, and we solve the problem

$$\min_{(x,\eta)} \tfrac{1}{2} x^T G x + x^T c + M \eta,$$

$$\text{subject to } \quad (a_i^T x - b_i) \leq \eta, \qquad i \in \mathcal{E},$$
$$-(a_i^T x - b_i) \leq \eta, \qquad i \in \mathcal{E}, \qquad (16.47)$$
$$b_i - a_i^T x \leq \eta, \qquad i \in \mathcal{I},$$
$$0 \leq \eta,$$

for some large positive value of M. It can be shown by applying the theory of exact penalty functions (see Chapter 17) that whenever there exist feasible points for the original problem (16.1), then for all M sufficiently large, the solution of (16.47) will have $\eta = 0$, with an x component that is a solution for (16.1).

Our strategy is to use some heuristic to choose a value of M and solve (16.47) by the usual means. If the solution we obtain has a positive value of η, we increase M and try again. Note that a feasible point is easy to obtain for the subproblem (16.47): We set $x = \tilde{x}$ (where, as before, $\tilde{x}$ is the user-supplied initial guess) and choose η large enough that all the constraints in (16.47) are satisfied. This approach is, in fact, an exact penalty method using the ℓ_∞ norm; see Chapter 17.

A variant of (16.47) that penalizes the ℓ_1 norm of the constraint violation rather than the ℓ_∞ norm is as follows:

$$\min_{(x,s,t,v)} \tfrac{1}{2}x^T G x + x^T c + M e_{\mathcal{E}}^T (s+t) + M e_{\mathcal{I}}^T v$$

$$\text{subject to} \quad a_i^T x - b_i + s_i - t_i = 0, \quad i \in \mathcal{E},$$
$$a_i^T x - b_i + v_i \geq 0, \quad i \in \mathcal{I}, \qquad (16.48)$$
$$s \geq 0, \ t \geq 0, \ v \geq 0.$$

Here, $e_{\mathcal{E}}$ is the vector $(1, 1, \ldots, 1)^T$ of length $|\mathcal{E}|$; similarly for $e_{\mathcal{I}}$. The slack variables s_i, t_i, and v_i soak up any infeasibility in the constraints.

In the following example we use subscripts on the vectors x and p to denote their components, and we use superscripts to indicate the iteration index. For example, x_1 denotes the first component, while x^4 denotes the fourth iterate of the vector x.

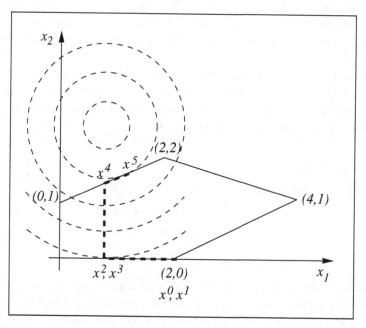

Figure 16.3   Iterates of the active-set method.

❑ **EXAMPLE 16.4**

We apply Algorithm 16.3 to the following simple 2-dimensional problem illustrated in Figure 16.3.

$$\min_{x} \ q(x) = (x_1 - 1)^2 + (x_2 - 2.5)^2 \tag{16.49a}$$

$$\text{subject to} \quad x_1 - 2x_2 + 2 \geq 0, \tag{16.49b}$$

$$-x_1 - 2x_2 + 6 \geq 0, \tag{16.49c}$$

$$-x_1 + 2x_2 + 2 \geq 0, \tag{16.49d}$$

$$x_1 \geq 0, \tag{16.49e}$$

$$x_2 \geq 0. \tag{16.49f}$$

We refer the constraints, in order, by indices 1 through 5. For this problem it is easy to determine a feasible initial point; say $x^0 = (2, 0)^T$. Constraints 3 and 5 are active at this point, and we set $\mathcal{W}_0 = \{3, 5\}$. (Note that we could just as validly have chosen $\mathcal{W}_0 = \{5\}$ or $\mathcal{W}_0 = \{3\}$ or even $\mathcal{W} = \emptyset$; each choice would lead the algorithm to perform somewhat differently.)

Since x^0 lies on a vertex of the feasible region, it is obviously a minimizer of the objective function q with respect to the working set $\mathcal{W}_0$; that is, the solution of (16.39) with $k = 0$ is $p = 0$. We can then use (16.42) to find the multipliers $\hat{\lambda}_3$ and $\hat{\lambda}_5$ associated with the active constraints. Substitution of the data from our problem into (16.42) yields

$$\begin{bmatrix} -1 \\ 2 \end{bmatrix} \hat{\lambda}_3 + \begin{bmatrix} 0 \\ 1 \end{bmatrix} \hat{\lambda}_5 = \begin{bmatrix} 2 \\ -5 \end{bmatrix},$$

which has the solution $(\hat{\lambda}_3, \hat{\lambda}_5) = (-2, -1)$.

We now remove constraint 3 from the working set, because it has the most negative multiplier, and set $\mathcal{W}_1 = \{5\}$. We begin iteration 1 by finding the solution of (16.39) for $k = 1$, which is $p^1 = (-1, 0)^T$. The step-length formula (16.41) yields $\alpha_1 = 1$, and the new iterate is $x^2 = (1, 0)^T$.

There are no blocking constraints, so that $\mathcal{W}_2 = \mathcal{W}_1 = \{5\}$, and we find at the start of iteration 2 that the solution of (16.39) is $p^2 = 0$. From (16.42) we deduce that the Lagrange multiplier for the lone working constraint is $\hat{\lambda}_5 = -5$, so we drop 5 from the working set to obtain $\mathcal{W}_3 = \emptyset$.

Iteration 3 starts by solving the unconstrained problem, to obtain the solution $p^3 = (0, 2.5)^T$. The formula (16.41) yields a step length of $\alpha_3 = 0.6$ and a new iterate $x^4 = (1, 1.5)^T$. There is a single blocking constraint (constraint 1), so we obtain $\mathcal{W}_4 = \{1\}$. The solution of (16.39) for $k = 4$ is then $p^4 = (0.4, 0.2)^T$, and the new step length is 1. There are no blocking constraints on this step, so the next working set is unchanged: $\mathcal{W}_5 = \{1\}$. The new iterate is $x^5 = (1.4, 1.7)^T$.

Finally, we solve (16.39) for $k = 5$ to obtain a solution $p^5 = 0$. The formula (16.42) yields a multiplier $\hat{\lambda}_1 = 0.8$, so we have found the solution. We set $x^* = (1.4, 1.7)^T$ and terminate.

$\square$

FURTHER REMARKS ON THE ACTIVE-SET METHOD

We noted above that there is flexibility in the choice of the initial working set and that each initial choice leads to a different iteration sequence. When the initial active constraints have independent gradients, as above, we can include them all in $\mathcal{W}_0$. Alternatively, we can select a subset. For instance, if in the example above we have chosen $\mathcal{W}_0 = \{3\}$, the first iterate would have yielded $p^0 = (0.2, 0.1)^T$ and a new iterate of $x^1 = (2.2, 0.1)^T$. If we had chosen $\mathcal{W}_0 = \{5\}$, we would have moved immediately to the new iterate $x^1 = (1, 0)^T$, without first performing the operation of dropping the index 3, as is done in the example. If we had selected $\mathcal{W}_0 = \emptyset$, we would have obtained $p^1 = (-1, 2.5)^T$, $\alpha_1 = \frac{2}{3}$, a new iterate of $x^1 = (\frac{4}{3}, \frac{5}{3})^T$, and a new working set of $\mathcal{W}_1 = \{1\}$. The solution x^* would have been found on the next iteration.

Even if the initial working set $\mathcal{W}_0$ coincides with the initial active set, the sets $\mathcal{W}_k$ and $\mathcal{A}(x^k)$ may differ at later iterations. For instance, when a particular step encounters more than one blocking constraint, just one of them is added to the working set, so the identification between $\mathcal{W}_k$ and $\mathcal{A}(x^k)$ is broken. Moreover, subsequent iterates differ in general according to what choice is made.

We require the constraint gradients in $\mathcal{W}_0$ to be linearly independent, and our strategy for modifying the working set ensures that this same property holds for all subsequent working sets $\mathcal{W}_k$. When we encounter a blocking constraint on a particular step, its constraint normal cannot be a linear combination of the normals a_i in the current working set (see Exercise 16.18). Hence, linear independence is maintained after the blocking constraint is added to the working set. On the other hand, deletion of an index from the working set cannot introduce linear dependence.

The strategy of removing the constraint corresponding to the most negative Lagrange multiplier often works well in practice but has the disadvantage that it is susceptible to the scaling of the constraints. (By multiplying constraint i by some factor $\beta > 0$ we do not change the geometry of the optimization problem, but we introduce a scaling of $1/\beta$ to the corresponding multiplier λ_i.) Choice of the most negative multiplier is analogous to Dantzig's original pivot rule for the simplex method in linear programming (see Chapter 13) and, as we noted there, strategies that are less sensitive to scaling often give better results. We do not discuss this advanced topic further.

We note that the strategy of adding or deleting at most one constraint at each iteration of the Algorithm 16.3 places a natural lower bound on the number of iterations needed to reach optimality. Suppose, for instance, that we have a problem in which m inequality constraints are active at the solution x^* but that we start from a point x^0 that is strictly feasible with respect to all the inequality constraints. In this case, the algorithm will need at least m iterations to move from x^0 to x^*. Even more iterations will be required if the algorithm adds some constraint j to the working set at some iteration, only to remove it at a later step.

FINITE TERMINATION OF ACTIVE-SET ALGORITHM ON STRICTLY CONVEX QPs

It is not difficult to show that, under certain assumptions, Algorithm 16.3 converges for strictly convex QPs, that is, it identifies the solution x^* in a finite number of iterations. This claim is certainly true if we assume that the method always takes a nonzero step length α_k whenever the direction p_k computed from (16.39) is nonzero. Our argument proceeds as follows:

- If the solution of (16.39) is $p_k = 0$, the current point x_k is the unique global minimizer of $q(\cdot)$ for the working set $\mathcal{W}_k$; see Theorem 16.6. If it is not the solution of the original problem (16.1) (that is, at least one of the Lagrange multipliers is negative), Theorems 16.5 and 16.6 together show that the step p_{k+1} computed after a constraint is dropped will be a strict decrease direction for $q(\cdot)$. Therefore, because of our assumption $\alpha_k > 0$, we have that the value of q is lower than $q(x_k)$ at all subsequent iterations. It follows that the algorithm can never return to the working set $\mathcal{W}_k$, because subsequent iterates have values of q that are lower than the global minimizer for this working set.

- The algorithm encounters an iterate k for which $p_k = 0$ solves (16.39) at least on every nth iteration. To demonstrate this claim, we note that for any k at which $p_k \neq 0$, either we have $\alpha_k = 1$ (in which case we reach the minimizer of q on the current working set $\mathcal{W}_k$, so that the next iteration will yield $p_{k+1} = 0$), or else a constraint is added to the working set $\mathcal{W}_k$. If the latter situation occurs repeatedly, then after at most n iterations the working set will contain n indices, which correspond to n linearly independent vectors. The solution of (16.39) will then be $p_k = 0$, since only the zero vector will satisfy the constraints (16.39b).

- Taken together, the two statements above indicate that the algorithm finds the global minimum of q on its current working set periodically (at least once every n iterations) and that, having done so, it never visits this particular working set again. It follows that, since there are only a finite number of possible working sets, the algorithm cannot iterate forever. Eventually, it encounters a minimizer for a current working set that satisfies optimality conditions for (16.1), and it terminates with a solution.

The assumption that we can always take a nonzero step along a nonzero descent direction p_k calculated from (16.39) guarantees that the algorithm does not undergo *cycling*. This term refers to the situation in which a sequence of consecutive iterations results in no movement in iterate x, while the working set $\mathcal{W}_k$ undergoes deletions and additions of indices and eventually repeats itself. That is, for some integers k and $l \geq 1$, we have that $x^k = x^{k+l}$ and $\mathcal{W}_k = \mathcal{W}_{k+l}$. At each iterate in the cycle, a constraint is dropped (as in Theorem 16.5), but a new constraint $i \notin \mathcal{W}_k$ is encountered immediately without any movement along the computed direction p. Procedures for handling degeneracy and cycling in quadratic programming are similar to those for linear programming discussed in Chapter 13; we do not discuss them here. Most QP implementations simply ignore the possibility of cycling.

UPDATING FACTORIZATIONS

We have seen that the step computation in the active-set method given in Algorithm 16.3 requires the solution of the equality-constrained subproblem (16.39). As mentioned at the beginning of this chapter, this computation amounts to solving the KKT system (16.5). Since the working set can change by just one index at every iteration, the KKT matrix differs in at most one row and one column from the previous iteration's KKT matrix. Indeed, G remains fixed, whereas the matrix A of constraint gradients corresponding to the current working set may change through addition and/or deletion of a single row.

It follows from this observation that we can compute the matrix factors needed to solve (16.39) at the current iteration by updating the factors computed at the previous iteration, rather than recomputing them from scratch. These updating techniques are crucial to the efficiency of active-set methods.

We limit our discussion to the case in which the step is computed with the null-space method (16.17)–(16.20). Suppose that A has m linearly independent rows and assume that the bases Y and Z are defined by means of a QR factorization of A (see Section 15.3 for details). Thus

$$A^T \Pi = Q \begin{bmatrix} R \\ 0 \end{bmatrix} = \begin{bmatrix} Q_1 & Q_2 \end{bmatrix} \begin{bmatrix} R \\ 0 \end{bmatrix} \tag{16.50}$$

(see (15.21)), where Π is a permutation matrix; R is square, upper triangular and nonsingular; $Q = \begin{bmatrix} Q_1 & Q_2 \end{bmatrix}$ is $n \times n$ orthogonal; and Q_1 and R both have m columns while Q_2 has $n - m$ columns. As noted in Chapter 15, we can choose Z to be simply the orthonormal matrix Q_2.

Suppose that one constraint is added to the working set at the next iteration, so that the new constraint matrix is $\bar{A}^T = \begin{bmatrix} A^T & a \end{bmatrix}$, where a is a column vector of length n such that $\bar{A}^T$ retains full column rank. As we now show, there is an economical way to update the Q and R factors in (16.50) to obtain new factors (and hence a new null-space basis matrix $\bar{Z}$, with $n - m - 1$ columns) for the expanded matrix $\bar{A}$. Note first that, since $Q_1 Q_1^T + Q_2 Q_2^T = I$, we have

$$\bar{A}^T \begin{bmatrix} \Pi & 0 \\ 0 & 1 \end{bmatrix} = \begin{bmatrix} A^T \Pi & a \end{bmatrix} = Q \begin{bmatrix} R & Q_1^T a \\ 0 & Q_2^T a \end{bmatrix}. \tag{16.51}$$

We can now define an orthogonal matrix $\hat{Q}$ that transforms the vector $Q_2^T a$ to a vector in which all elements except the first are zero. That is, we have

$$\hat{Q}(Q_2^T a) = \begin{bmatrix} \gamma \\ 0 \end{bmatrix},$$

where γ is a scalar. (Since $\hat{Q}$ is orthogonal, we have $\|Q_2^T a\| = |\gamma|$.) From (16.51) we now have

$$\bar{A}^T \begin{bmatrix} \Pi & 0 \\ 0 & 1 \end{bmatrix} = Q \begin{bmatrix} R & Q_1^T a \\ 0 & \hat{Q}^T \begin{bmatrix} \gamma \\ 0 \end{bmatrix} \end{bmatrix} = Q \begin{bmatrix} I & 0 \\ 0 & \hat{Q}^T \end{bmatrix} \begin{bmatrix} R & Q_1^T a \\ 0 & \gamma \\ 0 & 0 \end{bmatrix}.$$

This factorization has the form

$$\bar{A}^T \bar{\Pi} = \bar{Q} \begin{bmatrix} \bar{R} \\ 0 \end{bmatrix},$$

where

$$\bar{\Pi} = \begin{bmatrix} \Pi & 0 \\ 0 & 1 \end{bmatrix}, \quad \bar{Q} = Q \begin{bmatrix} I & 0 \\ 0 & \hat{Q}^T \end{bmatrix} = \begin{bmatrix} Q_1 & Q_2 \hat{Q}^T \end{bmatrix}, \quad \bar{R} = \begin{bmatrix} R & Q_1^T a \\ 0 & \gamma \end{bmatrix}.$$

We can therefore choose $\bar{Z}$ to be the last $n - m - 1$ columns of $Q_2 \hat{Q}^T$. If we know Z explicitly and need an explicit representation of $\bar{Z}$, we need to account for the cost of obtaining $\hat{Q}$ and the cost of forming the product $Q_2 \hat{Q}^T = Z \hat{Q}^T$. Because of the special structure of $\hat{Q}$, this cost is of order $n(n - m)$, compared to the cost of computing (16.50) from scratch, which is of order $n^2 m$. The updating strategy is less expensive, especially when the null space is small (that is, when $n - m \ll n$).

An updating technique can also be designed for the case in which a row is removed from A. This operation has the effect of deleting a column from R in (16.50), thus disturbing the upper triangular property of this matrix by introducing a number of nonzeros on the diagonal immediately below the main diagonal of the matrix. Upper triangularity can be restored by applying a sequence of plane rotations. These rotations introduce a number of inexpensive transformations into the first m columns of Q, and the updated null-space matrix is obtained by selecting the last $n - m + 1$ columns from this matrix after the transformations are complete. The new null-space basis in this case has the form

$$\bar{Z} = \begin{bmatrix} \bar{z} & Z \end{bmatrix}, \tag{16.52}$$

that is, the current matrix Z is augmented by a single column. The total cost of this operation varies with the location of the removed column in A but is in general cheaper than recomputing a QR factorization from scratch. For details of these procedures, see Gill et al. [124, Section 5].

We now consider the reduced Hessian. Because of the special form of (16.39), we have $h = 0$ in (16.5), and the step p_Y given in (16.18) is zero. Thus from (16.19), the null-space component p_Z is the solution of

$$(Z^T G Z) p_z = -Z^T g. \tag{16.53}$$

We can sometimes find ways of updating the factorization of the reduced Hessian $Z^T G Z$ after Z has changed. Suppose that we have the Cholesky factorization of the current reduced Hessian, written as

$$Z^T G Z = L L^T,$$

and that at the next step Z changes as in (16.52), gaining a column after deletion of a constraint. A series of inexpensive, elementary operations can be used to transform the Cholesky factor L into the new factor $\bar{L}$ for the new reduced Hessian $\bar{Z}^T G \bar{Z}$.

A variety of other simplifications are possible. For example, as discussed in Section 16.7, we can update the reduced gradient $Z^T g$ at the same time as we update Z to $\bar{Z}$.

16.6   INTERIOR-POINT METHODS

The interior-point approach can be applied to convex quadratic programs through a simple extension of the linear-programming algorithms described in Chapter 14. The resulting primal-dual algorithms are easy to describe and are quite efficient on many types of problems. Extensions of interior-point methods to nonconvex problems are discussed in Chapter 19.

For simplicity, we restrict our attention to convex quadratic programs with inequality constraints, which we write as follows:

$$\min_x \quad q(x) = \tfrac{1}{2} x^T G x + x^T c \tag{16.54a}$$

$$\text{subject to} \quad A x \geq b, \tag{16.54b}$$

where G is symmetric and positive semidefinite and where the $m \times n$ matrix A and right-hand side b are defined by

$$A = [a_i]_{i \in \mathcal{I}}, \quad b = [b_i]_{i \in \mathcal{I}}, \quad \mathcal{I} = \{1, 2, \ldots, m\}.$$

(If equality constraints are also present, they can be accommodated with simple extensions to the approaches described below.) Rewriting the KKT conditions (16.37) in this notation, we obtain

$$Gx - A^T \lambda + c = 0,$$
$$Ax - b \geq 0,$$
$$(Ax - b)_i \lambda_i = 0, \quad i = 1, 2, \ldots, m,$$
$$\lambda \geq 0.$$

By introducing the slack vector $y \geq 0$, we can rewrite these conditions as

$$Gx - A^T \lambda + c = 0, \tag{16.55a}$$
$$Ax - y - b = 0, \tag{16.55b}$$
$$y_i \lambda_i = 0, \quad i = 1, 2, \ldots, m, \tag{16.55c}$$
$$(y, \lambda) \geq 0. \tag{16.55d}$$

Since we assume that G is positive semidefinite, these KKT conditions are not only necessary but also sufficient (see Theorem 16.4), so we can solve the convex quadratic program (16.54) by finding solutions of the system (16.55).

Given a current iterate (x, y, λ) that satisfies $(y, \lambda) > 0$, we can define a complementarity measure μ by

$$\mu = \frac{y^T \lambda}{m}. \tag{16.56}$$

As in Chapter 14, we derive path-following, primal-dual methods by considering the *perturbed* KKT conditions given by

$$F(x, y, \lambda; \sigma\mu) = \begin{bmatrix} Gx - A^T \lambda + c \\ Ax - y - b \\ Y\Lambda e - \sigma\mu e \end{bmatrix} = 0, \tag{16.57}$$

where

$$Y = \mathrm{diag}(y_1, y_2, \ldots, y_m), \quad \Lambda = \mathrm{diag}(\lambda_1, \lambda_2, \ldots, \lambda_m), \quad e = (1, 1, \ldots, 1)^T,$$

and $\sigma \in [0, 1]$. The solutions of (16.57) for all positive values of σ and μ define the *central path*, which is a trajectory that leads to the solution of the quadratic program as $\sigma\mu$ tends to zero.

By fixing μ and applying Newton's method to (16.57), we obtain the linear system

$$\begin{bmatrix} G & 0 & -A^T \\ A & -I & 0 \\ 0 & \Lambda & Y \end{bmatrix} \begin{bmatrix} \Delta x \\ \Delta y \\ \Delta \lambda \end{bmatrix} = \begin{bmatrix} -r_d \\ -r_p \\ -\Lambda Y e + \sigma\mu e \end{bmatrix}, \tag{16.58}$$

where

$$r_d = Gx - A^T\lambda + c, \qquad r_p = Ax - y - b. \tag{16.59}$$

We obtain the next iterate by setting

$$(x^+, y^+, \lambda^+) = (x, y, \lambda) + \alpha(\Delta x, \Delta y, \Delta \lambda), \tag{16.60}$$

where α is chosen to retain the inequality $(y^+, \lambda^+) > 0$ and possibly to satisfy various other conditions.

In the rest of the chapter we discuss several enhancements of this primal-dual iteration that make it effective in practice.

SOLVING THE PRIMAL-DUAL SYSTEM

The major computational operation in the interior-point method is the solution of the system (16.58). The coefficient matrix in this system can be much more costly to factor than the matrix (14.9) arising in linear programming because of the presence of the Hessian matrix G. It is therefore important to exploit the structure of (16.58) by choosing a suitable direct factorization algorithm, or by choosing an appropriate preconditioner for an iterative solver.

As in Chapter 14, the system (16.58) may be restated in more compact forms. The "augmented system" form is

$$\begin{bmatrix} G & -A^T \\ A & \Lambda^{-1}\mathcal{Y} \end{bmatrix} \begin{bmatrix} \Delta x \\ \Delta \lambda \end{bmatrix} = \begin{bmatrix} -r_d \\ -r_p + (-y + \sigma\mu\Lambda^{-1}e) \end{bmatrix}. \tag{16.61}$$

After a simple transformation to symmetric form, a symmetric indefinite factorization scheme can be applied to the coefficient matrix in this system. The "normal equations" form (14.44a) is

$$(G + A^T\mathcal{Y}^{-1}\Lambda A)\Delta x = -r_d + A^T\mathcal{Y}^{-1}\Lambda[-r_p - y + \sigma\mu\Lambda^{-1}e], \tag{16.62}$$

which can be solved by means of a modified Cholesky algorithm. This approach is effective if the term $A^T(\mathcal{Y}^{-1}\Lambda)A$ is not too dense compared with G, and it has the advantage of being much smaller than (16.61) if there are many inequality constraints.

The projected CG method of Algorithm 16.2 can also be effective for solving the primal-dual system. We can rewrite (16.58) in the form

$$\begin{bmatrix} G & 0 & -A^T \\ 0 & \mathcal{Y}^{-1}\Lambda & I \\ A & -I & 0 \end{bmatrix} \begin{bmatrix} \Delta x \\ \Delta y \\ \Delta \lambda \end{bmatrix} = \begin{bmatrix} -r_d \\ -\Lambda e + \sigma\mu\mathcal{Y}^{-1}e \\ -r_p \end{bmatrix}, \tag{16.63}$$

and observe that these are the optimality conditions for an equality-constrained convex quadratic program of the form (16.3), in which the variable is $(\Delta x, \Delta y)$. Hence, we can make appropriate substitutions and solve this system using Algorithm 16.2. This approach may be useful for problems in which the direct factorization cannot be performed due to excessive memory demands. The projected CG method does not require that the matrix G be formed or factored; it requires only matrix-vector products.

STEP LENGTH SELECTION

We mentioned in Chapter 14 that interior-point methods for linear programming are more efficient if different step lengths α^{pri}, α^{dual} are used for the primal and dual variables. Equation (14.37) indicates that the greatest reduction in the residuals r_b and r_c is obtained by choosing the largest admissible primal and dual step lengths. The situation is different in quadratic programming. Suppose that we define the new iterate as

$$(x^+, y^+) = (x, y) + \alpha^{\text{pri}}(\Delta x, \Delta y), \qquad \lambda^+ = \lambda + \alpha^{\text{dual}}\Delta\lambda, \qquad (16.64)$$

where α^{pri} and α^{dual} are step lengths that ensure the positivity of (y^+, λ^+). By using (16.58) and (16.59), we see that the new residuals satisfy the following relations:

$$r_p^+ = (1 - \alpha^{\text{pri}})r_p, \qquad (16.65a)$$

$$r_d^+ = (1 - \alpha^{\text{dual}})r_d + (\alpha^{\text{pri}} - \alpha^{\text{dual}})G\Delta x. \qquad (16.65b)$$

If $\alpha^{\text{pri}} = \alpha^{\text{dual}} = \alpha$ then both residuals decrease linearly for all $\alpha \in (0, 1)$. For different step lengths, however, the dual residual r_d^+ may increase for certain choices of α^{pri}, α^{dual}, possibly causing divergence of the interior-point iteration.

One option is to use equal step lengths, as in (16.60), and to set $\alpha = \min(\alpha_\tau^{\text{pri}}, \alpha_\tau^{\text{dual}})$, where

$$\alpha_\tau^{\text{pri}} = \max\{\alpha \in (0, 1] : y + \alpha\Delta y \geq (1 - \tau)y\}, \qquad (16.66a)$$

$$\alpha_\tau^{\text{dual}} = \max\{\alpha \in (0, 1] : \lambda + \alpha\Delta\lambda \geq (1 - \tau)\lambda\}; \qquad (16.66b)$$

the parameter $\tau \in (0, 1)$ controls how far we back off from the maximum step for which the conditions $y + \alpha\Delta y \geq 0$ and $\lambda + \alpha\Delta\lambda \geq 0$ are satisfied. Numerical experience has shown, however, that using different step lengths in the primal and dual variables often leads to faster convergence. One way to choose unequal step lengths is to select $(\alpha^{\text{pri}}, \alpha^{\text{dual}})$ so as to (approximately) minimize the optimality measure

$$\|Gx^+ - A^T\lambda^+ + c\|_2^2 + \|Ax^+ - y^+ - b\|_2^2 + (y^+)^T z^+,$$

subject to $0 \leq \alpha^{\text{pri}} \leq \alpha_\tau^{\text{pri}}$ and $0 \leq \alpha^{\text{dual}} \leq \alpha_\tau^{\text{dual}}$, where x^+, y^+, λ^+ are defined as a function of the step lengths through (16.64).

A PRACTICAL PRIMAL-DUAL METHOD

The most popular interior-point method for convex QP is based on Mehrotra's predictor-corrector, originally developed for linear programming (see Section 14.2). The extension to quadratic programming is straightforward, as we now show.

First, we compute an affine scaling step $(\Delta x^{\text{aff}}, \Delta y^{\text{aff}}, \Delta \lambda^{\text{aff}})$ by setting $\sigma = 0$ in (16.58). We improve upon this step by computing a corrector step, which is defined following the same reasoning that leads to (14.31). Next, we compute the centering parameter σ using (14.34). The total step is obtained by solving the following system (cf. (14.35)):

$$
\begin{bmatrix} G & 0 & -A^T \\ A & -I & 0 \\ 0 & \Lambda & \mathcal{Y} \end{bmatrix}
\begin{bmatrix} \Delta x \\ \Delta y \\ \Delta \lambda \end{bmatrix}
=
\begin{bmatrix} -r_d \\ -r_p \\ -\Lambda \mathcal{Y} e - \Delta \Lambda^{\text{aff}} \Delta \mathcal{Y}^{\text{aff}} e + \sigma \mu e \end{bmatrix}. \qquad (16.67)
$$

We now specify the algorithm. For simplicity, we will assume in our description that equal step lengths are used in the primal and dual variables though, as noted above, unequal step lengths can give slightly faster convergence.

Algorithm 16.4 (Predictor-Corrector Algorithm for QP).
Compute (x_0, y_0, λ_0) with $(y_0, \lambda_0) > 0$;
for $k = 0, 1, 2, \ldots$
    Set $(x, y, \lambda) = (x_k, y_k, \lambda_k)$ and solve (16.58) with $\sigma = 0$ for
        $(\Delta x^{\text{aff}}, \Delta y^{\text{aff}}, \Delta \lambda^{\text{aff}})$;
    Calculate $\mu = y^T \lambda / m$;
    Calculate $\hat{\alpha}_{\text{aff}} = \max\{\alpha \in (0, 1] \mid (y, \lambda) + \alpha(\Delta y^{\text{aff}}, \Delta \lambda^{\text{aff}}) \geq 0\}$;
    Calculate $\mu_{\text{aff}} = (y + \hat{\alpha}^{\text{aff}} \Delta y^{\text{aff}})^T (\lambda + \hat{\alpha}^{\text{aff}} \Delta \lambda^{\text{aff}}) / m$;
    Set centering parameter to $\sigma = (\mu_{\text{aff}} / \mu)^3$;
    Solve (16.67) for $(\Delta x, \Delta y, \Delta \lambda)$;
    Choose $\tau_k \in (0, 1)$ and set $\hat{\alpha} = \min(\alpha_{\tau_k}^{\text{pri}}, \alpha_{\tau_k}^{\text{dual}})$ (see (16.66));
    Set $(x_{k+1}, y_{k+1}, \lambda_{k+1}) = (x_k, y_k, \lambda_k) + \hat{\alpha}(\Delta x, \Delta y, \Delta \lambda)$;
end (for)

We can choose τ_k to approach 1 as the iterates approach the solution, to accelerate the convergence.

As for linear programming, efficiency and robustness of this approach is greatly enhanced if we choose a good starting point. This selection can be done in several ways. The following simple heuristic accepts an initial point $(\bar{x}, \bar{y}, \bar{\lambda})$ from the user and moves it far enough away from the boundary of the region $(y, \lambda) \geq 0$ to permit the algorithm to take long steps on early iterations. First, we compute the affine scaling step $(\Delta x^{\text{aff}}, \Delta y^{\text{aff}}, \Delta \lambda^{\text{aff}})$

from the user-supplied initial point $(\bar{x}, \bar{y}, \bar{\lambda})$, then set

$$y_0 = \max(1, |\bar{y} + \Delta y^{\text{aff}}|), \qquad \lambda_0 = \max(1, |\bar{\lambda} + \Delta \lambda^{\text{aff}}|), \qquad x_0 = \bar{x},$$

where the max and absolute values are applied component-wise.

We conclude this section by contrasting some of the properties of active-set and interior-point methods for convex quadratic programming. Active-set methods generally require a large number of steps in which each search direction is relatively inexpensive to compute, while interior-point methods take a smaller number of more expensive steps. Active-set methods are more complicated to implement, particularly if the procedures for updating matrix factorizations try to take advantage of sparsity or structure in G and A. By contrast, the nonzero structure of the matrix to be factored at each interior-point iteration remains the same at all iterations (though the numerical values change), so standard sparse factorization software can be used to obtain the steps. For particular sparsity structures (for example, bandedness in the matrices A and G), efficient customized solvers for the linear system arising at each interior-point iteration can be devised.

For very large problems, interior-point methods are often more efficient. However, when an estimate of the solution is available (a "warm start"), the active-set approach may converge rapidly in just a few iterations, particularly if the initial value of x is feasible. Interior-point methods are less able to exploit a warm start, though research efforts to improve their performance in this regard are ongoing.

16.7 THE GRADIENT PROJECTION METHOD

In the active-set method described in Section 16.5, the active set and working set change slowly, usually by a single index at each iteration. This method may thus require many iterations to converge on large-scale problems. For instance, if the starting point x^0 has no active constraints, while 200 constraints are active at the (nondegenerate) solution, then at least 200 iterations of the active-set method will be required to reach the solution.

The gradient projection method allows the active set to change rapidly from iteration to iteration. It is most efficient when the constraints are simple in form—in particular, when there are only bounds on the variables. Accordingly, we restrict our attention to the following bound-constrained problem:

$$\min_{x} \quad q(x) = \tfrac{1}{2}x^T G x + x^T c \qquad (16.68a)$$

$$\text{subject to} \quad l \le x \le u, \qquad (16.68b)$$

where G is symmetric and l and u are vectors of lower and upper bounds on the components of x. We do not make any positive definiteness assumptions on G in this section, because the gradient projection approach can be applied to both convex and nonconvex problems. The feasible region defined by (16.68b) is sometimes called a "box" because of its rectangular shape. Some components of x may lack an upper or a lower bound; we handle these cases formally by setting the appropriate components of l and u to $-\infty$ and $+\infty$, respectively.

Each iteration of the gradient projection algorithm consists of two stages. In the first stage, we search along the steepest descent direction from the current point x, that is, the direction $-g$, where $g = Gx + c$; see (16.6). Whenever a bound is encountered, the search direction is "bent" so that it stays feasible. We search along the resulting piecewise-linear path and locate the first local minimizer of q, which we denote by x^c and refer to as the *Cauchy point*, by analogy with our terminology of Chapter 4. The working set is now defined to be the set of bound constraints that are active at the Cauchy point, denoted by $\mathcal{A}(x^c)$. In the second stage of each gradient projection iteration, we explore the face of the feasible box on which the Cauchy point lies by solving a subproblem in which the active components x_i for $i \in \mathcal{A}(x^c)$ are fixed at the values x_i^c.

We describe the gradient projection method in detail in the rest of this section. Our convention in this section is to denote the iteration number by a superscript (that is, x^k) and use subscripts to denote the elements of a vector.

CAUCHY POINT COMPUTATION

We now derive an explicit expression for the piecewise-linear path obtained by projecting the steepest descent direction onto the feasible box, and outline the search procedure for identifying the first local minimum of q along this path.

The projection of an arbitrary point x onto the feasible region (16.68b) is defined as follows. The ith component is given by

$$P(x, l, u)_i = \begin{cases} l_i & \text{if} \quad x_i < l_i, \\ x_i & \text{if} \quad x_i \in [l_i, u_i], \\ u_i & \text{if} \quad x_i > u_i. \end{cases} \tag{16.69}$$

(We assume, without loss of generality, that $l_i < u_i$ for all i.) The piecewise-linear path $x(t)$ starting at the reference point x and obtained by projecting the steepest descent direction at x onto the feasible region (16.68b) is thus given by

$$x(t) = P(x - tg, l, u), \tag{16.70}$$

where $g = Gx + c$; see Figure 16.4.

The Cauchy point x^c, is defined as the first local minimizer of the univariate, piecewise-quadratic function $q(x(t))$, for $t \geq 0$. This minimizer is obtained by examining each of the line segments that make up $x(t)$. To perform this search, we need to determine the values of t at which the kinks in $x(t)$, or *breakpoints*, occur. We first identify the values of t for which each component reaches its bound along the chosen direction $-g$. These values $\bar{t}_i$ are given

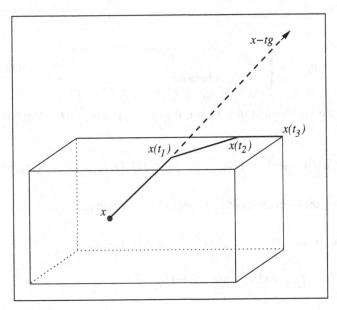

Figure 16.4   The piecewise-linear path $x(t)$, for an example in $\mathbb{R}^3$.

by the following explicit formulae:

$$\bar{t}_i = \begin{cases} (x_i - u_i)/g_i & \text{if } g_i < 0 \text{ and } u_i < +\infty, \\ (x_i - l_i)/g_i & \text{if } g_i > 0 \text{ and } l_i > -\infty, \\ \infty & \text{otherwise.} \end{cases} \tag{16.71}$$

The components of $x(t)$ for any t are therefore

$$x_i(t) = \begin{cases} x_i - tg_i & \text{if } t \le \bar{t}_i, \\ x_i - \bar{t}_i g_i & \text{otherwise.} \end{cases}$$

To search for the first local minimizer along $P(x - tg, l, u)$, we eliminate the duplicate values and zero values of $\bar{t}_i$ from the set $\{\bar{t}_1, \bar{t}_2, \ldots, \bar{t}_n\}$, to obtain a sorted, reduced set of breakpoints $\{t_1, t_2, \ldots, t_l\}$ with $0 < t_1 < t_2 < \cdots$. We now examine the intervals $[0, t_1]$, $[t_1, t_2]$, $[t_2, t_3]$, ... in turn. Suppose we have examined up to t_{j-1} and have not yet found a local minimizer. For the interval $[t_{j-1}, t_j]$, we have that

$$x(t) = x(t_{j-1}) + (\Delta t)p^{j-1},$$

where

$$\Delta t = t - t_{j-1} \in [0, t_j - t_{j-1}],$$

and

$$p_i^{j-1} = \begin{cases} -g_i & \text{if } t_{j-1} < \bar{t}_i, \\ 0 & \text{otherwise.} \end{cases} \tag{16.72}$$

We can then write the quadratic (16.68a) on the line segment $[x(t_{j-1}), x(t_j)]$ as follows:

$$q(x(t)) = c^T(x(t_{j-1}) + (\Delta t)p^{j-1}) + \tfrac{1}{2}(x(t_{j-1}) + (\Delta t)p^{j-1})^T G(x(t_{j-1}) + (\Delta t)p^{j-1}).$$

Expanding and grouping the coefficients of 1, Δt, and $(\Delta t)^2$, we find that

$$q(x(t)) = f_{j-1} + f'_{j-1}\Delta t + \tfrac{1}{2}f''_{j-1}(\Delta t)^2, \quad \Delta t \in [0, t_j - t_{j-1}], \tag{16.73}$$

where the coefficients f_{j-1}, f'_{j-1}, and f''_{j-1} are defined by

$$f_{j-1} \stackrel{\text{def}}{=} c^T x(t_{j-1}) + \tfrac{1}{2}x(t_{j-1})^T G x(t_{j-1}),$$
$$f'_{j-1} \stackrel{\text{def}}{=} c^T p^{j-1} + x(t_{j-1})^T G p^{j-1},$$
$$f''_{j-1} \stackrel{\text{def}}{=} (p^{j-1})^T G p^{j-1}.$$

Differentiating (16.73) with respect to Δt and equating to zero, we obtain $\Delta t^* = -f'_{j-1}/f''_{j-1}$. The following cases can occur. (i) If $f'_{j-1} > 0$ there is a local minimizer of $q(x(t))$ at $t = t_{j-1}$; else (ii) $\Delta t^* \in [0, t_j - t_{j-1})$ there is a minimizer at $t = t_{j-1} + \Delta t^*$; (iii) in all other cases we move on to the next interval $[t_j, t_{j+1}]$ and continue the search.

For the next search interval, we need to calculate the new direction p^j from (16.72), and we use this new value to calculate f_j, f'_j, and f''_j. Since p^j differs from p^{j-1} typically in just one component, computational savings can be made by updating these coefficients rather than computing them from scratch.

SUBSPACE MINIMIZATION

After the Cauchy point x^c has been computed, the components of x^c that are at their lower or upper bounds define the active set

$$\mathcal{A}(x^c) = \{i \mid x_i^c = l_i \text{ or } x_i^c = u_i\}.$$

In the second stage of the gradient projection iteration, we approximately solve the QP obtained by fixing the components x_i for $i \in \mathcal{A}(x^c)$ at the values x_i^c. The remaining components are determined from the subproblem

$$\min_{x} q(x) = \tfrac{1}{2}x^T G x + x^T c \tag{16.74a}$$

$$\text{subject to} \quad x_i = x_i^c, \ i \in \mathcal{A}(x^c), \tag{16.74b}$$

$$l_i \le x_i \le u_i, \ i \notin \mathcal{A}(x^c). \tag{16.74c}$$

It is not necessary to solve this problem exactly. Nor is it desirable in the large-dimensional case, because the subproblem may be almost as difficult as the original problem (16.68). In fact, to obtain global convergence of the gradient projection procedure, we require only that the approximate solution x^+ of (16.74) is feasible with respect to (16.68b) and has an objective function value no worse than that of x^c, that is, $q(x^+) \le q(x^c)$. A strategy that is intermediate between choosing $x^+ = x^c$ as the approximate solution (on the one hand) and solving (16.74) exactly (on the other hand) is to compute an approximate solution of (16.74) by using the conjugate gradient iteration described in Algorithm 16.1 or Algorithm 16.2. Note that for the equality constraints (16.74b), the Jacobian A and the null-space basis matrix Z have particularly simple forms. We could therefore apply conjugate gradient to the problem (16.74a), (16.74b) and terminate as soon as a bound $l \le x \le u$ is encountered. Alternatively, we could continue to iterate, temporarily ignoring the bounds and projecting the solution back onto the box constraints. The negative-curvature case can be handled as in Algorithm 7.2, the method for approximately solving possibly indefinite trust-region subproblems in unconstrained optimization.

We summarize the gradient projection algorithm for quadratic programming as follows.

Algorithm 16.5 (Gradient Projection Method for QP).
  Compute a feasible starting point x^0;
  for $k = 0, 1, 2, \ldots$
      if x^k satisfies the KKT conditions for (16.68)
          stop with solution $x^* = x^k$;
      Set $x = x^k$ and find the Cauchy point x^c;
      Find an approximate solution x^+ of (16.74) such that $q(x^+) \le q(x^c)$
          and x^+ is feasible;
      $x^{k+1} \leftarrow x^+$;
  end (for)

If the algorithm approaches a solution x^* at which the Lagrange multipliers associated with all the active bounds are nonzero (that is, strict complementarity holds), the active sets $\mathcal{A}(x^c)$ generated by the gradient projection algorithm are equal to the optimal active set for all k sufficiently large. That is, constraint indices do not repeatedly enter and leave the active set on successive iterations. When the problem is degenerate, the active set may not settle down at its optimal value. Various devices have been proposed to prevent this undesirable behavior from taking place.

While gradient projection methods can be applied in principle to problems with general linear constraints, significant computation may be required to perform the projection onto the feasible set in such cases. For example, if the constraint set is defined as $a_i^T x \geq b_i$, $i \in \mathcal{I}$, we must solve the following convex quadratic program to compute the projection of a given point $\bar{x}$ onto this set:

$$\max_x \|x - \bar{x}\|^2 \quad \text{subject to } a_i^T x \geq b_i \text{ for all } i \in \mathcal{I}.$$

The expense of solving this "projection subproblem" may approach the cost of solving the original quadratic program, so it is usually not economical to apply gradient projection to this case.

When we use duality to replace a strictly convex quadratic program with its dual (see Example 12.12), the gradient projection method may be useful in solving the bound-constrained dual problem, which is formulated in terms of the Lagrange multipliers λ as follows:

$$\max_\lambda \tilde{q}(\lambda) = -\frac{1}{2}(A^T \lambda - c)^T G^{-1}(A^T \lambda - c)^T + b^T \lambda, \quad \text{subject to } \lambda \geq 0.$$

(Note that the dual is conventionally written as a maximization problem; we can equivalently minimize $-\tilde{q}(\lambda)$ and note that this transformed problem is convex.) This approach is most useful when G has a simple form, for example, a diagonal or block-diagonal matrix.

16.8 PERSPECTIVES AND SOFTWARE

Active-set methods for convex quadratic programming are implemented in QPOPT [126], VE09 [142], BQPD [103], and QPA [148]. Several commercial interior-point solvers for QP are available, including CPLEX [172], XPRESS-MP [159] and MOSEK [5]. The code QPB [146] uses a two-phase interior-point method that can handle convex and nonconvex problems. OOPS [139] and OOQP [121] are object-oriented interior-point codes that allow the user to customize the linear algebra techniques to the particular structure of the data for an application. Some nonlinear programming interior-point packages, such as LOQO [294] and KNITRO [46], are also effective for convex and nonconvex quadratic programming.

The numerical comparison of active-set and interior-point methods for convex quadratic programming reported in [149] indicates that interior-point methods are generally much faster on large problems. If a warm start is required, however, active-set methods may be generally preferable. Although considerable research has been focused on improving the warm-start capabilities of interior-point methods, the full potential of such techniques is now yet known.

We have assumed in this chapter that all equality-constrained quadratic programs have linearly independent constraints, that is, the $m \times n$ constraint Jacobian matrix A has rank m.

If redundant constraints are present, they can be detected by forming a SVD or rank-revealing QR factorization of A^T, and then removed from the formulation. When A is larger, sparse Gaussian elimination techniques can be applied to A^T instead, but they are less reliable.

The KNITRO and OOPS software packages provide the option of solving the primal-dual equations (16.63) by means of the projected CG iteration of Algorithm 16.2.

We have not considered active-set methods for the case in which the Hessian matrix G is indefinite because these methods can be quite complicated to describe and it is not well understood how to adapt them to the large dimensional case. We make some comments here on the principal techniques.

Algorithm 16.3, the active-set method for convex QP, can be adapted to this indefinite case by modifying the computation of the search direction and step length in certain situations. To explain the need for the modification, we consider the computation of a step by a null-space method, that is, $p = Zp_z$, where p_z is given by (16.53). If the reduced Hessian $Z^T G Z$ is positive definite, then this step p points to the minimizer of the subproblem (16.39), and the logic of the iteration need not be changed. If $Z^T G Z$ has negative eigenvalues, however, p points only to a saddle point of (16.39) and is therefore not always a suitable step. Instead, we seek an alternative direction s_z that is a direction of *negative curvature* for $Z^T G Z$. We then have that

$$q(x + \alpha Z s_z) \to -\infty \quad \text{as } \alpha \to \infty. \tag{16.75}$$

Additionally, we change the sign of s_z if necessary to ensure that $Z s_z$ is a non-ascent direction for q at the current point x, that is, $\nabla q(x)^T Z s_z \le 0$. By moving along the direction $Z s_z$, we will encounter a constraint that can be added to the working set for the next iteration. (If we don't find such a constraint, the problem is unbounded.) If the reduced Hessian for the new working set is not positive definite, we repeat this process until enough constraints have been added to make the reduced Hessian positive definite. A difficulty with this general approach, however, is that if we allow the reduced Hessian to have several negative eigenvalues, it is difficult to make these methods efficient when the reduced Hessian changes from one working set to the next.

Inertia controlling methods are a practical class of algorithms for indefinite QP that never allow the reduced Hessian to have more than one negative eigenvalue. As in the convex case, there is a preliminary phase in which a feasible starting point x_0 is found. We place the additional demand on x_0 that it be either a vertex (in which case the reduced Hessian is the null matrix) or a constrained stationary point at which the reduced Hessian is positive definite. At each iteration, the algorithm will either add or remove a constraint from the working set. If a constraint is added, the reduced Hessian is of smaller dimension and must remain positive definite or be the null matrix. Therefore, an indefinite reduced Hessian can arise only when one of the constraints is removed from the working set, which happens only when the current point is a minimizer with respect to the current working set. In this case, we will choose the new search direction to be a direction of negative curvature for the reduced Hessian.

Various algorithms for indefinite QP differ in the way that indefiniteness is detected, in the computation of the negative curvature direction, and in the handling of the working set; see Fletcher [99] and Gill and Murray [126].

NOTES AND REFERENCES

The problem of determining whether a feasible point for a nonconvex QP (16.1) is a global minimizer is NP-hard (Murty and Kabadi [219]); so is the problem of determining whether a given point is a local minimizer (Vavasis [296, Theorem 5.1]). Various algorithms for convex QP with polynomial convexity are discussed in Nesterov and Nemirovskii [226].

The portfolio optimization problem was formulated by Markowitz [201].

For a discussion on the QMR, LSQR, and GMRES methods see, for example, [136, 272, 290]. The idea of using the projection (16.30) in the CG method dates back to at least Polyak [238]. The alternative (16.34), and its special case (16.32), are proposed in Coleman [64]. Although it can give rise to substantial rounding errors, they can be corrected by iterative refinement; see Gould et al. [143]. More recent studies on preconditioning of the projected CG method include Keller et al. [176] and Lukšan and Vlček [196].

For further discussion on the gradient projection method see, for example, Conn, Gould, and Toint [70] and Burke and Moré [44].

In some areas of application, the KKT matrix (16.7) not only is sparse but also contains special structure. For instance, the quadratic programs that arise in many control problems have banded matrices G and A (see Wright [315]), which can be exploited by interior-point methods via a suitable symmetric reordering of K. When active-set methods are applied to this problem, however, the advantages of bandedness and sparsity are lost after just a few updates of the factorization.

Further details of interior-point methods for convex quadratic programming can be found in Wright [316] and Vanderbei [293]. The first inertia-controlling method for indefinite quadratic programming was proposed by Fletcher [99]. See also Gill et al. [129] and Gould [142] for a discussion of methods for general quadratic programming.

✎ EXERCISES

✎ **16.1**

(a) Solve the following quadratic program and illustrate it geometrically.

$$\min f(x) = 2x_1 + 3x_2 + 4x_1^2 + 2x_1 x_2 + x_2^2,$$
$$\text{subject to } x_1 - x_2 \geq 0, \quad x_1 + x_2 \leq 4, \quad x_1 \leq 3.$$

(b) If the objective function is redefined as $q(x) = -f(x)$, does the problem have a finite minimum? Are there local minimizers?

✎ **16.2** The problem of finding the shortest distance from a point x_0 to the hyperplane $\{x \mid Ax = b\}$, where A has full row rank, can be formulated as the quadratic program

$$\min \tfrac{1}{2}(x - x_0)^T (x - x_0) \text{ subject to } Ax = b.$$

Show that the optimal multiplier is

$$\lambda^* = (AA^T)^{-1}(b - Ax_0)$$

and that the solution is

$$x^* = x_0 + A^T (AA^T)^{-1}(b - Ax_0).$$

Show that in the special case in which A is a row vector, the shortest distance from x_0 to the solution set of $Ax = b$ is $|b - Ax_0|/\|A\|_2$.

✎ **16.3** Use Theorem 12.1 to verify that the first-order necessary conditions for (16.3) are given by (16.4).

✎ **16.4** Suppose that G is positive semidefinite in (16.1) and that x^* satisfies the KKT conditions (16.37) for some $\lambda_i^*, i \in \mathcal{A}(x^*)$. Suppose in addition that second-order sufficient conditions are satisfied, that is, $Z^T GZ$ is positive definite where the columns of Z span the null space of the active constraint Jacobian matrix. Show that x^* is in fact the unique global solution for (16.1), that is, $q(x) > q(x^*)$ for all feasible x with $x \neq x^*$.

✎ **16.5** Verify that the inverse of the KKT matrix is given by (16.16).

✎ **16.6** Use Theorem 12.6 to show that if the conditions of Lemma 16.1 hold, then the second-order sufficient conditions for (16.3) are satisfied by the vector pair (x^*, λ^*) that satisfies (16.4).

✎ **16.7** Consider (16.3) and suppose that the projected Hessian matrix $Z^T GZ$ has a negative eigenvalue; that is, $u^T Z^T GZu < 0$ for some vector u. Show that if there exists any vector pair (x^*, λ^*) that satisfies (16.4), then the point x^* is only a stationary point of (16.3) and not a local minimizer. (Hint: Consider the function $q(x^* + \alpha Zu)$ for $\alpha \neq 0$, and use an expansion like that in the proof of Theorem 16.2.)

✎ **16.8** By using the QR factorization and a permutation matrix, show that for a full-rank $m \times n$ matrix A (with $m < n$) one can find an orthogonal matrix Q and an $m \times m$ upper triangular matrix $\hat{U}$ such that $AQ = \begin{bmatrix} 0 & \hat{U} \end{bmatrix}$. (Hint: Start by applying the standard QR factorization to A^T.)

✎ **16.9** Verify that the first-order conditions for optimality of (16.1) are equivalent to (16.37) when we make use of the active-set definition (16.36).

✐   **16.10** For each of the alternative choices of initial working set $\mathcal{W}_0$ in the example (16.49) (that is, $\mathcal{W}_0 = \{3\}$, $\mathcal{W}_0 = \{5\}$, and $\mathcal{W}_0 = \emptyset$) work through the first two iterations of Algorithm 16.3.

✐   **16.11** Program Algorithm 16.3, and use it to solve the problem

$$\min \quad x_1^2 + 2x_2^2 - 2x_1 - 6x_2 - 2x_1 x_2$$

$$\text{subject to} \quad \tfrac{1}{2}x_1 + \tfrac{1}{2}x_2 \le 1, \quad -x_1 + 2x_2 \le 2, \quad x_1, x_2 \ge 0.$$

Choose three initial starting points: one in the interior of the feasible region, one at a vertex, and one at a non-vertex point on the boundary of the feasible region.

✐   **16.12** Show that the operator P defined by (16.27) is independent of the choice of null-space basis Z. (Hint: First show that any null-space basis Z can be written as $Z = QB$ where Q is an orthogonal basis and B is a nonsingular matrix.)

✐   **16.13**

(a) Show that the the computation of the preconditioned residual g^+ in (16.28d) can be performed with (16.29) or (16.30).

(b) Show that we can also perform this computation by solving the system (16.32).

(c) Verify (16.33).

✐   **16.14**

(a) Show that if $Z^T G Z$ is positive definite, then the denominator in (16.28a) is nonzero.

(b) Show that if $Z^T r = r_z \ne 0$ and $Z^T H Z$ is positive definite, then the denominator in (16.28e) is nonzero.

✐   **16.15** Consider problem (16.3), and assume that A has full row rank and that Z is a basis for the null space of A. Prove that there are no finite solutions if $Z^T G Z$ has negative eigenvalues.

✐   **16.16**

(a) Assume that $A \ne 0$. Show that the KKT matrix (16.7) is indefinite.

(b) Prove that if the KKT matrix (16.7) is nonsingular, then A must have full rank.

✐   **16.17** Consider the quadratic program

$$\max \quad 6x_1 + 4x_2 - 13 - x_1^2 - x_2^2,$$

$$\text{subject to} \quad x_1 + x_2 \le 3, \quad x_1 \ge 0, \quad x_2 \ge 0. \tag{16.76}$$

First solve it graphically, and then use your program implementing the active-set method given in Algorithm 16.3.

✎   **16.18** Using (16.39) and (16.41), explain briefly why the gradient of each blocking constraint cannot be a linear combination of the constraint gradients in the current working set $\mathcal{W}_k$.

✎   **16.19** Let W be an $n \times n$ symmetric matrix, and suppose that Z is of dimension $n \times t$. Suppose that $Z^T W Z$ is positive definite and that $\bar{Z}$ is obtained by removing a column from Z. Show that $\bar{Z}^T W \bar{Z}$ is positive definite.

✎   **16.20** Find a null-space basis matrix Z for the equality-constrained problem defined by (16.74a), (16.74b).

✎   **16.21** Write down KKT conditions for the following convex quadratic program with mixed equality and inequality constraints:

$$\min \ q(x) = \tfrac{1}{2} x^T G x + x^T c \quad \text{subject to} \quad Ax \geq b, \quad \bar{A}x = \bar{b},$$

where G is symmetric and positive semidefinite. Use these conditions to derive an analogue of the generic primal-dual step (16.58) for this problem.

✎   **16.22** Explain why for a bound-constrained problems the number of possible active sets is at most 3^n.

✎   **16.23**

(a) Show that the primal-dual system (16.58) can be solved using the augmented system (16.61) or the normal equations (16.62). Describe in detail how all the components $(\Delta x, \Delta y, \Delta \lambda)$ are computed.

(b) Verify (16.65).

✎   **16.24** Program Algorithm 16.4 and use it to solve problem (16.76). Set all initial variables to be the vector $e = (1, 1, \dots, 1)^T$.

✎   **16.25** Let $\bar{x} \in R^n$ be given, and let x^* be the solution of the projection problem

$$\min \ \|x - \bar{x}\|^2 \quad \text{subject to} \quad l \leq x \leq u. \qquad\qquad (16.77)$$

For simplicity, assume that $-\infty < l_i < u_i < \infty$ for all $i = 1, 2, \dots, n$. Show that the solution of this problem coincides with the projection formula given by (16.69) that is, show that $x^* = P(\bar{x}, l, u)$. (*Hint*: Note that the problem is separable.)

✏ **16.26** Consider the bound-constrained quadratic problem (16.68) with

$$
G = \begin{bmatrix} 4 & 1 \\ 1 & 2 \end{bmatrix}, \quad c = \begin{bmatrix} -1 \\ 1 \end{bmatrix}, \quad l = \begin{bmatrix} 0 \\ 0 \end{bmatrix}, \quad \text{and } u = \begin{bmatrix} 5 \\ 3 \end{bmatrix}. \tag{16.78}
$$

Suppose $x^0 = (0, 2)^T$. Find $\bar{t}_1, \bar{t}_2, t_1, t_2, p^1, p^2$ and $x(t_1), x(t_2)$. Find the minimizer of $q(x(t))$.

✏ **16.27** Consider the search for the one dimensional minimizer of the function $q(x(t))$ defined by (16.73). There are 9 possible cases since f, f', f'' can each be positive, negative, or zero. For each case, determine the location of the minimizer. Verify that the rules described in Section 16.7 hold.

CHAPTER 17

Penalty and Augmented Lagrangian Methods

Some important methods for constrained optimization replace the original problem by a sequence of subproblems in which the constraints are represented by terms added to the objective. In this chapter we describe three approaches of this type. The *quadratic penalty* method adds a multiple of the square of the violation of each constraint to the objective. Because of its simplicity and intuitive appeal, this approach is used often in practice, although it has some important disadvantages. In *nonsmooth exact penalty* methods, a single unconstrained problem (rather than a sequence) takes the place of the original constrained problem. Using these penalty functions, we can often find a solution by performing a single unconstrained minimization, but the nonsmoothness may create complications. A popular function of this type is the ℓ_1 penalty function. A different kind of exact penalty approach is the *method of multipliers* or *augmented Lagrangian method*, in which explicit Lagrange multiplier estimates are used to avoid the ill-conditioning that is inherent in the quadratic penalty function.

A somewhat related approach is used in the *log-barrier method*, in which logarithmic terms prevent feasible iterates from moving too close to the boundary of the feasible region. This approach forms part of the foundation for interior-point methods for nonlinear programming and we discuss it further in Chapter 19.

17.1 THE QUADRATIC PENALTY METHOD

MOTIVATION

Let us consider replacing a constrained optimization problem by a single function consisting of

- the original objective of the constrained optimization problem, *plus*

- one additional term for each constraint, which is positive when the current point x violates that constraint and zero otherwise.

Most approaches define a *sequence* of such penalty functions, in which the penalty terms for the constraint violations are multiplied by a positive coefficient. By making this coefficient larger, we penalize constraint violations more severely, thereby forcing the minimizer of the penalty function closer to the feasible region for the constrained problem.

The simplest penalty function of this type is the *quadratic penalty function*, in which the penalty terms are the squares of the constraint violations. We describe this approach first in the context of the equality-constrained problem

$$\min_x \ f(x) \quad \text{subject to } c_i(x) = 0, \quad i \in \mathcal{E}, \tag{17.1}$$

which is a special case of (12.1). The quadratic penalty function $Q(x; \mu)$ for this formulation is

$$Q(x; \mu) \overset{\text{def}}{=} f(x) + \frac{\mu}{2} \sum_{i \in \mathcal{E}} c_i^2(x), \tag{17.2}$$

where $\mu > 0$ is the *penalty parameter*. By driving μ to ∞, we penalize the constraint violations with increasing severity. It makes good intuitive sense to consider a sequence of values $\{\mu_k\}$ with $\mu_k \uparrow \infty$ as $k \to \infty$, and to seek the approximate minimizer x_k of $Q(x; \mu_k)$ for each k. Because the penalty terms in (17.2) are smooth, we can use techniques from unconstrained optimization to search for x_k. In searching for x_k, we can use the minimizers x_{k-1}, x_{k-2}, etc., of $Q(\cdot; \mu)$ for smaller values of μ to construct an initial guess. For suitable choices of the sequence $\{\mu_k\}$ and the initial guesses, just a few steps of unconstrained minimization may be needed for each μ_k.

❑ EXAMPLE 17.1

Consider the problem (12.9) from Chapter 12, that is,

$$\min x_1 + x_2 \quad \text{subject to } x_1^2 + x_2^2 - 2 = 0, \tag{17.3}$$

for which the solution is $(-1, -1)^T$ and the quadratic penalty function is

$$Q(x; \mu) = x_1 + x_2 + \frac{\mu}{2} \left(x_1^2 + x_2^2 - 2 \right)^2. \tag{17.4}$$

We plot the contours of this function in Figures 17.1 and 17.2. In Figure 17.1 we have $\mu = 1$, and we observe a minimizer of Q near the point $(-1.1, -1.1)^T$. (There is also a local maximizer near $x = (0.3, 0.3)^T$.) In Figure 17.2 we have $\mu = 10$, so points that do not lie on the feasible circle defined by $x_1^2 + x_2^2 = 2$ suffer a much greater penalty than in the first figure—the "trough" of low values of Q is clearly evident. The minimizer in this figure is much closer to the solution $(-1, -1)^T$ of the problem (17.3). A local maximum lies near $(0, 0)^T$, and Q goes rapidly to ∞ outside the circle $x_1^2 + x_2^2 = 2$.   ❑

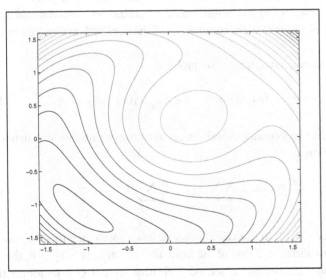

Figure 17.1   Contours of $Q(x; \mu)$ from (17.4) for $\mu = 1$, contour spacing 0.5.

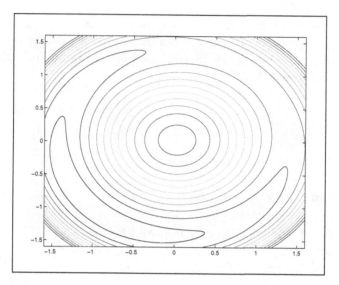

Figure 17.2 Contours of $Q(x; \mu)$ from (17.4) for $\mu = 10$, contour spacing 2.

The situation is not always so benign as in Example 17.1. For a given value of the penalty parameter μ, the penalty function may be unbounded below even if the original constrained problem has a unique solution. Consider for example

$$\min \; -5x_1^2 + x_2^2 \quad \text{subject to } x_1 = 1, \tag{17.5}$$

whose solution is $(1, 0)^T$. The penalty function is unbounded for any $\mu < 10$. For such values of μ, the iterates generated by an unconstrained minimization method would usually diverge. This deficiency is, unfortunately, common to all the penalty functions discussed in this chapter.

For the general constrained optimization problem

$$\min_x \; f(x) \quad \text{subject to } c_i(x) = 0, \quad i \in \mathcal{E}, \quad c_i(x) \geq 0, \quad i \in \mathcal{I}, \tag{17.6}$$

which contains inequality constraints as well as equality constraints, we can define the quadratic penalty function as

$$Q(x; \mu) \stackrel{\text{def}}{=} f(x) + \frac{\mu}{2} \sum_{i \in \mathcal{E}} c_i^2(x) + \frac{\mu}{2} \sum_{i \in \mathcal{I}} \left([c_i(x)]^- \right)^2, \tag{17.7}$$

where $[y]^-$ denotes $\max(-y, 0)$. In this case, Q may be less smooth than the objective and constraint functions. For instance, if one of the inequality constraints is $x_1 \geq 0$, then the function $\min(0, x_1)^2$ has a discontinuous second derivative, so that Q is no longer twice continuously differentiable.

ALGORITHMIC FRAMEWORK

A general framework for algorithms based on the quadratic penalty function (17.2) can be specified as follows.

Framework 17.1 (Quadratic Penalty Method).
> Given $\mu_0 > 0$, a nonnegative sequence $\{\tau_k\}$ with $\tau_k \to 0$, and a starting point x_0^s;
> for $k = 0, 1, 2, \ldots$
> > Find an approximate minimizer x_k of $Q(\cdot; \mu_k)$, starting at x_k^s,
> > > and terminating when $\|\nabla_x Q(x; \mu_k)\| \leq \tau_k$;
> > **if** final convergence test satisfied
> > > **stop** with approximate solution x_k;
> > **end (if)**
> > Choose new penalty parameter $\mu_{k+1} > \mu_k$;
> > Choose new starting point x_{k+1}^s;
> **end (for)**

The parameter sequence $\{\mu_k\}$ can be chosen adaptively, based on the difficulty of minimizing the penalty function at each iteration. When minimization of $Q(x; \mu_k)$ proves to be expensive for some k, we choose μ_{k+1} to be only modestly larger than μ_k; for instance $\mu_{k+1} = 1.5\mu_k$. If we find the approximate minimizer of $Q(x; \mu_k)$ cheaply, we could try a more ambitious increase, for instance $\mu_{k+1} = 10\mu_k$. The convergence theory for Framework 17.1 allows wide latitude in the choice of nonnegative tolerances τ_k; it requires only that $\tau_k \to 0$, to ensure that the minimization is carried out more accurately as the iterations progress.

There is no guarantee that the stop test $\|\nabla_x Q(x; \mu_k)\| \leq \tau_k$ will be satisfied because, as discussed above, the iterates may move away from the feasible region when the penalty parameter is not large enough. A practical implementation must include safeguards that increase the penalty parameter (and possibly restore the initial point) when the constraint violation is not decreasing rapidly enough, or when the iterates appear to be diverging.

When only equality constraints are present, $Q(x; \mu_k)$ is smooth, so the algorithms for unconstrained minimization described in the first chapters of the book can be used to identify the approximate solution x_k. However, the minimization of $Q(x; \mu_k)$ becomes more difficult to perform as μ_k becomes large, unless we use special techniques to calculate the search directions. For one thing, the Hessian $\nabla_{xx}^2 Q(x; \mu_k)$ becomes arbitrarily ill conditioned near the minimizer. This property alone is enough to make many unconstrained minimization algorithms such as quasi-Newton and conjugate gradient perform poorly. Newton's method, on the other hand, is not sensitive to ill conditioning of the Hessian, but it, too, may encounter difficulties for large μ_k for two other reasons. First, ill conditioning of $\nabla_{xx}^2 Q(x; \mu_k)$ might be expected to cause numerical problems when we solve the linear equations to calculate the Newton step. We discuss this issue below, and show that these effects are not severe and that a reformulation of the Newton equations is possible. Second, even when x is close to the minimizer of $Q(\cdot; \mu_k)$, the quadratic Taylor series approximation to $Q(x; \mu_k)$ about x is a reasonable approximation of the true function only in a small neighborhood of x. This property can be seen in Figure 17.2, where the contours of Q near the minimizer have a "banana" shape, rather than the elliptical shape that characterizes quadratic functions. Since Newton's method is based on the quadratic model, the steps that it generates may not make rapid progress toward the minimizer of $Q(x; \mu_k)$. This difficulty can be lessened by a judicious choice of the starting point x^s_{k+1}, or by setting $x^s_{k+1} = x_k$ and choosing μ_{k+1} to be only modestly larger than μ_k.

CONVERGENCE OF THE QUADRATIC PENALTY METHOD

We describe some convergence properties of the quadratic penalty method in the following two theorems. We restrict our attention to the equality-constrained problem (17.1), for which the quadratic penalty function is defined by (17.2).

For the first result we assume that the penalty function $Q(x; \mu_k)$ has a (finite) minimizer for each value of μ_k.

Theorem 17.1.

Suppose that each x_k is the exact global minimizer of $Q(x; \mu_k)$ defined by (17.2) in Framework 17.1 above, and that $\mu_k \uparrow \infty$. Then every limit point x^* of the sequence $\{x_k\}$ is a global solution of the problem (17.1).

PROOF.   Let $\bar{x}$ be a global solution of (17.1), that is,

$$f(\bar{x}) \le f(x) \quad \text{for all } x \text{ with } c_i(x) = 0, \quad i \in \mathcal{E}.$$

Since x_k minimizes $Q(\cdot; \mu_k)$ for each k, we have that $Q(x_k; \mu_k) \le Q(\bar{x}; \mu_k)$, which leads to the inequality

$$f(x_k) + \frac{\mu_k}{2} \sum_{i \in \mathcal{E}} c_i^2(x_k) \le f(\bar{x}) + \frac{\mu_k}{2} \sum_{i \in \mathcal{E}} c_i^2(\bar{x}) = f(\bar{x}). \tag{17.8}$$

By rearranging this expression, we obtain

$$\sum_{i \in \mathcal{E}} c_i^2(x_k) \le \frac{2}{\mu_k} [f(\bar{x}) - f(x_k)]. \tag{17.9}$$

Suppose that x^* is a limit point of $\{x_k\}$, so that there is an infinite subsequence $\mathcal{K}$ such that

$$\lim_{k \in \mathcal{K}} x_k = x^*.$$

By taking the limit as $k \to \infty$, $k \in \mathcal{K}$, on both sides of (17.9), we obtain

$$\sum_{i \in \mathcal{E}} c_i^2(x^*) = \lim_{k \in \mathcal{K}} \sum_{i \in \mathcal{E}} c_i^2(x_k) \leq \lim_{k \in \mathcal{K}} \frac{2}{\mu_k}[f(\bar{x}) - f(x_k)] = 0,$$

where the last equality follows from $\mu_k \uparrow \infty$. Therefore, we have that $c_i(x^*) = 0$ for all $i \in \mathcal{E}$, so that x^* is feasible. Moreover, by taking the limit as $k \to \infty$ for $k \in \mathcal{K}$ in (17.8), we have by nonnegativity of μ_k and of each $c_i(x_k)^2$ that

$$f(x^*) \leq f(x^*) + \lim_{k \in \mathcal{K}} \frac{\mu_k}{2} \sum_{i \in \mathcal{E}} c_i^2(x_k) \leq f(\bar{x}).$$

Since x^* is a feasible point whose objective value is no larger than that of the global solution $\bar{x}$, we conclude that x^*, too, is a global solution, as claimed. □

Since this result requires us to find the *global* minimizer for each subproblem, this desirable property of convergence to the global solution of (17.1) cannot be attained in general. The next result concerns convergence properties of the sequence $\{x_k\}$ when we allow inexact (but increasingly accurate) minimizations of $Q(\cdot; \mu_k)$. In contrast to Theorem 17.1, it shows that the sequence may be attracted to infeasible points, or to any KKT point (that is, a point satisfying first-order necessary conditions; see (12.34)), rather than to a minimizer. It also shows that the quantities $\mu_k c_i(x_k)$ may be used as estimates of the Lagrange multipliers λ_i^* in certain circumstances. This observation is important for the analysis of augmented Lagrangian methods in Section 17.3.

To establish the result we will make the (optimistic) assumption that the stop test $\|\nabla_x Q(x; \mu_k)\| \leq \tau_k$ is satisfied for all k.

Theorem 17.2.

Suppose that the tolerances and penalty parameters in Framework 17.1 satisfy $\tau_k \to 0$ and $\mu_k \uparrow \infty$. Then if a limit point x^* of the sequence $\{x_k\}$ is infeasible, it is a stationary point of the function $\|c(x)\|^2$. On the other hand, if a limit point x^* is feasible and the constraint gradients $\nabla c_i(x^*)$ are linearly independent, then x^* is a KKT point for the problem (17.1). For such points, we have for any infinite subsequence $\mathcal{K}$ such that $\lim_{k \in \mathcal{K}} x_k = x^*$ that

$$\lim_{k \in \mathcal{K}} -\mu_k c_i(x_k) = \lambda_i^*, \quad \text{for all } i \in \mathcal{E}, \tag{17.10}$$

where λ^* is the multiplier vector that satisfies the KKT conditions (12.34) for the equality-constrained problem (17.1).

PROOF.   By differentiating $Q(x; \mu_k)$ in (17.2), we obtain

$$\nabla_x Q(x_k; \mu_k) = \nabla f(x_k) + \sum_{i \in \mathcal{E}} \mu_k c_i(x_k) \nabla c_i(x_k), \tag{17.11}$$

so from the termination criterion for Framework 17.1, we have that

$$\left\| \nabla f(x_k) + \sum_{i\in\mathcal{E}} \mu_k c_i(x_k) \nabla c_i(x_k) \right\| \leq \tau_k. \tag{17.12}$$

By rearranging this expression (and in particular using the inequality $\|a\| - \|b\| \leq \|a+b\|$), we obtain

$$\left\| \sum_{i\in\mathcal{E}} c_i(x_k) \nabla c_i(x_k) \right\| \leq \frac{1}{\mu_k} \left[\tau_k + \|\nabla f(x_k)\| \right]. \tag{17.13}$$

Let x^* be a limit point of the sequence of iterates. Then there is a subsequence $\mathcal{K}$ such that $\lim_{k\in\mathcal{K}} x_k = x^*$. When we take limits as $k \to \infty$ for $k \in \mathcal{K}$, the bracketed term on the right-hand-side approaches $\|\nabla f(x^*)\|$, so because $\mu_k \uparrow \infty$, the right-hand-side approaches zero. From the corresponding limit on the left-hand-side, we obtain

$$\sum_{i\in\mathcal{E}} c_i(x^*) \nabla c_i(x^*) = 0. \tag{17.14}$$

We can have $c_i(x^*) \neq 0$ (if the constraint gradients $\nabla c_i(x^*)$ are dependent), but in this case (17.14) implies that x^* is a stationary point of the function $\|c(x)\|^2$.

If, on the other hand, the constraint gradients $\nabla c_i(x^*)$ are linearly independent at a limit point x^*, we have from (17.14) that $c_i(x^*) = 0$ for all $i \in \mathcal{E}$, so x^* is feasible. Hence, the second KKT condition (12.34b) is satisfied. We need to check the first KKT condition (12.34a) as well, and to show that the limit (17.10) holds.

By using $A(x)$ to denote the matrix of constraint gradients (also known as the Jacobian), that is,

$$A(x)^T = [\nabla c_i(x)]_{i\in\mathcal{E}}, \tag{17.15}$$

and λ_k to denote the vector $-\mu_k c(x_k)$, we have as in (17.12) that

$$A(x_k)^T \lambda_k = \nabla f(x_k) - \nabla_x Q(x_k; \mu_k), \quad \|\nabla_x Q(x_k; \mu_k)\| \leq \tau_k. \tag{17.16}$$

For all $k \in \mathcal{K}$ sufficiently large, the matrix $A(x_k)$ has full row rank, so that $A(x_k)A(x_k)^T$ is nonsingular. By multiplying (17.16) by $A(x_k)$ and rearranging, we have that

$$\lambda_k = \left[A(x_k)A(x_k)^T \right]^{-1} A(x_k) \left[\nabla f(x_k) - \nabla_x Q(x_k; \mu_k) \right].$$

Hence by taking the limit as $k \in \mathcal{K}$ goes to ∞, we find that

$$\lim_{k\in\mathcal{K}} \lambda_k = \lambda^* = \left[A(x^*)A(x^*)^T \right]^{-1} A(x^*)\nabla f(x^*).$$

By taking limits in (17.12), we conclude that

$$\nabla f(x^*) - A(x^*)^T \lambda^* = 0, \tag{17.17}$$

so that λ^* satisfies the first KKT condition (12.34a) for (17.1). Hence, x^* is a KKT point for (17.1), with unique Lagrange multiplier vector λ^*.   □

It is reassuring that, if a limit point x^* is not feasible, it is at least a stationary point for the function $\|c(x)\|^2$. Newton-type algorithms can always be attracted to infeasible points of this type. (We see the same effect in Chapter 11, in our discussion of methods for nonlinear equations that use the sum-of-squares merit function $\|r(x)\|^2$.) Such methods cannot be guaranteed to find a root, and can be attracted to a stationary point or minimizer of the merit function. In the case in which the nonlinear program (17.1) is infeasible, we often observe convergence of the quadratic-penalty method to stationary points or minimizers of $\|c(x)\|^2$.

ILL CONDITIONING AND REFORMULATIONS

We now examine the nature of the ill conditioning in the Hessian $\nabla_{xx}^2 Q(x; \mu_k)$. An understanding of the properties of this matrix, and the similar Hessians that arise in other penalty and barrier methods, is essential in choosing effective algorithms for the minimization problem and for the linear algebra calculations at each iteration.

The Hessian is given by the formula

$$\nabla_{xx}^2 Q(x; \mu_k) = \nabla^2 f(x) + \sum_{i \in \mathcal{E}} \mu_k c_i(x) \nabla^2 c_i(x) + \mu_k A(x)^T A(x), \tag{17.18}$$

where we have used the definition (17.15) of $A(x)$. When x is close to the minimizer of $Q(\cdot; \mu_k)$ and the conditions of Theorem 17.2 are satisfied, we have from (17.10) that the sum of the first two terms on the right-hand-side of (17.18) is approximately equal to the Hessian of the Lagrangian function defined in (12.33). To be specific, we have

$$\nabla_{xx}^2 Q(x; \mu_k) \approx \nabla_{xx}^2 \mathcal{L}(x, \lambda^*) + \mu_k A(x)^T A(x), \tag{17.19}$$

when x is close to the minimizer of $Q(\cdot; \mu_k)$. We see from this expression that $\nabla_{xx}^2 Q(x; \mu_k)$ is approximately equal to the sum of

- a matrix whose elements are independent of μ_k (the Lagrangian term), and

- a matrix of rank $|\mathcal{E}|$ whose nonzero eigenvalues are of order μ_k (the second term on the right-hand side of (17.19)).

The number of constraints $|\mathcal{E}|$ is usually smaller than n. In this case, the last term in (17.19) is singular. The overall matrix has some of its eigenvalues approaching a constant, while others are of order μ_k. Since μ_k is approaching ∞, the increasing ill conditioning of $\nabla^2_{xx} Q(x; \mu_k)$ is apparent.

One consequence of the ill conditioning is possible inaccuracy in the calculation of the Newton step for $Q(x; \mu_k)$, which is obtained by solving the following system:

$$\nabla^2_{xx} Q(x; \mu_k) p = -\nabla_x Q(x; \mu_k). \tag{17.20}$$

In general, the poor conditioning of this system will lead to significant errors in the computed value of p, regardless of the computational technique used to solve (17.20). For the same reason, iterative methods can be expected to perform poorly unless accompanied by a preconditioning strategy that removes the systematic ill conditioning.

There is an alternative formulation of the equations (17.20) that avoids the ill conditioning due to the final term in (17.18). By introducing a new variable vector ζ defined by $\zeta = \mu A(x) p$, we see that the vector p that solves (17.20) also satisfies the following system:

$$\begin{bmatrix} \nabla^2 f(x) + \displaystyle\sum_{i \in \mathcal{E}} \mu_k c_i(x) \nabla^2 c_i(x) & A(x)^T \\ A(x) & -(1/\mu_k) I \end{bmatrix} \begin{bmatrix} p \\ \zeta \end{bmatrix} = \begin{bmatrix} -\nabla_x Q(x; \mu_k) \\ 0 \end{bmatrix}. \tag{17.21}$$

When x is not too far from the solution x^*, the coefficient matrix in this system does not have large singular values (of order μ_k), so the system (17.21) can be viewed as a well conditioned reformulation of (17.20). We note, however, that neither system may yield a good search direction p because the coefficients $\mu_k c_i(x)$ in the summation term of the upper left block of (17.21) may be poor approximations to the Lagrange multipliers $-\lambda_i^*$, even when x is quite close to the minimizer x_k of $Q(x; \mu_k)$. This fact may cause the quadratic model on which p is based to be an inadequate model of $Q(\cdot; \mu_k)$, so the Newton step may be intrinsically an unsuitable search direction. We discussed possible remedies for this difficulty above, in our comments following Framework 17.1.

To compute the step via (17.21) involves the solution of a linear system of dimension $n + |\mathcal{E}|$ rather than the system of dimension n given by (17.19). A similar system must be solved to calculate the sequential quadratic programming (SQP) step (18.6), which is derived in Chapter 18. In fact, when μ_k is large, (17.21) can be viewed as a regularization of the SQP step (18.6) in which the term $-(1/\mu_k) I$ helps to ensure that the iteration matrix is nonsingular even when the Jacobian $A(x)$ is rank deficient. On the other hand, when μ_k is small, (17.21) shows that the step computed by the quadratic penalty method does not closely satisfy the linearization of the constraints. This situation is undesirable because the steps may not make significant progress toward the feasible region, resulting in inefficient global behavior. Moreover, if $\{\mu_k\}$ does not approach ∞ rapidly enough, we lose the possibility of a superlinear rate that occurs when the linearization is exact; see Chapter 18.

To conclude, the formulation (17.21) allows us to view the quadratic penalty method either as the application of unconstrained minimization to the penalty function $Q(\cdot; \mu_k)$ or as a variation on the SQP methods discussed in Chapter 18.

17.2 NONSMOOTH PENALTY FUNCTIONS

Some penalty functions are *exact*, which means that, for certain choices of their penalty parameters, a single minimization with respect to x can yield the exact solution of the nonlinear programming problem. This property is desirable because it makes the performance of penalty methods less dependent on the strategy for updating the penalty parameter. The quadratic penalty function of Section 17.1 is not exact because its minimizer is generally not the same as the solution of the nonlinear program for any positive value of μ. In this section we discuss *nonsmooth* exact penalty functions, which have proved to be useful in a number of practical contexts.

A popular nonsmooth penalty function for the general nonlinear programming problem (17.6) is the ℓ_1 *penalty function* defined by

$$\phi_1(x; \mu) = f(x) + \mu \sum_{i \in \mathcal{E}} |c_i(x)| + \mu \sum_{i \in \mathcal{I}} [c_i(x)]^-, \tag{17.22}$$

where we use again the notation $[y]^- = \max\{0, -y\}$. Its name derives from the fact that the penalty term is μ times the ℓ_1 norm of the constraint violation. Note that $\phi_1(x; \mu)$ is not differentiable at some x, because of the presence of the absolute value and $[\cdot]^-$ functions.

The following result establishes the *exactness* of the ℓ_1 penalty function. For a proof see [165, Theorem 4.4].

Theorem 17.3.

Suppose that x^* is a strict local solution of the nonlinear programming problem (17.6) at which the first-order necessary conditions of Theorem 12.1 are satisfied, with Lagrange multipliers $\lambda_i^*, i \in \mathcal{E} \cup \mathcal{I}$. Then x^* is a local minimizer of $\phi_1(x; \mu)$ for all $\mu > \mu^*$, where

$$\mu^* = \|\lambda^*\|_\infty = \max_{i \in \mathcal{E} \cup \mathcal{I}} |\lambda_i^*|. \tag{17.23}$$

If, in addition, the second-order sufficient conditions of Theorem 12.6 hold and $\mu > \mu^*$, then x^* is a strict local minimizer of $\phi_1(x; \mu)$.

Loosely speaking, at a solution of the nonlinear program x^*, any move into the infeasible region is penalized sharply enough that it produces an increase in the penalty function to a value greater than $\phi_1(x^*; \mu) = f(x^*)$, thereby forcing the minimizer of $\phi_1(\cdot; \mu)$ to lie at x^*.

❏ **EXAMPLE 17.2**

Consider the following problem in one variable:

$$\min x \qquad \text{subject to} \quad x \geq 1, \tag{17.24}$$

whose solution is $x^* = 1$. We have that

$$\phi_1(x; \mu) = x + \mu[x - 1]^- = \begin{cases} (1 - \mu)x + \mu & \text{if } x \leq 1, \\ x & \text{if } x > 1. \end{cases} \tag{17.25}$$

As can be seen in Figure 17.3, the penalty function has a minimizer at $x^* = 1$ when $\mu > 1$, but is a monotone increasing function when $\mu < 1$.

❏

Since penalty methods work by minimizing the penalty function directly, we need to characterize stationary points of ϕ_1. Even though ϕ_1 is not differentiable, it has a directional derivative $D(\phi_1(x; \mu); p)$ along any direction; see (A.51) and the example following this definition.

Definition 17.1.

A point $\hat{x} \in R^n$ is a stationary point *for the penalty function* $\phi_1(x; \mu)$ *if*

$$D(\phi_1(\hat{x}; \mu); p) \geq 0, \tag{17.26}$$

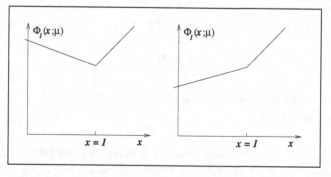

Figure 17.3 Penalty function for problem (17.24) with $\mu > 1$ (left) and $\mu < 1$ (right).

for all $p \in R^n$. Similarly, $\hat{x}$ is a stationary point of the measure of infeasibility

$$h(x) = \sum_{i\in\mathcal{E}} |c_i(x)| + \sum_{i\in\mathcal{I}} [c_i(x)]^- \qquad (17.27)$$

if $D(h(\hat{x}); p) \geq 0$ for all $p \in R^n$. If a point is infeasible for (17.6) but stationary with respect to the infeasibility measure h, we say that it is an infeasible stationary point.

For the function in Example 17.2, we have for $x^* = 1$ that

$$D(\phi_1(x^*; \mu); p) = \begin{cases} p & \text{if } p \geq 0 \\ (1-\mu)p & \text{if } p < 0; \end{cases}$$

it follows that when $\mu > 1$, we have $D(\phi_1(x^*; \mu); p) \geq 0$ for all $p \in R$.

The following result complements Theorem 17.3 by showing that stationary points of $\phi_1(x; \mu)$ correspond to KKT points of the constrained optimization problem (17.6) under certain assumptions.

Theorem 17.4.

Suppose that $\hat{x}$ is a stationary point of the penalty function $\phi_1(x; \mu)$ for all μ greater than a certain threshold $\hat{\mu} > 0$. Then, if $\hat{x}$ is feasible for the nonlinear program (17.6), it satisfies the KKT conditions (12.34) for (17.6). Is $\hat{x}$ is not feasible for (17.6), it is an infeasible stationary point.

PROOF. Suppose first that $\hat{x}$ is feasible. We have from (A.51) and the definition (17.22) of ϕ_1 that

$$D(\phi_1(\hat{x}; \mu); p) = \nabla f(\hat{x})^T p + \mu \sum_{i\in\mathcal{E}} |\nabla c_i(\hat{x})^T p| + \mu \sum_{i\in\mathcal{I}\cap\mathcal{A}(\hat{x})} [\nabla c_i(\hat{x})^T p]^-, \qquad (17.28)$$

where the active set $\mathcal{A}(\hat{x})$ is defined in Definition 12.1. (We leave verification of (17.28) as an exercise.) Consider any direction p in the linearized feasible direction set $\mathcal{F}(\hat{x})$ of Definition 12.3. By the properties of $\mathcal{F}(\hat{x})$, we have

$$|\nabla c_i(\hat{x})^T p| + \sum_{i\in\mathcal{I}\cap\mathcal{A}(\hat{x})} [\nabla c_i(\hat{x})^T p]^- = 0,$$

so that by the stationarity assumption on $\phi_1(\hat{x}; \mu)$, we have

$$0 \leq D(\phi_1(\hat{x}; \mu); p) = \nabla f(\hat{x})^T p, \quad \text{for all } p \in \mathcal{F}(\hat{x}).$$

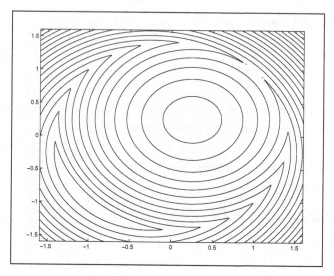

Figure 17.4   Contours of $\phi_1(x; \mu)$ from (17.3) for $\mu = 2$, contour spacing 0.5.

We can now apply Farkas' Lemma (Lemma 12.4) to deduce that

$$\nabla f(\hat{x}) = \sum_{i \in A(\hat{x})} \hat{\lambda}_i \nabla c_i(\hat{x}),$$

for some coefficients $\hat{\lambda}_i$ with $\hat{\lambda}_i \geq 0$ for all $i \in \mathcal{I} \cap A(\hat{x})$. As we noted earlier (see Theorem 12.1 and (12.35)), this expression implies that the KKT conditions (12.34) hold, as claimed.

We leave the second part of the proof (concerning infeasible $\hat{x}$) as an exercise.   □

□ **EXAMPLE 17.3**

Consider again problem (17.3), for which the ℓ_1 penalty function is

$$\phi_1(x; \mu) = x_1 + x_2 + \mu \left| x_1^2 + x_2^2 - 2 \right|. \tag{17.29}$$

Figure 17.4 plots the function $\phi_1(x; 2)$, whose minimizer is the solution $x^* = (-1, -1)^T$ of (17.3). In fact, following Theorem 17.3, we find that for all $\mu > |\lambda^*| = 0.5$, the minimizer of $\phi_1(x; \mu)$ coincides with x^*. The sharp corners on the contours indicate nonsmoothness along the boundary of the circle defined by $x_1^2 + x_2^2 = 2$.   □

These results provide the motivation for an algorithmic framework based on the ℓ_1 penalty function, which we now present.

Framework 17.2 (Classical ℓ_1 Penalty Method).
  Given $\mu_0 > 0$, tolerance $\tau > 0$, starting point x_0^s;
  for $k = 0, 1, 2, \ldots$
        Find an approximate minimizer x_k of $\phi_1(x; \mu_k)$, starting at x_k^s;
        if $h(x_k) \leq \tau$
              stop with approximate solution x_k;
        end (if)
        Choose new penalty parameter $\mu_{k+1} > \mu_k$;
        Choose new starting point x_{k+1}^s;
  end (for)

The minimization of $\phi_1(x; \mu_k)$ is made difficult by the nonsmoothness of the function. Nevertheless, as we discuss below, it is well understood how to compute minimization steps using a smooth model of $\phi_1(x; \mu_k)$, in a way that resembles SQP methods.

The simplest scheme for updating the penalty parameter μ_k is to increase it by a constant multiple (say 5 or 10), if the current value produces a minimizer that is not feasible to within the tolerance τ. This scheme sometimes works well in practice, but can also be inefficient. If the initial penalty parameter μ_0 is too small, many cycles of Framework 17.2 may be needed to determine an appropriate value. In addition, the iterates may move away from the solution x^* in these initial cycles, in which case the minimization of $\phi_1(x; \mu_k)$ should be terminated early and x_k^s should possibly be reset to a previous iterate. If, on the other hand, μ_k is excessively large, the penalty function will be difficult to minimize, possibly requiring a large number of iterations. We return to the issue of selecting the penalty parameter below.

A PRACTICAL ℓ_1 PENALTY METHOD

As noted already, $\phi_1(x; \mu)$ is nonsmooth—its gradient is not defined at any x for which $c_i(x) = 0$ for some $i \in \mathcal{E} \cup \mathcal{I}$. Rather than using techniques for nondifferentiable optimization, such as bundle methods [170], we prefer techniques that take account of the special nature of the nondifferentiabilities in this function. As in the algorithms for unconstrained optimization discussed in the first part of this book, we obtain a step toward the minimizer of $\phi_1(x; \mu)$ by forming a simplified model of this function and seeking the minimizer of this model. Here, the model can be defined by linearizing the constraints c_i and replacing the nonlinear programming objective f by a quadratic function, as follows:

$$q(p; \mu) = f(x) + \nabla f(x)^T p + \tfrac{1}{2} p^T W p + \mu \sum_{i \in \mathcal{E}} |c_i(x) + \nabla c_i(x)^T p| +$$

$$\mu \sum_{i \in \mathcal{I}} [c_i(x) + \nabla c_i(x)^T p]^-, \tag{17.30}$$

where W is a symmetric matrix which usually contains second derivative information about f and c_i, $i \in \mathcal{E} \cup \mathcal{I}$. The model $q(p; \mu)$ is not smooth, but we can formulate the problem of minimizing q as a smooth quadratic programming problem by introducing artificial variables r_i, s_i, and t_i, as follows:

$$\min_{p,r,s,t} \quad f(x) + \tfrac{1}{2} p^T W p + \nabla f(x)^T p + \mu \sum_{i \in \mathcal{E}} (r_i + s_i) + \mu \sum_{i \in \mathcal{I}} t_i$$

$$\text{subject to} \quad \nabla c_i(x)^T p + c_i(x) = r_i - s_i, \quad i \in \mathcal{E}$$

$$\nabla c_i(x)^T p + c_i(x) \geq -t_i, \quad i \in \mathcal{I} \tag{17.31}$$

$$r, \ s, \ t \geq 0.$$

This subproblem can be solved with a standard quadratic programming solver. Even after addition of a "box-shaped" trust region constraint of the form $\|p\|_\infty \leq \Delta$, it remains a quadratic program. This approach to minimizing ϕ_1 is closely related to sequential quadratic programming (SQP) and will be discussed further in Chapter 18.

The strategy for choosing and updating the penalty parameter μ_k is crucial to the practical success of the iteration. We mentioned that a simple (but not always effective) approach is to choose an initial value and increase it repeatedly until feasibility is attained. In some variants of the approach, the penalty parameter is chosen at every iteration so that $\mu_k > \|\lambda_k\|_\infty$, where λ_k is an estimate of the Lagrange multipliers computed at x_k. We base this strategy on Theorem 17.2, which suggests that in a neighborhood of a solution x^*, a good choice would be to set μ_k modestly larger than $\|\lambda^*\|_\infty$. This strategy is not always successful, as the multiplier estimates may be inaccurate and may in any case not provide a good appropriate value of μ_k far from the solution.

The difficulties of choosing appropriate values of μ_k caused nonsmooth penalty methods to fall out of favor during the 1990s and stimulated the development of filter methods, which do not require the choice of a penalty parameter (see Section 15.4). In recent years, however, there has been a resurgence of interest in penalty methods, in part because of their ability to handle degenerate problems. New approaches for updating the penalty parameter appear to have largely overcome the difficulties associated with choosing μ_k, at least for some particular implementations (see Algorithm 18.5).

Careful consideration should also be given to the choice of starting point x_{k+1}^s for the minimization of $\phi_1(x; \mu_{k+1})$. If the penalty parameter μ_k for the present cycle is appropriate, in the sense that the algorithm made progress toward feasibility, then we can set x_{k+1}^s to be the minimizer x_k of $\phi_1(x; \mu_k)$ obtained on this cycle. Otherwise, we may want to restore the initial point from an earlier cycle.

A GENERAL CLASS OF NONSMOOTH PENALTY METHODS

Exact nonsmooth penalty functions can be defined in terms of norms other than the ℓ_1 norm. We can write

$$\phi(x; \mu) = f(x) + \mu \|c_{\mathcal{E}}(x)\| + \mu \|[c_{\mathcal{I}}(x)]^-\|, \qquad (17.32)$$

where $\| \cdot \|$ is any vector norm, and all the equality and inequality constraints have been grouped in the vector functions $c_{\mathcal{E}}$ and $c_{\mathcal{I}}$, respectively. Framework 17.2 applies to any of these penalty functions; we simply redefine the measure of infeasibility as $h(x) = \|c_{\mathcal{E}}(x)\| + \|[c_{\mathcal{I}}(x)]^-\|$. The most common norms used in practice are the ℓ_1, ℓ_∞ and ℓ_2 (not squared). It is easy to find a reformulation similar to (17.31) for the ℓ_∞ norm.

The theoretical properties described for the ℓ_1 function extend to the general class (17.32). In Theorem 17.3, we replace the inequality (17.23) by

$$\mu^* = \|\lambda^*\|_D, \qquad (17.33)$$

where $\| \cdot \|_D$ is the dual norm of $\| \cdot \|$, defined in (A.6). Theorem 17.4 applies without modification.

We show now that penalty functions of the type considered so far in this chapter *must* be nonsmooth to be exact. For simplicity, we restrict our attention to the case when there is a single equality constraint $c_1(x) = 0$, and consider a penalty function of the form

$$\phi(x; \mu) = f(x) + \mu h(c_1(x)), \qquad (17.34)$$

where $h : \mathbb{R} \to \mathbb{R}$ is a function satisfying the properties $h(y) \geq 0$ for all $y \in \mathbb{R}$ and $h(0) = 0$. Suppose for contradiction that h is continuously differentiable. Since h has a minimizer at zero, we have from Theorem 2.2 that $\nabla h(0) = 0$. If x^* is a local solution of the problem (17.6), we have $c_1(x^*) = 0$ and therefore $\nabla h(c_1(x^*)) = 0$. If x^* is a local minimizer of $\phi(x; \mu)$, we therefore have

$$0 = \nabla \phi(x^*; \mu) = \nabla f(x^*) + \mu \nabla c_1(x^*) \nabla h(c_1(x^*)) = \nabla f(x^*).$$

However, it is not generally true that the gradient of f vanishes at the solution of a *constrained* optimization problem, so our original assumption that h is continuously differentiable must be incorrect, and $\phi(\cdot; \mu)$ cannot be smooth.

Nonsmooth penalty functions are also used as *merit functions* in methods that compute steps by some other mechanism. For further details see the general discussion of Section 15.4 and the concrete implementations given in Chapters 18 and 19.

17.3 AUGMENTED LAGRANGIAN METHOD: EQUALITY CONSTRAINTS

We now discuss an approach known as the *method of multipliers* or the *augmented Lagrangian method*. This algorithm is related to the quadratic penalty algorithm of Section 17.1, but it reduces the possibility of ill conditioning by introducing explicit Lagrange multiplier estimates into the function to be minimized, which is known as the augmented Lagrangian function. In contrast to the penalty functions discussed in Section 17.2, the augmented Lagrangian function largely preserves smoothness, and implementations can be constructed from standard software for unconstrained or bound-constrained optimization.

In this section we use superscripts (usually k and $k + 1$) on the Lagrange multiplier estimates to denote iteration index, and subscripts (usually i) to denote the component indices of the vector λ. For all other variables we use subscripts for the iteration index, as usual.

MOTIVATION AND ALGORITHMIC FRAMEWORK

We consider first the equality-constrained problem (17.1). The quadratic penalty function $Q(x; \mu)$ defined by (17.2) penalizes constraint violations by squaring the infeasibilities and scaling them by $\mu/2$. As we see from Theorem 17.2, however, the approximate minimizers x_k of $Q(x; \mu_k)$ do not quite satisfy the feasibility conditions $c_i(x) = 0, i \in \mathcal{E}$. Instead, they are perturbed (see (17.10)) so that

$$c_i(x_k) \approx -\lambda_i^*/\mu_k, \quad \text{for all } i \in \mathcal{E}. \tag{17.35}$$

To be sure, we have $c_i(x_k) \to 0$ as $\mu_k \uparrow \infty$, but one may ask whether we can alter the function $Q(x; \mu_k)$ to avoid this systematic perturbation—that is, to make the approximate minimizers more nearly satisfy the equality constraints $c_i(x) = 0$, even for moderate values of μ_k.

The augmented Lagrangian function $\mathcal{L}_A(x, \lambda; \mu)$ achieves this goal by including an explicit estimate of the Lagrange multipliers λ, based on the estimate (17.35), in the objective. From the definition

$$\mathcal{L}_A(x, \lambda; \mu) \stackrel{\text{def}}{=} f(x) - \sum_{i \in \mathcal{E}} \lambda_i c_i(x) + \frac{\mu}{2} \sum_{i \in \mathcal{E}} c_i^2(x), \tag{17.36}$$

we see that the augmented Lagrangian differs from the (standard) Lagrangian (12.33) for (17.1) by the presence of the squared terms, while it differs from the quadratic penalty function (17.2) in the presence of the summation term involving λ. In this sense, it is a combination of the Lagrangian function and the quadratic penalty function.

We now design an algorithm that fixes the penalty parameter μ to some value $\mu_k > 0$ at its kth iteration (as in Frameworks 17.1 and 17.2), fixes λ at the current estimate λ^k, and performs minimization with respect to x. Using x_k to denote the approximate minimizer of $\mathcal{L}_A(x, \lambda^k; \mu_k)$, we have by the optimality conditions for unconstrained minimization (Theorem 2.2) that

$$0 \approx \nabla_x \mathcal{L}_A(x_k, \lambda^k; \mu_k) = \nabla f(x_k) - \sum_{i \in \mathcal{E}} [\lambda_i^k - \mu_k c_i(x_k)] \nabla c_i(x_k). \qquad (17.37)$$

By comparing with the optimality condition (17.17) for (17.1), we can deduce that

$$\lambda_i^* \approx \lambda_i^k - \mu_k c_i(x_k), \qquad \text{for all } i \in \mathcal{E}. \qquad (17.38)$$

By rearranging this expression, we have that

$$c_i(x_k) \approx -\frac{1}{\mu_k}(\lambda_i^* - \lambda_i^k), \qquad \text{for all } i \in \mathcal{E},$$

so we conclude that if λ^k is close to the optimal multiplier vector λ^*, the infeasibility in x_k will be much smaller than $(1/\mu_k)$, rather than being proportional to $(1/\mu_k)$ as in (17.35). The relation (17.38) immediately suggests a formula for improving our current estimate λ^k of the Lagrange multiplier vector, using the approximate minimizer x_k just calculated: We can set

$$\lambda_i^{k+1} = \lambda_i^k - \mu_k c_i(x_k), \qquad \text{for all } i \in \mathcal{E}. \qquad (17.39)$$

This discussion motivates the following algorithmic framework.

Framework 17.3 (Augmented Lagrangian Method-Equality Constraints).
  Given $\mu_0 > 0$, tolerance $\tau_0 > 0$, starting points x_0^s and λ^0;
  for $k = 0, 1, 2, \ldots$
      Find an approximate minimizer x_k of $\mathcal{L}_A(\cdot, \lambda^k; \mu_k)$, starting at x_k^s,
          and terminating when $\|\nabla_x \mathcal{L}_A(x_k, \lambda^k; \mu_k)\| \leq \tau_k$;
      if a convergence test for (17.1) is satisfied
          stop with approximate solution x_k;
      end (if)
      Update Lagrange multipliers using (17.39) to obtain λ^{k+1};
      Choose new penalty parameter $\mu_{k+1} \geq \mu_k$;
      Set starting point for the next iteration to $x_{k+1}^s = x_k$;
      Select tolerance τ_{k+1};
  end (for)

We show below that convergence of this method can be assured without increasing μ indefinitely. Ill conditioning is therefore less of a problem than in Framework 17.1, so the choice of starting point x_{k+1}^s in Framework 17.3 is less critical. (In Framework 17.3 we

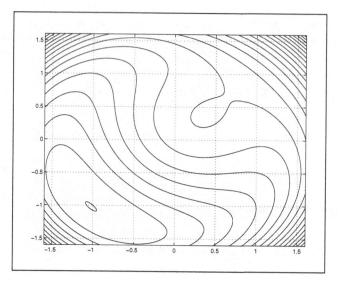

Figure 17.5 Contours of $\mathcal{L}_A(x, \lambda; \mu)$ from (17.40) for $\lambda = -0.4$ and $\mu = 1$, contour spacing 0.5.

simply start the search at iteration $k + 1$ from the previous approximate minimizer x_k.) The tolerance τ_k could be chosen to depend on the infeasibility $\sum_{i \in \mathcal{E}} |c(x_k)|$, and the penalty parameter μ may be increased if the reduction in this infeasibility measure is insufficient at the present iteration.

❏ EXAMPLE 17.4

Consider again problem (17.3), for which the augmented Lagrangian is

$$\mathcal{L}_A(x, \lambda; \mu) = x_1 + x_2 - \lambda(x_1^2 + x_2^2 - 2) + \frac{\mu}{2}(x_1^2 + x_2^2 - 2)^2. \qquad (17.40)$$

The solution of (17.3) is $x^* = (-1, -1)^T$ and the optimal Lagrange multiplier is $\lambda^* = -0.5$.

Suppose that at iterate k we have $\mu_k = 1$ (as in Figure 17.1), while the current multiplier estimate is $\lambda^k = -0.4$. Figure 17.5 plots the function $\mathcal{L}_A(x, -0.4; 1)$. Note that the spacing of the contours indicates that the conditioning of this problem is similar to that of the quadratic penalty function $Q(x; 1)$ illustrated in Figure 17.1. However, the minimizing value of $x_k \approx (-1.02, -1.02)^T$ is much closer to the solution $x^* = (-1, -1)^T$ than is the minimizing value of $Q(x; 1)$, which is approximately $(-1.1, -1.1)^T$. This example shows that the inclusion of the Lagrange multiplier term in the function $\mathcal{L}_A(x, \lambda; \mu)$ can result in a significant improvement over the quadratic penalty method, as a way to reformulate the constrained optimization problem (17.1).

❑

PROPERTIES OF THE AUGMENTED LAGRANGIAN

We now prove two results that justify the use of the augmented Lagrangian function and the method of multipliers for equality-constrained problems.

The first result validates the approach of Framework 17.3 by showing that when we have knowledge of the exact Lagrange multiplier vector λ^*, the solution x^* of (17.1) is a strict minimizer of $\mathcal{L}_A(x, \lambda^*; \mu)$ for all μ sufficiently large. Although we do not know λ^* exactly in practice, the result and its proof suggest that we can obtain a good estimate of x^* by minimizing $\mathcal{L}_A(x, \lambda; \mu)$ even when μ is not particularly large, provided that λ is a reasonably good estimate of λ^*.

Theorem 17.5.

Let x^* be a local solution of (17.1) at which the LICQ is satisfied (that is, the gradients $\nabla c_i(x^*)$, $i \in \mathcal{E}$, are linearly independent vectors), and the second-order sufficient conditions specified in Theorem 12.6 are satisfied for $\lambda = \lambda^*$. Then there is a threshold value $\bar{\mu}$ such that for all $\mu \geq \bar{\mu}$, x^* is a strict local minimizer of $\mathcal{L}_A(x, \lambda^*; \mu)$.

PROOF.   We prove the result by showing that x^* satisfies the second-order sufficient conditions to be a strict local minimizer of $\mathcal{L}_A(x, \lambda^*; \mu)$ (see Theorem 2.4) for all μ sufficiently large; that is,

$$\nabla_x \mathcal{L}_A(x^*, \lambda^*; \mu) = 0, \quad \nabla^2_{xx} \mathcal{L}_A(x^*, \lambda^*; \mu) \text{ positive definite.} \tag{17.41}$$

Because x^* is a local solution for (17.1) at which LICQ is satisfied, we can apply Theorem 12.1 to deduce that $\nabla_x \mathcal{L}(x^*, \lambda^*) = 0$ and $c_i(x^*) = 0$ for all $i \in \mathcal{E}$, so that

$$\nabla_x \mathcal{L}_A(x^*, \lambda^*; \mu) = \nabla f(x^*) - \sum_{i \in \mathcal{E}} [\lambda_i^* - \mu c_i(x^*)] \nabla c_i(x^*)$$

$$= \nabla f(x^*) - \sum_{i \in \mathcal{E}} \lambda_i^* \nabla c_i(x^*) = \nabla_x \mathcal{L}(x^*, \lambda^*) = 0,$$

verifying the first part of (17.41), independently of μ.

For the second part of (17.41), we define A to be the constraint gradient matrix in (17.15) evaluated at x^*, and write

$$\nabla^2_{xx} \mathcal{L}_A(x^*, \lambda^*; \mu) = \nabla^2_{xx} \mathcal{L}(x^*, \lambda^*) + \mu A^T A.$$

If the claim in (17.41) were not true, then for each integer $k \geq 1$, we could choose a vector w_k with $\|w_k\| = 1$ such that

$$0 \geq w_k^T \nabla^2_{xx} \mathcal{L}_A(x^*, \lambda^*; k) w_k = w_k^T \nabla^2_{xx} \mathcal{L}(x^*, \lambda^*) w_k + k \|A w_k\|_2^2, \tag{17.42}$$

and therefore

$$\|Aw_k\|_2^2 \leq -(1/k)w_k^T\nabla_{xx}^2\mathcal{L}(x^*,\lambda^*)w_k \rightarrow 0, \quad \text{as } k \rightarrow \infty. \tag{17.43}$$

Since the vectors $\{w_k\}$ lie in a compact set (the surface of the unit sphere), they have an accumulation point w. The limit (17.43) implies that $Aw = 0$. Moreover, by rearranging (17.42), we have that

$$w_k^T\nabla_{xx}^2\mathcal{L}(x^*,\lambda^*)w_k \leq -k\|Aw_k\|_2^2 \leq 0,$$

so by taking limits we have $w^T\nabla_{xx}^2\mathcal{L}(x^*,\lambda^*)w \leq 0$. However, this inequality contradicts the second-order conditions in Theorem 12.6 which, when applied to (17.1), state that we must have $w^T\nabla_{xx}^2\mathcal{L}(x^*,\lambda^*)w > 0$ for all nonzero vectors w with $Aw = 0$. Hence, the second part of (17.41) holds for all μ sufficiently large. $\qquad\square$

The second result, given by Bertsekas [19, Proposition 4.2.3], describes the more realistic situation of $\lambda \neq \lambda^*$. It gives conditions under which there is a minimizer of $\mathcal{L}_A(x,\lambda;\mu)$ that lies close to x^* and gives error bounds on both x_k and the updated multiplier estimate λ^{k+1} obtained from solving the subproblem at iteration k.

Theorem 17.6.

Suppose that the assumptions of Theorem 17.5 are satisfied at x^* and λ^* and let $\bar{\mu}$ be chosen as in that theorem. Then there exist positive scalars δ, ϵ, and M such that the following claims hold:

(a) *For all λ^k and μ_k satisfying*

$$\|\lambda^k - \lambda^*\| \leq \mu_k\delta, \qquad \mu_k \geq \bar{\mu}, \tag{17.44}$$

the problem

$$\min_x \mathcal{L}_A(x,\lambda^k;\mu_k) \quad \text{subject to } \|x - x^*\| \leq \epsilon$$

has a unique solution x_k. Moreover, we have

$$\|x_k - x^*\| \leq M\|\lambda^k - \lambda^*\|/\mu_k. \tag{17.45}$$

(b) *For all λ^k and μ_k that satisfy (17.44), we have*

$$\|\lambda^{k+1} - \lambda^*\| \leq M\|\lambda^k - \lambda^*\|/\mu_k, \tag{17.46}$$

where λ^{k+1} is given by the formula (17.39).

(c) For all λ^k and μ_k that satisfy (17.44), the matrix $\nabla^2_{xx} \mathcal{L}_A(x_k, \lambda^k; \mu_k)$ is positive definite and the constraint gradients $\nabla c_i(x_k)$, $i \in \mathcal{E}$, are linearly independent.

This theorem illustrates some salient properties of the augmented Lagrangian approach. The bound (17.45) shows that x_k will be close to x^* if λ_k is accurate *or* if the penalty parameter μ_k is large. Hence, this approach gives us two ways of improving the accuracy of x_k, whereas the quadratic penalty approach gives us only one option: increasing μ_k. The bound (17.46) states that, locally, we can ensure an improvement in the accuracy of the multipliers by choosing a sufficiently large value of μ_k. The final observation of the theorem shows that second-order sufficient conditions for unconstrained minimization (see Theorem 2.4) are also satisfied for the kth subproblem under the given conditions, so one can expect good performance by applying standard unconstrained minimization techniques.

17.4 PRACTICAL AUGMENTED LAGRANGIAN METHODS

In this section we discuss practical augmented Lagrangian procedures, in particular, procedures for handling inequality constraints. We discuss three approaches based, respectively, on bound-constrained, linearly constrained, and unconstrained formulations. The first two are the basis of the successful nonlinear programming codes LANCELOT [72] and MINOS [218].

BOUND-CONSTRAINED FORMULATION

Given the general nonlinear program (17.6), we can convert it to a problem with equality constraints and bound constraints by introducing slack variables s_i and replacing the general inequalities $c_i(x) \geq 0$, $i \in \mathcal{I}$, by

$$c_i(x) - s_i = 0, \quad s_i \geq 0, \quad \text{for all } i \in \mathcal{I}. \tag{17.47}$$

Bound constraints, $l \leq x \leq u$, need not be transformed. By reformulating in this way, we can write the nonlinear program as follows:

$$\min_{x \in \mathbb{R}^n} f(x) \quad \text{subject to} \quad c_i(x) = 0, \ i = 1, 2, \ldots, m, \ l \leq x \leq u. \tag{17.48}$$

(The slacks s_i have been incorporated into the vector x and the constraint functions c_i have been redefined accordingly. We have numbered the constraints consecutively with $i = 1, 2, \ldots, m$ and in the discussion below we gather them into the vector function $c : \mathbb{R}^n \to \mathbb{R}^m$.) Some of the components of the lower bound vector l may be set to $-\infty$, signifying that there is no lower bound on the components of x in question; similarly for u.

The bound-constrained Lagrangian (BCL) approach incorporates only the equality constraints from (17.48) into the augmented Lagrangian, that is,

$$\mathcal{L}_A(x, \lambda; \mu) = f(x) - \sum_{i=1}^{m} \lambda_i c_i(x) + \frac{\mu}{2} \sum_{i=1}^{m} c_i^2(x). \tag{17.49}$$

The bound constraints are enforced explicitly in the subproblem, which has the form

$$\min_{x} \mathcal{L}_A(x, \lambda; \mu) \quad \text{subject to } l \le x \le u. \tag{17.50}$$

After this problem has been solved approximately, the multipliers λ and the penalty parameter μ are updated and the process is repeated.

An efficient technique for solving the nonlinear program with bound constraints (17.50) (for fixed μ and λ) is the (nonlinear) gradient projection method discussed in Section 18.6. By specializing the KKT conditions (12.34) to the problem (17.50), we find that the first-order necessary condition for x to be a solution of (17.50) is that

$$x - P(x - \nabla_x \mathcal{L}_A(x, \lambda; \mu), l, u) = 0, \tag{17.51}$$

where $P(g, l, u)$ is the projection of the vector $g \in \mathbb{R}^n$ onto the rectangular box $[l, u]$ defined as follows

$$P(g, l, u)_i = \begin{cases} l_i & \text{if } g_i \le l_i, \\ g_i & \text{if } g_i \in (l_i, u_i), \\ u_i & \text{if } g_i \ge u_i, \end{cases} \quad \text{for all } i = 1, 2, \ldots, n. \tag{17.52}$$

We are now ready to describe the algorithm implemented in the LANCELOT software package.

Algorithm 17.4 (Bound-Constrained Lagrangian Method).
   Choose an initial point x_0 and initial multipliers λ^0;
   Choose convergence tolerances η_* and ω_*;
   Set $\mu_0 = 10$, $\omega_0 = 1/\mu_0$, and $\eta_0 = 1/\mu_0^{0.1}$;
   for $k = 0, 1, 2, \ldots$
      Find an approximate solution x_k of the subproblem (17.50) such that

$$\left\| x_k - P\left(x_k - \nabla_x \mathcal{L}_A(x_k, \lambda^k; \mu_k), l, u\right) \right\| \le \omega_k;$$

      if $\|c(x_k)\| \le \eta_k$
         (* test for convergence *)
         if $\|c(x_k)\| \le \eta_*$ and $\left\| x_k - P\left(x_k - \nabla_x \mathcal{L}_A(x_k, \lambda^k; \mu_k), l, u\right) \right\| \le \omega_*$
            stop with approximate solution x_k;

   end (if)
   (* update multipliers, tighten tolerances *)
    $\lambda^{k+1} = \lambda^k - \mu_k c(x_k);$
    $\mu_{k+1} = \mu_k;$
    $\eta_{k+1} = \eta_k / \mu_{k+1}^{0.9};$
    $\omega_{k+1} = \omega_k / \mu_{k+1};$
  else
   (* increase penalty parameter, tighten tolerances *)
    $\lambda^{k+1} = \lambda^k;$
    $\mu_{k+1} = 100\mu_k;$
    $\eta_{k+1} = 1/\mu_{k+1}^{0.1};$
    $\omega_{k+1} = 1/\mu_{k+1};$
  end (if)
 end (for)

The main branch in the algorithm occurs after problem (17.50) has been solved approximately, when the algorithm tests to see if the constraints have decreased sufficiently, as measured by the condition

$$\|c(x_k)\| \le \eta_k. \tag{17.53}$$

If this condition holds, the penalty parameter is not changed for the next iteration because the current value of μ_k is producing an acceptable level of constraint violation. The Lagrange multiplier estimates are updated according to the formula (17.39) and the tolerances ω_k and η_k are tightened in advance of the next iteration. If, on the other hand, (17.53) does not hold, then we increase the penalty parameter to ensure that the next subproblem will place more emphasis on decreasing the constraint violations. The Lagrange multiplier estimates are not updated in this case; the focus is on improving feasibility.

  The constants 0.1, 0.9, and 100 appearing in Algorithm 17.4 are to some extent arbitrary; other values can be used without compromising theoretical convergence properties. LANCELOT uses the gradient projection method with trust regions (see (18.61)) to solve the bound-constrained nonlinear subproblem (17.50). In this context, the gradient projection method constructs a quadratic model of the augmented Lagrangian $\mathcal{L}_A$ and computes a step d by approximately solving the trust region problem

$$\min_d \tfrac{1}{2} d^T \left[\nabla_{xx}^2 \mathcal{L}(x_k, \lambda^k) + \mu_k A_k^T A_k \right] d + \nabla_x \mathcal{L}_A(x_k, \lambda^k; \mu_k)^T d \tag{17.54}$$

$$\text{subject to } l \le x_k + d \le u, \qquad \|d\|_\infty \le \Delta,$$

where $A_k = A(x_k)$ and Δ is a trust region radius. (We can formulate the trust-region constraint by means of the bounds $-\Delta e \le d \le \Delta e$, where $e = (1, 1, \ldots, 1)^T$.) Each iteration of the algorithm for solving this subproblem proceeds in two stages. First, a projected gradient line search is performed to determine which components of d should be set at one of their bounds. Second, a conjugate gradient iteration minimizes (17.54) with respect to the free components of d—those not at one of their bounds. Importantly, this algorithm does not require the factorizations of a KKT matrix or of the constraint Jacobian A_k. The conjugate gradient iteration only requires matrix-vector products, a feature that makes LANCELOT suitable for large problems.

The Hessian of the Lagrangian $\nabla^2_{xx} \mathcal{L}(x_k, \lambda^k)$ in (17.54) can be replaced by a quasi-Newton approximation based on the BFGS or SR1 updating formulas. LANCELOT is designed to take advantage of partially separable structure in the objective function and constraints, either in the evaluation of the Hessian of the Lagrangian or in the quasi-Newton updates (see Section 7.4).

LINEARLY CONSTRAINED FORMULATION

The principal idea behind *linearly constrained Lagrangian* (LCL) methods is to generate a step by minimizing the Lagrangian (or augmented Lagrangian) subject to linearizations of the constraints. If we use the formulation (17.48) of the nonlinear programming problem, the subproblem used in the LCL approach takes the form

$$\min_x \quad F_k(x) \tag{17.55a}$$

$$\text{subject to} \quad c(x_k) + A_k(x - x_k) = 0, \quad l \le x \le u. \tag{17.55b}$$

There are several possible choices for $F_k(x)$. Early LCL methods defined

$$F_k(x) = f(x) - \sum_{i=1}^{m} \lambda_i^k \bar{c}_i^k(x), \tag{17.56}$$

where λ^k is the current Lagrange multiplier estimate and $\bar{c}_i^k(x)$ is the difference between $c_i(x)$ and its linearization at x_k, that is,

$$\bar{c}_i^k(x) = c_i(x) - c_i(x_k) - \nabla c_i(x_k)^T(x - x_k). \tag{17.57}$$

One can show that as x_k converges to a solution x^*, the Lagrange multiplier associated with the equality constraint in (17.55b) converges to the optimal multiplier. Therefore, one can set λ^k in (17.56) to be the Lagrange multiplier for the equality constraint in (17.55b) from the previous iteration.

Current LCL methods define F_k to be the augmented Lagrangian function

$$F_k(x) = f(x) - \sum_{i=1}^{m} \lambda_i^k \bar{c}_i^k(x) + \frac{\mu}{2} \sum_{i=1}^{m} [\bar{c}_i^k(x)]^2. \tag{17.58}$$

This definition of F_k appears to yield more reliable convergence from remote starting points than does (17.56), in practice.

There is a notable similarity between (17.58) and the augmented Lagrangian (17.36), the difference being that the original constraints $c_i(x)$ have been replaced by the functions $\bar{c}_i^k(x)$, which capture only the "second-order and above" terms of c_i. The subproblem (17.55) differs from the augmented Lagrangian subproblem in that the new x is required to satisfy exactly a linearization of the equality constraints, while the linear part of each constraint is factored out of the objective via the use of $\bar{c}_i^k$ in place of c_i. A procedure similar to the one in Algorithm 17.4 can be used for updating the penalty parameter μ and for adjusting the tolerances that govern the accuracy of the solution of the subproblem.

Since $\bar{c}_i^k(x)$ has zero gradient at $x = x_k$, we have that $\nabla F_k(x_k) = \nabla f(x_k)$, where F_k is defined by either (17.56) or (17.58). We can also show that the Hessian of F_k is closely related to the Hessians of the Lagrangian or augmented Lagrangian functions for (17.1). Because of these properties, the subproblem (17.55) is similar to the SQP subproblems described in Chapter 18, with the quadratic objective in SQP being replaced by a nonlinear objective in LCL.

The well known code MINOS [218] uses the nonlinear model function (17.58) and solves the subproblem via a reduced gradient method that employs quasi-Newton approximations to the reduced Hessian of F_k. A fairly accurate solution of the subproblem is computed in MINOS to try to ensure that the Lagrange multiplier estimates for the equality constraint in (17.55b) (subsequently used in (17.58)) are of good quality. As a result, MINOS typically requires more evaluations of the objective f and constraint functions c_i (and their gradients) in total than SQP methods or interior-point methods. The total number of subproblems (17.55) that are solved in the course of the algorithm is, however, sometimes smaller than in other approaches.

UNCONSTRAINED FORMULATION

We can obtain an unconstrained form of the augmented Lagrangian subproblem for inequality-constrained problems by using a derivation based on the proximal point approach. Supposing for simplicity that the problem has no equality constraints ($\mathcal{E} = \emptyset$), we can write the problem (17.6) equivalently as an unconstrained optimization problem:

$$\min_{x \in \mathbb{R}^n} F(x), \tag{17.59}$$

where

$$F(x) = \max_{\lambda \geq 0} \left\{ f(x) - \sum_{i \in \mathcal{I}} \lambda_i c_i(x) \right\} = \begin{cases} f(x) & \text{if } x \text{ is feasible,} \\ \infty & \text{otherwise.} \end{cases} \tag{17.60}$$

To verify these expressions for F, consider first the case of x infeasible, that is, $c_i(x) < 0$ for some i. We can then choose λ_i arbitrarily large and positive while setting $\lambda_j = 0$ for all

$j \neq i$, to verify that $F(x)$ is infinite in this case. If x is feasible, we have $c_i(x) \geq 0$ for all $i \in \mathcal{I}$, so the maximum is attained at $\lambda = 0$, and $F(x) = f(x)$ in this case. By combining (17.59) with (17.60), we have

$$\min_{x \in \mathbb{R}^n} F(x) = \min_{x \text{ feasible}} f(x), \tag{17.61}$$

which is simply the original inequality-constrained problem. It is not practical to minimize F directly, however, since this function is not smooth—it jumps from a finite value to an infinite value as x crosses the boundary of the feasible set.

We can make this approach more practical by replacing F by a smooth approximation $\hat{F}(x; \lambda^k, \mu_k)$ which depends on the penalty parameter μ_k and Lagrange multiplier estimate λ^k. This approximation is defined as follows:

$$\hat{F}(x; \lambda^k, \mu_k) = \max_{\lambda \geq 0} \left\{ f(x) - \sum_{i \in \mathcal{I}} \lambda_i c_i(x) - \frac{1}{2\mu_k} \sum_{i \in \mathcal{I}} (\lambda_i - \lambda_i^k)^2 \right\}. \tag{17.62}$$

The final term in this expression applies a penalty for any move of λ away from the previous estimate λ^k; it encourages the new maximizer λ to stay *proximal* to the previous estimate λ^k. Since (17.62) represents a bound-constrained quadratic problem in λ, separable in the individual components λ_i, we can perform the maximization explicitly, to obtain

$$\lambda_i = \begin{cases} 0 & \text{if } -c_i(x) + \lambda_i^k/\mu_k \leq 0; \\ \lambda_i^k - \mu_k c_i(x) & \text{otherwise.} \end{cases} \tag{17.63}$$

By substituting these values in (17.62), we find that

$$\hat{F}(x; \lambda^k, \mu_k) = f(x) + \sum_{i \in \mathcal{I}} \psi(c_i(x), \lambda_i^k; \mu_k), \tag{17.64}$$

where the function ψ of three scalar arguments is defined as follows:

$$\psi(t, \sigma; \mu) \stackrel{\text{def}}{=} \begin{cases} -\sigma t + \dfrac{\mu}{2} t^2 & \text{if } t - \sigma/\mu \leq 0, \\ -\dfrac{1}{2\mu} \sigma^2 & \text{otherwise,} \end{cases} \tag{17.65}$$

Hence, we can obtain the new iterate x_k by minimizing $\hat{F}(x; \lambda^k, \mu_k)$ with respect to x, and use the formula (17.63) to obtain the updated Lagrange multiplier estimates λ^{k+1}. By comparing with Framework 17.3, we see that F plays the role of $\mathcal{L}_A$ and that the scheme just described extends the augmented Lagrangian methods for equality constraints neatly to the inequality-constrained case. Unlike the bound-constrained and linearly constrained formulations, however, this unconstrained formulation is not the basis of any widely used software packages, so its practical properties have not been tested.

17.5   PERSPECTIVES AND SOFTWARE

The quadratic penalty approach is often used by practitioners when the number of constraints is small. In fact, minimization of $Q(x; \mu)$ is sometimes performed for just one large value of μ. Unless μ is chosen wisely (with the benefit of experience with the underlying application), the resulting solution may not be very accurate. Since the main software packages for constrained optimization do not implement a quadratic penalty approach, little attention has been paid to techniques for updating the penalty parameter, adjusting the tolerances τ_k, and choosing the starting points x_k^s for each iteration. (See Gould [141] for a discussion of these issues.)

Despite the intuitive appeal and simplicity of the quadratic penalty method of Framework 17.1, the augmented Lagrangian method of Sections 17.3 and 17.4 is generally preferred. The subproblems are in general no more difficult to solve, and the introduction of multiplier estimates reduces the likelihood that large values of μ will be needed to obtain good feasibility and accuracy, thereby avoiding ill conditioning of the subproblem. The quadratic penalty approach remains, however, an important mechanism for regularizing other algorithms such as sequential quadratic programming (SQP) methods, as we mention at the end of Section 17.1.

A general-purpose ℓ_1 penalty method was developed by Fletcher in the 1980's. It is known as the $S\ell_1QP$ method because it has features in common with SQP methods. More recently, an ℓ_1 penalty method that uses linear programming subproblems has been implemented as part of the KNITRO [46] software package. These two methods are discussed in Section 18.5.

The ℓ_1 penalty function has received significant attention in recent years. It has been successfully used to treat difficult problems, such as mathematical programs with complementarity constraints (MPCCs), in which the constraints do not satisfy standard constraint qualifications [274]. By including these problematic constraints as a penalty term, rather than linearizing them exactly, and treating the remaining constraints using other techniques such as SQP or interior-point, it is possible to extend the range of applicability of these other approaches. See [8] for an active-set method and [16, 191] for interior-point methods for MPCCs. The SNOPT software package uses an ℓ_1 penalty approach within an SQP method as a safeguard strategy in case the quadratic model appears to be infeasible or unbounded or to have unbounded multipliers.

Augmented Lagrangian methods have been popular for many years because, in part, of their simplicity. The MINOS and LANCELOT packages rank among the best implementations of augmented Lagrangian methods. Both are suitable for large-scale nonlinear programming problems. At a general level, the linearly constrained Lagrangian (LCL) of MINOS and the bound-constrained Lagrangian (BCL) method of LANCELOT have important features in common. They differ significantly, however, in the formulation of the step-computation subproblems and in the techniques used to solve these subproblems. MINOS follows a reduced-space approach to handle linearized constraints and employs a (dense) quasi-Newton approximation to the Hessian of the Lagrangian. As a result, MINOS is most successful for problems with relatively few degrees of freedom. LANCELOT, on the other hand, is more effective when there are relatively few constraints. As indicated in Section 17.4, LANCELOT does not require a factorization of the constraint Jacobian matrix A, again enhancing its suitability for very large problems, and provides a variety of Hessian approximation options and preconditioners. The PENNON software package [184] is based on an augmented Lagrangian approach and has the advantage of permitting semi-definite matrix constraints.

A weakness of both the bound-constrained and unconstrained Lagrangian methods is that they complicate constraints by squaring them in (17.49); progress in feasibility is only achieved through the minimization of the augmented Lagrangian. In contrast, the LCL formulation (17.55) promotes steady progress toward feasibility by performing a Newton-like step on the constraints. Not surprisingly, numerical experience has shown an advantage of MINOS over LANCELOT for problems with linear constraints.

Smooth exact penalty functions have been constructed from the augmented Lagrangian functions of Section 17.3, but these are considerably more complicated. As an example, we mention the function of Fletcher for equality-constrained problems, defined as follows:

$$\phi_F(x; \mu) = f(x) - \lambda(x)^T c(x) + \frac{\mu}{2} \sum_{i \in \mathcal{E}} c_i(x)^2. \tag{17.66}$$

The Lagrange multiplier estimates $\lambda(x)$ are defined explicitly in terms of x via the least-squares estimate, defined as

$$\lambda(x) = [A(x)A(x)^T]^{-1} A(x) \nabla f(x). \tag{17.67}$$

The function ϕ_F is differentiable and exact, though the threshold value μ^* defining the exactness property is not as easy to specify as for the nonsmooth ℓ_1 penalty function. Drawbacks of the penalty function ϕ_F include the cost of evaluating $\lambda(x)$ via (17.67), the fact that $\lambda(x)$ is not uniquely defined when $A(x)$ does not have full rank, and the observation that estimates of λ may be poor when $A(x)$ is nearly singular.

NOTES AND REFERENCES

The quadratic penalty function was first proposed by Courant [81]. Gould [140] addresses the issue of stable determination of the Newton step for $Q(x; \mu_k)$. His formula (2.2) differs from our formula (17.20) in the right-hand-side, but both systems give rise to the same p component.

The augmented Lagrangian method was proposed by Hestenes [167] and Powell [240]. In the early days it was known as the "method of multipliers." A key reference in this area is Bertsekas [18]. Chapters 1–3 of that book contain a thorough motivation of the method that outlines its connections to other approaches. Other introductory discussions are given by Fletcher [101, Section 12.2], and Polak [236, Section 2.8]. The extension to inequality constraints in the unconstrained formulation was described by Rockafellar [269] and Powell [243].

Linearly constrained Lagrangian methods were proposed by Robinson [266] and Rosen and Kreuser [271]. The MINOS implementation is due to Murtagh and Saunders [218] and the LANCELOT implementation due to Conn, Gould and Toint [72]. We have followed Friedlander and Saunders [114] in our use of the terms "linearly constrained Lagrangian" and "bound-constrained Lagrangian."

✐ EXERCISES

✐   17.1

(a) Write an equality-constrained problem which has a local solution and for which the quadratic penalty function Q is unbounded for *any* value of the penalty parameter.

(b) Write a problem with a single inequality constraint that has the same unboundedness property.

✐   **17.2** Draw the contour lines of the quadratic penalty function Q for problem (17.5) corresponding to $\mu = 1$. Find the stationary points of Q.

✐   **17.3** Minimize the quadratic penalty function for problem (17.3) for $\mu_k = 1, 10, 100, 1000$ using an unconstrained minimization algorithm. Set $\tau_k = 1/\mu_k$ in Framework 17.1, and choose the starting point x_{k+1}^s for each minimization to be the solution for the previous value of the penalty parameter. Report the approximate solution of each penalty function.

✐   **17.4** For $z \in \mathbb{R}$, show that the function $\min(0, z)^2$ has a discontinuous second derivative at $z = 0$. (It follows that quadratic penalty function (17.7) may not have continuous second derivatives even when f and c_i, $i \in \mathcal{E} \cup \mathcal{I}$, in (17.6) are all twice continuously differentiable.)

✐   **17.5** Write a quadratic program similar to (17.31) for the case when the norm in (17.32) is the infinity norm.

✐   **17.6** Suppose that a nonlinear program has a minimizer x^* with Lagrange multiplier vector λ^*. One can show (Fletcher [101, Theorem 14.3.2]) that the function $\phi_1(x; \mu)$ does not have a local minimizer at x^* unless $\mu > \|\lambda^*\|_\infty$. Verify that this observation holds for Example 17.1.

✐   **17.7** Verify (17.28).

✐   **17.8** Prove the second part of Theorem 17.4. That is, if $\hat{x}$ is a stationary point of $\phi_1(x; \mu)$ for all μ sufficiently large, but $\hat{x}$ is infeasible for problem (17.6), then $\hat{x}$ is an infeasible stationary point. (Hint: Use the fact that $D(\phi_1(\hat{x}; \mu); p) = \nabla f(\hat{x})^T p + \mu D(h(\hat{x}); p)$, where h is defined in (17.27).)

✏ **17.9** Verify that the KKT conditions for the bound-constrained problem

$$\min_{x \in R^n} \phi(x) \quad \text{subject to } l \le x \le u$$

are equivalent to the compactly stated condition

$$x - P(x - \nabla\phi(x), l, u) = 0,$$

where the projection operator P onto the rectangular box $[l, u]$ is defined in (17.52).

✏ **17.10** Calculate the gradient and Hessian of the LCL objective functions $F_k(x)$ defined by (17.56) and (17.58). Evaluate these quantities at $x = x_k$.

✏ **17.11** Show that the function $\psi(t, \sigma; \mu)$ defined in (17.65) has a discontinuity in its second derivative with respect to t when $t = \sigma/\mu$. Assuming that $c_i : R^n \to R$ is twice continuously differentiable, write down the second partial derivative matrix of $\psi(c_i(x), \lambda_i; \mu)$ with respect to x for the two cases $c_i(x) < \lambda_i/\mu$ and $c_i(x) \ge a\lambda_i/\mu$.

✏ **17.12** Verify that the multipliers λ_i, $i \in \mathcal{I}$ defined in (17.63) are indeed those that attain the maximum in (17.62), and that the equality (17.64) holds. Hint: Use the fact that KKT conditions for the problem

$$\max \phi(x) \quad \text{subject to } x \ge 0$$

indicate that at a stationary point, we either have $x_i = 0$ and $[\nabla\phi(x)]_i \le 0$, or $x_i > 0$ and $[\nabla\phi(x)]_i = 0$.

CHAPTER 18

Sequential Quadratic Programming

One of the most effective methods for nonlinearly constrained optimization generates steps by solving quadratic subproblems. This sequential quadratic programming (SQP) approach can be used both in line search and trust-region frameworks, and is appropriate for small or large problems. Unlike linearly constrained Lagrangian methods (Chapter 17), which are effective when most of the constraints are linear, SQP methods show their strength when solving problems with significant nonlinearities in the constraints.

All the methods considered in this chapter are active-set methods; a more descriptive title for this chapter would perhaps be "Active-Set Methods for Nonlinear Programming."

In Chapter 14 we study interior-point methods for nonlinear programming, a competing approach for handling inequality-constrained problems.

There are two types of active-set SQP methods. In the IQP approach, a general inequality-constrained quadratic program is solved at each iteration, with the twin goals of computing a step and generating an estimate of the optimal active set. EQP methods decouple these computations. They first compute an estimate of the optimal active set, then solve an equality-constrained quadratic program to find the step. In this chapter we study both IQP and EQP methods.

Our development of SQP methods proceeds in two stages. First, we consider local methods that motivate the SQP approach and allow us to introduce the step computation techniques in a simple setting. Second, we consider practical line search and trust-region methods that achieve convergence from remote starting points. Throughout the chapter we give consideration to the algorithmic demands of solving large problems.

18.1 LOCAL SQP METHOD

We begin by considering the equality-constrained problem

$$\min f(x) \tag{18.1a}$$

$$\text{subject to } c(x) = 0, \tag{18.1b}$$

where $f : \mathbb{R}^n \to \mathbb{R}$ and $c : \mathbb{R}^n \to \mathbb{R}^m$ are smooth functions. The idea behind the SQP approach is to model (18.1) at the current iterate x_k by a quadratic programming subproblem, then use the minimizer of this subproblem to define a new iterate x_{k+1}. The challenge is to design the quadratic subproblem so that it yields a good step for the nonlinear optimization problem. Perhaps the simplest derivation of SQP methods, which we present now, views them as an application of Newton's method to the KKT optimality conditions for (18.1).

From (12.33), we know that the Lagrangian function for this problem is $\mathcal{L}(x, \lambda) = f(x) - \lambda^T c(x)$. We use $A(x)$ to denote the Jacobian matrix of the constraints, that is,

$$A(x)^T = [\nabla c_1(x), \nabla c_2(x), \dots, \nabla c_m(x)], \tag{18.2}$$

where $c_i(x)$ is the ith component of the vector $c(x)$. The first-order (KKT) conditions (12.34) of the equality-constrained problem (18.1) can be written as a system of $n + m$ equations in the $n + m$ unknowns x and λ:

$$F(x, \lambda) = \begin{bmatrix} \nabla f(x) - A(x)^T \lambda \\ c(x) \end{bmatrix} = 0. \tag{18.3}$$

Any solution (x^*, λ^*) of the equality-constrained problem (18.1) for which $A(x^*)$ has full rank satisfies (18.3). One approach that suggests itself is to solve the nonlinear equations (18.3) by using Newton's method, as described in Chapter 11.

The Jacobian of (18.3) with respect to x and λ is given by

$$F'(x, \lambda) = \begin{bmatrix} \nabla_{xx}^2 \mathcal{L}(x, \lambda) & -A(x)^T \\ A(x) & 0 \end{bmatrix}. \tag{18.4}$$

The Newton step from the iterate (x_k, λ_k) is thus given by

$$\begin{bmatrix} x_{k+1} \\ \lambda_{k+1} \end{bmatrix} = \begin{bmatrix} x_k \\ \lambda_k \end{bmatrix} + \begin{bmatrix} p_k \\ p_\lambda \end{bmatrix}, \tag{18.5}$$

where p_k and p_λ solve the Newton–KKT system

$$\begin{bmatrix} \nabla_{xx}^2 \mathcal{L}_k & -A_k^T \\ A_k & 0 \end{bmatrix} \begin{bmatrix} p_k \\ p_\lambda \end{bmatrix} = \begin{bmatrix} -\nabla f_k + A_k^T \lambda_k \\ -c_k \end{bmatrix}. \tag{18.6}$$

This Newton iteration is well defined when the KKT matrix in (18.6) is nonsingular. We saw in Chapter 16 that this matrix is nonsingular if the following assumption holds at $(x, \lambda) = (x_k, \lambda_k)$.

Assumptions 18.1.

(a) *The constraint Jacobian $A(x)$ has full row rank;*

(b) *The matrix $\nabla_{xx}^2 \mathcal{L}(x, \lambda)$ is positive definite on the tangent space of the constraints, that is, $d^T \nabla_{xx}^2 \mathcal{L}(x, \lambda)d > 0$ for all $d \neq 0$ such that $A(x)d = 0$.*

The first assumption is the linear independence constraint qualification discussed in Chapter 12 (see Definition 12.4), which we assume throughout this chapter. The second condition holds whenever (x, λ) is close to the optimum (x^*, λ^*) and the second-order sufficient condition is satisfied at the solution (see Theorem 12.6). The Newton iteration (18.5), (18.6) can be shown to be quadratically convergent under these assumptions (see Theorem 18.4) and constitutes an excellent algorithm for solving equality-constrained problems, provided that the starting point is close enough to x^*.

SQP FRAMEWORK

There is an alternative way to view the iteration (18.5), (18.6). Suppose that at the iterate (x_k, λ_k) we model problem (18.1) using the quadratic program

$$\min_p \quad f_k + \nabla f_k^T p + \tfrac{1}{2} p^T \nabla_{xx}^2 \mathcal{L}_k p \tag{18.7a}$$

$$\text{subject to} \quad A_k p + c_k = 0. \tag{18.7b}$$

If Assumptions 18.1 hold, this problem has a unique solution (p_k, l_k) that satisfies

$$\nabla^2_{xx} \mathcal{L}_k p_k + \nabla f_k - A_k^T l_k = 0, \tag{18.8a}$$

$$A_k p_k + c_k = 0. \tag{18.8b}$$

The vectors p_k and l_k can be identified with the solution of the Newton equations (18.6). If we subtract $A_k^T \lambda_k$ from both sides of the first equation in (18.6), we obtain

$$\begin{bmatrix} \nabla^2_{xx} \mathcal{L}_k & -A_k^T \\ A_k & 0 \end{bmatrix} \begin{bmatrix} p_k \\ \lambda_{k+1} \end{bmatrix} = \begin{bmatrix} -\nabla f_k \\ -c_k \end{bmatrix}. \tag{18.9}$$

Hence, by nonsingularity of the coefficient matrix, we have that $\lambda_{k+1} = l_k$ and that p_k solves (18.7) and (18.6).

The new iterate (x_{k+1}, λ_{k+1}) can therefore be defined either as the solution of the quadratic program (18.7) or as the iterate generated by Newton's method (18.5), (18.6) applied to the optimality conditions of the problem. Both viewpoints are useful. The Newton point of view facilitates the analysis, whereas the SQP framework enables us to derive practical algorithms and to extend the technique to the inequality-constrained case.

We now state the SQP method in its simplest form.

Algorithm 18.1 (Local SQP Algorithm for solving (18.1)).
   Choose an initial pair (x_0, λ_0); set $k \leftarrow 0$;
   repeat until a convergence test is satisfied
            Evaluate $f_k, \nabla f_k, \nabla^2_{xx} \mathcal{L}_k, c_k$, and A_k;
            Solve (18.7) to obtain p_k and l_k;
            Set $x_{k+1} \leftarrow x_k + p_k$ and $\lambda_{k+1} \leftarrow l_k$;
   end (repeat)

We note in passing that, in the objective (18.7a) of the quadratic program, we could replace the linear term $\nabla f_k^T p$ by $\nabla_x \mathcal{L}(x_k, \lambda_k)^T p$, since the constraint (18.7b) makes the two choices equivalent. In this case, (18.7a) is a quadratic approximation of the Lagrangian function. This fact provides a motivation for our choice of the quadratic model (18.7): We first replace the nonlinear program (18.1) by the problem of minimizing the Lagrangian subject to the equality constraints (18.1b), then make a quadratic approximation to the Lagrangian and a linear approximation to the constraints to obtain (18.7).

INEQUALITY CONSTRAINTS

The SQP framework can be extended easily to the general nonlinear programming problem

$$\min f(x) \tag{18.10a}$$

$$\text{subject to } c_i(x) = 0, \quad i \in \mathcal{E}, \tag{18.10b}$$

$$c_i(x) \geq 0, \quad i \in \mathcal{I}. \tag{18.10c}$$

To model this problem we now linearize both the inequality and equality constraints to obtain

$$\min_{p} \quad f_k + \nabla f_k^T p + \tfrac{1}{2} p^T \nabla_{xx}^2 \mathcal{L}_k p \tag{18.11a}$$

$$\text{subject to } \quad \nabla c_i(x_k)^T p + c_i(x_k) = 0, \quad i \in \mathcal{E}, \tag{18.11b}$$

$$\nabla c_i(x_k)^T p + c_i(x_k) \geq 0, \quad i \in \mathcal{I}. \tag{18.11c}$$

We can use one of the algorithms for quadratic programming described in Chapter 16 to solve this problem. The new iterate is given by $(x_k + p_k, \lambda_{k+1})$ where p_k and λ_{k+1} are the solution and the corresponding Lagrange multiplier of (18.11). A local SQP method for (18.10) is thus given by Algorithm 18.1 with the modification that the step is computed from (18.11).

In this IQP approach the set of active constraints $\mathcal{A}_k$ at the solution of (18.11) constitutes our guess of the active set at the solution of the nonlinear program. If the SQP method is able to correctly identify this optimal active set (and not change its guess at a subsequent iteration) then it will act like a Newton method for equality-constrained optimization and will converge rapidly. The following result gives conditions under which this desirable behavior takes place. Recall that strict complementarity is said to hold at a solution pair (x^*, λ^*) if there is no index $i \in \mathcal{I}$ such that $\lambda_i^* = c_i(x^*) = 0$.

Theorem 18.1 (Robinson [267]).

Suppose that x^* is a local solution of (18.10) at which the KKT conditions are satisfied for some λ^*. Suppose, too, that the linear independence constraint qualification (LICQ) (Definition 12.4), the strict complementarity condition (Definition 12.5), and the second-order sufficient conditions (Theorem 12.6) hold at (x^*, λ^*). Then if (x_k, λ_k) is sufficiently close to (x^*, λ^*), there is a local solution of the subproblem (18.11) whose active set $\mathcal{A}_k$ is the same as the active set $\mathcal{A}(x^*)$ of the nonlinear program (18.10) at x^*.

It is also remarkable that, far from the solution, the SQP approach is usually able to improve the estimate of the active set and guide the iterates toward a solution; see Section 18.7.

18.2   PREVIEW OF PRACTICAL SQP METHODS

IQP AND EQP

There are two ways of designing SQP methods for solving the general nonlinear programming problem (18.10). The first is the approach just described, which solves at every iteration the quadratic subprogram (18.11), taking the active set at the solution of this subproblem as a guess of the optimal active set. This approach is referred to as the IQP (inequality-constrained QP) approach; it has proved to be quite successful in practice. Its main drawback is the expense of solving the general quadratic program (18.11), which can be high when the problem is large. As the iterates of the SQP method converge to the solution, however, solving the quadratic subproblem becomes economical if we use information from the previous iteration to make a good guess of the optimal solution of the current subproblem. This *warm-start* strategy is described below.

The second approach selects a subset of constraints at each iteration to be the so-called working set, and solves only equality-constrained subproblems of the form (18.7), where the constraints in the working sets are imposed as equalities and all other constraints are ignored. The working set is updated at every iteration by rules based on Lagrange multiplier estimates, or by solving an auxiliary subproblem. This EQP (equality-constrained QP) approach has the advantage that the equality-constrained quadratic subproblems are less expensive to solve than (18.11) in the large-scale case.

An example of an EQP method is the sequential linear-quadratic programming (SLQP) method discussed in Section 18.5. This approach constructs a linear program by omitting the quadratic term $p^T \nabla_{xx}^2 \mathcal{L}_k p$ from (18.11a) and adding a trust-region constraint $\|p\|_\infty \leq \Delta_k$ to the subproblem. The active set of the resulting linear programming subproblem is taken to be the working set for the current iteration. The method then fixes the constraints in the working set and solves an equality-constrained quadratic program (with the term $p^T \nabla_{xx}^2 \mathcal{L}_k p$ reinserted) to obtain the SQP step. Another successful EQP method is the gradient projection method described in Section 16.7 in the context of bound constrained quadratic programs. In this method, the working set is determined by minimizing a quadratic model along the path obtained by projecting the steepest descent direction onto the feasible region.

ENFORCING CONVERGENCE

To be practical, an SQP method must be able to converge from remote starting points and on nonconvex problems. We now outline how the local SQP strategy can be adapted to meet these goals.

We begin by drawing an analogy with unconstrained optimization. In its simplest form, the Newton iteration for minimizing a function f takes a step to the minimizer of the quadratic model

$$m_k(p) = f_k + \nabla f_k^T p + \tfrac{1}{2} p^T \nabla^2 f_k p.$$

This framework is useful near the solution, where the Hessian $\nabla^2 f(x_k)$ is normally positive definite and the quadratic model has a well defined minimizer. When x_k is not close to the solution, however, the model function m_k may not be convex. Trust-region methods ensure that the new iterate is always well defined and useful by restricting the candidate step p_k

to some neighborhood of the origin. Line search methods modify the Hessian in $m_k(p)$ to make it positive definite (possibly replacing it by a quasi-Newton approximation B_k), to ensure that p_k is a descent direction for the objective function f.

Similar strategies are used to globalize SQP methods. If $\nabla_{xx}^2 \mathcal{L}_k$ is positive definite on the tangent space of the active constraints, the quadratic subproblem (18.7) has a unique solution. When $\nabla_{xx}^2 \mathcal{L}_k$ does not have this property, line search methods either replace it by a positive definite approximation B_k or modify $\nabla_{xx}^2 \mathcal{L}_k$ directly during the process of matrix factorization. In all these cases, the subproblem (18.7) becomes well defined, but the modifications may introduce unwanted distortions in the model.

Trust-region SQP methods add a constraint to the subproblem, limiting the step to a region within which the model (18.7) is considered reliable. These methods are able to handle indefinite Hessians $\nabla_{xx}^2 \mathcal{L}_k$. The inclusion of the trust region may, however, cause the subproblem to become infeasible, and the procedures for handling this situation complicate the algorithms and increase their computational cost. Due to these tradeoffs, neither of the two SQP approaches—line search or trust-region—is currently regarded as clearly superior to the other.

The technique used to accept or reject steps also impacts the efficiency of SQP methods. In unconstrained optimization, the merit function is simply the objective f, and it remains fixed throughout the minimization procedure. For constrained problems, we use devices such as a merit function or a filter (see Section 15.4). The parameters or entries used in these devices must be updated in a way that is compatible with the step produced by the SQP method.

18.3   ALGORITHMIC DEVELOPMENT

In this section we expand on the ideas of the previous section and describe various ingredients needed to produce practical SQP algorithms. We focus on techniques for ensuring that the subproblems are always feasible, on alternative choices for the Hessian of the quadratic model, and on step-acceptance mechanisms.

HANDLING INCONSISTENT LINEARIZATIONS

A possible difficulty with SQP methods is that the linearizations (18.11b), (18.11c) of the nonlinear constraints may give rise to an infeasible subproblem. Consider, for example, the case in which $n = 1$ and the constraints are $x \le 1$ and $x^2 \ge 4$. When we linearize these constraints at $x_k = 1$, we obtain the inequalities

$$-p \ge 0 \quad \text{and} \quad 2p - 3 \ge 0,$$

which are inconsistent.

To overcome this difficulty, we can reformulate the nonlinear program (18.10) as the ℓ_1 penalty problem

$$\min_{x,v,w,t} \quad f(x) + \mu \sum_{i \in \mathcal{E}} (v_i + w_i) + \mu \sum_{i \in \mathcal{I}} t_i \tag{18.12a}$$

$$\text{subject to} \quad c_i(x) = v_i - w_i, \quad i \in \mathcal{E}, \tag{18.12b}$$

$$c_i(x) \geq -t_i, \quad i \in \mathcal{I}, \tag{18.12c}$$

$$v, \, w, \, t \geq 0, \tag{18.12d}$$

for some positive choice of the penalty parameter μ. The quadratic subproblem (18.11) associated with (18.12) is always feasible. As discussed in Chapter 17, if the nonlinear problem (18.10) has a solution x^* that satisfies certain regularity assumptions, and if the penalty parameter μ is sufficiently large, then x^* (along with $v_i^* = w_i^* = 0$, $i \in \mathcal{E}$ and $t_i^* = 0, i \in \mathcal{I}$) is a solution of the penalty problem (18.12). If, on the other hand, there is no feasible solution to the nonlinear problem and μ is large enough, then the penalty problem (18.12) usually determines a stationary point of the infeasibility measure. The choice of μ has been discussed in Chapter 17 and is considered again in Section 18.5. The SNOPT software package [127] uses the formulation (18.12), which is sometimes called the *elastic mode*, to deal with inconsistencies of the linearized constraints.

Other procedures for relaxing the constraints are presented in Section 18.5 in the context of trust-region methods.

FULL QUASI-NEWTON APPROXIMATIONS

The Hessian of the Lagrangian $\nabla_{xx}^2 \mathcal{L}(x_k, \lambda_k)$ is made up of second derivatives of the objective function and constraints. In some applications, this information is not easy to compute, so it is useful to consider replacing the Hessian $\nabla_{xx}^2 \mathcal{L}(x_k, \lambda_k)$ in (18.11a) by a quasi-Newton approximation. Since the BFGS and SR1 formulae have proved to be successful in the context of unconstrained optimization, we can employ them here as well.

The update for B_k that results from the step from iterate k to iterate $k + 1$ makes use of the vectors s_k and y_k defined as follows:

$$s_k = x_{k+1} - x_k, \qquad y_k = \nabla_x \mathcal{L}(x_{k+1}, \lambda_{k+1}) - \nabla_x \mathcal{L}(x_k, \lambda_{k+1}). \tag{18.13}$$

We compute the new approximation B_{k+1} using the BFGS or SR1 formulae given, respectively, by (6.19) and (6.24). We can view this process as the application of quasi-Newton updating to the case in which the objective function is given by the Lagrangian $\mathcal{L}(x, \lambda)$ (with λ fixed). This viewpoint immediately reveals the strengths and weaknesses of this approach.

If $\nabla_{xx}^2 \mathcal{L}$ is positive definite in the region where the minimization takes place, then BFGS quasi-Newton approximations B_k will reflect some of the curvature information of the problem, and the iteration will converge robustly and rapidly, just as in the unconstrained BFGS method. If, however, $\nabla_{xx}^2 \mathcal{L}$ contains negative eigenvalues, then the BFGS approach of approximating it with a positive definite matrix may be problematic. BFGS updating requires that s_k and y_k satisfy the curvature condition $s_k^T y_k > 0$, which may not hold when s_k and y_k are defined by (18.13), even when the iterates are close to the solution.

To overcome this difficulty, we could *skip* the BFGS update if the condition

$$s_k^T y_k \geq \theta s_k^T B_k s_k \tag{18.14}$$

is not satisfied, where θ is a positive parameter (10^{-2}, say). This strategy may, on occasion, yield poor performance or even failure, so it cannot be regarded as adequate for general-purpose algorithms.

A more effective modification ensures that the update is always well defined by modifying the definition of y_k.

Procedure 18.2 (Damped BFGS Updating).
  Given: symmetric and positive definite matrix B_k;
  Define s_k and y_k as in (18.13) and set

$$r_k = \theta_k y_k + (1 - \theta_k) B_k s_k,$$

where the scalar θ_k is defined as

$$\theta_k = \begin{cases} 1 & \text{if } s_k^T y_k \geq 0.2 s_k^T B_k s_k, \\ (0.8 s_k^T B_k s_k)/(s_k^T B_k s_k - s_k^T y_k) & \text{if } s_k^T y_k < 0.2 s_k^T B_k s_k; \end{cases} \tag{18.15}$$

Update B_k as follows:

$$B_{k+1} = B_k - \frac{B_k s_k s_k^T B_k}{s_k^T B_k s_k} + \frac{r_k r_k^T}{s_k^T r_k}. \tag{18.16}$$

The formula (18.16) is simply the standard BFGS update formula, with y_k replaced by r_k. It guarantees that B_{k+1} is positive definite, since it is easy to show that when $\theta_k \neq 1$ we have

$$s_k^T r_k = 0.2 s_k^T B_k s_k > 0. \tag{18.17}$$

To gain more insight into this strategy, note that the choice $\theta_k = 0$ gives $B_{k+1} = B_k$, while $\theta_k = 1$ gives the (possibly indefinite) matrix produced by the unmodified BFGS update. A value $\theta_k \in (0, 1)$ thus produces a matrix that interpolates the current approximation B_k and the one produced by the unmodified BFGS formula. The choice of θ_k ensures that the new approximation stays close enough to the current approximation B_k to ensure positive definiteness.

Damped BFGS updating often works well but it, too, can behave poorly on difficult problems. It still fails to address the underlying problem that the Lagrangian Hessian may not be positive definite. For this reason, SR1 updating may be more appropriate, and is indeed a good choice for trust-region SQP methods. An SR1 approximation to the Hessian of the Lagrangian is obtained by applying formula (6.24) with s_k and y_k defined by (18.13), using the safeguards described in Chapter 6. Line search methods cannot, however, accept indefinite Hessian approximations and would therefore need to modify the SR1 formula, possibly by adding a sufficiently large multiple of the identity matrix; see the discussion around (19.25).

All quasi-Newton approximations B_k discussed above are dense $n \times n$ matrices that can be expensive to store and manipulate in the large-scale case. Limited-memory updating is useful in this context and is often implemented in software packages. (See (19.29) for an implementation of limited-memory BFGS in a constrained optimization algorithm.)

REDUCED-HESSIAN QUASI-NEWTON APPROXIMATIONS

When we examine the KKT system (18.9) for the equality-constrained problem (18.1), we see that the part of the step p_k in the range space of A_k^T is completely determined by the second block row $A_k p_k = -c_k$. The Lagrangian Hessian $\nabla_{xx}^2 \mathcal{L}_k$ affects only the part of p_k in the orthogonal subspace, namely, the null space of A_k. It is reasonable, therefore, to consider quasi-Newton methods that find approximations to only that part of $\nabla_{xx}^2 \mathcal{L}_k$ that affects the component of p_k in the null space of A_k. In this section, we consider quasi-Newton methods based on these reduced-Hessian approximations. Our focus is on equality-constrained problems in this section, as existing SQP methods for the full problem (18.10) use reduced-Hessian approaches only after an equality-constrained subproblem has been generated.

To derive reduced-Hessian methods, we consider solution of the step equations (18.9) by means of the null space approach of Section 16.2. In that section, we defined matrices Y_k and Z_k whose columns span the range space of A_k^T and the null space of A_k, respectively. By writing

$$p_k = Y_k p_Y + Z_k p_Z, \tag{18.18}$$

and substituting into (18.9), we obtain the following system to be solved for p_Y and p_Z:

$$(A_k Y_k) p_Y = -c_k, \tag{18.19a}$$

$$\left(Z_k^T \nabla_{xx}^2 \mathcal{L}_k Z_k \right) p_Z = -Z_k^T \nabla_{xx}^2 \mathcal{L}_k Y_k p_Y - Z_k^T \nabla f_k. \tag{18.19b}$$

From the first block of equations in (18.9) we see that the Lagrange multipliers λ_{k+1}, which are sometimes called QP multipliers, can be obtained by solving

$$(A_k Y_k)^T \lambda_{k+1} = Y_k^T (\nabla f_k + \nabla_{xx}^2 \mathcal{L}_k p_k). \tag{18.20}$$

We can avoid computation of the Hessian $\nabla_{xx}^2 \mathcal{L}_k$ by introducing several approximations in the null-space approach. First, we delete the term involving p_k from the right-hand-side of (18.20), thereby decoupling the computations of p_k and λ_{k+1} and eliminating the need for $\nabla_{xx}^2 \mathcal{L}_k$ in this term. This simplification can be justified by observing that p_k converges to zero as we approach the solution, whereas ∇f_k normally does not. Therefore, the multipliers computed in this manner will be good estimates of the QP multipliers near the solution. More specifically, if we choose $Y_k = A_k^T$ (which is a valid choice for Y_k when A_k has full row rank; see (15.16)), we obtain

$$\hat{\lambda}_{k+1} = (A_k A_k^T)^{-1} A_k \nabla f_k. \tag{18.21}$$

These are called the *least-squares multipliers* because they can also be derived by solving the problem

$$\min_{\lambda} \|\nabla_x \mathcal{L}(x_k, \lambda)\|_2^2 = \|\nabla f_k - A_k^T \lambda\|_2^2. \tag{18.22}$$

This observation shows that the least-squares multipliers are useful even when the current iterate is far from the solution, because they seek to satisfy the first-order optimality condition in (18.3) as closely as possible. Conceptually, the use of least-squares multipliers transforms the SQP method from a primal-dual iteration in x and λ to a purely primal iteration in the x variable alone.

Our second simplification of the null-space approach is to remove the cross term $Z_k^T \nabla_{xx}^2 \mathcal{L}_k Y_k p_Y$ in (18.19b), thereby yielding the simpler system

$$(Z_k^T \nabla_{xx}^2 \mathcal{L}_k Z_k) p_Z = -Z_k^T \nabla f_k. \tag{18.23}$$

This approach has the advantage that it needs to approximate only the matrix $Z_k^T \nabla_{xx}^2 \mathcal{L}_k Z_k$, not the $(n - m) \times m$ cross-term matrix $Z_k^T \nabla_{xx}^2 \mathcal{L}_k Y_k$, which is a relatively large matrix when $m \gg n - m$. Dropping the cross term is justified when $Z_k^T \nabla_{xx}^2 \mathcal{L}_k Z_k$ is replaced by a quasi-Newton approximation because the normal component p_Y usually converges to zero faster than the tangential component p_Z, thereby making (18.23) a good approximation of (18.19b).

Having dispensed with the partial Hessian $Z_k^T \nabla_{xx}^2 \mathcal{L}_k Y_k$, we discuss how to approximate the remaining part $Z_k^T \nabla_{xx}^2 \mathcal{L}_k Z_k$. Suppose we have just taken a step $\alpha_k p_k = x_{k+1} - x_k = \alpha_k Z_k p_Z + \alpha_k Y_k p_Y$. By Taylor's theorem, writing $\nabla_{xx}^2 \mathcal{L}_{k+1} = \nabla_{xx}^2 \mathcal{L}(x_{k+1}, \lambda_{k+1})$, we have

$$\nabla_{xx}^2 \mathcal{L}_{k+1} \alpha_k p_k \approx \nabla_x \mathcal{L}(x_k + \alpha_k p_k, \lambda_{k+1}) - \nabla_x \mathcal{L}(x_k, \lambda_{k+1}).$$

By premultiplying by Z_k^T, we have

$$Z_k^T \nabla_{xx}^2 \mathcal{L}_{k+1} Z_k \alpha_k p_Z \tag{18.24}$$
$$\approx -Z_k^T \nabla_{xx}^2 \mathcal{L}_{k+1} Y_k \alpha_k p_Y + Z_k^T \left[\nabla_x \mathcal{L}(x_k + \alpha_k p_k, \lambda_{k+1}) - \nabla_x \mathcal{L}(x_k, \lambda_{k+1}) \right].$$

If we drop the cross term $Z_k^T \nabla_{xx}^2 \mathcal{L}_{k+1} Y_k \alpha_k p_Y$ (using the rationale discussed earlier), we see that the secant equation for M_k can be defined by

$$M_{k+1} s_k = y_k, \tag{18.25}$$

where s_k and y_k are given by

$$s_k = \alpha_k p_z, \qquad y_k = Z_k^T \left[\nabla_x \mathcal{L}(x_k + \alpha_k p_k, \lambda_{k+1}) - \nabla_x \mathcal{L}(x_k, \lambda_{k+1}) \right]. \tag{18.26}$$

We then apply the BFGS or SR1 formulae, using these definitions for the correction vectors s_k and y_k, to define the new approximation M_{k+1}. An advantage of this reduced-Hessian approach, compared to full-Hessian quasi-Newton approximations, is that the reduced Hessian is much more likely to be positive definite, even when the current iterate is some distance from the solution. When using the BFGS formula, the safeguarding mechanism discussed above will be required less often in line search implementations.

MERIT FUNCTIONS

SQP methods often use a merit function to decide whether a trial step should be accepted. In line search methods, the merit function controls the size of the step; in trust-region methods it determines whether the step is accepted or rejected and whether the trust-region radius should be adjusted. A variety of merit functions have been used in SQP methods, including nonsmooth penalty functions and augmented Lagrangians. We limit our discussion to exact, nonsmooth merit functions typified by the ℓ_1 merit function discussed in Chapters 15 and 17.

For the purpose of step computation and evaluation of a merit function, inequality constraints $c(x) \geq 0$ are often converted to the form

$$\bar{c}(x, s) = c(x) - s = 0,$$

where $s \geq 0$ is a vector of slacks. (The condition $s \geq 0$ is typically not monitored by the merit function.) Therefore, in the discussion that follows we assume that all constraints are in the form of equalities, and we focus our attention on problem (18.1).

The ℓ_1 merit function for (18.1) takes the form

$$\phi_1(x; \mu) = f(x) + \mu \|c(x)\|_1. \tag{18.27}$$

In a line search method, a step $\alpha_k p_k$ will be accepted if the following sufficient decrease condition holds:

$$\phi_1(x_k + \alpha_k p_k; \mu_k) \leq \phi_1(x_k, \mu_k) + \eta \alpha_k D(\phi_1(x_k; \mu); p_k), \qquad \eta \in (0, 1), \tag{18.28}$$

where $D(\phi_1(x_k; \mu); p_k)$ denotes the directional derivative of ϕ_1 in the direction p_k. This requirement is analogous to the Armijo condition (3.4) for unconstrained optimization provided that p_k is a descent direction, that is, $D(\phi_1(x_k; \mu); p_k) < 0$. This descent condition holds if the penalty parameter μ is chosen sufficiently large, as we show in the following result.

Theorem 18.2.
  Let p_k and λ_{k+1} be generated by the SQP iteration (18.9). Then the directional derivative of ϕ_1 in the direction p_k satisfies

$$D(\phi_1(x_k; \mu); p_k) = \nabla f_k^T p_k - \mu \|c_k\|_1. \tag{18.29}$$

Moreover, we have that

$$D(\phi_1(x_k; \mu); p_k) \leq -p_k^T \nabla_{xx}^2 \mathcal{L}_k p_k - (\mu - \|\lambda_{k+1}\|_\infty) \|c_k\|_1. \tag{18.30}$$

PROOF.   By applying Taylor's theorem (see (2.5)) to f and $c_i, i = 1, 2, \ldots, m$, we obtain

$$\phi_1(x_k + \alpha p; \mu) - \phi_1(x_k; \mu) = f(x_k + \alpha p) - f_k + \mu \|c(x_k + \alpha p)\|_1 - \mu \|c_k\|_1$$
$$\leq \alpha \nabla f_k^T p + \gamma \alpha^2 \|p\|^2 + \mu \|c_k + \alpha A_k p\|_1 - \mu \|c_k\|_1,$$

where the positive constant γ bounds the second-derivative terms in f and c. If $p = p_k$ is given by (18.9), we have that $A_k p_k = -c_k$, so for $\alpha \leq 1$ we have that

$$\phi_1(x_k + \alpha p_k; \mu) - \phi_1(x_k; \mu) \leq \alpha [\nabla f_k^T p_k - \mu \|c_k\|_1] + \alpha^2 \gamma \|p_k\|^2.$$

By arguing similarly, we also obtain the following lower bound:

$$\phi_1(x_k + \alpha p_k; \mu) - \phi_1(x_k; \mu) \geq \alpha [\nabla f_k^T p_k - \mu \|c_k\|_1] - \alpha^2 \gamma \|p_k\|^2.$$

Taking limits, we conclude that the directional derivative of ϕ_1 in the direction p_k is given by

$$D(\phi_1(x_k; \mu); p_k) = \nabla f_k^T p_k - \mu \|c_k\|_1, \tag{18.31}$$

which proves (18.29). The fact that p_k satisfies the first equation in (18.9) implies that

$$D(\phi_1(x_k; \mu); p_k) = -p_k^T \nabla_{xx}^2 \mathcal{L}_k p_k + p_k^T A_k^T \lambda_{k+1} - \mu \|c_k\|_1.$$

From the second equation in (18.9), we can replace the term $p_k^T A_k^T \lambda_{k+1}$ in this expression by $-c_k^T \lambda_{k+1}$. By making this substitution in the expression above and invoking the inequality

$$-c_k^T \lambda_{k+1} \leq \|c_k\|_1 \|\lambda_{k+1}\|_\infty,$$

we obtain (18.30). $\qquad\qquad\qquad\qquad\qquad\qquad\qquad\qquad\qquad\qquad\qquad$ □

It follows from (18.30) that p_k will be a descent direction for ϕ_1 if $p_k \neq 0$, $\nabla_{xx}^2 \mathcal{L}_k$ is positive definite and

$$\mu > \|\lambda_{k+1}\|_\infty. \tag{18.32}$$

(A more detailed analysis shows that this assumption on $\nabla_{xx}^2 \mathcal{L}_k$ can be relaxed; we need only the reduced Hessian $Z_k^T \nabla_{xx}^2 \mathcal{L}_k Z_k$ to be positive definite.)

One strategy for choosing the new value of the penalty parameter μ in $\phi_1(x; \mu)$ at every iteration is to increase the previous value, if necessary, so as to satisfy (18.32), with some margin. It has been observed, however, that this strategy may select inappropriate values of μ and often interferes with the progress of the iteration.

An alternative approach, based on (18.29), is to require that the directional derivative be sufficiently negative in the sense that

$$D(\phi_1(x_k; \mu); p_k) = \nabla f_k^T p_k - \mu \|c_k\|_1 \leq -\rho \mu \|c_k\|_1,$$

for some $\rho \in (0, 1)$. This inequality holds if

$$\mu \geq \frac{\nabla f_k^T p_k}{(1 - \rho)\|c_k\|_1}. \tag{18.33}$$

This choice is not dependent on the Lagrange multipliers and performs adequately in practice.

A more effective strategy for choosing μ, which is appropriate both in the line search and trust-region contexts, considers the effect of the step on a model of the merit function. We define a (piecewise) quadratic model of ϕ_1 by

$$q_\mu(p) = f_k + \nabla f_k^T p + \frac{\sigma}{2} p^T \nabla_{xx}^2 \mathcal{L}_k p + \mu m(p), \tag{18.34}$$

where

$$m(p) = \|c_k + A_k p\|_1,$$

and σ is a parameter to be defined below. After computing a step p_k, we choose the penalty parameter μ large enough that

$$q_\mu(0) - q_\mu(p_k) \geq \rho \mu [m(0) - m(p_k)], \tag{18.35}$$

for some parameter $\rho \in (0, 1)$. It follows from (18.34) and (18.7b) that inequality (18.35) is satisfied for

$$\mu \geq \frac{\nabla f_k^T p_k + (\sigma/2) p_k^T \nabla_{xx}^2 \mathcal{L}_k p_k}{(1 - \rho)\|c_k\|_1}. \tag{18.36}$$

If the value of μ from the previous iteration of the SQP method satisfies (18.36), it is left unchanged. Otherwise, μ is increased so that it satisfies this inequality with some margin.

The constant σ is used to handle the case in which the Hessian $\nabla^2_{xx}\mathcal{L}_k$ is not positive definite. We define σ as

$$\sigma = \begin{cases} 1 & \text{if } p_k^T \nabla^2_{xx}\mathcal{L}_k p_k > 0, \\ 0 & \text{otherwise.} \end{cases} \tag{18.37}$$

It is easy to verify that, if μ satisfies (18.36), this choice of σ ensures that $D(\phi_1(x_k; \mu); p_k) \leq -\rho\mu\|c_k\|_1$, so that p_k is a descent direction for the merit function ϕ_1. This conclusion is not always valid if $\sigma = 1$ and $p_k^T \nabla^2_{xx}\mathcal{L}_k p_k < 0$. By comparing (18.33) and (18.36) we see that, when $\sigma > 0$, the strategy based on (18.35) selects a larger penalty parameter, thus placing more weight on the reduction of the constraints. This property is advantageous if the step p_k decreases the constraints but increases the objective, for in this case the step has a better chance of being accepted by the merit function.

SECOND-ORDER CORRECTION

In Chapter 15, we showed by means of Example 15.4 that many merit functions can impede progress of an optimization algorithm, a phenomenon known as the Maratos effect. We now show that the step analyzed in that example is, in fact, produced by an SQP method.

❑ **EXAMPLE 18.1**   (EXAMPLE 15.4, REVISITED)

Consider problem (15.34). At the iterate $x_k = (\cos\theta, \sin\theta)^T$, let us compute a search direction p_k by solving the SQP subproblem (18.7) with $\nabla^2_{xx}\mathcal{L}_k$ replaced by $\nabla^2_{xx}\mathcal{L}(x^*, \lambda^*) = I$. Since

$$f_k = -\cos\theta, \quad \nabla f_k = \begin{bmatrix} 4\cos\theta - 1 \\ 4\sin\theta \end{bmatrix}, \quad A_k^T = \begin{bmatrix} 2\cos\theta \\ 2\sin\theta \end{bmatrix},$$

the quadratic subproblem (18.7) takes the form

$$\min_p \quad (4\cos\theta - 1)p_1 + 4\sin\theta p_2 + \frac{1}{2}p_1^2 + \frac{1}{2}p_2^2$$
$$\text{subject to} \quad p_2 + \cot\theta p_1 = 0.$$

By solving this subproblem, we obtain the direction

$$p_k = \begin{bmatrix} \sin^2\theta \\ -\sin\theta\cos\theta \end{bmatrix}, \tag{18.38}$$

which coincides with (15.35).   ❑

We mentioned in Section 15.4 that the difficulties associated with the Maratos effect can be overcome by means of a *second-order correction*. There are various ways of applying this technique; we describe one possible implementation next.

Suppose that the SQP method has computed a step p_k from (18.11). If this step yields an increase in the merit function ϕ_1, a possible cause is that our linear approximations to the constraints are not sufficiently accurate. To overcome this deficiency, we could re-solve (18.11) with the linear terms $c_i(x_k) + \nabla c_i(x_k)^T p$ replaced by quadratic approximations,

$$c_i(x_k) + \nabla c_i(x_k)^T p + \tfrac{1}{2} p^T \nabla^2 c_i(x_k) p. \tag{18.39}$$

However, even if the Hessians of the constraints are individually available, the resulting quadratically constrained subproblem may be too difficult to solve. Instead, we evaluate the constraint values at the new point $x_k + p_k$ and make use of the following approximations. By Taylor's theorem, we have

$$c_i(x_k + p_k) \approx c_i(x_k) + \nabla c_i(x_k)^T p_k + \tfrac{1}{2} p_k^T \nabla^2 c_i(x_k) p_k. \tag{18.40}$$

Assuming that the (still unknown) second-order step p will not be too different from p_k, we can approximate the last term in (18.39) as follows:

$$p^T \nabla^2 c_i(x_k) p = p_k^T \nabla^2 c_i(x_k) p_k. \tag{18.41}$$

By making this substitution in (18.39) and using (18.40), we obtain the second-order correction subproblem

$$\min_p \quad \nabla f_k^T p + \tfrac{1}{2} p^T \nabla_{xx}^2 \mathcal{L}_k p$$
$$\text{subject to} \quad \nabla c_i(x_k)^T p + d_i = 0, \quad i \in \mathcal{E},$$
$$\nabla c_i(x_k)^T p + d_i \geq 0, \quad i \in \mathcal{I},$$

where

$$d_i = c_i(x_k + p_k) - \nabla c_i(x_k)^T p_k, \quad i \in \mathcal{E} \cup \mathcal{I}.$$

The second-order correction step requires evaluation of the constraints $c_i(x_k + p_k)$ for $i \in \mathcal{E} \cup \mathcal{I}$, and therefore it is preferable not to apply it every time the merit function increases. One strategy is to use it only if the increase in the merit function is accompanied by an increase in the constraint norm.

It can be shown that when the step p_k is generated by the SQP method (18.11) then, near a solution satisfying second-order sufficient conditions, the algorithm above takes either the full step p_k or the corrected step $p_k + \hat{p}_k$. The merit function does not interfere with the iteration, so superlinear convergence is attained, as in the local algorithm.

18.4 A PRACTICAL LINE SEARCH SQP METHOD

From the discussion in the previous section, we can see that there is a wide variety of line search SQP methods that differ in the way the Hessian approximation is computed, in the step acceptance mechanism, and in other algorithmic features. We now incorporate some of these ideas into a concrete, practical SQP algorithm for solving the nonlinear programming problem (18.10). To keep the description simple, we will not include a mechanism such as (18.12) to ensure the feasibility of the subproblem, or a second-order correction step. Rather, the search direction is obtained simply by solving the subproblem (18.11). We also assume that the quadratic program (18.11) is convex, so that we can solve it by means of the active-set method for quadratic programming (Algorithm 16.3) described in Chapter 16.

Algorithm 18.3 (Line Search SQP Algorithm).
  Choose parameters $\eta \in (0, 0.5)$, $\tau \in (0, 1)$, and an initial pair (x_0, λ_0);
  Evaluate f_0, ∇f_0, c_0, A_0;
  If a quasi-Newton approximation is used, choose an initial $n \times n$ symmetric positive definite Hessian approximation B_0, otherwise compute $\nabla^2_{xx} \mathcal{L}_0$;
  repeat until a convergence test is satisfied
      Compute p_k by solving (18.11); let $\hat{\lambda}$ be the corresponding multiplier;
      Set $p_\lambda \leftarrow \hat{\lambda} - \lambda_k$;
      Choose μ_k to satisfy (18.36) with $\sigma = 1$;
      Set $\alpha_k \leftarrow 1$;
      while $\phi_1(x_k + \alpha_k p_k; \mu_k) > \phi_1(x_k; \mu_k) + \eta \alpha_k D_1(\phi(x_k; \mu_k)p_k)$
          Reset $\alpha_k \leftarrow \tau_\alpha \alpha_k$ for some $\tau_\alpha \in (0, \tau]$;
      end (while)
      Set $x_{k+1} \leftarrow x_k + \alpha_k p_k$ and $\lambda_{k+1} \leftarrow \lambda_k + \alpha_k p_\lambda$;
      Evaluate f_{k+1}, ∇f_{k+1}, c_{k+1}, A_{k+1}, (and possibly $\nabla^2_{xx} \mathcal{L}_{k+1}$);
      If a quasi-Newton approximation is used, set
          $s_k \leftarrow \alpha_k p_k$ and $y_k \leftarrow \nabla_x \mathcal{L}(x_{k+1}, \lambda_{k+1}) - \nabla_x \mathcal{L}(x_k, \lambda_{k+1})$,
      and obtain B_{k+1} by updating B_k using a quasi-Newton formula;
  end (repeat)

We can achieve significant savings in the solution of the quadratic subproblem by warm-start procedures. For example, we can initialize the working set for each QP subproblem to be the final active set from the previous SQP iteration.

We have not given particulars of the quasi-Newton approximation in Algorithm 18.3. We could use, for example, a limited-memory BFGS approach that is suitable for large-scale problems. If we use an exact Hessian $\nabla^2_{xx} \mathcal{L}_k$, we assume that it is modified as necessary to be positive definite on the null space of the equality constraints.

Instead of a merit function, we could employ a filter (see Section 15.4) in the inner "while" loop to determine the steplength α_k. As discussed in Section 15.4, a feasibility restoration phase is invoked if a trial steplength generated by the backtracking line search is smaller than a given threshold. Regardless of whether a merit function or filter are used, a mechanism such as second-order correction can be incorporated to overcome the Maratos effect.

18.5   TRUST-REGION SQP METHODS

Trust-region SQP methods have several attractive properties. Among them are the facts that they do not require the Hessian matrix $\nabla_{xx}^2 \mathcal{L}_k$ in (18.11) to be positive definite, they control the quality of the steps even in the presence of Hessian and Jacobian singularities, and they provide a mechanism for enforcing global convergence. Some implementations follow an IQP approach and solve an inequality-constrained subproblem, while others follow an EQP approach.

The simplest way to formulate a trust-region SQP method is to add a trust-region constraint to subproblem (18.11), as follows:

$$\min_p \ f_k + \nabla f_k^T p + \tfrac{1}{2} p^T \nabla_{xx}^2 \mathcal{L}_k p \tag{18.43a}$$

$$\text{subject to} \ \ \nabla c_i(x_k)^T p + c_i(x_k) = 0, \quad i \in \mathcal{E}, \tag{18.43b}$$

$$\nabla c_i(x_k)^T p + c_i(x_k) \geq 0, \quad i \in \mathcal{I}, \tag{18.43c}$$

$$\|p\| \leq \Delta_k. \tag{18.43d}$$

Even if the constraints (18.43b), (18.43c) are compatible, this problem may not always have a solution because of the trust-region constraint (18.43d). We illustrate this fact in Figure 18.1 for a problem that contains only one equality constraint whose linearization is represented by the solid line. In this example, any step p that satisfies the linearized constraint must lie outside the trust region, which is indicated by the circle of radius Δ_k. As we see from this example, a consistent system of equalities and inequalities may not have a solution if we restrict the norm of the solution.

To resolve the possible conflict between the linear constraints (18.43b), (18.43c) and the trust-region constraint (18.43d), it is not appropriate simply to increase Δ_k until the set of steps p satisfying the linear constraints intersects the trust region. This approach would defeat the purpose of using the trust region in the first place as a way to define a region within which we trust the model (18.43a)–(18.43c) to accurately reflect the behavior of the objective and constraint functions. Analytically, it would harm the convergence properties of the algorithm.

A more appropriate viewpoint is that there is no reason to satisfy the linearized constraints exactly at every step; rather, we should aim to improve the feasibility of these constraints at each step and to satisfy them exactly only if the trust-region constraint permits it. This point of view is the basis of the three classes of methods discussed in this section: relaxation methods, penalty methods, and filter methods.

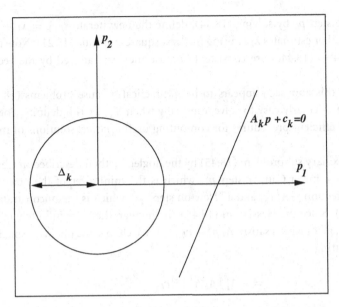

Figure 18.1 Inconsistent constraints in trust-region model.

A RELAXATION METHOD FOR EQUALITY-CONSTRAINED OPTIMIZATION

We describe this method in the context of the equality-constrained optimization problem (18.1); its extension to general nonlinear programs is deferred to Chapter 19 because it makes use of interior-point techniques. (Active-set extensions of the relaxation approach have been proposed, but have not been fully explored.)

At the iterate x_k, we compute the SQP step by solving the subproblem

$$\min_{p} \quad f_k + \nabla f_k^T p + \frac{1}{2} p^T \nabla_{xx}^2 \mathcal{L}_k p \tag{18.44a}$$

$$\text{subject to} \quad A_k p + c_k = r_k, \tag{18.44b}$$

$$\|p\|_2 \leq \Delta_k. \tag{18.44c}$$

The choice of the relaxation vector r_k requires careful consideration, as it impacts the efficiency of the method. Our goal is to choose r_k as the smallest vector such that (18.44b), (18.44c) are consistent for some reduced value of trust-region radius Δ_k. To do so, we first solve the subproblem

$$\min_{v} \quad \|A_k v + c_k\|_2^2 \tag{18.45a}$$

$$\text{subject to} \quad \|v\|_2 \leq 0.8\Delta_k. \tag{18.45b}$$

Denoting the solution of this subproblem by v_k, we define

$$r_k = A_k v_k + c_k. \tag{18.46}$$

We now compute the step p_k by solving (18.44), define the new iterate $x_{k+1} = x_k + p_k$, and obtain new multiplier estimates λ_{k+1} using the least squares formula (18.21). Note that the constraints (18.44b), (18.44c) are consistent because they are satisfied by the vector $p = v_k$.

At first glance, this approach appears to be impractical because problems (18.44) and (18.45) are not particularly easy to solve, especially when $\nabla^2_{xx} \mathcal{L}_k$ is indefinite. Fortunately, we can design efficient procedures for computing useful *inexact* solutions of these problems.

We solve the auxiliary subproblem (18.45) by the dogleg method described in Chapter 4. This method requires a Cauchy step p^U, which is the minimizer of the objective (18.45a) along the direction $-A_k^T c_k$, and a "Newton step" p^B, which is the unconstrained minimizer of (18.45a). Since the Hessian in (18.45a) is singular, there are infinitely many possible choices of p^B, all of which satisfy $A_k p^B + c_k = 0$. We choose the one with smallest Euclidean norm by setting

$$p^B = -A_k^T [A_k A_k^T]^{-1} c_k.$$

We now take v_k to be the minimizer of (18.45a) along the path defined by p^U, p^B, and the formula (4.16).

The preferred technique for computing an approximate solution p_k of (18.44) is the projected conjugate gradient method of Algorithm 16.2. We apply this algorithm to the equality-constrained quadratic program (18.44a)–(18.44b), monitoring satisfaction of the trust-region constraint (18.44c) and stopping if the boundary of this region is reached or if negative curvature is detected; see Section 7.1. Algorithm 16.2 requires a feasible starting point, which may be chosen as v_k.

A merit function that fits well with this approach is the nonsmooth ℓ_2 function $\phi_2(x; \mu) = f(x) + \mu \|c(x)\|_2$. We model it by means of the function

$$q_\mu(p) = f_k + \nabla f_k^T p + \frac{1}{2} p^T \nabla^2_{xx} \mathcal{L}_k p + \mu m(p), \qquad (18.47)$$

where

$$m(p) = \|c_k + A_k p\|_2,$$

(see (18.34)). We choose the penalty parameter large enough that inequality (18.35) is satisfied. To judge the acceptability of a step p_k, we monitor the ratio

$$\rho_k = \frac{\text{ared}_k}{\text{pred}_k} = \frac{\phi_2(x_k, \mu) - \phi_2(x_k + p_k, \mu)}{q_\mu(0) - q_\mu(p_k)}. \qquad (18.48)$$

We can now give a description of this trust-region SQP method for the equality-constrained optimization problem (18.1).

Algorithm 18.4 (Byrd–Omojokun Trust-Region SQP Method).
  Choose constants $\epsilon > 0$ and $\eta, \gamma \in (0, 1)$;
  Choose starting point x_0, initial trust region $\Delta_0 > 0$;
  for $k = 0, 1, 2, \ldots$
      Compute $f_k, c_k, \nabla f_k, A_k$;
      Compute multiplier estimates $\hat{\lambda}_k$ by (18.21);
      if $\|\nabla f_k - A_k^T \hat{\lambda}_k\|_\infty < \epsilon$ **and** $\|c_k\|_\infty < \epsilon$
          stop with approximate solution x_k;
      Solve normal subproblem (18.45) for v_k and compute r_k from (18.46);
      Compute $\nabla_{xx}^2 \mathcal{L}_k$ or a quasi-Newton approximation;
      Compute p_k by applying the projected CG method to (18.44);
      Choose μ_k to satisfy (18.35);
      Compute $\rho_k = \text{ared}_k / \text{pred}_k$;
      if $\rho_k > \eta$
          Set $x_{k+1} = x_k + p_k$;
          Choose Δ_{k+1} to satisfy $\Delta_{k+1} \geq \Delta_k$;
      else
          Set $x_{k+1} = x_k$;
          Choose Δ_{k+1} to satisfy $\Delta_{k+1} \leq \gamma \|p_k\|$;
  end (for).

A second-order correction can be added to avoid the Maratos effect. Beyond the cost of evaluating the objective function f and constraints c, the main costs of this algorithm lie in the projected CG iteration, which requires products of the Hessian $\nabla_{xx}^2 \mathcal{L}_k$ with vectors, and in the factorization and backsolves with the projection matrix (16.32); see Section 16.3.

Sℓ_1QP (SEQUENTIAL ℓ_1 QUADRATIC PROGRAMMING)

In this approach we move the linearized constraints (18.43b), (18.43c) into the objective of the quadratic program, in the form of an ℓ_1 penalty term, to obtain the following subproblem:

$$\min_p \quad q_\mu(p) \stackrel{\text{def}}{=} f_k + \nabla f_k^T p + \frac{1}{2} p^T \nabla_{xx}^2 \mathcal{L}_k p + \mu \sum_{i \in \mathcal{E}} |c_i(x_k) + \nabla c_i(x_k)^T p|$$

$$+\mu \sum_{i \in \mathcal{I}} [c_i(x_k) + \nabla c_i(x_k)^T p]^- \tag{18.49}$$

subject to $\|p\|_\infty \leq \Delta_k$,

for some penalty parameter μ, where we use the notation $[y]^- = \max\{0, -y\}$. Introducing slack variables v, w, t, we can reformulate this problem as follows:

$$\min_{p,v,w,t} \quad f_k + \nabla f_k^T p + \frac{1}{2} p^T \nabla_{xx}^2 \mathcal{L}_k p + \mu \sum_{i \in \mathcal{E}} (v_i + w_i) + \mu \sum_{i \in \mathcal{I}} t_i \quad (18.50a)$$

$$\text{s.t.} \quad \nabla c_i(x_k)^T p + c_i(x_k) = v_i - w_i, \quad i \in \mathcal{E}, \quad (18.50b)$$

$$\nabla c_i(x_k)^T p + c_i(x_k) \geq -t_i, \quad i \in \mathcal{I}, \quad (18.50c)$$

$$v, w, t \geq 0, \quad (18.50d)$$

$$\|p\|_\infty \leq \Delta_k. \quad (18.50e)$$

This formulation is simply a linearization of the elastic-mode formulation (18.12) with the addition of a trust-region constraint.

The constraints of this problem are always consistent. Since the trust region has been defined using the ℓ_∞ norm, (18.50) is a smooth quadratic program that can be solved by means of a quadratic programming algorithm. Warm-start strategies can significantly reduce the solution time of (18.50) and are invariably used in practical implementations.

It is natural to use the ℓ_1 merit function

$$\phi_1(x; \mu) = f(x) + \mu \sum_{i \in \mathcal{E}} |c_i(x)| + \mu \sum_{i \in \mathcal{I}} [c_i(x)]^- \quad (18.51)$$

to determine step acceptance. In fact, the function q_μ defined in (18.49) can be viewed as a model of $\phi_1(x, \mu)$ at x_k in which we approximate each constraint function c_i by its linearization, and replace f by a quadratic function whose curvature term includes information from both objective and constraints.

After computing the step p_k from (18.50), we determine the ratio ρ_k via (18.48), using the merit function ϕ_1 and defining q_μ by (18.49). The step is accepted or rejected according to standard trust-region rules, as implemented in Algorithm 18.4. A second-order correction step can be added to prevent the occurence of the Maratos effect.

The $S\ell_1 QP$ approach has several attractive properties. Not only does the formulation (18.49) overcome the possible inconsistency among the linearized constraints, but it also ensures that the trust-region constraint can always be satisfied. Further, the matrix $\nabla_{xx}^2 \mathcal{L}_k$ can be used without modification in subproblem (18.50) or else can be replaced by a quasi-Newton approximation. There is no requirement for it to be positive definite.

This choice of the penalty parameter μ plays an important role in the efficiency of this method. Unlike the SQP methods described above, which use a penalty function only to determine the acceptability of a trial point, the step p_k of the $S\ell_1 QP$ algorithm depends on μ. Values of μ that are too small can lead the algorithm away from the solution (Section 17.2), while excessively large values can result in slow progress. To obtain good practical performance over a range of applications, the value of μ must be chosen carefully at each iteration; see Algorithm 18.5 below.

SEQUENTIAL LINEAR-QUADRATIC PROGRAMMING (SLQP)

The SQP methods discussed above require the solution of a general (inequality-constrained) quadratic problem at each iteration. The cost of solving this subproblem imposes a limit on the size of problems that can be solved in practice. In addition, the incorporation of (indefinite) second derivative information in SQP methods has proved to be difficult [147].

The sequential linear-quadratic programming (SLQP) method attempts to overcome these concerns by computing the step in two stages, each of which scales well with the number of variables. First, a linear program (LP) is solved to identify a working set $\mathcal{W}$. Second, there is an equality-constrained quadratic programming (EQP) phase in which the constraints in the working set $\mathcal{W}$ are imposed as equalities. The total step of the algorithm is a combination of the steps obtained in the linear programming and equality-constrained phases, as we now discuss.

In the LP phase, we would like to solve the problem

$$\min_{p} \quad f_k + \nabla f_k^T p \tag{18.52a}$$

$$\text{subject to} \quad c_i(x_k) + \nabla c_i(x_k)^T p = 0, \quad i \in \mathcal{E}, \tag{18.52b}$$

$$c_i(x_k) + \nabla c_i(x_k)^T p \geq 0, \quad i \in \mathcal{I}, \tag{18.52c}$$

$$\|p\|_\infty \leq \Delta_k^{\text{LP}}, \tag{18.52d}$$

which differs from the standard SQP subproblem (18.43) only in that the second-order term in the objective has been omitted and that an ℓ_∞ norm is used to define the trust region. Since the constraints of (18.52) may be inconsistent, we solve instead the ℓ_1 penalty reformulation of (18.52) defined by

$$\min_{p} \quad l_\mu(p) \overset{\text{def}}{=} f_k + \nabla f_k^T p + \mu \sum_{i \in \mathcal{E}} |c_i(x_k) + \nabla c_i(x_k)^T p|$$

$$+ \mu \sum_{i \in \mathcal{I}} [c_i(x_k) + \nabla c_i(x_k)^T p]^- \tag{18.53a}$$

$$\text{subject to} \quad \|p\|_\infty \leq \Delta_k^{\text{LP}}. \tag{18.53b}$$

By introducing slack variables as in (18.50), we can reformulate (18.53) as an LP. The solution of (18.53), which we denote by p^{LP}, is computed by the simplex method (Chapter 13). From this solution we obtain the following explicit estimate of the optimal active set:

$$\mathcal{A}_k(p^{\text{LP}}) = \{i \in \mathcal{E} \mid c_i(x_k) + \nabla c_i(x_k)^T p^{\text{LP}} = 0\} \cup \{i \in \mathcal{I} \mid c_i(x_k) + \nabla c_i(x_k)^T p^{\text{LP}} = 0\}.$$

Likewise, we define the set $\mathcal{V}_k$ of violated constraints as

$$\mathcal{V}_k(p^{\text{LP}}) = \{i \in \mathcal{E} \mid c_i(x_k) + \nabla c_i(x_k)^T p^{\text{LP}} \neq 0\} \cup \{i \in \mathcal{I} \mid c_i(x_k) + \nabla c_i(x_k)^T p^{\text{LP}} < 0\}.$$

We define the working set $\mathcal{W}_k$ as some linearly independent subset of the active set $\mathcal{A}_k(p^{\text{LP}})$. To ensure that the algorithm makes progress on the penalty function ϕ_1, we define the *Cauchy step*,

$$p^{\text{C}} = \alpha^{\text{LP}} p^{\text{LP}}, \tag{18.54}$$

where $\alpha^{\text{LP}} \in (0, 1]$ is a steplength that provides sufficient decrease in the model q_μ defined in (18.49).

Given the working set $\mathcal{W}_k$, we now solve an equality-constrained quadratic program (EQP) treating the constraints in $\mathcal{W}_k$ as equalities and ignoring all others. We thus obtain the subproblem

$$\min_{p} \quad f_k + \tfrac{1}{2} p^T \nabla^2_{xx} \mathcal{L}_k p + \left(\nabla f_k + \mu_k \sum_{i \in \mathcal{V}_k} \gamma_i \nabla c_i(x_k) \right)^T p \tag{18.55a}$$

$$\text{subject to} \quad c_i(x_k) + \nabla c_i(x_k)^T p = 0, \quad i \in \mathcal{E} \cap \mathcal{W}_k, \tag{18.55b}$$

$$c_i(x_k) + \nabla c_i(x_k)^T p = 0, \quad i \in \mathcal{I} \cap \mathcal{W}_k, \tag{18.55c}$$

$$\|p\|_2 \leq \Delta_k, \tag{18.55d}$$

where γ_i is the algebraic sign of the i-th violated constraint. Note that the trust region (18.55d) is spherical, and that Δ_k is distinct from the trust-region radius Δ_k^{LP} used in (18.53b). Problem (18.55) is solved for the vector p^{Q} by applying the projected conjugated gradient procedure of Algorithm 16.2, handling the trust-region constraint by Steihaug's strategy (Algorithm 7.2). The total step p_k of the SLQP method is given by

$$p_k = p^{\text{C}} + \alpha^{\text{Q}}(p^{\text{Q}} - p^{\text{C}}),$$

where $\alpha^{\text{Q}} \in [0, 1]$ is a steplength that approximately minimizes the model q_μ defined in (18.49).

The trust-region radius Δ_k for the EQP phase is updated using standard trust-region update strategies. The choice of radius Δ_{k+1}^{LP} for the LP phase is more delicate, since it influences our guess of the optimal active set. The value of Δ_{k+1}^{LP} should be set to be a little larger than the total step p_k, subject to some other restrictions [49]. The multiplier estimates λ_k used in the Hessian $\nabla^2_{xx} \mathcal{L}_k$ are least squares estimates (18.21) using the working set $\mathcal{W}_k$, and modified so that $\lambda_i \geq 0$ for $i \in \mathcal{I}$.

An appealing feature of the SLQP algorithm is that established techniques for solving large-scale versions of the LP and EQP subproblems are readily available. High quality LP software is capable of solving problems with very large numbers of variables and constraints, while the solution of the EQP subproblem can be performed efficiently using the projected conjugate gradient method.

A TECHNIQUE FOR UPDATING THE PENALTY PARAMETER

We have mentioned that penalty methods such as $S\ell_1 QP$ and SLQP can be sensitive to the choice of the penalty parameter μ. We now discuss a procedure for choosing μ that has proved to be effective in practice and is supported by global convergence guarantees. The goal is to choose μ small enough to avoid an unnecessary imbalance in the merit function, but large enough to cause the step to make sufficient progress in linearized feasibility at each iteration. We present this procedure in the context of the $S\ell_1 QP$ method and then describe its extension to the SLQP approach.

We define a piecewise linear model of constraint violation at a point x_k by

$$m_k(p) = \sum_{i \in \mathcal{E}} |c_i(x_k) + \nabla c_i(x_k)^T p| + \sum_{i \in \mathcal{I}} [c_i(x_k) + \nabla c_i(x_k)^T p]^-, \tag{18.56}$$

so that the objective of the SQP subproblem (18.49) can be written as

$$q_\mu(p) = f_k + \nabla f_k^T p + \frac{1}{2} p^T \nabla_{xx}^2 \mathcal{L}_k p + \mu m_k(p). \tag{18.57}$$

We begin by solving the QP subproblem (18.49) (or equivalently, (18.50)) using the previous value μ_{k-1} of the penalty parameter. If the constraints (18.50b), (18.50c) are satisfied with the slack variables v_i, w_i, t_i all equal to zero (that is, $m_k(p_k) = 0$), then the current value of μ is adequate, and we set $\mu_k = \mu_{k-1}$. This is the felicitous case in which we can achieve linearized feasibility with a step p_k that is no longer in norm than the trust-region radius.

If $m_k(p) > 0$, on the other hand, it may be appropriate to increase the penalty parameter. The question is: by how much? To obtain a reference value, we re-solve the QP (18.49) using an "infinite" value of μ, by which we mean that the objective function in (18.49) is replaced by $m_k(p)$. After computing the new step, which we denote by p_∞, two outcomes are possible. If $m_k(p_\infty) = 0$, meaning that the linearized constraints are feasible within the trust region, we choose $\mu_k > \mu_{k-1}$ such that $m_k(p_k) = 0$. Otherwise, if $m_k(p_\infty) > 0$, we choose $\mu_k \geq \mu_{k-1}$ such that the reduction in m_k caused by the step p_k is at least a fraction of the (optimal) reduction given by p_∞.

The selection of $\mu_k > \mu_{k-1}$ is achieved in all cases by successively increasing the current trial value of μ (by a factor of 10, say) and re-solving the quadratic program (18.49). To describe this strategy more precisely, we write the solution of the QP problem (18.49) as $p(\mu)$ to stress its dependence on the penalty parameter. Likewise, p_∞ denotes the minimizer of $m_k(p)$ subject to the trust-region constraint (18.50e). The following algorithm describes the selection of the penalty parameter μ_k and the computation of the $S\ell_1 QP$ step p_k.

Algorithm 18.5 (Penalty Update and Step Computation).
  Initial data: x_k, $\mu_{k-1} > 0$, $\Delta_k > 0$, and parameters $\epsilon_1, \epsilon_2 \in (0, 1)$.

Solve the subproblem (18.50) with $\mu = \mu_{k-1}$ to obtain $p(\mu_{k-1})$;
if $m_k(p(\mu_{k-1})) = 0$
  Set $\mu^+ \leftarrow \mu_{k-1}$;
else
  Compute p_∞;
  if $m_k(p_\infty) = 0$
      Find $\mu^+ > \mu_{k-1}$ such that $m_k(p(\mu^+)) = 0$;
  else
      Find $\mu^+ \geq \mu_{k-1}$ such that
$$m_k(0) - m_k(p(\mu^+)) \geq \epsilon_1[m_k(0) - m_k(p_\infty)];$$
  end(if)
end(if)
  Increase μ^+ if necessary to satisfy
$$q_{\mu^+}(0) - q_{\mu^+}(p(\mu^+)) \geq \epsilon_2 \mu^+ [m_k(0) - m_k(p(\mu^+))];$$
Set $\mu_k \leftarrow \mu^+$ and $p_k \leftarrow p(\mu^+)$.

(Note that the inequality in the penultimate line is the same as condition (18.35).) Although Algorithm 18.5 requires the solution of some additional quadratic programs, we hope to reduce the total number of iterations (and the total number of QP solves) by identifying an appropriate penalty parameter value more quickly than rules based on feasibility monitoring (see Framework 17.2).

Numerical experience indicates that these savings occur when an adaptation of Algorithm 18.5 is used in the SLQP method. This adaptation is obtained simply by setting $\nabla_{xx}^2 \mathcal{L}_k = 0$ in the definition (18.49) of q_μ and applying Algorithm 18.5 to determine μ and to compute the LP step p^{LP}. The extra LP solves required by Algorithm 18.5 in this case are typically inexpensive, requiring relatively few simplex iterations, because we can use warm-start information from LPs solved earlier, with different values of the penalty parameter.

18.6   NONLINEAR GRADIENT PROJECTION

In Section 16.7, we discussed the gradient projection method for bound constrained quadratic programming. It is not difficult to extend this method to the problem

$$\min f(x) \qquad \text{subject to} \quad l \leq x \leq u, \tag{18.58}$$

where f is a nonlinear function and l and u are vectors of lower and upper bounds, respectively.

We begin by describing a line search approach. At the current iterate x_k, we form the quadratic model

$$q_k(x) = f_k + \nabla f_k^T (x - x_k) + \tfrac{1}{2}(x - x_k)^T B_k (x - x_k), \qquad (18.59)$$

where B_k is a positive definite approximation to $\nabla^2 f(x_k)$. We then use the gradient projection method for quadratic programming (Algorithm 16.5) to find an approximate solution $\hat{x}$ of the subproblem

$$\min q_k(x) \qquad \text{subject to} \quad l \leq x \leq u. \qquad (18.60)$$

The search direction is defined as $p_k = \hat{x} - x_k$ and the new iterate is given by $x_{k+1} = x_k + \alpha_k p_k$, where the steplength α_k is chosen to satisfy

$$f(x_k + \alpha_k p_k) \leq f(x_k) + \eta \alpha_k \nabla f_k^T p_k$$

for some parameter $\eta \in (0, 1)$.

To see that the search direction p_k is indeed a descent direction for the objective function, we use the properties of Algorithm 16.5, as discussed in Section 16.7. Recall that this method searches along a piecewise linear path—the projected steepest descent path—for the Cauchy point x^c, which minimizes q_k along this path. It then identifies the components of x that are at their bounds and holds these components constant while performing an unconstrained minimization of q_k over the remaining components to obtain the approximate solution $\hat{x}$ of the subproblem (18.60).

The Cauchy point x^c satisfies $q_k(x^c) < q_k(x_k)$ if the projected gradient is nonzero. Since Algorithm 16.5 produces a subproblem solution $\hat{x}$ with $q_k(\hat{x}) \leq q_k(x^c)$, we have

$$f_k = q_k(x_k) > q_k(x^c) \geq q_k(\hat{x}) = f_k + \nabla f_k^T p_k + \tfrac{1}{2} p_k^T B_k p_k.$$

This inequality implies that $\nabla f_k^T p_k < 0$, since B_k is assumed to be positive definite.

We now consider a trust-region gradient projection method for solving (18.58). We begin by forming the quadratic model (18.59), but since there is no requirement for q_k to be convex, we can define B_k to be the Hessian $\nabla^2 f(x_k)$ or a quasi-Newton approximation obtained from the BFGS or SR1 formulas. The step p_k is obtained by solving the subproblem

$$\min q_k(x) \qquad \text{subject to} \quad \{l \leq x \leq u, \quad \|x - x_k\|_\infty \leq \Delta_k\}, \qquad (18.61)$$

for some $\Delta_k > 0$. This problem can be posed as a bound-constrained quadratic program as follows:

$$\min q_k(x) \qquad \text{subject to} \quad \max(l, x_k - \Delta_k e) \leq x \leq \min(u, x_k + \Delta_k e),$$

where $e = (1, 1, \ldots, 1)^T$. Algorithm 16.5 can be used to solve this subproblem. The step p_k is accepted or rejected following standard trust-region strategies, and the radius Δ_k is updated according to the agreement between the change in f and the change in q_k produced by the step p_k; see Chapter 4.

The two gradient projection methods just outlined require solution of an inequality-constrained quadratic subproblem at each iteration, and so are formally IQP methods. They can, however, be viewed also as EQP methods because of their use of Algorithm 16.5 in solving the subproblem. This algorithm first identifies a working set (by finding the Cauchy point) and then solves an equality-constrained subproblem (by fixing the working-set constraints at their bounds). For large problems, it is efficient to perform the subspace minimization (16.74) by using the conjugate gradient method. A preconditioner is sometimes needed to make this approach practical; the most popular choice is the incomplete (and modified) Cholesky factorization outlined in Algorithm 7.3.

The gradient projection approach can be extended in principle to more general (linear or convex) constraints. Practical implementations are however limited to the bound constrained problem (18.58) because of the high cost of computing projections onto general constraint sets.

18.7 CONVERGENCE ANALYSIS

Numerical experience has shown that the SQP and SLQP methods discussed in this chapter often converge to a solution from remote starting points. Hence, there has been considerable interest in understanding what drives the iterates toward a solution and what can cause the algorithms to fail. These global convergence studies have been valuable in improving the design and implementation of algorithms.

Some early results make strong assumptions, such as boundedness of multipliers, well posedness of the subproblem (18.11), and regularity of constraint Jacobians. More recent studies relax many of these assumptions with the goal of understanding both the successful and unsuccessful outcomes of the iteration. We now state a classical global convergence result that gives conditions under which a standard SQP algorithm always identifies a KKT point of the nonlinear program.

Consider an SQP method that computes a search direction p_k by solving the quadratic program (18.11). We assume that the Hessian $\nabla_{xx}^2 \mathcal{L}_k$ is replaced in (18.11a) by some symmetric and positive definite approximation B_k. The new iterate is defined as $x_{k+1} + \alpha_k p_k$, where α_k is computed by a backtracking line search, starting from the unit steplength, and terminating when

$$\phi_1(x_k + \alpha_k p_k; \mu) \le \phi_1(x_k; \mu) - \eta \alpha_k (q_\mu(0) - q_\mu(p_k)),$$

where $\eta \in (0, 1)$, with ϕ_1 defined as in (18.51) and q_μ defined as in (18.49). To establish the convergence result, we assume that each quadratic program (18.11) is feasible and determines a bounded solution p_k. We also assume that the penalty parameter μ is fixed for all k and sufficiently large.

Theorem 18.3.

Suppose that the SQP algorithm just described is applied to the nonlinear program (18.10). Suppose that the sequences $\{x_k\}$ and $\{x_k + p_k\}$ are contained in a closed, bounded, convex region of $\mathbb{R}^n$ in which f and c_i have continuous first derivatives. Suppose that the matrices B_k and multipliers are bounded and that μ satisfies $\mu \geq \|\lambda_k\|_\infty + \rho$ for all k, where ρ is a positive constant. Then all limit points of the sequence $\{x_k\}$ are KKT points of the nonlinear program (18.10).

The conclusions of the theorem are quite satisfactory, but the assumptions are somewhat restrictive. For example, the condition that the sequence $\{x_k + p_k\}$ stays within in a bounded set rules out the case in which the Hessians B_k or constraint Jacobians become ill conditioned. Global convergence results that are established under more realistic conditions are surveyed by Conn, Gould, and Toint [74]. An example of a result of this type is Theorem 19.2. Although this theorem is established for a nonlinear interior-point method, similar results can be established for trust-region SQP methods.

RATE OF CONVERGENCE

We now derive conditions that guarantee the local convergence of SQP methods, as well as conditions that ensure a superlinear rate of convergence. For simplicity, we limit our discussion to Algorithm 18.1 for equality-constrained optimization, and consider both exact Hessian and quasi-Newton versions. The results presented here can be applied to algorithms for inequality-constrained problems once the active set has settled at its final optimal value (see Theorem 18.1).

We begin by listing a set of assumptions on the problem that will be useful in this section.

Assumptions 18.2.

The point x^* is a local solution of problem (18.1) at which the following conditions hold.

(a) The functions f and c are twice differentiable in a neighborhood of x^* with Lipschitz continuous second derivatives.

(b) The linear independence constraint qualification (Definition 12.4) holds at x^*. This condition implies that the KKT conditions (12.34) are satisfied for some vector of multipliers λ^*.

(c) The second-order sufficient conditions (Theorem 12.6) hold at (x^*, λ^*).

We consider first an SQP method that uses exact second derivatives.

Theorem 18.4.

Suppose that Assumptions 18.2 hold. Then, if (x_0, λ_0) is sufficiently close to (x^*, λ^*), the pairs (x_k, λ_k) generated by Algorithm 18.1 converge quadratically to (x^*, λ^*).

The proof follows directly from Theorem 11.2, since we know that Algorithm 18.1 is equivalent to Newton's method applied to the nonlinear system $F(x, \lambda) = 0$, where F is defined by (18.3).

We turn now to quasi-Newton variants of Algorithm 18.1, in which the Lagrangian Hessian $\nabla^2_{xx} \mathcal{L}(x_k, \lambda_k)$ is replaced by a quasi-Newton approximation B_k. We discussed in Section 18.3 algorithms that used approximations to the full Hessian, and also reduced-Hessian methods that maintained approximations to the projected Hessian $Z_k^T \nabla^2_{xx} \mathcal{L}(x_k, \lambda_k) Z_k$. As in the earlier discussion, we take Z_k to be the $n \times (n - m)$ matrix whose columns span the null space of A_k, assuming in addition that the columns of Z_k are orthornormal; see (15.22).

If we multiply the first block row of the KKT system (18.9) by Z_k, we obtain

$$Z_k^T \nabla^2_{xx} \mathcal{L}_k p_k = -Z_k^T \nabla f_k. \tag{18.62}$$

This equation, together with the second block row $A_k p_k = -c_k$ of (18.9), is sufficient to determine fully the value of p_k when x_k and λ_k are not too far from their optimal values. In other words, only the projection of the Hessian $Z_k^T \nabla^2_{xx} \mathcal{L}_k$ is significant; the remainder of $\nabla^2_{xx} \mathcal{L}_k$ (its projection onto the range space of A_k^T) does not play a role in determinining p_k.

By multiplying (18.62) by Z_k, and defining the following matrix P_k, which projects onto the null space of A_k:

$$P_k = I - A_k^T \left[A_k A_k^T \right]^{-1} A_k = Z_k Z_k^T,$$

we can rewrite (18.62) equivalently as follows:

$$P_k \nabla^2_{xx} \mathcal{L}_k p_k = -P_k \nabla f_k.$$

The discussion above, together with Theorem 18.4, suggests that a quasi-Newton method will be locally convergent if the quasi-Newton matrix B_k is chosen so that $P_k B_k$ is a reasonable approximation of $P_k \nabla^2_{xx} \mathcal{L}_k$, and that it will be superlinearly convergent if $P_k B_k$ approximates $P_k \nabla^2_{xx} \mathcal{L}_k$ well. To make the second statement more precise, we present a result that can be viewed as an extension of characterization of superlinear convergence (Theorem 3.6) to the equality-constrained case. In the following discussion, $\nabla^2_{xx} \mathcal{L}_*$ denotes $\nabla^2_{xx} \mathcal{L}(x^*, \lambda^*)$.

Theorem 18.5.

Suppose that Assumptions 18.2 hold and that the iterates x_k generated by Algorithm 18.1 with quasi-Newton approximate Hessians B_k converge to x^*. Then x_k converges superlinearly if and only if the Hessian approximation B_k satisfies

$$\lim_{k \to \infty} \frac{\| P_k (B_k - \nabla^2_{xx} \mathcal{L}_*)(x_{k+1} - x_k) \|}{\| x_{k+1} - x_k \|} = 0. \tag{18.63}$$

We can apply this result to the quasi-Newton updating schemes discussed earlier in this chapter, beginning with the full BFGS approximation based on (18.13). To guarantee that the BFGS approximation is always well defined, we make the (strong) assumption that the Hessian of the Lagrangian is positive definite at the solution.

Theorem 18.6.

Suppose that Assumptions 18.2 hold. Assume also that $\nabla_{xx}^2 \mathcal{L}_*$ and B_0 are symmetric and positive definite. If $\|x_0 - x^*\|$ and $\|B_0 - \nabla_{xx}^2 \mathcal{L}_*\|$ are sufficiently small, the iterates x_k generated by Algorithm 18.1 with BFGS Hessian approximations B_k defined by (18.13) and (18.16) (with $r_k = s_k$) satisfy the limit (18.63). Therefore, the iterates x_k converge superlinearly to x^*.

For the damped BFGS updating strategy given in Procedure 18.2, we can show that the rate of convergence is R-superlinear (not the usual Q-superlinear rate; see the Appendix).

We now consider reduced-Hessian SQP methods that update an approximation M_k to $Z_k^T \nabla_{xx}^2 \mathcal{L}_k Z_k$. From the definition of P_k, we see that $Z_k M_k Z_k^T$ can be considered as an approximation to the *two-sided* projection $P_k \nabla_{xx}^2 \mathcal{L}_k P_k$. Since reduced-Hessian methods do not approximate the one-sided projection $P_k \nabla_{xx}^2 \mathcal{L}_k$, we cannot expect (18.63) to hold. For these methods, we can state a condition for superlinear convergence by writing (18.63) as

$$\lim_{k \to \infty} \left[\frac{P_k (B_k - \nabla_{xx}^2 \mathcal{L}_*) P_k (x_{k+1} - x_k)}{\|x_{k+1} - x_k\|} \right.$$
$$\left. + \frac{P_k (B_k - \nabla_{xx}^2 \mathcal{L}_*)(I - P_k)(x_{k+1} - x_k)}{\|x_{k+1} - x_k\|} \right] = 0, \tag{18.64}$$

and defining $B_k = Z_k M_k Z_k^T$. The following result shows that it is necessary only for the first term in (18.64) to go to zero to obtain a weaker form of superlinear convergence, namely, two-step superlinear convergence.

Theorem 18.7.

Suppose that Assumption 18.2(a) holds and that the matrices B_k are bounded. Assume also that the iterates x_k generated by Algorithm 18.1 with approximate Hessians B_k converge to x^*, and that

$$\lim_{k \to \infty} \frac{\|P_k (B_k - \nabla_{xx}^2 \mathcal{L}_*) P_k (x_{k+1} - x_k)\|}{\|x_{k+1} - x_k\|} = 0. \tag{18.65}$$

Then the sequence $\{x_k\}$ converges to x^* two-step superlinearly, that is,

$$\lim_{k \to \infty} \frac{\|x_{k+2} - x^*\|}{\|x_k - x^*\|} = 0.$$

In a reduced-Hessian method that uses BFGS updating, the iteration is $x_{k+1} = x_k + Y_k p_Y + Z_k p_Z$, where p_Y and p_Z are given by (18.19a), (18.23) (with $(Z_k^T \nabla_{xx}^2 \mathcal{L}_k Z_k)$ replaced by M_k). The reduced-Hessian approximation M_k is updated by the BFGS formula using the correction vectors (18.26), and the initial approximation M_0 is symmetric and positive definite. If we make the assumption that the null space bases Z_k used to define the correction vectors (18.26) vary smoothly, then we can apply Theorem 18.7 to show that x_k converges two-step superlinearly.

18.8   PERSPECTIVES AND SOFTWARE

SQP methods are most efficient if the number of active constraints is nearly as large as the number of variables, that is, if the number of free variables is relatively small. They require few evaluations of the functions, in comparison with augmented Lagrangian methods, and can be more robust on badly scaled problems than the nonlinear interior-point methods described in the next chapter. It is not known at present whether the IQP or EQP approach will prove to be more effective for large problems. Current reasearch focuses on widening the class of problems that can be solved with SQP and SLQP approaches.

Two established SQP software packages are SNOPT [128] and FILTERSQP [105]. The former code follows a line search approach, while the latter implements a trust-region strategy using a filter for step acceptance. The SLQP approach of Section 18.5 is implemented in KNITRO/ACTIVE [49]. All three packages include mechanisms to ensure that the subproblems are always feasible and to guard against rank-deficient constraint Jacobians. SNOPT uses the penalty (or elastic) mode (18.12), which is invoked if the SQP subproblem is infeasible or if the Lagrange multiplier estimates become very large in norm. FILTERSQP includes a feasibility restoration phase that, in addition to promoting convergence, provides rapid identification of convergence to infeasible points. KNITRO/ACTIVE implements a penalty method using the update strategy of Algorithm 18.5.

There is no established implementation of the $S\ell_1 QP$ approach, but prototype implementations have shown promise. The CONOPT [9] package implements a generalized reduced gradient method as well as an SQP method.

Quasi-Newton approximations to the Hessian of the Lagrangian $\nabla_{xx}^2 \mathcal{L}_k$ are often used in practice. BFGS updating is generally less effective for constrained problems than in the unconstrained case because of the requirement of maintaining a positive definite approximation to an underlying matrix that often does not have this property. Nevertheless, the BFGS and limited-memory BFGS approximations implemented in SNOPT and KNITRO perform adequately in practice. KNITRO also offers an SR1 option that may be more effective than the BFGS option, but the question of how best to implement full quasi-Newton approximations for constrained optimization requires further investigation. The RSQP package [13] implements an SQP method that maintains a quasi-Newton approximation to the reduced Hessian.

The Maratos effect, if left unattended, can significantly slow optimization algorithms that use nonsmooth merit functions or filters. However, selective application of second-order correction steps adequately resolves the difficulties in practice.

Trust-region implementations of the gradient projection method include TRON [192] and LANCELOT [72]. Both codes use a conjugate gradient iteration to perform the subspace minimization and apply an incomplete Cholesky preconditioner. Gradient projection methods in which the Hessian approximation is defined by limited-memory BFGS updating are implemented in LBFGS-B [322] and BLMVM [17]. The properties of limited-memory BFGS matrices can be exploited to perform the projected gradient search and subspace minimization efficiently. SPG [23] implements the gradient projection method using a nonmonotone line search.

NOTES AND REFERENCES

SQP methods were first proposed in 1963 by Wilson [306] and were developed in the 1970s by Garcia-Palomares and Mangasarian [117], Han [163, 164], and Powell [247, 250, 249], among others. Trust-region variants are studied by Vardi [295], Celis, Dennis, and Tapia [56], and Byrd, Schnabel, and Shultz [55]. See Boggs and Tolle [33] and Gould, Orban, and Toint [147] for literature surveys.

The SLQP approach was proposed by Fletcher and Sainz de la Maza [108] and was further developed by Chin and Fletcher [59] and Byrd et al. [49]. The latter paper discusses how to update the LP trust region and many other details of implementation. The technique for updating the penalty parameter implemented in Algorithm 18.5 is discussed in [49, 47]. The $S\ell_1 QP$ method was proposed by Fletcher; see [101] for a complete discussion of this method.

Some analysis shows that several—but not all—of the good properties of BFGS updating are preserved by damped BFGS updating. Numerical experiments exposing the weakness of the approach are reported by Powell [254]. Second-order correction strategies were proposed by Coleman and Conn [65], Fletcher [100], Gabay [116], and Mayne and Polak [204]. The watchdog technique was proposed by Chamberlain et al. [57] and other nonmonotone strategies are described by Bonnans et al. [36]. For a comprehensive discussion of second-order correction and nonmonotone techniques, see the book by Conn, Gould, and Toint [74].

Two filter SQP algorithms are described by Fletcher and Leyffer [105] and Fletcher, Leyffer, and Toint [106]. It is not yet known whether the filter strategy has advantages over merit functions. Both approaches are undergoing development and improved implementations can be expected in the future. Theorem 18.3 is proved by Powell [252] and Theorem 18.5 by Boggs, Tolle, and Wang [34].

✐ EXERCISES

✐   **18.1** Show that in the quadratic program (18.7) we can replace the linear term $\nabla f_k^T p$ by $\nabla_x \mathcal{L}(x_k, \lambda_k)^T p$ without changing the solution.

✐ **18.2** Prove Theorem 18.4.

✐ **18.3** Write a program that implements Algorithm 18.1. Use it to solve the problem

$$\min e^{x_1 x_2 x_3 x_4 x_5} - \tfrac{1}{2}(x_1^3 + x_2^3 + 1)^2 \qquad (18.66)$$
$$\text{subject to} \quad x_1^2 + x_2^2 + x_3^2 + x_4^2 + x_5^2 - 10 = 0, \qquad (18.67)$$
$$x_2 x_3 - 5 x_4 x_5 = 0, \qquad (18.68)$$
$$x_1^3 + x_2^3 + 1 = 0. \qquad (18.69)$$

Use the starting point $x_0 = (-1.71, 1.59, 1.82, -0.763, -0.763)^T$. The solution is $x^* = (-1.8, 1.7, 1.9, -0.8, -0.8)^T$.

✐ **18.4** Show that the damped BFGS updating satisfies (18.17).

✐ **18.5** Consider the constraint $x_1^2 + x_2^2 = 1$. Write the linearized constraints (18.7b) at the following points: $(0, 0)^T$, $(0, 1)^T$, $(0.1, 0.02)^T$, $-(0.1, 0.02)^T$.

✐ **18.6** Prove Theorem 18.2 for the case in which the merit function is given by $\phi(x; \mu) = f(x) + \mu\|c(x)\|_q$, where $q > 0$. Use this lemma to show that the condition that ensures descent is given by $\mu > \|\lambda_{k+1}\|_r$, where $r > 0$ satisfies $r^{-1} + q^{-1} = 1$.

✐ **18.7** Write a program that implements the reduced-Hessian method given by (18.18), (18.19a), (18.21), (18.23). Use your program to solve the problem given in Exercise 18.3.

✐ **18.8** Show that the constraints (18.50b)–(18.50e) are always consistent.

✐ **18.9** Show that the feasibility problem (18.45a)–(18.45b) always has a solution v_k lying in the range space of A_k^T. Hint: First show that if the trust-region constraint (18.45b) is active, v_k lies in the range space of A_k^T. Next, show that if the trust region is inactive, the minimum-norm solution of (18.45a) lies in the range space of A_k^T.

Interior-Point Methods for Nonlinear Programming

Interior-point (or barrier) methods have proved to be as successful for nonlinear optimization as for linear programming, and together with active-set SQP methods, they are currently considered the most powerful algorithms for large-scale nonlinear programming. Some of the key ideas, such as primal-dual steps, carry over directly from the linear programming case, but several important new challenges arise. These include the treatment of nonconvexity, the strategy for updating the barrier parameter in the presence of nonlinearities, and the need to ensure progress toward the solution. In this chapter we describe two classes of interior-point methods that have proved effective in practice.

The methods in the first class can be viewed as direct extensions of interior-point methods for linear and quadratic programming. They use line searches to enforce convergence and employ direct linear algebra (that is, matrix factorizations) to compute steps. The methods in the second class use a quadratic model to define the step and incorporate a trust-region constraint to provide stability. These two approaches, which coincide asymptotically, have similarities with line search and trust-region SQP methods.

Barrier methods for nonlinear optimization were developed in the 1960s but fell out of favor for almost two decades. The success of interior-point methods for linear programming stimulated renewed interest in them for the nonlinear case. By the late 1990s, a new generation of methods and software for nonlinear programming had emerged. Numerical experience indicates that interior-point methods are often faster than active-set SQP methods on large problems, particularly when the number of free variables is large. They may not yet be as robust, but significant advances are still being made in their design and implementation. The terms "interior-point methods" and "barrier methods" are now used interchangeably.

In Chapters 14 and 16 we discussed interior-point methods for linear and quadratic programming. It is not essential that the reader study those chapters before reading this one, although doing so will give a better perspective. The first part of this chapter assumes familiarity primarily with the KKT conditions and Newton's method, and the second part of the chapter relies on concepts from sequential quadratic programming presented in Chapter 18.

The problem under consideration in this chapter is written as follows:

$$\min_{x,s} \; f(x) \tag{19.1a}$$

$$\text{subject to} \qquad c_E(x) = 0, \tag{19.1b}$$

$$c_I(x) - s = 0, \tag{19.1c}$$

$$s \geq 0. \tag{19.1d}$$

The vector $c_I(x)$ is formed from the scalar functions $c_i(x)$, $i \in \mathcal{I}$, and similarly for $c_E(x)$. Note that we have transformed the inequalities $c_i(x) \geq 0$ into equalities by the introduction of a vector s of slack variables. We use l to denote the number of equality constraints (that is, the dimension of the vector c_E) and m to denote the number of inequality constraints (the dimension of c_I).

19.1 TWO INTERPRETATIONS

Interior-point methods can be seen as continuation methods or as barrier methods. We discuss both derivations, starting with the continuation approach.

The KKT conditions (12.1) for the nonlinear program (19.1) can be written as

$$\nabla f(x) - A_E^T(x)y - A_I^T(x)z = 0, \tag{19.2a}$$

$$Sz - \mu e = 0, \tag{19.2b}$$

$$c_E(x) = 0, \qquad (19.2c)$$
$$c_I(x) - s = 0, \qquad (19.2d)$$

with $\mu = 0$, together with

$$s \geq 0, \qquad z \geq 0. \qquad (19.3)$$

Here $A_E(x)$ and $A_I(x)$ are the Jacobian matrices of the functions c_E and c_I, respectively, and y and z are their Lagrange multipliers. We define S and Z to be the diagonal matrices whose diagonal entries are given by the vectors s and z, respectively, and let $e = (1, 1, \ldots, 1)^T$.

Equation (19.2b), with $\mu = 0$, and the bounds (19.3) introduce into the problem the combinatorial aspect of determining the optimal active set, illustrated in Example 15.1. We circumvent this difficulty by letting μ be strictly positive, thus forcing the variables s and z to take positive values. The homotopy (or continuation) approach consists of (approximately) solving the *perturbed KKT conditions* (19.2) for a sequence of positive parameters $\{\mu_k\}$ that converges to zero, while maintaining $s, z > 0$. The hope is that, in the limit, we will obtain a point that satisfies the KKT conditions for the nonlinear program (19.1). Furthermore, by requiring the iterates to decrease a merit function (or to be acceptable to a filter), the iteration is likely to converge to a minimizer, not simply a KKT point.

The homotopy approach is justified locally. In a neighborhood of a solution (x^*, s^*, y^*, z^*) that satisfies the linear independence constraint qualification (LICQ) (Definition 12.4), the strict complementarity condition (Definition 12.5), and the second-order sufficient conditions (Theorem 12.6), we have that for all sufficiently small positive values of μ, the system (19.2) has a locally unique solution, which we denote by $(x(\mu), s(\mu), y(\mu), z(\mu))$. The trajectory described by these points is called the *primal-dual central path*, and it converges to (x^*, s^*, y^*, z^*) as $\mu \to 0$.

The second derivation of interior-point methods associates with (19.1) the barrier problem

$$\min_{x,s} \ f(x) - \mu \sum_{i=1}^{m} \log s_i \qquad (19.4a)$$
$$\text{subject to} \qquad c_E(x) = 0, \qquad (19.4b)$$
$$c_I(x) - s = 0, \qquad (19.4c)$$

where μ is a positive parameter and $\log(\cdot)$ denotes the natural logarithm function. One need not include the inequality $s \geq 0$ in (19.4) because minimization of the barrier term $-\mu \sum_{i=1}^{m} \log s_i$ in (19.4a) prevents the components of s from becoming too close to zero. (Recall that $(-\log t) \to \infty$ as $t \downarrow 0$.) Problem (19.4) also avoids the combinatorial aspect of nonlinear programs, but its solution does not coincide with that of (19.1) for $\mu > 0$. The barrier approach consists of finding (approximate) solutions of the barrier problem (19.4) for a sequence of positive barrier parameters $\{\mu_k\}$ that converges to zero.

To compare the homotopy and barrier approaches, we write the KKT conditions for (19.4) as follows:

$$\nabla f(x) - A_{\scriptscriptstyle E}^{\,T}(x)y - A_{\scriptscriptstyle I}^{\,T}(x)z = 0, \tag{19.5a}$$

$$-\mu S^{-1}e + z = 0, \tag{19.5b}$$

$$c_{\scriptscriptstyle E}(x) = 0, \tag{19.5c}$$

$$c_{\scriptscriptstyle I}(x) - s = 0. \tag{19.5d}$$

Note that they differ from (19.2) only in the second equation, which becomes quite nonlinear near the solution as $s \to 0$. It is advantageous for Newton's method to transform the rational equation (19.5b) into a quadratic equation. We do so by multiplying this equation by S, a procedure that does not change the solution of (19.5) because the diagonal elements of S are positive. After this transformation, the KKT conditions for the barrier problem coincide with the perturbed KKT system (19.2).

The term "interior point" derives from the fact that early barrier methods [98] did not use slacks and assumed that the initial point x_0 is feasible with respect to the inequality constraints $c_i(x) \geq 0$, $i \in \mathcal{I}$. These methods used the barrier function

$$f(x) - \mu \sum_{i \in \mathcal{I}} \log c_i(x)$$

to prevent the iterates from leaving the feasible region defined by the inequalities. (We discuss this barrier function further in Section 19.6.) Most modern interior-point methods are infeasible (they can start from any initial point x_0) and remain interior only with respect to the constraints $s \geq 0, z \geq 0$. However, they can be designed so that once they generate a feasible iterate, all subsequent iterates remain feasible with respect to the inequalities.

In the next sections we will see that the homotopy and barrier interpretations are both useful. The homotopy view gives rise to the definition of the primal-dual direction, whereas the barrier view is vital in the design of globally convergent iterations.

19.2   A BASIC INTERIOR-POINT ALGORITHM

Applying Newton's method to the nonlinear system (19.2), in the variables x, s, y, z, we obtain

$$\begin{bmatrix} \nabla_{xx}^2 \mathcal{L} & 0 & -A_{\scriptscriptstyle E}^{\,T}(x) & -A_{\scriptscriptstyle I}^{\,T}(x) \\ 0 & Z & 0 & S \\ A_{\scriptscriptstyle E}(x) & 0 & 0 & 0 \\ A_{\scriptscriptstyle I}(x) & -I & 0 & 0 \end{bmatrix} \begin{bmatrix} p_x \\ p_s \\ p_y \\ p_z \end{bmatrix} = - \begin{bmatrix} \nabla f(x) - A_{\scriptscriptstyle E}^{\,T}(x)y - A_{\scriptscriptstyle I}^{\,T}(x)z \\ Sz - \mu e \\ c_{\scriptscriptstyle E}(x) \\ c_{\scriptscriptstyle I}(x) - s \end{bmatrix}, \tag{19.6}$$

where $\mathcal{L}$ denotes the Lagrangian for (19.1a)–(19.1c):

$$\mathcal{L}(x, s, y, z) = f(x) - y^T c_E(x) - z^T (c_I(x) - s).$$ (19.7)

The system (19.6) is called the *primal-dual system* (in contrast with the primal system discussed in Section 19.3). After the step $p = (p_x, p_s, p_y, p_z)$ has been determined, we compute the new iterate (x^+, s^+, y^+, z^+) as

$$x^+ = x + \alpha_s^{\max} p_x, \quad s^+ = s + \alpha_s^{\max} p_s,$$ (19.8a)
$$y^+ = y + \alpha_z^{\max} p_y, \quad z^+ = z + \alpha_z^{\max} p_z,$$ (19.8b)

where

$$\alpha_s^{\max} = \max\{\alpha \in (0, 1] : s + \alpha p_s \ge (1 - \tau)s\},$$ (19.9a)
$$\alpha_z^{\max} = \max\{\alpha \in (0, 1] : z + \alpha p_z \ge (1 - \tau)z\},$$ (19.9b)

with $\tau \in (0, 1)$. (A typical value of τ is 0.995.) The condition (19.9), called the *fraction to the boundary rule*, prevents the variables s and z from approaching their lower bounds of 0 too quickly.

This simple iteration provides the basis of modern interior-point methods, though various modifications are needed to cope with nonconvexities and nonlinearities. The other major ingredient is the procedure for choosing the sequence of parameters $\{\mu_k\}$, which from now on we will call the *barrier parameters*. In the approach studied by Fiacco and McCormick [98], the barrier parameter μ is held fixed for a series of iterations until the KKT conditions (19.2) are satisfied to some accuracy. An alternative approach is to update the barrier parameter at each iteration. Both approaches have their merits and are discussed in Section 19.3.

The primal-dual matrix in (19.6) remains nonsingular as the iteration converges to a solution that satisfies the second-order sufficiency conditions and strict complementarity. More specifically, if x^* is a solution point for which strict complementarity holds, then for every index i either s_i or z_i remains bounded away from zero as the iterates approach x^*, ensuring that the second block row of the primal-dual matrix (19.6) has full row rank. Therefore, the interior-point approach does not, in itself, give rise to ill conditioning or singularity. This fact allows us to establish a fast (superlinear) rate of convergence; see Section 19.8.

We summarize the discussion by describing a concrete implementation of this basic interior-point method. We use the following error function, which is based on the perturbed KKT system (19.2):

$$E(x, s, y, z; \mu) = \max \left\{ \| \nabla f(x) - A_E(x)^T y - A_I(x)^T z \|, \| Sz - \mu e \|, \right.$$
$$\left. \| c_E(x) \|, \| c_I(x) - s \| \right\},$$ (19.10)

for some vector norm $\| \cdot \|$.

Algorithm 19.1 (Basic Interior-Point Algorithm).
Choose x_0 and $s_0 > 0$, and compute initial values for the multipliers y_0 and $z_0 > 0$.
Select an initial barrier parameter $\mu_0 > 0$ and parameters $\sigma, \tau \in (0, 1)$. Set $k \leftarrow 0$.

> **repeat** until a stopping test for the nonlinear program (19.1) is satisfied
>> **repeat** until $E(x_k, s_k, y_k, z_k; \mu_k) \leq \mu_k$
>>> Solve (19.6) to obtain the search direction $p = (p_x, p_s, p_y, p_z)$;
>>> Compute $\alpha_s^{\max}, \alpha_z^{\max}$ using (19.9);
>>> Compute $(x_{k+1}, s_{k+1}, y_{k+1}, z_{k+1})$ using (19.8);
>>> Set $\mu_{k+1} \leftarrow \mu_k$ and $k \leftarrow k + 1$;
>> **end**
>> Choose $\mu_k \in (0, \sigma\mu_k)$;
> **end**

An algorithm that updates the barrier parameter μ_k at every iteration is easily obtained from Algorithm 19.1 by removing the requirement that the KKT conditions be satisfied for each μ_k (the inner "repeat" loop) and by using a dynamic rule for updating μ_k in the penultimate line.

The following theorem provides a theoretical foundation for interior-point methods that compute only approximate solutions of the barrier problem.

Theorem 19.1.

Suppose that Algorithm 19.1 generates an infinite sequence of iterates $\{x_k\}$ and that $\{\mu_k\} \to 0$ (that is, that the algorithm does not loop infinitely in the inner "repeat" statement). Suppose that f and c are continuously differentiable functions. Then all limit points $\hat{x}$ of $\{x_k\}$ are feasible. Furthermore, if any limit point $\hat{x}$ of $\{x_k\}$ satisfies the linear independence constraint qualification (LICQ), then the first-order optimality conditions of the problem (19.1) hold at $\hat{x}$.

PROOF. For simplicity, we prove the result for the case in which the nonlinear program (19.1) contains only inequality constraints, leaving the extension of the result as an exercise. For ease of notation, we denote the inequality constraints c_i by c. Let $\hat{x}$ be a limit point of the sequence $\{x_k\}$, and let $\{x_{k_i}\}$ be a convergent subsequence, namely, $\{x_{k_i}\} \to \hat{x}$. Since $\mu_k \to 0$, the error E given by (19.10) converges to zero, so we have $(c_{k_i} - s_{k_i}) \to 0$. By continuity of c, this fact implies that $\hat{c} \stackrel{\text{def}}{=} c(\hat{x}) \geq 0$ (that is, $\hat{x}$ is feasible) and $s_{k_i} \to \hat{s} = \hat{c}$.

Now suppose that the linear independence constraint qualification holds at $\hat{x}$, and consider the set of active indices

$$\mathcal{A} = \{i : \hat{c}_i = 0\}.$$

For $i \notin \mathcal{A}$, we have $\hat{c}_i > 0$ and $\hat{s}_i > 0$, and thus by the complementarity condition (19.2b), we have that $[z_{k_l}]_i \to 0$. From this fact and $\nabla f_{k_l} - A_{k_l}^T z_{k_l} \to 0$, we deduce that

$$\nabla f_{k_l} - \sum_{i \in \mathcal{A}} [z_{k_l}]_i \nabla c_i(x_{k_l}) \to 0. \tag{19.11}$$

By the constraint qualification hypothesis, the vectors $\{\nabla \hat{c}_i : i \in \mathcal{A}\}$ are linearly independent. Hence, by (19.11) and continuity of $\nabla f(\cdot)$ and $\nabla c_{(i)}(\cdot)$, $i \in \mathcal{A}$, the positive sequence $\{z_{k_l}\}$ converges to some value $\hat{z} \geq 0$. Taking the limit in (19.11), we have that

$$\nabla f(\hat{x}) = \sum_{i \in \mathcal{A}} \hat{z}_i \nabla c_i(\hat{x}).$$

We also have that $\hat{c}^T \hat{z} = 0$, completing the proof. □

Practical interior-point algorithms fall into two categories. The first builds on Algorithm 19.1, adding a line search and features to control the rate of decrease in the slacks s and multipliers z, and introducing modifications in the primal-dual sytem when negative curvature is encountered. The second category of algorithms, presented in Section 19.5, computes steps by minimizing a quadratic model of (19.4), subject to a trust-region constraint. The two approaches share many features described in the next section.

19.3 ALGORITHMIC DEVELOPMENT

We now discuss a series of modifications and extensions of Algorithm 19.1 that enable it to solve nonconvex nonlinear problems, starting from any initial estimate.

Often, the primal-dual system (19.6) is rewritten in the symmetric form

$$\begin{bmatrix} \nabla_{xx}^2 \mathcal{L} & 0 & A_E^T(x) & A_I^T(x) \\ 0 & \Sigma & 0 & -I \\ A_E(x) & 0 & 0 & 0 \\ A_I(x) & -I & 0 & 0 \end{bmatrix} \begin{bmatrix} p_x \\ p_s \\ -p_y \\ -p_z \end{bmatrix} = - \begin{bmatrix} \nabla f(x) - A_E^T(x)y - A_I^T(x)z \\ z - \mu S^{-1} e \\ c_E(x) \\ c_I(x) - s \end{bmatrix}, \tag{19.12}$$

where

$$\Sigma = S^{-1} Z. \tag{19.13}$$

This formulation permits the use of a symmetric linear equations solver, which reduces the computational work of each iteration.

PRIMAL VS. PRIMAL-DUAL SYSTEM

If we apply Newton's method directly to the optimality conditions (19.5) of the barrier problem (instead of transforming to (19.5b) first) and then symmetrize the iteration matrix, we obtain the system (19.12) but with Σ given by

$$\Sigma = \mu S^{-2}. \tag{19.14}$$

This is often called the *primal* system, in contrast with the *primal-dual* system arising from (19.13). (This nomenclature owes more to the historical development of interior-point methods than to the concept of primal-dual iterations.) Whereas in the primal-dual choice (19.13) the vector z can be seen as a general multiplier estimate, the primal term (19.14) is obtained by making the specific selection $Z = \mu S^{-1}$; we return to this choice of multipliers in Section 19.6.

Even though the systems (19.2) and (19.5) are equivalent, Newton's method applied to them will generally produce different iterates, and there are reasons for preferring the primal-dual system. Note that (19.2b) has the advantage that its derivatives are bounded as any slack variables approach zero; such is not the case with (19.5b). Moreover, analysis of the primal step as well as computational experience has shown that, under some circumstances, the primal step (19.12), (19.14) tends to produce poor steps that violate the bounds $s > 0$ and $z > 0$ significantly, resulting in slow progress; see Section 19.6.

SOLVING THE PRIMAL-DUAL SYSTEM

Apart from the cost of evaluating the problem functions and their derivatives, the work of the interior-point iteration is dominated by the solution of the primal-dual system (19.12), (19.13). An efficient linear solver, using either sparse factorization or iterative techniques, is therefore essential for fast solution of large problems.

The symmetric matrix in (19.12) has the familiar form of a KKT matrix (cf. (16.7), (18.6)), and the linear system can be solved by the approaches described in Chapter 16. We can first reduce the system by eliminating p_s using the second equation in (19.6), giving

$$\begin{bmatrix} \nabla_{xx}^2 \mathcal{L} & A_E^T(x) & A_I^T(x) \\ A_E(x) & 0 & 0 \\ A_I(x) & 0 & -\Sigma^{-1} \end{bmatrix} \begin{bmatrix} p_x \\ -p_y \\ -p_z \end{bmatrix} = - \begin{bmatrix} \nabla f(x) - A_E^T(x)y - A_I^T(x)z \\ c_E(x) \\ c_I(x) - \mu Z^{-1} e \end{bmatrix}. \tag{19.15}$$

This system can be factored by using a symmetric indefinite factorization; see (16.12). If we denote the coefficient matrix in (19.15) by K, this factorization computes $P^T K P = L B L^T$, where L is lower triangular and B is block diagonal, with blocks of size 1×1 or 2×2. P is a matrix of row and column permutations that seeks a compromise between the goals of preserving sparsity and ensuring numerical stability; see (3.51) and the discussion that follows.

The system (19.15) can be reduced further by eliminating p_z using the last equation, to obtain the condensed coefficient matrix

$$
\begin{bmatrix} \nabla_{xx}^2 \mathcal{L} + A_{\scriptscriptstyle I}^T \Sigma A_{\scriptscriptstyle I} & A_{\scriptscriptstyle E}^T(x) \\ A_{\scriptscriptstyle E}(x) & 0 \end{bmatrix}, \tag{19.16}
$$

which is much smaller than (19.12) when the number of inequality constraints is large. Although significant fill-in can arise from the term $A_{\scriptscriptstyle I}^T \Sigma A_{\scriptscriptstyle I}$, it is tolerable in many applications. A particularly favorable case, in which $A_{\scriptscriptstyle I}^T \Sigma A_{\scriptscriptstyle I}$ is diagonal, arises when the inequality constraints are simple bounds.

The primal-dual system in any of the symmetric forms (19.12), (19.15), (19.16) is ill conditioned because, by (19.13), some of the elements of Σ diverge to ∞, while others converge to zero as $\mu \rightarrow 0$. Nevertheless, because of the special form in which this ill conditioning arises, the direction computed by a stable direct factorization method is usually accurate. Damaging errors result only when the slacks s or multipliers z become very close to zero (or when the Hessian $\nabla_{xx}^2 \mathcal{L}$ or the Jacobian matrix $A_{\scriptscriptstyle E}$ is almost rank deficient). For this reason, direct factorization techniques are considered the most reliable techniques for computing steps in interior-point methods.

Iterative linear algebra techniques can also be used for the step computation. Ill conditioning is a grave concern in this context, and preconditioners that cluster the eigenvalues of Σ must be used. Fortunately, such preconditioners are easy to construct. For example, let us introduce the change of variables $\tilde{p}_s = S^{-1} p_s$ in the system (19.12), and multiply the second equation in (19.12) by S, transforming the term Σ into $S \Sigma S$. As $\mu \rightarrow 0$ (and assuming that $SZ \approx \mu I$) we have from (19.13) that all the elements of $S \Sigma S$ cluster around μI. Other scalings can be used as well. The change of variables $\tilde{p}_s = \Sigma^{1/2} p_s$ provides the perfect preconditioner, while $\tilde{p}_s = \sqrt{\mu} S^{-1} p_s$ transforms Σ to $S \Sigma S / \mu$, which converges to I as $\mu \rightarrow 0$.

We can apply an iterative method to one of the symmetric indefinite systems (19.12), (19.15), or (19.16). The conjugate gradient method is not appropriate (except as explained below) because it is designed for positive definite systems, but we can use GMRES, QMR, or LSQR (see [136]). In addition to employing preconditioning that removes the ill conditioning caused by the barrier approach, as discussed above, we need to deal with possible ill conditioning caused by the Hessian $\nabla_{xx}^2 \mathcal{L}$ or the Jacobian matrices $A_{\scriptscriptstyle E}$ and $A_{\scriptscriptstyle I}$. General-purpose preconditioners are difficult to find in this context, and the success of an iterative method hinges on the use of problem-specific or structured preconditioners.

An effective alternative is to use a null-space approach to solve the primal-dual system and apply the CG method in the (positive definite) reduced space. As explained in Section 16.3, we can do this by applying the *projected CG iteration* of Algorithm 16.2 using a so-called constraint preconditioner. In the context of the system (19.12) the preconditioner has the form

$$
\begin{bmatrix}
G & 0 & A_{\mathrm{E}}^T(x) & A_{\mathrm{I}}^T(x) \\
0 & T & 0 & -I \\
A_{\mathrm{E}}(x) & 0 & 0 & 0 \\
A_{\mathrm{I}}(x) & -I & 0 & 0
\end{bmatrix}, \tag{19.17}
$$

where G is a sparse matrix that is positive definite on the null space of the constraints and T is a diagonal matrix that equals or approximates Σ. This preconditioner keeps the Jacobian information of A_{E} and A_{I} intact and thereby removes any ill conditioning present in these matrices.

UPDATING THE BARRIER PARAMETER

The sequence of barrier parameters $\{\mu_k\}$ must converge to zero so that, in the limit, we recover the solution of the nonlinear programming problem (19.1). If μ_k is decreased too slowly, a large number of iterations will be required for convergence; but if it is decreased too quickly, some of the slacks s or multipliers z may approach zero prematurely, slowing progress of the iteration. We now describe several techniques for updating μ_k that have proved to be effective in practice.

The strategy implemented in Algorithm 19.1, which we call the *Fiacco–McCormick approach*, fixes the barrier parameter until the perturbed KKT conditions (19.2) are satisfied to some accuracy. Then the barrier parameter is decreased by the rule

$$
\mu_{k+1} = \sigma_k \mu_k, \quad \text{with} \quad \sigma_k \in (0, 1). \tag{19.18}
$$

Some early implementations of interior-point methods chose σ_k to be a constant (for example, $\sigma_k = 0.2$). It is, however, preferable to let σ_k take on two or more values (for example, 0.2 and 0.1), choosing smaller values when the most recent iterations make significant progress toward the solution. Furthermore, by letting $\sigma_k \to 0$ near the solution, and letting the parameter τ in (19.9) converge to 1, a superlinear rate of convergence can be obtained.

The Fiacco–McCormick approach works well on many problems, but it can be sensitive to the choice of the initial point, the initial barrier parameter value, and the scaling of the problem.

Adaptive strategies for updating the barrier parameter are more robust in difficult situations. These strategies, unlike the Fiacco–McCormick approach, vary μ at every iteration depending on the progress of the algorithm. Most such strategies are based on complementarity, as in the linear programming case (see Framework 14.1), and have the form

$$
\mu_{k+1} = \sigma_k \frac{s_k^T z_k}{m}, \tag{19.19}
$$

which allows μ_k to reflect the scale of the problem. One choice of σ_k, implemented in the LOQO package [294], is based on the deviation of the smallest complementarity product $[s_k]_i [z_k]_i$ from the average:

$$\sigma_k = 0.1 \min \left(0.05 \frac{1 - \xi_k}{\xi_k}, 2 \right)^3, \quad \text{where} \quad \xi_k = \frac{\min_i [s_k]_i [z_k]_i}{(s^k)^T z^k / m}. \qquad (19.20)$$

Here $[s_k]_i$ denotes the ith component of the iterate s_k, and similarly for $[z_k]_i$. When $\xi_k \approx 1$ (all the individual products are near to their average), the barrier parameter is decreased aggressively.

Predictor or probing strategies (see Section 14.2) can also be used to determine the parameter σ_k in (19.19). We calculate a predictor (affine scaling) direction

$$(\Delta x^{\text{aff}}, \Delta s^{\text{aff}}, \Delta y^{\text{aff}}, \Delta z^{\text{aff}})$$

by setting $\mu = 0$ in (19.12). We probe this direction by finding α_p^{aff} and α_d^{aff} to be the longest step lengths that can be taken along the affine scaling direction before violating the nonnegativity conditions $(s, z) \geq 0$. Explicit formulas for these step lengths are given by (19.9) with $\tau = 1$. We then define μ_{aff} to be the value of complementarity along the (shortened) affine scaling step, that is,

$$\mu_{\text{aff}} = (s_k + \alpha_s^{\text{aff}} \Delta s^{\text{aff}})^T (z_k + \alpha_z^{\text{aff}} \Delta z^{\text{aff}}) / m, \qquad (19.21)$$

and define σ_k as follows:

$$\sigma_k = \left(\frac{\mu_{\text{aff}}}{s_k^T z_k / m} \right)^3. \qquad (19.22)$$

This heuristic choice of σ_k was proposed for linear programming problems (see (14.34)) and also works well for nonlinear programs.

HANDLING NONCONVEXITY AND SINGULARITY

The direction defined by the primal-dual system (19.12) is not always productive because it seeks to locate only KKT points; it can move toward a maximizer or other stationary points. In Chapter 18 we have seen that the Newton step (18.9) for the equality-constrained problem (18.1) can be guaranteed to be a descent direction for a large class of merit functions—and to be a productive direction for a filter—if the Hessian W is positive definite on the tangent space of the constraints. The reason is that, in this case, the step can be interpreted as the minimization of a convex model in the reduced space obtained by eliminating the linearized constraints.

For the primal-dual system (19.12), the step p is a descent direction if the matrix

$$
\begin{bmatrix}
\nabla_{xx}^2 \mathcal{L} & 0 \\
0 & \Sigma
\end{bmatrix}
\tag{19.23}
$$

is positive definite on the null space of the constraint matrix

$$
\begin{bmatrix}
A_E(x) & 0 \\
A_I(x) & -I
\end{bmatrix}.
$$

Lemma 16.3 states that this positive definiteness condition holds if the inertia of the primal-dual matrix in (19.12) is given by

$$
(n + m, l + m, 0),
\tag{19.24}
$$

in other words, if this matrix has exactly $n + m$ positive, $l + m$ negative, and no zero eigenvalues. (Recall that l and m denote the number of equality and inequality constraints, respectively.) As discussed in Section 3.4, the inertia can be obtained from the symmetric-indefinite factorization of (19.12).

If the primal-dual matrix does not have the desired inertia, we can modify it as follows. Note that the diagonal matrix Σ is positive definite by construction but $\nabla_{xx}^2 \mathcal{L}$ can be indefinite. Therefore, we can replace the latter matrix by $\nabla_{xx}^2 \mathcal{L} + \delta I$, where $\delta > 0$ is sufficiently large to ensure that the inertia is given by (19.24). The size of this modification is not known beforehand, but we can try successively larger values of δ until the desired inertia is obtained.

We must also guard against singularity of the primal-dual matrix caused by the rank deficiency of A_E (the matrix $[A_I \ -I]$ always has full rank). We do so by including a regularization parameter $\gamma \geq 0$, in addition to the modification term δI, and work with the modified primal-dual matrix

$$
\begin{bmatrix}
\nabla_{xx}^2 \mathcal{L} + \delta I & 0 & A_E(x)^T & A_I(x)^T \\
0 & \Sigma & 0 & -I \\
A_E(x) & 0 & -\gamma I & 0 \\
A_I(x) & -I & 0 & 0
\end{bmatrix}.
\tag{19.25}
$$

A procedure for selecting γ and δ is given in Algorithm B.1 in Appendix B. It is invoked at every iteration of the interior-point method to enforce the inertia condition (19.24) and to guarantee nonsingularity. Other matrix modifications to ensure positive definiteness have been discussed in Chapter 3 in the context of unconstrained minimization.

STEP ACCEPTANCE: MERIT FUNCTIONS AND FILTERS

The role of the merit function or filter is to determine whether a step is productive and should be accepted. Since interior-point methods can be seen as methods for solving the barrier problem (19.4), it is appropriate to define the merit function ϕ or filter in terms of barrier functions. We may use, for example, an exact merit function of the form

$$\phi_v(x, s) = f(x) - \mu \sum_{i=1}^{m} \log s_i + v\|c_{\varepsilon}(x)\| + v\|c_{\iota}(x) - s\|, \qquad (19.26)$$

where the norm is chosen, say, to be the ℓ_1 or the ℓ_2 norm (unsquared). The penalty parameter $v > 0$ can be updated by using the strategies described in Chapter 18.

In a line search method, after the step p has been computed and the maximum step lengths (19.9) have been determined, we perform a backtracking line search that computes the step lengths

$$\alpha_s \in (0, \alpha_s^{\max}], \qquad \alpha_z \in (0, \alpha_z^{\max}], \qquad (19.27)$$

providing sufficient decrease of the merit function or ensuring acceptability by the filter. The new iterate is then defined as

$$x^+ = x + \alpha_s p_x, \quad s^+ = s + \alpha_s p_s, \qquad (19.28a)$$
$$y^+ = y + \alpha_z p_y, \quad z^+ = z + \alpha_z p_z. \qquad (19.28b)$$

When defining a filter (see Section 15.4) the pairs of the filter are formed, on the one hand, by the values of the barrier function $f(x) - \mu \sum_{i=1}^{m} \log s_i$ and, on the other hand, by the constraint violations $\|(c_{\varepsilon}(x), c_{\iota}(x) - s)\|$. A step will be accepted if it is not dominated by any element in the filter. Under certain circumstances, if the step is not accepted by the filter, instead of reducing the step length α_s in (19.8a), a feasibility restoration phase is invoked; see the Notes and References at the end of the chapter.

QUASI-NEWTON APPROXIMATIONS

A quasi-Newton version of the primal-dual step is obtained by replacing $\nabla_{xx}^2 \mathcal{L}$ in (19.12) by a quasi-Newton approximation B. We can use the BFGS (6.19) or SR1 (6.24) update formulas described in Chapter 6 to define B, or we can follow a limited-memory BFGS approach (see Chapter 7). It is important to approximate the Hessian of the Lagrangian of the nonlinear program, not the Hessian of the barrier function, which is highly ill conditioned and changes rapidly.

The correction pairs used by the quasi-Newton updating formula are denoted here by $(\Delta x, \Delta l)$, replacing the notation (s, y) of Chapter 6. After computing a step from (x, s, y, z)

to (x^+, s^+, y^+, z^+), we define

$$\Delta l = \nabla_x \mathcal{L}(x^+, s^+, y^+, z^+) - \nabla_x \mathcal{L}(x, s^+, y^+, z^+),$$
$$\Delta x = x^+ - x.$$

To ensure that the BFGS method generates a positive definite matrix, one can skip or damp the update; see (18.14) and (18.15). SR1 updating must be safeguarded to avoid unboundedness, as discussed in Section 6.2, and may also need to be modified so that the inertia of the primal-dual matrix is given by (19.24). This modification can be performed by means of Algorithm B.1.

The quasi-Newton matrices B generated in this manner are dense $n \times n$ matrices. For large problems, limited-memory updating is desirable. One option is to implement a limited-memory BFGS method by using the compact representations described in Section 7.2. Here B has the form

$$B = \xi I + W M W^T, \tag{19.29}$$

where $\xi > 0$ is a scaling factor, W is an $n \times 2\hat{m}$ matrix, M is a $2\hat{m} \times 2\hat{m}$ symmetric and nonsingular matrix, and $\hat{m}$ denotes the number of correction pairs saved in the limited-memory updating procedure. The matrices W and M are formed by using the vectors $\{\Delta l^k\}$ and $\{\Delta x^k\}$ accumulated in the last $\hat{m}$ iterations. Since the limited-memory matrix B is positive definite, and assuming A_E has full rank, the primal-dual matrix is nonsingular, and we can compute the solution to (19.12) by inverting the coefficient matrix using the Sherman–Morrison–Woodbury formula (see Exercise 19.14).

FEASIBLE INTERIOR-POINT METHODS

In many applications, it is desirable for all of the iterates generated by an optimization algorithm to be feasible with respect to some or all of the *inequality* constraints. For example, the objective function may be defined only when some of the constraints are satisfied, making this feature essential.

Interior-point methods provide a natural framework for deriving feasible algorithms. If the current iterate x satisfies $c_i(x) > 0$, then it is easy to adapt the primal-dual iteration (19.12) so that feasibility is preserved. After computing the step p, we let $x^+ = x + p_x$, redefine the slacks as

$$s^+ \leftarrow c_I(x^+), \tag{19.30}$$

and test whether the point (x^+, s^+) is acceptable for the merit function ϕ. If so, we define this point to be the new iterate; otherwise we reject the step p and compute a new, shorter trial step. In a line search algorithm we backtrack, and in a trust-region method we compute a new step with a reduced trust-region bound. This strategy is justified by the fact that if at a trial point we have that $c_i(x^+) \leq 0$ for some inequality constraint, the value of the merit function is $+\infty$, and we reject the trial point. We will also reject steps $x + p_x$ that are too close to the boundary of the feasible region because such steps increase the barrier term $-\mu \sum_{i \in \mathcal{I}} \log(s_i)$ in the merit function (19.26).

Making the substitution (19.30) has the effect of replacing $\log(s_i)$ with $\log(c_i(x))$ in the merit function, a technique reminiscent of the classical primal log-barrier approach discussed in Section 19.6.

19.4 A LINE SEARCH INTERIOR-POINT METHOD

We now give a more detailed description of a line search interior-point method. We denote by $D\phi(x, s; p)$ the directional derivative of the merit function ϕ_ν at (x, s) in the direction p. The stopping conditions are based on the error function (19.10).

Algorithm 19.2 (Line Search Interior-Point Algorithm).
Choose x_0 and $s_0 > 0$, and compute initial values for the multipliers y_0 and $z_0 > 0$. If a quasi-Newton approach is used, choose an $n \times n$ symmetric and positive definite initial matrix B_0. Select an initial barrier parameter $\mu > 0$, parameters $\eta, \sigma \in (0, 1)$, and tolerances ϵ_μ and ϵ_{TOL}. Set $k \leftarrow 0$.

repeat until $E(x_k, s_k, y_k, z_k; 0) \leq \epsilon_{\text{TOL}}$
    repeat until $E(x_k, s_k, y_k, z_k; \mu) \leq \epsilon_\mu$
        Compute the primal-dual direction $p = (p_x, p_s, p_y, p_z)$ from
            (19.12), where the coefficient matrix is modified as in
            (19.25), if necessary;
        Compute $\alpha_s^{\max}, \alpha_z^{\max}$ using (19.9); Set $p_w = (p_x, p_s)$;
        Compute step lengths α_s, α_z satisfying both (19.27) and
            $\phi_\nu(x_k + \alpha_s p_x, s_k + \alpha_s p_s) \leq \phi_\nu(x_k, s_k) + \eta \alpha_s D\phi_\nu(x_k, s_k; p_w)$;
        Compute $(x_{k+1}, s_{k+1}, y_{k+1}, z_{k+1})$ using (19.28);
        if a quasi-Newton approach is used
            update the approximation B_k;
        Set $k \leftarrow k + 1$;
    end
    Set $\mu \leftarrow \sigma \mu$ and update ϵ_μ;
end

The barrier tolerance can be defined, for example, as $\epsilon_\mu = \mu$, as in Algorithm 19.1. An adaptive strategy that updates the barrier parameter μ at every step is easily implemented in this framework. If the merit function can cause the Maratos effect (see Section 15.4), a second-order correction or a nonmonotone strategy should be implemented. An alternative to using a merit function is to employ a filter mechanism to perform the line search.

We will see in Section 19.7 that Algorithm 19.2 must be safeguarded to ensure global convergence.

19.5   A TRUST-REGION INTERIOR-POINT METHOD

We now consider an interior-point method that uses trust regions to promote convergence. As in the unconstrained case, the trust-region formulation allows great freedom in the choice of the Hessian and provides a mechanism for coping with Jacobian and Hessian singularities. The price to pay for this flexibility is a more complex iteration than in the line search approach.

The interior-point method described below is asymptotically equivalent to the line search method discussed in Section 19.4, but differs significantly in two respects. First, it is not fully a primal-dual method in the sense that it first computes a step in the variables (x, s) and then updates the estimates for the multipliers, as opposed to the approach of Algorithm 19.1, in which primal and dual variables are computed simultaneously. Second, the trust-region method uses a scaling of the variables that discourages moves toward the boundary of the feasible region. This causes the algorithm to generate steps that can be different from, and enjoy more favorable convergence properties than, those produced by a line search method.

We first describe a trust-region algorithm for finding approximate solutions of a fixed barrier problem. We then present a complete interior-point method in which the barrier parameter is driven to zero.

AN ALGORITHM FOR SOLVING THE BARRIER PROBLEM

The barrier problem (19.4) is an equality-constrained optimization problem and can be solved by using a sequential quadratic programming method with trust regions. A straightforward application of SQP techniques to the barrier problem leads, however, to inefficient steps that tend to violate the positivity of the slack variables and are frequently cut short by the trust-region constraint. To overcome this problem, we design an SQP method tailored to the structure of barrier problems.

At the iterate (x, s), and for a given barrier parameter μ, we first compute Lagrange multiplier estimates (y, z) and then compute a step $p = (p_x, p_s)$ that approximately solves the subproblem

$$\min_{p_x, p_s} \quad \nabla f^T p_x + \frac{1}{2} p_x^T \nabla_{xx}^2 \mathcal{L} p_x - \mu e^T S^{-1} p_s + \frac{1}{2} p_s^T \Sigma p_s \tag{19.31a}$$

$$\text{subject to} \quad A_{\scriptscriptstyle E}(x) p_x + c_{\scriptscriptstyle E}(x) = r_{\scriptscriptstyle E}, \tag{19.31b}$$

$$A_{\scriptscriptstyle I}(x) p_x - p_s + (c_{\scriptscriptstyle I}(x) - s) = r_{\scriptscriptstyle I}, \tag{19.31c}$$

$$\|(p_x, S^{-1} p_s)\|_2 \leq \Delta, \tag{19.31d}$$

$$p_s \geq -\tau s. \tag{19.31e}$$

Here Σ is the primal-dual matrix (19.13), and the scalar $\tau \in (0, 1)$ is chosen close to 1 (for example, 0.995). The inequality (19.31e) plays the same role as the fraction to the boundary rule (19.9). Ideally, we would like to set $r = (r_E, r_I) = 0$, but since this can cause the constraints (19.31b)–(19.31d) to be incompatible or to give a step p that makes little progress toward feasibility, we choose the parameter r by an auxiliary computation, as in Algorithm 18.4.

We motivate the choice of the objective (19.31a) by noting that the first-order optimality conditions of (19.31a)–(19.31c) are given by (19.2) (with the second block of equations scaled by S^{-1}). Thus the step computed from the subproblem (19.31) is related to the primal-dual line search step in the same way as the SQP and Newton–Lagrange steps of Section 18.1.

The trust-region constraint (19.31d) guarantees that the problem (19.31) has a finite solution even when $\nabla_{xx}^2 \mathcal{L}(x, s, y, z)$ is not positive definite, and therefore this Hessian need never be modified. In addition, the trust-region formulation ensures that adequate progress is made at every iteration. To justify the scaling S^{-1} used in (19.31d), we note that the shape of the trust region must take into account the requirement that the slacks not approach zero prematurely. The scaling S^{-1} serves this purpose because it restricts those components i of the step vector p_s for which s_i is close to its lower bound of zero. As we see below, it also plays an important role in the choice of the relaxation vectors r_E and r_I.

We outline this SQP trust-region approach as follows. The stopping condition is defined in terms of the error function E given by (19.10), and the merit function ϕ_ν can be defined as in (19.26) using the 2-norm, $\| \cdot \|_2$.

Algorithm 19.3 (Trust-Region Algorithm for Barrier Problems).

Input parameters: $\mu > 0$, x_0, $s_0 > 0$, ϵ_μ, and $\Delta_0 > 0$. Compute Lagrange multiplier estimates y_0 and $z_0 > 0$. Set $k \leftarrow 0$.

repeat until $E(x_k, s_k, y_k, z_k; \mu) \leq \epsilon_\mu$
    Compute $p = (p_x, p_s)$ by approximately solving (19.31).
    if p provides sufficient decrease in the merit function ϕ_ν
        Set $x_{k+1} \leftarrow x_k + p_x$, $s_{k+1} \leftarrow s_k + p_s$;
        Compute new multiplier estimates $y_{k+1}, z_{k+1} > 0$
            and set $\Delta_{k+1} \geq \Delta_k$;
    else
        Define $x_{k+1} \leftarrow x_k$, $s_{k+1} \leftarrow s_k$, and set $\Delta_{k+1} < \Delta_k$;
    end
    Set $k \leftarrow k + 1$;
end (repeat)

Algorithm 19.3 is applied for a fixed value of the barrier parameter μ. A complete interior-point algorithm driven by a sequence $\{\mu_k\} \rightarrow 0$ is described below. First, we discuss how to find an approximate solution of the subproblem (19.31), along with Lagrange multiplier estimates (y_{k+1}, z_{k+1}).

STEP COMPUTATION

The subproblem (19.31a)–(19.31e) is difficult to minimize exactly because of the presence of the nonlinear constraint (19.31d) and the bounds (19.31e). An important observation is that we can compute useful *inexact* solutions, at moderate cost. Since this approach scales up well with the number of variables and constraints, it provides a framework for developing practical interior-point methods for large-scale optimization.

The first step in the solution process is to make a change of variables that transforms the trust-region constraint (19.31d) into a ball. By defining

$$\tilde{p} = \begin{bmatrix} p_x \\ \tilde{p}_s \end{bmatrix} = \begin{bmatrix} p_x \\ S^{-1} p_s \end{bmatrix}, \tag{19.32}$$

we can write problem (19.31) as

$$\min_{p_x, \tilde{p}_s} \quad \nabla f^T p_x + \frac{1}{2} p_x^T \nabla_{xx}^2 \mathcal{L} p_x - \mu e^T \tilde{p}_s + \frac{1}{2} \tilde{p}_s^T S \Sigma S \tilde{p}_s \tag{19.33a}$$

$$\text{subject to} \quad A_{\text{E}}(x) p_x + c_{\text{E}}(x) = r_{\text{E}}, \tag{19.33b}$$

$$A_{\text{I}}(x) p_x - S \tilde{p}_s + (c_{\text{I}}(x) - s) = r_{\text{I}}, \tag{19.33c}$$

$$\|(p_x, \tilde{p}_s)\|_2 \le \Delta, \tag{19.33d}$$

$$\tilde{p}_s \ge -\tau e. \tag{19.33e}$$

To compute the vectors r_{E} and r_{I}, we proceed as in Section 18.5 and formulate the following *normal subproblem* in the variable $v = (v_x, v_s)$:

$$\min_{v} \quad \|A_{\text{E}}(x) v_x + c_{\text{E}}(x)\|_2^2 + \|A_{\text{I}}(x) v_x - S v_s + (c_{\text{I}}(x) - s)\|_2^2$$

$$\tag{19.34a}$$

$$\text{subject to} \quad \|(v_x, v_s)\|_2 \le 0.8\Delta, \tag{19.34b}$$

$$v_s \ge -(\tau/2)e. \tag{19.34c}$$

If we ignore (19.34c), this problem has the standard form of a trust-region problem, and we can compute an approximate solution by using the techniques discussed in Chapter 4, such as the dogleg method. If the solution violates the bounds (19.34c), we can backtrack so that these bounds are satisfied.

Having solved (19.34), we define the vectors r_{E} and r_{I} in (19.33b)–(19.33c) to be the residuals in the normal step computation, namely,

$$r_{\text{E}} = A_{\text{E}}(x) v_x + c_{\text{E}}(x), \qquad r_{\text{I}} = A_{\text{I}}(x) v_x - S v_s + (c_{\text{I}}(x) - s). \tag{19.35}$$

We are now ready to compute an approximate solution $\tilde{d}$ of the subproblem (19.33). By (19.35), the vector v is a particular solution of the linear constraints (19.33b)–(19.33c). We can then solve the equality-constrained quadratic program (19.33a)–(19.33c) by using the projected conjugate gradient iteration given in Algorithm 16.2. We terminate the projected CG iteration by Steihaug's rules: During the solution by CG we monitor the satisfaction of the trust-region constraint (19.33d) and stop if the boundary of this region is reached, if negative curvature is detected, or if an approximate solution is obtained. If the solution given by the projected CG iteration does not satisfy the bounds (19.33e), we backtrack so that they are satisfied. After the step $(p_x, \tilde{p}_s)$ has been computed, we recover p from (19.32).

As discussed in Section 16.3, every iteration of the projected CG iteration requires the solution of a linear system in order to perform the projection operation. For the quadratic program (19.33a)–(19.33c) this projection matrix is given by

$$
\begin{bmatrix} I & \hat{A}^T \\ \hat{A} & 0 \end{bmatrix}, \quad \text{with} \quad \hat{A} = \begin{bmatrix} A_{\mathrm{E}}(x) & 0 \\ A_{\mathrm{I}}(x) & -S \end{bmatrix}.
\tag{19.36}
$$

Thus, although this trust-region approach still requires the solution of an augmented system, the matrix (19.36) is simpler than the primal-dual matrix (19.12). In particular, the Hessian $\nabla_{xx}^2 \mathcal{L}$ need never be factored because the CG approach requires only products of this matrix with vectors.

We mentioned in Section 19.3 that the term $S\Sigma S$ in (19.33a) has a much tighter distribution of eigenvalues than Σ. Therefore the CG method will normally not be adversely affected by ill conditioning and is a viable approach for solving the quadratic program (19.33a)–(19.33c).

LAGRANGE MULTIPLIERS ESTIMATES AND STEP ACCEPTANCE

At an iterate (x, s), we choose (y, z) to be the least-squares multipliers (see (18.21)) corresponding to (19.33a)–(19.33c). We obtain the formula

$$
\begin{bmatrix} y \\ z \end{bmatrix} = \left(\hat{A}\hat{A}^T \right)^{-1} \hat{A} \begin{bmatrix} \nabla f(x) \\ -\mu e \end{bmatrix},
\tag{19.37}
$$

where $\hat{A}$ is given by (19.36) The multiplier estimates z obtained in this manner may not always be positive; to enforce positivity, we may redefine them as

$$
z_i \leftarrow \min(10^{-3}, \mu/s_i), \qquad i = 1, 2, \ldots, m.
\tag{19.38}
$$

The quantity μ/s_i is called the ith primal multiplier estimate because if all components of z were defined by (19.38), then Σ would reduce to the primal choice, (19.14).

As is standard in trust-region methods, the step p is accepted if

$$
\mathrm{ared}(p) \geq \eta \, \mathrm{pred}(p),
\tag{19.39}
$$

where

$$\text{ared}(p) = \phi_\nu(x, s) - \phi_\nu(x + p_x, s + p_s) \tag{19.40}$$

and where η is a constant in $(0, 1)$ (say, $\eta = 10^{-8}$). The predicted reduction is defined as

$$\text{pred}(p) = q_\nu(0) - q_\nu(p), \tag{19.41}$$

where q_ν is defined as

$$q_\nu(p) = \nabla f^T p_x + \frac{1}{2} p_x^T \nabla_{xx}^2 \mathcal{L} p_x - \mu e^T S^{-1} p_s + \frac{1}{2} p_s^T \Sigma p_s + \nu m(p),$$

and

$$m(p) = \left\| \begin{bmatrix} A_E(x) p_x + c_E(x) \\ A_I(x) p_x - p_s + c_I(x) - s \end{bmatrix} \right\|_2.$$

To determine an appropriate value of the penalty parameter ν, we require that ν be large enough that

$$\text{pred}(p) \geq \rho \nu(m(0) - m(p)), \tag{19.42}$$

for some parameter $\rho \in (0, 1)$. This is the same as condition (18.35) used in Section 18.5, and the value of ν can be computed by the procedure described in that section.

DESCRIPTION OF A TRUST-REGION INTERIOR-POINT METHOD

We now present a more detailed description of the trust-region interior-point algorithm for solving the nonlinear programming problem (19.1). For concreteness we follow the Fiacco–McCormick strategy for updating the barrier parameter. The stopping conditions are stated, once more, in terms of the error function E defined by (19.10). In a quasi-Newton approach, the Hessian $\nabla_{xx}^2 \mathcal{L}$ is replaced by a symmetric approximation.

Algorithm 19.4 (Trust-Region Interior-Point Algorithm).
Choose a value for the parameters $\eta > 0$, $\tau \in (0, 1)$, $\sigma \in (0, 1)$, and $\zeta \in (0, 1)$, and select the stopping tolerances ϵ_μ and ϵ_{TOL}. If a quasi-Newton approach is used, select an $n \times n$ symmetric initial matrix B_0. Choose initial values for $\mu > 0$, $x_0, s_0 > 0$, and Δ_0. Set $k \leftarrow 0$.

  repeat until $E(x_k, s_k, y_k, z_k; 0) \leq \epsilon_{\text{TOL}}$
    repeat until $E(x_k, s_k, y_k, z_k; \mu) \leq \epsilon_\mu$
      Compute Lagrange multipliers from (19.37)–(19.38);

> Compute $\nabla_{xx}^2 \mathcal{L}(x_k, s_k, y_k, z_k)$ or upate a quasi-Newton
>   approximation B_k, and define Σ_k by (19.13);
> Compute the normal step $v_k = (v_x, v_s)$;
> Compute $\tilde{p}_k$ by applying the projected CG method to (19.33);
> Obtain the total step p_k from (19.32);
> Update v_k to satisfy (19.42);
> Compute $\text{pred}_k(p_k)$ by (19.41) and $\text{ared}_k(p_k)$ by (19.40);
> **if** $\text{ared}_k(p_k) \geq \eta \, \text{pred}_k(p_k)$
>   Set $x_{k+1} \leftarrow x_k + p_x$, $s_{k+1} \leftarrow s_k + p_s$;
>   Choose $\Delta_{k+1} \geq \Delta_k$;
> **else**
>   set $x_{k+1} = x_k$, $s_{k+1} = s_k$; and choose $\Delta_{k+1} < \Delta_k$;
> **endif**
>   Set $k \leftarrow k + 1$;
> **end**
> Set $\mu \leftarrow \sigma \mu$ and update ϵ_μ;
> **end**

The merit function (19.26) can reject steps that make good progress toward a solution: the Maratos effect discussed in Chapter 18. This deficiency can be overcome by selective application of a second-order correction step; see Section 15.4.

Algorithm 19.4 can easily be modified to implement an adaptive barrier update strategy. The barrier stop tolerance can be defined as $\epsilon_\mu = \mu$. Algorithm 19.4 is the basis of the KNITRO/CG method [50], which implements both exact Hessian and quasi-Newton options.

19.6 THE PRIMAL LOG-BARRIER METHOD

Prior to the introduction of primal-dual interior methods, barrier methods worked in the space of primal variables x. As in the quadratic penalty function approach of Chapter 17, the goal was to solve nonlinear programming problems by unconstrained minimization applied to a parametric sequence of functions.

Primal barrier methods are more easily described in the context of inequality-constrained problems of the form

$$\min_x \; f(x) \quad \text{subject to } c(x) \geq 0. \tag{19.43}$$

The log-barrier function is defined by

$$P(x; \mu) = f(x) - \mu \sum_{i \in \mathcal{I}} \log c_i(x), \tag{19.44}$$

where $\mu > 0$. One can show that the minimizers of $P(x; \mu)$, which we denote by $x(\mu)$, approach a solution of (19.43) as $\mu \downarrow 0$, under certain conditions; see, for example, [111]. The trajectory $\mathcal{C}_p$ defined by

$$\mathcal{C}_p \stackrel{\text{def}}{=} \{x(\mu) \mid \mu > 0\} \tag{19.45}$$

is often referred to as the *primal central path*.

Since the minimizer $x(\mu)$ of $P(x; \mu)$ lies in the strictly feasible set $\{x \mid c(x) > 0\}$ (where no constraints are active), we can in principle search for it by using any of the unconstrained minimization algorithms described in the first part of this book. These methods need to be modified, as explained in the discussion following equation (19.30), so that they reject steps that leave the feasible region or are too close to the constraint boundaries.

One way to obtain an estimate of the Lagrange multipliers is based on differentiating P to obtain

$$\nabla_x P(x; \mu) = \nabla f(x) - \sum_{i \in \mathcal{I}} \frac{\mu}{c_i(x)} \nabla c_i(x). \tag{19.46}$$

When x is close to the minimizer $x(\mu)$ and μ is small, we see from Theorem 12.1 that the optimal Lagrange multipliers z_i^*, $i \in \mathcal{I}$, can be estimated as follows:

$$z_i^* \approx \mu/c_i(x), \quad i \in \mathcal{I}. \tag{19.47}$$

A general framework for algorithms based on the primal log-barrier function (19.44) can be specified as follows.

Framework 19.5 (Unconstrained Primal Barrier Method).
  Given $\mu_0 > 0$, a sequence $\{\tau_k\}$ with $\tau_k \to 0$, and a starting point x_0^s;
  for $k = 0, 1, 2, \ldots$
      Find an approximate minimizer x_k of $P(\cdot; \mu_k)$, starting at x_k^s,
          and terminating when $\|\nabla P(x_k; \mu_k)\| \le \tau_k$;
      Compute Lagrange multipliers z_k by (19.47);
      if final convergence test satisfied
          stop with approximate solution x_k;
      Choose new penalty parameter $\mu_{k+1} < \mu_k$;
      Choose new starting point x_{k+1}^s;
  end (for)

The primal barrier approach was first proposed by Frisch [115] in the 1950s and was analyzed and popularized by Fiacco and McCormick [98] in the late 1960s. It fell out of favor after the introduction of SQP methods and has not regained its popularity because it suffers from several drawbacks compared to primal-dual interior-point methods. The most important drawback is that the minimizer $x(\mu)$ becomes more and more difficult to find as $\mu \downarrow 0$ because of the nonlinearity of the function $P(x; \mu)$

❏ **EXAMPLE 19.1**

Consider the problem

$$\min (x_1 + 0.5)^2 + (x_2 - 0.5)^2 \quad \text{subject to} \quad x_1 \in [0, 1], \quad x_2 \in [0, 1], \quad (19.48)$$

for which the primal barrier function is

$$P(x; \mu) = (x_1 + 0.5)^2 + (x_2 - 0.5)^2 \qquad\qquad (19.49)$$
$$- \mu \left[\log x_1 + \log(1 - x_1) + \log x_2 + \log(1 - x_2) \right].$$

Contours of this function for the value $\mu = 0.01$ are plotted in Figure 19.1. The elongated nature of the contours indicates bad scaling, which causes poor performance of unconstrained optimization methods such as quasi-Newton, steepest descent, and conjugate gradient. Newton's method is insensitive to the poor scaling, but the nonelliptical property—the contours in Figure 19.1 are almost straight along the left edge while being circular along the right edge—indicates that the quadratic approximation on which Newton's method is based does not capture well the behavior of the barrier function. Hence, Newton's method, too, may not show rapid convergence to the minimizer of (19.49) except in a small neighborhood of this point. ❏

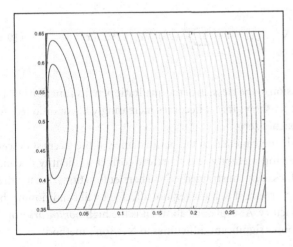

Figure 19.1
Contours of $P(x; \mu)$ from
(19.49) for $\mu = 0.01$

To lessen this nonlinearity, we can proceed as in (17.21) and introduce additional variables. Defining $z_i = \mu/c_i(x)$, we rewrite the stationarity condition (19.46) as

$$\nabla f(x) - \sum_{i \in \mathcal{I}} z_i \nabla c_i(x) = 0, \qquad (19.50a)$$

$$C(x)z - \mu e = 0, \qquad (19.50b)$$

where $C(x) = \text{diag}(c_1(x), c_2(x), \ldots, c_m(x))$. Note that this system is equivalent to the perturbed KKT conditions (19.2) for problem (19.43) if, in addition, we introduce slacks as in (19.2d). Finally, if we apply Newton's method in the variables (x, s, z) and temporarily ignore the bounds $s, z \geq 0$, we arrive at the primal-dual formulation. Thus, with hindsight, we can transform the primal log-barrier approach into the primal-dual line search approach of Section 19.4 or into the trust-region algorithm of Section 19.5.

Other drawbacks of the classical primal barrier approach are that it requires a feasible initial point, which can be difficult to find in many cases, and that the incorporation of equality constraints in a primal function is problematic. (A formulation in which the equality constraints are replaced by quadratic penalties suffers from the shortcomings of quadratic penalty functions discussed in Section 17.1.)

The shortcomings of the primal barrier approach were attributed for many years to the ill conditioning of the Hessian of the barrier function P. Note that

$$\nabla_{xx}^2 P(x; \mu) = \nabla^2 f(x) - \sum_{i \in \mathcal{I}} \frac{\mu}{c_i(x)} \nabla^2 c_i(x) + \sum_{i \in \mathcal{I}} \frac{\mu}{c_i^2(x)} \nabla c_i(x) \nabla c_i(x)^T. \qquad (19.51)$$

By substituting (19.47) into (19.51) and using the definition (12.33) of the Lagrangian $\mathcal{L}(x, z)$, we find that

$$\nabla_{xx}^2 P(x; \mu) \approx \nabla_{xx}^2 \mathcal{L}(x, z^*) + \sum_{i \in \mathcal{I}} \frac{1}{\mu} (z_i^*)^2 \nabla c_i(x) \nabla c_i(x)^T. \qquad (19.52)$$

Note the similarity of this expression to the Hessian of the quadratic penalty function (17.19). Analysis of the matrix $\nabla_{xx}^2 P(x; \mu)$ shows that it becomes increasingly ill conditioned near the minimizer $x(\mu)$, as μ approaches zero.

This ill conditioning will be detrimental to the performance of the steepest descent, conjugate gradient, or quasi-Newton methods. It is therefore correct to identify ill conditioning as a source of the difficulties of unconstrained primal barrier functions that use these unconstrained methods. Newton's method is, however, not affected by ill conditioning, but its performance is still not satisfactory. As explained above, it is the high *nonlinearity* of the primal barrier function P that poses significant difficulties to Newton's method.

19.7 GLOBAL CONVERGENCE PROPERTIES

We now study some global convergence properties of the primal-dual interior-point methods described in Sections 19.4 and 19.5. Theorem 19.1 provides the starting point for the analysis. It gives conditions under which limit points of the iterates generated by the interior-point methods are KKT points for the nonlinear problem. Theorem 19.1 relies on the assumption that the perturbed KKT conditions (19.2) can be satisfied (to a certain accuracy) for every value of μ_k. In this section we study conditions under which this assumption holds, that is, conditions that guarantee that our algorithms can find stationary points of the barrier problem (19.4).

We begin with a surprising observation. Whereas the line search primal-dual approach is the basis of globally convergent interior-point algorithms for linear and quadratic programming, it is not guaranteed to be successful for nonlinear programming, even for nondegenerate problems.

FAILURE OF THE LINE SEARCH APPROACH

We have seen in Chapter 11 that line search Newton iterations for nonlinear equations can fail when the Jacobian loses rank. We now discuss a different kind of failure specific to interior-point methods. It is caused by the lack of coordination between the step computation and the imposition of the bounds.

❏ **EXAMPLE 19.2**   (WÄCHTER AND BIEGLER [299])

Consider the problem

$$\min x \tag{19.53a}$$

$$\text{subject to} \quad c_1(x) - s \overset{\text{def}}{=} x^2 - s_1 - 1 = 0, \tag{19.53b}$$

$$c_2(x) - s \overset{\text{def}}{=} x - s_2 - \tfrac{1}{2} = 0, \tag{19.53c}$$

$$s_1 \geq 0, \ s_2 \geq 0. \tag{19.53d}$$

Note that the Jacobian of the equality constraints (19.53b)–(19.53c) with respect to (x, s) has full rank everywhere. Let us apply a line search interior-point method of the form (19.6)–(19.9), starting from an initial point $x^{(0)}$ such that $(s_1^{(0)}, s_2^{(0)}) > 0$, and $c_1(x^{(0)}) - s^{(0)} \geq 0$. (In this example, we use superscripts to denote iteration indices.) Figure 19.2 illustrates the feasible region (the dotted segment of the parabola) and the initial point, all projected onto the x-s_1 plane. The primal-dual step, which satisfies the linearization of the constraints (19.53b)–(19.53c), leads from $x^{(0)}$ to the tangent to the parabola. Here p_1 and p_2 are examples of possible steps satisfying the linearization of (19.53b)–(19.53c). The new iterate $x^{(1)}$ therefore lies between $x^{(0)}$ and this tangent, but since s_1 must remain positive, $x^{(1)}$ will

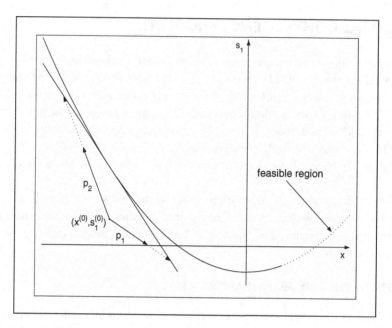

Figure 19.2 Problem (19.53) projected onto the x-s_1 plane.

lie above the horizontal axis. Thus, from any starting point above the x-axis and to the left of the parabola, namely, in the region

$$\{(x, s_1, s_2) \;:\; x^2 - s_1 - 1 \geq 0, \; s_1 \geq 0\}, \tag{19.54}$$

the new iterate will remain in this region. The argument can now be repeated to show that the iterates $\{x^{(k)}\}$ never leave the region (19.54) and therefore never become feasible.

This convergence failure affects any method that generates directions that satisfy the linearization of the constraints (19.53b)–(19.53c) and that enforces the bounds (19.53d) by the fraction to the boundary rule (19.8). The merit function can only restrict the step length further and is therefore incapable of resolving the difficulties. The strategy for updating μ is also irrelevant because the argument given above makes use only of the linearizations of the constraints.   ❑

These difficulties can be observed when practical line-search codes are applied to the problem (19.53). For a wide range of starting points in the region (19.54), the interior-point iteration converges to points of the form $(-\beta, 0, 0)$, with $\beta > 0$. In other words, the iterates can converge to an infeasible, non-optimal point on the boundary of the set $\{(x_1, s_1, s_2) \;:\; s_1 \geq 0, \; s_2 \geq 0\}$, a situation that barrier methods are supposed to prevent. Furthermore, such limit points are not stationary for a feasibility measure (see Definition 17.1).

Failures of this type are rare in practice, but they highlight a theoretical deficiency of the algorithmic class (19.6)–(19.9) that may manifest itself more often as inefficient behavior than as outright convergence failure.

MODIFIED LINE SEARCH METHODS

To remedy this problem, as well as the inefficiencies caused by Hessian and constraint Jacobian singularities, we must modify the search direction of the line search interior-point iteration in some circumstances. One option is to use penalizations of the constraints [147]. Such penalty-barrier methods have been investigated only recently and mature implementations have not yet emerged.

An approach that has been successful in practice is to monitor the step lengths α_s, α_z in (19.28); if they are smaller than a given threshold, then we replace the primal-dual step by a step that guarantees progress in feasibility and, preferably, improvement in optimality, too. In a filter method, when the step lengths are very small, we can invoke the feasibility restoration phase (see Section 15.4), which is designed to generate a new iterate that reduces the infeasibility. A different approach, which assumes that a trust-region algorithm is at hand, is to replace the primal-dual step by a trust-region step, such as that produced by Algorithm 19.4.

Safeguarding the primal-dual step when the step lengths are very small is justified theoretically because, when line search iterations converge to non-stationary points, the step lengths α_s, α_z converge to zero. From a practical perspective, however, this strategy is not totally satisfactory because it attempts to react when bad steps are generated, rather than trying to prevent them. It also requires the choice of a heuristic to determine when a step length is too small. As we discuss next, the trust-region approach always generates productive steps and needs no safeguarding.

GLOBAL CONVERGENCE OF THE TRUST-REGION APPROACH

The interior-point trust-region method specified in Algorithm 19.4 has favorable global convergence properties, which we now discuss. For simplicity, we present the analysis in the context of inequality-constrained problems of the form (19.43). We first study the solution of the barrier problem (19.4) for a fixed value of μ, and then consider the complete algorithm.

In the result that follows, B_k denotes the Hessian $\nabla^2_{xx}\mathcal{L}_k$ or a quasi-Newton approximation to it. We use the measure of infeasibility $h(x) = \|[c(x)]^-\|$, where $[y] = \max\{0, -y\}$. This measure vanishes if and only if x is feasible for problem (19.43). Note that $h(x)^2$ is differentiable and its gradient is

$$\nabla[h(x)^2] = 2A(x)c(x)^-.$$

We say that a sequence $\{x_k\}$ is *asymptotically feasible* if $c(x_k)^- \to 0$. To apply Algorithm 19.4 to a fixed barrier problem, we dispense with the outer "repeat" loop.

Theorem 19.2.

  Suppose that Algorithm 19.4 is applied to the barrier problem (19.4), that is, μ is fixed and the inner "repeat" loop is executed with $\epsilon_\mu = 0$. Suppose that the sequence $\{f_k\}$ is bounded below and the sequences $\{\nabla f_k\}$, $\{c_k\}$, $\{A_k\}$, and $\{B_k\}$ are bounded. Then one of the following three situations occurs:

  (i) *The sequence $\{x_k\}$ is not asymptotically feasible. In this case, the iterates approach stationarity of the measure of infeasibility $h(x) = \|c(x)^-\|$, meaning that $A_k c_k^- \to 0$, and the penalty parameters ν_k tend to infinity.*

  (ii) *The sequence $\{x_k\}$ is asymptotically feasible, but the sequence $\{(c_k, A_k)\}$ has a limit point $(\bar{\gamma}, \bar{A})$ failing the linear independence constraint qualification. In this situation also, the penalty parameters ν_k tend to infinity.*

  (iii) *The sequence $\{x_k\}$ is asymptotically feasible, and all limit points of the sequence $\{(c_k, A_k)\}$ satisfy the linear independence constraint qualification. In this case, the penalty parameter ν_k is constant and $c_k > 0$ for all large indices k, and the stationarity conditions of problem (19.4) are satisfied in the limit.*

  This theorem is proved in [48], where it is assumed, for simplicity, that Σ is given by the primal choice (19.14). The theorem accounts for two situations in which the KKT conditions may not be satisfied in the limit, both of which are of interest. Outcome (i) is a case in which, in the limit, there is no direction that improves feasibility to first order. This outcome cannot be ruled out because finding a feasible point is a problem that a local method cannot always solve without a good starting point. (Note that we do not assume that the constraint Jacobian A_k has full rank.)

  In considering outcome (ii), we must keep in mind that in some cases the solution to problem (19.43) is a point where the linear independence constraint qualification fails and that is not a KKT point. Outcome (iii) is the most desirable outcome and can be monitored in practice by observing, for example, the behavior of the penalty parameter ν_k.

  We now study the complete interior-point method given in Algorithm 19.4 applied to the nonlinear programming problem (19.43). By combining Theorems 19.1 and 19.2 we see that the following outcomes can occur:

  - For some barrier parameter μ generated by the algorithm, either the inequality $\|c_k - s_k\| \le \epsilon_\mu$ is never satisfied, in which case the stationarity condition for minimizing $h(x)$ is satisfied in the limit, or else $(c_k - s_k) \to 0$, in which case the sequence $\{(c_k, A_k)\}$ has a limit point $(\bar{c}, \bar{A})$ failing the linear independence constraint qualification;

  - At each outer iteration of Algorithm 19.4 the inner stop test $E(x_k, s_k, y_k, z_k; \mu) \le \epsilon_\mu$ is satisfied. Then all limit points of the iteration sequence are feasible. Furthermore, if any limit point $\hat{x}$ satisfies the linear independence constraint qualification, the first-order necessary conditions for problem (19.43) hold at $\hat{x}$.

19.8 SUPERLINEAR CONVERGENCE

We can implement primal-dual interior-point methods so that they converge quickly near the solution. All is needed is that we carefully control the decrease in the barrier parameter μ and the inner convergence tolerance ϵ_μ, and let the parameter τ in (19.9) converge to 1 sufficiently rapidly. We now describe strategies for updating these parameters in the context of the line search iteration discussed in Section 19.4; these strategies extend easily to the trust-region method of Section 19.5.

In the discussion that follows, we assume that the merit function or filter is inactive. This assumption is realistic because with a careful implementation (which may include second-order correction steps or other features), we can ensure that, near a solution, all the steps generated by the primal-dual method are acceptable to the merit function or filter. We denote the primal-dual iterates by

$$v = (x, s, y, z) \tag{19.55}$$

and define the full primal-dual step (without backtracking) by

$$v^+ = v + p, \tag{19.56}$$

where p is the solution of (19.12). To establish local convergence results, we assume that the iterates converge to a solution point satisfying certain regularity assumptions.

Assumptions 19.1.

(a) v^* is a solution of the nonlinear program (19.1) for which the first-order KKT conditions are satisfied.

(b) The Hessian matrices $\nabla^2 f(x)$ and $\nabla^2 c_i(x)$, $i \in \mathcal{E} \cup \mathcal{I}$, are locally Lipschitz continuous at v^*.

(c) The linear independence constraint qualification (LICQ) (Definition 12.4), the strict complementarity condition (Definition 12.5), and the second-order sufficient conditions (Theorem 12.6) hold at v^*.

We assume that v is an iterate at which the inner stop test $E(v, \mu) \le \epsilon_\mu$ is satisfied, so that the barrier parameter is decreased from μ to μ^+. We now study how to control the parameters in Algorithm 19.2 so that the following three properties hold in a neighborhood of v^*:

1. The iterate v^+ satisfies the fraction to the boundary rule (19.9), that is, $\alpha_s^{\max} = \alpha_z^{\max} = 1$.

2. The inner stop test is satisfied at v^+, that is, $E(v^+; \mu^+) \leq \epsilon_{\mu^+}$.

3. The sequence of iterates (19.56) converge superlinearly to v^*.

We can achieve these three goals by letting

$$\epsilon_\mu = \theta\mu \qquad \text{and} \qquad \epsilon_{\mu^+} = \theta\mu^+, \tag{19.57}$$

for $\theta > 0$, and setting the other parameters as follows:

$$\mu^+ = \mu^{1+\delta}, \quad \delta \in (0, 1); \qquad\qquad \tau = 1 - \mu^\beta, \quad \beta > \delta. \tag{19.58}$$

There are other practical ways of controlling the parameters of the algorithm. For example, we may prefer to determine the change in μ from the reduction achieved in the KKT conditions of the nonlinear program, as measured by the function E. The three results mentioned above can be established if the convergence tolerance ϵ_μ is defined as in (19.57) and if we replace μ by $E(v; 0)$ in the right-hand sides of the definitions (19.58) of μ^+ and τ.

There is a limit to how fast we can decrease μ and still be able to satisfy the inner stop test after just one iteration (condition 2). One can show that there is no point in decreasing μ at a faster than quadratic rate, since the overall convergence cannot be faster than quadratic. Not suprising, if τ is constant and $\mu^+ = \sigma\mu$, with $\sigma \in (0, 1)$, then the interior-point algorithm is only linearly convergent.

Although it is desirable to implement interior-point methods so that they achieve a superlinear rate of convergence, this rate is typically observed only in the last few iterations in practice.

19.9   PERSPECTIVES AND SOFTWARE

Software packages that implement nonlinear interior-point methods are widely available. Line search implementations include LOQO [294], KNITRO/DIRECT [303], IPOPT [301], and BARNLP [21], and for convex problems, MOSEK [5]. The trust-region algorithm discussed in Section 19.5 has been implemented in KNITRO/CG [50]. These interior-point packages have proved to be strong competitors of the leading active-set and augmented Lagrangian packages, such as MINOS [218], SNOPT [128], LANCELOT [72], FILTERSQP [105], and KNITRO/ACTIVE [49]. At present, interior-point and active-set methods appear to be the most promising approaches, while augmented Lagrangian methods seem to be less efficient. The KNITRO package provides crossover from interior-point to active-set modes [46].

Interior-point methods show their strength in large-scale applications, where they often (but not always) outperform active-set methods. In interior-point methods, the linear system to be solved at every iteration has the same block structure, so effort can be focused on exploiting this structure. Both direct factorization techniques and projected CG methods are available, allowing the user to solve many types of applications efficiently. On the other hand, interior-point methods, unlike active-set methods, consider all the constraints at each iteration, even if they are irrelevant to the solution. As a result, the cost of the primal-dual iteration can be excessive in some applications.

One of the main weaknesses of interior-point methods is their sensitivity to the choice of the initial point, the scaling of the problem, and the update strategy for the barrier parameter μ. If the iterates approach the boundary of the feasible region prematurely, interior-point methods may have difficulty escaping it, and convergence can be slow. The availability of adaptive strategies for updating μ is, however, beginning to lessen this sensitivity, and more robust implementations can be expected in the coming years.

Although the description of the line search algorithm in Section 19.4 is fairly complete, various details of implementation (such as second-order corrections, iterative refinement, and resetting of parameters) are needed to obtain a robust code. Our description of the trust-region method of Algorithm 19.4 leaves some important details unspecified, particularly concerning the procedure for computing approximate solutions of the normal and tangential subproblems; see [50] for further discussion. The KNITRO/CG implementation of this trust-region algorithm uses a projected CG iteration in the computation of the step, which allows the method to work even when only Hessian–vector products are available, not the Hessian itself.

Filters and merit functions have each been used to globalize interior-point methods. Although some studies have shown that merit functions restrict the progress of the iteration unduly [298], recent developments in penalty update procedures (see Chapter 18) have altered the picture, and it is currently unclear whether filter globalization approaches are preferable.

NOTES AND REFERENCES

The development of modern nonlinear interior-point methods was influenced by the success of interior-point methods for linear and quadratic programming. The concept of primal-dual steps arises from the homotopy formulation given in Section 19.1, which is an extension of the systems (14.13) and (16.57) for linear and quadratic programming. Although the primal barrier methods of Section 19.6 predate primal-dual methods by at least 15 years, they played a limited role in their development.

There is a vast literature on nonlinear interior-point methods. We refer the reader to the surveys by Forsgren, Gill, and Wright [111] and Gould, Orban, and Toint [147] for a comprehensive list of references. The latter paper also compares and contrasts interior-point methods with other nonlinear optimization methods. For an analysis of interior-point methods that use filter globalization see, for example, Ulbrich, Ulbrich, and Vicente [291] and Wächter and Biegler [300]. The book by Conn, Gould, and Toint [74] gives a thorough presentation of several interior-point methods.

Primal barrier methods were originally proposed by Frisch [115] and were analyzed in an authoritative book by Fiacco and McCormick [98]. The term "interior-point method" and the concept of the primal central path C_p appear to have originated in this book. Nesterov and Nemirovskii [226] propose and analyze several families of barrier methods and establish polynomial-time complexity results for very general classes of problems such as semidefinite and second-order cone programming. For a discussion of the history of barrier function methods, see Nash [221].

✎ **EXERCISES**

✎ **19.1** Consider the nonlinear program

$$\min\ f(x)\ \text{subject to}\ c_{\mathcal{E}}(x) = 0,\quad c_{\mathcal{I}}(x) \geq 0. \tag{19.59}$$

(a) Write down the KKT conditions of (19.1) and (19.59), and establish a one-to-one correspondence between KKT points of these problems (despite the different numbers of variables and multipliers).

(b) The multipliers z correspond to the *equality* constraints (19.1c) and should therefore be unsigned. Nonetheless, argue that (19.2) with $\mu = 0$ together with (19.3) can be seen as the KKT conditions of problem (19.1). Moreover, argue that the multipliers z in (19.2) can be seen as the multipliers of the inequalities $c_{\mathcal{I}}$ in (19.59).

(c) Suppose $\bar{x}$ is feasible for (19.59). Show that LICQ holds at $\bar{x}$ for (19.59) if and only if LICQ holds at $(\bar{x}, \bar{s})$ for (19.1), with $\bar{s} = c_{\mathcal{I}}(\bar{x})$.

(d) Repeat part (c) assuming that the MFCQ condition holds (see Definition 12.6) instead of LICQ.

✎ **19.2** This question concerns Algorithm 19.1.

(a) Extend the proof of Theorem 19.1 to the general nonlinear program (19.1).

(b) Show that the theorem still holds if the condition $E(x_k, s_k, y_k, z_k) \leq \mu_k$ is replaced by $E(x_k, s_k, y_k, z_k) \leq \epsilon_{\mu_k}$, for any sequence ϵ_{μ_k} that converges to 0 as $\mu_k \to 0$.

(c) Suppose that in Algorithm 19.1 the new iterate $(x_{k+1}, s_{k+1}, y_{k+1}, z_{k+1})$ is obtained by *any* means. What conditions are required on this iterate so that Theorem 19.1 holds?

✎ **19.3** Consider the nonlinear system of equations (11.1). Show that Newton's method (11.6) is invariant to scalings of the equations. More precisely, show that the Newton step p does not change if each component of r is multiplied by a nonzero constant.

19.4 Consider the system

$$x_1 + x_2 - 2 = 0,$$
$$x_1 x_2 - 2x_2^2 + 1 = 0.$$

Find all the solutions to this system. Show that if the first equation is multiplied by x_2, the solutions do not change but the Newton step taken from $(1, -1)$ will not be the same as that for the original system.

19.5 Let (x, s, y, z) be a primal-dual solution that satisfies the LICQ and strict complementarity conditions.

(a) Give conditions on $\nabla^2_{xx} c_L(x, s, y, z)$ that ensure that the primal-dual matrix in (19.6) is nonsingular.

(b) Show that some diagonal elements of Σ tend to infinity and others tend to zero when $\mu \to 0$. Can you characterize each case? Consider the cases in which Σ is defined by (19.13) and (19.14).

(c) Argue that the matrix in (19.6) is not ill conditioned under the assumptions of this problem.

19.6

(a) Introduce the change of variables $\tilde{p}_s = S^{-1} p_s$ in (19.12), and show that the $(2, 2)$ block of the primal-dual matrix has a cluster of eigenvalues around 0 when $\mu \to 0$.

(b) Analyze the eigenvalue distribution of the $(2, 2)$ block if the change of variables is given by $\tilde{p}_s = \Sigma^{1/2} p_s$ or $\tilde{p}_s = \sqrt{\mu} S^{-1} p_s$.

(c) Let $\gamma > 0$ be the smallest eigenvalue of $\nabla^2_{xx} c_L$. Describe a change of variables for which all the eigenvalues of the $(2, 2)$ block converge to γ as $\mu \to 0$.

19.7 Program the simple interior-point method Algorithm 19.1 and apply it to the problem (18.69). Use the same starting point as in that problem. Try different values for the parameter σ.

19.8

(a) Compute the minimum-norm solution of the system of equations defined by (19.35). (This system defines the Newton component in the dogleg method used to find an approximate solution to (19.34).) Show that the computation of the Newton component can use the factorization of the augmented matrix defined in (19.36).

(b) Compute the unconstrained minimizer of the quadratic in (19.34a) along the steepest descent direction, starting from $v = 0$. (This minimizer defines the Cauchy component in the dogleg method used to find an approximate solution to (19.34).)

(c) The dogleg step is a combination of the Newton and Cauchy steps from parts (a) and (b). Show that the dogleg step is in the range space of $\hat{A}^T$.

19.9

(a) If the normal subproblem (19.34a)–(19.34c) is solved by using the dogleg method, show that the solution v is in the range space of matrix $\hat{A}^T$ defined in (19.36).

(b) After the normal step v is obtained, we define the residual vectors r_E and r_I as in (19.35) and $w = \tilde{p} - v$. Show that (19.33) becomes a quadratic program with circular trust-region constraint and bound constraint in the variables w.

(c) Show that the solution w of the problem derived in part (b) is orthogonal to the normal step v, that is, that $w^T v = 0$.

19.10 Verify that the least-squares multiplier formula (18.21) corresponding to (19.33a)–(19.33c) is given by (19.37).

19.11

(a) Write the primal-dual system (19.6) for problem (19.53), considering s_1, s_2 as slacks and denoting the multipliers of (19.53b), (19.53c) by z_1, z_2. (You should get a system of five equations with five unknowns.) Show that the matrix of the system is singular at any iterate of the form $(x, 0, 0)$.

(b) Show that if the starting point in Example (19.53) lies in the region (19.54), the interior-point step leads to a point on the tangent line to the parabola, as illustrated in Figure 19.2. (More specifically, show that the tangent line never lies to the left of the parabola.)

(c) Let $x^{(0)} = -2$, $s_1^{(0)} = 1$, $s_2^{(0)} = 1$, let $z_1^{(0)} = z_2^{(0)} = 1$, and let $\mu = 0$. Compute the full Newton step based on the system in part (a). Truncate, if necessary, to satisfy a fraction to the boundary rule with $\tau = 1$. Verify that the new iterate is still in the region (19.54).

(d) Let us the consider the behavior of an SQP method. For the initial point in (c), show that the linearized constraints of problem (18.56) (don't forget the constraints $s_1 \geq 0$, $s_2 \geq 0$) are inconsistent. Therefore, the SQP subproblem (18.11) is inconsistent, and a relaxation of the constraint of the SQP subproblem must be performed.

19.12 Consider the following problem in a single variable x:

$$\min x \quad \text{subject to } x \geq 0, \ 1 - x \geq 0.$$

(a) Write the primal barrier function $P(x; \mu)$ associated with this problem.

(b) Plot the barrier function for different values of μ.

(c) Characterize the minimizers of the barrier function as a function of μ and consider the limit as μ goes to 0.

19.13 Consider the scalar minimization problem

$$\min_x \frac{1}{1+x^2}, \quad \text{subject to } x \geq 1.$$

Write down $P(x; \mu)$ for this problem, and show that $P(x; \mu)$ is unbounded below for any positive value of μ. (See Powell [242] and M. Wright [313].)

19.14

The goal of this exercise is to describe an efficient implementation of the limited-memory BFGS version of the interior-point method using the compact representation (19.29). First we decompose the primal-dual matrix as

$$\begin{bmatrix} \xi I & 0 & A_E^T & A_I^T \\ 0 & \Sigma & 0 & I \\ A_E & 0 & 0 & 0 \\ A_I & I & 0 & 0 \end{bmatrix} + \begin{bmatrix} W \\ 0 \\ 0 \\ 0 \end{bmatrix} \begin{bmatrix} MW^T & 0 & 0 & 0 \end{bmatrix}. \qquad (19.60)$$

Use the Sherman–Morrison–Woodbury formula to express the inverse (19.60). Then show that the primal-dual step (19.12) requires the solution of systems of the form $Cv = b$, where C is the left matrix in (19.60) and v and b are certain vectors.

Background Material

A.1 ELEMENTS OF LINEAR ALGEBRA

VECTORS AND MATRICES

In this book we work exclusively with vectors and matrices whose components are real numbers. Vectors are usually denoted by lowercase roman characters, and matrices by uppercase roman characters. The space of real vectors of length n is denoted by $\mathbb{R}^n$, while the space of real $m \times n$ matrices is denoted by $\mathbb{R}^{m \times n}$.

Given a vector $x \in \mathbb{R}^n$, we use x_i to denote its ith component. We invariably assume that x is a *column* vector, that is,

$$x = \begin{bmatrix} x_1 \\ x_2 \\ \vdots \\ x_n \end{bmatrix}.$$

The transpose of x, denoted by x^T is the row vector

$$x^T = \begin{bmatrix} x_1 & x_2 & \cdots & x_n \end{bmatrix},$$

and is often also written with parentheses as $x = (x_1, x_2, \ldots, x_n)$. We write $x \geq 0$ to indicate componentwise nonnegativity, that is, $x_i \geq 0$ for all $i = 1, 2, \ldots, n$, while $x > 0$ indicates that $x_i > 0$ for all $i = 1, 2, \ldots, n$.

Given $x \in \mathbb{R}^n$ and $y \in \mathbb{R}^n$, the standard inner product is $x^T y = \sum_{i=1}^n x_i y_i$.

Given a matrix $A \in \mathbb{R}^{m \times n}$, we specify its components by double subscripts as A_{ij}, for $i = 1, 2, \ldots, m$ and $j = 1, 2, \ldots, n$. The transpose of A, denoted by A^T, is the $n \times m$ matrix whose components are A_{ji}. The matrix A is said to be *square* if $m = n$. A square matrix is *symmetric* if $A = A^T$.

A square matrix A is *positive definite* if there is a positive scalar α such that

$$x^T A x \geq \alpha x^T x, \qquad \text{for all } x \in \mathbb{R}^n. \tag{A.1}$$

It is *positive semidefinite* if

$$x^T A x \geq 0, \qquad \text{for all } x \in \mathbb{R}^n.$$

We can recognize that a symmetric matrix is positive definite by computing its eigenvalues and verifying that they are all positive, or by performing a Cholesky factorization. Both techniques are discussed further in later sections.

The diagonal of the matrix $A \in \mathbb{R}^{m \times n}$ consists of the elements A_{ii}, for $i = 1, 2, \ldots \min(m, n)$. The matrix $A \in \mathbb{R}^{m \times n}$ is *lower triangular* if $A_{ij} = 0$ whenever $i < j$; that is, all elements above the diagonal are zero. It is *upper triangular* if $A_{ij} = 0$ whenever $i > j$; that is, all elements below the diagonal are zero. A is *diagonal* if $A_{ij} = 0$ whenever $i \neq j$.

The *identity matrix*, denoted by I, is the square diagonal matrix whose diagonal elements are all 1.

A square $n \times n$ matrix A is *nonsingular* if for any vector $b \in \mathbb{R}^n$, there exists $x \in \mathbb{R}^n$ such that $Ax = b$. For nonsingular matrices A, there exists a unique $n \times n$ matrix B such that $AB = BA = I$. We denote B by A^{-1} and call it the *inverse* of A. It is not hard to show that the inverse of A^T is the transpose of A^{-1}.

A square matrix Q is *orthogonal* if it has the property that $QQ^T = Q^T Q = I$. In other words, the inverse of an orthogonal matrix is its transpose.

NORMS

For a vector $x \in \mathbb{R}^n$, we define the following norms:

$$\|x\|_1 \stackrel{\text{def}}{=} \sum_{i=1}^n |x_i|, \tag{A.2a}$$

$$\|x\|_2 \stackrel{\text{def}}{=} \left(\sum_{i=1}^n x_i^2 \right)^{1/2} = (x^T x)^{1/2}, \tag{A.2b}$$

$$\|x\|_\infty \stackrel{\text{def}}{=} \max_{i=1,\dots,n} |x_i|. \tag{A.2c}$$

The norm $\| \cdot \|_2$ is often called the *Euclidean norm*. We sometimes refer to $\| \cdot \|_1$ as the ℓ_1 norm and to $\| \cdot \|_\infty$ as the ℓ_∞ norm. All these norms measure the length of the vector in some sense, and they are equivalent in the sense that each one is bounded above and below by a multiple of the other. To be precise, we have for all $x \in \mathbb{R}^n$ that

$$\|x\|_\infty \le \|x\|_2 \le \sqrt{n} \|x\|_\infty, \qquad \|x\|_\infty \le \|x\|_1 \le n\|x\|_\infty, \tag{A.3}$$

and so on. In general, a norm is any mapping $\| \cdot \|$ from $\mathbb{R}^n$ to the nonnegative real numbers that satisfies the following properties:

$$\|x + z\| \le \|x\| + \|z\|, \qquad \text{for all } x, z \in \mathbb{R}^n; \tag{A.4a}$$

$$\|x\| = 0 \Rightarrow x = 0; \tag{A.4b}$$

$$\|\alpha x\| = |\alpha| \|x\|, \qquad \text{for all } \alpha \in \mathbb{R} \text{ and } x \in \mathbb{R}^n. \tag{A.4c}$$

Equality holds in (A.4a) if and only if one of the vectors x and z is a nonnegative scalar multiple of the other.

Another interesting property that holds for the Euclidean norm $\| \cdot \| = \| \cdot \|_2$ is the Cauchy–Schwarz inequality, which states that

$$\left| x^T z \right| \le \|x\| \|z\|, \tag{A.5}$$

with equality if and only if one of these vectors is a nonnegative multiple of the other. We can prove this result as follows:

$$0 \le \|\alpha x + z\|^2 = \alpha^2 \|x\|^2 + 2\alpha x^T z + \|z\|^2.$$

The right-hand-side is a convex function of α, and it satisfies the required nonnegativity property only if there exist fewer than 2 distinct real roots, that is,

$$(2x^T z)^2 \le 4\|x\|^2 \|z\|^2,$$

proving (A.5). Equality occurs when the quadratic α has exactly one real root (that is, $|x^T z| = \|x\| \|z\|$) and when $\alpha x + z = 0$ for some α, as claimed.

Any norm $\| \cdot \|$ has a *dual norm* $\| \cdot \|_D$ defined by

$$\|x\|_D = \max_{\|y\|=1} x^T y. \tag{A.6}$$

It is easy to show that the norms $\| \cdot \|_1$ and $\| \cdot \|_\infty$ are duals of each other, and that the Euclidean norm is its own dual.

We can derive definitions for certain matrix norms from these vector norm definitions. If we let $\| \cdot \|$ be generic notation for the three norms listed in (A.2), we define the corresponding matrix norm as

$$\|A\| \stackrel{\text{def}}{=} \sup_{x \neq 0} \frac{\|Ax\|}{\|x\|}. \tag{A.7}$$

The matrix norms defined in this way are said to be *consistent* with the vector norms (A.2). Explicit formulae for these norms are as follows:

$$\|A\|_1 = \max_{j=1,\ldots,n} \sum_{i=1}^{m} |A_{ij}|, \tag{A.8a}$$

$$\|A\|_2 = \text{largest eigenvalue of } (A^T A)^{1/2}, \tag{A.8b}$$

$$\|A\|_\infty = \max_{i=1,\ldots,m} \sum_{j=1}^{n} |A_{ij}|. \tag{A.8c}$$

The Frobenius norm $\|A\|_F$ of the matrix A is defined by

$$\|A\|_F = \left(\sum_{i=1}^{m} \sum_{j=1}^{n} A_{ij}^2 \right)^{1/2}. \tag{A.9}$$

This norm is useful for many purposes, but it is not consistent with any vector norm. Once again, these various matrix norms are equivalent with each other in a sense similar to (A.3).

For the Euclidean norm $\| \cdot \| = \| \cdot \|_2$, the following property holds:

$$\|AB\| \leq \|A\| \|B\|, \tag{A.10}$$

for all matrices A and B with consistent dimensions.

The *condition number* of a nonsingular matrix is defined as

$$\kappa(A) = \|A\| \|A^{-1}\|, \tag{A.11}$$

where any matrix norm can be used in the definition. Different norms can by the use of a subscript—$\kappa_1(\cdot)$, $\kappa_2(\cdot)$, and $\kappa_\infty(\cdot)$, respectively—with κ denoting κ_2 by default.

Norms also have a meaning for scalar, vector, and matrix-valued functions that are defined on a particular domain. In these cases, we can define Hilbert spaces of functions for which the inner product and norm are defined in terms of an integral over the domain. We omit details, since all the development of this book takes place in the space $\mathbb{R}^n$, though many of the algorithms can be extended to more general Hilbert spaces. However, we mention for purposes of the analysis of Newton-like methods that the following inequality holds for functions of the type that we consider in this book:

$$\left\| \int_a^b F(t) \right\| \leq \int_a^b \| F(t) \| \, dt, \tag{A.12}$$

where F is a continuous scalar-, vector-, or matrix-valued function on the interval $[a, b]$.

SUBSPACES

Given the Euclidean space $\mathbb{R}^n$, the subset $\mathcal{S} \subset \mathbb{R}^n$ is a *subspace of* $\mathbb{R}^n$ if the following property holds: If x and y are any two elements of $\mathcal{S}$, then

$$\alpha x + \beta y \in \mathcal{S}, \qquad \text{for all } \alpha, \beta \in \mathbb{R}.$$

For instance, $\mathcal{S}$ is a subspace of $\mathbb{R}^2$ if it consists of (i) the whole space $\mathbb{R}^n$; (ii) any line passing through the origin; (iii) the origin alone; or (iv) the empty set.

Given any set of vectors $a_i \in \mathbb{R}^n$, $i = 1, 2, \ldots, m$, the set

$$\mathcal{S} = \left\{ w \in \mathbb{R}^n \mid a_i^T w = 0, \; i = 1, 2, \ldots, m \right\} \tag{A.13}$$

is a subspace. However, the set

$$\left\{ w \in \mathbb{R}^n \mid a_i^T w \geq 0, \; i = 1, 2, \ldots, m \right\} \tag{A.14}$$

is not in general a subspace. For example, if we have $n = 2, m = 1$, and $a_1 = (1, 0)^T$, this set would consist of all vectors $(w_1, w_2)^T$ with $w_1 \geq 0$, but then given two vectors $x = (1, 0)^T$ and $y = (2, 3)$ in this set, it is easy to choose multiples α and β such that $\alpha x + \beta y$ has a negative first component, and so lies outside the set.

Sets of the forms (A.13) and (A.14) arise in the discussion of second-order optimality conditions for constrained optimization.

A set of vectors $\{s_1, s_2, \ldots, s_m\}$ in $\mathbb{R}^n$ is called a *linearly independent set* if there are no real numbers $\alpha_1, \alpha_2, \ldots, \alpha_m$ such that

$$\alpha_1 s_2 + \alpha_2 s_2 + \cdots + \alpha_m s_m = 0,$$

unless we make the trivial choice $\alpha_1 = \alpha_2 = \cdots = \alpha_m = 0$. Another way to define linear independence is to say that none of the vectors $s_1, s_2, \ldots, s_m$ can be written as a linear combination of the other vectors in this set. If in fact we have $s_i \in S$ for all $i = 1, 2, \ldots, m$, we say that $\{s_1, s_2, \ldots, s_m\}$ is a *spanning set* for S if *any* vector $s \in S$ can be written as

$$s = \alpha_1 s_2 + \alpha_2 s_2 + \cdots + \alpha_m s_m,$$

for some particular choice of the coefficients $\alpha_1, \alpha_2, \ldots, \alpha_m$.

If the vectors $s_1, s_2, \ldots, s_m$ are both linearly independent and a spanning set for S, we call them a *basis* of S. In this case, m (the number of elements in the basis) is referred to as the *dimension* of S, and denoted by $\dim(S)$. Note that there are many ways to choose a basis of S in general, but that all bases contain the same number of vectors.

If A is any real matrix, the *null space* is the subspace

$$\text{Null}(A) = \{w \mid Aw = 0\},$$

while the *range space* is

$$\text{Range}(A) = \{w \mid w = Av \ \text{ for some vector } v\}.$$

The *fundamental theorem of linear algebra* states that

$$\text{Null}(A) \oplus \text{Range}(A^T) = \mathbb{R}^n,$$

where n is the number of columns in A. (Here, "$\oplus$" denotes the direct sum of two sets: $A \oplus B = \{x + y \mid x \in A, y \in B\}$.)

When A is square ($n \times n$) and nonsingular, we have $\text{Null}A = \text{Null}A^T = \{0\}$ and $\text{Range}A = \text{Range}A^T = \mathbb{R}^n$. In this case, the columns of A form a basis of $\mathbb{R}^n$, as do the columns of A^T.

EIGENVALUES, EIGENVECTORS, AND THE SINGULAR-VALUE DECOMPOSITION

A scalar value λ is an *eigenvalue* of the $n \times n$ matrix A if there is a nonzero vector q such that

$$Aq = \lambda q.$$

The vector q is called an *eigenvector* of A. The matrix A is nonsingular if none of its eigenvalues are zero. The eigenvalues of symmetric matrices are all real numbers, while nonsymmetric matrices may have imaginary eigenvalues. If the matrix is positive definite as well as symmetric, its eigenvalues are all *positive* real numbers.

All matrices A (not necessarily square) can be decomposed as a product of three matrices with special properties. When $A \in \mathbb{R}^{m \times n}$ with $m > n$, (that is, A has more rows than columns), this *singular-value decomposition (SVD)* has the form

$$A = U \begin{bmatrix} S \\ 0 \end{bmatrix} V^T, \tag{A.15}$$

where U and V are orthogonal matrices of dimension $m \times m$ and $n \times n$, respectively, and S is an $n \times n$ diagonal matrix with diagonal elements $\sigma_i, i = 1, 2, \ldots, n$, that satisfy

$$\sigma_1 \geq \sigma_2 \geq \cdots \geq \sigma_n \geq 0.$$

These diagonal values are called the singular values of A. We can define the condition number (A.11) of the $m \times n$ (possibly nonsquare) matrix A to be σ_1 / σ_n. (This definition is identical to $\kappa_2(A)$ when A happens to be square and nonsingular.)

When $m \leq n$ (the number of columns is at least equal to the number of rows), the SVD has the form

$$A = U \begin{bmatrix} S & 0 \end{bmatrix} V^T,$$

where again U and V are orthogonal of dimension $m \times m$ and $n \times n$, respectively, while S is $m \times m$ diagonal with nonnegative diagonal elements $\sigma_1 \geq \sigma_2 \geq \cdots \geq \sigma_m$.

When A is symmetric, its n real eigenvalues $\lambda_1, \lambda_2, \ldots, \lambda_n$ and their associated eigenvectors $q_1, q_2, \ldots, q_n$ can be used to write a *spectral decomposition* of A as follows:

$$A = \sum_{i=1}^{n} \lambda_i q_i q_i^T.$$

This decomposition can be restated in matrix form by defining

$$\Lambda = \text{diag}(\lambda_1, \lambda_2, \cdots, \lambda_n), \qquad Q = [q_1 \,|\, q_2 \,|\, \ldots \,|\, q_n],$$

and writing

$$A = Q \Lambda Q^T. \tag{A.16}$$

In fact, when A is positive definite as well as symmetric, this decomposition is identical to the singular-value decomposition (A.15), where we define $U = V = Q$ and $S = \Lambda$. Note that the singular values σ_i and the eigenvalues λ_i coincide in this case.

In the case of the Euclidean norm (A.8b), we have for symmetric positive definite matrices A that the singular values and eigenvalues of A coincide, and that

$$\|A\| = \sigma_1(A) = \text{largest eigenvalue of } A,$$
$$\|A^{-1}\| = \sigma_n(A)^{-1} = \text{inverse of smallest eigenvalue of } A.$$

Hence, we have for all $x \in \mathbb{R}^n$ that

$$\sigma_n(A)\|x\|^2 = \|x\|^2/\|A^{-1}\| \le x^T A x \le \|A\|\|x\|^2 = \sigma_1(A)\|x\|^2.$$

For an orthogonal matrix Q, we have for the Euclidean norm that

$$\|Qx\| = \|x\|,$$

and that all the singular values of this matrix are equal to 1.

DETERMINANT AND TRACE

The trace of an $n \times n$ matrix A is defined by

$$\text{trace}(A) = \sum_{i=1}^{n} A_{ii}. \tag{A.17}$$

If the eigenvalues of A are denoted by $\lambda_1, \lambda_2, \ldots, \lambda_n$, it can be shown that

$$\text{trace}(A) = \sum_{i=1}^{n} \lambda_i, \tag{A.18}$$

that is, the trace of the matrix is the sum of its eigenvalues.

The determinant of an $n \times n$ matrix A, denoted by $\det A$, is the product of its eigenvalues; that is,

$$\det A = \prod_{i=1}^{n} \lambda_i. \tag{A.19}$$

The determinant has several appealing (and revealing) properties. For instance,

$\det A = 0$ if and only if A is singular;

$\det AB = (\det A)(\det B)$;

$\det A^{-1} = 1/\det A$.

Recall that any orthogonal matrix A has the property that $QQ^T = Q^T Q = I$, so that $Q^{-1} = Q^T$. It follows from the property of the determinant that $\det Q = \det Q^T = \pm 1$. The properties above are used in the analysis of Chapter 6.

MATRIX FACTORIZATIONS: CHOLESKY, LU, QR

Matrix factorizations are important both in the design of algorithms and in their analysis. One such factorization is the singular-value decomposition defined above in (A.15). Here we define the other important factorizations.

All the factorization algorithms described below make use of *permutation matrices*. Suppose that we wish to exchange the first and fourth rows of a matrix A. We can perform this operation by premultiplying A by a permutation matrix P, which is constructed by interchanging the first and fourth rows of an identity matrix that contains the same number of rows as A. Suppose, for example, that A is a 5×5 matrix. The appropriate choice of P would be

$$
P = \begin{bmatrix}
0 & 0 & 0 & 1 & 0 \\
0 & 1 & 0 & 0 & 0 \\
0 & 0 & 1 & 0 & 0 \\
1 & 0 & 0 & 0 & 0 \\
0 & 0 & 0 & 0 & 1
\end{bmatrix}.
$$

A similar technique is used to to find a permutation matrix P that exchanges columns of a matrix.

The LU factorization of a matrix $A \in \mathbb{R}^{n \times n}$ is defined as

$$
PA = LU, \tag{A.20}
$$

where

P is an $n \times n$ permutation matrix (that is, it is obtained by rearranging the rows of the $n \times n$ identity matrix),

L is unit lower triangular (that is, lower triangular with diagonal elements equal to 1, and

U is upper triangular.

This factorization can be used to solve a linear system of the form $Ax = b$ efficiently by the following three-step process:

form $\tilde{b} = Pb$ by permuting the elements of b;

solve $Lz = \tilde{b}$ by performing triangular forward-substitution, to obtain the vector z;

solve $Ux = z$ by performing triangular back-substitution, to obtain the solution vector x.

The factorization (A.20) can be found by using Gaussian elimination with row partial pivoting, an algorithm that requires approximately $2n^3/3$ floating-point operations when A is dense. Standard software that implements this algorithm (notably, LAPACK [7]) is readily available. The method can be stated as follows.

Algorithm A.1 (Gaussian Elimination with Row Partial Pivoting).
  Given $A \in \mathbb{R}^{n \times n}$;
  Set $P \leftarrow I, L \leftarrow 0$;
  for $i = 1, 2, \ldots, n$
          find the index $j \in \{i, i + 1, \ldots, n\}$ such that $|A_{ji}| = \max_{k=i,i+1,\ldots,n} |A_{ki}|$;
          if $A_{ij} = 0$
                  stop; (* matrix A is singular *)
          if $i \neq j$
                  swap rows i and j of matrices A and L;
          (* elimination step *)
          $L_{ii} \leftarrow 1$;
          for $k = i + 1, i + 2, \ldots, n$
                  $L_{ki} \leftarrow A_{ki}/A_{ii}$;
                  for $l = i + 1, i + 2, \ldots, n$
                          $A_{kl} \leftarrow A_{kl} - L_{ki} A_{il}$;
                  end (for)
          end (if)
  end (for)
  $U \leftarrow$ upper triangular part of A.

Variants of the basic algorithm allow for rearrangement of the columns as well as the rows during the factorization, but these do not add to the practical stability properties of the algorithm. Column pivoting may, however, improve the performance of Gaussian elimination when the matrix A is sparse. by ensuring that the factors L and U are also reasonably sparse.

Gaussian elimination can be applied also to the case in which A is not square. When A is $m \times n$, with $m > n$, the standard row pivoting algorithm produces a factorization of the form (A.20), where $L \in \mathbb{R}^{m \times n}$ is unit lower triangular and $U \in \mathbb{R}^{n \times n}$ is upper triangular. When $m < n$, we can find an LU factorization of A^T rather than A, that is, we obtain

$$PA^T = \begin{bmatrix} L_1 \\ L_2 \end{bmatrix} U, \tag{A.21}$$

where L_1 is $m \times m$ (square) unit lower triangular, U is $m \times m$ upper triangular, and L_2 is a general $(n - m) \times m$ matrix. If A has full row rank, we can use this factorization to calculate its null space explicitly as the space spanned by the columns of the matrix

$$M = P^T \begin{bmatrix} L_1^{-T} L_2^T \\ -I \end{bmatrix} U^{-T}. \tag{A.22}$$

It is easy to check that M has dimensions $n \times (n - m)$ and that $AM = 0$.

When $A \in \mathbb{R}^{n \times n}$ is symmetric positive definite, it is possible to compute a similar but more specialized factorization at about half the cost—about $n^3/3$ operations. This factorization, known as the Cholesky factorization, produces a matrix L such that

$$A = LL^T. \tag{A.23}$$

(If we require L to have positive diagonal elements, it is uniquely defined by this formula.) The algorithm can be specified as follows.

Algorithm A.2 (Cholesky Factorization).
   Given $A \in \mathbb{R}^{n \times n}$ symmetric positive definite;
   for $i = 1, 2, \ldots, n$;
      $L_{ii} \leftarrow \sqrt{A_{ii}}$;
      for $j = i + 1, i + 2, \ldots, n$
         $L_{ji} \leftarrow A_{ji}/L_{ii}$;
         for $k = i + 1, i + 2, \ldots, j$
            $A_{jk} \leftarrow A_{jk} - L_{ji}L_{ki}$;
         end (for)
      end (for)
   end (for)

Note that this algorithm references only the lower triangular elements of A; in fact, it is only necessary to store these elements in any case, since by symmetry they are simply duplicated in the upper triangular positions.

Unlike the case of Gaussian elimination, the Cholesky algorithm can produce a valid factorization of a symmetric positive definite matrix without swapping any rows or columns. However, symmetric permutation (that is, reordering the rows and columns in the same way) can be used to improve the sparsity of the factor L. In this case, the algorithm produces a permutation of the form

$$P^T A P = LL^T$$

for some permutation matrix P.

The Cholesky factorization can be used to compute solutions of the system $Ax = b$ by performing triangular forward- and back-substitutions with L and L^T, respectively, as in the case of L and U factors produced by Gaussian elimination.

The Cholesky factorization can also be used to verify positive definiteness of a symmetric matrix A. If Algorithm A.2 runs to completion with all L_{ii} values well defined and positive, then A is positive definite.

Another useful factorization of rectangular matrices $A \in \mathbb{R}^{m \times n}$ has the form

$$AP = QR, \tag{A.24}$$

where

  P is an $n \times n$ permutation matrix,

  A is $m \times m$ orthogonal, and

  R is $m \times n$ upper triangular.

In the case of a square matrix $m = n$, this factorization can be used to compute solutions of linear systems of the form $Ax = b$ via the following procedure:

  set $\tilde{b} = Q^T b$;

  solve $Rz = \tilde{b}$ for z by performing back-substitution;

  set $x = P^T z$ by rearranging the elements of x.

For a dense matrix A, the cost of computing the QR factorization is about $4m^2 n/3$ operations. In the case of a square matrix, the operation count is about twice as high as for an LU factorization via Gaussian elimination. Moreover, it is more difficult to maintain sparsity in a QR factorization than in an LU factorization.

Algorithms to perform QR factorization are almost as simple as algorithms for Gaussian elimination and for Cholesky factorization. The most widely used algorithms work by applying a sequence of special orthogonal matrices to A, known either as Householder transformations or Givens rotations, depending on the algorithm. We omit the details, and refer instead to Golub and Van Loan [136, Chapter 5] for a complete description.

In the case of a rectangular matrix A with $m < n$, we can use the QR factorization of A^T to find a matrix whose columns span the null space of A. To be specific, we write

$$A^T P = QR = \begin{bmatrix} Q_1 & Q_2 \end{bmatrix} R,$$

where Q_1 consists of the first m columns of Q, and Q_2 contains the last $n - m$ columns. It is easy to show that columns of the matrix Q_2 span the null space of A. This procedure yields a more satisfactory basis matrix for the null space than the Gaussian elimination procedure (A.22), because the columns of Q_2 are orthogonal to each other and have unit length. It may be more expensive to compute, however, particularly in the case in which A is sparse.

When A has full column rank, we can make an identification between the R factor in (A.24) and the Cholesky factorization. By multiplying the formula (A.24) by its transpose, we obtain

$$P^T A^T A P = R^T Q^T Q R = R^T R,$$

and by comparison with (A.23), we see that R^T is simply the Cholesky factor of the symmetric positive definite matrix $P^T A^T A P$. Recalling that L is uniquely defined when we restrict its diagonal elements to be positive, this observation implies that R is also uniquely defined for a given choice of permutation matrix P, provided that we enforce positiveness of the diagonals of R. Note, too, that since we can rearrange (A.24) to read $A P R^{-1} = Q$, we can conclude that Q is also uniquely defined under these conditions.

Note that by definition of the Euclidean norm and the property (A.10), and the fact that the Euclidean norms of the matrices P and Q in (A.24) are both 1, we have that

$$\|A\| = \|Q R P^T\| \le \|Q\| \|R\| \|P^T\| = \|R\|,$$

while

$$\|R\| = \|Q^T A P\| \le \|Q^T\| \|A\| \|P\| = \|A\|.$$

We conclude from these two inequalities that $\|A\| = \|R\|$. When A is square, we have by a similar argument that $\|A^{-1}\| = \|R^{-1}\|$. Hence the Euclidean-norm condition number of A can be estimated by substituting R for A in the expression (A.11). This observation is significant because various techniques are available for estimating the condition number of triangular matrices R; see Golub and Van Loan [136, pp. 128–130] for a discussion.

SYMMETRIC INDEFINITE FACTORIZATION

When matrix A is symmetric but indefinite, Algorithm A.2 will break down by trying to take the square root of a negative number. We can however produce a factorization, similar to the Cholesky factorization, of the form

$$P A P^T = L B L^T, \tag{A.25}$$

where L is unit lower triangular, B is a block diagonal matrix with blocks of dimension 1 or 2, and P is a permutation matrix. The first step of this symmetric indefinite factorization proceeds as follows. We identify a submatrix E of A that is suitable to be used as a pivot block. The precise criteria that can be used to choose E are described below, but we note here that E is either a single diagonal element of A (a 1×1 pivot block), or else the 2×2 block consisting of two diagonal elements of A (say, a_{ii} and a_{jj}) along with the corresponding off-diagonal elements (that is, a_{ij} and a_{ji}). In either case, E must be nonsingular. We then find a permutation matrix P_1 that makes E a leading principal submatrix of A, that is,

$$P_1 A P_1 = \begin{bmatrix} E & C^T \\ C & H \end{bmatrix}, \tag{A.26}$$

and then perform a block factorization on this rearranged matrix, using E as the pivot block, to obtain

$$P_1 A P_1^T = \begin{bmatrix} I & 0 \\ CE^{-1} & I \end{bmatrix} \begin{bmatrix} E & 0 \\ 0 & H - CE^{-1}C^T \end{bmatrix} \begin{bmatrix} I & E^{-1}C^T \\ 0 & I \end{bmatrix}.$$

The next step of the factorization consists in applying exactly the same process to $H - CE^{-1}C^T$, known as the *remaining matrix* or the *Schur complement*, which has dimension either $(n-1) \times (n-1)$ or $(n-2) \times (n-2)$. We now apply the same procedure recursively, terminating with the factorization (A.25). Here P is defined as a product of the permutation matrices from each step of the factorization, and B contains the pivot blocks E on its diagonal.

The symmetric indefinite factorization requires approximately $n^3/3$ floating-point operations—the same as the cost of the Cholesky factorization of a positive definite matrix—but to this count we must add the cost of identifying suitable pivot blocks E and of performing the permutations, which can be considerable. There are various strategies for determining the pivot blocks, which have an important effect on both the cost of the factorization and its numerical properties. Ideally, our strategy for choosing E at each step of the factorization procedure should be inexpensive, should lead to at most modest growth in the elements of the remaining matrix at each step of the factorization, and should avoid excessive fill-in (that is, L should not be too much more dense than A).

A well-known strategy, due to Bunch and Parlett [43], searches the whole remaining matrix and identifies the largest-magnitude diagonal and largest-magnitude off-diagonal elements, denoting their respective magnitudes by ξ_{dia} and ξ_{off}. If the diagonal element whose magnitude is ξ_{dia} is selected to be a 1×1 pivot block, the element growth in the remaining matrix is bounded by the ratio $\xi_{\text{dia}}/\xi_{\text{off}}$. If this growth rate is acceptable, we choose this diagonal element to be the pivot block. Otherwise, we select the off-diagonal element whose magnitude is ξ_{off} (a_{ij}, say), and choose E to be the 2×2 submatrix that includes this element, that is,

$$E = \begin{bmatrix} a_{ii} & a_{ij} \\ a_{ij} & a_{jj} \end{bmatrix}.$$

This pivoting strategy of Bunch and Parlett is numerically stable and guarantees to yield a matrix L whose maximum element is bounded by 2.781. Its drawback is that the evaluation of ξ_{dia} and ξ_{off} at each iteration requires many comparisons between floating-point numbers to be performed: $O(n^3)$ in total during the overall factorization. Since each comparison costs roughly the same as an arithmetic operation, this overhead is not insignificant.

The more economical pivoting strategy of Bunch and Kaufman [42] searches at most two columns of the working matrix at each stage and requires just $O(n^2)$ comparisons in total. Its rationale and details are somewhat tricky, and we refer the interested reader to the original paper [42] or to Golub and Van Loan [136, Section 4.4] for details. Unfortunately, this algorithm can give rise to arbitrarily large elements in the lower triangular factor L, making it unsuitable for use with a modified Cholesky strategy.

The bounded Bunch–Kaufman strategy is essentially a compromise between the Bunch–Parlett and Bunch–Kaufman strategies. It monitors the sizes of elements in L, accepting the (inexpensive) Bunch–Kaufman choice of pivot block when it yields only modest element growth, but searching further for an acceptable pivot when this growth is excessive. Its total cost is usually similar to that of Bunch–Kaufman, but in the worst case it can approach the cost of Bunch–Parlett.

So far, we have ignored the effect of the choice of pivot block E on the sparsity of the final L factor. This consideration is important when the matrix to be factored is large and sparse, since it greatly affects both the CPU time and the amount of storage required by the algorithm. Algorithms that modify the strategies above to take account of sparsity have been proposed by Duff et al. [97], Duff and Reid [95], and Fourer and Mehrotra [113].

SHERMAN–MORRISON–WOODBURY FORMULA

If the square nonsingular matrix A undergoes a rank-one update to become

$$\bar{A} = A + ab^T,$$

where $a, b \in \mathbb{R}^n$, then if $\bar{A}$ is nonsingular, we have

$$\bar{A}^{-1} = A^{-1} - \frac{A^{-1}ab^T A^{-1}}{1 + b^T A^{-1}a}. \tag{A.27}$$

It is easy to verify this formula: Simply multiply the definitions of $\bar{A}$ and $\bar{A}^{-1}$ together and check that they produce the identity.

This formula can be extended to higher-rank updates. Let U and V be matrices in $\mathbb{R}^{n \times p}$ for some p between 1 and n. If we define

$$\hat{A} = A + UV^T,$$

then $\hat{A}$ is nonsingular if and only if $(I + V^T A^{-1} U)$ is nonsingular, and in this case we have

$$\hat{A}^{-1} = A^{-1} - A^{-1}U(I + V^T A^{-1}U)^{-1}V^T A^{-1}. \tag{A.28}$$

We can use this formula to solve linear systems of the form $\bar{A}x = d$. Since

$$x = \hat{A}^{-1}d = A^{-1}d - A^{-1}U(I + V^T A^{-1}U)^{-1}V^T A^{-1}d,$$

we see that x can be found by solving $p+1$ linear systems with the matrix A (to obtain $A^{-1}d$ and $A^{-1}U$), inverting the $p \times p$ matrix $I + V^T A^{-1}U$, and performing some elementary matrix algebra. Inversion of the $p \times p$ matrix $I + V^T A^{-1}U$ is inexpensive when $p \ll n$.

INTERLACING EIGENVALUE THEOREM

The following result is proved for example in Golub and Van Loan [136, Theorem 8.1.8].

Theorem A.1 (Interlacing Eigenvalue Theorem).
Let $A \in \mathbb{R}^{n \times n}$ be a symmetric matrix with eigenvalues $\lambda_1, \lambda_2, \ldots, \lambda_n$ satisfying

$$\lambda_1 \geq \lambda_2 \geq \cdots \geq \lambda_n,$$

and let $z \in \mathbb{R}^n$ be a vector with $\|z\| = 1$, and $\alpha \in \mathbb{R}$ be a scalar. Then if we denote the eigenvalues of $A + \alpha z z^T$ by $\xi_1, \xi_2, \ldots, \xi_n$ (in decreasing order), we have for $\alpha > 0$ that

$$\xi_1 \geq \lambda_1 \geq \xi_2 \geq \lambda_2 \geq \xi_3 \geq \cdots \geq \xi_n \geq \lambda_n,$$

with

$$\sum_{i=1}^{n} \xi_i - \lambda_i = \alpha. \tag{A.29}$$

If $\alpha < 0$, we have that

$$\lambda_1 \geq \xi_1 \geq \lambda_2 \geq \xi_2 \geq \lambda_3 \geq \cdots \geq \lambda_n \geq \xi_n,$$

where the relationship (A.29) is again satisfied.

Informally stated, the eigenvalues of the modified matrix "interlace" the eigenvalues of the original matrix, with nonnegative adjustments if the coefficient α is positive, and nonpositive adjustments if α is negative. The total magnitude of the adjustments equals α, whose magnitude is identical to the Euclidean norm $\|\alpha z z^T\|_2$ of the modification.

ERROR ANALYSIS AND FLOATING-POINT ARITHMETIC

In most of this book our algorithms and analysis deal with real numbers. Modern digital computers, however, cannot store or compute with general real numbers. Instead, they work with a subset known as *floating-point numbers*. Any quantities that are stored on the computer, whether they are read directly from a file or program or arise as the intermediate result of a computation, must be approximated by a floating-point number. In general, then, the numbers that are produced by practical computation differ from those that would be produced if the arithmetic were exact. Of course, we try to perform our computations in such a way that these differences are as tiny as possible.

Discussion of errors requires us to distinguish between *absolute error* and *relative error*. If x is some exact quantity (scalar, vector, matrix) and $\tilde{x}$ is its approximate value, the absolute error is the norm of the difference, namely, $\|x - \tilde{x}\|$. (In general, any of the norms (A.2a), (A.2b), and (A.2c) can be used in this definition.) The relative error is the ratio of the absolute error to the size of the exact quantity, that is,

$$\frac{\|x - \tilde{x}\|}{\|x\|}.$$

When this ratio is significantly less than one, we can replace the denominator by the size of the approximate quantity—that is, $\|\tilde{x}\|$—without affecting its value very much.

Most computations associated with optimization algorithms are performed in double-precision arithmetic. Double-precision numbers are stored in words of length 64 bits. Most of these bits (say t) are devoted to storing the *fractional part*, while the remainder encode the *exponent e* and other information, such as the sign of the number, or an indication of whether it is zero or "undefined." Typically, the fractional part has the form

$$.d_1 d_2 \ldots d_t,$$

where each $d_i, i = 1, 2, \ldots, t$, is either zero or one. (In some systems d_1 is implicitly assumed to be 1 and is not stored.) The value of the floating-point number is then

$$\sum_{i=1}^{t} d_i 2^{-i} \times 2^e.$$

The value 2^{-t-1} is known as *unit roundoff* and is denoted by $\mathbf{u}$. Any real number whose absolute value lies in the range $[2^L, 2^U]$ (where L and U are lower and upper bounds on the value of the exponent e) can be approximated to within a relative accuracy of $\mathbf{u}$ by a floating-point number, that is,

$$\mathrm{fl}(x) = x(1 + \epsilon), \qquad \text{where } |\epsilon| \leq \mathbf{u}, \tag{A.30}$$

where $\mathrm{fl}(\cdot)$ denotes floating-point approximation. The value of $\mathbf{u}$ for double-precision IEEE arithmetic is about 1.1×10^{-16}. In other words, if the real number x and its floating-point approximation are both written as base-10 numbers (the usual fashion), they agree to at least 15 digits.

For further information on floating-point computations, see Overton [233], Golub and Van Loan [136, Section 2.4], and Higham [169].

When an arithmetic operation is performed with one or two floating-point numbers, the result must also be stored as a floating-point number. This process introduces a small *roundoff error*, whose size can be quantified in terms of the size of the arguments. If x and y are two floating-point numbers, we have that

$$|\mathrm{fl}(x * y) - x * y| \le \mathbf{u}|x * y|, \tag{A.31}$$

where $*$ denotes any of the operations $+, -, \times, \div$.

Although the error in a single floating-point operation appears benign, more significant errors may occur when the arguments x and y are floating-point approximations of two *real* numbers, or when a sequence of computations are performed in succession. Suppose, for instance, that x and y are large real numbers whose values are very similar. When we store them in a computer, we approximate them with floating-point numbers $\mathrm{fl}(x)$ and $\mathrm{fl}(y)$ that satisfy

$$\mathrm{fl}(x) = x + \epsilon_x, \quad \mathrm{fl}(y) = y + \epsilon_y, \quad \text{where } |\epsilon_x| \le \mathbf{u}|x|, |\epsilon_y| \le \mathbf{u}|y|.$$

If we take the difference of the two stored numbers, we obtain a final result $\mathrm{fl}(\mathrm{fl}(x) - \mathrm{fl}(y))$ that satisfies

$$\mathrm{fl}(\mathrm{fl}(x) - \mathrm{fl}(y)) = (\mathrm{fl}(x) - \mathrm{fl}(y))(1 + \epsilon_{xy}), \quad \text{where } |\epsilon_{xy}| \le \mathbf{u}.$$

By combining these expressions, we find that the difference between this result and the true value $x - y$ may be as large as

$$\epsilon_x + \epsilon_y + \epsilon_{xy},$$

which is bounded by $\mathbf{u}(|x| + |y| + |x - y|)$. Hence, since x and y are large and close together, the relative error is approximately $2\mathbf{u}|x|/|x - y|$, which may be quite large, since $|x| \gg |x - y|$.

This phenomenon is known as *cancellation*. It can also be explained (less formally) by noting that if both x and y are accurate to k digits, and if they agree in the first $\bar{k}$ digits, then their difference will contain only about $k - \bar{k}$ significant digits—the first $\bar{k}$ digits cancel each other out. This observation is the reason for the well-known adage of numerical computing—that one should avoid taking the difference of two similar numbers if at all possible.

CONDITIONING AND STABILITY

Conditioning and *stability* are two terms that are used frequently in connection with numerical computations. Unfortunately, their meaning sometimes varies from author to author, but the general definitions below are widely accepted, and we adhere to them in this book.

Conditioning is a property of the numerical problem at hand (whether it is a linear algebra problem, an optimization problem, a differential equations problem, or whatever). A problem is said to be *well conditioned* if its solution is not affected greatly by small perturbations to the data that define the problem. Otherwise, it is said to be *ill conditioned*. A simple example is given by the following 2×2 system of linear equations:

$$\begin{bmatrix} 1 & 2 \\ 1 & 1 \end{bmatrix} \begin{bmatrix} x_1 \\ x_2 \end{bmatrix} = \begin{bmatrix} 3 \\ 2 \end{bmatrix}.$$

By computing the inverse of the coefficient matrix, we find that the solution is simply

$$\begin{bmatrix} x_1 \\ x_2 \end{bmatrix} = \begin{bmatrix} -1 & 2 \\ 1 & -1 \end{bmatrix} \begin{bmatrix} 3 \\ 2 \end{bmatrix} = \begin{bmatrix} 1 \\ 1 \end{bmatrix}.$$

If we replace the first right-hand-side element by 3.00001, the solution becomes $(x_1, x_2)^T = (0.99999, 1.00001)^T$, which is only slightly different from its exact value $(1, 1)^T$. We would note similar insensitivity if we were to perturb the other elements of the right-hand-side or elements of the coefficient matrix. We conclude that this problem is well conditioned. On the other hand, the problem

$$\begin{bmatrix} 1.00001 & 1 \\ 1 & 1 \end{bmatrix} \begin{bmatrix} x_1 \\ x_2 \end{bmatrix} = \begin{bmatrix} 2.00001 \\ 2 \end{bmatrix}$$

is ill conditioned. Its exact solution is $x = (1, 1)^T$, but if we change the first element of the right-hand-side from 2.00001 to 2, the solution would change drastically to $x = (0, 2)^T$.

For general square linear systems $Ax = b$ where $A \in \mathbb{R}^{n \times n}$, the condition number of the matrix (defined in (A.11)) can be used to quantify the conditioning. Specifically, if we perturb A to $\tilde{A}$ and b to $\tilde{b}$ and take $\tilde{x}$ to be the solution of the perturbed system $\tilde{A}\tilde{x} = \tilde{b}$, it can be shown that

$$\frac{\|x - \tilde{x}\|}{\|x\|} \approx \kappa(A) \left[\frac{\|A - \tilde{A}\|}{\|A\|} + \frac{\|b - \tilde{b}\|}{\|b\|} \right]$$

(see, for instance, Golub and Van Loan [136, Section 2.7]). Hence, a large condition number $\kappa(A)$ indicates that the problem $Ax = b$ is ill conditioned, while a modest value indicates well conditioning.

Note that the concept of conditioning has nothing to do with the particular algorithm that is used to solve the problem, only with the numerical problem itself.

Stability, on the other hand, is a property of the algorithm. An algorithm is stable if it is guaranteed to produce accurate answers to all well-conditioned problems in its class, even when floating-point arithmetic is used.

As an example, consider again the linear equations $Ax = b$. We can show that Algorithm A.1, in combination with triangular substitution, yields a computed solution $\tilde{x}$ whose relative error is approximately

$$\frac{\|x - \tilde{x}\|}{\|x\|} \approx \kappa(A)\frac{\text{growth}(A)}{\|A\|}\mathbf{u}, \tag{A.32}$$

where $\text{growth}(A)$ is the size of the largest element that arises in A during execution of Algorithm A.1. In the worst case, we can show that $\text{growth}(A)/\|A\|$ may be around 2^{n-1}, which indicates that Algorithm A.1 is an unstable algorithm, since even for modest n (say, $n = 200$), the right-hand-side of (A.32) may be large even when $\kappa(A)$ is modest. In practice, however, large growth factors are rarely observed, so we conclude that Algorithm A.1 is stable for all practical purposes.

Gaussian elimination without pivoting, on the other hand, is definitely unstable. If we omit the possible exchange of rows in Algorithm A.1, the algorithm will fail to produce a factorization even of some well-conditioned matrices, such as

$$A = \begin{bmatrix} 0 & 1 \\ 1 & 2 \end{bmatrix}.$$

For systems $Ax = b$ in which A is symmetric positive definite, the Cholesky factorization in combination with triangular substitution constitutes a stable algorithm for producing a solution x.

A.2   ELEMENTS OF ANALYSIS, GEOMETRY, TOPOLOGY

SEQUENCES

Suppose that $\{x_k\}$ is a sequence of points belonging to $\mathbb{R}^n$. We say that a sequence $\{x_k\}$ *converges* to some point x, written $\lim_{k\to\infty} x_k = x$, if for any $\epsilon > 0$, there is an index K such that

$$\|x_k - x\| \le \epsilon, \quad \text{for all } k \ge K.$$

For example, the sequence $\{x_k\}$ defined by $x_k = (1 - 2^{-k}, 1/k^2)^T$ converges to $(1, 0)^T$.

Given a index set $\mathcal{S} \subset \{1, 2, 3, \ldots\}$, we can define a *subsequence* of $\{t_k\}$ corresponding to $\mathcal{S}$, and denote it by $\{t_k\}_{k \in \mathcal{S}}$.

We say that $\hat{x} \in \mathbb{R}^n$ is an *accumulation point* or *limit point* for $\{x_k\}$ if there is an infinite set of indices $k_1, k_2, k_3, \ldots$ such that the subsequence $\{x_{k_i}\}_{i=1,2,3,\ldots}$ converges to $\hat{x}$; that is,

$$\lim_{i \to \infty} x_{k_i} = \hat{x}.$$

Alternatively, we say that for any $\epsilon > 0$ and all positive integers K, we have

$$\|x_k - x\| \le \epsilon, \quad \text{for some } k \ge K.$$

An example is given by the sequence

$$\begin{bmatrix} 1 \\ 1 \end{bmatrix}, \begin{bmatrix} 1/2 \\ 1/2 \end{bmatrix}, \begin{bmatrix} 1 \\ 1 \end{bmatrix}, \begin{bmatrix} 1/4 \\ 1/4 \end{bmatrix}, \begin{bmatrix} 1 \\ 1 \end{bmatrix}, \begin{bmatrix} 1/8 \\ 1/8 \end{bmatrix}, \ldots, \tag{A.33}$$

which has exactly two limit points: $\hat{x} = (0, 0)^T$ and $\hat{x} = (1, 1)^T$. A sequence can even have an infinite number of limit points. An example is the sequence $x_k = \sin k$, for which every point in the interval $[-1, 1]$ is a limit point. A sequence converges if and only if it has exactly one limit point.

A sequence is said to be a *Cauchy sequence* if for any $\epsilon > 0$, there exists an integer K such that $\|x_k - x_l\| \le \epsilon$ for all indices $k \ge K$ and $l \ge K$. A sequence converges if and only if it is a Cauchy sequence.

We now consider *scalar sequences* $\{t_k\}$, that is, $t_k \in \mathbb{R}$ for all k. This sequence is said to be *bounded above* if there exists a scalar u such that $t_k \le u$ for all k, and bounded below if there is a scalar v with $t_k \ge v$ for all k. The sequence $\{t_k\}$ is said to be *nondecreasing* if $t_{k+1} \ge t_k$ for all k, and *nonincreasing* if $t_{k+1} \le t_k$ for all k. If $\{t_k\}$ is nondecreasing and *bounded above*, then it converges, that is, $\lim_{k \to \infty} t_k = t$ for some scalar t. Similarly, if $\{t_k\}$ is nonincreasing and bounded below, it converges.

We define the *supremum* of the scalar sequence $\{t_k\}$ as the smallest real number u such that $t_k \le u$ for all $k = 1, 2, 3, \ldots$, and denote it by $\sup\{t_k\}$. The *infimum*, denoted by $\inf\{t_k\}$, is the largest real number v such that $v \le t_k$ for all $k = 1, 2, 3, \ldots$. We can now define the sequence of suprema as $\{u_i\}$, where

$$u_i \stackrel{\text{def}}{=} \sup\{t_k \,|\, k \ge i\}.$$

Clearly, $\{u_i\}$ is a nonincreasing sequence. If bounded below, it converges to a finite number $\bar{u}$, which we call the "lim sup" of $\{t_k\}$, denoted by $\limsup t_k$. Similarly, we can denote the sequence of infima by $\{v_i\}$, where

$$v_i \stackrel{\text{def}}{=} \inf\{t_k \,|\, k \ge i\},$$

which is nondecreasing. If $\{v_i\}$ is bounded above, it converges to a point $\bar{v}$ which we call the "lim inf" of $\{t_k\}$, denoted by $\liminf t_k$. As an example, the sequence $1, \frac{1}{2}, 1, \frac{1}{4}, 1, \frac{1}{8}, \ldots$ has a lim inf of 0 and a lim sup of 1.

RATES OF CONVERGENCE

One of the key measures of performance of an algorithm is its rate of convergence. Here, we define the terminology associated with different types of convergence.

Let $\{x_k\}$ be a sequence in $\mathbb{R}^n$ that converges to x^*. We say that the convergence is *Q-linear* if there is a constant $r \in (0, 1)$ such that

$$\frac{\|x_{k+1} - x^*\|}{\|x_k - x^*\|} \leq r, \qquad \text{for all } k \text{ sufficiently large.} \tag{A.34}$$

This means that the distance to the solution x^* decreases at each iteration by at least a constant factor bounded away from 1. For example, the sequence $1 + (0.5)^k$ converges Q-linearly to 1, with rate $r = 0.5$. The prefix "Q" stands for "quotient," because this type of convergence is defined in terms of the quotient of successive errors.

The convergence is said to be *Q-superlinear* if

$$\lim_{k \to \infty} \frac{\|x_{k+1} - x^*\|}{\|x_k - x^*\|} = 0.$$

For example, the sequence $1 + k^{-k}$ converges superlinearly to 1. (Prove this statement!) *Q-quadratic* convergence, an even more rapid convergence rate, is obtained if

$$\frac{\|x_{k+1} - x^*\|}{\|x_k - x^*\|^2} \leq M, \qquad \text{for all } k \text{ sufficiently large,}$$

where M is a positive constant, not necessarily less than 1. An example is the sequence $1 + (0.5)^{2^k}$.

The speed of convergence depends on r and (more weakly) on M, whose values depend not only on the algorithm but also on the properties of the particular problem. Regardless of these values, however, a quadratically convergent sequence will always eventually converge faster than a linearly convergent sequence.

Obviously, any sequence that converges Q-quadratically also converges Q-super-linearly, and any sequence that converges Q-superlinearly also converges Q-linearly. We can also define higher rates of convergence (cubic, quartic, and so on), but these are less interesting in practical terms. In general, we say that the Q-order of convergence is p (with $p > 1$) if there is a positive constant M such that

$$\frac{\|x_{k+1} - x^*\|}{\|x_k - x^*\|^p} \leq M, \qquad \text{for all } k \text{ sufficiently large.}$$

Quasi-Newton methods for unconstrained optimization typically converge Q-superlinearly, whereas Newton's method converges Q-quadratically under appropriate assumptions. In contrast, steepest descent algorithms converge only at a Q-linear rate, and when the problem is ill-conditioned the convergence constant r in (A.34) is close to 1.

In the book, we omit the letter Q and simply talk about superlinear convergence, quadratic convergence, and so on.

A slightly weaker form of convergence, characterized by the prefix "R" (for "root"), is concerned with the overall rate of decrease in the error, rather than the decrease over each individual step of the algorithm. We say that convergence is *R-linear* if there is a sequence of nonnegative scalars $\{v_k\}$ such that

$$\|x_k - x^*\| \leq v_k \text{ for all } k, \text{ and } \{v_k\} \text{ converges Q-linearly to zero.}$$

The sequence $\{\|x_k - x^*\|\}$ is said to be *dominated* by $\{v_k\}$. For instance, the sequence

$$x_k = \begin{cases} 1 + (0.5)^k, & k \text{ even,} \\ 1, & k \text{ odd,} \end{cases} \tag{A.35}$$

(the first few iterates are 2, 1, 1.25, 1, 1.03125, 1, ...) converges R-linearly to 1, because we have $(1 + (0.5)^k) - 1| = (0.)^k$, and the sequence $\{(0.5)^k\}$ converges Q-linearly to zero. Likewise, we say that $\{x_k\}$ converges R-superlinearly to x^* if $\{\|x_k - x^*\|\}$ is dominated by a sequence of scalars converging Q-superlinearly to zero, and $\{x_k\}$ converges R-quadratically to x^* if $\{\|x_k - x^*\|\}$ is dominated by a sequence converging Q-quadratically to zero.

Note that in the R-linear sequence (A.35), the error actually increases at every second iteration! Such behavior occurs even in sequences whose R-rate of convergence is arbitrarily high, but it cannot occur for Q-linear sequences, which insist on a decrease at every step k, for k sufficiently large.

For an extensive discussion of convergence rates see Ortega and Rheinboldt [230].

TOPOLOGY OF THE EUCLIDEAN SPACE $\mathbb{R}^n$

The set $\mathcal{F}$ is *bounded* if there is some real number $M > 0$ such that

$$\|x\| \leq M, \quad \text{for all } x \in \mathcal{F}.$$

A subset $\mathcal{F} \subset \mathbb{R}^n$ is *open* if for every $x \in \mathcal{F}$, we can find a positive number $\epsilon > 0$ such that the ball of radius ϵ around x is contained in $\mathcal{F}$; that is,

$$\{y \in \mathbb{R}^n \mid \|y - x\| \leq \epsilon\} \subset \mathcal{F}.$$

The set $\mathcal{F}$ is *closed* if for all possible sequences of points $\{x_k\}$ in $\mathcal{F}$, all limit points of $\{x_k\}$ are elements of $\mathcal{F}$. For instance, the set $\mathcal{F} = (0, 1) \cup (2, 10)$ is an open subset of $\mathbb{R}$, while

$\mathcal{F} = [0, 1] \cup [2, 5]$ is a closed subset of $\mathbf{R}$. The set $\mathcal{F} = (0, 1]$ is a subset of $\mathbf{R}$ that is neither open nor closed.

The *interior* of a set $\mathcal{F}$, denoted by int $\mathcal{F}$, is the largest open set contained in $\mathcal{F}$. The *closure* of $\mathcal{F}$, denoted by cl $\mathcal{F}$, is the smallest closed set containing $\mathcal{F}$. In other words, we have

$$x \in \text{cl} \mathcal{F} \quad \text{if } \lim_{k \to \infty} x_k = x \text{ for some sequence } \{x_k\} \text{ of points in } \mathcal{F}.$$

If $\mathcal{F} = (-1, 1] \cup [2, 4)$, then

$$\text{cl} \mathcal{F} = [-1, 1] \cup [2, 4], \quad \text{int } \mathcal{F} = (-1, 1) \cup (2, 4).$$

Note that if $\mathcal{F}$ is open, then int $\mathcal{F} = \mathcal{F}$, while if $\mathcal{F}$ is closed, then cl $\mathcal{F} = \mathcal{F}$.

We note the following facts about open and closed sets. The union of finitely many closed sets is closed, while any intersection of closed sets is closed. The intersection of finitely many open sets is open, while any union of open sets is open.

The set $\mathcal{F}$ is *compact* if every sequence $\{x^k\}$ of points in $\mathcal{F}$ has at least one limit point, and all such limit points are in $\mathcal{F}$. (This definition is equivalent to the more formal one involving covers of $\mathcal{F}$.) The following is a central result in topology:

$$\mathcal{F} \in \mathbf{R}^n \text{ is closed and bounded} \Rightarrow \mathcal{F} \text{ is compact.}$$

Given a point $x \in \mathbf{R}^n$, we call $\mathcal{N} \in \mathbf{R}^n$ a *neighborhood of* x if it is an open set containing x. An especially useful neighborhood is the *open ball of radius* ϵ *around* x, which is denoted by $\mathbf{B}(x, \epsilon)$; that is,

$$\mathbf{B}(x, \epsilon) = \{y \mid \|y - x\| < \epsilon\}.$$

Given a set $\mathcal{F} \subset \mathbf{R}^n$, we say that $\mathcal{N}$ is a *neighborhood of* $\mathcal{F}$ if there is $\epsilon > 0$ such that

$$\cup_{x \in \mathcal{F}} \mathbf{B}(x, \epsilon) \subset \mathcal{N}.$$

CONVEX SETS IN $\mathbf{R}^n$

A *convex combination* of a finite set of vectors $\{x_1, x_2, \ldots, x_m\}$ in $\mathbf{R}^m$ is any vector x of the form

$$x = \sum_{i=1}^{m} \alpha_i x_i, \quad \text{where } \sum_{i=1}^{m} \alpha_i = 1, \quad \text{and } \alpha_i \geq 0 \text{ for all } i = 1, 2, \ldots, m.$$

The *convex hull* of $\{x_1, x_2, \ldots, x_m\}$ is the set of all convex combinations of these vectors.

A *cone* is a set $\mathcal{F}$ with the property that for all $x \in \mathcal{F}$ we have

$$x \in \mathcal{F} \Rightarrow \alpha x \in \mathcal{F}, \quad \text{for all } \alpha > 0. \tag{A.36}$$

For instance, the set $\mathcal{F} \subset \mathbf{R}^2$ defined by

$$\{(x_1, x_2)^T \mid x_1 > 0, \; x_2 \geq 0\}$$

is a cone in $\mathbf{R}^2$. Note that cones are not necessarily convex. For example, the set $\{(x_1, x_2)^T \mid x_1 \geq 0 \text{ or } x_2 \geq 0\}$, which encompasses three quarters of the two-dimensional plane, is a cone.

The *cone generated by* $\{x_1, x_2, \ldots, x_m\}$ is the set of all vectors x of the form

$$x = \sum_{i=1}^{m} \alpha_i x_i, \quad \text{where } \alpha_i \geq 0 \text{ for all } i = 1, 2, \ldots, m.$$

Note that all cones of this form are convex.

Finally, we define the affine hull and relative interior of a set. An *affine set* in $\mathbf{R}^n$ is a the set of all vectors $\{x\} \oplus S$, where $x \in \mathbf{R}^n$ and S is a subspace of $\mathbf{R}^n$. Given $\mathcal{F} \subset \mathbf{R}^n$, the *affine hull* of $\mathcal{F}$ (denoted by aff $\mathcal{F}$) is the smallest affine set containing $\mathcal{F}$. For instance, when $\mathcal{F}$ is the "ice-cream cone" defined in three dimensions as

$$\Omega = \left\{ x \in \mathbf{R}^3 \mid x_3 \geq 2\sqrt{x_1^2 + x_2^2} \right\} \tag{A.37}$$

(see Figure A.1), we have aff $\mathcal{F} = \mathbf{R}^3$. If $\mathcal{F}$ is the set of two isolated points $\mathcal{F} = \{(1, 0, 0)^T, (0, 2, 0)^T\}$, we have

$$\text{aff } \mathcal{F} = \{(1, 0, 0)^T + \alpha(-1, 2, 0)^T \mid \text{for all } \alpha \in \mathbf{R}\}.$$

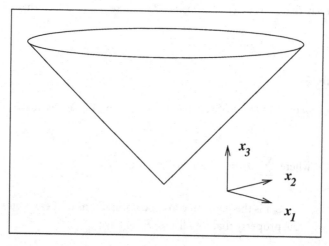

Figure A.1   "Ice-cream cone" set.

The *relative interior* ri $\mathcal{F}$ of the set $\mathcal{F}$ is its *interior relative to* aff $\mathcal{F}$. If $x \in \mathcal{F}$, then $x \in$ ri $\mathcal{F}$ if there is an $\epsilon > 0$ such that

$$(x + \epsilon B) \cap \text{aff} \, \mathcal{F} \subset \mathcal{F}.$$

Referring again to the ice-cream cone (A.37), we have that

$$\text{ri} \, \mathcal{F} = \left\{ x \in \mathbb{R}^3 \,\middle|\, x_3 > 2\sqrt{x_1^2 + x_2^2} \right\}.$$

For the set of two isolated points $\mathcal{F} = \{(1, 0, 0)^T, (0, 2, 0)^T\}$, we have ri $\mathcal{F} = \emptyset$. For the set $\mathcal{F}$ defined by

$$\mathcal{F} \overset{\text{def}}{=} \{x \in \mathbb{R}^3 \,|\, x_1 \in [0, 1], \ x_2 \in [0, 1], \ x_3 = 0\},$$

we have that

$$\text{aff} \, \mathcal{F} = \mathbb{R} \times \mathbb{R} \times \{0\}, \quad \text{ri} \, \mathcal{F} = \{x \in \mathbb{R}^3 \,|\, x_1 \in (0, 1), \ x_2 \in (0, 1), \ x_3 = 0\}.$$

CONTINUITY AND LIMITS

Let f be a function that maps some domain $\mathcal{D} \subset \mathbb{R}^n$ to the space $\mathbb{R}^m$. For some point $x_0 \in \text{cl}\mathcal{D}$, we write

$$\lim_{x \to x_0} f(x) = f_0 \tag{A.38}$$

(spoken "the limit of $f(x)$ as x approaches x_0 is f_0") if for all $\epsilon > 0$, there is a value $\delta > 0$ such that

$$\|x - x_0\| < \delta \text{ and } x \in \mathcal{D} \implies \|f(x) - f_0\| < \epsilon.$$

We say that f is *continuous* at x_0 if $x_0 \in \mathcal{D}$ and the expression (A.38) holds with $f_0 = f(x_0)$. We say that f is continuous on its domain $\mathcal{D}$ if f is continuous for all $x_0 \in \mathcal{D}$.

An example is provided by the function

$$f(x) = \begin{cases} -x & \text{if } x \in [-1, 1], x \neq 0, \\ 5 & \text{for all other } x \in [-10, 10]. \end{cases} \tag{A.39}$$

This function is defined on the domain $[-10, 10]$ and is continuous at all points of the domain except the points $x = 0$, $x = 1$, and $x = -1$. At $x = 0$, the expression (A.38) holds with $f_0 = 0$, but the function is not continuous at this point because $f_0 \neq f(0) = 5$. At

$x = -1$, the limit (A.38) is not defined, because the function values in the neighborhood of this point are close to both 5 and -1, depending on whether x is slightly smaller or slightly larger than -1. Hence, the function is certainly not continuous at this point. The same comments apply to the point $x = 1$.

In the special case of $n = 1$ (that is, the argument of f is a real scalar), we can also define the *one-sided limit*. Given $x_0 \in clD$, We write

$$\lim_{x \downarrow x_0} f(x) = f_0 \tag{A.40}$$

(spoken "the limit of $f(x)$ as x approaches x_0 from above is f_0") if for all $\epsilon > 0$, there is a value $\delta > 0$ such that

$$x_0 < x < x_0 + \delta \quad \text{and} \quad x \in \mathcal{D} \quad \Rightarrow \quad \|f(x) - f_0\| < \epsilon.$$

Similarly, we write

$$\lim_{x \uparrow x_0} f(x) = f_0 \tag{A.41}$$

(spoken "the limit of $f(x)$ as x approaches x_0 from below is f_0") if for all $\epsilon > 0$, there is a value $\delta > 0$ such that

$$x_0 - \delta < x < x_0 \quad \text{and} \quad x \in \mathcal{D} \quad \Rightarrow \quad \|f(x) - f_0\| < \epsilon.$$

For the function defined in (A.39), we have that

$$\lim_{x \downarrow 1} f(x) = 5, \quad \lim_{x \uparrow 1} f(x) = 1.$$

Considering again the general case of $f : \mathcal{D} \to \mathbb{R}^m$ where $\mathcal{D} \subset \mathbb{R}^n$ for general m and n. The function f is said to be *Lipschitz continuous* on some set $\mathcal{N} \subset \mathcal{D}$ if there is a constant $L > 0$ such that

$$\|f(x_1) - f(x_0)\| \leq L\|x_1 - x_0\|, \quad \text{for all } x_0, x_1 \in \mathcal{N}. \tag{A.42}$$

(L is called the *Lipschitz constant*.) The function f is *locally Lipschitz continuous* at a point $\bar{x} \in \text{int}\mathcal{D}$ if there is some neighborhood $\mathcal{N}$ of $\bar{x}$ with $\mathcal{N} \subset \mathcal{D}$ such that the property (A.42) holds for some $L > 0$.

If g and h are two functions mapping $\mathcal{D} \subset \mathbb{R}^n$ to $\mathbb{R}^m$, Lipschitz continuous on a set $\mathcal{N} \subset \mathcal{D}$, their sum $g + h$ is also Lipschitz continuous, with Lipschitz constant equal to the sum of the Lipschitz constants for g and h individually. If g and h are two functions mapping $\mathcal{D} \subset \mathbb{R}^n$ to $\mathbb{R}$, the product gh is Lipschitz continuous on a set $\mathcal{N} \subset \mathcal{D}$ if both g and h are Lipschitz continuous on $\mathcal{N}$ and both are bounded on $\mathcal{N}$ (that is, there is $M > 0$

such that $|g(x)| \leq M$ and $|h(x)| \leq M$ for all $x \in \mathcal{N}$). We prove this claim via a sequence of elementary inequalities, for arbitrary $x_0, x_1 \in \mathcal{N}$:

$$
\begin{aligned}
&|g(x_0)h(x_0) - g(x_1)h(x_1)| \\
&\leq |g(x_0)h(x_0) - g(x_1)h(x_0)| + |g(x_1)h(x_0) - g(x_1)h(x_1)| \\
&= |h(x_0)|\, |g(x_0) - g(x_1)| + |g(x_1)|\, |h(x_0) - h(x_1)| \\
&\leq 2ML\|x_0 - x_1\|,
\end{aligned}
\tag{A.43}
$$

where L is an upper bound on the Lipschitz constant for both g and h.

DERIVATIVES

Let $\phi : \mathbb{R} \to \mathbb{R}$ be a real-valued function of a real variable (sometimes known as a *univariate* function). The first derivative $\phi'(\alpha)$ is defined by

$$
\frac{d\phi}{d\alpha} = \phi'(\alpha) \overset{\text{def}}{=} \lim_{\epsilon \to 0} \frac{\phi(\alpha + \epsilon) - \phi(\alpha)}{\epsilon}.
\tag{A.44}
$$

The second derivative is obtained by substituting ϕ by ϕ' in this same formula; that is,

$$
\frac{d^2\phi}{d\alpha^2} = \phi''(\alpha) \overset{\text{def}}{=} \lim_{\epsilon \to 0} \frac{\phi'(\alpha + \epsilon) - \phi'(\alpha)}{\epsilon}.
\tag{A.45}
$$

Suppose now that α in turn depends on another quantity β (we denote this dependence by writing $\alpha = \alpha(\beta)$). We can use the *chain rule* to calculate the derivative of ϕ with respect to β:

$$
\frac{d\phi(\alpha(\beta))}{d\beta} = \frac{d\phi}{d\alpha}\frac{d\alpha}{d\beta} = \phi'(\alpha)\alpha'(\beta).
\tag{A.46}
$$

Consider now the function $f : \mathbb{R}^n \to \mathbb{R}$, which is a real-valued function of n independent variables. We typically gather the variables into a vector $x = (x_1, x_2, \ldots, x_n)^T$. We say that f is differentiable at x if there exists a vector $g \in \mathbb{R}^n$ such that

$$
\lim_{y \to 0} \frac{f(x + y) - f(x) - g^T y}{\|y\|} = 0,
\tag{A.47}
$$

where $\| \cdot \|$ is any vector norm of y. (This type of differentiability is known as *Frechet differentiability*.) If g satisfying (A.47) exists, we call it the *gradient* of f at x, and denote it by $\nabla f(x)$, written componentwise as

$$\nabla f(x) = \begin{bmatrix} \dfrac{\partial f}{\partial x_1} \\ \vdots \\ \dfrac{\partial f}{\partial x_n} \end{bmatrix}. \tag{A.48}$$

Here, $\partial f/\partial x_i$ represents the partial derivative of f with respect to x_i. By setting $y = \epsilon e_i$ in (A.47), where e_i is the vector in $\mathbb{R}^n$ consisting all all zeros, except for a 1 in position i, we obtain

$$
\begin{aligned}
\frac{\partial f}{\partial x_i} &\stackrel{\text{def}}{=} \lim_{\epsilon \to 0} \frac{f(x_1, \ldots, x_{i-1}, x_i + \epsilon, x_{i+1}, \ldots, x_n) - f(x_1, \ldots, x_{i-1}, x_i, x_{i+1}, \ldots, x_n)}{\epsilon} \\
&= \frac{f(x + \epsilon e_i) - f(x)}{\epsilon}.
\end{aligned}
$$

A gradient with respect to only a subset of the unknowns can be expressed by means of a subscript on the symbol ∇. Thus for the function of two vector variables $f(z, t)$, we use $\nabla_z f(z, t)$ to denote the gradient with respect to z (holding t constant).

The matrix of second partial derivatives of f is known as the *Hessian*, and is defined as

$$\nabla^2 f(x) = \begin{bmatrix} \dfrac{\partial^2 f}{\partial x_1^2} & \dfrac{\partial^2 f}{\partial x_1 \partial x_2} & \cdots & \dfrac{\partial^2 f}{\partial x_1 \partial x_n} \\ \dfrac{\partial^2 f}{\partial x_2 \partial x_1} & \dfrac{\partial^2 f}{\partial x_2^2} & \cdots & \dfrac{\partial^2 f}{\partial x_2 \partial x_n} \\ \vdots & \vdots & & \vdots \\ \dfrac{\partial^2 f}{\partial x_n \partial x_1} & \dfrac{\partial^2 f}{\partial x_n \partial x_2} & \cdots & \dfrac{\partial^2 f}{\partial x_n^2} \end{bmatrix}.$$

We say that f is *differentiable* on a domain $\mathcal{D}$ if $\nabla f(x)$ exists for all $x \in \mathcal{D}$, and *continuously differentiable* if $\nabla f(x)$ is a continuous functions of x. Similarly, f is *twice differentiable* on $\mathcal{D}$ if $\nabla^2 f(x)$ exists for all $x \in \mathcal{D}$ and *twice continuously differentiable* if $\nabla^2 f(x)$ is continuous on $\mathcal{D}$. Note that when f is twice continuously differentiable, the Hessian is a symmetric matrix, since

$$\frac{\partial^2 f}{\partial x_i \partial x_j} = \frac{\partial^2 f}{\partial x_j \partial x_i}, \quad \text{for all } i, j = 1, 2, \ldots, n.$$

When f is a vector valued function that is $f : \mathbb{R}^n \to \mathbb{R}^m$ (See Chapters 10 and 11), we define $\nabla f(x)$ to be the $n \times m$ matrix whose ith column is $\nabla f_i(x)$, that is, the gradient of

f_i with respect to x. Often, for notational convenience, we prefer to work with the transpose of his matrix, which has dimensions $m \times n$. This matrix is called the *Jacobian* and is often denoted by $J(x)$. Specifically, the (i, j) element of $J(x)$ is $\partial f_i(x)/\partial x_j$.

When the vector x in turn depends on another vector t (that is, $x = x(t)$), we can extend the chain rule (A.46) for the univariate function. Defining

$$h(t) = f(x(t)), \tag{A.49}$$

we have

$$\nabla h(t) = \sum_{i=1}^{n} \frac{\partial f}{\partial x_i} \nabla x_i(t) = \nabla x(t) \nabla f(x(t)). \tag{A.50}$$

❏ **EXAMPLE A.1**

Let $f : \mathbb{R}^2 \to \mathbb{R}$ be defined by $f(x_1, x_2) = x_1^2 + x_1 x_2$, where $x_1 = \sin t_1 + t_2^2$ and $x_2 = (t_1 + t_2)^2$. Defining $h(t)$ as in (A.49), the chain rule (A.50) yields

$$\nabla h(t)$$

$$= \sum_{i=1}^{n} \frac{\partial f}{\partial x_i} \nabla x_i(t)$$

$$= (2x_1 + x_2) \begin{bmatrix} \cos t_1 \\ 2t_2 \end{bmatrix} + x_1 \begin{bmatrix} 2(t_1 + t_2) \\ 2(t_1 + t_2) \end{bmatrix}$$

$$= \left(2 \left(\sin t_1 + t_2^2 \right) + (t_1 + t_2)^2 \right) \begin{bmatrix} \cos t_1 \\ 2t_2 \end{bmatrix} + \left(\sin t_1 + t_2^2 \right) \begin{bmatrix} 2(t_1 + t_2) \\ 2(t_1 + t_2) \end{bmatrix}.$$

If, on the other hand, we substitute directly for x into the definition of f, we obtain

$$h(t) = f(x(t)) = \left(\sin t_1 + t_2^2 \right)^2 + \left(\sin t_1 + t_2^2 \right) (t_1 + t_2)^2.$$

The reader should verify that the gradient of this expression is identical to the one obtained above by applying the chain rule.   ❏

Special cases of the chain rule can be derived when $x(t)$ in (A.50) is a linear function of t, say $x(t) = Ct$. We then have $\nabla x(t) = C^T$, so that

$$\nabla h(t) = C^T \nabla f(Ct).$$

In the case in which f is a scalar function, we can differentiate twice using the chain rule to obtain

$$\nabla^2 h(t) = C^T \nabla^2 f(Ct) C.$$

(The proof of this statement is left as an exercise.)

DIRECTIONAL DERIVATIVES

The *directional derivative* of a function $f : \mathbb{R}^n \to \mathbb{R}$ in the direction p is given by

$$D(f(x); p) \overset{\text{def}}{=} \lim_{\epsilon \to 0} \frac{f(x + \epsilon p) - f(x)}{\epsilon}. \tag{A.51}$$

The directional derivative may be well defined even when f is not continuously differentiable; in fact, it is most useful in such situations. Consider for instance the ℓ_1 norm function $f(x) = \|x\|_1$. We have from the definition (A.51) that

$$D(\|x\|_1; p) = \lim_{\epsilon \to 0} \frac{\|x + \epsilon p\|_1 - \|x\|_1}{\epsilon} = \lim_{\epsilon \to 0} \frac{\sum_{i=1}^n |x_i + \epsilon p_i| - \sum_{i=1}^n |x_i|}{\epsilon}.$$

If $x_i > 0$, we have $|x_i + \epsilon p_i| = |x_i| + \epsilon p_i$ for all ϵ sufficiently small. If $x_i < 0$, we have $|x_i + \epsilon p_i| = |x_i| - \epsilon p_i$, while if $x_i = 0$, we have $|x_i + \epsilon p_i| = \epsilon |p_i|$. Therefore, we have

$$D(\|x\|_1; p) = \sum_{i | x_i < 0} -p_i + \sum_{i | x_i > 0} p_i + \sum_{i | x_i = 0} |p_i|,$$

so the directional derivative of this function exists for any x and p. The first derivative $\nabla f(x)$ does *not* exist, however, whenever any of the components of x are zero.

When f is in fact continuously differentiable in a neighborhood of x, we have

$$D(f(x); p) = \nabla f(x)^T p.$$

To verify this formula, we define the function

$$\phi(\alpha) = f(x + \alpha p) = f(y(\alpha)), \tag{A.52}$$

where $y(\alpha) = x + \alpha p$. Note that

$$\lim_{\epsilon \to 0} \frac{f(x + \epsilon p) - f(x)}{\epsilon} = \lim_{\epsilon \to 0} \frac{\phi(\epsilon) - \phi(0)}{\epsilon} = \phi'(0).$$

By applying the chain rule (A.50) to $f(y(\alpha))$, we obtain

$$\phi'(\alpha) = \sum_{i=1}^{n} \frac{\partial f(y(\alpha))}{\partial y_i} \nabla y_i(\alpha) \tag{A.53}$$

$$= \sum_{i=1}^{n} \frac{\partial f(y(\alpha))}{\partial y_i} p_i = \nabla f(y(\alpha))^T p = \nabla f(x + \alpha p)^T p.$$

We obtain (A.51) by setting $\alpha = 0$ and comparing the last two expressions.

MEAN VALUE THEOREM

We now recall the mean value theorem for univariate functions. Given a continuously differentiable function $\phi : \mathbb{R} \rightarrow \mathbb{R}$ and two real numbers α_0 and α_1 that satisfy $\alpha_1 > \alpha_0$, we have that

$$\phi(\alpha_1) = \phi(\alpha_0) + \phi'(\xi)(\alpha_1 - \alpha_0) \tag{A.54}$$

for some $\xi \in (\alpha_0, \alpha_1)$. An extension of this result to a multivariate function $f : \mathbb{R}^n \rightarrow \mathbb{R}$ is that for any vector p we have

$$f(x + p) = f(x) + \nabla f(x + \alpha p)^T p, \tag{A.55}$$

for some $\alpha \in (0, 1)$. (This result can be proved by defining $\phi(\alpha) = f(x + \alpha p)$, $\alpha_0 = 0$, and $\alpha_1 = 1$ and applying the chain rule, as above.)

❑ EXAMPLE A.2

Consider $f : \mathbb{R}^2 \rightarrow \mathbb{R}$ defined by $f(x) = x_1^3 + 3x_1 x_2^2$, and let $x = (0, 0)^T$ and $p = (1, 2)^T$. It is easy to verify that $f(x) = 0$ and $f(x + p) = 13$. Since

$$\nabla f(x + \alpha p) = \begin{bmatrix} 3(x_1 + \alpha p_1)^2 + 3(x_2 + \alpha p_2)^2 \\ 6(x_1 + \alpha p_1)(x_2 + \alpha p_2) \end{bmatrix} = \begin{bmatrix} 15\alpha^2 \\ 12\alpha^2 \end{bmatrix},$$

we have that $\nabla f(x + \alpha p)^T p = 39\alpha^2$. Hence the relation (A.55) holds when we set $\alpha = 1/\sqrt{13}$, which lies in the open interval $(0, 1)$, as claimed.   ❑

An alternative expression to (A.55) can be stated for twice differentiable functions: We have

$$f(x + p) = f(x) + \nabla f(x)^T p + \frac{1}{2} p^T \nabla^2 f(x + \alpha p)^T p, \qquad (\text{A.56})$$

for some $\alpha \in (0, 1)$. In fact, this expression is one form of Taylor's theorem, Theorem 2.1 in Chapter 2, to which we refer throughout the book.

The extension of (A.55) to a vector-valued function $r : \mathbb{R}^n \to \mathbb{R}^m$ for $m > 1$ is not immediate. There is in general no scalar α such that the natural extension of (A.55) is satisfied. However, the following result is often a useful analog. As in (10.3), we denote the Jacobian of $r(x)$, by $J(x)$, where $J(x)$ is the $m \times n$ matrix whose (j, i) entry is $\partial r_j / \partial x_i$, for $j = 1, 2, \dots, m$ and $i = 1, 2, \dots, n$, and asssume that $J(x)$ is defined and continuous on the domain of interest. Given x and p, we then have

$$r(x + p) - r(x) = \int_0^1 J(x + \alpha p) p \, d\alpha. \qquad (\text{A.57})$$

When p is sufficiently small in norm, we can approximate the right-hand side of this expression adequately by $J(x)p$, that is,

$$r(x + p) - r(x) \approx J(x)p.$$

If J is Lipschitz continuous in the vicinity of x and $x + p$ with Lipschitz constant L, we can use (A.12) to estimate the error in this approximation as follows:

$$\begin{aligned} \|r(x + p) - r(x) - J(x)p\| &= \left\| \int_0^1 [J(x + \alpha p) - J(x)] p \, d\alpha \right\| \\ &\leq \int_0^1 \|J(x + \alpha p) - J(x)\| \, \|p\| \, d\alpha \\ &\leq \int_0^1 L\alpha \|p\|^2 \, d\alpha = \tfrac{1}{2} L \|p\|^2. \end{aligned}$$

IMPLICIT FUNCTION THEOREM

The implicit function theorem lies behind a number of important results in local convergence theory of optimization algorithms and in the characterization of optimality (see Chapter 12). Our statement of this result is based on Lang [187, p. 131] and Bertsekas [19, Proposition A.25].

Theorem A.2 (Implicit Function Theorem).

Let $h : \mathbf{R}^n \times \mathbf{R}^m \to \mathbf{R}^n$ be a function such that

(i) $h(z^*, 0) = 0$ for some $z^* \in \mathbf{R}^n$,

(ii) the function $h(\cdot, \cdot)$ is continuously differentiable in some neighborhood of $(z^*, 0)$, and

(iii) $\nabla_z h(z, t)$ is nonsingular at the point $(z, t) = (z^*, 0)$.

Then there exist open sets $\mathcal{N}_z \subset \mathbf{R}^n$ and $\mathcal{N}_t \subset \mathbf{R}^m$ containing z^* and 0, respectively, and a continuous function $z : \mathcal{N}_t \to \mathcal{N}_z$ such that $z^* = z(0)$ and $h(z(t), t) = 0$ for all $t \in \mathcal{N}_t$. Further, $z(t)$ is uniquely defined. Finally, if h is p times continuously differentiable with respect to both its arguments for some $p > 0$, then $z(t)$ is also p times continuously differentiable with respect to t, and we have

$$\nabla z(t) = -\nabla_t h(z(t), t)[\nabla_z h(z(t), t)]^{-1}$$

for all $t \in \mathcal{N}_t$.

This theorem is frequently applied to parametrized systems of linear equations, in which z is obtained as the solution of

$$M(t)z = g(t),$$

where $M(\cdot) \in \mathbf{R}^{n \times n}$ has $M(0)$ nonsingular, and $g(\cdot) \in \mathbf{R}^n$. To apply the theorem, we define

$$h(z, t) = M(t)z - g(t).$$

If $M(\cdot)$ and $g(\cdot)$ are continuously differentiable in some neighborhood of 0, the theorem implies that $z(t) = M(t)^{-1}g(t)$ is a continuous function of t in some neighborhood of 0.

ORDER NOTATION

In much of our analysis we are concerned with how the members of a sequence behave *eventually*, that is, when we get far enough along in the sequence. For instance, we might ask whether the elements of the sequence are bounded, or whether they are similar in size to the elements of a corresponding sequence, or whether they are decreasing and, if so, how rapidly. *Order notation* is useful shorthand to use when questions like these are being examined. It saves us defining many constants that clutter up the argument and the analysis.

We will use three varieties of order notation: $O(\cdot)$, $o(\cdot)$, and $\Omega(\cdot)$. Given two nonnegative infinite sequences of scalars $\{\eta_k\}$ and $\{\nu_k\}$, we write

$$\eta_k = O(\nu_k)$$

if there is a positive constant C such that

$$|\eta_k| \le C|\nu_k|$$

for all k sufficiently large. We write

$$\eta_k = o(\nu_k)$$

if the sequence of ratios $\{\eta_k/\nu_k\}$ approaches zero, that is,

$$\lim_{k \to \infty} \frac{\eta_k}{\nu_k} = 0.$$

Finally, we write

$$\eta_k = \Omega(\nu_k)$$

if there are two constants C_0 and C_1 with $0 < C_0 \le C_1 < \infty$ such that

$$C_0|\nu_k| \le |\eta_k| \le C_1|\nu_k|,$$

that is, the corresponding elements of both sequences stay in the same ballpark for all k. This definition is equivalent to saying that $\eta_k = O(\nu_k)$ and $\nu_k = O(\eta_k)$.

The same notation is often used in the context of quantities that depend continuously on each other as well. For instance, if $\eta(\cdot)$ is a function that maps $\mathbb{R}$ to $\mathbb{R}$, we write

$$\eta(\nu) = O(\nu)$$

if there is a constant C such that $|\eta(\nu)| \le C|\nu|$ for all $\nu \in \mathbb{R}$. (Typically, we are interested only in values of ν that are either very large or very close to zero; this should be clear from the context. Similarly, we use

$$\eta(\nu) = o(\nu) \tag{A.58}$$

to indicate that the ratio $\eta(\nu)/\nu$ approaches zero either as $\nu \to 0$ or $\nu \to \infty$. (Again, the precise meaning should be clear from the context.)

As a slight variant on the definitions above, we write

$$\eta_k = O(1)$$

to indicate that there is a constant C such that $|\eta_k| \le C$ for all k, while

$$\eta_k = o(1)$$

indicates that $\lim_{k \to \infty} \eta_k = 0$. We sometimes use vector and matrix quantities as arguments, and in these cases the definitions above are intended to apply to the norms of these quantities. For instance, if $f : \mathbb{R}^n \to \mathbb{R}^n$, we write $f(x) = O(\|x\|)$ if there is a constant $C > 0$ such that $\|f(x)\| \le C\|x\|$ for all x in the domain of f. Typically, as above, we are interested only in some subdomain of f, usually a small neighborhood of 0. As before, the precise meaning should be clear from the context.

ROOT-FINDING FOR SCALAR EQUATIONS

In Chapter 11 we discussed methods for finding solutions of nonlinear systems of equations $F(x) = 0$, where $F : \mathbb{R}^n \to \mathbb{R}^n$. Here we discuss briefly the case of scalar equations ($n = 1$), for which the algorithm is easy to illustrate. Scalar root-finding is needed in the trust-region algorithms of Chapter 4, for instance. Of course, the general theorems of Chapter 11 can be applied to derive rigorous convergence results for this special case.

The basic step of Newton's method (Algorithm Newton of Chapter 11) in the scalar case is simply

$$p_k = -F(x_k)/F'(x_k), \qquad x_{k+1} \leftarrow x_k + p_k \tag{A.59}$$

(cf. (11.6)). Graphically, such a step involves taking the tangent to the graph of F at the point x_k and taking the next iterate to be the intersection of this tangent with the x axis (see Figure A.2). Clearly, if the function F is nearly linear, the tangent will be quite a good approximation to F itself, so the Newton iterate will be quite close to the true root of F.

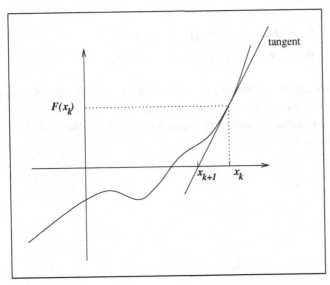

Figure A.2   One step of Newton's method for a scalar equation.

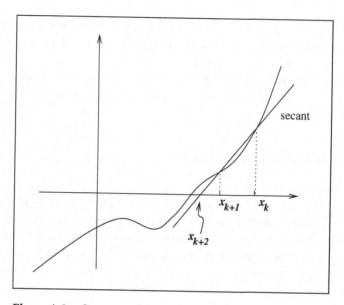

Figure A.3 One step of the secant method for a scalar equation.

The secant method for scalar equations can be viewed as the specialization of Broyden's method to the case of $n = 1$. The issues are simpler in this case, however, since the secant equation (11.27) completely determines the value of the 1×1 approximate Hessian B_k. That is, we do not need to apply extra conditions to ensure that B_k is fully determined. By combining (11.24) with (11.27), we find that the secant method for the case of $n = 1$ is defined by

$$B_k = (F(x_k) - F(x_{k-1}))/(x_k - x_{k-1}), \tag{A.60a}$$

$$p_k = -F(x_k)/B_k, \qquad x_{k+1} = x_k + p_k. \tag{A.60b}$$

By illustrating this algorithm, we see the origin of the term "secant." B_k approximates the slope of the function at x_k by taking the secant through the points $(x_{k-1}, F(x_{k-1})$ and $(x_k, F(x_k))$, and x_{k+1} is obtained by finding the intersection of this secant with the x axis. The method is illustrated in Figure A.3.

APPENDIX B

A Regularization Procedure

The following algorithm chooses parameters δ, γ that guarantee that the regularized primal-dual matrix (19.25) is nonsingular and satisfies the inertia condition (19.24). The algorithm assumes that, at the beginning of the interior-point iteration, δ_{old} has been initialized to zero.

Algorithm B.1 (Inertia Correction and Regularization).

    Given the current barrier parameter μ, constants $\eta > 0$ and $\beta < 1$, and the perturbation δ_{old} used in the previous interior-point iteration.

Factor (19.25) with $\delta = \gamma = 0$.
if (19.25) is nonsingular and its inertia is $(n + m, l + m, 0)$
        compute the primal-dual step; **stop;**
if (19.25) has zero eigenvalues
        set $\gamma \leftarrow 10^{-8} \eta \mu^{\beta}$;
if $\delta_{old} = 0$
        set $\delta \leftarrow 10^{-4}$;

else
        set $\delta \leftarrow \delta_{old}/2$;
repeat
        Factor the modified matrix (19.25);
        if the inertia is $(n + m, l + m, 0)$
                Set $\delta_{old} \leftarrow \delta$;
                Compute the primal-dual step (19.12) using the coefficient
                        matrix (19.25); stop;
        else
                Set $\delta \leftarrow 10\delta$;
end (repeat)

This algorithm has been adapted from a more elaborate procedure described by Wächter and Biegler [301]. All constants used in the algorithm are arbitrary; we have provided typical choices. The algorithm aims to avoid unnecessarily large modifications δI of $\nabla^2_{xx} \mathcal{L}$ while trying to minimize the number of matrix factorizations. Excessive modifications degrade the performance of the algorithm because they erase the second derivative information contained in $\nabla^2_{xx} \mathcal{L}$, and cause the step to take on steepest-descent like characteristics. The first trial value $(\delta = \delta_{old}/2)$ is based on the previous modification δ_{old} because the minimum perturbation δ required to achieve the desired inertia will often not vary much from one interior-point iteration to the next.

The heuristics implemented in Algorithm B.1 provide an alternative to those employed in Algorithm 7.3, which were presented in the context of unconstrained optimization. We emphasize, however, that all of these are indeed heuristics and may not always provide adequate safeguards.

References

[1] R. K. AHUJA, T. L. MAGNANTI, AND J. B. ORLIN, *Network Flows: Theory, Algorithms, and Applications*, Prentice-Hall, Englewood Cliffs, N.J., 1993.

[2] H. AKAIKE, *On a successive transformation of probability distribution and its application to the analysis of the optimum gradient method*, Annals of the Institute of Statistical Mathematics, 11 (1959), pp. 1–17.

[3] M. AL-BAALI, *Descent property and global convergence of the Fletcher-Reeves method with inexact line search*, I.M.A. Journal on Numerical Analysis, 5 (1985), pp. 121–124.

[4] E. D. ANDERSEN AND K. D. ANDERSEN, *Presolving in linear programming*, Mathematical Programming, 71 (1995), pp. 221–245.

[5] ——, *The MOSEK interior point optimizer for linear programming: an implementation of the homogeneous algorithm*, in High Performance Optimization, T. T. H. Frenk, K. Roos and S. Zhang, eds., Kluwer Academic Publishers, 2000, pp. 197–232.

[6] E. D. ANDERSEN, J. GONDZIO, C. MÉSZÁROS, AND X. XU, *Implementation of interior-point methods for large scale linear programming*, in Interior Point Methods in Mathematical Programming, T. Terlaky, ed., Kluwer, 1996, ch. 6, pp. 189–252.

[7] E. ANDERSON, Z. BAI, C. BISCHOF, J. DEMMEL, J. DONGARRA, J. DU CROZ, A. GREENBAUM, S. HAMMARLING, A. MCKENNEY, S. OSTROUCHOV, AND D. SORENSEN, *LAPACK User's Guide*, SIAM, Philadelphia, 1992.

[8] M. ANITESCU, *On solving mathematical programs with complementarity constraints as nonlinear programs*, SIAM Journal on Optimization, 15 (2005), pp. 1203–1236.

[9] ARKI CONSULTING AND DEVELOPMENT A/S, *CONOPT version 3*, 2004.

[10] B. M. AVERICK, R. G. CARTER, J. J. MORÉ, AND G. XUE, *The MINPACK-2 test problem collection*, Preprint MCS–P153–0692, Argonne National Laboratory, 1992.

[11] P. BAPTIST AND J. STOER, *On the relation between quadratic termination and convergence properties of minimization algorithms, Part II: Applications*, Numerische Mathematik, 28 (1977), pp. 367–392.

[12] R. H. BARTELS AND G. H. GOLUB, *The simplex method of linear programming using LU decomposition*, Communications of the ACM, 12 (1969), pp. 266–268.

[13] R. BARTLETT AND L. BIEGLER, *rSQP++: An object-oriented framework for successive quadratic programming*, in Large-Scale PDE-Constrained Optimization, L. T. Biegler, O. Ghattas, M. Heinkenschloss, and B. van Bloemen Waanders, eds., vol. 30, New York, 2003, Springer-Verlag, pp. 316–330. Lecture Notes in Computational Science and Engineering.

[14] M. BAZARAA, H. SHERALI, AND C. SHETTY, *Nonlinear Programming, Theory and Applications.*, John Wiley & Sons, New York, second ed., 1993.

[15] A. BEN-TAL AND A. NEMIROVSKI, *Lectures on Modern Convex Optimization: Analysis, Algorithms, and Engineering Applications*, MPS-SIAM Series on Optimization, SIAM, 2001.

[16] H. Y. BENSON, A. SEN, D. F. SHANNO, AND R. J. VANDERBEI, *Interior-point algorithms, penalty methods and equilibrium problems*, Technical Report ORFE-03-02, Operations Research and Financial Engineering, Princeton University, 2003.

[17] S. BENSON AND J. MORÉ, *A limited-memory variable-metric algorithm for bound constrained minimization*, Numerical Analysis Report P909-0901, ANL, Argonne, IL, USA, 2001.

[18] D. P. BERTSEKAS, *Constrained Optimization and Lagrange Multiplier Methods*, Academic Press, New York, 1982.

[19] ——, *Nonlinear Programming*, Athena Scientific, Belmont, MA, second ed., 1999.

[20] M. BERZ, C. BISCHOF, C. F. CORLISS, AND A. GRIEWANK, eds., *Computational Differentiation: Techniques, Applications, and Tools*, SIAM Publications, Philadelphia, PA, 1996.

[21] J. BETTS, S. K. ELDERSVELD, P. D. FRANK, AND J. G. LEWIS, *An interior-point nonlinear programming algorithm for large scale optimization*, Technical report MCT TECH-003, Mathematics and Computing Technology, The Boeing Company, P.O. Box 3707, Seattle, WA 98124-2207, 2000.

[22] J. R. BIRGE AND F. LOUVEAUX, *Introduction to Stochastic Programming*, Springer-Verlag, New York, 1997.

[23] E. G. BIRGIN, J. M. MARTINEZ, AND M. RAYDAN, *Algorithm 813: SPG software for convex-constrained optimization*, ACM Transactions on Mathematical Software, 27 (2001), pp. 340–349.

[24] C. BISCHOF, A. BOUARICHA, P. KHADEMI, AND J. J. MORÉ, *Computing gradients in large-scale optimization using automatic differentiation*, INFORMS Journal on Computing, 9 (1997), pp. 185–194.

[25] C. BISCHOF, A. CARLE, P. KHADEMI, AND A. MAUER, *ADIFOR 2.0: Automatic differentiation of FORTRAN 77 programs*, IEEE Computational Science & Engineering, 3 (1996), pp. 18–32.

[26] C. BISCHOF, G. CORLISS, AND A. GRIEWANK, *Structured second- and higher-order derivatives through univariate Taylor series*, Optimization Methods and Software, 2 (1993), pp. 211–232.

[27] C. BISCHOF, P. KHADEMI, A. BOUARICHA, AND A. CARLE, *Efficient computation of gradients and Jacobians by transparent exploitation of sparsity in automatic differentiation*, Optimization Methods and Software, 7 (1996), pp. 1–39.

[28] C. BISCHOF, L. ROH, AND A. MAUER, *ADIC: An extensible automatic differentiation tool for ANSI-C,* Software—Practice and Experience, 27 (1997), pp. 1427–1456.

[29] Å. BJÖRCK, *Numerical Methods for Least Squares Problems,* SIAM Publications, Philadelphia, PA, 1996.

[30] P. T. BOGGS, R. H. BYRD, AND R. B. SCHNABEL, *A stable and efficient algorithm for nonlinear orthogonal distance regression,* SIAM Journal on Scientific and Statistical Computing, 8 (1987), pp. 1052–1078.

[31] P. T. BOGGS, J. R. DONALDSON, R. H. BYRD, AND R. B. SCHNABEL, *ODRPACK—Software for weighted orthogonal distance regression,* ACM Transactions on Mathematical Software, 15 (1981), pp. 348–364.

[32] P. T. BOGGS AND J. W. TOLLE, *Convergence properties of a class of rank-two updates,* SIAM Journal on Optimization, 4 (1994), pp. 262–287.

[33] ———, *Sequential quadratic programming,* Acta Numerica, 4 (1996), pp. 1–51.

[34] P. T. BOGGS, J. W. TOLLE, AND P. WANG, *On the local convergence of quasi-Newton methods for constrained optimization,* SIAM Journal on Control and Optimization, 20 (1982), pp. 161–171.

[35] I. BONGARTZ, A. R. CONN, N. I. M. GOULD, AND P. L. TOINT, *CUTE: Constrained and unconstrained testing environment,* Research Report, IBM T.J. Watson Research Center, Yorktown Heights, NY, 1993.

[36] J. F. BONNANS, E. R. PANIER, A. L. TITS, AND J. L. ZHOU, *Avoiding the Maratos effect by means of a nonmonotone line search. II. Inequality constrained problems — feasible iterates,* SIAM Journal on Numerical Analysis, 29 (1992), pp. 1187–1202.

[37] S. BOYD, L. EL GHAOUI, E. FERON, AND V. BALAKRISHNAN, *Linear Matrix Inequalities in Systems and Control Theory,* SIAM Publications, Phildelphia, 1994.

[38] S. BOYD AND L. VANDENBERGHE, *Convex Optimization,* Cambridge University Press, Cambridge, 2003.

[39] R. P. BRENT, *Algorithms for minimization without derivatives,* Prentice Hall, Englewood Cliffs, NJ, 1973.

[40] H. M. BÜCKER, G. F. CORLISS, P. D. HOVLAND, U. NAUMANN, AND B. NORRIS, eds., *Automatic Differentiation: Applications, Theory, and Implementations,* vol. 50 of Lecture Notes in Computational Science and Engineering, Springer, New York, 2005.

[41] R. BULIRSCH AND J. STOER, *Introduction to Numerical Analysis,* Springer-Verlag, New York, 1980.

[42] J. R. BUNCH AND L. KAUFMAN, *Some stable methods for calculating inertia and solving symmetric linear systems,* Mathematics of Computation, 31 (1977), pp. 163–179.

[43] J. R. BUNCH AND B. N. PARLETT, *Direct methods for solving symmetric indefinite systems of linear equations,* SIAM Journal on Numerical Analysis, 8 (1971), pp. 639–655.

[44] J. V. BURKE AND J. J. MORÉ, *Exposing constraints,* SIAM Journal on Optimization, 4 (1994), pp. 573–595.

[45] W. BURMEISTER, *Die konvergenzordnung des Fletcher-Powell algorithmus,* Zeitschrift für Angewandte Mathematik und Mechanik, 53 (1973), pp. 693–699.

[46] R. BYRD, J. NOCEDAL, AND R. WALTZ, *Knitro: An integrated package for nonlinear optimization,* Technical Report 18, Optimization Technology Center, Evanston, IL, June 2005.

[47] R. BYRD, J. NOCEDAL, AND R. A. WALTZ, *Steering exact penalty methods,* Technical Report OTC 2004/07, Optimization Technology Center, Northwestern University, Evanston, IL, USA, April 2004.

[48] R. H. BYRD, J.-C. GILBERT, AND J. NOCEDAL, *A trust region method based on interior point techniques for nonlinear programming,* Mathematical Programming, 89 (2000), pp. 149–185.

[49] R. H. BYRD, N. I. M. GOULD, J. NOCEDAL, AND R. A. WALTZ, *An algorithm for nonlinear optimization using linear programming and equality constrained subproblems*, Mathematical Programming, Series B, 100 (2004), pp. 27–48.

[50] R. H. BYRD, M. E. HRIBAR, AND J. NOCEDAL, *An interior point method for large scale nonlinear programming*, SIAM Journal on Optimization, 9 (1999), pp. 877–900.

[51] R. H. BYRD, H. F. KHALFAN, AND R. B. SCHNABEL, *Analysis of a symmetric rank-one trust region method*, SIAM Journal on Optimization, 6 (1996), pp. 1025–1039.

[52] R. H. BYRD, J. NOCEDAL, AND R. B. SCHNABEL, *Representations of quasi-Newton matrices and their use in limited-memory methods*, Mathematical Programming, Series A, 63 (1994), pp. 129–156.

[53] R. H. BYRD, J. NOCEDAL, AND Y. YUAN, *Global convergence of a class of quasi-Newton methods on convex problems*, SIAM Journal on Numerical Analysis, 24 (1987), pp. 1171–1190.

[54] R. H. BYRD, R. B. SCHNABEL, AND G. A. SCHULTZ, *Approximate solution of the trust regions problem by minimization over two-dimensional subspaces*, Mathematical Programming, 40 (1988), pp. 247–263.

[55] R. H. BYRD, R. B. SCHNABEL, AND G. A. SHULTZ, *A trust region algorithm for nonlinearly constrained optimization*, SIAM Journal on Numerical Analysis, 24 (1987), pp. 1152–1170.

[56] M. R. CELIS, J. E. DENNIS, AND R. A. TAPIA, *A trust region strategy for nonlinear equality constrained optimization*, in Numerical Optimization, P. T. Boggs, R. H. Byrd, and R. B. Schnabel, eds., SIAM, 1985, pp. 71–82.

[57] R. CHAMBERLAIN, C. LEMARECHAL, H. C. PEDERSEN, AND M. J. D. POWELL, *The watchdog technique for forcing convergence in algorithms for constrained optimization*, Mathematical Programming, 16 (1982), pp. 1–17.

[58] S. H. CHENG AND N. J. HIGHAM, *A modified Cholesky algorithm based on a symmetric indefinite factorization*, SIAM Journal of Matrix Analysis and Applications, 19 (1998), pp. 1097–1100.

[59] C. M. CHIN AND R. FLETCHER, *On the global convergence of an SLP-filter algorithm that takes EQP steps*, Mathematical Programming, Series A, 96 (2003), pp. 161–177.

[60] T. D. CHOI AND C. T. KELLEY, *Superlinear convergence and implicit filtering*, SIAM Journal on Optimization, 10 (2000), pp. 1149–1162.

[61] V. CHVÁTAL, *Linear Programming*, W. H. Freeman and Company, New York, 1983.

[62] F. H. CLARKE, *Optimization and Nonsmooth Analysis*, John Wiley & Sons, New York, 1983 (Reprinted by SIAM Publications, 1990).

[63] A. COHEN, *Rate of convergence of several conjugate gradient algorithms*, SIAM Journal on Numerical Analysis, 9 (1972), pp. 248–259.

[64] T. F. COLEMAN, *Linearly constrained optimization and projected preconditioned conjugate gradients*, in Proceedings of the Fifth SIAM Conference on Applied Linear Algebra, J. Lewis, ed., Philadelphia, USA, 1994, SIAM, pp. 118–122.

[65] T. F. COLEMAN AND A. R. CONN, *Non-linear programming via an exact penalty-function: Asymptotic analysis*, Mathematical Programming, 24 (1982), pp. 123–136.

[66] T. F. COLEMAN, B. GARBOW, AND J. J. MORÉ, *Software for estimating sparse Jacobian matrices*, ACM Transactions on Mathematical Software, 10 (1984), pp. 329–345.

[67] ———, *Software for estimating sparse Hessian matrices*, ACM Transactions on Mathematical Software, 11 (1985), pp. 363–377.

[68] T. F. COLEMAN AND J. J. MORÉ, *Estimation of sparse Jacobian matrices and graph coloring problems*, SIAM Journal on Numerical Analysis, 20 (1983), pp. 187–209.

[69] ———, *Estimation of sparse Hessian matrices and graph coloring problems*, Mathematical Programming, 28 (1984), pp. 243–270.

[70] A. R. CONN, N. I. M. GOULD, AND P. L. TOINT, *Testing a class of algorithms for solving minimization problems with simple bounds on the variables*, Mathematics of Computation, 50 (1988), pp. 399–430.

[71] ——, *Convergence of quasi-Newton matrices generated by the symmetric rank one update*, Mathematical Programming, 50 (1991), pp. 177–195.

[72] ——, *LANCELOT: a FORTRAN package for large-scale nonlinear optimization (Release A)*, no. 17 in Springer Series in Computational Mathematics, Springer-Verlag, New York, 1992.

[73] ——, *Numerical experiments with the LANCELOT package (Release A) for large-scale nonlinear optimization*, Report 92/16, Department of Mathematics, University of Namur, Belgium, 1992.

[74] A. R. CONN, N. I. M. GOULD, AND P. L. TOINT, *Trust-Region Methods*, MPS-SIAM Series on Optimization, SIAM, 2000.

[75] A. R. CONN, K. SCHEINBERG, AND P. L. TOINT, *On the convergence of derivative-free methods for unconstrained optimization*, in Approximation Theory and Optimization: Tributes to M. J. D. Powell, A. Iserles and M. Buhmann, eds., Cambridge University Press, Cambridge, UK, 1997, pp. 83–108.

[76] ——, *Recent progress in unconstrained nonlinear optimization without derivatives*, Mathematical Programming, Series B, 79 (1997), pp. 397–414.

[77] W. J. COOK, W. H. CUNNINGHAM, W. R. PULLEYBLANK, AND A. SCHRIJVER, *Combinatorial Optimization*, John Wiley & Sons, New York, 1997.

[78] B. F. CORLISS AND L. B. RALL, *An introduction to automatic differentiation*, in Computational Differentiation: Techniques, Applications, and Tools, M. Berz, C. Bischof, G. F. Corliss, and A. Griewank, eds., SIAM Publications, Philadelphia, PA, 1996, ch. 1.

[79] T. H. CORMEN, C. E. LEISSERSON, AND R. L. RIVEST, *Introduction to Algorithms*, MIT Press, 1990.

[80] R. W. COTTLE, J.-S. PANG, AND R. E. STONE, *The Linear Complementarity Problem*, Academic Press, San Diego, 1992.

[81] R. COURANT, *Variational methods for the solution of problems with equilibrium and vibration*, Bulletin of the American Mathematical Society, 49 (1943), pp. 1–23.

[82] H. P. CROWDER AND P. WOLFE, *Linear convergence of the conjugate gradient method*, IBM Journal of Research and Development, 16 (1972), pp. 431–433.

[83] A. CURTIS, M. J. D. POWELL, AND J. REID, *On the estimation of sparse Jacobian matrices*, Journal of the Institute of Mathematics and its Applications, 13 (1974), pp. 117–120.

[84] J. CZYZYK, S. MEHROTRA, M. WAGNER, AND S. J. WRIGHT, *PCx: An interior-point code for linear programming*, Optimization Methods and Software, 11/12 (1999), pp. 397–430.

[85] Y. DAI AND Y. YUAN, *A nonlinear conjugate gradient method with a strong global convergence property*, SIAM Journal on Optimization, 10 (1999), pp. 177–182.

[86] G. B. DANTZIG, *Linear Programming and Extensions*, Princeton University Press, Princeton, NJ, 1963.

[87] W. C. DAVIDON, *Variable metric method for minimization*, Technical Report ANL–5990 (revised), Argonne National Laboratory, Argonne, IL, 1959.

[88] ——, *Variable metric method for minimization*, SIAM Journal on Optimization, 1 (1991), pp. 1–17.

[89] R. S. DEMBO, S. C. EISENSTAT, AND T. STEIHAUG, *Inexact Newton methods*, SIAM Journal on Numerical Analysis, 19 (1982), pp. 400–408.

[90] J. E. DENNIS, D. M. GAY, AND R. E. WELSCH, *Algorithm 573 — NL2SOL, An adaptive nonlinear least-squares algorithm*, ACM Transactions on Mathematical Software, 7 (1981), pp. 348–368.

[91] J. E. DENNIS AND J. J. MORÉ, *Quasi-Newton methods, motivation and theory*, SIAM Review, 19 (1977), pp. 46–89.

[92] J. E. DENNIS AND R. B. SCHNABEL, *Numerical Methods for Unconstrained Optimization and Nonlinear Equations*, Prentice-Hall, Englewood Cliffs, NJ, 1983. Reprinted by SIAM Publications, 1993.

[93] J. E. DENNIS AND R. B. SCHNABEL, *A view of unconstrained optimization*, in Optimization, vol. 1 of Handbooks in Operations Research and Management, Elsevier Science Publishers, Amsterdam, The Netherlands, 1989, pp. 1–72.

[94] I. I. DIKIN, *Iterative solution of problems of linear and quadratic programming*, Soviet Mathematics-Doklady, 8 (1967), pp. 674–675.

[95] I. S. DUFF AND J. K. REID, *The multifrontal solution of indefinite sparse symmetric linear equations*, ACM Transactions on Mathematical Software, 9 (1983), pp. 302–325.

[96] I. S. DUFF AND J. K. REID, *The design of MA48: A code for the direct solution of sparse unsymmetric linear systems of equations*, ACM Transactions on Mathematical Software, 22 (1996), pp. 187–226.

[97] I. S. DUFF, J. K. REID, N. MUNKSGAARD, AND H. B. NEILSEN, *Direct solution of sets of linear equations whose matrix is sparse symmetric and indefinite*, Journal of the Institute of Mathematics and its Applications, 23 (1979), pp. 235–250.

[98] A. V. FIACCO AND G. P. MCCORMICK, *Nonlinear Programming: Sequential Unconstrained Minimization Techniques*, John Wiley & Sons, New York, N.Y., 1968. Reprinted by SIAM Publications, 1990.

[99] R. FLETCHER, *A general quadratic programming algorithm*, Journal of the Institute of Mathematics and its Applications, 7 (1971), pp. 76–91.

[100] ———, *Second order corrections for non-differentiable optimization*, in Numerical Analysis, D. Griffiths, ed., Springer Verlag, 1982, pp. 85–114. Proceedings Dundee 1981.

[101] ———, *Practical Methods of Optimization*, John Wiley & Sons, New York, second ed., 1987.

[102] ———, *An optimal positive definite update for sparse Hessian matrices*, SIAM Journal on Optimization, 5 (1995), pp. 192–218.

[103] ———, *Stable reduced hessian updates for indefinite quadratic programming*, Mathematical Programming, 87 (2000), pp. 251–264.

[104] R. FLETCHER, A. GROTHEY, AND S. LEYFFER, *Computing sparse Hessian and Jacobian approximations with optimal hereditary properties*, technical report, Department of Mathematics, University of Dundee, 1996.

[105] R. FLETCHER AND S. LEYFFER, *Nonlinear programming without a penalty function*, Mathematical Programming, Series A, 91 (2002), pp. 239–269.

[106] R. FLETCHER, S. LEYFFER, AND P. L. TOINT, *On the global convergence of an SLP-filter algorithm*, Numerical Analysis Report NA/183, Dundee University, Dundee, Scotland, UK, 1999.

[107] R. FLETCHER AND C. M. REEVES, *Function minimization by conjugate gradients*, Computer Journal, 7 (1964), pp. 149–154.

[108] R. FLETCHER AND E. SAINZ DE LA MAZA, *Nonlinear programming and nonsmooth optimization by successive linear programming*, Mathematical Programming, 43 (1989), pp. 235–256.

[109] C. FLOUDAS AND P. PARDALOS, eds., *Recent Advances in Global Optimization*, Princeton University Press, Princeton, NJ, 1992.

[110] J. J. H. FORREST AND J. A. TOMLIN, *Updated triangular factors of the basis to maintain sparsity in the product form simplex method*, Mathematical Programming, 2 (1972), pp. 263–278.

[111] A. FORSGREN, P. E. GILL, AND M. H. WRIGHT, *Interior methods for nonlinear optimization*, SIAM Review, 44 (2003), pp. 525–597.

[112] R. FOURER, D. M. GAY, AND B. W. KERNIGHAN, *AMPL: A Modeling Language for Mathematical Programming*, The Scientific Press, South San Francisco, CA, 1993.

[113] R. FOURER AND S. MEHROTRA, *Solving symmetric indefinite systems in an interior-point method for linear programming*, Mathematical Programming, 62 (1993), pp. 15–39.

[114] M. P. FRIEDLANDER AND M. A. SAUNDERS, *A globally convergent linearly constrained Lagrangian method for nonlinear optimization*, SIAM Journal on Optimization, 15 (2005), pp. 863–897.

[115] K. R. FRISCH, *The logarithmic potential method of convex programming*, Technical Report, University Institute of Economics, Oslo, Norway, 1955.

[116] D. GABAY, *Reduced quasi-Newton methods with feasibility improvement for nonlinearly constrained optimization*, Mathematical Programming Studies, 16 (1982), pp. 18–44.

[117] U. M. GARCIA-PALOMARES AND O. L. MANGASARIAN, *Superlinearly convergent quasi-Newton methods for nonlinearly constrained optimization problems*, Mathematical Programming, 11 (1976), pp. 1–13.

[118] D. M. GAY, *More AD of nonlinear AMPL models: computing Hessian information and exploiting partial separability*, in Computational Differentiation: Techniques, Applications, and Tools, M. Berz, C. Bischof, G. F. Corliss, and A. Griewank, eds., SIAM Publications, Philadelphia, PA, 1996, pp. 173–184.

[119] R.-P. GE AND M. J. D. POWELL, *The convergence of variable metric matrices in unconstrained optimization*, Mathematical Programming, 27 (1983), pp. 123–143.

[120] A. H. GEBREMEDHIN, F. MANNE, AND A. POTHEN, *What color is your Jacobian? Graph coloring for computing derivatives*, SIAM Review, 47 (2005), pp. 629–705.

[121] E. M. GERTZ AND S. J. WRIGHT, *Object-oriented software for quadratic programming*, ACM Transactions on Mathematical Software, 29 (2003), pp. 58–81.

[122] J. GILBERT AND C. LEMARÉCHAL, *Some numerical experiments with variable-storage quasi-Newton algorithms*, Mathematical Programming, Series B, 45 (1989), pp. 407–435.

[123] J. GILBERT AND J. NOCEDAL, *Global convergence properties of conjugate gradient methods for optimization*, SIAM Journal on Optimization, 2 (1992), pp. 21–42.

[124] P. E. GILL, G. H. GOLUB, W. MURRAY, AND M. A. SAUNDERS, *Methods for modifying matrix factorizations*, Mathematics of Computation, 28 (1974), pp. 505–535.

[125] P. E. GILL AND M. W. LEONARD, *Limited-memory reduced-Hessian methods for unconstrained optimization*, SIAM Journal on Optimization, 14 (2003), pp. 380–401.

[126] P. E. GILL AND W. MURRAY, *Numerically stable methods for quadratic programming*, Mathematical Programming, 14 (1978), pp. 349–372.

[127] P. E. GILL, W. MURRAY, AND M. A. SAUNDERS, *User's guide for SNOPT (Version 5.3): A FORTRAN package for large-scale nonlinear programming*, Technical Report NA 97-4, Department of Mathematics, University of California, San Diego, 1997.

[128] ——, *SNOPT: An SQP algorithm for large-scale constrained optimization*, SIAM Journal on Optimization, 12 (2002), pp. 979–1006.

[129] P. E. GILL, W. MURRAY, M. A. SAUNDERS, AND M. H. WRIGHT, *User's guide for SOL/QPSOL*, Technical Report SOL84-6, Department of Operations Research, Stanford University, Stanford, California, 1984.

[130] P. E. GILL, W. MURRAY, AND M. H. WRIGHT, *Practical Optimization*, Academic Press, 1981.

[131] ——, *Numerical Linear Algebra and Optimization, Vol. 1*, Addison Wesley, Redwood City, California, 1991.

[132] D. GOLDFARB, *Curvilinear path steplength algorithms for minimization which use directions of negative curvature*, Mathematical Programming, 18 (1980), pp. 31–40.

[133] D. GOLDFARB AND J. FORREST, *Steepest edge simplex algorithms for linear programming*, Mathematical Programming, 57 (1992), pp. 341–374.

[134] D. GOLDFARB AND J. K. REID, *A practicable steepest-edge simplex algorithm*, Mathematical Programming, 12 (1977), pp. 361–373.

[135] G. GOLUB AND D. O'LEARY, *Some history of the conjugate gradient methods and the Lanczos algorithms: 1948–1976*, SIAM Review, 31 (1989), pp. 50–100.

[136] G. H. GOLUB AND C. F. VAN LOAN, *Matrix Computations*, The Johns Hopkins University Press, Baltimore, third ed., 1996.

[137] J. GONDZIO, *HOPDM (version 2.12): A fast LP solver based on a primal-dual interior point method*, European Journal of Operations Research, 85 (1995), pp. 221–225.

[138] ———, *Multiple centrality corrections in a primal-dual method for linear programming*, Computational Optimization and Applications, 6 (1996), pp. 137–156.

[139] J. GONDZIO AND A. GROTHEY, *Parallel interior point solver for structured quadratic programs: Application to financial planning problems*, Technical Report MS-03-001, School of Mathematics, University of Edinburgh, Scotland, 2003.

[140] N. I. M. GOULD, *On the accurate determination of search directions for simple differentiable penalty functions*, I.M.A. Journal on Numerical Analysis, 6 (1986), pp. 357–372.

[141] ———, *On the convergence of a sequential penalty function method for constrained minimization*, SIAM Journal on Numerical Analysis, 26 (1989), pp. 107–128.

[142] ———, *An algorithm for large scale quadratic programming*, I.M.A. Journal on Numerical Analysis, 11 (1991), pp. 299–324.

[143] N. I. M. GOULD, M. E. HRIBAR, AND J. NOCEDAL, *On the solution of equality constrained quadratic problems arising in optimization*, SIAM Journal on Scientific Computing, 23 (2001), pp. 1375–1394.

[144] N.I.M. GOULD, S. LEYFFER, AND P. L. TOINT, *A multidimensional filter algorithm for nonlinear equations and nonlinear least squares*, SIAM Journal on Optimization, 15 (2004), pp. 17–38.

[145] N. I. M. GOULD, S. LUCIDI, M. ROMA, AND P. L. TOINT, *Solving the trust-region subproblem using the Lanczos method.* SIAM Journal on Optimization, 9 (1999), pp. 504–525.

[146] N. I. M. GOULD, D. ORBAN, AND P. L. TOINT, GALAHAD—*a library of thread-safe Fortran 90 packages for large-scale nonlinear optimization*, ACM Transactions on Mathematical Software, 29 (2003), pp. 353–372.

[147] N. I. M. GOULD, D. ORBAN, AND P. L. TOINT, *Numerical methods for large-scale nonlinear optimization*, Acta Numerica, 14 (2005), pp. 299–361.

[148] N. I. M. GOULD AND P. L. TOINT, *An iterative working-set method for large-scale non-convex quadratic programming*, Applied Numerical Mathematics, 43 (2002), pp. 109–128.

[149] ———, *Numerical methods for large-scale non-convex quadratic programming*, in Trends in Industrial and Applied Mathematics, A. H. Siddiqi and M. Kočvara, eds., Dordrecht, The Netherlands, 2002, Kluwer Academic Publishers, pp. 149–179.

[150] A. GRIEWANK, *Achieving logarithmic growth of temporal and spatial complexity in reverse automatic differentiation*, Optimization Methods and Software, 1 (1992), pp. 35–54.

[151] ———, *Automatic directional differentiation of nonsmooth composite functions*, in Seventh French-German Conference on Optimization, 1994.

[152] A. GRIEWANK, *Evaluating Derivatives: Principles and Techniques of Automatic Differentiation*, vol. 19 of Frontiers in Applied Mathematics, SIAM, 2000.

[153] A. GRIEWANK AND G. F. CORLISS, eds., *Automatic Differentition of Algorithms*, SIAM Publications, Philadelphia, Penn., 1991.

[154] A. GRIEWANK, D. JUEDES, AND J. UTKE, *ADOL-C, A package for the automatic differentiation of algorithms written in C/C++*, ACM Transactions on Mathematical Software, 22 (1996), pp. 131–167.

[155] A. GRIEWANK AND P. L. TOINT, *Local convergence analysis of partitioned quasi-Newton updates*, Numerische Mathematik, 39 (1982), pp. 429–448.

[156] ———, *Partitioned variable metric updates for large structured optimization problems*, Numerische Mathematik, 39 (1982), pp. 119–137.

[157] J. GRIMM, L. POTTIER, AND N. ROSTAING-SCHMIDT, *Optimal time and minimum space time product for reversing a certain class of programs*, in Computational Differentiation, Techniques, Applications, and Tools, M. Berz, C. Bischof, G. Corliss, and A. Griewank, eds., SIAM, Philadelphia, 1996, pp. 95–106.

[158] L. GRIPPO, F. LAMPARIELLO, AND S. LUCIDI, *A nonmonotone line search technique for Newton's method*, SIAM Journal on Numerical Analysis, 23 (1986), pp. 707–716.

[159] C. GUÉRET, C. PRINS, AND M. SEVAUX, *Applications of optimization with Xpress-MP*, Dash Optimization, 2002.

[160] W. W. HAGER, *Minimizing a quadratic over a sphere*, SIAM Journal on Optimization, 12 (2001), pp. 188–208.

[161] W. W. HAGER AND H. ZHANG, *A new conjugate gradient method with guaranteed descent and an efficient line search*, SIAM Journal on Optimization, 16 (2005), pp. 170–192.

[162] ———, *A survey of nonlinear conjugate gradient methods*. To appear in the Pacific Journal of Optimization, 2005.

[163] S. P. HAN, *Superlinearly convergent variable metric algorithms for general nonlinear programming problems*, Mathematical Programming, 11 (1976), pp. 263–282.

[164] ———, *A globally convergent method for nonlinear programming*, Journal of Optimization Theory and Applications, 22 (1977), pp. 297–309.

[165] S. P. HAN AND O. L. MANGASARIAN, *Exact penalty functions in nonlinear programming*, Mathematical Programming, 17 (1979), pp. 251–269.

[166] HARWELL SUBROUTINE LIBRARY, *A catalogue of subroutines (release 13)*, AERE Harwell Laboratory, Harwell, Oxfordshire, England, 1998.

[167] M. R. HESTENES, *Multiplier and gradient methods*, Journal of Optimization Theory and Applications, 4 (1969), pp. 303–320.

[168] M. R. HESTENES AND E. STIEFEL, *Methods of conjugate gradients for solving linear systems*, Journal of Research of the National Bureau of Standards, 49 (1952), pp. 409–436.

[169] N. J. HIGHAM, *Accuracy and Stability of Numerical Algorithms*, SIAM Publications, Philadelphia, 1996.

[170] J.-B. HIRIART-URRUTY AND C. LEMARECHAL, *Convex Analysis and Minimization Algorithms*, Springer-Verlag, Berlin, New York, 1993.

[171] P. HOUGH, T. KOLDA, AND V. TORCZON, *Asynchronous parallel pattern search for nonlinear optimization*, SIAM Journal on Optimization, 23 (2001), pp. 134–156.

[172] ILOG CPLEX 8.0, *User's Manual*, ILOG SA, Gentilly, France, 2002.

[173] D. JONES, C. PERTTUNEN, AND B. STUCKMAN, *Lipschitzian optimization without the Lipschitz constant*, Journal of Optimization Theory and Applications, 79 (1993), pp. 157–181.

[174] P. KALL AND S. W. WALLACE, *Stochastic Programming*, John Wiley & Sons, New York, 1994.

[175] N. KARMARKAR, *A new polynomial-time algorithm for linear programming*, Combinatorics, 4 (1984), pp. 373–395.

[176] C. KELLER, N. I. M. GOULD, AND A. J. WATHEN, *Constraint preconditioning for indefinite linear systems*, SIAM Journal on Matrix Analysis and Applications, 21 (2000), pp. 1300–1317.

[177] C. T. KELLEY, *Iterative Methods for Linear and Nonlinear Equations*, SIAM Publications, Philadelphia, PA, 1995.

[178] C. T. KELLEY, *Detection and remediation of stagnation in the Nelder-Mead algorithm using a sufficient decrease condition*, SIAM Journal on Optimization, 10 (1999), pp. 43–55.

[179] ——, *Iterative Methods for Optimization*, no. 18 in Frontiers in Applied Mathematics, SIAM Publications, Philadelphia, PA, 1999.

[180] L. G. KHACHIYAN, *A polynomial algorithm in linear programming*, Soviet Mathematics Doklady, 20 (1979), pp. 191–194.

[181] H. F. KHALFAN, R. H. BYRD, AND R. B. SCHNABEL, *A theoretical and experimental study of the symmetric rank one update*, SIAM Journal on Optimization, 3 (1993), pp. 1–24.

[182] V. KLEE AND G. J. MINTY, *How good is the simplex algorithm?* in Inequalities, O. Shisha, ed., Academic Press, New York, 1972, pp. 159–175.

[183] T. G. KOLDA, R. M. LEWIS, AND V. TORCZON, *Optimization by direct search: New perspectives on some classical and modern methods*, SIAM Review, 45 (2003), pp. 385–482.

[184] M. KOČVARA AND M. STINGL, *PENNON, a code for nonconvex nonlinear and semidefinite programming*, Optimization Methods and Software, 18 (2003), pp. 317–333.

[185] H. W. KUHN AND A. W. TUCKER, *Nonlinear programming*, in Proceedings of the Second Berkeley Symposium on Mathematical Statistics and Probability, J. Neyman, ed., Berkeley, CA, 1951, University of California Press, pp. 481–492.

[186] J. W. LAGARIAS, J. A. REEDS, M. H. WRIGHT, AND P. E. WRIGHT, *Convergence properties of the Nelder-Mead simplex algorithm in low dimensions*, SIAM Journal on Optimization, 9 (1998), pp. 112–147.

[187] S. LANG, *Real Analysis*, Addison-Wesley, Reading, MA, second ed., 1983.

[188] C. L. LAWSON AND R. J. HANSON, *Solving Least Squares Problems*, Prentice-Hall, Englewood Cliffs, NJ, 1974.

[189] C. LEMARÉCHAL, *A view of line searches*, in Optimization and Optimal Control, W. Oettli and J. Stoer, eds., no. 30 in Lecture Notes in Control and Information Science, Springer-Verlag, 1981, pp. 59–78.

[190] K. LEVENBERG, *A method for the solution of certain non-linear problems in least squares*, Quarterly of Applied Mathematics, 2 (1944), pp. 164–168.

[191] S. LEYFFER, G. LOPEZ-CALVA, AND J. NOCEDAL, *Interior methods for mathematical programs with complementarity constraints*, technical report 8, Optimization Technology Center, Northwestern University, Evanston, IL, 2004.

[192] C. LIN AND J. MORÉ, *Newton's method for large bound-constrained optimization problems*, SIAM Journal on Optimization, 9 (1999), pp. 1100–1127.

[193] C. LIN AND J. J. MORÉ, *Incomplete Cholesky factorizations with limited memory*, SIAM Journal on Scientific Computing, 21 (1999), pp. 24–45.

[194] D. C. LIU AND J. NOCEDAL, *On the limited-memory BFGS method for large scale optimization*, Mathematical Programming, 45 (1989), pp. 503–528.

[195] D. LUENBERGER, *Introduction to Linear and Nonlinear Programming*, Addison Wesley, second ed., 1984.

[196] L. LUKŠAN AND J. VLČEK, *Indefinitely preconditioned inexact Newton method for large sparse equality constrained nonlinear programming problems*, Numerical Linear Algebra with Applications, 5 (1998), pp. 219–247.

[197] *Macsyma User's Guide*, second ed., 1996.

[198] O. L. MANGASARIAN, *Nonlinear Programming*, McGraw-Hill, New York, 1969. Reprinted by SIAM Publications, 1995.

[199] N. MARATOS, *Exact penalty function algorithms for finite dimensional and control optimization problems*, PhD thesis, University of London, 1978.

[200] M. MARAZZI AND J. NOCEDAL, *Wedge trust region methods for derivative free optimization*, Mathematical Programming, Series A, 91 (2002), pp. 289–305.

[201] H. M. MARKOWITZ, *Portfolio selection*, Journal of Finance, 8 (1952), pp. 77–91.

[202] ———, *The elimination form of the inverse and its application to linear programming*, Management Science, 3 (1957), pp. 255–269.

[203] D. W. MARQUARDT, *An algorithm for least squares estimation of non-linear parameters*, SIAM Journal, 11 (1963), pp. 431–441.

[204] D. Q. MAYNE AND E. POLAK, *A superlinearly convergent algorithm for constrained optimization problems*, Mathematical Programming Studies, 16 (1982), pp. 45–61.

[205] L. MCLINDEN, *An analogue of Moreau's proximation theorem, with applications to the nonlinear complementarity problem*, Pacific Journal of Mathematics, 88 (1980), pp. 101–161.

[206] N. MEGIDDO, *Pathways to the optimal set in linear programming*, in Progress in Mathematical Programming: Interior-Point and Related Methods, N. Megiddo, ed., Springer-Verlag, New York, NY, 1989, ch. 8, pp. 131–158.

[207] S. MEHROTRA, *On the implementation of a primal-dual interior point method*, SIAM Journal on Optimization, 2 (1992), pp. 575–601.

[208] S. MIZUNO, M. TODD, AND Y. YE, *On adaptive step primal-dual interior-point algorithms for linear programming*, Mathematics of Operations Research, 18 (1993), pp. 964–981.

[209] J. L. MORALES AND J. NOCEDAL, *Automatic preconditioning by limited memory quasi-newton updating*, SIAM Journal on Optimization, 10 (2000), pp. 1079–1096.

[210] J. J. MORÉ, *The Levenberg-Marquardt algorithm: Implementation and theory*, in Lecture Notes in Mathematics, No. 630–Numerical Analysis, G. Watson, ed., Springer-Verlag, 1978, pp. 105–116.

[211] ———, *Recent developments in algorithms and software for trust region methods*, in Mathematical Programming: The State of the Art, Springer-Verlag, Berlin, 1983, pp. 258–287.

[212] ———, *A collection of nonlinear model problems*, in Computational Solution of Nonlinear Systems of Equations, vol. 26 of Lectures in Applied Mathematics, American Mathematical Society, Providence, RI, 1990, pp. 723–762.

[213] J. J. MORÉ AND D. C. SORENSEN, *On the use of directions of negative curvature in a modified Newton method*, Mathematical Programming, 16 (1979), pp. 1–20.

[214] ———, *Computing a trust region step*, SIAM Journal on Scientific and Statistical Computing, 4 (1983), pp. 553–572.

[215] ———, *Newton's method*, in Studies in Numerical Analysis, vol. 24 of MAA Studies in Mathematics, The Mathematical Association of America, 1984, pp. 29–82.

[216] J. J. MORÉ AND D. J. THUENTE, *Line search algorithms with guaranteed sufficient decrease*, ACM Transactions on Mathematical Software, 20 (1994), pp. 286–307.

[217] J. J. MORÉ AND S. J. WRIGHT, *Optimization Software Guide*, SIAM Publications, Philadelphia, 1993.

[218] B. A. MURTAGH AND M. A. SAUNDERS, *MINOS 5.1 User's guide*, Technical Report SOL-83-20R, Stanford University, 1987.

[219] K. G. MURTY AND S. N. KABADI, *Some NP-complete problems in quadratic and nonlinear programming*, Mathematical Programming, 19 (1987), pp. 200–212.

[220] S. G. NASH, *Newton-type minimization via the Lanczos method*, SIAM Journal on Numerical Analysis, 21 (1984), pp. 553–572.

[221] ———, *SUMT (Revisited)*, Operations Research, 46 (1998), pp. 763–775.

[222] U. NAUMANN, *Optimal accumulation of Jacobian matrices by elimination methods on the dual computational graph*, Mathematical Programming, 99 (2004), pp. 399–421.

[223] J. A. NELDER AND R. MEAD, *A simplex method for function minimization*, The Computer Journal, 8 (1965), pp. 308–313.

[224] G. L. NEMHAUSER AND L. A. WOLSEY, *Integer and Combinatorial Optimization*, John Wiley & Sons, New York, 1988.

[225] A. S. NEMIROVSKII AND D. B. YUDIN, *Problem complexity and method efficiency*, John Wiley & Sons, New York, 1983.

[226] Y. E. NESTEROV AND A. S. NEMIROVSKII, *Interior Point Polynomial Methods in Convex Programming*, SIAM Publications, Philadelphia, 1994.

[227] G. N. NEWSAM AND J. D. RAMSDELL, *Estimation of sparse Jacobian matrices*, SIAM Journal on Algebraic and Discrete Methods, 4 (1983), pp. 404–418.

[228] J. NOCEDAL, *Updating quasi-Newton matrices with limited storage*, Mathematics of Computation, 35 (1980), pp. 773–782.

[229] ———, *Theory of algorithms for unconstrained optimization*, Acta Numerica, 1 (1992), pp. 199–242.

[230] J. M. ORTEGA AND W. C. RHEINBOLDT, *Iterative solution of nonlinear equations in several variables*, Academic Press, New York and London, 1970.

[231] M. R. OSBORNE, *Nonlinear least squares—the Levenberg algorithm revisited*, Journal of the Australian Mathematical Society, Series B, 19 (1976), pp. 343–357.

[232] ———, *Finite Algorithms in Optimization and Data Analysis*, John Wiley & Sons, New York, 1985.

[233] M. L. OVERTON, *Numerical Computing with IEEE Floating Point Arithmetic*, SIAM, Philadelphia, PA, 2001.

[234] C. C. PAIGE AND M. A. SAUNDERS, *LSQR: An algorithm for sparse linear equations and sparse least squares*, ACM Transactions on Mathematical Software, 8 (1982), pp. 43–71.

[235] C. H. PAPADIMITRIOU AND K. STEIGLITZ, *Combinatorial Optimization: Algorithms and Complexity*, Prentice Hall, Englewood Cliffs, NJ, 1982.

[236] E. POLAK, *Optimization: Algorithms and Consistent Approximations*, no. 124 in Applied Mathematical Sciences, Springer, 1997.

[237] E. POLAK AND G. RIBIÈRE, *Note sur la convergence de méthodes de directions conjuguées*, Revue Française d'Informatique et de Recherche Opérationnelle, 16 (1969), pp. 35–43.

[238] B. T. POLYAK, *The conjugate gradient method in extremal problems*, U.S.S.R. Computational Mathematics and Mathematical Physics, 9 (1969), pp. 94–112.

[239] M. J. D. POWELL, *An efficient method for finding the minimum of a function of several variables without calculating derivatives*, Computer Journal, 91 (1964), pp. 155–162.

[240] ———, *A method for nonlinear constraints in minimization problems*, in Optimization, R. Fletcher, ed., Academic Press, New York, NY, 1969, pp. 283–298.

[241] ———, *A hybrid method for nonlinear equations*, in Numerical Methods for Nonlinear Algebraic Equations, P. Rabinowitz, ed., Gordon & Breach, London, 1970, pp. 87–114.

[242] ———, *Problems related to unconstrained optimization*, in Numerical Methods for Unconstrained Optimization, W. Murray, ed., Academic Press, 1972, pp. 29–55.

[243] ———, *On search directions for minimization algorithms*, Mathematical Programming, 4 (1973), pp. 193–201.

[244] ———, *Convergence properties of a class of minimization algorithms*, in Nonlinear Programming 2, O. L. Mangasarian, R. R. Meyer, and S. M. Robinson, eds., Academic Press, New York, 1975, pp. 1–27.

[245] ———, *Some convergence properties of the conjugate gradient method*, Mathematical Programming, 11 (1976), pp. 42–49.

[246] ———, *Some global convergence properties of a variable metric algorithm for minimization without exact line searches*, in Nonlinear Programming, SIAM-AMS Proceedings, Vol. IX, R. W. Cottle and C. E. Lemke, eds., SIAM Publications, 1976, pp. 53–72.

[247] ———, *A fast algorithm for nonlinearly constrained optimization calculations*, in Numerical Analysis Dundee 1977, G. A. Watson, ed., Springer Verlag, Berlin, 1977, pp. 144–157.

[248] ———, *Restart procedures for the conjugate gradient method*, Mathematical Programming, 12 (1977), pp. 241–254.

[249] ———, *Algorithms for nonlinear constraints that use Lagrangian functions*, Mathematical Programming, 14 (1978), pp. 224–248.

[250] ———, *The convergence of variable metric methods for nonlinearly constrained optimization calculations*, in Nonlinear Programming 3, Academic Press, New York and London, 1978, pp. 27–63.

[251] ———, *On the rate of convergence of variable metric algorithms for unconstrained optimization*, Technical Report DAMTP 1983/NA7, Department of Applied Mathematics and Theoretical Physics, Cambridge University, 1983.

[252] ———, *Variable metric methods for constrained optimization*, in Mathematical Programming: The State of the Art, Bonn, 1982, Springer-Verlag, Berlin, 1983, pp. 288–311.

[253] ———, *Nonconvex minimization calculations and the conjugate gradient method*, Lecture Notes in Mathematics, 1066 (1984), pp. 122–141.

[254] ———, *The performance of two subroutines for constrained optimization on some difficult test problems*, in Numerical Optimization, P. T. Boggs, R. H. Byrd, and R. B. Schnabel, eds., SIAM Publications, Philadelphia, 1984.

[255] ———, *Convergence properties of algorithms for nonlinear optimization*, SIAM Review, 28 (1986), pp. 487–500.

[256] ———, *Direct search algorithms for optimization calculations*, Acta Numerica, 7 (1998), pp. 287–336.

[257] ———, *UOBYQA: unconstrained optimization by quadratic approximation*, Mathematical Programming, Series B, 92 (2002), pp. 555–582.

[258] ———, *On trust-region methods for unconstrained minimization without derivatives*, Mathematical Programming, 97 (2003), pp. 605–623.

[259] ———, *Least Frobenius norm updating of quadratic models that satisfy interpolation conditions*, Mathematical Programming, 100 (2004), pp. 183–215.

[260] ———, *The NEWUOA software for unconstrained optimization without derivatives*, Numerical Analysis Report DAMPT 2004/NA05, University of Cambridge, Cambridge, UK, 2004.

[261] M. J. D. POWELL AND P. L. TOINT, *On the estimation of sparse Hessian matrices*, SIAM Journal on Numerical Analysis, 16 (1979), pp. 1060–1074.

[262] R. L. RARDIN, *Optimization in Operations Research*, Prentice Hall, Englewood Cliffs, NJ, 1998.

[263] F. RENDL AND H. WOLKOWICZ, *A semidefinite framework for trust region subproblems with applications to large scale minimization*, Mathematical Programming, 77 (1997), pp. 273–299.

[264] J. M. RESTREPO, G. K. LEAF, AND A. GRIEWANK, *Circumventing storage limitations in variational data assimilation studies*, SIAM Journal on Scientific Computing, 19 (1998), pp. 1586–1605.

[265] K. RITTER, *On the rate of superlinear convergence of a class of variable metric methods*, Numerische Mathematik, 35 (1980), pp. 293–313.

[266] S. M. ROBINSON, *A quadratically convergent algorithm for general nonlinear programming problems*, Mathematical Programming, 3 (1972), pp. 145–156.

[267] ———, *Perturbed Kuhn-Tucker points and rates of convergence for a class of nonlinear-programming algorithms*, Mathematical Programming, 7 (1974), pp. 1–16.

[268] ———, *Generalized equations and their solutions. Part II: Applications to nonlinear programming*, Mathematical Programming Study, 19 (1982), pp. 200–221.

[269] R. T. ROCKAFELLAR, *The multiplier method of Hestenes and Powell applied to convex programming*, Journal of Optimization Theory and Applications, 12 (1973), pp. 555–562.

[270] ———, *Lagrange multipliers and optimality*, SIAM Review, 35 (1993), pp. 183–238.

[271] J. B. ROSEN AND J. KREUSER, *A gradient projection algorithm for nonlinear constraints*, in Numerical Methods for Non-Linear Optimization, F. A. Lootsma, ed., Academic Press, London and New York, 1972, pp. 297–300.

[272] ———, *Iterative Methods for Sparse Linear Systems*, SIAM Publications, Philadelphia, PA, second ed., 2003.

[273] Y. SAAD AND M. SCHULTZ, *GMRES: A generalized minimal residual algorithm for solving non-symmetric linear systems*, SIAM Journal on Scientific and Statistical Computing, 7 (1986), pp. 856–869.

[274] H. SCHEEL AND S. SCHOLTES, *Mathematical programs with complementarity constraints: Stationarity, optimality and sensitivity*, Mathematics of Operations Research, 25 (2000), pp. 1–22.

[275] T. SCHLICK, *Modified Cholesky factorizations for sparse preconditioners*, SIAM Journal on Scientific Computing, 14 (1993), pp. 424–445.

[276] R. B. SCHNABEL AND E. ESKOW, *A new modified Cholesky factorization*, SIAM Journal on Scientific Computing, 11 (1991), pp. 1136–1158.

[277] R. B. SCHNABEL AND P. D. FRANK, *Tensor methods for nonlinear equations*, SIAM Journal on Numerical Analysis, 21 (1984), pp. 815–843.

[278] G. SCHULLER, *On the order of convergence of certain quasi-Newton methods*, Numerische Mathematik, 23 (1974), pp. 181–192.

[279] G. A. SCHULTZ, R. B. SCHNABEL, AND R. H. BYRD, *A family of trust-region-based algorithms for unconstrained minimization with strong global convergence properties*, SIAM Journal on Numerical Analysis, 22 (1985), pp. 47–67.

[280] G. A. F. SEBER AND C. J. WILD, *Nonlinear Regression*, John Wiley & Sons, New York, 1989.

[281] T. STEIHAUG, *The conjugate gradient method and trust regions in large scale optimization*, SIAM Journal on Numerical Analysis, 20 (1983), pp. 626–637.

[282] J. STOER, *On the relation between quadratic termination and convergence properties of minimization algorithms. Part I: Theory*, Numerische Mathematik, 28 (1977), pp. 343–366.

[283] K. TANABE, *Centered Newton method for mathematical programming*, in System Modeling and Optimization: Proceedings of the 13th IFIP conference, vol. 113 of Lecture Notes in Control and Information Systems, Berlin, 1988, Springer-Verlag, pp. 197–206.

[284] M. J. TODD, *Potential reduction methods in mathematical programming*, Mathematical Programming, Series B, 76 (1997), pp. 3–45.

[285] ———, *Semidefinite optimization*, Acta Numerica, 10 (2001), pp. 515–560.

[286] ———, *Detecting infeasibility in infeasible-interior-point methods for optimization*, in Foundations of Computational Mathematics, Minneapolis, 2002, F. Cucker, R. DeVore, P. Olver, and E. Suli, eds., Cambridge University Press, Cambridge, 2004, pp. 157–192.

[287] M. J. TODD AND Y. YE, *A centered projective algorithm for linear programming*, Mathematics of Operations Research, 15 (1990), pp. 508–529.

[288] P. L. TOINT, *On sparse and symmetric matrix updating subject to a linear equation*, Mathematics of Computation, 31 (1977), pp. 954–961.

[289] ———, *Towards an efficient sparsity exploiting Newton method for minimization*, in Sparse Matrices and Their Uses, Academic Press, New York, 1981, pp. 57–87.

[290] L. TREFETHEN AND D. BAU, *Numerical Linear Algebra*, SIAM, Philadelphia, PA, 1997.

[291] M. ULBRICH, S. ULBRICH, AND L. N. VICENTE, *A globally convergence primal-dual interior-point filter method for nonlinear programming*, Mathematical Programming, Series B, 100 (2004), pp. 379–410.

[292] L. VANDENBERGHE AND S. BOYD, *Semidefinite programming*, SIAM Review, 38 (1996), pp. 49–95.

[293] R. J. VANDERBEI, *Linear Programming: Foundations and Extensions*, Springer Verlag, New York, second ed., 2001.

[294] R. J. VANDERBEI AND D. F. SHANNO, *An interior point algorithm for nonconvex nonlinear programming*, Computational Optimization and Applications, 13 (1999), pp. 231–252.

[295] A. VARDI, *A trust region algorithm for equality constrained minimization: convergence properties and implementation*, SIAM Journal of Numerical Analysis, 22 (1985), pp. 575–591.

[296] S. A. VAVASIS, *Quadratic programming is NP*, Information Processing Letters, 36 (1990), pp. 73–77.

[297] ———, *Nonlinear Optimization*, Oxford University Press, New York and Oxford, 1991.

[298] A. WÄCHTER, *An interior point algorithm for large-scale nonlinear optimization with applications in process engineering*, PhD thesis, Department of Chemical Engineering, Carnegie Mellon University, Pittsburgh, PA, USA, 2002.

[299] A. WÄCHTER AND L. T. BIEGLER, *Failure of global convergence for a class of interior point methods for nonlinear programming*, Mathematical Programming, 88 (2000), pp. 565–574.

[300] ———, *Line search filter methods for nonlinear programming: Motivation and global convergence*, SIAM Journal on Optimization, 16 (2005), pp. 1–31.

[301] ———, *On the implementation of an interior-point filter line-search algorithm for large-scale nonlinear programming*, Mathematical Programming, 106 (2006), pp. 25–57.

[302] H. WALKER, *Implementation of the GMRES method using Householder transformations*, SIAM Journal on Scientific and Statistical Computing, 9 (1989), pp. 815–825.

[303] R. A. WALTZ, J. L. MORALES, J. NOCEDAL, AND D. ORBAN, *An interior algorithm for nonlinear optimization that combines line search and trust region steps*, Tech. Rep. 2003-6, Optimization Technology Center, Northwestern University, Evanston, IL, USA, June 2003.

[304] WATERLOO MAPLE SOFTWARE, INC, *Maple V software package*, 1994.

[305] L. T. WATSON, *Numerical linear algebra aspects of globally convergent homotopy methods*, SIAM Review, 28 (1986), pp. 529–545.

[306] R. B. WILSON, *A simplicial algorithm for concave programming*, PhD thesis, Graduate School of Business Administration, Harvard University, 1963.

[307] D. WINFIELD, *Function and functional optimization by interpolation in data tables*, PhD thesis, Harvard University, Cambridge, USA, 1969.

[308] W. L. WINSTON, *Operations Research*, Wadsworth Publishing Co., third ed., 1997.

[309] P. WOLFE, *A duality theorem for nonlinear programming*, Quarterly of Applied Mathematics, 19 (1961), pp. 239–244.

[310] ———, *The composite simplex algorithm*, SIAM Review, 7 (1965), pp. 42–54.

[311] S. WOLFRAM, *The Mathematica Book*, Cambridge University Press and Wolfram Media, Inc., third ed., 1996.

[312] L. A. WOLSEY, *Integer Programming*, Wiley–Interscience Series in Discrete Mathematics and Optimization, John Wiley & Sons, New York, NY, 1998.

[313] M. H. WRIGHT, *Interior methods for constrained optimization*, in Acta Numerica 1992, Cambridge University Press, Cambridge, 1992, pp. 341–407.

[314] ———, *Direct search methods: Once scorned, now respectable*, in Numerical Analysis 1995 (Proceedings of the 1995 Dundee Biennial Conference in Numerical Analysis), Addison Wesley Longman, 1996, pp. 191–208.

[315] S. J. WRIGHT, *Applying new optimization algorithms to model predictive control*, in Chemical Process Control-V, J. C. Kantor, ed., CACHE, 1997.

[316] ———, *Primal-Dual Interior-Point Methods*, SIAM Publications, Philadelphia, PA, 1997.

[317] ———, *Modified Cholesky factorizations in interior-point algorithms for linear programming*, SIAM Journal on Optimization, 9 (1999), pp. 1159–1191.

[318] S. J. WRIGHT AND J. N. HOLT, *An inexact Levenberg-Marquardt method for large sparse nonlinear least squares problems*, Journal of the Australian Mathematical Society, Series B, 26 (1985), pp. 387–403.

[319] E. A. YILDIRIM AND S. J. WRIGHT, *Warm-start strategies in interior-point methods for linear programming*, SIAM Journal on Optimization, 12 (2002), pp. 782–810.

[320] Y. YUAN, *On the truncated conjugate-gradient method*, Mathematical Programming, Series A, 87 (2000), pp. 561–573.

[321] Y. ZHANG, *Solving large-scale linear programs with interior-point methods under the Matlab environment*, Optimization Methods and Software, 10 (1998), pp. 1–31.

[322] C. ZHU, R. H. BYRD, P. LU, AND J. NOCEDAL, *Algorithm 778: L-BFGS-B, FORTRAN subroutines for large scale bound constrained optimization*, ACM Transactions on Mathematical Software, 23 (1997), pp. 550–560.

Index

Printed in the United States
By Bookmasters